Evolution Science and Ethics in the Third Millennium

Robert Cliquet · Dragana Avramov

Evolution Science and Ethics in the Third Millennium

Challenges and Choices for Humankind

Springer

Robert Cliquet
Anthropology and Social Biology
Ghent University
Ghent
Belgium

and

Population and Family Study Centre
(CBGS), Flemish Scientific Institute
Brussels
Belgium

Dragana Avramov
Population and Social Policy Consultants
(PSPC)
Brussels
Belgium

ISBN 978-3-319-73089-9 ISBN 978-3-319-73090-5 (eBook)
https://doi.org/10.1007/978-3-319-73090-5

Library of Congress Control Number: 2017962031

Printed on acid-free paper

This Springer imprint is published by Springer Nature
The registered company is Springer International Publishing AG
The registered company address is: Gewerbestrasse 11, 6330 Cham, Switzerland

For the children of our children,
Una, Lore, Joren, Brent, Arthur, Emiel,
Sander and Thomas

Preface

This book aims to revitalise and substantiate the idea that evolution science can, in the context of further progressing modernisation, provide a rational and robust framework for elaborating the main guidelines of a universal morality for the future. It also traces pathways for long-term biological evolution and cultural development and adaptation of humanity, in the direction of ecological sustainability of our planetary environment.

It argues that the traditional theistic belief systems are, in many respects, no longer well adapted to the needs of the novel environment of modernity and its exigencies for further human evolution and cultural development and adaptation, on the one hand; on the other hand, most secular ideologies only deal with humanity's present and future in a rather fragmented way and with a short-term perspective. Hence, the need for another framework to rethink values and norms that would safely guide the human species in making choices throughout new subsequent stages of biological evolution and cultural development.

The backbone of the general approach in this book is evolution science, and in particular bio-anthropology, which is understood here as the study of the biological origin, present and future of humanity. Bio-anthropology investigates the biology of the hominins, and its interaction with human societies, cultures and value systems, from a long-term evolutionary perspective. The rationale of the discourse in this book is based on the interaction of the long-term hominisation process with the fast changing environment of modernisation.

Building upon the theoretical framework of a former book by one of the authors ('Biosocial interactions in modernisation') ethical aspects are discussed for each of the major biosocial sources of human variation: individual variation, inter-personal variation, inter-group variation and inter-generational variation.

This book about evolutionary ethics is typically an interdisciplinary work. It should be of interest to a variety of human behavioural and social sciences, such as biological and cultural anthropology, biology, sociology, psychology, ethics, philosophy and theology. Although it is mainly addressed to scholars and students in social and behavioural sciences, it targets also lay people, since it deals in a holistic way with long-term challenges for the human species. Policy makers may find issues and reflections that can help them better understand from where we are coming and inspire them to take action or orient their future policy direction in a longer-term perspective.

In addressing different ideologies, faiths and philosophical systems based on scientific study and personal reflections, the authors' aims are not to argue or try to demonstrate the primacy or superiority of one or other belief system or faith. This book's concern regarding the origin, evolution and long-term future of humanity relates to the human species as a whole and not to the interests or prerogatives of a particular group or population.

The authors' striving to make a synthesis of such a complex subject as shaping the future of humankind is underpinned by the huge and ever-increasing body of scientific literature on various aspects of a possible, probable and desirable future. This case is documented in some detail, referring the reader to a quite substantial body of literature. Most footnote references are limited to one or a few examples, whilst all of the consulted literature has been included in the bibliography.

Robert Cliquet
Dragana Avramov

Contents

1 Setting the Stage for Reflecting on a Universal Morality

Abstract

Dealing with the ethical challenges of humankind at the turn of the twenty-first century, and safely guiding the human species through new subsequent stages of biological evolution and adaptation and cultural development, requires rethinking of our values and norms in a longer-term perspective and at the planetary level. Therefore, this chapter starts by discussing the meaning of evolution and presenting an overview of the major stages of the development of evolution science—the Darwinian revolution, the Modern Evolutionary Synthesis, the molecular-genetic revolution and the Second Darwinian Revolution. Next, the two major developmental processes, the hominisation process and the modernisation process, are addressed, which are considered by the authors to be of pivotal importance for the future of human morality. Finally, by confronting the hominisation and modernisation processes, this chapter sets the stage for revealing the necessary changes in values and norms in view of adapting to further progressing modernisation and evolving toward higher levels of hominisation.

1.1 Evolution Science

Humans have always, and everywhere, raised questions about the origin and meaning of life—in particular human life—as well as about its causes.[1] Since the eighteenth century a process developed through which the traditional vision—the supernaturally, spiritually evoked creationism—was gradually ousted by the

[1]Sproul (1979), Leeming (2009).

R. Cliquet and D. Avramov, *Evolution Science and Ethics in the Third Millennium*,
https://doi.org/10.1007/978-3-319-73090-5_1

scientific view—the naturally and materialistically based evolutionism, and its explanatory mechanism, the present-day Modern Evolutionary Synthesis.[2]

The scientific, philosophical, ethical and societal repercussions of evolutionism was so strong, broad and profound that it may rightly be considered as the most fundamental intellectual revolution in human history.[3] The replacement of the creationist vision by the evolutionist view and its explanation completely changed the way in which the human could think about his origin, development, meaning and future: it provoked a transition from a static towards a dynamic worldview. It replaced the spiritual conception of life by a materialistic one; the supernatural origin of life and humankind was replaced by natural determinants; the anthropocentric cosmic vision, which was already undermined in its spatial dimension by Nicolaus Copernicus's theory in the sixteenth century and the discoveries of Galileo Galilei in the seventeenth century, was now also fundamentally altered in its temporal and biological dimensions.[4] As John Stewart[5] states:

> For the first time humans have a powerful, science-based story that explains where they have come from, and their place in the unfolding of the universe.

The emergence and development of evolution science, with its consecutive Darwinian revolution, Modern Evolutionary Synthesis, molecular revolution and Second Darwinian Revolution, provides us with a unifying framework, and a solid, naturalistic basis for rethinking values and norms to secure a safe and progressive future ontogenetic development[6] and phylogenetic evolution[7] of humankind.

1.1.1 The Darwinian Revolution

In the middle of the nineteenth century modern evolutionary theory, nowadays usually referred to as Darwinism, was independently developed by two scholars, namely Charles Darwin (1809–1882) and Arthur Wallace (1823–1913). It was Darwin's extensive treatise of 1859 *On the Origin of Species by Means of Natural Selection of the Preservation of Favoured Races in the Struggle for Life* that forced the definitive breakthrough of the evolutionary paradigm and its major explanatory mechanism, natural selection. From a cultural historical point of view, this monumental treatise rightly has been considered as the most important book ever written by one author.

When speaking about evolutionism, one has to distinguish between evolution as phenomenon and the evolutionary mechanism as the explanation of that phenomenon. In general terms, evolution can be described as the occurrence of changes

[2]Bowler (1984), Larson (2006).
[3]Mayr (1978).
[4]Gingerich (1993).
[5]Stewart (2008).
[6]Ontogeny: the development of an organism within its own lifetime from conception to death.
[7]Phylogeny: the evolutionary development and history of a species or larger groups of related organisms as they change through time.

in the genetic composition and structure of populations. Darwin and Wallace's theory about the explanation of evolution concentrated on the process of natural selection.

The essence of the discovery of natural selection to explain the origin of new species consists of recognising the following facts and their interrelationships:

- species are characterised by individual genetic variations;
- species have the potential for an exponential population increase, but remain demographically stationary due to the limited capacity of their environment, inferring that in nature there is a constant competition ('struggle for existence');
- from the confrontation of the variability principle and the reproductive competition it was deduced that some genetic variants adapted and survived better than others ('survival of the fittest').

These interrelationships can be represented by the following chained formulae[8,9]:

$$GRI + LR \rightarrow SE + V \rightarrow NS + T \rightarrow BI$$

1.1.2 The Modern Evolutionary Synthesis

The evolution theory as an explanatory mechanism is nowadays much more developed and refined. In the course of the first half of the twentieth century, several other evolutionary mechanisms were discovered, not instead of natural selection, but complementary to the Darwin-Wallace explanation, namely mutation,[10] genetic drift,[11] genetic migration,[12] and assortative mating.[13] In the same period, or even somewhat earlier, Mendelian genetics,[14] biometrical or quantitative genetics,[15] and population genetics developed. A crucial milestone in the development of the Modern Evolutionary Synthesis was the formulation of the Hardy-Weinberg Law[16]

[8]Huxley and Flew, quoted in Oldroyd (1980, 118–119).

[9]GRI: geometrical ratio of increase; LR: limited resources; SE: struggle for existence; V: variation; NS: natural selection; T: time; BI: biological improvement.

[10]Morgan (1903), De Vries (1904).

[11]Genetic drift: change of allele frequencies of monogenes as a result of the accumulation of random fluctuations in the intergenerational transmission of alleles in small populations (Wright 1929); see also Chap. 2, Sect. 2.1.4.

[12]Genetic migration: transfer of genes from one population to a genetically different one; see also Chap. 2, Sect. 2.1.3.

[13]Assortative mating: deviation of partner choice from a random mating pattern (Fisher 1918; Wright 1921, 1922); see also Chap. 2, Sect. 2.1.5.

[14]Mendel (1865), Correns (1900), De Vries (1900).

[15]Biometrical or quantitative genetics: a branch of genetics that deals with biological characteristics that show a continuous variation (Galton 1889; Pearson 1896; Kearsey and Pooni 1998).

[16]Hardy (1908), Weinberg (1908).

of 1908 that shows in a simple equation the incidence and intergenerational transmission of genes and genotypes in populations: this forms the basis of the mathematical modelling of the whole evolutionary process.

Around the 1930s, the theory of Darwin and Wallace had evolved to a much more comprehensive synthesis, also completely mathematically formalised, in which all known evolutionary determinants—mutation, selection, genetic drift, gene flow, and partner choice—were analysed in their mutual relations and interactions.[17] Somewhat later, the Modern Evolutionary Synthesis was further elaborated and illustrated in a number of impressive comprehensive evolutionary treatises.[18]

In recent years several authors have, on the basis of mathematical models and computer simulations, developed theories about the self-organising capacities of complex systems.[19] Self-organising potentials of complex systems have probably played an important role in the origin of life. If those theories applied to all living organisms, biological evolution would be driven by three complementary and interacting mechanisms: chance events (mutation, genetic drift), selection, and self-organisation of complex systems.[20] With the development of evolution science, humanity now possesses the intellectual capacity to consciously self-organise its future evolution, assuming it desires and decides to do so.

Due to the continuous progress in evolutionary biology, several scholars have in recent decades argued for a further shift in the denomination of the evolutionary paradigm, replacing 'Modern Evolutionary Synthesis' (MES) by 'Extended Evolutionary Synthesis' (EES). The aim is to integrate in the evolution synthesis not only primary evolutionary mechanisms but also more complex secondary evolutionary causes, and in particular organism- and ecology-centred evolutionary concepts or developments, such as developmental evolutionary processes,[21] epigenetics,[22] genomic evolution,[23] phenotypic plasticity,[24] evolvability,[25] self-organization,[26] evolutionary adaptive landscapes,[27] evolutionary capacitance,[28] ecological inheritance through niche construction,[29] multilevel selection,[30] etc. However, in our view, the continually developing insights in more complex evolutionary processes fits very well into the basic MES and does not reflect a fundamental evolutionary paradigm change.

[17]Chetverikov (1927), Fisher (1930), Wright (1931), Haldane (1932).
[18]Dobzhansky (1937), Mayr (1942), Huxley (1942), Simpson (1944).
[19]Prigogine and Stengers (1984), Kauffman (1993, 1995).
[20]Depew and Weber (1995).
[21]Gould (1977), Gilbert et al. (1996).
[22]Newman and Muller (2001), Jablonka and Lamb (2005).
[23]Quayle and Bullock (2006), Wray (2010).
[24]Pigliucci (2001), West-Eberhard (2003).
[25]Wagner and Altenberg (1996), Kirschner and Gerhart (1998).
[26]Kauffman (1993), Johnson and Lam (2010).
[27]Gavrilets (1997), Svensson and Calsbeek (2012).
[28]Rutherford and Lindquist (1998), Bergman and Siegal (2003).
[29]Odling-Smee (2003), Abouheif et al. (2014).
[30]Wilson (2010), Gardner (2015).

1.1.3 The Molecular-Genetic Revolution

The discovery of the molecular structure of genes (*deoxyribonucleic acid, DNA*) in 1953[31] and the subsequent explosive development of molecular genetics[32] enabled understanding, at the molecular level, of the duplication process of genes and the enormous genetic variability that exists in nature. It confirmed at the molecular level the knowledge of fields developed earlier in genetics at the cellular, individual, family and population level, and it opened the prospect of correcting deleterious genes and enhancing desirable genetic variants in plants, animals and humans. Recently, molecular geneticists successfully created a first form of 'synthetic life' by assembling a working bacterial genome from raw nucleotides.[33]

In the domain of evolution science, molecular genetics has given a considerable new impetus for evolutionary studies, especially for the study of human evolution. The ingenious analytical methods developed on the basis of the knowledge about the nucleotide sequences in the DNA molecules of humans and other primates have, at the molecular level, confirmed or even refined the insights into the evolution, diversification and migration of hominins that were developed earlier on the basis of paleontological, anatomical and archaeological data.[34]

Molecular genetics opens promising prospects for obtaining a deeper and more profound understanding of the development of many aspects of morality, to the extent that moral behaviour is linked to a multitude of personality characteristics, and the latter are partly influenced by genes.[35]

1.1.4 The Second Darwinian Revolution

In the 1960s and 1970s evolutionary biology made a great leap forward with the development of a number of refined or new concepts and theories about the biological evolution of altruism, sex relations, and sociality in general: inclusive fitness,[36] kin selection,[37] reciprocity selection,[38] group selection,[39] evolutionary

[31]Watson and Crick (1953).
[32]Strachan and Read (2010).
[33]Venter (2013).
[34]Cavalli-Sforza et al. (1994), Relethford (2001), DeSalle and Ian Tattersall (2008), Fairbanks (2010).
[35]For instance Benjamin et al. (2002), Noblett and Coccaro (2005), Canli (2008).
[36]Inclusive fitness: the sum of the number of offspring an individual produces and the number of offspring of his relatives that results from his altruistic behaviour (Hamilton 1964); see also Chap. 2, Sect. 2.1.2.5.
[37]Kin selection: the evolutionary mechanism through which inclusive fitness of an individual is being achieved (Maynard Smith 1964); see also Chap. 2, Sect. 2.1.2.5.
[38]Reciprocity selection: the evolutionary mechanism through which genes are selected thanks to altruistic behaviour between non-relatives (Trivers 1971; Alexander 1987; Nowak and Sigmund 1998; 2005); see also Chap. 2, Sect. 2.1.2.6.
[39]Group selection: the evolutionary mechanism through which natural selection produces differences in reproductive fitness between groups (Maynard Smith 1964; Alexander and Borgia 1978; Sober and Wilson 1998); see also Chap. 2, Sect. 2.1.2.8.

stable strategy,[40] Red Queen theory,[41] the Machiavellian hypothesis,[42] selfish gene theory,[43] evolution of sex theory,[44] evolutionary game theory,[45] the Handicap Principle,[46] biocultural co-evolution,[47] etc. The innovation was so striking and fundamental, especially for the understanding of many aspects of the evolution of sociality, that Steven Gangestad and Jeffry Simpson[48] refer to it as "the theoretical reawakening" of the evolutionary sciences in the 1960s and early 1970s. Some authors already now refer to that period as the Second Darwinian Revolution,[49] which is probably the only fundamental novelty in the study of biosocial life since the Darwinian revolution of the 19th century.[50]

In 1974 many of the new concepts and theories about the biological evolution of sociality were synthesised by Michael T. Ghiselin,[51] but the public breakthrough of the evolutionary study of sociality came in 1975 with Edward O. Wilson's[52] epoch-making oeuvre on the behaviour of social species: *Sociobiology: The New Synthesis*. Wilson defined sociobiology as the systematic study of the biological evolution of social behaviour, in which knowledge from ethology,[53] ecology and genetics is incorporated in order to show how social species adapt to the environment by evolution.

[40]Evolutionary stable strategy: a strategy that cannot be invaded by any alternative strategy in a population (Maynard Smith and Price 1973); see also Chap. 2, Sect. 2.1.2.8.
[41]Red Queen Theory: organisms that live in coevolved interactions with other evolving organisms in a changing environment, must constantly evolve (Van Valen 1973); see also Chap. 5, Sect. 5.3.2.3.
[42]Machiavellian Hypothesis: the increase in brain size during human evolution evolved due to intense social competition in which increasingly sophisticated 'Machiavellian' strategies were used as a means to achieve higher social and reproductive success (Alexander 1974; Humphrey 1976; De Waal 1982; Byrne and Whiten 1988); see also this Chap., Sect. 1.2.
[43]Selfish gene theory: evolution occurs through the differential reproduction of competing genes, the more successful forms of which survive at the detriment of alternative ones. Richard Dawkins (1976) coined in this respect the term 'selfish gene'; see also Chap. 2, Sect. 2.2.4.4.
[44]Ghiselin (1974), Maynard Smith (1978), Daly and Wilson (1978).
[45]Evolutionary game theory: application of game theory to the evolution of living organisms (Maynard Smith and Price 1973; Maynard Smith 1982; Gintis 2000; Barash 2003); see also Chap. 2, Sect. 2.1.2.6.
[46]The handicap principle: living beings display their biological superiority through costly morphological or behavioural signals, showing their ability to squander wastefully some of their natural resources. (Zahavi and Zahavi 1997).
[47]Biocultural co-evolution: the feedback-causal relationship between biological evolution and cultural change, resulting in an acceleration of both processes (For instance, Washburn 1959, 1960; Cavalli-Sforza and Feldman 1981; Lumsden and Wilson 1981; Durham 1991; Boyd and Richerson 1985; Gintis 2011); see also this Chap., Sect. 1.2 and Chap. 2, Sect. 2.1.
[48]Gangestad and Simpson (2007, 435).
[49]Wright (1994), Horgan (1995), Machalek and Martin (2004).
[50]See also Gardner (2013, 104).
[51]Ghiselin (1974).
[52]Wilson (1975).
[53]Ethology: the study of (comparative) animal behaviour, usually with a focus on behaviour as an evolutionarily adaptive phenomenon.

From the 1970s onward the new biological concepts and theories concerning the evolution of social life have also been applied to human social behaviour.[54] Richard Alexander[55] did pioneering work in generating ingenious hypotheses and theories with regard to various issues such as the evolution of morality in intergroup competition, the relationship between biological evolution and culture, human parental investment and nepotism, and scenario building, consciousness, and human communication. Furthermore, in 1978, Edward O. Wilson clarified his ideas on human social evolution in *On Human Nature*.[56]

In addition to theoretical work, including mathematical modelling of the evolution of social behaviour, and empirical investigations on populations in various stages of cultural development—hunter/gatherer, agrarian and industrial stages—valuable new insights were also gained by applying game theory and experiments[57] to evolutionary theory.[58] An interesting finding is that the results of well-controlled laboratory game experiments correspond well to the behaviour of people in natural settings.[59]

Specifically, human-oriented sociobiological theoretical work and empirical research concern a broad variety of issues, such as individual drives or traits,[60] sex

[54]For instance, Gregory et al. (1978), Chagnon and Irons (1979), Bowles and Gintis (2011), Voland (2013).
[55]Alexander (1975, 1979, 1987).
[56]Wilson (1978), see also Wilson's recent book on 'The Social Conquest of Earth' (2012).
[57]For instance, prisoner dilemma game (Axelrod and Hamilton 1981; Axelrod 1986); public goods game (Yamagishi 1986; Fehr and Gächter 2000; 2002); dictator game (Kahneman et al. 1986); gift exchange game (Fehr et al. 1993); trust game (Berg et al. 1995); ultimatum game (Henrich 2000).
[58]For overviews of evolutionary game experiments see, amongst others, Maynard Smith (1982), Gintis (2000), Barash (2003), Bowles and Gintis (2011).
[59]Bowles and Gintis (2011, 39ff).
[60]For instance, nepotism (Alexander 1979; Bellow 2004); dominance (Omark et al. 1980); jealousy (Daly et al. 1982); cheating behaviour (Trivers 1974); cheating detection (Cosmides and Tooby 1992); self-deception (Trivers 2000; 2011); suicide (Mascaro et al. 2001); menopause (Peccei 1995); senescence (Hamilton 1966).

relations,[61] reproductive behaviour,[62] social relations in general,[63] and, last but not least, morality.[64]

Parallel to the post WWII developments in evolutionary theory about the evolution of social behaviour was the transformation of bio-anthropology from a mainly descriptive to a more explanatory science, mainly through the application of the Modern Evolutionary Synthesis. A major result of this transformation was the development of the theory about biocultural co-evolution,[65] based on the salient parallelism between the biological evolution of the hominins[66] and the development of human culture. The general idea is that genetic evolution and cultural change permanently interact and mutually reinforce each other and that cultural phenomena are under the same evolutionary pressures—mutation, selection, drift, and migration—as genetic traits.[67]

The Second Darwinian Revolution also underpinned the emergence of evolutionary psychology, although much of what is published in this field is in fact sociobiology.[68] An evolutionary approach to social and cultural behaviour requires in-depth study at the individual level of the way in which the brain functions in order to create social and cultural adaptations. Indeed, evolutionary psychology studies the evolved human psychological mechanisms regulating individual

[61]For instance, mating behaviour (Daly and Wilson 1978; Buss 1994; 2007; Miller 2000); kinship systems (Van den Berghe 1979); monogamy (Melotti 1980; Fisher 1992; De La Croix and Mariani 2015); incest avoidance and incest taboo (Van den Berghe 1980; Wolf 1995); cuckoldry and mate guarding (Hiatt 1989); polygyny (Borgerhoff Mulder 1990); sexual attractiveness (Gangestad and Thornhill 1997).

[62]For instance, parental investment and sexual selection (Trivers 1972); sex ratio and male surmortality (Trivers and Willard 1973); parent-offspring conflict (Trivers 1974); sexual dimorphism and reproductive strategies (Daly and Wilson 1978); paternal confidence (Gaulin and Schlegel 1980); paternity security and avunculate (Greene 1980); infanticide (Dickemann 1979); child abuse (Lenington 1981); hidden ovulation (Daniels 1983); birth spacing (Blurton Jones 1987); adoption (Silk 1990); rape (Thornhill and Palmer 2000); demographic transition (Borgerhoff Mulder 1998).

[63]For instance, biopolitics (Somit 1976); food sharing (Isaac 1978); evolution of cooperation (Axelrod and Hamilton 1981); cooperation and international politics (Axelrod 1984); ostracism (Gruter and Masters 1986); in-group/out-group relations, xenophobia and racism (Reynolds et al. 1987); aggression and war (Shaw and Wong 1989; Van der Dennen 1995); life history theory (Hill 1993); wary cooperation theory (Alford and Hibbing 2004).

[64]For instance, Campbell (1975), Stent (1980), Alexander (1987), Wilson (1993), Wright (1994), Hauser (2006), Krebs (2011), Boehm (2012).

[65]For instance, Washburn (1959), Cavalli-Sforza and Feldman (1981), Lumsden and Wilson (1981), Durham (1991), Boyd and Richerson (1985), Gintis (2011).

[66]Hominins: the various human-like species that evolved in the course of the hominisation process, ultimately resulting in the emergence of the present species *Homo sapiens sapiens*.

[67]Cavalli-Sforza and Feldman (1981), Boyd and Richerson (1985), Cziko (1995), Mesoudi (2011; 2016).

[68]In many quarters, the terminological shift from sociobiology to evolutionary psychology has probably more to do with political correctness than scientific scrupulousness, such as the desire to avoid association with a field that has been accused of biological determinism and reductionism, racism, sexism, etc. (Silverman 2003; Webster 2007) or the fact that, particularly in the United States, the more individual-oriented psychology is politically more fashionable than the more socially oriented sociobiology.

behaviour resulting in social and cultural dynamics.[69] Evolutionary psychology endeavours to produce a synthesis of evolutionary biology and psychology,[70] focusing on the cognitive-mental level as the mediator between social reality and individual behaviour. It seeks to identify specific cognitive mechanisms that were designed to solve specific adaptive problems in the environment in which they emerged. A fundamental proposition of evolutionary psychology is that the mechanisms of our social cognition were adaptations to the hunter-gatherer culture of the Pleistocene past. This implies that our brain, with its basic mental content-specific cognitive mechanisms, is not adapted to the present environment of high population density and social complexity that we experience in modern culture.[71] Today the neuro-cognitive behavioural sciences contribute significantly to evolutionary psychological theory. The processing of social cognition appears to occur via specific neuro-cognitive processes and sections of the brain.[72]

The Second Darwinian Revolution is not only influencing psychology. Sociobiology has also given a boost to behavioural ecology.[73] This is of great importance for the study of the evolution of social behaviour, because that discipline usually examines biosocial relations under pressure from environmental conditions and the availability of resources. The sociobiological approach is also, albeit slowly, being introduced to sociology.[74] Other offshoots of the recent evolutionary revival are bioeconomics,[75] sociobiological psychiatry,[76] and evolutionary political science.[77] Last but not least, there is a revival of evolutionary ethics.[78]

The broad diversity of the specialised—and sometimes competing—research fields in the realm of the biosocial sciences, each with their distinct conceptual and methodological approaches, that in recent decades developed around the evolution of social behaviour, might be interpreted as a sign of the weakness or even the fallacy of the evolutionary approach. However, as Kevin N. Laland and Gillian R. Brown[79] pertinently concluded in their comparative overview of various evolutionary perspectives on human behaviour, the recent explosive development of new concepts, theories, observations, and experiments about the evolutionary background of (human) social behaviour is complementary rather than incompatible.

[69]Tooby and Cosmides (1990), Wright (1994), Gangestad and Simpson (2007).
[70]Buss (1999), Barrett et al. (2002).
[71]Dunbar (2007).
[72]Bechara (2002).
[73]Borgerhoff Mulder and Schacht (2012).
[74]For instance, Ducros (1981), Crippen (1994), Niedenzu et al. (2008), Turner et al. (2015).
[75]For instance, Koslowsky (1999), Landa and Ghiselin (1999), Hodgson (2007).
[76]For instance, Rancour-Laferriere (1985), Stevens and Price (1996), McGuire and Troisi (1998).
[77]For instance, Rubin (2002), Alford and Hibbing (2004), Fowler and Schreiber (2008).
[78]Campbell (1979), Alexander (1987), Nitecki and Nitecki (1993), Farber (1994), Katz (2000), Hinde (2002), Boniolo and De Anna (2006), Høgh-Olesen (2010), Krebs (2011), Boehm (2012).
[79]Laland and Brown (2002, 317).

1.2 The Hominisation Process

The evolution of humankind is the result of the hominisation process that took place over a period of six to seven million years during which a prehominin anthropoid was transformed, over a series of successive hominin waves and radiations[80]—*Australopithecus/Homo habilis/Homo ergaster/Homo erectus/Archaic Homo sapiens*—to the present-day *Homo sapiens sapiens* (Fig. 1.1). This major evolutionary transformation[81] was initiated by the acquisition of bipedalism and was mainly characterised by a gradual, though substantial increase in brain capacity and the associated development of unusual levels of novelty[82] in the domains of nutrition (shift from mainly vegetarianism to omnivorism), complex social life (including cooperative breeding[83]), language and culture.

The study of the parallelism between the biological evolution of the hominins and the development of human culture during the hominisation process has resulted in several anthropological theories of biocultural co-evolution.[84] The hominisation process was not only accompanied by the emergence of the specific human type of culture—euculture[85] as opposed to protoculture[86] of some animal species—but also by a gradual increase in complexity of that culture and of the speed with which successive cultural phases followed each other. The success of the hominins in the Pleistocene was due to the cumulative cultural change that much more rapidly

[80]It appears more and more clearly that each major stage of the hominisation process, and in particular the earliest stage, was characterised by the development of several variants. For instance, the Australopithecus/Paranthropus stage included variants such as the *Sahelanthropus Tchadensis, Adripithecus ramidus, A. anamensis, A. afarensis, A. bahrelghazali, A. africanus, A. garhi, A. sediba, A. deyiremeda, A. prometheus, A. naledi, Kenyanthropus platyops, P. aethiopicus, P. boisei,* and *P. robustus* (Stringer 2012; Stringer and Andrews 2012; Tattersall 2012; Berger et al. 2015).

[81]Maynard Smith and Szathmáry (1995).

[82]Flinn and Coe (2007, 340), Antón and Snodgrass (2012), Isler et al. (2012).

[83]Hrdy (2011).

[84]For instance, Washburn (1959), Lumsden and Wilson (1981), Boyd and Richerson (1985), Durham (1991).

[85]Euculture: specifically human culture, showing high complexity, depending on intentional and symbolic behaviour.

[86]Protoculture: rudimentary and non-symbolic forms of intergenerationally transmitting learned behaviour among non-human primates.

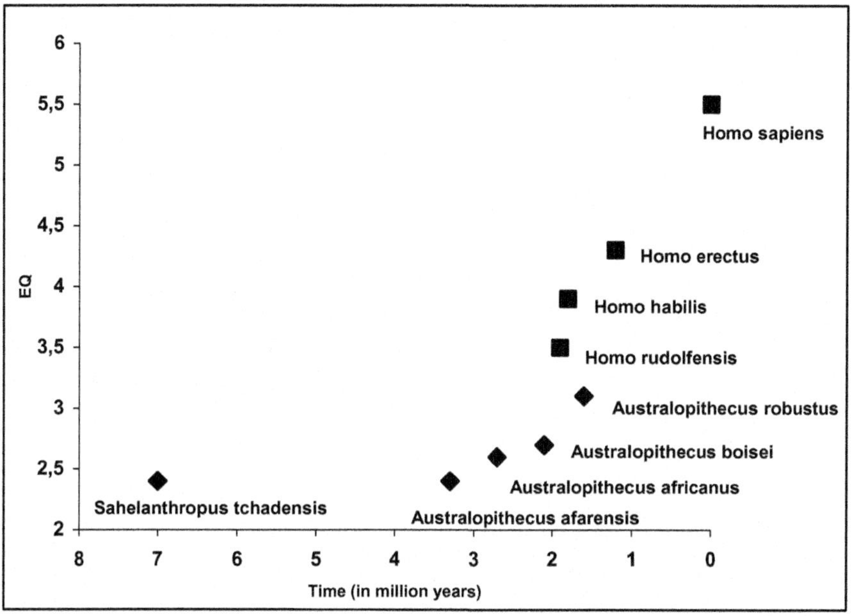

Fig. 1.1 The hominisation process (Cliquet 2010, 4) *Legend* EQ = encephalisation quotient is the ratio of the actual brain mass to the expected brain mass of a typical species that size

results in complex adaptations[87] than natural selection is able to produce genetic adaptations.[88]

Biocultural co-evolution appears to reflect a neurological-cultural associative-causal relationship and a feedback-causal relationship. The hominin brain not only produced culture, but the adaptive advantage of this production also increased the selective pressure on the brain and favoured the accumulation and dissemination of mutations allowing for the brain's further growth.[89] This increasing encephalisation,[90] in turn, stimulated the further increase in cultural

[87]Adaptation is a concept that may have two different meanings, namely (phylo)genetic adaptation and ontogenetic adaptation. The first refers to a process through which a genetically determined or influenced feature spreads in a population by means of natural selection and thanks to which this feature succeeds in contributing to the survival and reproduction of its carriers; the second relates to physiological or behavioural changes during the ontogenetic development of individuals as adjustment to environmental living conditions, but are not genetically transmitted to subsequent generations. The concept of adaptation is applicable to biological as well as cultural traits. Whereas (phylo)genetic adaptations are intergenerationally transmitted through genes, vertically from parents to children, cultural adaptations can, vertically as well as horizontally, be spread through cultural learning processes. G.G. Williams (1966, 159) called an adaptation a "design for survival".

[88]Richerson and Boyd (2005, 146).

[89]Flinn and Coe (2007).

[90]Encephalisation: the tendency of the human evolutionary lineage toward larger brains through evolutionary time.

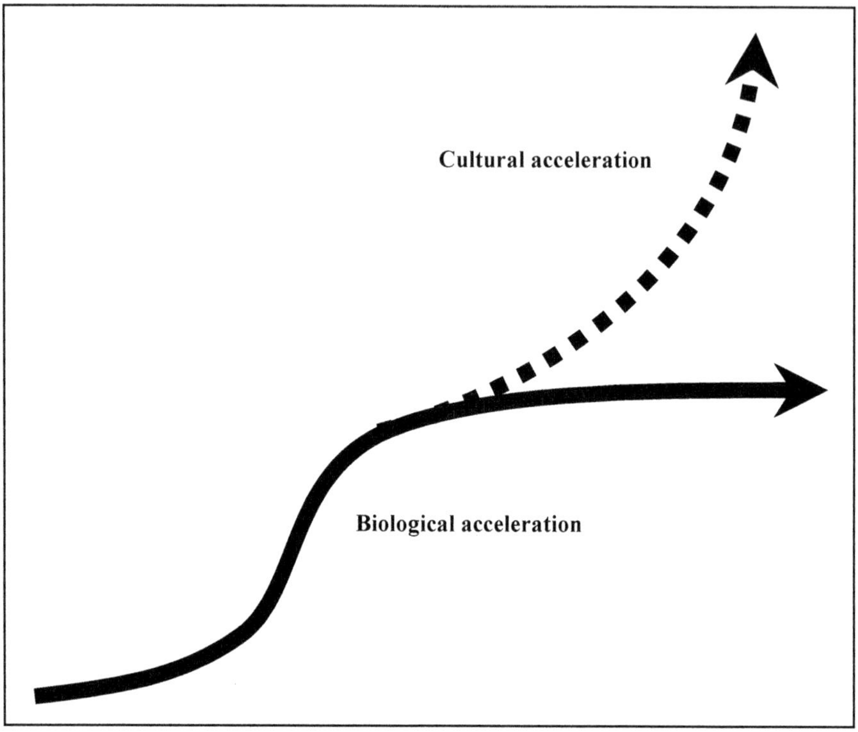

Fig. 1.2 Biocultural co-evolution among the hominins (Cliquet and Thienpont 2002, 620)

innovation and complexity.[91] The biocultural co-evolutionary system moreover implies that today's human genome is partially a product of the culture that developed in the course of hominisation. In other words, *Homo sapiens sapiens* is the result—obviously unconscious—of a process of *autopoiesis*.[92] As Peter A. Corning[93] formulated it:

> In a very real sense, our species invented itself.

During the last phase of hominisation—the *Homo sapiens sapiens* stage—a divergence appears between the evolutionary tempo of individual neurological progress on the one hand and that of socio-cultural complexity on the other hand. Encephalisation, as measured by the size of the brain, seems to have stopped, whilst cultural acceleration has continued to progress, especially since the advent of the

[91]Not only the human brain, but several other biological characteristics of the human species are the result of the biocultural co-evolutionary process. The most salient example is the anatomy and physiology of human speech and facial communication (Cliquet and Thienpont 2002, 600; Gintis and Helbing 2015, 17), but also the dexterity of the human hands is a good example. Obviously, the same applies to many essential components of human culture, sociality and morality.
[92]Autopoesis: 'αὐτοποίησις' = self-creation in Greek.
[93]Corning (2014, 242).

agricultural phase of human history, and even more since the Industrial Revolution (Fig. 1.2). The paradoxical divergence between the apparently stagnating individual encephalisation of *Homo sapiens sapiens* and the remarkable cultural growth in recent millennia can be explained by the transition from an individual level increase in neurological capacity to a biosocial type of encephalisation. Just as brain growth during hominisation was characterised by an exponential increase in the number of multiple interconnected neurons, resulting in an exponential enhancement of the associative capacity of the individual human brain, the more recent phases of cultural development in human history have been made possible by an exponential increase in the number of multiple interconnected individuals in demographically growing human societies. Hence, individual level encephalisation has been complemented by biosocial interconnectivity, resulting in an exponential increase in the overall capacity of growing and evolving human societies. This 'biosocial encephalisation' obviously applies only to cultural forms that can be developed via social mechanisms, such as technology and social organisation. It does not apply to cultural expressions that remain dependent upon individual creativity.[94]

There are two major evolutionary biological features of *Homo sapiens sapiens* that made our species strongly dependent upon socio-cultural structures and processes for its development and survival. These features were, on the one hand, the shift from programmed behaviour based on fixed instincts and inherited action patterns toward a conscious control of behaviour through the development of the large brain hemispheres, and, on the other hand, the relatively short human pregnancy duration which caused women to give birth prematurely, before the baby's brain had fully matured. Moreover, both the biosocial dependency of the human children and adolescents and the interdependency of adults increased and became more prolonged as human culture and society became more complex.

Initially, the evolutionary explanations for the enlargement of the brain in the course of the hominisation process strongly referred to the natural environment and/or human technology: the increasing brain capacity of the hominins was thought to result from changing selective pressures to demands emanating from the environment[95] or in response to tool making.[96] In recent decades, the causal explanation for the hominin brain increase has shifted to the exigencies of the increasingly complex social life.[97] In this context, theories about several specific explanations for the quick evolvement of the hominin brain have been developed, such as pressures for the development of social intelligence,[98] the need for dealing with social deception and manipulation ('Machiavellian intelligence'),[99] the

[94]Cliquet and Thienpont (2002, 623).
[95]Clutton-Brock and Harvey (1980).
[96]Oakley (1959), Washburn (1959, 1960).
[97]For instance, Etkin (1963), Dunbar (2003), Gamble et al. (2014), Gintis (2014).
[98]Humphrey (1976).
[99]Whiten and Byrne (1988).

exigencies of social cooperation,[100] the challenges of warfare,[101] and the exigencies of mate choice.[102]

The apparent stagnation of brain expansion in human evolution since the appearance of *Homo sapiens sapiens* does not mean that the evolution of humankind has reached an endpoint. Human genetic evolution is still on-going, with important genetic changes in the recent past.[103] Contrary to the long-held view that natural selection has ceased to affect humans, recent molecular-genetic studies show, in contrast, that the advent of the agricultural and industrial cultural eras has provoked an acceleration in genetic changes in *Homo sapiens sapiens.*[104]

Although many detailed technical questions about the emergence, evolution and specificities of the hominins, and in particular the present *Homo sapiens sapiens*, remain to be resolved,[105] the development and accumulation of knowledge in various sciences, such as evolution science, geology and cosmology, genetics, including population genetics, behavioural genetics and molecular genetics, palaeontology and archaeology, primatology, bio-anthropology and cultural anthropology, sociobiology and evolutionary psychology, and neurology, has resulted in the fact that the origin, evolution, and specificity of humankind is no longer a mystery. Knowledge about the hominisation process allows us to distinguish future alternative directions in which humanity could evolve and to make rational choices about the trajectory to opt for.

1.3 The Modernisation Process

In bio-anthropology, it is now generally accepted that the specific human genome emerged as an adaptation to natural and socio-cultural living conditions that existed in the Pleistocene era, the Environment of Evolutionary Adaptedness (EEA),[106] sometimes called the Environment of Ancestral Adaptedness (EAA), which does not mean that no further genetic change would have occurred in the recent past, particularly since the invention of agriculture,[107] or is currently occurring.

With the onset of the scientific-industrial cultural stage, the human species entered a phase in its evolution and history that, in many respects, is a developmental threshold. Mankind has created an evolutionary novel environment of technologically advanced societies inhabited by anonymous millions, very different from the primordial environment of small face-to-face bands of hunter-gatherers, resulting in a shift from largely kin-based to largely non-kin-based social

[100]Moll and Tomasello (2007), Brosnan et al. (2010), McNally et al. (2012).
[101]Alexander (1989).
[102]Miller (2000).
[103]Evans et al. (2005), Hawks et al. (2007), Williamson et al. (2007), Hawks (2016).
[104]Armelagos and Harper (2005), Cochran and Harpending (2009), Byars et al. (2010).
[105]For instance, Barash (2012), Weaver (2012).
[106]Bowlby (1969).
[107]Cochran and Harpending (2009).

networks.[108] Modernity created a macro-cosmos, very different from life in small groups, the 'micro-cosmos' to which we are biologically adapted.[109] Nevertheless, Edward O. Wilson[110] considers modernity only as "a mosaic of cultural hypertrophies of archaic behavioural adaptations" to survival and reproduction in hunter-gatherer populations.

Cultural changes such as the development of science with its technological and societal applications, the emergence of the market economy, the shift towards industrial modes of economic production, the creation and use of new energy sources, and last but not least, the emergence and the implementation of the normative framework of the Enlightenment,[111] provoked a giant leap generally labelled as the modernisation process. Of course, modernisation has roots in a much longer historical process in which Steven Pinker[112] distinguishes three main stages: (1) the pacification process as a consequence of state organisation at the emergence of the agricultural era; (2) the civilising process in Renaissance times; and (3) the humanisation revolution in the wake of the Enlightenment.

The most fundamental feature and key determinant of modernisation consists, in the authors' view, of the development of science—both 'low' and 'high' science[113]—which brought not only a more thorough knowledge and understanding of reality, but which made possible more effective ways of intervening in life, society and the environment. As Jacques Monod[114] wrote in his well-known book *Le hazard et la néccesité*:

> Les sociétés modernes sont construites sur la science. Elles lui doivent leur richesse, leur puissance et la certitude que des richesses et des pouvoirs bien plus grands encore seront demain, s'il le veut, accessibles à l'Homme.[115]

This interventionist nature of modern culture fundamentally changed humankind's biosocial and biocultural relationships, not only bringing about new challenges but also offering new opportunities for the future. The achievements of science and technology increasingly allow humans to master biosocial and environmental processes. In combination with the humanistic principles of the Enlightenment, modernisation may result in a sustained enhancement of the quality of life, provided that the sources of biosocial variation and ecological equilibria are understood and maladaptive practices are countered.

[108]Newson and Richerson (2009) argue with their 'kin influence hypothesis' that the leading causal variable of modernity is a marked change in social network structure leading to a lower ratio of relatives to non-relatives.
[109]Hayek (1979), Richerson and Boyd (2005, 230).
[110]Wilson (1978, 89).
[111]Enlightenment: intellectual movement in Europe in the eighteenth century that emphasized the use of reason and the scientific method. It advanced ideals such as individual liberty, progress, tolerance, fraternity, constitutional government, and separation of church and state (Habermas 1978, 2; Heilbroner 1995, 58; Bruce 2002, 2).
[112]Pinker (2011, 56, 59, 129).
[113]McCloskey (2016).
[114]Monod (1970, 185).
[115]"Modern societies are built on science. They owe their wealth, power and the certainty that, tomorrow, still far greater wealth and power will, if he wants so, be accessible to the Human."

Contrary to some doom- or scare-mongering individuals and groups who are of the view that modernity leads to a degeneration, de-humanisation or even the end of humankind,[116] there is ample evidence that modernity has considerably improved material and spiritual living conditions, and consequently also the potential for morality of the human species. Modernity succeeded in freeing the human from the highly restrictive and oppressive cages of extended kinship and agrarian society, thus allowing for a richer fulfilment of human-specific individual and social potentialities.[117]

Indeed, never before in human history have so many people, at least in privileged parts of the world, lived decent lives, free from poverty, hunger, illness, disaster, misfortune, suppression, hard labour, drudgery, war, ignorance and superstition.[118] Nevertheless, this positive view of the achievements of modernity does not blind the authors to the many shortcomings and derailments of modern culture, with its new technogenic risks (ABC warfare[119] and GNR derailments[120]), its dysgenic consequences, its failure to avoid non-contagious diseases, its poor mastering of exorbitant social inequalities, especially between nations, its persistent in-group oriented tribalism and tense inter-group relations, its belated population growth control, its improper use and abuse of natural resources, its decimation of biodiversity, its pollution of the environment and anthropogenic climate change, and, last but not least, its incapacity to adequately master biological drives such as greed, aggressiveness, envy, jealousy, spite and hatred, some of which might have had evolutionary advantages in the EEA, but currently endanger our future development and evolution.

The persistence of this mixture of old and new challenges to a safe and further progressing future is largely due to our inability to adapt swiftly and sufficiently to the novel environment of modernity, and we continue to behave as we did in the past, driven by our basic instincts of selfishness, greed, and multiplication that were adaptive in pre-modern living conditions.

Indeed, humanity has not had sufficient time to adapt to its novel environment of modernity. As Nick Bostrom[121] states:

> Human nature is in an evolutionary disequilibrium; our evolved dispositions are not adapted to the contemporary fitness landscape and do not maximize the inclusive fitness of current individuals.

In addition, modern societies still function partly on the basis of values and norms that emerged and were adaptive in pre-scientific living conditions, but are no longer adapted to the novel environment of modernity. In this respect, Marc Hauser[122] rightly pointed out that many norms prescribed by traditional law or

[116]For instance, Chauchard (1959, 41), Kass (2002), Sandel (2007).
[117]Maryanski and Turner (1992), Veenhoven (2005).
[118]See also Kurzweil (2005, 396, 408).
[119]ABC: atomic, biological and chemical weapons (see, for instance, Croddy et al. 2004).
[120]GNR: advanced technologies of the genetics, nanotechnology and robotics revolution (see Joy 2000; Mulhall 2002; Kurzweil 2005).
[121]Bostrom (2004, 339), see also Richerson and Boyd (2005, 230).
[122]Hauser (2006, 423).

religion, or both, for instance topics in the field of bioethics such as contraception, abortion and euthanasia, clash with some of our innate moral intuitions that influence many of our decisions.

Hence, it is not surprising that many people are seriously worried about the future of modern culture that, despite its magnificent achievements, faces considerable challenges in multiple domains. As Robert Heilbroner saliently observes in his book *Visions of the Future,* the distant past evoked feelings of resignation and yesterday was characterised by optimism, today gives rise to anxiety.[123]

The characteristics of the modernisation process allow us to distinguish between various possible future orientations and to a rational choice about the desirable direction to go in.

1.4 Confronting Hominisation with Modernisation

The divergence between the human genetic predispositions, that are still largely adapted to the ancient Pleistocene environment in which humans emerged and the present evolutionarily novel environment of modernity, requires cultural adaptation via technological interventions, on the one hand, and value changes, on the other hand, because the natural evolutionary mechanism is unable to produce the necessary genetic adaptations quickly enough.[124]

Many challenges confronting modern societies today, or which can be anticipated in the near future, may result from the fact that the human body—particularly the human brain with its psychological mechanisms that specifically evolved as adaptations to Pleistocene living conditions—is in many respects no longer well adapted to the powerful process of increasingly rapid cultural change that started with the appearance of agriculture some 10,000–12,000 years ago, and that accelerated tremendously with the emergence of modern culture some 400–500 years ago. This resulted in a significant discrepancy between the original Environment of Evolutionary Adaptedness (EEA) and the evolutionary novel environment created by science. It is important to keep in mind that the EEA era covered 95% of the total time of existence of *Homo sapiens sapiens.* If the major former hominin stage in human evolution—the *Homo erectus* stage—is included in the calculation, the EEA era of existence accounts for up to 99%.

Although there exists today considerable between-country diversity in the degree or stage of achievement of science- and technology-driven modernisation, it can be observed that modernisation is seizing virtually all nations and cultures on the planet, perhaps with the exception of some remote and isolated hunter-gatherer populations. The processes that advanced industrial societies have experienced are,

[123]Heilbroner (1995, 95).

[124]However, not only biological evolution and modern cultural development evolve at different speeds, also the technological and social dimensions of modernity evolve asynchronously and unevenly. As Glover (1984, 186) stated: "Our present wave of problems exists because modern physical technology has come too early in our social development."

therefore, also of importance to developing countries, many of which are experiencing a much more rapid modernisation than Western countries ever did.

Through the confrontation of the hominisation and modernisation processes, this book endeavours to deal with the necessary value and norm changes needed to adapt to the opportunities and challenges of a further progressing modernisation process and to evolve to higher levels of hominisation, being well aware that it is not possible—in the short run—to fundamentally change the human genetic predispositions which we inherited from Pleistocene adaptive processes. Notwithstanding the many obvious advantages modernity provides for, it may, as Bjørn Grinde[125] points out, in a number of respects, be an unnatural environment that causes people to behave abnormally.

Modernisation entails a matured and further progressing cultural stage, based on the application of science and technology allowing the optimal development of human-specific potentialities and enhancing quality of life. Looking forward requires reflection about ways to overcome present-day threats of biodiversity extinction, depletion of natural resources, overpopulation, underdevelopment, overconsumption, environmental pollution, anthropogenic climate change, ABC-weaponry threats and neo-colonial exploitation.

The mediating role of technology is having a transformative role for shared values and norms. However, forward-looking choices to be made are not technical/technological in nature. They are ethical.

The authors share Milton Rokeach's[126] understanding of values as enduring beliefs about specific modes of conduct or end states of existence that are personally or socially preferable. Norms are behavioural rules according to which one ought to behave. Morality refers to values and principles of conduct held by individuals and/or groups. The ensemble of values and norms in a culture constitutes its ethics, a domain of human thought that deals with good and bad.[127]

Modern societies, notwithstanding their evolving ideologically pluralistic features and growing tolerance towards outliers and anomaly, continue to be strongly polarised along lines of religious and secular ideological struggles. As a result, many new ethical and political challenges remain unresolved or are inadequately addressed. In this book it is argued that theistic belief systems are, in many respects, no longer well adapted to the requirements of the novel environment. They do not trace pathways for managing future human evolution and cultural adaptation and development. Theistic beliefs tend to be more oriented towards buffering change than managing it. Most secular ideologies usually deal with humanity's present controversies by piecemeal development and address forward-looking in a too fragmented, conflicting and short-term perspective. They are usually limited to meeting the needs of specific population sub-groups and rarely address possible changes as they relate to the human species.

[125]Grinde (1996).
[126]Rokeach (1973).
[127]Broom (2004; 2006), Adams (2005).

Dealing with the ethical challenges of humankind at the turn of the twenty-first century, and safely guiding the human species through new subsequent stages of biological evolution and adaptation and cultural development, requires rethinking of our values and norms in a longer-term perspective and at the planetary level. By confronting the hominisation and modernisation processes, later chapters reveal the necessary changes in values and norms in consideration of adapting to a further progressing modernisation and evolving toward higher levels of hominisation.

1.5 The Time Dimension: The Third Millennium

Futurologists[128] focus their attention mostly on a period of between five and fifty years. The reason is that the immediate future (less than five years) belongs to the domain of daily care: government-terms rarely exceed that time, at least in democratically ruled countries. The period over fifty years is also usually disregarded because it is expected that so many changes will occur that long-term prediction and planning are too uncertain.[129]

Evolution scientists use an immensely broader time perspective than futurologists. Bio-anthropologists, who study the origin and the present and future evolution of the hominins, take a much longer-term perspective. Towards the past, this includes the full history of the hominins, extended over a period of several million years; for the present, this includes the study of the specificity, variability, and changeability of the currently living members of the hominin tribe, *Homo sapiens sapiens;* for the future, bio-anthropologists are interested in the further evolution of humankind, a time span that also can include many thousands, if not millions of years.

This book looks at ecological, biological and cultural developments in the coming decades up to the end of twenty-first century without losing sight of a longer-term evolutionary perspective of this millennium.

[128]Futurology: study that deals with future possibilities based on current trends.
[129]Cornish (1977).

2 Origin and Evolution of Morality

Abstract
This chapter starts with a discussion of the role of evolutionary mechanisms in the development of predispositions to moral behaviour and the development of moral values and norms. Next, the evolutionary background of morality is dealt with. The major stages of evolutionary ethics as a scientific discipline are reviewed and the biological bases and causes of morality are discussed. Then an overview is given of the major stages in the evolution and historical development of morality and the content of moral systems. Finally, the genetic and neurological determinants of moral behaviour are addressed.

2.1 Evolutionary Mechanisms Producing Predispositions to Morality

Evolution science includes not only knowledge about how life, including human life, evolved on our planet in the course of time, but it also includes knowledge about the basic mechanisms that allow life to evolve—mutation, various forms of selection, migration, drift and partner choice.

Genes producing the capacity for developing innate moral sentiments and moral reasoning, as well as cultural processes resulting in moral codes that have a life-sustaining or life-reproducing effect, are subject to mutation and various forms of selection, and may also be influenced by chance fluctuations and migration. In other words, all of the known basic evolutionary mechanisms are, or may be, involved in the evolution of the two enabling components—biological and cultural —of human morality.[1]

[1]Cavalli-Sforza and Feldman (1981), McKenzie Alexander (2007).

R. Cliquet and D. Avramov, *Evolution Science and Ethics in the Third Millennium*,
https://doi.org/10.1007/978-3-319-73090-5_2

Biological and cultural factors involved in the production of human morality are, in other words, not different from other biological or cultural characteristics related to life-sustaining or life-reproducing processes. This view is founded on the Dual Inheritance Theory (DIT)[2] which explains the biocultural co-evolution of human behaviour as a product result of two different and interacting evolutionary processes: genetic evolution and cultural change. Many ethical choices are, just like cultural innovations in general and physical biological features, dependent on the evolutionary mechanism, and specifically Darwinian selection.[3] Human morality evolved to contribute solving adaptive problems and achieving adaptive goals.[4] Moral systems are, in fact, cultural instruments that serve the same goal as biological organ systems, namely to promote ontogenetic and phylogenetic adaptation.[5]

The present scientific insight into the evolutionary process explains why a morally 'blind' mechanism was able to produce a purposefully oriented human morality with strongly universal, objective moral standards.[6]

2.1.1 Mutation

2.1.1.1 Genetic Mutation

Genetic mutation is a change in the chemical structure—the DNA—of a gene or a group of genes. Mutations are at the root of genetic variability and, consequently, form the basic condition for possible changes in the genetic composition of a population. Neutral mutations are evolutionarily neither advantageous nor disadvantageous; they can be responsible for some genetic variation between individuals. Harmful mutations can also be present in a population: this is either because they have been newly introduced into the gene pool[7] or because they have not been completely eliminated by selection and, hence, were transmitted from earlier generations. Even beneficial mutations can be responsible for a certain amount of genetic variation, because the effect of selection that is responsible for their distribution in the population is not yet completed. Favourable genetic variants can spread in a gene pool, because they had a relative selective advantage in former generations due to environmental circumstances. In humans some less favourable mutations can thrive, because they have sufficient survival value in society's sheltered cultural or economical conditions, or are even fostered by such conditions.

Genetic mutations are responsible for the neurological, hormonal, and other biological changes hominins underwent in the course of the hominisation process, so as to make several parts of the human brain more susceptible to develop or experience moral sentiments and to be able to produce moral ideas.

[2]Boyd and Richerson (1985), Durham (1991), Henrich and McElreath (2007).
[3]Bajema (1978), Alexander (1979), Cavalli-Sforza and Feldman (1981), Lumsden and Wilson (1981), Boyd and Richerson (1985).
[4]See also Krebs (2011, 257).
[5]See also Ruse (1999, 241).
[6]See also Talbott (2015, 707).
[7]Gene pool: the whole of the genes present in a reproductive community or a population.

2.1.1.2 Cultural Mutants

Cultural analogues of genetic mutations have been given several names, the best known of which is Richard Dawkins' 'meme' concept.[8] In the domain of values and norms, new ideas are about how to change individual life and re-organise social life —the relations between the sexes and age groups, between social classes, between in-groups and out-groups—in order to adapt to new environmental or social challenges. They also need to include ideas about how to relate the human species to the evolving planetary biosphere, the planet itself, and the cosmos to which we belong. Well-known examples of complex moral innovations of great social importance were the in-group transcendent norms of early Christianity and Enlightenment humanism.

An important difference between genetic mutations and cultural innovations is that the first are, as far as is known, purely chance phenomena, meaning that they are probabilistic rather than deterministic in nature, whilst the second are more often the consequence of non-chance events. Inventions are often the result of a conscious search operation, which means that chance might also be involved. The non-chance, consciously or unconsciously directional, transformative or reconstructive character of cultural innovations[9] has important consequences for the frequency of cultural mutants as well as for their spatial dispersion.[10] Whereas biological evolution functions on the basis of Darwinian principles, cultural change occurs on the basis of both Darwinism and Lamarckism. The term Lamarckism refers to Lamarck's[11] theory of the biological inheritance of acquired characteristics. Biology has refuted the Lamarckian principle[12] in favour of Darwinian selection. Furthermore, the recent development of epigenetics—the study of the systems and processes by which genes' expression can be altered and transmitted to next generation(s) without changes in the DNA[13]—has raised the question whether this is a form of Lamarckism which would require a fundamental revision of the Modern Evolutionary Synthesis.[14] Epigenetic processes do not alter genes, they only regulate their expression. So, epigenetic phenomena are, just as ordinary genetic phenomena, subject to selection processes for their long-term transmission and maintenance in the gene pool. Epigenetics cannot be considered as Lamarckism. However, the transfer of culturally acquired competences by means of learning applies very well to intergenerationally transmitted cultural features.[15]

[8]Dawkins (1976, 206).

[9]For a discussion of the different types of cultural innovation and their implications for the evolution of culture, see, for instance, Acerbi and Mesoudi (2015).

[10]Henrich et al. (2008).

[11]Lamarck (1809).

[12]Heard and Martienssen (2014), Penny (2015, 2016). See also our discussion of Extended Evolutionary Synthesis (EES) in Chap. 1, p. 4.

[13]For instance, Allis et al. (2015), Giuliani et al. (2015).

[14]For instance, Skinner (2015).

[15]For instance, Gould (1980, 84): "Human cultural evolution, in strong opposition to our biological history, is Lamarckian in character. What we learn in one generation, we transmit directly by teaching and writing."

Another important difference between genetic and cultural mutants is that the former can only be transmitted via biological parents, whilst the latter can also be transmitted—vertically as well as horizontally and obliquely—by non-parents. This is the reason why cultural change can progress so much faster than biological evolution.

Hence, at the mutational level, one can already see that co-evolution between biological and cultural mutants may be present to a certain degree: moral ideas cannot be produced if the required brain capacity to produce them and the sensitivity to accept them are not available. Conversely, cultural innovations, also in the domain of morality, may change the direction or strength of selective processes on biological predispositions facilitating moral behaviour.

2.1.2 Selection

In general terms, the evolutionary concept of selection refers to the differential reproduction of genes and memes. Positive selection results in the preservation or even increase of particular genetic or cultural variants, whilst negative selection leads to a decrease, and eventually the elimination, of the targeted genes or memes.

In the evolutionary discourse about the effects of selection, two different but interrelated concepts are used: reproductive fitness and adaptation/maladaptation. Reproductive fitness, sometimes called Darwinian fitness or genetic fitness, refers to the degree that a genetic or cultural variant is differentially reproduced and transmitted to subsequent generations. It is, in fact, a quantitative indicator for the degree to which a trait is intergenerationally decreased, preserved or increased. The concept of adaptation/maladaptation refers to the degree to which an organism succeeds or fails, in a long-term perspective, in ontogenetically developing and phylogenetically evolving well-adjusted characteristics in its (changing) environment.

Hence, selection has a quantitative (reproduction) and a qualitative (adaptation) dimension. In the evolutionary evaluation of biological or cultural changes, often only the quantitative aspect of the change and its short-term dimension are considered, whereas the qualitative aspect and long-term dimension are left out of scope. A classical example is the demographic transition where the shift from high to low fertility is often said to be a maladaptive trait,[16] because it would violate the maximisation of inclusive fitness principle.[17]

Considering both the quantitative and qualitative aspects of evolutionary change is particularly important for the human species with its enormous environmentally intervening potentialities. Not only the numerical reproduction of human genes or individuals but also the degree to which the ontogenetic and phylogenetic

[16]See, for instance, Hill (1984), Vining (1986), Pérusse (1993), Borgerhoff Mulder (1998), Richerson and Boyd (2005, 149, 169); see also Chap. 8, Sect. 8.2.2.1.

[17]Maximisation of inclusive fitness: theory which asserts that humans, like other organisms, developed through natural selection evolved behavioural tendencies in order to maximise their genetic representation in future generations in the context of constraints set by the environment and their phylogenetic past (Hamilton 1964).

development of human-specific potentialities is being achieved, should be duly taken into consideration. Moreover, reproductive fitness should be evaluated in combination and interaction with the degree to which the human succeeds in mastering its environment and achieving long-term ecological sustainability, resulting in what people would usually consider as enhancing quality of life, well-being, welfare, and happiness.[18]

In evolutionary biology several forms of selection are distinguished, e.g. natural selection, sexual selection, kin selection, group selection, social selection. In the domain of culture we speak of cultural selection. Biological and cultural forms of selection resulted in the formulation of a universal selection theory[19] or a holistic Darwinian theory.[20] This means that Darwin's theory and explanatory mechanism —the presence of blind variation and the selective retention effect thereupon—is not only applicable to biological phenomena but also to cultural, technological and societal phenomena, and not least to moral phenomena.

2.1.2.1 Natural Selection

Natural selection is the well-known mechanism through which genes are differentially transmitted to subsequent generations due to the interaction with the natural environment.

The selection intensity against a genetic variant determines the reproductive fitness of that variant. Reproductive fitness can be achieved via various biodemographic mechanisms, namely differential mating, differential fertility, differential mortality, and differences in generation length.[21]

Whereas single step selection can change the frequency of individual genes or even complete genomes,[22] it is only cumulative selection that can manufacture biological complexity, eventually also resulting in the emergence and evolution of different races and species.[23]

Natural selection has not only been observed in numerous empirical and experimental investigations on a wide variety of species,[24] but it has also been developed in a mathematical theory of selection, expressed by what is now known as The Price Equation, in which intergenerational change is partly ascribed to the action of selection and partly due to other genetic mechanisms or environmental factors that influence the transmission of particular traits.[25]

[18]See also Sartorius (2003, 171) who evaluates 'fitness' not only in terms of relative numbers of genes or individuals, but also in terms of increasing control over Earth's resources, resulting in differential growth in wealth and power.

[19]Universal selection theory: application of Darwinian selection beyond biological processes in order to explain evolutionary processes in a wide variety of other domains (see, for instance, Dawkins 1983; Cziko 1995).

[20]Campbell (1965), Corning (2005).

[21]For instance, Cavalli-Sforza and Bodmer (2013).

[22]For instance, Gintis (2014).

[23]For instance, Dawkins (1986).

[24]For instance, Dawkins (2010).

[25]Price (1970, 1972, 1995), Hamilton (1975), Maynard Smith and Price (1973); see also Frank (1995; 1997; 1998; 2012), Grafen (2000), Gardner (2008), Harman (2011).

2.1.2.2 Cultural Selection

The concept of cultural selection can have two different meanings: it can refer to selection of biological characteristics due to cultural factors, but it can also refer to selective processes pertaining to cultural variables themselves, meaning that one cultural variant is more likely to be intergenerationally reproduced than alternative variants.[26] In the latter case, such differential cultural inheritance can be achieved through purely culturally transmitted mechanisms (learning, coercion), and/or through biodemographic mechanisms (differential mating, fertility, or mortality of the carriers of the cultural traits).[27]

The transgenerational transmission of cultural traits is obviously much more complicated than that of genetic traits, because memes can be transferred through many more people than through parents. Indeed, whereas genes can intergenerationally only be inherited vertically (i.e. through biological parents), memes can also be transmitted horizontally or diagonally, i.e. through other persons or groups. Cultural heritage can even come from ancestors living many generations ago; it can have different effects according to the kind of trait, the type of heir, or the change of its contents in the process. Moreover, socio-cultural phenomena may be subject to such complex developmental changes, both within and between generations. Some researchers raise the question as to whether the current evolutionary mechanistic toolkit and its mathematics can suffice for measuring and fully explaining intergenerational cultural dynamics.[28]

In an analogy to the genetic fitness concept, the term cultural fitness has been devised, aiming to measure the overall transgenerational influence, resulting from the combination of the transmission mechanisms and reproduction intensity. However, cultural fitness is an indicator that, because of its complexity, is much more difficult to quantify than genetic fitness.[29]

In the context of this discourse, it is necessary to reflect more extensively on the concept of selection of cultural and, more particularly, moral ideas. The application of the evolutionary and population genetic framework on cultural variables is less commonly known in ethics and social sciences in general.

Indeed, cultural and in particular moral standards may, much like biological characteristics, be promoted or eliminated by selection. They can be positively selected when they have socially, culturally, or biologically, adaptively advantageous effects. In general, it can be stated that ethical ideas, like genetic variants, are under selective pressure.[30] The prevalence of moral values and norms that promote social, cultural or biological adaptation will increase, whilst moral variants that unfavourably influence societal functioning, or impede the ontogenetic development and genetic survival, will decrease or be eliminated, often together with their inventors or carriers. This applies mainly to values and norms that are, directly or

[26]For instance, Stephen (1882), Keller (1915), Cavalli-Sforza and Feldman (1981), Mesoudi (2011, 64, 79; 2016).

[27]For instance, Henrich (2004).

[28]Laor and Jablonka (2013), Claidiere et al. (2014).

[29]Claidière and Andrè (2012, 15), El Mouden et al. (2014, 233).

[30]See, for example, Chudek and Henrich (2011, 224).

indirectly, of vital importance either for societal or biological intergenerational continuity.

This does not mean that all values and norms that spread or are maintained optimise inclusive fitness or are adaptive. In the same way as some biological characteristics, some values and norms can be neutral; others can even have unfavourable effects—be maladaptive—but, for a variety of reasons, nevertheless temporarily or partially survive.

With regard to biological characteristics, the presence or even persistence of maladaptive traits can be due to a variety of causes. They can be due to unfavourable mutations, which increase morbidity or mortality through which reproduction is partially reduced. Some unfavourable genetic variants may not be immediately eliminated ontogenetically or early in the life course or appear only after the peak of reproductive life. Some harmful genetic variants can have favourable effects in heterozygote combinations, in interaction with other genes, or in particular environmental conditions. Some favourable genetic variants may have become unfavourable due to changing environmental or cultural living conditions.[31]

Some of the biological mechanisms producing maladaptiveness are also applicable to maladaptive secular or religious norms, because compliance with such norms can also produce higher levels of morbidity or mortality, reducing reproductive fitness. Maladaptive cultural traits can nevertheless also persist, because they are compensated or protected by other cultural or even biological characteristics. In pre-scientific cultural stages, when no adequate insights existed into the causes of evolution, many cultural practices spread: some were founded on lucid insight, however, most emanated from superstition and resulted in inefficiency, if not in misery.[32] As William J. Talbott[33] formulated so well, moral systems often include a 'hodgepodge' of social norms, including many norms that were not directly involved in the solution of their biosocial problems, because in prehistorical times—one could even say in pre-Darwinian times—when those systems emerged, people did not understand well their norms as means of solving their problems. Some of these maladaptive behavioural patterns can spread or maintain themselves temporarily, because selection against them is weak or the population fails to respond adequately, or their elimination is biologically or culturally compensated by other factors.[34]

The degree of maladaptation must be evaluated at different levels of organisation —individual, population, generation—and according to time dimensions—short

[31]McGuire et al. (1997, 30), Nesse and Williams (1997, 15).

[32]Many examples of relatively or potentially maladaptive cultural practices are known: drug abuse, alcoholism, smoking, necrophage customs, mutilations (e.g. ritual male circumcision, female genital mutilation, female foot-binding), human sacrifice, celibacy, infanticide, nutritional customs resulting in qualitative starvation, slavery, torture, wife-beatings, rape, witchcraft, environmental pollution, homicide, suicide, dysfunctional food and health care practices that increase infant mortality, reduce life-expectancy and/or lower personal productivity (Cavalli-Sforza and Feldman 1981; Barkow 1989; Edgerton 1992).

[33]Talbott (2015, 700).

[34]For instance, Barkow (1989).

versus long-term—and must be situated in the total ecological context in which the phenomenon prevails. Thus, it is not impossible that particular cultural practices, such as ritual mutilations, which have maladaptive effects at the individual level, may have had some advantageous effects at the population level in earlier cultural eras, for instance through population control, initiation into adulthood or preparation for war.[35]

Nevertheless, a fact remains that many moral codes are in line with the biological predispositions of the human species.[36] This is especially the case where the genetic programming of particular biological predispositions is weak or incomplete, so that moral rules are necessary to complement or strengthen the effects of genetic factors in order to guarantee survival or reproduction. Well-known and evolutionarily well-established examples are moral codes for childcare, incest avoidance, in-group favouritism, altruism, reciprocal altruism and mutualism, sexual attraction/love, reproductive behaviour, and elderly care.

However, the different means and tools of biological evolution and culture, and in particular the different speed at which they can change, results in genetic fitness and cultural fitness not always being positively correlated.[37] They can be antagonistic and provoke serious biosocial stress as can be observed in modernisation.

2.1.2.3 Social Selection

Social selection is a concept often used in social biology, referring to social factors producing reproductive differentials of biological characteristics.[38] Social selection can operate at the group level, for instance, in the case of social class characteristics related to genetic factors and influencing reproductive behaviour[39]; but it can be limited to situations whereby fitness is influenced by the behaviour of other individuals.[40] In this sense, social selection differs from other forms of selection, because it tends to be more reciprocal with several individuals selecting traits in each other.[41]

Social selection is to be distinguished from social assortment[42] through which biological characteristics are differentially distributed over different social groups in a population.

The concept of social selection is often used as a synonym for cultural selection in the sense of selection of biological characteristics due to cultural factors.

---

[35]Kardong (2010, 153).
[36]Durham (1991), Ayala (2009), Teehan (2010), Mouden et al. (2014, 235).
[37]El Mouden et al. (2014, 235).
[38]Broca (1872), Fisher (1930; 1958), Nesse (2009).
[39]Schwidetzky (1950), Retherford and Sewell (1988), Lynn and Harvey (2008).
[40]For instance, in the case of hereditary diseases, mate finding and fertility may be changed considerably by the presence of affected family members (Yokoyama 1983).
[41]Krebs (2011, 60).
[42]Scheidt (1925), Schwidetzky (1950), West-Eberhard (1979). The concept of social selection is often confused with the term social (as)sortment. In some cases the term social selection is used to refer to assortative processes (e.g. Montagu 1950, 331; Strickland and Shetty 1998, 8; Blane et al. 2008), in other cases the concept of sortment in fact refers to selective processes (e.g. in the book 'The Sorting Society' edited by Skene and Thompson 2001).

L. Krzywicki,[43] for instance, uses it in the domain of morality as the elimination of individuals who do not obey particular moral norms, resulting in the fixation of those norms. Another example of a moral factor involved in social selection is Christopher Boehm's theory that group punishment, through aggressive suppression of free-riding deviants, could have changed the human potential for behaving altruistically.[44]

2.1.2.4 Sexual Selection

The specific sexual dimorphic features of a species are the result of sexual selection.[45] Charles Darwin, who initially developed the theory of sexual selection,[46] defined it as an evolutionary mechanism through which individuals acquire, via selection pressures on some of their sex-specific traits, reproductive advantages over other individuals of the same sex, and transmit their characteristics to their descendants of the same sex. Darwin distinguished two complementary components of sexual selection: competition within one sex over members of the other sex, and differential choice by members of one sex for members of the other sex. In 1915 Ronald A. Fisher[47] produced a first genetic analysis of mate selection, developing the concept of runaway sexual selection, a positive feedback mechanism between the selection of favourable traits and the higher mating success for such traits that explains the development of costly male epigamic traits as indicators of the presence of good genes.

Sexual selection is a form of natural selection, but whereas the latter operates through competition for survival, sexual selection operates through competition for reproduction.[48] After Darwin/Fisher, sexual selection fell somewhat into oblivion, but with the Second Darwinian Revolution it revived strongly with theoretical and empirical contributions by several eminent scholars.[49]

Many biological and behavioural differences between the sexes, in particular those relating to reproductive strategies, can be explained by the sexually differentiated relative parental investment in offspring.[50] The strongly investing sex in offspring produces fewer descendants than the weakly investing sex. The strong investors will, consequently, develop a qualitative or K-strategy in order to ensure that each offspring produced has maximal opportunities for survival. The reproductive success of the weakly investing sex, in contrast, will be favoured by producing as many offspring as possible: it will develop a quantitative or r-strategy.[51]

[43] Krzywicki (1951), quoted in Urbanek (1993, 328).
[44] Boehm (2014).
[45] For instance, Hutchinson (1959), Campbell (1972), Cronin (1993), Miller (2000).
[46] Darwin (1859; 1871).
[47] Fisher (1915); see also Fisher (1930).
[48] Miller (2000, 8).
[49] Williams (1966), Trivers (1972), Ghiselin (1974), Zahavi (1975), Maynard Smith (1978), Symons (1979), West-Eberhard (1979), O'Donald (1980), Lande (1981), Kirkpatrick (1982), Pomiankowski (1987), Bulmer (1989), Cronin (1993), Andersson (1994), Buss (1994), Miller (2000).
[50] Trivers (1972).
[51] Daly and Wilson (1978).

These different reproductive strategies lead to different mating strategies. The less investing sex—usually the male sex—tries to get and inseminate as many partners as possible, leading to intra-sexual competition ('male-male competition'). The more investing sex runs more risks. It will consequently be more selective ('female choice'),[52] since its interest is to get partners who are likely to guarantee a high chance of survival for their offspring. Together, male-male competition on the one hand, and female choice on the other, result in an enlargement of the sexual dimorphism of characteristics that fulfil functions in partner choice and adaptation. This double sexual selection leads to sexual differences in morphology, physiology and psychology.

In the course of hominisation, the sexual dimorphism for secondary sexual characteristics gradually decreased, leading—among other characteristics—to the so-called feminisation process of the human male.[53] As men's parental investment in larger-brained and longer-maturing offspring increased, the mating strategy of females and males evolved towards the establishment of more enduring relations and courtship investments, associated with a reduction in male-male competition, decreasing male physical robustness and behavioural aggressiveness, but enhancing the requirements of cooperation and sociability. In particular, the sex role differentiation within enduring family units that developed with group hunting, and in later cultural stages, with other sex differentiated activities (in which the more mobile males got separated from the food gathering and child caring females) required behavioural inclinations as well as normative agreements for mastering sexual competition. Whereas in prehominin species the sexual rank order is enforced by the physical presence of the dominant male(s), in human societies—where adult males and females perform spatially separated activities—biological control systems not only have to change but also have to be partly supplemented by socio-cultural regulations.

The hominisation process was not only characterised by changes in male sexual biology and psychology. Also the hominin female underwent important changes. While the male transformation mainly concerned secondary sex characteristics, the evolutionary changes of the female sex also concerned primary sexual features including concealed ovulation, large breasts, orgasm, multiple erogenous zones, frontal intercourse linked to face-to-face interaction accompanying bipedalism, and menopause.[54]

The changes in sexual dimorphism pertained not only to morphological traits, but also, given the decreasing genetic programming of behaviour during the hominisation process, pertain to the evolution of innate moral sentiments and cultural codes regulating sexual behaviour. Hence, morality has not only evolved through the process of natural selection, but because moral virtues such as altruism,

[52]Beware! The 'female choice' principle does not exclude the existence of ordinary, naturally or socially induced selective processes within the female sex, interacting with and reinforcing the effects of the specific sexual selective processes (see, e.g. Vaillancourt 2013).

[53]Brace (1973), Armelagos and Van Gerven (1980), Hall (1982), Cliquet (1984), Ghesquiere et al. (1985), Steerneman et al. (1992), Plavcan and van Schaik (1997), Cieri et al. (2014).

[54]For instance, Lancaster (1985).

kindness, generosity and cooperativity are sexually attractive, it is likely that their evolution was influence by sexual selection as well.[55] Many human virtues may have evolved in both sexes through mutual mate choice for features such as altruism, kindness, empathy, magnanimity, conscientiousness, agreeableness honesty, self-control, courtship generosity, fidelity, heroism, and parenting abilities.[56]

2.1.2.5 Kin Selection

Of fundamental importance for the understanding of the evolution of social behaviour, in particular of human moral behaviour, are biological concepts and theories regarding the evolution of altruism.

Due to the many different definitions or interpretations that have been given to this concept, it is necessary to specify how it is usually applied in evolutionary biology. Originally coined by Auguste Comte in the nineteenth century as the antithesis to egoism, the term altruism is sometimes used as an ethical principle and sometimes as a behavioural act. In the latter case, it can strongly vary from modest forms of helping behaviour towards others to extreme forms of self-sacrificing deeds for others. In evolutionary theory, the concept of altruism is used in its behavioural meaning, independently of its moral intention that can be neutral, selfless or even selfish. It is defined as behaviour that, ultimately, reduces the reproductive fitness of a cooperating individual compared to the reproductive fitness of individuals who behave selfishly.[57]

Darwin's theory of natural selection states that genetically influenced characteristics in a population maintain or spread themselves by means of a relatively higher reproduction of the carriers of the relevant genes. This concept of Darwinian or reproductive fitness explains the maintenance or spreading of virtually all genes that promote adaptive characteristics, whether they are of a morphological, physiological or psychological nature. However, there is one exception: genes determining or influencing biological characteristics that produce—in interaction with environmental factors—altruistic behaviour. Such behaviour implies that the genes of the altruist, who helps others or sacrifices himself for others, will be more or less reduced in the gene pool of the next generation(s), and will eventually become completely eliminated by natural selection. Nevertheless, altruistic behaviour exists in all social species and clearly appears to have great adaptive advantages. For a long time altruistic behaviour remained a paradox in Darwinian evolutionary theory.

Darwin was aware of the fact that behavioural traits that lead to reproductive self-sacrifice cannot be maintained evolutionarily by means of differential reproductive success. With his characteristic discernment, Darwin hypothesised that,

[55] Cela-Conde (1987), Miller (2000), Nesse (2007), Cela-Conde et al. (2010), Phillips et al. (2010).
[56] Miller (2000, 292; 2008, 219); see also Tang and Ye (2016).
[57] Altruism can, obviously, include any kind of benefit to other individuals at some cost for the altruist—'behavioural altruism', as distinguished from 'reproductive altruism' by Clavien and Chapuisat (2013, 128)—but the insertion of the genetic endowment for the capacity of such behaviour in the individual genome and its spread in the population is dependent upon several evolutionary mechanisms (see also Clavien 2010).

on the basis of the genetic relationship between reproducing and non-reproducing individuals, some kind of family selection occurred whereby such altruistic characteristics nevertheless could spread, but he did not elaborate this idea.

In the 1930s, Ronald Fisher[58] and J.B.S. Haldane[59] suggested that genes which favour altruistic behaviour could be selected for if the beneficiaries would be genetically sufficiently related to the altruist, so that such genes could nevertheless increase.

The definitive breakthrough of this idea occurred in the 1960s when brilliant contributions by William Hamilton[60] developed a fully-fledged theory about kin selection. Hamilton proved, in a formal mathematical way, that the altruistic behaviour of an individual who lowers his/her personal reproductive fitness can be positively selected for when his altruistic behaviour increases the reproductive fitness of related persons to such a degree that his genes are increased in the next generation via related beneficiaries. Hamilton showed that natural selection promotes altruistic behaviour between relatives when the benefit for the recipient relative, weighted for the degree of relatedness,[61] is larger than the cost for the altruist himself. In 1975 Hamilton,[62] building upon the work of George Price,[63] showed that evolution of altruism between relatives is a specific form of group selection. (See also Sect. 2.1.2.8.)

Hamilton developed the inclusive fitness concept by referring to the degree to which genes are transferred to the next generation thanks to the ordinary reproductive fitness of an individual and the fitness of his/her relatives, which is the result of his/her altruistic behaviour. Thus, Hamilton amended the classical theory of natural selection by broadening the concept of fitness from reproductive fitness to inclusive fitness. In this way he gave an explanation for the possible genetic transmission of altruistic behavioural characteristics by means of differential reproductive behaviour of individuals.

John Maynard Smith[64] introduced the term 'kin selection' to Hamilton's theory on inclusive fitness. The kin selection concept is not limited to the effect an altruist has on the reproductive behaviour of relatives. It also includes the reproductive success of the altruist himself. Hamilton's theory also implies that the degree to which the reproductive behaviour of an individual is associated with nepotistic behaviour towards his/her own offspring, the principle of inclusive fitness, extends nepotism to other close relatives.[65]

The kin selection and inclusive fitness concepts are very important for studies embedded in evolutionary theory. They offer a coherent explanation for the genetic transmission of characteristics that promote social behaviour transcending the

[58]Fisher (1930).
[59]Haldane (1932).
[60]Hamilton (1963; 1964).
[61]See Wright (1922).
[62]Hamilton (1975, 141).
[63]Price (1972).
[64]Maynard Smith (1964).
[65]Alexander (1979).

parent-offspring bond. They are empirically well documented in the zoological and cultural anthropological and sociological literature.[66] However, some aspects of the mathematical way of analysis have been challenged.[67] Indeed, the inclusive fitness of a trait is very difficult to measure either in the field or in experiments.[68] Moreover, the fitness of a gene always involves the whole genome in which it is imbedded, and more particularly the other genes with which it interacts.[69]

2.1.2.6 Reciprocity Selection

Building upon Hamilton's theory of kin selection, Robert Trivers[70] developed, in his paper on *The Evolution of Reciprocal Altruism*, an evolutionary model explaining the occurrence of altruistic behaviour between non-relatives, thus extending the evolutionary theory of altruism from kin to non-kin. Reciprocal altruistic behaviour changes not only the social relations in a population, but can also influence, via differential reproductive effects, the genetic composition of the population. This occurs in the first place regarding the transmission of predispositions for reciprocal altruism, and in the second place for all other traits that are linked to reciprocity. Therefore, the term used here, reciprocity selection,[71] is analogous to the term kin selection.

The idea of the importance of reciprocity for the evolution of social life had already been advanced by earlier authors—among others Charles Darwin himself and especially Peter Kropotkin[72] for whom mutual aid represented an important element for survival and progressive social development. However, it is only with Robert Trivers' contribution of 1971 that this question was dealt with in a more extensive and thorough way, and that a theory was elaborated which coherently links up with the present-day evolutionary synthesis. Moreover, Trivers' contribution represented the point of departure for a large number of other scholars, such as Richard Alexander, Mary West-Eberhard, Robert Axelrod, William Hamilton, Lee Alan Dugatkin, David S. Wilson, Elliot Sober, Martin A. Nowak, Karl Sigmund, Herbert Gintis, Samuel Bowles,[73] to further elaborate or refine the evolutionary study of reciprocity, also partly on the basis of economic game experiments,[74] and to develop it into a genuine evolutionary theory of cooperation. (See also Sect. 1.2.8.) Countless contributions have meanwhile been produced,

[66]For an overview, see Dugatkin (2006, 123–141); see also, Essock-Vitale and McGuire (1980), Boehm (1999), Fry (2006).
[67]Nowak et al. (2010), Allen et al. (2013).
[68]Nowak and Highfield (2011, 109).
[69]Gintis (2014, 494).
[70]Trivers (1971).
[71]See also Boorman and Levitt (1973), Hamilton (1977).
[72]Kropotkin (1902); see also West-Eberhard (1975).
[73]Alexander (1974), West-Eberhard (1975), Axelrod and Hamilton (1981), Axelrod (1984; 2001), Dugatkin et al. (1992), Wilson and Sober (1994), Nowak and Sigmund (1998; 2005), Gintis (2000), Bowles and Gintis (2011).
[74]For example, Axelrod and Hamilton (1981), Henrich et al. (2001; 2004; 2010), Fehr et al. (2002; 2003; 2004). For general overviews of game theory in evolutionary processes, see Maynard Smith (1982), Gintis (2000), Barash (2003), Bowles and Gintis (2011).

further refining or specifying the population genetic mechanisms that favour the origin and persistence of intraspecific cooperation in humans and other species.[75] Recently, Martin A. Nowak[76] distinguished and compared five possible mechanisms for the evolution of cooperation—kin selection,[77] direct reciprocity,[78] indirect reciprocity,[79] network reciprocity,[80] and group selection.[81]

In this context, special mention is deserved for the thesis of Herbert Gintis[82] who proposed and modelled a form of prosocial behaviour that he calls strong reciprocity; he introduced the epithet *Homo reciprocans* for this type of behaviour.[83] Mainly based on game experiments, Gintis argues that people behave prosocially and punish anti-social behaviour to the detriment of themselves, even if the likelihood of future interactions is negligible. Gintis et al.[84] take into account different sources of knowledge, such as the demographic, ecological and social living conditions prevailing in Pleistocene times, empirical data on the moral values and norms of current-day hunter-gatherers and present modern populations, and the results of economic game experiments in well controlled laboratory conditions. They conclude that simple reciprocal interactions, let alone kin selection,[85] would not suffice to explain the origin of the specifically human morality in which moral values are treated as ends in themselves, rather than to just promote the interests of their proponents. The efforts that crime victims make to ensure that offenders are punished, or the selfless morally inspired actions many people get engaged in with the aim of changing society's norms and policies, are mentioned as examples of prevailing forms of strong reciprocity in everyday social life.

The theory about strong reciprocity has been subject to criticism,[86] not because selfless cooperation or altruistic punishment is being denied, but because it is explained by means of group selection instead of one of the evolutionary mechanisms that focus on individual selection. It has also been challenged, because there is no strong empirical evidence from field studies that uncoordinated costly material punishment is used in small societies, except in the regulation of sexual conflict.[87] Moreover, some evolutionary game experiments show that other forms of

[75]For an overview, see for instance, Lehmann and Keller (2006), Corning (2008), Bowles and Gintis (2011).
[76]Nowak (2006).
[77]Hamilton (1964).
[78]Trivers (1971), Axelrod (1984).
[79]Alexander (1987, 93, 94) defined indirect reciprocity as "reciprocity occurring in the presence of interested audiences—groups of people who continually evaluate the members of their society as possible future interactants from whom they would like to gain more than they lose"; see also Boyd and Richerson (1989), Nowak and Sigmund (1998; 2005).
[80]Nowak and May (1992), Ohtsuki et al. (2006).
[81]Wilson (1975), Wade (1978), Wilson and Sober (1994), Landa and Wilson (2008).
[82]Gintis (2000); see also Fehr et al. (2002), Henrich et al. (2004), Gintis et al. (2008).
[83]Gintis (2000, 251).
[84]Gintis et al. (2008); see also Gächter and Herrmann (2006), Bowles and Gintis (2011).
[85]See also Tomasello (2009, 52).
[86]For instance, Burnham and Johnson (2005), Trivers (2006, 79ff), Nowak and Highfield (2011, 224ff), Yamagishi et al. (2012).
[87]Guala (2012).

behaviour, such as cooperative success[88] or reward,[89] are more effective than punishment in promoting cooperative behaviour.

Notwithstanding the differences about the ultimate explanations for the type of selective processes which resulted in our capacity for cooperative altruistic behaviour, it must be acknowledged that evolutionary game experiments about reciprocity have considerably enriched our knowledge and insight about social interactions. They have also confirmed experimentally knowledge which was available via other research procedures—evolutionary theory, computer simulations, and above all, empirical observations in natural settings. Relevant findings of experimental games are: status quo is preferred over any alternative strategy; people cooperate even in one-shot game experiments; people favour short-term gains that entail long-term losses; cooperation increases strongly in stable groups; people cooperate more with partners than strangers; people are about twice as averse to taking losses as to enjoying an equal level of gains and are displeased when subjected to relative deprivation; free-riders[90] evoke negative emotions; altruistic punishment and reputation enhances cooperation; people respond strongly to increased costs of punishment.[91]

Hence, biological predispositions and cultural practices to detect, expose and punish selfish cheaters and free-riders are related to the mechanisms of kin selection and reciprocity selection, resulting in sensitivity to the needs of others. Such predispositions and practices are a way of establishing (reciprocal) altruistic behaviour.[92]

In recent years, many researchers have continued, mainly by means of mathematical modelling and evolutionary game experiments, to further explore and nuance reciprocity theory by identifying specific factors and conditions under which various forms of reciprocal altruism—strong reciprocity, indirect reciprocity—can be selected for and explain the evolution of human sociality.[93]

In particular, the recent progress in the study of reciprocity behaviour has led to a shift in focus from altruistic to mutualistic cooperation.[94] In sociobiological theory altruism and mutualism are often distinguished, whereby altruism is behaviour that involves a cost to the actor's survival or reproduction, while mutualism usually refers merely to (selfish) cooperation and mutual benefit for the actors involved.[95]

[88]Dreber et al. (2008).
[89]Rand et al. (2009).
[90]Free-riders: people with exploitative motives (Delton and Krasnow 2015, 23).
[91]Gächter and Herrmann (2006, 300); see also Gintis (2000, 245ff), Bowles and Gintis (2011), Nowak and Highfield (2011).
[92]For instance, Fehr and Gächter (2000), Herrmann et al. (2008), Shimao and Nakamaru (2013).
[93]André (2010), Bravo (2010), Fehr and Schneider (2010), Iwagami and Masuda (2010), Saavedra et al. (2010), Smead (2010), Barta et al. (2011), Chiang et al. (2011), Delton et al. (2011), Krupp et al. (2011), Pena et al. (2011), Sigmund (2012), Vollan (2012), Berger (2013), DeScioli and Krishna (2013), Jaeggi and Gurven (2013), Phelps (2013), Suzuki and Kimura (2013), Sylwester and Roberts (2013).
[94]Baumard et al. (2013), Forber and Smead (2015), Tomasello et al. (2012).
[95]For instance, Wenegrat (1990, 24), Joyce (2006, 13), Bowles and Gintis (2011, 2), Baumard et al. (2013, 61).

Mutualism is a term that is not only used for intraspecific forms of reciprocal altruism but also for interspecific forms of cooperation.[96]

Mutually beneficial social behaviour is not only less costly than altruistic behaviour, but it is also less sensitive to cheating. An important difference between reciprocal altruistic behaviour and mutualism is that the first mainly relies on partner control, whereas the latter is based on assortative partner choice that is, as such, a form of social selection.[97] Mutualistic cooperation would therefore be evolutionarily more stable and facilitate more varied and complex prosocial behaviour.[98]

The evolutionary theory about reciprocal altruism/mutualism is a pertinent example of the emergence and evolution of a biological predisposition that is related to a universally prevailing cultural precept called the Golden Rule that is not only in line with but also reinforces the biological predisposition.[99]

2.1.2.7 Coercive Selection

In socially hierarchical societies, social cooperation can be induced by yet another mechanism, namely the forcing of group members to such behaviour. Daniel Krebs[100] speaks in such cases about forced altruism. Indeed, the importance of social coercion in the development of helping behaviour should not be underestimated, especially not in the human species. Social coercion can be used in two opposite ways: strengthening the social position of dominant individuals leading to increased forms of social inequalities, or weakening the potential abuse of rulers, cheaters and free-riders. The latter proximately results in the strengthening of cooperation and social cohesion, and ultimately enhances the reproductive fitness of the punishers and/or diminishes the fitness of cheaters and free-riders.

Pierre Van den Berghe[101] is of the view that social coercion is to be considered one of the three major foundations—together with kin selection and reciprocity selection—of human social life. Although coercion and deference to coercion are ubiquitous in social animals in the form of dominance hierarchies,[102] it is a behavioural pattern that is specific to the human species. Social coercion of conspecifics,[103] resulting in intra-species parasitism, is a unique phenomenon for the human species. In the animal kingdom slavery prevails among some ant species, but the enslaving only concerns other species, not the same one.[104] In the relatively egalitarian foraging hunter-gatherer societies social coercion, through moralistic aggression, kept alpha-male types from dominating group life.[105] With the

[96]For instance, Bergstrom, in Hammerstein (2003, 241ff).
[97]Eshel and Cavalli-Sforza (1982), Baumard et al. (2013, 61).
[98]Forber and Smead (2015, 414).
[99]For instance, Wilkins and Thurner (2010), Nowak and Highfield (2011, 273), Goodman (2014).
[100]Krebs (1983, 65); see also Richerson et al. (2003, 373) who use the term coercive dominance.
[101]Van den Berghe (1979, 15, 16).
[102]Clutton-Brock and Parker (1995).
[103]Conspecifics: organisms belonging to the same species.
[104]Wilson (1975).
[105]Boehm (1999, 207).

development of the socially strongly stratified societies in the agrarian and early industrial eras social coercion developed very strongly and widely.

Although social coercion as a factor in community formation differs from kin selection and reciprocity selection, it may have as an ultimate effect that it allows for a higher reproductive fitness, either for the dominant individuals or for the group as a whole. Therefore, the term used here is coercive selection.

Punishment—or menace of punishment—is an important means to enforce socially desirable behaviour, and in particular cooperative behaviour in society. This is the reason why scholars started using the concept of altruistic punishment,[106] a somewhat antithetical term, because punishing cheaters and free riders is ultimately a selfish rather than an altruistic deed.

There already exists an impressive body of theoretical,[107] empirical,[108] and experimental research[109] showing that punishment of non-cooperators, free-riders and cheaters has not only a deterrent effect, but it promotes cooperative behaviour. Punishment also sustains large-scale cooperation in intergroup conflict and warfare.[110] Altruistic punishment occurs even when there is a high cost for the punishers,[111] although other factors and conditions, such as cost-to-impact ratio,[112] reputation,[113] gossip,[114] trust,[115] sympathy,[116] reward treatment,[117] inequity aversion,[118] spite,[119] envy,[120] and antisocial punishment,[121] may be involved.

Punishment of selfish behaviour seems to be present in all human cultures.[122] It is even found in extant mobile hunter-gatherer cultures—similar to our Palaeolithic ancestors—where they help to maintain a relatively egalitarian society.[123] Punishment is even common in animal societies where it is used to invigorate dominance, repress or restrain cheating, control offspring or sexual partners, and strengthen cooperative behaviour.[124] Punishment suggests the possibility of the evolution of 'moralistic' strategies in which punishers punish not only reluctant cooperators but also others who fail to cooperate and even those who fail to punish

[106]Fehr and Gächter (2002), Boyd et al. (2003), de Quervain et al. (2004), Fowler (2005).

[107]For instance, Frey and Rusch (2012), Guala (2012), Shimao and Nakamaru (2013).

[108]For instance, Boehm (1993), Henrich et al. (2001).

[109]For instance, Gürerk et al. (2006), Pedersen et al. (2013), Przepiorka and Diekmann (2013).

[110]Mathew and Boyd (2011), Gneezy and Fessler (2012).

[111]Henrich et al. (2006), Hauert et al. (2007).

[112]Egas and Riedl (2008).

[113]For instance, Rockenbach and Milinski (2006), Dos Santos et al. (2013), Kroupa (2014).

[114]Kroupa (2014).

[115]For instance, Balliet and Van Lange (2013).

[116]Ye et al. (2011).

[117]Choi and Ahn (2013).

[118]Hetzer (2013), Bone and Raihani (2015).

[119]Nakamaru and Iwasa (2007), Jensen (2010).

[120]Pedersen et al. (2013), Bone and Raihani (2015).

[121]Hermann et al. (2008), Rand et al. (2010), Powers et al. (2012).

[122]Fehr and Gächter (2000), Herrmann et al. (2008), Bowles and Gintis (2011).

[123]Boehm (1999, 249).

[124]Clutton-Brock and Parker (1995), Frank (1995), Boyd et al. (2003).

non-cooperators.[125] People take pleasure from punishing norm violations as can be seen in the brain activity associated with punishing defectors.[126] Nevertheless, humans prefer pool punishment to peer punishment.[127] Altruistic punishment is of particular importance in large societies of non-related people where kin selection and reciprocal selection perform less well,[128] but where institutional sanctioning, either in the form of beliefs in moralising Gods[129] or by secular authorities,[130] became a major coercive force.[131] Overall, social coercion in the form of altruistic punishment is an important explanatory factor for understanding the evolution of human cooperation and the rationale of ethics.

2.1.2.8 Group Selection

Group selection is an evolutionary mechanism through which natural selection produces differences in reproductive fitness between groups.

In the past decades this type of selection has been the subject of a fairly sharp scientific controversy.[132] This was probably due to an insufficient distinction between within-group selection, which has rightly been rejected as the explanation for the evolution of altruistic behaviour, and between-group selection that can be a powerful mechanism of selection between groups or populations and lead to the replacement of less successful groups by more successful ones.[133] It may also be related to an insufficient distinction between genuine altruistic behaviour that involves a fitness cost for the altruists and selfish cooperative group behaviour that nevertheless benefits all group members.[134] It is not impossible that the controversy was also partly fuelled by subtle, though often unconscious, differences in ideological views, more particularly concerning the primacy of views about individualism and collectivism.

In 1970, Michael E. Price[135] showed, with what is now known as the Price Equation, that altruism could spread via covariance between group benefit and the frequency of altruists in the group. Price's Equation is composed of a term that measures the contribution of within-group selection and a term that targets between-group selection by measuring the covariance between a character and its

[125] Richerson and Boyd (1997).

[126] de Quervain et al. (2004), Camerer et al. (2005), Strobel et al. (2011).

[127] Traulsen et al. (2012).

[128] Boyd et al. (2003), Nakamaru and Iwasa (2006), Marlowe et al. (2008).

[129] See also the discussion on 'supernatural punishment' in Chap. 3, Sect. 3.6.1.2.

[130] Yilmaz and Bahçekapili (2016).

[131] For instance, Marlowe et al. (2008), Baldassarri and Grossman (2011), Kuemmerli (2011), Isakov and Rand (2012).

[132] Wynne-Edwards (1962), Williams (1966), Wade (1978), Trivers (1985) Dugatkin and Reeve (1994), Wilson and Sober (1994), Nesse (1994) Sober and Wilson (1998), Borrello (2005), Boyd and Richerson (2007), Corning (2008), Leigh (2010).

[133] This form of between-group selection (Maynard Smith 1964; Leigh 1983) has been referred to by West et al. (2007) as 'old' group selection and by Molleman et al. (2013) as 'replacement' group selection.

[134] Andre and Morin (2011, 2538).

[135] Price (1970; 1972); see also Hamilton (1975), Frank (1995; 2012), Sober and Wilson (1998), Traulsen et al. (2005).

fitness, thus mathematically formalising the multilevel character of natural selection.

Regarding within-group selection, it is beyond doubt that Darwin's theory of natural selection concerns the individual—or even the gene—level of organisation. Darwin only hypothesised for human morality that selection operates at the population level. Probably, the abuse of the theory of individually oriented selection, that was made in the so-called social-Darwinist discourse,[136] partly caused the essence of Darwin's theory about the level at which selection operates to fade away somewhat. The idea spread that selection, particularly for the transmission of altruistic behaviour, operates for the good of the community.[137] However, as an explanatory mechanism for altruism, within-group selection raises a fundamental theoretical problem which was addressed by several leading scholars who initiated what is now called the Second Darwinian Revolution:[138] genetic mutations producing altruistic, i.e. self-sacrificing, behaviour are out-selected in a population consisting of selfish behaving individuals. Within-group selection based on altruistic behaviour is not an evolutionarily stable strategy, because mutations that promote selfish rather than altruistic behaviour will increase through natural selection.

However, genes for altruistic behaviour can be positively selected through group selection in situations where a population is divided into several subgroups.[139] An altruistic gene that is being selected against within its group will nevertheless increase in frequency when its group grows more strongly than other groups. This is thanks to the action of the carriers of the altruistic gene displaying behaviours beneficial to their group, for instance due to stronger cooperative actions or moralistic policing of cheats and shirkers.[140] Due to what is called Simpson's paradox,[141] the share of group beneficial traits in the total population can increase in frequency. Individual competition within groups and group competition between groups can have opposite effects: selection on the lower level (within groups) favours selfish individuals, whereas selection on the higher level (between groups) favours cooperating individuals.

Several scholars have drawn attention to the possible reinforcing role of social assortment[142] through which cooperative individuals tend to group and act together, thus increasing the likelihood of selection at the group level. In fact, this is

---

[136]See, for instance, the discussions in Hofstadter (1944), Jones (1980), De Tarde (1984), Tort (1992).

[137]For instance, Hauser (2006, 359).

[138]Hamilton (1964; 1996) Maynard Smith (1964; 1982; 1989), Williams (1966), Price (1970), Trivers (1971; 1985), Maynard Smith and Price (1973), Wilson (1975; 1978), Dawkins (1976), Axelrod and Hamilton (1981), Axelrod (1984; 2001).

[139]Haldane (1932), Wright (1945), Williams and Williams (1957), Maynard Smith (1964), Price (1970; 1972), Hamilton (1975); see the discussion in Sober and Wilson (1998, 55–100); see also Bowles and Gintis (2011, 46ff).

[140]Boehm (1996; 1999, 205ff).

[141]Simpson (1951); see also the discussion in Sober and Wilson (1999, 23–26).

[142]For instance, Eshel and Cavalli-Sforza (1982), Wilson and Dugatkin (1997), Wilson and Sober (1999, 135–142), Forber and Smead (2015).

Hamilton's kin selection mechanism, whereby its coefficient of relatedness is replaced by a comparable coefficient for altruistic assortative interaction between nonrelated people.[143] Also, group selection can occur or be intensified through the differential production of individuals migrating to other groups, a phenomenon that Lucas Molleman et al.[144] called contagion group selection.

The group selection controversy has also been fuelled, because altruistic individual behaviour and selfish cooperative group behaviour—the latter sometimes being called mutualism—have not been sufficiently distinguished. Group selection can also occur through a form of selection which John Maynard Smith[145] called synergistic selection and as the evolutionary model Peter Corning developed as The Synergism Hypothesis.[146] Indeed, selfish functional cooperation between members of a group, independently of their degree of genetic relationship, or their altruistic or reciprocal behaviour, might provide mutual advantages—proximately economic and ultimately reproductive—for the cooperating partners who may become a unit of selection.

Group selection has in all probability been of particular importance in the evolution of the hominins, because they possess a dual inheritance system—genetic and cultural. Dual inheritance theory (DIT)[147] predicts that, due to the nature of cultural inheritance, it may be an important force in the evolution of both genetically based cooperative predispositions and culturally developed moral codes systems favouring cooperative actions, which is an example of biocultural co-evolution.

Indeed, group selection can apply to cultural traits as well as to genetic traits. For cultural traits, the concept of cultural group selection is used.[148] Charles Darwin[149] advanced the idea that cultural factors can operate at the group level, because the Lamarckian transmission of acquired characteristics bypasses the biological inheritance system. J.B.S. Haldane[150] was the first to view the population as a whole as a unit of selection. Also, Ernst Mayr[151] argued that human cultural groups, as wholes, could serve as the target of selection. A more elaborated development of this idea is found in the recent work of Robert Boyd and Peter Richerson, David Sloan Wilson and Elliot Sober, Christopher Boehm, Joseph Henrich, and Samuel Bowles and Herbert Gintis.[152] These authors argue that the ultrasocial scale of organisation of the human species requires, in addition to kin selection and

[143]Gintis (2000, 271).
[144]Molleman et al. (2013).
[145]Maynard Smith (1982; 1984; 1989), Maynard Smith and Szathmáry (1995; 1999).
[146]Corning (1983; 1996; 1997; 2005; 2008); see also Corning and Szathmáry (2015).
[147]For instance, Cavalli-Sforza and Feldman (1981), Durham (1991), Henrich and McElreath (2007).
[148]Campbell (1965), Soltis et al. (1995), Henrich (2004).
[149]Darwin (1871, 147).
[150]Haldane (1932).
[151]Mayr (1988).
[152]Boyd and Richerson (1990; 2005; 2007), Richerson and Boyd (1997; 2005), Wilson and Sober (1994), Wilson (1997; 2002), Sober and Wilson (1998), Boehm (1999), Henrich (2004), Bowles and Gintis (2011), Richerson et al. (2016).

reciprocity, cultural selection theories as explanatory models for large human populations with institutions such as moral systems and religions that can be considered as biologically and culturally evolved adaptations, which allow groups to act as singular units rather than as a collection of individuals.[153] However, cultural group selection may, in turn, elicit the further spreading of prosocial genes by purely within-group selection processes.[154]

Indeed, as group size increases, the genetically based predisposition to nepotism or reciprocity may not suffice to further enhance sociality, therefore cultural institutions are required to promote or enforce the application of cooperation in large populations.[155] Herbert Gintis and Dirk Helbing[156] label the need for extra-familial socialisation institutions as support for altruistic forms of prosociality "the fundamental theorem of sociology".

The evolution of the hominins in the harsh Palaeolithic era[157] in which those vulnerable bipedal and slow maturing creatures succeeded in surviving and evolving, thanks to the development of social life, strongly stimulated the development of altruistic features and decreased within-group competition, resulting in what Samuel Bowles[158] calls reproductive levelling. Christopher Boehm[159] has argued that the shift from hierarchical to relatively egalitarian social relations among early hunter-gatherer bands of anatomically modern humans—or even of *Homo erectus* in the Palaeolithic—strongly favoured the selection of genes for altruism by tilting the balance from within-group selection to between-group selection, thus further supporting the development and spreading of altruistic or cooperative traits, which was a thesis already advanced by J.B.S. Haldane in the 1930s.[160] Oddly enough, this shift intensified inter-group warfare because of the spread of altruistic genes that enhance the capacity for patriotic self-sacrifice; in turn, warfare further supported, via between-group selection, the spread of biological traits and moral codes for altruistic behaviour. Between-group selection supports altruistic genes favouring cooperation, because groups with many cooperators will have a reproductive and demographic advantage over groups with few cooperators.[161] Cooperative groups will also be more successful in inter-group conflicts in which reputation, reciprocation, and retribution play such an important role.[162]

In conclusion, at the end of the twentieth century, consideration of group selection re-emerged as an important component of a multilevel theory of evolution,

[153]Wilson (2002).
[154]Henrich et al. (2003, 462).
[155]Henrich and Boyd (2001, 208), Wilkins and Thurner (2010, 635), Bowles and Gintis (2011, 93ff).
[156]Gintis and Helbing (2015, 20).
[157]Harms (2000), Baschetti (2007, 243).
[158]Bowles (2006).
[159]Boehm (1999, 220ff).
[160]See also Choi and Bowles (2007), Smirnov et al. (2007), Lehmann and Feldman (2008), Bowles (2009), Ginges and Atran (2011), Gneezy and Fessler (2011), Saaksvuori et al. (2011), Halevy et al. (2012), Konrad and Morath (2012), Rusch (2014).
[161]Henrich and Boyd (2001, 85), Bowles (2006; 2009).
[162]Boyd and Richerson (2007, 224), Bowles and Gintis (2011, 76).

more particularly for humans.[163] Natural selection is to be considered as a hierarchical process in which groups, as higher units of the biological hierarchy, are the vehicles of selection for individuals, just as individuals are vehicles of selection for genes.[164]

John Teehan[165] refers to the whole of the different evolutionary mechanisms that promote altruistic behaviour and result in intensive social cooperation among genetically related and non-related group members—kin selection, direct and indirect reciprocity, coercion, and cultural group selection—as the human specific moral grammar.

2.1.3 Migration

Genetic migration occurs when a genetically different section of a population leaves a population and joins another population. Genetic migration can occur on a small scale, at the level of individual (mate) exchange, or on a large scale, as a massive population invasion. It can occur as a single, non-recurring population move, or as a continuous gene flow between two or more populations. It can be of a uni- or a bi-directional nature. It can be merely of a deterministic nature, assuming an infinite population size, so that no random elements resulting in drift are included; in contrast, it can also be subject to stochastic forces resulting in complex interactions between migration and drift.[166] Genetic migration may change the genetic make-up of either the sending or the receiving population. Genetic immigration leads to an increased heterozygosity and increased genetic variability within the invaded population, and to a decrease in between-population variance. It lowers the ratio of the variance between populations to the total (between and within) variance.[167]

Migration is a mechanism that also can contribute to the geographical spread or social dissemination of cultural traits. An important difference between genetic and cultural migration is that the latter is not limited to the intergenerational transmission from parents to descendants, but can also occur horizontally via non-related persons.[168]

2.1.4 Genetic and Cultural Drift

Chance events of a different nature to mutations may also influence the presence or disappearance of biological or cultural characteristics that fulfil a role in the development of moral sentiments, ideas or behaviours.

[163]Borrello (2005), Traulsen and Nowak (2006).
[164]Sober (1994), Sober and Wilson (1998), Wilson and Wilson (2007; 2008), Field (2008), Landa and Wilson (2008).
[165]Teehan (2010, 41); see also Hauser (2006, 300).
[166]For instance, Piazza (1990).
[167]Wright (1965).
[168]Cavalli-Sforza and Feldman (1981), Mesoudi (2011, 81).

In genetics, a well-known phenomenon is gene drift, also referred to as the *Sewall-Wright-effect*,[169] which can change allele frequencies of monogenes as a result of the accumulation of random fluctuations in the intergenerational transmission of alleles in small populations. The formation of each new generation is, in fact, a sampling process of the gametes available in the gene pool of the population. The smaller the population, the greater is the risk that allele frequencies in the next generation will deviate from those of the preceding one. Therefore this evolutionary mechanism has become less important in modern societies, which are usually composed of many millions of individuals.

In the cultural realm, cultural features can, just like genetic traits, be subject to random fluctuations as a result of sampling phenomena.[170] In the same way as its genetic analogue, cultural drift may have caused intergenerational changes in small populations. Cultural drift may also have occurred as a consequence of demographic catastrophes producing so-called demographic bottlenecks and founder effects, or as a result of migration of small population fractions.

Whereas cultural drift may have been a (minor) source of cultural variability in the past, when reproductive communities were very small and easily subject to random sampling phenomena, it has become irrelevant in huge modern populations.

2.1.5 Partner Choice

So far this chapter has discussed evolutionary factors that influence the genetic or cultural composition of a population, factors that change the relative prevalence of the different genes or memes in a population. In addition to evolutionary mechanisms that change the genetic or cultural composition of a population, there is also a mechanism through which the genetic or cultural structure of a population can be changed, i.e. the way in which genes present in sex cells or memes are combined in offspring through mate choice.

Partner choice may or may not occur at random, meaning that people choose their partner either by chance or on the basis of similarity (= positive assortative mating) or dissimilarity (= negative assortative mating) in some of their characteristics. In positive assortative mating more identical genes will be combined in offspring or identical memes transferred to individuals and the genetic or cultural differences between individuals in the population will increase. In the case of negative assortative mating the opposite occurs.[171] A special case of genetic assortative mating concerns the positive or negative choice of blood relatives. A positive assortative mating for blood relatives leads to inbreeding, whereas a negative choice results in outbreeding. Inbreeding is a genetic consequence of biologically consanguineous mating, resulting in offspring with a higher than

---

[169]Wright (1931).

[170]Koerper and Stickel (1980), Bentley et al. (2004), Mesoudi (2011, 76).

[171]For instance, Mascie-Taylor and Boyce (1988).

random risk of carrying a double dose of all the genes that were present in a single dose in the common ancestor.[172]

The type of assortative mating is important for individuals and populations, because it can more or less visualise particular characteristics in the population structure. Indirectly, it can also favourably or unfavourably influence some selective processes and influence the population composition.

In an analogy to the assortative mating for biological characteristics and its effect on the population variance of those characteristics, it can be assumed that this mechanism can also influence the distribution and variance of moral ideas and behaviour. Homogamy is well known to exist for a variety of cultural characteristics—educational level, leisure activities, and especially religious, moral or political convictions.

2.2 Evolutionary Background of Morality

The evolutionary mechanisms through which biological and cultural traits can be transmitted and disseminated have now been explained, it is possible to elaborate on the evolutionary background of biological and cultural phenomena related to values and norms, i.e. ideas about what is and what should be right or wrong.

Values and norms can be expressed through three channels: moral rules, religious rules and laws. The latter are usually developed on the basis of the first two. In this chapter, the discussion is limited to the evolutionary background of morality.

The concept of morality refers to a system of attitudes, standards, and/or behaviour by which humans evaluate certain forms of behaviour as good and other forms of behaviour as bad; morals are usually taken to refer to rules about what people ought to do and what they ought not to do, in particular regarding interpersonal and social relations.[173]

Evolutionists usually address morality as a form of behaviour—behavioural morality—as opposed to abstract morality as it is usually developed by moral philosophers and theologians.[174] To the degree that biological predispositions to moral behaviour (as well as cultural values and norms) are important for the realisation of ontogenetic and/or phylogenetic processes, they are subject to natural selection, and thus will influence the reproductive fitness of individuals,[175] and the social relations between individuals and groups of individuals.[176]

[172]For instance, Cavalli-Sforza and Bodmer (1999, 341).
[173]For instance, Irons (1996, 1), Hinde (2002, 3), Krebs (2011, 27).
[174]Gintis (2015, 216).
[175]Allchin (2009, 599).
[176]Krebs (2011, 27).

2.2.1 A Brief Review of Evolutionary Ethics Theory

Evolutionary ethics is the study field that looks at how morality is in one way or another related to the biological evolution of humankind, and consequently to its present biological constitution. It has already a long history and is underpinned by extensive literature.[177]

When discussing the relationships between biological evolution and morality, it is important to acknowledge that evolutionary ethics deals with three different issues, namely (1) to what degree do human moral sentiments as biological drives have an evolutionary foundation and, consequently, a genetic and neurological basis; (2) to what degree does biological evolution, via natural selection (or other evolutionary mechanisms), also contribute to the development of moral values and norms embedded in culture; and (3) can moral values and norms in turn change the evolutionary course?

The term evolutionary ethics covers a broad variety of theories and approaches about the relations between biological evolution and moral issues,[178] although some authors limit it to the attempt to derive specific ethical principles from evolutionary theory.[179] Two major approaches are usually distinguished, namely descriptive and prescriptive.[180]

The descriptive approach endeavours to explain the evolutionary causal origin and/or functional maintenance of the biological dispositions or capacities for moral behaviour and/or of cultural values and norms, resulting in a biological theory of the ethical systems as the evolutionary product of natural selection or some other evolutionary mechanism.[181] This is an approach to evolutionary ethics that Florian Von Schilcher and Neil Tennant identify as evolution of ethics.[182]

The prescriptive (or imperative) approach attempts to derive substantive moral principles or moral guidance from the facts of biology/evolution and to provide a meta-ethical justification for morality on the basis of evolution science. This is in addition to its capacity to explain moral behaviour.[183] Evolutionary ethics can contribute both to normative or substantive ethics, which deals with what one should do and to metaethics, which examines why one ought to do what.[184] Von Schilcher and Tennant identify this approach as evolutionary ethics sensu stricto, namely ethics from evolution.

[177]Referencing this extensive literature would take too much space. A broad selection of books and articles, going from Darwin (1871), up to Krebs (2011), Boehm (2012), and Voland (2013) is included in the bibliography at the end of this book.
[178]Farber (1998), Teehan (2006).
[179]For instance, Schloss (2004, 1).
[180]Murphy (1982), Richards (1986), Maienschein and Ruse (1999), Woolcock (1999).
[181]For instance, Darwin (1871), Alexander (1987), Wright (1994), De Waal (1996), Katz (2000), Levy (2004), Joyce (2006), Krebs (2011), Gibson and Lawson (2015).
[182]Von Schilcher and Tennant (1984, 160); see also Tennant (1983, 290).
[183]For instance, Spencer (1892), Stephen (1882–1907), Keith (1946), Huxley (1957), Cattell (1972), Wilson (1978), Richards (1987), Rachels (1990), Dennett (1995), Mataré (1999), Hinde (2002).
[184]Ruse, in Boniolo and De Anna (2006, 13), James (2011, 117).

However, not all of the relevant literature fits exactly into one of those two categories. Many authors develop a discourse of a more general nature, without explicitly taking a position concerning their precise approach. Most acknowledge the importance of knowledge about the evolutionary background of moral innate dispositions and cultural values and norms. They also hold the view that such knowledge about human nature has to be taken seriously when considering how to pursue desirable social or political goals and how to devise or choose between ethical norms. This stand can be observed even among those who categorically reject a prescriptive approach—for whatever reason.[185] Many contributions deal with methodological issues, such as the applicability or otherwise of the naturalistic fallacy.[186] A number of publications compare the evolutionary ethics approach to other, mostly religious, approaches.[187]

Another issue related to the implications of evolution science and ethics is refuting beliefs and ideas that contradict the scientific knowledge acquired by evolutionary science or by one of the basic disciplines on which it is based.[188]

When consulting and evaluating the older literature on evolutionary ethics from a present-day perspective, one should take into account that the scientific knowledge base in those days was much more limited. For instance, Darwin, Spencer, Haeckel and many others wrote their innovating treatises before the development of Mendelian genetics, population genetics, behavioural genetics and molecular genetics. Furthermore, the meaning of many concepts used might have been somewhat different from their current contents. The language used often mirrored the cultural and ideological context of those times, for instance the word race was often used for what is now considered to be the human species. Indeed, scientists, even when they try to approach human and societal phenomena in an 'objective', if not an aloof way, may be partially influenced by the cultural climate of their times, the character of their era, with its era-specific features including prejudices as well as achievements. This must have been strongly present in the era of the emerging novel way to look at human phenomena from a scientific point of view. Today it is often still partially the case, because many scientists are still under the spell of traditional religious convictions or are advocates of modern secular ideologies.

Therefore, the authors strive to read the historical 'classics' benevolently, taking into account those limitations and trying to assess what constituted the innovative aspects of the evolutionary ethical approach relative to the then dominant religious-philosophical views.

[185]For instance, Huxley (1894), Dewey (1898; 1922), Chauchard (1959), Simpson (1964), Kitcher (1985), Hughes (1986), Williams (1988), Knapp (1989), Paridis and Williams (1989), Farber (1994), Campbell (1996), Woodcock (1999), Singer (1999), Ehrlich (2000), Stenmark (2001), Clavien and El Bez (2007), Ayala (2009), Purdom and Lisle (2009), Harris (2010), Illies (2010).
[186]For instance, Richards (1986), Ruse (1994), Teehan and DiCarlo (2004), Curry (2006), Walter (2006), Rottschaefer (2007).
[187]For instance, Knapp (1989), Williams (1996), Stenmark (2001), Clayton and Schloss (2004), Pope (2007).
[188]See Singer (1999, 16).

Paul L. Farber[189] rightly distinguishes roughly three major episodes in the development of evolutionary ethics theory: (1) the Darwinian and Spencerian beginnings of evolutionary ethics as a consequence of Darwin's and Wallace's evolutionary theory of natural selection in the nineteenth century; (2) the new evolutionary ethics following the development of the Neo-Darwinian Evolutionary Synthesis in the 1930–1940s; and (3) the latest revision of the evolutionary ethics theory inspired by the Second Darwinian Revolution in the 1960–1970s.

Regarding the relationship between evolutionary ethics and traditional ethics, Michael Bradie[190] distinguishes three major approaches: (1) evolutionary ethics is a competitor of traditional ethics;[191] (2) evolutionary ethics is complementary to traditional ethics;[192] and (3) evolutionary ethics is a successor discipline to traditional ethics.[193]

2.2.1.1 The Is/Ought Question and the Naturalistic Fallacy

As stated in the introductory chapter, this book approaches global morality from a biological evolutionary perspective, well adapted to the opportunities and challenges of the scientifically driven modernisation process. This implies that values and norms will be related, justified, and possibly derived from facts. This immediately raises the eternal philosophical and logical 'Is/Ought' question and the problem of the naturalistic fallacy.

Attempting to define the ultimate goal of value and norm systems in terms of ontogenetic development and phylogenetic evolution implies the rejection of the philosophically unbridgeable transition from fact to value—the notorious 'Is/Ought' controversy—and sinning against David Hume's[194] 'guillotine' which states that one cannot make a normative claim based on facts about the world, or George E. Moore's[195] 'naturalistic fallacy' which states that it is impossible to define what is 'good' in terms of one or more natural properties. Many philosophers and ethicists continue to maintain that a logical transition from fact to value is impossible. Science—the study of what 'is'—would not be able to make inferences about ethics—the prescription of the 'ought'.[196] Moore's thesis about the naturalistic fallacy caused "generations of philosophers to either ignore or ridicule discoveries in the biological sciences".[197]

---

189 Farber (1994, 7).

190 Bradie (1994, 7).

191 For instance, Spencer (1879), Huxley (1894).

192 For instance, Westermarck (1906), Campbell (1979).

193 For instance, Wilson (1998), Ruse (1999).

194 Hume (1739, 521). NB. Hume's brief discourse on the is-ought relationship is less categorical than is usually reported in the philosophical literature. Hume mainly argued that the is-ought affirmation "should be observed and explained), and at the same time that a reason should be given, for what seems altogether inconceivable, how this new relation can be a deduction from others, which are entirely different from it."

195 Moore (1903).

196 For instance, Sidgwick (1876), Dewey (1898), Moore (1903), Flew (1967), Ruse (1979), Kitcher (1994), Woolcock (1999), Singer (1999, 2002), van der Steen (1999), Elqayam and Evans (2011), Pruss (2013).

197 Hauser (2006, 3).

Opposition to evolutionary ethics, and in particular its prescriptive approach—ethics from evolution—comes not only from philosophical proponents of the naturalistic fallacy but also from opponents of a rational, scientific approach to ethical matters in general, such as most proponents of theological[198] and other speculative-metaphysical ethical theories, and proponents of intuitionist[199] and emotivist[200] ethical theories.

Even a number of renowned evolutionists, notwithstanding the fact that they consider evolutionary knowledge of extreme importance for understanding the origin and evolution of values and norms, hold the view that evolutionary knowledge cannot serve as the foundation for a universal ethics, or prescribe desirable moral goals, or determine in which direction values and norms should be modified in the future.[201]

Indeed, a simple transition from facts, and especially from isolated or static facts, to values is not always possible. Facts can often be interpreted in multiple or contradictory ways. For instance, it is not because the human species has experienced during the largest part of its existence phenomena such as disease and hunger that these phenomena must be considered as natural, or as good, and consequently that they ought to be preserved. Indeed, as Garrett Hardin[202] stated:

> … the past is not a necessary guide to the future.

The natural fallacy has too often been used as a handy instrument to preserve existing abuses, mainly regarding social inequalities and inequities, with the aim of safeguarding the position of privileged groups in society.

Whilst a simple transition from fact to value is not always possible, a total rejection of the 'Is-Ought' transition is also not useful or reasonable. Here one risks lapsing into the opposite error, namely to commit the anti-naturalistic fallacy,[203] sometimes called the cultural fallacy,[204] or even the philosophical fallacy.[205] In this respect the renowned Edward O. Wilson[206] pertinently stated:

> the naturalistic fallacy is itself a fallacy.

Steven Pinker[207] rightly argues that the naturalistic fallacy leads quickly to its converse, the moralistic fallacy, meaning that 'Ought' implies 'Is', rather than 'Is' implies 'Ought'. Margaret Mead's[208] superficial and prejudiced, though much

[198]For instance, Zagzebski (2004).
[199]For instance, Audi (2004).
[200]For instance, Stevenson (1963).
[201]For instance, Wallace (1889), Huxley (1894), Simpson (1949), Dobzhansky (1967), Dawkins (1976), Barash (1977), Alexander (1987, 2005), Miller (1999), Ehrlich (2000), Allchin (2009).
[202]Hardin (1977, 115).
[203]Casebeer (2003), Walter (2006).
[204]Petrinovich (1995, 24).
[205]Baschetti (2007, 4).
[206]Wilson (1998, 250).
[207]Pinker (2002, 162); see also Ridley (1996, 257).
[208]Mead (1928).

admired study of the sexual behaviour of the Samoans[209] is a striking example of reverse naturalistic fallacy.

Moreover, if ethics could not be based on facts, what is left for justifying moral principles? Supranatural revelations as described in the traditional organised religions? Intuitive feelings of about what is good? Arbitrarily chosen culturally constructed values and norms? From whatever source ethics derives its values and norms, they are facts—religious, emotional, cultural, but in any case, facts. Consequently, as Robert J. Richards[210] concludes:

> …either the naturalistic fallacy is no fallacy, or no ethical system can be justified.

The authors argue that the mistake made by those who deny the possibility of an 'Is/Ought' transition is embedded in the formulation of the problem itself, especially in the understanding of the concept 'Is'. Applied to life processes, too often this concept is considered to be a static or chaotic situation, whilst the 'Ought' is supposed to be dynamic and ordering. However, life is anything but a static or chaotic phenomenon. Life is—intra- as well as intergenerationally—an evolving homeostatic phenomenon, which is very vulnerable to outside conditions.[211] Essentially it is a generic process, not only ontogenetically but also phylogenetically; it is also an ordered, goal-oriented phenomenon.[212] The realisation of this ordered genesis—onto and phylo—requires an operating system to successfully achieve its completion.

In the human species the ontogenetic and phylogenetic development is no longer completely genetically programmed and does not occur automatically or autonomously. Human life cannot ontogenetically develop itself or evolve phylogenetically when all of the species-specific building blocks—physical, organic and socio-cultural—and the programmes that combine these components into functional structures and processes are not available. The onto- as well as the phylogenetic development of the human requires (in addition to molecular, morphological and physiological building blocks) cultural value and norm systems that need to interfere and mediate where genetic programming no longer suffices to achieve the generic processes. This discourse is in line with the argumentation developed by many present sociobiologists and evolutionary psychologists who turn the naturalistic fallacy on its head: ethical behaviour is a consequence of our genetic heritage resulting from natural selection that adapts us to our environments.[213] Indeed, a large part of the living conditions, which are culturally induced for the realisation of ontogenetic and phylogenetic programmes, are determined by the genetic specificity of the human species. Almost all important ethical and ideological challenges turn out to be biosocial challenges as well. As Alex Walter[214] stated:

[209]Freeman (1983).
[210]Richards (1986, 286).
[211]Cannon (1939).
[212]See also de Waal (2014, 201).
[213]Ruse and Wilson (1986, 430, 431), Ayala (2009, 12); see also the discussion on "evolutionary anti-realism" in James (2011, 168ff).
[214]Walter (2006, 34, 35).

> We must recognize that while not all natural facts are relevant to ethical or moral discourse, all facts that are relevant to ethical and moral discourse will nonetheless be natural facts.

Hence, the naturalistic fallacy sets no obstacles to the existence of objective moral values.[215] On the contrary, as John Teehan and Christopher DiCarlo[216] argue, the naturalistic fallacy is the basis rather than the obstacle to evolutionary ethics. However, this does not mean that, within certain boundaries, no choices have to be made. Moral dilemmas require decisions about what to do.[217]

Since the ontogenetic as well as the phylogenetic development of the human depends partially on socio-cultural living conditions, largely determined by values and norms, as well as by varying environmental living conditions, human ontogeny as well as phylogeny can be subject to some variation. This issue is addressed in Chaps. 5–8, which discuss the possible alternatives in the future ontogenetic and phylogenetic development of the human species and their ethical implications in a further modernising environment. Essential values and norms, resulting from the human-specific biogram,[218] needed to guarantee the ontogenetic development and phylogenetic evolution. They have to be related, confronted and justified with respect to the specific ecological and socio-cultural environment in which they have to be realised and to which they have to adapt. This choice of the specific ecological and socio-cultural environment in which humankind has to develop and evolve considerably narrows the normative options that can theoretically be conceived.

The authors approach the relation between facts and values, between 'Is' and 'Ought', from a scientific naturalistic perspective—a typical consequentialist approach—based on the premise that science provides the best available empirical knowledge for understanding and prescribing morality.[219]

As Simon Young[220] stated:

> Naturalistic ethics are the only conceivably valid kind, because humanity is a part of nature and thus limited by her laws.

Indeed, since moral codes of vital importance are ultimately the result of the interaction between human nature and the physical and cultural environment in which it evolved or evolves, there is, as Robert Hinde stated, "no need to seek for any other source for oughts."[221] Consequently, the authors are of the view that science, and particularly evolutionary science, must be the primary source for

---

[215] Curry (2006, 243), Rottschaefer (2007, 397).
[216] Teehan and DiCarlo (2004, 43).
[217] Dewey (1922).
[218] Biogram or biogrammar: the overall genetic determined biological programme of an organism that predisposes it to behave in distinctive ways (Tiger and Fox 1971).
[219] Rottschaefer (1998, 16; 2007, 374), Kurtz and Koepsell (2007), Stenger (2009).
[220] Young (2006, 203).
[221] Hinde (2002, x).

constructing a new, universal ethic.[222] In a recent contribution, Gregory Gorelik and Todd K. Shackelford[223] advanced the evolutionary awareness concept for such an evolutionarily informed ethical framework.

2.2.1.2 The Darwinian and Spencerian Beginnings of Evolutionary Ethics

The development of evolutionary ethics[224] began shortly after the publication of Charles Darwin's monumental book *On the Origin of Species by Natural Selection*. However, the idea that human nature includes some passions that constitute the basis of human morality could already be found in David Hume's famous *A Treatise of Human Nature,*[225] and may go even back to Greek antiquity with the writings of Plato and Aristotle.[226]

Charles Darwin[227] himself and also Alfred R. Wallace[228] elaborated views on the evolution of ethical behaviour. Darwin addressed the evolution of the moral sense in Chaps. 4 and 5 of *The Descent of Man*. Darwin explained how moral predispositions, as evolutionary selected innate traits, could have evolved by the mechanism of natural selection. Although Darwin did not make explicit and systematic attempts to derive substantive moral principles from man's evolutionary past, in many places his book includes evolutionary inspired hints concerning specific ethical norms, for instance on sociality, slavery, and reproductive behaviour.

Herbert Spencer[229] was the first to elaborate a full-fledged theory of ethics based on evolution, although he was more Lamarckist than Darwinist in his approach. His evolutionary views about the struggle for existence and survival of the fittest in society aimed at a socially progressive evolution, based on mutual aid rather than competition. This part of his work is usually shadowed by the fact that his writings have largely been interpreted and used to support a *laissez-faire* socio-economic philosophy.[230]

---

[222] For instance, Loye (1999).

[223] Gorelik and Shackelford (2014, 784).

[224] Williams (1893, 2), Ruse (1999), Thompson (1999).

[225] Hume (1739, 522ff).

[226] See, for instance, Curry (2006).

[227] Darwin (1871).

[228] Wallace (1900; 1905).

[229] Spencer (1862; 1864; 1892).

[230] The writings of authors such as Spencer (1851; 1862; 1864; 1879; 1892), Sumner (1883; 1914), and many others have resulted in the so-called Social-Darwinist school of thought in which the principles of Darwinian evolutionary theory were transferred and applied to the analysis of social order and social structure. Social Darwinists attempted to give a naturalistic account of ethical values based on the theory of evolution. The term social Darwinism was first used in the 1880s in Europe—in all probability it was the Gautier (1880) who invented the term and employed it in a pejorative sense to refer to theories that saw social laws as extensions of natural laws (Tort 1992). However, with time, the social Darwinist discourse evolved from what is now considered traditional social Darwinism in the second half of the nineteenth century, in which ideas about individual economic competition were used to justify laissez-faire economic policies, to several variants of collective social Darwinism in the early decades of the twentieth century. These included a militarist or imperialist social Darwinism (Fiske 1874; Strong 1885) and a racialist

In the nineteenth century several other scholars elaborated on the relationship between evolutionary theory and ethics.[231] Among the most well known was Thomas Huxley[232] who vigorously rejected the idea that evolution could serve as a foundation for ethics. Although a strong advocate of evolution, he was not in favour of an evolutionary ethics:

> Cosmic evolution may teach us how the good and the evil tendencies of man may have come about; but, in itself, it is incompetent to furnish any better reason why that we call good is preferable to what we call evil than we had before.... Let us understand, once and for all, that the ethical process of society depends, not on imitating the cosmic process, still less in running away from it, but in combating it.

In contrast, in the nineteenth century the less well known Leslie Stephen[233] produced the most developed elaboration of the Darwinian position on ethics. Just as Darwin, he believed that the hominisation process was characterised by an overall progress in human moral evolution.[234]

In addition, outside the English-speaking world several scholars developed ideas about evolutionary ethics. The sixty submissions to a contest for a prize in Germany in response to the question "*Was lernen wir aus den Prinzipiën der Deszendenztheorie für die innerpolitische Entwicklung und Gezetzgebung der Staaten?*"[235] provide an idea about the enormous boost that Darwin's revolutionary theory gave to the development of ethically oriented biosocial writings.

In nineteenth century Germany, Ernst Haeckel[236] was the most prominent among these scholars.[237] Like Darwin, Haeckel was of the view that the evolutionary process promoted cooperation and interdependency between congeners as an instrument for survival. For Haeckel, the basic unit of evolution is not the individual, but society as a whole. Although Haeckel did not produce a specific publication on evolutionary ethics, several of his works include ideas or even whole

social Darwinism (de Gobineau 1853–1855; Chamberlain 1911) that used natural selection as an argument for the superiority of particular nations or races (cf. Hofstadter 1944; Jones 1980). Given this contentious history, the expression 'social Darwinism' can cover different meanings and is consequently often misunderstood, misused and abused. The recent developments in evolutionary theory, more particularly regarding the evolution of social behaviour, prompted Clavien (2015, 730) to state that one should rather speak about *Pro*-social Darwinism.

[231]For an overview see Williams (1893), who provided not only an overview of the most prominent nineteenth century theoreticians on evolutionary ethics, but also deals with all major concepts and issues in this field, and also discusses a number of practical ethical implications of the application of evolution to ethics, arguing that the traditional ideologies on which moral action was based need replacement with a "newer and higher system... founded on the solid rock of scientific Truth".

[232]Huxley (1894, 80, 83).

[233]Stephen (1882–1907).

[234]See Farber (1994).

[235]Schallmayer (1910): "What do we learn from the principles of evolutionary theory about the internal political development and legislation of states?" (authors' translation).

[236]Haeckel (1866; 1868).

[237]See, for instance, also Rée (1877) with his interesting '*Der Ursprung der moralischen Empfindungen*' in which he develops ideas that point in the direction of group selection.

chapters about the significance of evolution for morality.[238] Haeckel has in recent years been linked to anti-Semitism and Nazi ideology by Christian conservative authors such as Daniel Gasman[239] and Richard Weikart.[240] They chose to ignore that Haeckel praised Jews as being important contributors to the German culture "who have always stood bravely for enlightenment and freedom against the forces of reaction, inexhaustible opponents, as often as needed, against the obscurantists". They choose to forget that the Nazis outlawed immediately the German Monist League and completely rejected the Haeckelian teachings.[241]

In France at the turn of the twentieth century, Jean-Louis de Lanessan[242] wrote an interesting treatise on natural ethics, based on a comparative biological approach and a cross-cultural approach, which he opposed to the beliefs in the traditional (Mediterranean) religions.

2.2.1.3 The New Evolutionary Ethics Following the Development of the Neo-Darwinian Evolutionary Synthesis

The emergence of the Neo-Darwinian evolutionary synthesis in the 1930–1940s[243] gave a second boost to evolutionary ethics with the writings of a broad group of eminent evolutionary scholars.[244]

Whereas Julian Huxley, Conrad H. Waddington and Raymond Cattell thought that evolution science would allow the derivation of ethical values and norms from the study of evolution, Arthur Keith, Theodosius Dobzhansky and George G.

---

[238]See Heie (2004).

[239]Gasman (1998).

[240]Weikart (2004; 2009).

[241]Christian conservatives, such as Richard Weikart, are eager to link simplistically, erroneously and maliciously evolutionary theory and in particular Darwinism to Nazism and racism, and in particular anti-Semitism. In his recent books "*From Darwin to Hitler*" and "*Hitler's Ethic: The Nazi Pursuit of Evolutionary Progress*", Weikart (2004; 2009) examined the supposed links between evolutionary ethics and Nazi ideology. He argues that there are strong similarities between evolutionary ethics and Hitler's worldview. Of course, Nazi theorists used evolutionary theory, just as they used religious ('Gott mit Uns') and socialist ('National-Sozialism') ideas to support and justify their ethnocentric in-group ideology aimed at obtaining power, resources and territories. What Weikart does not mention is that the evolutionarily inspired eugenics of the Nazis had nothing to do with eugenics: on the contrary, the Nazi so-called 'Rassenhygiene' was a dysgenic policy. Also their euthanasia programme had nothing to do with eugenics, let alone with evolutionary ethics; idem for their racial ideology that was absolutely deceptive, because the so-called 'Arian race' is not a biological entity. It is understandable that *bona fide* Darwinists are furious about the cheap and misleading—but ideologically (and probably also commercially) interesting—tactic among critics of Darwin's theory to draw a link to a criminal ideology such as Nazism. Furthermore, Weikart chooses to ignore the roots of Nazi ideology to be found in old time Christian apology, First World War consequences and economic crises (e.g. Richards 2010).

[242]De Lanessan (1908).

[243]Fisher (1930), Wright (1931), Haldane (1932), Dobzhansky (1937), Huxley (1942), Mayr (1942), Simpson (1944).

[244]Haldane (1924), Huxley (1927; 1957; 1964), Leake and Romanell (1950), Keith (1946), Waddington (1941; 1960), Dobzhansky (1956; 1962; 1967), Simpson (1949; 1964; 1969), and Cattell (1933; 1944; 1950; 1972).

Simpson acknowledged that evolution was important for understanding the emergence and evolution of morality, but one ought not to look to evolution for its justification. Simpson wrote:

> The evolutionary process in itself is nonethical—there simply is no point in considering whether it is good, bad, a mixture of the two, or neither[245]

Julian S. Huxley,[246] one of the main architects of the new evolutionary synthesis, was also one of the main scholars who attempted to derive a foundation of morality from the study of evolution. Rooted in evolutionary science in the same line of thought as Herbert Spencer, Huxley advocated evolutionary progress, allowing increased control over and independence of the environment, as a measure for human action. Huxley[247] was well aware of the fact that there is no purpose in evolution, only a direction, a trajectory—the line of evolutionary progress that can serve as a guide in formulating man's purpose for the future:

> If we wish to work towards a purpose for the future of man, we must formulate that purpose ourselves. Purposes in life are made, not found.

Huxley wanted to establish a scientifically based, rational and agnostic evolutionary humanism, as an alternative to the obsolete religious ideologies and the modern ideologies such as Marxism, existentialism and liberalism.[248]

Another renowned scholar to be mentioned is the psychologist and behavioural geneticist Raymond B. Cattell.[249] His remarkable *A New Morality from Science: Beyondism* is one of the very few treatises that deals in great detail and in a concrete way with the foundations and derivations of an evolutionary ethics. Although the authors disagree with one of his major primary moral aims—"the maintenance of an ethos or atmosphere defined as 'cooperative competition' among groups"—they consider Cattell's *Beyondism* as one of the most important and relevant contributions to the elaboration of a scientifically founded evolutionary ethics.

2.2.1.4 The Latest Revision of the Evolutionary Ethics Theory Inspired by the Second Darwinian Revolution

In recent decades a third wave of renewed and explosive interest in matters of evolutionary ethics can be observed.[250] It is mainly a result of the Second Darwinian Revolution, as reflected in fields such as sociobiology,[251] behavioural

[245]Simpson (1964, 143).
[246]Huxley (1927; 1957; 1964).
[247]Huxley (1942, 576); for a discussion of the concept of evolutionary progress, see also Corning (1983), Nitecki (1988), Ruse (1996), Zarandi (2003).
[248]See Knapp (1989), Farber (1994), Ruse (1999), Phillips (2007).
[249]Cattell (1933; 1944; 1950; 1972).
[250]Murphy (1982), Alexander (1987), Nitecki and Nitecki (1993), Bradie (1994), Wright (1994), Petrinovich (1995), Thompson (1995), Ridley (1996), Arnhart (1998), Farber (1998), Maienschein and Ruse (1999), Mataré (1999), Katz (2000), Broom (2004), Joyce (2006), Sinnott-Armstrong (2008), Høgh-Olesen (2010), James (2011), Krebs (2011).
[251]Hamilton (1964; 1996), Trivers (1971; 1985), Ghiselin (1974), Campbell (1975), Wilson (1975; 1978), Dawkins (1976), Alexander (1978), Daly and Wilson (1978), Chagnon and Irons (1979),

ecology,[252] and evolutionary psychology,[253] but also in evolutionary anthropology,[254] cultural anthropology,[255] the evolutionary study of religion,[256] neurology and cognitive science,[257] molecular genetics and other fields of biomedical research,[258] and cosmology.[259]

The sociobiologist Edward O. Wilson[260] has strongly contributed to igniting or reviving the discussion on the reciprocal relationship between evolutionary theory and ethics. His outstanding *Sociobiology: The New Synthesis* gives an overview of the evolutionary-biological mechanisms that lie at the basis of the evolution of social behaviour, illustrated in chapters about all major social species, including the human species. Following partly justified criticisms on his last, brief chapter on the human species, Wilson wrote in a much more nuanced and detailed way about specific human sociobiological issues in his book *On Human Nature*. In his view, the capacity for and drive towards moral behaviour is, just as morphological and physiological characteristics, the result of evolutionary processes, and particularly natural selection. Wilson argues, in particular in *On Human Nature*, that knowledge about evolution and biology in general must serve as a foundation and guide for moral action. His discourse mainly concentrates on the analysis of the evolutionary emergence of morality and some problems humans experience in modernity because of their Pleistocene heritage, on the one hand, and the failures of traditional religious belief systems and modern secular ideologies, on the other hand, to reconcile our genetic heritage with the opportunities and challenges of modernity. Following Spencer's and Huxley's thinking, Wilson argues that evolution is characterised by a progressive development towards increased complexity. Contrary to Raymond Cattell's *Beyondism*, Wilson's works only occasionally include hints about the concrete contents of a future ethics, well adapted to the novel environment of modernity—cf. his cardinal values of the survival of human genes

Cronin (1993), de Waal (1996), Cronk et al. (2000), Nesse (2001), Corning (2005), Bowles and Gintis (2011), Voland (2013).

[252]Borgerhoff Mulder and Schacht (2012).

[253]Barkow et al. (1992), Wright (1994), Baron-Cohen (1997), Simpson and Kenrick (1997), Buss (1999; 2007), Barrett et al. (2002), Crawford and Salmon (2004), Gangestad and Simpson (2007), Crawford and Krebs (2008), Dunbar and Barrett (2009), Krebs (2011).

[254]For instance, Durham (1991), Dunbar (2004), Barash (2012).

[255]For instance, Boehm (1999; 2012), Cronk (1999), Richerson and Boyd (2005).

[256]Crippen and Machalek (1989), Guthrie (1993), Gauchet (1997), Boyer (2001), Pyysiäinen (2001), Atran (2002), Bruce (2002), Edis (2002), Wilson (2002), Barrett (2004), Broom (2004), Clayton and Schloss (2004), Everitt (2004), Harris (2004, 2010), Baril (2006), Dawkins (2006), Dennett (2007), Bulbulia et al. (2008), Stenger (2008), Voland and Schiefenhövel (2009), Wright (2009), Haught (2010), Pinxten (2010), Teehan (2010), Bellah (2011), Cunningham (2011), Shermer (2011), Stenger (2012).

[257]For instance, Guthrie (1993), Baron-Cohen (1997), d'Aquili and Newberg (1999), Newberg et al. (2001), Greene (2003), Harris (2004; 2010), McNamara (2006), Gangestad and Simpson (2007), Mendez (2009), Verplaetse et al. (2009).

[258]For instance, Hughes (2004), Mulhall (2002), Bostrom and Roache (2008), Buchanan (2011), Venter (2013).

[259]Kauffman (1995), Hawking and Mlodinow (2010), Stenger (2011), Krauss (2012).

[260]Wilson (1975; 1978; 1998); see also Ruse and Wilson (1985; 1986).

in the form of a common pool over generations, the favouring of diversity in the gene pool, and the promotion of universal human rights.

Wilson's[261] provocative statement that

> the time has come for ethics to be removed temporarily from the hand of the philosophers and biologicized

obviously provoked angry territorial reactions from many quarters.

In recent decades, many other renowned scholars from different fields have contributed extensively to the current discourse on evolutionary ethics. Examples are sociobiologist Richard Alexander,[262] biologists Francisco J. Ayala[263] and Robert Hinde,[264] historian Robert J. Richards,[265] evolutionary anthropologist Christopher H. Boehm,[266] philosophers Michael Ruse[267] and Peter Singer,[268] neurologist Joshua D. Greene,[269] and evolutionary psychologist Dennis L. Krebs.[270] Moreover, the bibliography of the present treatise includes some fifty books and a plurality of journal articles that have been written on evolutionary ethics since 1975.

It will not surprise the reader that, just as in the past, the recent literature on evolutionary ethics continues including many contributions in which the authors, for a variety of reasons (philosophical, cultural, but mainly religious) question or oppose the propositions of the third wave of evolutionary ethics.[271] For instance, Anthony O'Hear argues in his *Beyond evolution: Human Nature and the Limits of Evolutionary Explanation*[272] that evolutionary science cannot explain in a deterministic way every aspect of human nature, in particular its moral sense, because human consciousness has the freedom to make its choices autonomically. He overlooks that human consciousness is also the outcome of the evolutionary process and can be the cause and instrument of further human evolution and moral growth.

The opposition to an evolution-based ethics is part of a more general negative attitude towards evolution science, genetics, and biosocial science in general. Many social scientists, philosophers and ethicists, policy makers, and national or international governmental bodies often deny the importance and relevance of evolutionary biological, and more particular genetic factors that influence or even drive individual or societal processes. In United Nations, European Union or Council of Europe contexts some fundamentally important sociobiological issues or causes of societal challenges are simply not addressed or are concealed. Topics such as

261 Wilson (1975, 562).
262 Alexander (1979; 1987).
263 Ayala (1987; 1995; 2007; 2009).
264 Hinde (1974; 2002).
265 Richards (1986; 1987; 1993; 2008).
266 Boehm (1996; 1999; 2012).
267 Ruse and Wilson (1985), Ruse (1986; 1989; 1996; 1993; 1999).
268 Singer (1981; 1984; 1993; 1999; 2002; 2009).
269 Greene (2003; 2009).
270 Krebs (2011).
271 For instance, Peters (1999), Woolcock (1999), Pava (2009).
272 O'Hear (1997).

genetic causes of individual or group diversity, biological predispositions to group conflict, high fertility and population growth as a cause of persistent economic and cultural underdevelopment or environmental and climatic disaster, dysgenic effects of medical interventions or demographic trends, and eugenic measures for improving quality of life of future generations are usually considered to be sensitive issues and consequently politically incorrect. Ignoring clusters of such causes relevant at the global level inevitably results in lukewarm effects of fragmented policies.

From the very beginning of the emergence of ideas about the impact of evolutionary theory on the origin and development of morality, evolutionary debunking arguments have arisen, expressing the worry that morality might be discredited because of the finding that moral senses and even moral codes are the result of evolutionary forces. In recent decades, these worries seem even to have increased.[273] For the evolutionary scientist, these worries appear to be highly unjustified, because instead of debunking, the opposite may be expected. The evolutionary basis of morality, together with culturally evolved moral senses and/or moral codes, precisely demonstrates the gigantic importance of morality for human development, survival and evolution.

2.2.1.5 Continuity, Refinement, but Persisting Discordance About Evolutionary Ethics

There is a strong continuity throughout the three waves of the development of evolutionary ethics up to date, both regarding the descriptive/explanatory approach and the prescriptive approach. A gradual increase is observed in the scientific evidence that moral instincts and moral values have a strong, albeit not a unique, evolutionary foundation, as well as in the facts supporting the possibility of deriving values and norms from the study of hominin evolution. There has been a considerable increase in the current scientific knowledge base about human evolution (in particular biocultural co-evolution) and the evolution of social behaviour (with the discoveries of mechanisms such as kin selection and reciprocity selection), genetics (population genetics, behavioural genetics and molecular genetics) and neurology. It forms a very solid scientific background, not only for understanding the biological origin and evolution of innate moral sentiments and cultural values but also for devising norms for future action.

Regarding the descriptive/explanatory approach, there is definitely continuity in the analysis of the origin and development of human morality from Charles Darwin and Ernst Haeckel, then Julian Huxley and C. Waddington, to Edward Wilson, Richard Alexander, Michael Ruse, and Dennis L. Krebs and Christopher Boehm.[274]

[273]For the discussion of evolutionary debunking arguments (EDAs) see, for instance, Kahane (2011) and Millhouse et al. (2015).

[274]Darwin (1871), Haeckel (1870), Huxley (1957), Waddington (1960), Wilson (1978), Alexander (1987), Ruse (1986; 1999), Krebs (2011), Boehm (2012).

However, in the prescriptive approach there is also a consistency, going from Hebert Spencer and Leslie Stephen, then Julian Huxley and Raymond Cattell, to Edward Wilson, and Robert Richards.[275]

Notwithstanding the continuity of evolutionary ethics theory throughout its subsequent waves, there are also important shifts in substance and methodology. The most striking is probably the shift from the traditional emphasis of evolutionary ethics as a competitive ethic to the newer approach of a cooperative ethic.[276] In this regard, Michael Ruse[277] refers to the traditional account as evolutionary ethics and the newer account as Darwinian ethics. The recent change in the contents (and methodology) of evolutionary ethics is largely the result of the explosive development of sociobiology and more particularly of the discovery or reconsideration of evolutionary mechanisms such as kin selection, reciprocity selection and group selection. However, sociobiologically or bioethically related research in scientific fields such as primatology,[278] neurology[279] and bioeconomic game experiments[280] are also contributing. Contrary to Maurizio Meloni,[281] who perceives the recent developments in all of those fields as a transformation from "a biologization of morality into a moralization of biology", the authors consider those developments to be a continuous process characterised by an ever further deepening of the insights into the evolutionary foundation of morality.

From the recent literature, it also appears that a lot of effort has been made to clarify the logical and philosophical questions about the naturalistic fallacy.[282] In addition, scholars in the domain of evolutionary ethics are much more aware of the complex interactions between biological-evolutionary factors and socio-cultural and ecological living conditions and determinants of values and norms.

However, notwithstanding this continuity and progress, both in the methodology and the descriptive/explanatory and prescriptive approaches, the scientific community continues to remain divided as to whether the study of evolution can only explain the biological origin and evolution of moral innate dispositions and values and rules or can also serve as the scientific basis for reflecting and devising values and norms for the future development and evolution of humankind.

Evolution science continues to provoke uneasiness, if not outright opposition in many quarters where people have difficulties in accepting a non-purposive, non-design approach to key philosophical questions.[283] Some continue to believe

[275]Spencer (1862, 1892), Stephen (1882–1907), Huxley (1957; 1964), Cattell (1972), Richards (1987).
[276]Ruse (1989), Mizzoni (2002), Meloni (2013).
[277]Ruse (1986, 207).
[278]De Waal (1996, 2006), Kappeler and van Schaik (2007).
[279]Moll et al. (2003), Gazzaniga (2005), Tancredi (2005), Glannon (2007), Verplaetse et al. (2009), Churchland (2011), Marazziti et al. (2013), Seung (2013), Shoemaker (2012), Alvaro-Gonzalez (2014).
[280]Maynard Smith (1982), Gintis (2000), Bowles and Gintis (2011), Barash (2012).
[281]Meloni (2013, 85).
[282]For instance, Richards (1986), Ruse (1986), Teehan and DiCarlo (2004), Curry (2006), Rottschaefer (2007), de Waal (2014), Kitcher (2014).
[283]See discussion in Vannelli (2001, 31).

that morality is essentially a cultural phenomenon and not a biological adaptation as a consequence of natural selection.[284] Others argue that moral systems are an adaptation, comparable to essential biological characteristics.[285]

Paul L. Farber,[286] in his analysis of the history of evolutionary ethics, pertinently observes that:

> humans read values into nature, rather than discover them there.

Indeed, it is striking that people with very different ideological or political orientations made use of evolutionary theory to support their views and programmes. Contrary to what is often thought that only advocates of conservative or right wing liberalism with its '*laissez-faire*' socio-economic doctrine[287] looked for a justification in evolutionary theory, progressives and left-wing advocates such as Marxists[288] and feminists[289] have also sought to base their revolutionary views on evolution science.[290]

Finally, looking at the extensive literature on evolutionary ethics that has been produced since Darwin's era, one is struck by the salient contrast between the large number of theoretical and methodological contributions, elaborating on the evolutionary origins and development of moral sentiments and ethical values and norms or how to achieve desirable values and norms, on the one hand, and the limited number of treatises that deal with the concrete implications of evolutionary theory for practical ethical applications on the other hand.[291] For instance, Raymond Cattell's 'Beyondism' is one of the very few works that includes both a theoretical approach and practical applications.

2.2.2 Biological Bases of Morality: Natural Needs and Drives

David Hume[292] already argued in the eighteenth century that moral values are the expressions of natural human drives aiming at the common good of society. Recent progress in evolutionary biology, particularly sociobiology and evolutionary psychology, but also in fields such as neuroscience and evolutionary or economic game experiments, espouses the Humean approach to ethics and philosophy.[293]

In the course of the second half of the twentieth century, scholars from different human disciplines have tried to list and classify natural human needs and drives and some have also attempted to derive values and norms from those predispositions.

[284]For instance, Sahlins (1976), Ayala (2009).
[285]For instance, Ruse (1995).
[286]Farber (1994, 9, 85).
[287]Spencer (1862; 1892), Sumner (1883; 1914), Carnegie (1920), Arnhart (2010).
[288]For instance, Woltmann (1899).
[289]For instance, Bebel (1879), Gowaty (1997), Hrdy (1999), Vandermassen (2005).
[290]See Richards (1986), Farber (1994).
[291]See also Clavien (2015, 731).
[292]Hume (1739, 522ff).
[293]See Curry (2006), Hauser (2006).

Various sources of information have been or are being used in this venture: cultural anthropological studies about human universals,[294] comparative zoological studies,[295] bioanthropological, sociobiological and psychological studies,[296] as well as studies using or referring to evolutionary science.[297]

In the authors' view, there are three major biological bases of morality: individual ontogenetic development, sociality, and reproduction; the latter being the basic condition for long-term biological evolution and adaptation.

2.2.2.1 Individual Ontogenetic Development

Most human needs and drives relate initially to the requirements of the individual ontogenetic development: drive for survival in general, thermoregulation, sleep, respiration, need for nutrition, physical safety, health care, family care, social bonding, mobility, resource acquisition, and habitat and territorial security. However, there is also the need for language development, creative and aesthetic pleasure, and—last but not least—intellectual understanding (including spirituality, religiosity and morality). Needs and drives deriving from individual ontogenetic development induced not only the innate sense of having the opportunity for individual self-realisation that includes aspirations such as survival, reproduction, freedom, welfare, well-being, and happiness, but may also result in the desire to implement basic values and norms usually grouped in modern times under the heading of human rights.[298] The issue of individual ontogenetic development will be detailed further in Chap. 6.

2.2.2.2 Sociality

These are the human needs for living in groups. The recent advances in sociobiology have greatly enlarged the knowledge of biological needs for sociality: nepotism (altruism toward closest kin); reciprocal altruism between non-relatives; mutualism; in-group favouritism and out-group enmity; male bonding; social dominance and submissive forms of behaviour; egalitarianism; punishment of criminals, cheaters and free-riders. It will not come as a surprise that many scholars consider human supersociality as the primary source of our moral nature.[299] The problems of human social relations are discussed further in Chap. 7.

2.2.2.3 Reproduction

Several human needs and drives are related to the process of reproduction, or, ultimately, the intergenerational transmission of genes (and memes): development

[294]For instance, Murdock (1945), Malinovsky (1960), Parsons (1964), Hockett (1973), Brown (1991).
[295]For instance, Tiger and Fox (1971), Wilson (1975), de Waal (1996).
[296]For instance, Montagu (1955), Bowlby (1969), Van den Berghe (1975), Doyal and Gough (1991), Maslow (1954; 1999).
[297]For instance, Cattell (1972), Wilson (1978), Alexander (1987), Arnhart (1998), Corning (2000; 2010; 2011), Bowles and Gintis (2011), Krebs (2011).
[298]Krebs (2011, 209, 210).
[299]For instance, Fowers (2015, 4).

of sexual identity; sexual mating; love; female choice; male-male competition; sexual jealousy; incest avoidance; childbearing, childcare and education. Some of the needs and drives listed above under ontogenetic development and group relations are also relevant for intergenerational processes, e.g. practically everything that appertains to social behaviour and particularly social status/hierarchy seeking, and resource and territorial acquisition. The challenges related to intergenerational replacement are further addressed in Chap. 8.

2.2.2.4 Competition Between Natural Needs and Drives

The natural needs and drives of humans not only allow the identification of basic values and norms in order to allow for a harmonious ontogenetic and phylogenetic development, but they can also be the source of serious and difficult moral controversies. In the first place, this is because some needs may be in competition or conflict with other needs; and in the second place, because some needs of some individuals or groups may be in competition with some needs of other individuals or groups. Moreover, since individuals may show variation in the degree of their needs and drives, some individuals may experience needs, drives and desires which are detrimental to the harmonious development of other individuals or even whole societies, e.g. the desire to cheat, steal, dominate, pester, torture, rape, enslave, or kill conspecifics. Finally, in many circumstances, particularly in times of scarcity of resources or territories, or in situations of strong social inequities, intra- as well as intergenerational needs may raise feelings such as jealousy, envy, spite, disgust, hate, anger, rage, grudge, revenge, or aggressivity leading to all kinds of (social) competition and even to social strife and revolution, if not war.

In conclusion, contrary to the view of those scholars that science, and in particular evolutionary science, cannot assign either meaning or purpose to the world it explores,[300] the authors are of the opinion that evolution explains not only our biology, but can tell us a lot about what is good, right, or moral—providing that a comprehensive view is adopted, taking into account all relevant biological knowledge, and that this knowledge is considered in the context of the cultural opportunities modernity offers and the ecological constraints our planet imposes.

2.2.3 Evolutionary Causes of Human Morality

As already mentioned, human morality includes two different but interrelated phenomena:

(1) The biological (neurological) basis of innate moral sentiments and the capacity for reasoning, also in the domain of morality, ultimately determined by genes;
(2) The cultural values and norms (ethical codes), resulting from cultural creative processes and social learning and internalisation, but ultimately influenced or

[300]See, for instance, Miller (1999, 280), Wade (2009, 5).

driven by innate moral sentiments on the one hand, and the capacity for moral reasoning and judging on the other hand.

When discussing the evolutionary causes of human morality, both components of morality—the neurological basis (the genes) and the cultural codes (the memes) and their interactions—have to be considered, because both are subject to the same evolutionary mechanisms, namely mutation (genetic mutations of genes/cultural changes of memes), selection (natural selection of particular alleles/cultural selection of particular memes), chance fluctuations ('drift') (genetic drift of alleles/cultural drift of memes), migration (gene flow/cultural dispersion of memes).[301] (See Sect. 2.1).

The essential proposition here is that human morality (in both of its components) is the result and—as will be argued below—partly the cause of the hominisation process. Whether human morality is a direct biological adaptation to the hominisation process[302] or merely an exaptation (namely an indirect consequence of man's specific intellectual abilities)[303] does not really matter. Direct adaptation or indirect exaptation is considered by many to be plausible for the explanation of the existence of a neurological basis of human moral sentiments. However, at first sight adaptation or exaptation may be less evident for the origin of values and norms that are traditionally believed to be the product of culture and disassociated from biological evolution. Of course, values and norms are cultural products and there is a considerable diversity in values and norms between cultures and within cultures. By stating that values and norms are a biological adaptation (or exaptation) the authors mean that they are phenomena deriving from certain biological needs, but whose actual content can vary within a certain range, in interaction with socio-cultural and environmental living conditions. Successful value and norm systems fulfilled for the socialising hominins the same survival functions as their somatic organ systems. Without those guiding and control systems, ontogenetic development—as well as further phylogenetic evolution of the socially dependent, large brain-steered hominins—was impossible. Among such organisms behaviour is no longer exclusively genetically programmed, but consequently requires cultural intervention to guarantee survival and further evolution.

The question is now which biological changes in the course of the hominisation process caused the development of human morality, both of innate moral sentiments and of cultural value and norm systems? Three, though mutually interrelated, biological-evolutionary changes, are distinguishable:

(1) The shift from instinctive to conscious behaviour in hominin evolution;
(2) The increasing prematurity of hominin offspring during the hominisation process;
(3) The development of human sociality beyond the stage of the family.

[301]For instance, Cavalli-Sforza and Feldman (1981), Lumsden and Wilson (1981), Boyd and Richerson (1985), Durham (1991), McKenzie Alexander (2007), Baumard (2016).
[302]For instance, Wilson (1978).
[303]For instance, Prinz (2008, 368), Ayala (2009; 2010, 9019).

All three of those biological-evolutionary changes during the hominisation process involve both the Darwinian evolution of biological entities (genes, neurons, hormones) and the combined Darwinian/Lamarckian change of cultural features (including values and norms), both of which merged in a biocultural co-evolutionary integration.

2.2.3.1 The Shift from Instinctive to Conscious Behaviour

The evolution of the hominins, starting from a pre-hominin primate, over a series of successive hominin waves, to the present *Homo sapiens sapiens* was not only characterised by the development of a bipedal gait, but mainly by the increase of the large brain hemispheres. This resulted in a gradual shift from a predominantly automatically programmed instinctual behaviour toward a predominantly conscious control of behaviour.[304] The shift was characterised by an increasingly incomplete genetic programming of behaviour, necessitating an ever increasing degree of learned behaviour. In turn, this required the development of more complex forms of social organisation whereby cooperation and task division increased in importance and transcended the boundaries of close blood relationship. Such social behaviour requires the development of cultural regulatory systems, namely value and norm systems, in order to efficiently direct the ontogenetic development of the human. The increasingly more complex sociality of successive hominins was itself a major factor in the selective pressures for higher brain capacities—probably the most important case of biosocial co-evolutionary interaction.

2.2.3.2 The Increasing Prematurity of Hominin New-Borns and Enduring Dependency of Infants and Juveniles

The hominin brain growth with its shift from instinctive towards conscious control of behaviour during the hominisation process was associated with a prolongation of almost all life course phases—infancy, adolescence, adulthood, and old age. The only exception in this prolongation trend concerns the duration of human pregnancy. Compared to the duration of the great apes, the human pregnancy should last 21 months. In order to allow the big-brained human new-born to pass through the female pelvis at birth,[305] and to keep the foetal metabolic demands in balance with the mother's metabolic capacity,[306] evolution's adaptive strategy consisted, among others, of the premature birth of the human new-born. Hence, the human postnatal growth and development during infancy and adolescence became much more strongly dependent upon socio-cultural values, norms and structures for childcare. Moreover, the sociobiological prolonged infantile and juvenile dependency of the human child and adolescent increased and prolonged as human culture and society became more complex. Since humans are born so helpless, and infants and juveniles are dependent for so long, it takes many years of intensive learning and socialisation for the individual to independently survive, grow, and become a

[304] Jerison (1973).
[305] Portman (1944), Leutenegger (1982).
[306] Dunsworth et al. (2012).

functional member of a group. Even the motivation to learn and socialise must be stimulated by means of value and norm systems. The individual no longer knows instinctively what and how to teach his/her offspring in many domains that are important for survival. In modern culture, with its extensive educational requirements and rising standards for social and cultural performance, the care and oversight of infants, adolescents and even young adults transcends by far the role of parents and other kin and involves many more non-related adults. This brings us to the next crucial biological change in the hominisation process.

2.2.3.3 The Development of Human Sociality Extending Beyond the Family

The primary social bond in humans consists of the mother-child bond.[307] In the course of the hominisation process, with its increasing needs for raising large-brained offspring, more enduring relations between mating partners/parents emerged, extending the mother-child bond to the nuclear family unit. Also, broader kinship relations had a supporting effect on childcare and survival. Family and more general kinship relations were biosocial structures that emerged in response to the two specific characteristics of the human brain, mentioned above.[308]

However, human sociality largely transcends the family/kinship level, by forming groups of people including less closely related or even non-related individuals, resulting in larger population units such as tribes and nations. Those larger population units fulfil many more biosocial functions, not only in regulating the relations between group members but also in securing and/or expanding resource and territory acquisitions, particularly in defending against or competing with other population units.[309] Ultimately, these larger population units, including more distantly related individuals, also increased the survival and development opportunities of descendants—the intergenerational transmission of genes (and memes) of group members.[310] Human sociality is one of the most specific singularities in the hominisation process.[311]

The development of more complex biosocial units, such as enduring couples, larger kinship groups, and population units including less closely related individuals, in which prosocial predispositions, cooperation and division of labour, also between the sexes, increased in importance and transcended close kinship bonds, induced the selection and acquisition of innate moral sentiments that facilitate such behaviour, as well as the development of moral reasoning leading to abstract ideas about morality. Both the innate moral sentiments and the capacity for moral reasoning resulted in the establishment of moral rules that culturally regulate the social relations not only within but also between groups.[312]

[307] For instance, Rossi and Rossi (1990).

[308] Van den Berghe (1979), Filsinger (1988), Salmon and Shackleford (2007).

[309] Alexander (1979; 1987), Diamond (1997; 2005).

[310] Sanderson (2001), Wilson (2012).

[311] Wilson (2012, 45).

[312] For a detailed discussion of the role of moral intuitions and moral reasoning in human evolution, see Chaps. 16 (203ff) and 17 (217ff) in Krebs (2011); see also Wilson (1993), Hauser (2006), Bowles and Gintis (2011), Boehm (2012).

Innate moral intuitions as well as rationally designed social value and norm systems not only need to be able to control, or keep within reasonable limits, selfish-oriented biological drives, such as dominance, aggressiveness, laziness, avarice, envy, pride, lust, greed, cheat, cowardliness, rage and anger, spite and revenge,[313] but also to foster cooperative and other socially important attributes, such as altruism, reciprocity and mutualism, generosity, compassion, equality, fairness, honesty, and deference.[314] Research on very young children shows that prosocial inclinations and preferences are intuitive and widespread, occurring in the form of helping behaviour, sensitivity to fairness, indirect reciprocity and cooperative interaction.[315] Also, studies on adults show that social cooperation occurs on an intuitive basis and is perceived as emotionally rewarding.[316] Comparative studies across the primate order show that human prosocial sentiments have a solid evolutionary foundation.[317]

Biological control systems had to be complemented by socio-cultural ordering systems in the big-brained hominins. Group hunting, defence against or attacking other human groups required a totally different balance between competitive and cooperative drives and actions than a solitary way of life. Whereas morality is largely irrelevant to solitary organisms, it is a *conditio sine qua non* for developing and sustaining social life and becomes more complex as the size of the population increases.[318]

Human morality, grounded in moral intuitive emotions and in the capacity for moral reasoning, became an evolutionarily advantageous and indispensible instrument for the survival and the successful propagation of the large-brained, long-maturing, strongly socialised hominins living in groups, which were at a high risk of extinction due to frequently occurring severe environmental crises and belligerently competing with other groups.

As Dennis L. Krebs[319] concluded in his magnificent *The Origins of Morality—An Evolutionary Account*:

> The function of conceptions of morality is to induce individuals to uphold the social orders of their groups by constraining their selfish urges and biases, upholding relationships, promoting group harmony, resolving conflicts of interest in effective ways, dealing effectively with those who violate the rules, and fostering their interests in ways that, if everyone adopted them, would produce a better life for all.

[313] Campbell (1975), Wilson (1993, 12), Krebs (2011, 92).
[314] Krebs (2011, 201ff).
[315] For instance, Warneken and Tomasello (2006; 2007; 2013), Hamlin et al. (2011), Schmidt and Sommerville (2011), Hepach et al. (2012), House et al. (2012; 2013), Kato-Shimizu et al. (2013), Sebastian-Enesco et al. (2013), Cortes Barragan and Dweck (2014), Kuhlmeier et al. (2014).
[316] For instance, Rand et al. (2012), Aknin et al. (2013), Zaki and Mitchell (2013), Keltner et al. (2014).
[317] For instance, de Waal (1996), Silk and House (2011), Grueter et al. (2012).
[318] Rossano (2010, 174ff), Wilson (1993, 2).
[319] Krebs (2011, 27).

2.2.4 Major Stages in the Evolution and Historical Development of Morality and Content of Moral Systems

Paraphrasing Theodosius Dobzhansky's well-known catchphrase '*Nothing in biology makes sense except in the light of evolution*',[320] Joshua M. Tybur pertinently noted that '*Nothing in morality makes sense except in the light of evolution*'.[321]

Indeed, in recent decades, particularly since the emergence of the Second Darwinian Revolution—with its discovery of evolutionary mechanisms such as kin selection, reciprocity selection, coercive selection, group selection, and gene-culture co-evolutionary processes—a considerable body of research consisting of theoretical analyses (including mathematically modelled evolutionary studies, empirical investigations, and evolutionary game experiments) has enormously enlarged our understanding of the origin and evolution of phenomena such as altruism, cooperation, sociality, and morality.

Also, in recent years there has been an ongoing interest in evolutionary ethics as can be witnessed from the continuing flow of scientific books and articles about the origin and evolution of morality.[322]

In this discourse about the major stages in the evolution and historical development of morality the following issues are addressed:

(1) The major evolutionary changes in the dispositions of moral behaviour;
(2) The major stages in the historical development of morality;
(3) The temporal changes in the contents of moral systems.

2.2.4.1 The Major Evolutionary Changes in the Dispositions of Moral Senses, Moral Learning and Moral Reasoning

The concept of morality is usually defined as views and/or behaviour by which humans evaluate certain forms of behaviour as good and other forms of behaviour as bad, in particular regarding interpersonal and social relations. However, in order to consider the evolutionary antecedents and the evolutionary trajectory of human moral sentiments and behaviour, the concept of morality needs to be understood in a somewhat broader sense. It has to include precursor forms of social behaviours out of which typical human moral behaviour evolved.

As is well known, people have very different views of how morality emerged in the evolution and history of humankind. Religious people usually believe that morality is absolute, has a divine origin and is imposed by the supernatural power

[320]Dobzhansky (1973, 125).

[321]Tybur (2012, 35); see also Richerson and Boyd (2005, 237).

[322]For instance, Ayala (2010), Cartwright (2010), Komter (2010), Manner and Gowdy (2010), Bowles and Gintis (2011), Brosnan 2011), Krebs (2011), Sussman and Cloninger (2011), Boehm (2012), Gaitan Torres (2012), Gamble et al. (2014), Gintis (2014), Hodgson (2014), Kitcher (2014).

(s) in which they believe.[323] They often feel offended by the idea that human morality might have evolved from a so-called 'monkey morality'.[324] In contrast, most scientists think that morality emerged and evolved as a consequence of the interaction of biological needs, ecological constraints and socio-cultural living conditions. However, the scientific community is still quite divided about the relative share and interactions of biological and cultural factors in the development of moral principles. Bio-anthropologists, sociobiologists, and evolutionary psychologists and other students of evolution,[325] as well as social scientists and philosophers, who have a good knowledge of evolution science,[326] usually take a comprehensive view on the origin and evolution of morality. They acknowledge the biological origin and predispositions of moral sentiments and the capacity for moral reasoning, albeit being fully aware of the normative variability that can be produced by diverse ecological and socio-cultural living conditions. In the light of scientific knowledge there is no more room for a simplistic 'moral nativism'.[327]

Behavioural research on various groups of mammals, and in particular primates, shows that many of the emotional sentiments and cognitive abilities that are foundational for developing human morality, precede the hominisation process. Human morality is grounded in our mammalian prehistory, particularly the mother-child bond. However, in many group living non-human primates, many more innate dispositions for 'norm-related characteristics', such as attachment and bonding, cooperation and mutual aid, sympathy and empathy, direct and indirect reciprocity, altruism and reciprocal altruism, righteousness, fairness, consolation, conflict anticipation, conflict resolution and peace-making, deception and deception detection, cheating and free-riding curbing, and group loyalty community concern are already present to a certain extent. This occurs within groups due to tensions between individual interests and collective interests, or between groups in order to manage or master intergroup conflicts.[328]

In the course of the hominisation process, the strongly increasing need for prosocial forms of behaviour induced selective processes. Through these processes the development of moral sentiments and the capacity for moral learning and reasoning became embedded in the human genome and made our brain function accordingly. In addition

[323]For instance, Haught (2000).

[324]For instance, Koukl (2012). There are, obviously, other objections to the facts or views that morality is grounded in our biology; in particular, objections against the supposed deterministic, and hence unalterable, nature of evolutionary biological factors, or the misapprehension that the involvement of evolutionary biological factors in moral behaviour implies that moral principles must be genetically encoded. For a discussion of these fallacies, see, for instance, Hauser (2006, 420).

[325]For instance, Darwin (1871), Westermarck (1906), Keith (1946), Waddington (1960), Cattell (1972), Campbell (1975), Alexander (1987), Wright (1994), Katz (2000), Hinde (2002), Hauser (2006), Joyce (2006), Sinnott-Armstrong (2008), James (2011), Krebs (2011), Boehm (2012).

[326]For instance, Singer (1981), Richards (1987), Maienschein and Ruse (1999), Richerson and Boyd (2005), Verplaetse (2009).

[327]See, for instance, the discussion on moral nativism by Jesse Prinz, Susan Swyer, and Valerie Tiberius in Sinnott-Armstrong (2008, 367–439).

[328]De Waal (1996, 2006), Flack and de Waal (2000), Joyce (2006), Verbeek (2006), Bekoff and Pierce (2009), Silk et al. (2013), Van Schaik (2016).

to the basic emotions—pleasure, surprise, pain, fear, disgust, anger—the hominisation process (with its increasingly strong development of prosocial forms of behaviour) evoked specific social emotions such as love, pride, embarrassment, guilt, shame, envy, and jealousy.[329] Moral behaviour had a positive effect on reproductive fitness, because human groups with many altruists were more successful in dealing with the challenges of survival, adaptation, well-being, and competition with other human groups. Such group successes could even compensate for the fitness losses of altruistic individuals.[330] Moreover, biological and cultural selective processes induced people to cooperate not only for self-interests and the well-being of others, but even for promoting social values and norms for their own sake.[331]

Christopher Boehm[332] developed a theory in his recent book *Moral Origins—The Evolution of Virtue, Altruism and Shame* that the evolutionary origin of morality lies in the gradual development of a self-regulating conscience—the aptitude, faculty, intuition or judgment in distinguishing right from wrong. In his view, conscience evolution began with the social control or even suppression of alpha males, free-riders, cheaters and other socially deviant individuals, resulting in the social selection of genes that enabled people to internalise their group's moral rules, ultimately leading to more egalitarian, cooperative and altruistic group relations. Boehm situates this egalitarian revolution around 250,000 years ago, with the emergence of large game group hunting and meat sharing[333] among archaic types of *Homo sapiens*, a period in hominin evolution that lies close to the appearance of anatomically modern humans some 200,000 to 300,000 years ago.[334]

In another recent book, *The Origin of Morality,* Dennis L. Krebs[335] discusses the evolution of four primitive prosocial dispositions—deference, self-control, altruism and cooperation—as essential precursors to the development of the human moral senses and the capacity for moral learning and reasoning. In addition, he elaborates on the evolution of the uniquely human prosocial dispositions—deference to abstract ideas and distant rulers; suppression of selfish and aggressive urges; development of complex, indirect and equitable forms of cooperation; egalitarianism; strong reciprocity and indirect reciprocity; development of in-group favouritism ('tribal instincts').

The concept of moral sense goes back to Lord Shaftesbury who, in his *Characteristics of Men, Manners, Opinions, Times* of 1711, considered it as the origin of our moral ideas.[336] However, this concept has acquired or is commonly being used in a more specific meaning, namely as the innate ability to discern between good and evil, between right and wrong.[337]

---

[329]Bowles and Gintis (2011, 186ff).

[330]Gintis et al. (2008).

[331]Bowles and Gintis (2011, 1).

[332]Boehm (2012, 315–341).

[333]See also the 'egalitarian meat-sharing hypothesis' of Mameli (2013, 922).

[334]White et al. (2003), McDougall et al. (2005), Hublin et al. (2017).

[335]Krebs (2011, 75–142, 163–186); see also Wilson (1993, 29–120).

[336]Shaftesbury (1711).

[337]For instance, Broom (2004, XIII), Wright (2004, 328), Dawkins (2006, 214), Hauser (2006, 36), Krebs (2011, 9).

Currently, most scholars use the umbrella of the moral sense concept to distinguish a broad range of moral predispositions. For instance, in addition to the above mentioned primitive and uniquely human prosocial dispositions, Dennis L. Krebs[338] distinguishes specific moral senses which he considers to be associated with the welfare of other individuals and groups and evoked by moral norms, namely conscience, moral sentiments about others (empathy), the sense of rights (self-control), the sense of moral obligation (duty), and the sense of justice (fairness).

It is needless to elaborate on the well-known fact that the strong increase in brain size and complexity during the hominisation process equipped humankind with a considerably enlarged capacity for learning; in particular for developing social and moral behaviour, not only through teaching but also through the processes of approval and disapproval, rewarding, and punishment.[339]

Finally, there is the salient human capacity of reasoning that allows for the development of conscious moral judgment and decision-making.[340] This is also the result of the selective processes that occurred in the course of the evolution of the hominins to solve adaptive problems. Moral reasoning is a topic that raises, in scientific quarters as well as among lay people, much less discord than the issue of innate moral sentiments.

2.2.4.2 Major Stages in the Historical Development of Morality

Several scholars have distinguished and classified hierarchically the successive stages of the biocultural evolution of morality that shows phylogenetically as well as culturally a gradual increase and progressive complexity.[341]

For instance, Gregory Peterson[342] distinguishes three stages in the evolving morality: (1) 'quasi-morality' in animals whereby genetically programmed instinctual drives enable organisms to influence the well-being of one another; (2) 'proto-morality' among social mammals, and in particular primates, whereby "animals begin to be able to rationally deliberate actions and their consequences, based on the acquisition of a variety of cognitive skills, including enhanced memory and planning abilities, the ability to map social relations and social hierarchies, some awareness of how one's actions affect others, and the ability to form goals and roughly weigh pains and pleasures"; and (3) 'genuine morality' requiring mental capabilities beyond those needed for proto-morality, namely enabling abstraction and symbolic expression, and resulting in being disposed to develop moral sentiments and the capacity for moral learning and reasoning.

As a result of reading and interpreting the literature on the evolution of morality among the hominins, the authors would propose a following classification, roughly distinguishing three major stages in the evolution of morality: (1) a biosocial stage; (2) a cultural-religious stage; and (3) a cultural-scientific stage.

---

[338] Krebs (2011, 206–212).
[339] Crittenden (1992), May and Clark (1996).
[340] Krebs (2011, 217–232).
[341] Shermer (2004, 26ff).
[342] Peterson (2000, 472–476).

The first one—the biosocial stage—concerns the very slow and gradual shift from a biologically based proto-morality of non-human primates in which social sentiments were developed, as listed above, to the biosocially based morality of the hominins in which emotional personality features further evolved and cognitive capabilities were increasingly engaged in moral judgment.[343] This process took many thousands of generations during which the successive waves of hominins gradually succeeded in largely controlling and inhibiting their innate wanton drives. Philip Kitcher[344] is probably correct in hypothesising that this evolution was linked to the evolution of our linguistic capacity that facilitated conversation and negotiation. The evolving hominin morality was characterised, among others, by changes in social relations from predominantly biologically determined dominance hierarchies to relatively more socially controlled egalitarian relations—the dominance of individuals being superseded by the dominance of the group, as Christopher Boehm puts it.[345]

The second stage—the cultural-religious stage—is a relatively recent stage that occurred some 8000–10,000 years ago, when human societies shifted from tribes to chiefdoms and then to states in the agrarian era. Organised religions started to gradually incorporate moral ideas and behaviours into divinely sanctioned moral values and norms systems.[346] In the course of the agrarian era, with its increased means of subsistence and larger population sizes, the moral systems founded on religion were characterised by the re-establishment of stronger, but this time largely culturally and economically based, dominance hierarchies. This resulted in strongly increased social inequalities between different groups of people—men versus women, masters versus slaves, higher versus lower social classes, dominating versus subjugated populations.[347] The satisfaction of innate egalitarian drives was, in a shrewd way, ideologically largely transferred to a heavenly and more righteous hereafter.[348]

In the third stage—the cultural-scientific stage—emerging in the wake of the scientific revolution and its associated Enlightenment, morality very gradually started to be more strongly influenced by scientific insights. The scientifically based stage is mainly characterised by two seemingly opposite features: (1) the extension of moral concerns to the self-actualisation of each individual—moral individualism; and (2) the extension of morality to the complete human species, and even to the biosphere and its planetary basis as a whole—moral universalism.[349] This third stage is still in a very early phase of development, with values and norms again changing to more egalitarian standards, transcending the population genetic, ethnic or ideological in-group borders, and enlarging moral concerns to the biosphere and the ecology of the globe.[350]

[343] See, for instance, Bouchard and Loehlin (2001), Greene (2003, 847), de Waal (2007), Previc (2009), Boehm (2012), Tomasello (2016).
[344] Kitcher (2006, 136).
[345] Boehm (1999; 2012).
[346] Knauft (2000).
[347] Lenski (1984), Maynard Smith et al. (2010).
[348] Talbott (2015, 702).
[349] Darwin (1871, 147), Wilson (1993, 191ff).
[350] See also Singer (1981), Wilson (1993, 191ff).

Given this early stage of development of the present scientifically grounded moral era, it is understandable that its main features—moral individualism and moral universalism—are opposed and combated by conservative religious groups and institutions. Moral individualism and moral universalism are also parasitised by several regressive maladaptive behavioural trends such as increased moral relativism, minimised self-control, weakened moral teaching and learning, decreasing personalised social responsibility and duty, increased free-riding.[351] However, these epiphenomena should not derail the pathways to the establishment of a better-adapted moral balance addressing a broad variety of needs and drives at individual, group, national, international and planetary levels. The technical and technological modernisation is occurring at such a high speed that a comprehensive moral adaptation to the novel environment of modernity is urgent in order to avoid pending human-made disasters. However, moral adaptation has up-to-date been a slow process.

As argued above, several basic evolutionary mechanisms, such as mutation, natural selection, sexual selection, kin selection, reciprocal selection, coercive selection, and group selection, as well as cultural selection and biocultural co-evolution, have contributed to shaping the emotional and cognitive attributes that made humans sensitive to and receptive for the components of human morality that are universally present in human populations. They have also contributed to shaping many life- and reproduction-sustaining moral codes universally present in successful cultures.[352]

Evolutionary scholars have developed empirically well-documented and plausible theories about the mechanisms and scenarios concerning the biological evolution of moral sentiments and capabilities for moral learning and reasoning during the long-lasting hunter-gatherer stage of hominin evolution. However, it is not yet evidenced whether the agrarian/pastoral stage of cultural evolution has also influenced—and whether the scientific stage is currently influencing—the human genetic make-up related to moral predispositions. Whilst there is evidence for many physical and mental traits during the on-going biological evolution in the agrarian/pastoral and industrial stages of cultural evolution, it can only be hypothesised that the same process applies to neurological and/or hormonal characteristics that are related to moral sentiments. In this area, it is unlikely that really important genetic changes have already occurred, because of the relatively short duration of these cultural stages, the multifactorial nature of the genetic determinants of moral sentiments and cognitive capacities, and above all, the relatively small behavioural changes between the late Pleistocene, the agrarian/pastoral and the early industrial cultural eras.[353] One can argue that the currently emerging moral stage in human evolution and history, based on scientific insights in the evolutionary process, might in the future have more substantial impacts. This is based on a premise that the direction of changes of the modern social, cultural and

[351]See also Wilson (1993, 246ff).

[352]See, for instance, Huxley (1957), Cattell (1972), Alexander (1987), Cela-Conde (1987), Katz (2000), Hinde (2002), Broom (2004), Richerson and Boyd (2005), Lindsay (2008), Sinnott-Armstrong (2008), Cartwright (2010), Teehan (2010), Krebs (2011).

[353]Wilson (1978, 89).

technological environment can now be more effectively and consciously shaped to further evolution.

2.2.4.3 Major Temporal Changes in the Contents of Moral Systems

Many ethical theorists have dealt with the question of whether humanity has made moral progress in its long evolution and history. Has humanity been morally improving over time?[354] Whereas there is unanimity about technological progress and quite some agreement about the development of culture, expert opinions differ about moral progress.

Overall, most authors acknowledge that the development of modernity, in comparison with previous cultural stages, has been accompanied by a considerable moral progress. Classical examples of improvements in moral values and norms are the development of democratic decision making and governance; and the abolition of oppression of women, child labour, slavery and the death penalty. There has been a general decrease in criminality; the recognition of individual rights; the acceptance of coexistence of diversity: ideological/religious, racial, ethnic and some aspects of sexuality.[355] There is evidence of the prevention or reduction of inter-group violence and aggression;[356] the extension of ethical concerns to other than human life forms and the physical environment on the planet.[357] These changes show how strongly altering living conditions (particularly improving them) can affect the scope of moral reasoning.[358]

Conservatively thinking people may consider some of the modern personal decision-making liberties in matters of birth, partnership, and death, that people have acquired in advanced democracies as signs of moral degradation—the right to birth control (contraception, sterilisation, abortion, medically assisted fertility), partnership (personal partner choice, consensual union, LAT-relations,[359] divorce), and death control (euthanasia, palliative care). However, these are clashes between the moral values embedded in earlier cultural stages and the evolving ones.

Some developments in the twentieth and twenty-first centuries—for instance, the (re)appearance of discriminatory ideologies have raised the question as to whether modernity is actually experiencing a dramatic moral regression instead of a further progression; for example Nazism, fascism, communism, and Islamist fundamentalism, the virtual globalisation of intergroup conflicts, the reappearance or resurgence of exploitative capitalism in the form of what is now called Neo-liberalism, the rise or revival of some forms of ego-centric behaviour at the cost of community development.[360]

[354]See, for instance, Kant (1790), Williams (1893, 466), Huxley (1942, 576), Stent (1969), Nisbet (1980; 2009), Corning (1983), Nitecki (1988), Ruse (1996), Zarandi (2003), Pinker (2011).
[355]For instance, Neuhaus (2009).
[356]For instance, Pinker (2011).
[357]For instance, Jamieson (2003).
[358]For instance, Sachdeva et al. (2011).
[359]LAT relations: living-apart-together.
[360]For instance, Neuhaus (2009).

The judgment of whether a new moral development is a progression or regression is usually based on purely ideological premises. The authors argue that a comprehensive and long-term evolutionary approach is an appropriate way to deal adequately with the question of adaptive versus maladaptive ethics and behaviour.

2.2.4.4 Selfishness and Selflessness in the Evolution of Morality

Evolutionary ethics has raised and dealt with several crucial sociobiological questions and biophilosophical controversies concerning the origin and evolution of morality: the natural/cultural determination of morality, the emotional/rational nature of morality, the Is/Ought relationship, the gene/individual/community level of selection, the individual/societal approach to morality, and last but not least the question of the relationship between selfishness and selflessness in moral behaviour.[361] At present, it is not an exaggeration to affirm that most of those controversies have largely been resolved or at least clarified.

Notwithstanding its high costs, altruistic behaviour has multiple advantageous effects at the individual as well at the group level, on health and well-being, on social cohesion and cooperation, and last but not least on intergenerational continuity.[362]

Although the controversy about selfishness/selflessness goes back to the earliest stages of evolution science, it strongly flared up with the emergence of the Second Darwinian Revolution through the basic works of William Hamilton, George Williams, and Robert Trivers[363] and the more contemplative early writings of Michael Ghiselin, Edward Wilson, and Richard Dawkins.[364] In particular, Dawkins' 1976 classic *The Selfish Gene* became a stumbling stone, because its provocative title made it subject to a lot of misunderstanding, if not coarse distortion.[365]

Apart from the conceptual confusion about the multiple meanings of the concept of altruism,[366] it is quite understandable why the sociobiological findings about the relation between selfishness and selflessness are so profoundly disturbing. The hominisation process was made possible thanks to the development of human eusociality that, in turn, was partly based on the evolution of genetic predispositions and partly on compensating or even reinforcing cultural values and norms promoting altruistic behaviour. One of the major functions of altruistic sentiments and altruism-focussed cultural values and norms consists of curtailing powerful self-oriented drives; hence, our innate as well as our learned need to advocate, justify, and/or display our altruistic nature and reputation. The evolutionary explanation of the selfish/altruistic dynamics is strongly at odds with the

[361] See also Harman (2013, 2014).

[362] Manner and Gowdy (2010), Lozada et al. (2011), Rand et al. (2012), Aknin et al. (2013), Keltner et al. (2014).

[363] Hamilton (1963; 1964; 1975), Williams (1966), Trivers (1972).

[364] Ghiselin (1974), Wilson (1975, 1978), Dawkins (1976).

[365] See, for instance, Midgley (1985); for a discussion of this issue, see also Tanghe (2013).

[366] Clavien and Chapuisat (2013).

widespread but obsolete view that altruism is a lofty moral virtue that is highly elevated above the allegedly sordid materialistic evolutionary explanations.

In the section on kin selection (Sect. 2.1.2.5) the discourse was started by defining altruism, irrespective of its intention, as behaviour that reduces the reproductive fitness of a cooperating individual compared to the fitness of individuals who behave selfishly. Sociobiology discovered evolutionary mechanisms—kin selection, reciprocity selection, and group selection—through which genetic predispositions for altruistic behaviour can nevertheless be biologically transmitted and promote the inclusive fitness of the altruists.

However, some scholars argue, mainly on the basis of their ideological beliefs, that 'genuine' or 'moral' or 'sacrificial' altruism—defined as love behaviour that has no compensating reproductive benefit at all—cannot be explained by any form of Darwinian selection and, hence, would be 'maladaptive'. Its existence can only be explained on the basis of religious (supernatural) elements.[367]

First of all, the idea that moral or genuine altruism is a kind of behaviour that would be a specific characteristic of religious believers, and in particular Christian believers, is not only a typical in-group prejudice, but also flagrantly in contradiction with real facts. Moral altruistic behaviour is well known to be equally well present and practiced among people of various non-religious convictions—atheists, humanists, socialists, communists, ecologists, etc.[368]

Moreover, the theological interpretation of the application of the evolutionary toolkit on moral altruism sounds quite narrow-minded and does not fully grasp the complexity and diversity of Darwinian selective processes. It largely ignores the phenomenon of variability in altruistic behaviour, which can indeed be merely instrumental or more or less intensively motivated, and completely overlooks the neurological basis of emotions that are at the basis of various forms and shades of altruistic behaviour.

In contrast, the authors suggest that in *Homo sapiens sapiens*, with its high cognitive potentialities and refined emotional characteristics, deeply altruistically motivated forms of behaviour—moral altruism—might have as high a probability of being positively selected as more rudimentary forms of altruistic behaviour. This can occur via several evolutionary mechanisms; for instance, social assortment, direct or indirect reciprocal selection, or group selection.

2.2.5 Biological Determinants of Moral Behaviour

2.2.5.1 Genetic Determinants of Moral Behaviour

The question as to what degree genetic factors are involved in the development of moral sentiments or behaviour is difficult to answer. Moral behaviour is a complex phenomenon that includes features such as pro- and antisocial behaviour, altruistic and reciprocal behaviour, cooperation and mutual aid, sympathy and empathy,

[367]For instance, Schloss (2013, 213), Pruss (2013, 339), Clayton (2013, 347).
[368]For instance, Hofmann et al. (2014).

compassion, conscientiousness, agreeableness, moral judgment, conflict resolution and peace-making, deception and deception detection, cheating and free-riding curbing, community concern. Many of those components are in one or another way related to cognition, emotional personality, and psychopathology.

Hence, it cannot be expected that there exists one single gene for moral sense or behaviour. As is the case for many other complex characteristics, it can be expected that many genes as well as environmental factors are involved.

There are several general arguments in favour of the hypothesis that the human capacity for moral sentiments and behaviour is partially genetically determined. However, there is still rare hard evidence of the detection of genes linked to particular moral sentiments or forms of moral behaviour.

The general arguments are as follows. First of all, there are the findings on altruistic and other forms of moral sentiments and behaviour among social animal species, for which the comparative studies allow the identification of evolutionary origins and trends (see Sect. 2.2.4.1). Second, there is the universality of morality in the human species. This is a strong, although in itself insufficient, indication for a genetic predisposition. Third, there is the striking and ubiquitous phenomenon of in-group morality.[369] Fourth, there is the overwhelming evidence of the role of genetic factors in cognition, emotional personality and psychopathology that are all involved in particular forms of moral sentiments or behaviour.[370] For instance, all major emotional personality factors—neuroticism, extraversion, openness to experience, agreeableness and conscientiousness—have been found to be linked to a broad variety of moral sentiments or forms of moral behaviour.[371] In the domain of cognition, low cognitive abilities have often been found to be associated with delinquent or criminal forms of behaviour.[372] In particular, the important role of genetic factors in the aetiology of sociopathy and/or psychopathy is well established.[373] People diagnosed with 'antisocial personality disorder' or 'psychopathy' show signs of moral insensitivity from an early age.[374] Finally, there are the findings on moral sentiments or behaviour related brain structures and functions which are likely to have a genetic basis, because they cannot be linked to environmental or life course accidents. (See Sect. 2.2.5.2.)

What types of direct genetic knowledge about moral sentiments or behaviour are available? As is the case for any form of emotion or behaviour, three types of information are directly relevant for the involvement of genetic factors on particular forms of moral behaviour: (1) pedigree analysis of monogenic or chromosomal variants which are clearly linked to particular, usually pathological, conditions; (2) heritability estimates measuring the degree to which within-population variance

[369]Hartung (1995)

[370]For instance, Fuller and Thompson (1978), Plomin (1989), Plomin et al. (2008).

[371]For instance, Costa and McCrae (1992), Ebstein (2006), Knafo and Israel (2009), Johnson et al. (2011).

[372]For instance, Hirschi and Hindelang (1977), Moffit (1993), Walsh and Ellis (2003).

[373]For instance, Jang (2005), Livesley and Jang (2008).

[374]Pinker (2008, 4).

in particular forms of moral sentiments or behaviour is determined by genetic factors, environmental factors, and genetic-environmental interaction and covariance; (3) molecular genetics of particular sentiments or forms of behaviour relevant to morality.

A well-known example of a monogenic factor is the MAO-A gene on the X-chromosome that controls the function of the enzyme monoamine oxidase that helps to break down important neurotransmitters. Men (who have only one X chromosome) with low MAO-A activity (MAO-A-L) demonstrate aggression with greater intensity and frequency, particularly when provoked or maltreated.[375]

Studies about the degree to which genetic and environmental factors, as well as gene-environment interaction, are involved in within-population variation of various forms of prosociality yield significant heritability estimates, ranging between 30 and 50%, for many components of moral sentiments of behaviour, including empathy, helping, cooperation, altruism, and trustworthiness.[376]

Also, heritability studies about antisocial behaviour show that approximately 50% of the phenotypic variance is due to genetic factors.[377] Many studies have been undertaken on genetic influences on individual differences in aggression and antisocial behaviour,[378] although in this domain heritability estimates show a broader variation, ranging from 35%[379] to as much as 80%.[380]

The interesting fact about heritability studies of components of moral behaviour is that they indicate the degree to which behavioural differences between individuals are influenced by genetic factors. At the same time the above mentioned heritability estimates tell us something about the degree of possible effects of environmental factors (and many investigations also have estimations about genetic-environmental interactions), indicating that there is ample room and necessity for moral behaviour being environmentally influenced. The newly developing field of sociogenomics,[381] in which the effects of environmental factors on the expression of alleles will also be considered, will certainly allow deeper and more nuanced insights into the complex relations and interactions between genes and environments.[382]

[375] Ellis (1991), Brunner et al. (1993), Buckholtz et al. (2008), Sjoberg et al. (2008), McDermott et al. (2009).

[376] Rushton et al. (1986), Eisenberg et al. (2002), Fehr and Fischbacher (2003), Rushton (2004), Scourfield et al. (2004), Penner et al. (2005), Knafo and Plomin (2006), Hur and Rushton (2007), O'Connor et al. (2007), Volbrecht et al. (2007), Gregory et al. (2009), Knafo et al. (2009), Knafo and Israel (2009).

[377] Grove (1990), Blair et al. (2005), Ferguson and Beaver (2009), Nordio (2012).

[378] Knafo and Israel (2006), Hur and Rushton (2007), Craig and Halton (2009).

[379] Van der Valk et al. (1998).

[380] Dionne et al. (2003).

[381] Sociogenomics: the study of the molecular basis of social life by means of identifying genes that are implicated in social evolution (for instance, Robinson et al. 2005).

[382] For instance, Robinson et al. (2005), Roberts and Jackson (2008), Slavich and Cole (2013).

Notwithstanding the overwhelming evidence about the importance of genetic factors in predicting pro/anti-social behaviour, little is yet known about the specific genes involved. As is the case with other complex behavioural traits, many genes with small effects may be involved and are difficult to unravel.[383] Also, it cannot be expected that direct relations between genes and components of moral behaviour can be detected. Genes do not code for moral codes, but for proteins that may—often together with or in interaction with other genes and/or environmental factors—influence aspects of body build and function, cognition and emotional personality. In turn, these can lead to the presence (or absence) of moral emotions or forms of behaviour which are considered as moral (or immoral).

In recent decades, molecular genetics research has expanded its radius of action in the domain of behavioural genetics, and a specific molecular personality genetics is emerging.[384] Given the substantial relations between personality traits and moral sentiments or forms of (im)moral behaviour, this domain opens promising perspectives to better understand the development of many aspects of morality. Although molecular personality genetics is still in its infancy, several interesting discoveries have already been made. They include genes that are involved in the control of several brain signalling molecules, such as the neurohormone oxytocin,[385]

[383]Plomin et al. (2008).

[384]For instance, Benjamin et al. (2002), Noblett and Coccaro (2005), Canli (2008).

[385]For a long time oxytocin has been well known for its role in reproductive behaviour, particularly in labour at childbirth and breastfeeding. In recent studies oxytocin is also related to several sexual forms of behaviour, such as orgasm (Lee et al. 2009), pair bonding (Walum et al. 2008), and maternity (Bakermans-Kranenburg and Van IJzendoorn 2008), which is reason some authors started called it 'the love hormone'. However, more broadly, molecular genetic studies identifying variations in specific genes have found them to be associated with individual differences in forms of prosocial behaviour such as altruism, empathy, emotional perception, generosity, reciprocity, trust, and moral judgment (e.g. Kosfeld et al. 2005; Zak et al. 2007; Campbell 2008; Carter et al. 2008; Donaldson and Young 2008; Heinrichs and Domes 2008; Israel et al. 2008; 2009; Ebstein et al. 2009; Rodrigues et al. 2009; Mikolajczak et al. 2010; Tost et al. 2010; Van Dijk and Feith 2010; Reuter et al. 2011; Poulin et al. 2012; Walter et al. 2012; Zak 2012; Feldman et al. 2013; Jiang et al. 2013). Such findings led Zak (2012) to call it, in a recent book addressed to a broader audience, 'the Moral Molecule'. In contrast, oxytocin deficit disorder (ODD) has been found to be linked to autism, sociopathy, psychopathy and narcissism (Ebstein et al. 2010).

and the neurotransmitters dopamine[386] and serotonin.[387] In recent years extensive reviews of the relevant literature have been published.[388]

2.2.5.2 Neurological Determinants of Moral Behaviour

This book has repeatedly distinguished two neurological pathways involved in the development of morality:

(1) The neurological basis determining an innate moral sense;
(2) The neurological capacity for moral learning and reasoning.

In the expert literature one will find differences in the weight that is being given to these two types of neurological processes involved in the development of moral behaviour, a discussion which goes back to David Hume's and Adam Smith's questions as to whether the foundation of morals is derived from reason or from sentiment.[389] For instance, Francisco J. Ayala[390] considers that the human moral evaluation of actions mainly result from human rationality, which allows us to anticipate the consequences of one's own actions, to make value judgments, and to choose between alternative courses of action. Although it must be admitted that many moral dilemmas are not instinctively and instantly addressed, but only after careful rational deliberation,[391] most authors now also stress the importance of

---

[386]Dopamine functions in the brain as a neurotransmitter that plays a major role in reward-driven learning. Variants of the dopamine receptor genes have been found to be associated with temperament dimensions such as novelty seeking, extraversion, reward, and ADHD (Ebstein et al. 1996; 2010; Benjamin et al. 1996; Okuyama et al. 2000; Faraone et al. 2001; Kluger et al. 2002; Schinka et al. 2002; Becker et al. 2005; Eichhammer et al. 2005; Lesch 2007; Kovacs et al. 2009), with increased risks of criminal behaviour, alcoholism, drug abuse, and antisocial personality disorder (Tahir et al. 2000; Rowe 2002), and with an increased risk of various psychopathologies (Beaver et al. 2013). The dopamine receptor genes are consequently thought to play a role in pro/anti-social behaviour (Eisenberg et al. 2007).

[387]Serotonin is a monoamine neurotransmitter with various functions, including the regulation of mood, appetite, and sleep. Serotonin also performs tasks in some cognitive functions, including memory and learning. Several investigations have established an association between variants of the serotonin transporter gene and neuroticism/harm avoidance (anxiety-related personality traits) (Lesch et al. 1996; Cohen et al. 2002; Tsai et al. 2002; Schinka et al. 2004; Sen et al. 2004; Willis-Owen et al. 2005; Crockett, et al. 2010). It has also been found that serotonin modulates striatal responses to fairness and retaliation in humans (Crockett et al. 2013). A number of studies have documented a statistically significant association between the serotonin transporter promoter region polymorphism and antisocial outcomes. For example, carriers of the low expressing alleles are at-risk of displaying ADHD symptoms (Cadoret et al. 2003), consuming large amounts of alcohol (Herman et al. 2003), and having childhood conduct disorder (Cadoret et al. 2003). The serotonergic system has also been identified as being potentially involved in the aetiology of extreme violence and serious aggression: lower levels of serotonin were found to correspond with greater involvement in acts of extreme violence and consequently can be a source in the development of antisocial behaviour (Lesch et al. 1996; Virkkunen et al. 1996; Moore et al. 2002; Retz et al. 2004; Hu et al. 2006).

[388]Knafo and Israel (2009), Ebstein et al. (2010), Israel et al. (2015).

[389]Hume (1739/1969, 509), Maynard Smith (1759); see also Wierzbicka (2007, 75).

[390]Ayala (2010, 9018); see also Thompson (1990).

[391]Blackburn (2001), quoted in Churchland (2011, 111).

emotional personality characteristics, such as empathy, in the evolution of moral codes.[392] Emotions can even be an important drive for seemingly irrational behaviour.[393]

Ever since Charles Darwin,[394] many evolutionary biologists have argued that human morality is partly founded on a natural moral sense, as a result of the hominisation process during which the human brain universally acquired the capacity to evoke moral behaviour and be receptive to moral codes, to intuitively develop moral judgements, to learn to behave morally, to reason about moral questions, and to develop moral codes.[395] This does not imply that moral codes would be neurologically programmed, but that a neurological biogram was selected, which allowed for facilitating the production and receptivity of moral feelings and ideas. In *Moral Minds* Marc Hauser[396] argues even that our moral faculty is equipped with an innate, universal moral grammar for building moral systems and making moral judgments. Or, as Mario F. Mendez[397] concluded:

> Humans have an innate moral sense....

The question now is which brain structures are involved in the production of evolved emotional and rational capabilities for moral behaviour? There is already reference above to the important roles some neurological chemicals play in the development and variability of moral behaviour.

In recent years, progress in neurobiology[398] has allowed exploration and understanding of the neural bases of some of the most distinctive moral behavioural attributes of the human species, such as human altruism,[399] reciprocity,[400] care,[401] charitable donation,[402] social cooperation,[403] social interaction,[404] social

[392]For instance, Hoffman (2000), Greene et al. (2001), Greene (2009) Hauser (2006) Joyce (2006), Haidt (2001, 2007), Haidt and Craig (2007), Young and Koenigs (2007), Richerson et al. (2010), Krebs (2011, 213), Shoemacher (2012), Prinz (2015).

[393]For instance, Frank (1988, 254).

[394]Darwin (1871, Chap. 4).

[395]For instance, Pugh (1976), Wilson (1978), Kieffer (1979), Peters (2003), Greene (2003), Tancredi (2005), Hauser (2006), Verplaetse et al. (2009), Krebs (2011).

[396]Hauser (2006); see, for instance, also Greene (2003), Tancredi (2005), Mikhail (2007), Dupoux and Jacob (2007), Teehan (2010, 41).

[397]Mendez (2009)

[398]Moll et al. (2002; 2003; 2005; 2008), Jean-Baptiste (2003), Baschetti (2007b), Killen and Smetana (2007), Miller (2008), Narvaez and Vaydich (2008), Sinnott-Armstrong (2008), Mendez (2009), Verplaetse et al. (2009), Marazziti et al. (2013), Alvaro-Gonzalez (2014), Darragh et al. (2015), Decety and Wheatley (2015).

[399]Tankersley et al. (2007), Mathur et al. (2010), Marsh et al. (2014), Sul et al. (2015), Hein et al. (2016).

[400]Van den Bos et al. (2009), Watanabe et al. (2014).

[401]Robertson et al. (2007).

[402]Moll et al. (2006).

[403]Rilling et al. (2002), Declerck, et al. (2013), Schroeder et al. (2013).

[404]Schilbach et al. (2006).

attachment,[405] social reward,[406] social decision-making,[407] empathy,[408] honesty,[409] moral sensitivity,[410] moral judgments,[411] distributional fairness[412] and justice,[413] equitable decision-making,[414] egalitarian behaviour,[415] moral cognition,[416] moral disgust,[417] distaste for inequality and unfair treatment,[418] moral motivation,[419] moral action,[420] moral attitude,[421] beauty and goodness,[422] and wisdom.[423] This research has crystallised into a specific research domain some have called 'neuroethics'.[424]

In particular, new imaging techniques are used to define the neuro-anatomy of moral behaviour, to explore the centres of the brain associated with emotion, motivation, and moral behaviour in general,[425] to determine what parts of the brain are used when solving moral dilemmas and passing moral judgments. Functional magnetic resonance imaging (fMRI) studies, along with other psychophysiological measurements in normal individuals,[426] research on psychopathy or chronic anti-social behaviour,[427] and investigations on patients with frontotemporal dementia (FTD),[428] or with brain lesions leading to sociopathy,[429] point to a neurobiology of

[405]Moll and Schulkin (2009), Lewis et al. (2012).
[406]Bhanji and Delgado (2014).
[407]Leben (2011), Rilling and Sanfey (2011).
[408]Gallese (2003), Seitz et al. (2006), Singer et al. (2006), Decety (2010), Decety and Porges (2011), Bernhardt and Singer (2012), Decety and Svetlova (2012), Ferrari (2014), Decety and Cowell (2015).
[409]Abe and Greene (2014).
[410]Moll et al. (2002, 2007), Robertson et al. (2007).
[411]Greene et al. (2001), Moll et al. (2002), Borg et al. (2006), Koenigs et al. (2007), Prehn et al. (2008), Greene (2009), Young et al. (2010), Parkinson et al. (2011), Ciaramelli, et al. (2012), Yoder and Decety (2014).
[412]Hauser (2006, 83).
[413]Robertson et al. (2007), Buckholtz and Marois (2012).
[414]Zaki and Mitchell (2011).
[415]Dawes et al. (2012).
[416]Sevinc and Spreng (2014).
[417]Moll et al. (2005), Yang et al. (2013).
[418]Camerer et al. (2005), Tricomi et al. (2010), Takahashi et al. (2012).
[419]Frimer and Walker (2008), Narvaez and Lapsley (2009).
[420]Narvaez and Vaydich (2008).
[421]Knoch et al. (2006), Luo et al. (2006).
[422]Zaidal and Nadal (2011).
[423]Meeks and Jeste (2009).
[424]Rees and Rose (2004), Gazzaniga (2005), Hubbeling (2011), Schirmann (2013), Clausen and Levy (2015), Decety and Wheatley (2015).
[425]Cunnigham (2010), Decety and Howard (2013).
[426]Greene et al. (2001), Moll et al. (2002; 2003), Decety and Cacioppo (2012).
[427]Hoptman (2003), Blair et al. (2005), Yang et al. (2009), De Oliveira-Souza, et al. (2008), Harenski et al. (2010).
[428]Neary et al. (1998), Mendez et al. (2005).
[429]Damasio et al. (1990), Eslinger et al. (1992), Anderson et al. (1999).

moral behaviour.[430] In fact, all of the major dimensions of morality are related to the activity of one or more brain areas. As Frans De Waal[431] stated:

> Morality is as firmly grounded in neurobiology as anything else we do or are.

Morality may be considered to be a social construct, but it would not exist without the brain. Whereas it is clear that there is no unique moral brain centre, there are many brain areas that are involved in the development of various forms of moral behaviour.[432]

Morality-related brain structures and functions are not immutable over the life course of an individual. Although humans develop moral sentiments at a very early age,[433] there is increasing evidence for epigenetic and lifelong effects of attachment, care giving, educational activities, and responsive parenting on brain functioning and emotional regulation.[434] Both behavioural genetic and developmental psychological research shows that environmental factors—internal as well as external—can influence brain functions related to moral behaviour.[435] Responsible parenting stimulates earlier conscience development.[436] Early experiences result in better or worse equipped brains for moral life.[437] Brain damage at a young age or even during adulthood can unfavourably influence moral development or behaviour.[438] The human brain possesses a neuroplasticity, namely the ability of neurons to develop new connections, resulting in a rewiring of the brain, even in cases of brain degeneration due to lesions.

To conclude, despite the predisposition of the human brain to develop an innate moral sense, moral learning and training remain of crucial importance.[439]

2.2.6 Why Variability in Moral Behaviour?

An important question that remains is how to explain, from an evolutionary point of view, the existence or persistence of a variation in personality characteristics and/or moral codes that can, in particular circumstances, result in diverse forms of (im)-moral behaviour?

As for all biological or cultural features that can influence survival, sociality or reproduction in a positive or negative way, there are several causal factors that have

---

[430]Relatively recent reviews of the neurological bases of moral behaviour can be found in Tancredi (2005), Hauser (2006), Moll et al. (2008), Narvaez and Vaydich (2008), Narvaez (2008), Mendez (2009), Ebstein et al. (2010), Churchland (2011).
[431]De Waal (1996, 217).
[432]Joyce (2006, 140), Young (2012), Greene (2015), Oliveira-Souza et al. (2015).
[433]Hauser (2006, 303), Allchin (2009), Narvaez (2010).
[434]For instance, Karen (1994), Narvaez (2014), Bankard (2015).
[435]Narvaez and Vaydich (2008), Knafo and Israel (2009), Cowell and Decety (2015), Crockett and Rini (2015).
[436]Kochanska (2002).
[437]Lewis et al. (2000), Siegel (2001), Narvaez and Vaydich (2008).
[438]Eslinger et al. (1992), Christen and Regard (2012).
[439]Tancredi (2005, 43), Narvaez and Vaydich (2008).

to be considered for the variability in predispositions to moral behaviour or the production of moral codes. In particular, both genetic and environmental factors and their interactions can influence (im)moral behaviour. These factors and their interactions can elicit different forms of selection, resulting in varying degrees of adaptation or maladaptation.

First of all, genetic mutants can produce personality characteristics that induce, in combinations with particular environmental circumstances or life events, well-adapted or maladapted normative behaviour. A classical example is the presence of psychopathy which may have (had) some advantages in particular living conditions, but overall it is subject to strong negative selection or frequency-dependent selection,[440] resulting in a very low prevalence at the population level.[441] Also, some morality-related memes can fail to meet the needs for a well-adapted ontogenetic, societal, or intergenerational development; hence, they are under pressure of partial negative selection, because they produce a higher morbidity or mortality, or because they appear less satisfactory from an existential point of view, and are consequently selected against and regress. For example, female genital mutilation, a custom related to male cuckoldry fears practiced in patriarchally dominated societies.[442] Once women become empowered to understand that this practice has unfavourable effects on their sexuality and health and that it is used to control and suppress their femininity, they will oppose it. The practice will regress or disappear, as is being witnessed in populations that are emancipating from their patriarchal domination.[443]

Some behavioural variants may be due to the presence of particular gene or genotypic combinations. For instance, many socially important biological characteristics that show a continuous variability are controlled by several genes—so-called polygenes or quantitative trait loci (QTLs)—the extreme variants of which can be stronger under positive or negative selection. Another example consists of allele combinations in heterozygous and homozygous genotypes that are subject to balancing selection.[444]

Some current genes and/or memes may have evolved in the harsh ecological and social living conditions of earlier cultural eras, where they had fitness enhancing effects. In current modern living conditions, they have lost their original advantage or have even become a disadvantage—they are the so-called evolutionary hangovers. For instance, many scholars believe that ADHD (attention deficit/hyperactivity disorder) which has a relatively high prevalence in human

[440]For instance, Avilés (2002) showed that freeloaders increase in frequency when rare, but are selected against when common due to the reduced productivity of the groups they overburden with their presence.

[441]For instance, Glenn et al. (2011), Boehm (2012, 28).

[442]Hrdy (1981, 184).

[443]For instance, Almroth et al. (2005).

[444]In balancing selection, a selective advantage exists in favour of the heterozygote genotypes, whereby the allelic variants of a gene in heterozygote combinations are favoured over their respective homozygote genotypes. Thus, both the alleles of a gene, in proportions dependent on the adaptive advantages of the heterozygotes compared to the homozygotes, are maintained in the gene pool. A well-known example is the sickle-cell polymorphism (e.g. O'Malley 2006).

populations (between 3 and 5%) was considered adaptive in prehistory, because the harsh living conditions required hypervigilance, rapid-scanning, quickness of movement, hyperactivity and response-readiness, and because women were more attracted to male risk takers.[445] ADHD is obviously maladaptive in modern living and learning conditions.

It is very important for the understanding of the evolution of morality that humans are endowed with both egocentric and socially oriented drives.[446] In matters of mutual competition, the three basic drives in this context—egoism, nepotism and altruism (extrafamilial generosity)—are clearly hierarchically ordered, self-interest being the strongest and altruism the weakest.[447] These drives may be expressed in different ways and intensities according to social, cultural or ecological contexts. On an evolutionary scale, such basic drives may result in the simultaneous co-existence of different behavioural strategies that compete with each other.[448] For example, variation in several emotional personality features, such as the big five dimensions of human personality (extraversion, neuroticism, openness to experience, conscientiousness, and agreeableness) can have different trade-offs between fitness benefits and costs depending upon the nature of the socio-ecological or socio-economic context.[449] A classical example is violent behaviour[450] that in conditions of inter-group conflict may be advantageous, but in peace times must be subdued (see also Chap. 7, Sect. 7.5).[451]

A complicating factor is that humans can and do alter or vary their moral behaviour according to the identity of their beneficiaries or opponents. Kin, friends, and biological, cultural, religious or political conspecifics are often more amiably treated than strangers, or socially, philosophically (religiously), politically, ethnically or genetically different (groups of) individuals.[452] Peter Richerson and Robert Boyd[453] developed their theory in this respect on the evolution of tribal instincts through cultural group selection.

Last but not least, various environmental factors[454]—biological, psychological, social, cultural, economic, ecological—can all seriously affect moral codes and/or forms of behaviour, in particular those regarding group related forms of behaviour. A striking example is the physiological effect of semi-starvation on the prevalence of social strife and crime.[455] Moreover, moral values may be conceptualised in

[445]See Hartmann (1995), Shelley-Tremblay and Rosén (1996), Jensen et al. (1997), Crawford and Salmon (2002), Hartmann and Palladino (2005), Williams and Taylor (2006), Glover (2011).
[446]For instance, Sibly and Curnow (2012).
[447]Alexander (1987, 139–142), Boehm (2012, 331).
[448]Thomas (1984), Barr and Quinsey (2004), Cesarini et al. (2010).
[449]Nettle (2006), Cesarini et al. (2010).
[450]Gottschalk and Ellis (2010).
[451]Hawley et al. (2007).
[452]For instance, Petrinovic et al. (1993), Hartung (1995), Bernhard et al. (2006), Mifune et al. (2010), DeScioli and Krishna (2013).
[453]Richerson and Boyd (2005, 192); Richerson et al. (2003, 368).
[454]Tancredi (2005), Fumagalli and Priori (2011), Krebs (2011, 89).
[455]Keys and Brozek (1951).

different ways. For example, the necessity to control fertility in modern culture is differentially implemented by freethinking people who apply a broad range of contraceptive methods; whereas many people follow the Roman Catholic Church which rejects the use of appliance methods, and only allows sexual abstinence or the use of so-called natural family planning methods.

2.2.7 Moral Ambiguity of the Evolutionary Mechanism

From the very beginning of the development of evolution science, eminent evolutionists have drawn attention to the amoral, ruthless character of the nature of the biological-evolutionary system (natural selection, mutation, genetic drift—"*le hazard et la nécessité*", in the words of Jacques Monod[456]) and the need to replace it by a more humane system of evolution.[457] Thomas Huxley's[458] concluding remark of his famous essay on *Evolution and Ethics* is well known:

> Let us understand, once and for all, that the ethical process of society depends, not on imitating the cosmic process, still less in running away from it, but in combating it.

However, the recent development of sociobiological insights into the biological-evolutionary origin of morality, and more particularly of altruistic behaviour, shows that the evolutionary mechanism is not only responsible for the 'red in tooth and claw'[459] character of nature with its selfishness, greed, and competition, but also for highly valued forms of behaviour and moral principles such as unselfishness, (reciprocal) altruism and mutualism, and cooperation.[460] Moreover, the biological evolutionary mechanism, with on the one hand its chance events (mutation, gene recombination, gene drift, developmental accidents) and on the other hand its deterministic process (natural selection), resulted in the construction of the human brain. Evolutionary mechanisms enabled life on this planet to attain a level of consciousness and cognitive ability that allows us, more particularly since the development of science, to decide about and direct our own future evolution and development. As Simon Young sums it up, we can be "designers of our own evolution and destiny".[461]

---

[456]Monod (1970, 135).

[457]For instance, in a letter to Hooker, Darwin wrote: "What a book a Devil's Chaplain might write on the clumsy, wasteful, blundering low and horridly cruel works of nature" (quoted in Dawkins 2003, 8). See also Galton (1883), Stephen (1893), Huxley (1894), Dawkins (2003).

[458]Huxley (1894, 83).

[459]Alfred Lord Tennyson's (1908) '*In Memoriam A. H. Hallam*' (1850):

> Who trusted God was love indeed
> And love Creation's final law
> Tho' Nature, red in tooth and claw
> With ravine, shriek'd against his creed.

[460]De Waal (1996, 5), Mysterud and Penn (2007, 291).

[461]Young (2006, 32).

The perceived moral ambiguity of the evolutionary process can nowadays be addressed effectively, with its ruthless mechanisms—mutation, selection—on the one hand, but its remarkable end products on the other hand[462]—origin and evolution of life, consciousness, sociability, love, altruism, morality. There is the need to control the natural processes (natural selection, spontaneous mutation) by culturally induced biological processes (culturally induced selection and mutation), without, however, impeding the evolutionary process itself.

These aspects will be addressed in the following chapters.

[462]See also Joyce (2006, 222).

3 Adaptive and Maladaptive Features of Religious Beliefs as Sources of Morality

Abstract

The aim of this chapter is to look at religion and religiosity as sources of morality from an evolutionary perspective. The evolutionary origins of religious beliefs are investigated, genetic and neurological factors involved in religious behaviour are reviewed, and adaptive advantages and disadvantages of religions in pre-modern and modern living conditions are evaluated. The discourse on the organised religions is mainly focused on the Mediterranean region—Judaism, Christianity and Islam—whose essential characteristics and historical developments are briefly described and evaluated from an evolutionary point of view. The doctrines of the Abrahamic religions, as revealed in their basic scriptures, raise some anthropological questions and paradoxes about religions as sources of morality. The core of the chapter is devoted to the discussion about (1) individual and social effects; (2) proximate and ultimate effects; and (3) effects in ancestral and modern living conditions of religions as sources of morality and guidance for behaviour. The closing section of this chapter deals with the relation between science and religion. Two major issues are addressed: (1) the (in)compatibility of science and religion; and (2) the persistence of (neo)creationist beliefs.

3.1 Introduction

Substantial proportions of populations in modern(ising) societies are religious, with churches maintaining much of their traditional societal power and influence. However, their omnipotence as institutions has been gradually eroded, especially in countries that have embraced humanistic worldviews, high standards of education of the whole population, and where citizens have a strong sense of social protection. Religions have had to seek compromise with the findings of science and the values and norms emanating from philosophical or political ideologies such as liberalism,

R. Cliquet and D. Avramov, *Evolution Science and Ethics in the Third Millennium*,
https://doi.org/10.1007/978-3-319-73090-5_3

socialism, feminism, ecologism, or humanism. A typical example is the ethical basis of gender relations and values and norms related to rights of women.

For several hundred years, the achievements of the natural and human sciences shattered and undermined the very philosophical foundations and moral authority of the traditional revealed religions and the personal beliefs of many people.[1]

Some scholars are puzzled by the persisting religiosity at times of significant achievements and applications of science. How is the persistence or even resurgence of religiosity to be explained? Indoctrination in early stages of the life course?[2] Family pressure? Social status maintenance? Attraction to rituals? In-group sense of belonging? Loyalty to the tribe or to an ancient tradition? Disappointment with scientific materialism? Assignment of responsibility to supranatural power—de-responsibilisation of humans? Or is there, in addition, a more profound biosocial drive that might be genetically and/or neurologically hardwired?

The persisting religiosity in modern societies has incited many scholars to conclude that the predisposition to religiosity belongs to the "most ubiquitous and powerful forces in human life",[3] or is to be considered "as an ineradicable part of nature".[4] Is this really so?

This book will argue that religions and religiosity are part of our biological evolutionary and cultural historical heritage, which are adaptations—or exaptations[5]—to pre-scientific living conditions. Those times were characterised by a total ignorance of the causes of natural and life processes, and by the experience of powerless suffering resulting from disease, famine, crime, war, and natural catastrophes. However, the functional advantages of religious beliefs may change in the future as the living conditions modify.

3.2 Notions and Concepts

Since many of the notions and concepts related to religious issues can have different meanings, the authors want to clarify how they understand and use them in this book. Some readers may not agree with the interpretation or use of some of those concepts, but at least the choices are made clear and explicit.

3.2.1 Religion

In the expert literature on religion many definitions have been given about this phenomenon. The essential feature of religion, which many scholars agree upon, is

[1]Gazzaniga (2005, 163).
[2]Shariff et al. (2016).
[3]Hamer (2005, 4).
[4]Wilson (1978, 169).
[5]Exaptation: shifts in the function of a trait during evolution (Gould and Vrba 1982).

the belief in a supernatural agent or power that created the universe, explains its existence and meaning, and often imposes a moral code according to which humans should behave.[6] Most religions prescribe how the world or the human should be, and provide an integrated set of worldviews and practices.[7] A common feature of all religious behaviour consists of the presence of elements that are not identifiable by the senses or can be verified by evidence.[8] Religions distinguish themselves by involving counter-intuitive traits.[9] Religions may include a variety of elements, such as prayer, rituals, magic, mysticism, myths, miracles, spirituality, or divine revelation, but beliefs in God and the afterlife are the most prevalent.[10] Jared Diamond[11] assigns five major attributes to religious systems: (1) belief in a postulated supernatural agent; (2) groups of people identifying themselves as sharing deeply held beliefs; (3) presence of costly or painful sacrifices as commitment signals to the group; (4) belief in gods and other postulated supernatural agents implying how people should behave; (5) supernatural agents can be induced by prayers, donations, and sacrifices to intervene on behalf of mortal petitioners.

The above components of religion are typical for the Mediterranean religions of Judaism, Christianity, and Islam. In his comparative study of religions, Rik Pinxten[12] proposes a more general definition:

> Religions can be understood as a particular way of dealing with the Whole, or of reaching for a wholeness which transcends any individual in time and space, even in substance. In religious activities and utterances people express their cognitive, emotional and evaluative relations vis-à-vis others, animals, plants, the earth and the celestial phenomena.

Indeed, religions can be theistic—including a god concept—or non-theistic; they can rely on natural spirits, dead ancestors, or supernatural forces.

This chapter approaches the study of religions as cultural instruments that underpin sociobiological functions to justify and impose values and norms that were developed in the pre-scientific era of human history and were indispensable for onto- and/or phylogenesis.[13]

3.2.2 God

In the scientific literature on religions as well as among religious believers, two major meanings of the concept of God are usually distinguished. Firstly, a personal God (=theism) is an anthropomorphic, omniscient, omnipotent, omnibenevolent, and merciful, supernatural being that created not only the universe but also life, including human life on our planet; this personal God imposes, through revelations

[6] For instance, Tremlin (2006, 5).
[7] Saroglou and Muñoz-García (2008, 97).
[8] Dennett (2007, 25), Dow (2008), Steadman and Palmer (2008, 6).
[9] Boyer (2001, 65), Teehan (2010, 54).
[10] Previc (2006, 501), Rossano (2010, 27).
[11] Diamond (2012, 329).
[12] Pinxten (2010, 55).
[13] Cliquet and Thienpont (2002, 601).

via messiahs and prophets, moral rules of conduct, and judges people thereupon in a hypothetical hereafter. Secondly, an impersonal God (=deism) is divine essence who created the universe, but does not intervene in people's personal lives.[14]

The authors are focused on the moral imperatives imposed by organised religions in recognition of God's magnanimity or in the name of a supernatural power that is believed to have created us and intends to judge us. The concept of God is used here in the sense that organised religions such as Judaism, Christianity and Islam conceived it in their basic scriptures—the Tanakh (Hebrew Bible or Old Testament), the Christian Bible (including the Old and New Testaments) and the Muslim Qur'an. The authors are well aware of the fact that within some of the organised religions there are theological developments or theological variants that try to interpret and adapt the religious doctrines, by taking into account (some of) the findings of sciences. However, none of the organised religions have succeeded so far to fully take into consideration and assimilate the scientific achievements of evolutionary science and to reconcile or integrate those achievements into their religious belief systems.

3.2.3 Religiosity

The concept of religiosity is intimately related to religion and refers to various aspects of religious activity, dedication, and belief. Religiosity is difficult to measure, because it can include various components related to beliefs and practices of organised religions, such as belief in God and worship attendance, as well as items related to non-organised religious behaviour such as prayer and meditation.[15]

National or international attitudinal surveys, including questions on religiosity and other philosophical/ideological/political convictions,[16] show that there exists, within many countries as well as between countries, a substantial diversity in individual religiosity. Worldwide, 85% of people experience some form of religious belief,[17] or report that religion is an important part of their daily lives.[18] However, there is a considerable within-country as well as between-country variation. For example, the results of the 2005–2007 wave of the World Value Study (WVS),[19] about the proportion of people identifying themselves as a religious person, varies between 24% in China and 98% in Mali.

[14]Stark (2001, 5), Graffin and Provine (2007, 294).
[15]Hill and Hood (1999), Hall et al. (2008).
[16]Although many surveys contain relevant and interesting data about those issues, it must be noted that many of them often approach the question of ideological diversity in a superficial and/or a prejudiced, lopsided, or unbalanced way. A striking example is the otherwise highly interesting European Value Study that includes detailed questions about religious practice and religiosity, but almost completely neglects to capture the characteristics and diversity of the views of non-religious people such as apatheists, agnostics, freethinkers, atheists and humanists.
[17]Zuckerman (2005, 16).
[18]Crabtree (2009).
[19]http://www.wvsevsdb.com/wvs/WVSAnalizeQuestion.jsp.

In particular, culturally and technologically advanced countries are characterised by more or less strong religious/ideological diversity amongst their populations. In developing countries this variation is much more limited, and may even be non-existent.[20] This lack of variation in individual religiosity, or the strongly skewed distribution of the religious-nonreligious attitudes in many countries, is not surprising. Religious education and indoctrination are strongly embedded in many countries in the educational systems of families, churches and schools. In contrast, information and education about non-religious ideologies, such as agnostic or atheist humanism, is virtually absent in schools in most countries. In many countries, until recently refute of religious doctrine was or is still today subject to stigmatisation, ostracism, or—worse—persecution.[21]

The theist-atheist gradient in modern(ising) societies consists of three major groups of people: (1) citizens who still adhere to one of the religious faiths and are churchgoers; (2) people who are believers but are not part of any congregation; and (3) non-religious people, agnostics, and atheists, often referred to as freethinking people. In addition, one can have a variation in moral convictions that cuts across the theist-atheist bipolarity. Varying proportions of religious people may still profess the moral teachings and follow the moral rules of their religious denomination, but many others may have become what could be called 'modernist' religious believers, i.e. people who in general continue to believe in the theist tenets of their religion, but who foster in many domains of life modern or secular values and norms: for instance they have premarital sex, use modern contraceptives, apply induced abortion, live in consensual unions, have small numbers of children, divorce and remarry, and apply euthanasia.

Individual differences in religiosity have been invariably found to be differentially associated with several basic personality traits[22] as well as to their cultural adaptations, namely values.[23] Although the associations between religion and basic personality factors are rather weak, religious people tend to show higher degrees of agreeableness and conscientiousness, but they are less inclined towards novelty/open-mindedness.[24] Religious people are also more sensitive to traditional and conservative values, and are less prone to values indicating openness to change and self-enhancement.[25] The fundamentalist variant of religiosity, characterised by authoritarian and dogmatic religious attitudes, beliefs, and practices, is particularly adverse to openness to experience.[26]

[20]For instance, Inglehart et al. (2004), Pickel and Müller (2009), Haller et al. (2009), Crabtree and Pelham (2009).
[21]Edgell et al. (2006).
[22]McCullough et al. (2003), Wink et al. (2007), Saroglou (2010), Kandler and Riemann (2015).
[23]Hills et al. (2004), Saroglou and Muñoz-García (2008), Saroglou (2010), Lynn et al. (2009).
[24]Personality traits are invariable patterns of responses to the exigencies of the environment, have a high heritability and are highly stable throughout adulthood. The five-factor model of personality distinguishes the factors 'neuroticism', 'novelty', 'conscientiousness', 'agreeableness' and 'extraversion' (Costa and McCrae 1992; 2008).
[25]Saroglou and Muñoz-García (2008), Saroglou (2010).
[26]Altemeyer and Hunsberger (2005).

Research in developed countries shows that there is a striking negative relation between degree of religiosity and educational level.[27] Also, population studies in Europe and the United States,[28] as well as internationally comparative studies,[29] point to a clear negative correlation between religiosity and cognitive ability. This means that the degree of religiosity decreases with increasing education and higher levels of cognitive ability.

Finally, comparative international studies show that religiosity is dependent upon the level of economic development: religiosity prevails most strongly among poorer nations where life-threatening, societal or physical risks are high. In contrast, secularisation has progressed most among the more wealthy in more developed countries.[30] On comprehensive indicators of welfare or well-being, the least religious countries perform better than the most religious: higher life expectancy, lower infant mortality, less crime, higher literacy, higher GDP, better child welfare, more economic equality, higher economic competitiveness, higher gender equality, better healthcare, more investment in education, higher rates of university enrolment, more access to telephone, TV and internet, better environmental protection, less corruption, more political stability, more assistance to poorer nations.[31]

3.2.4 Spirituality

The concept of spirituality has also been given different meanings: "the presence of a relationship with a Higher Power that affects the way in which one operates in the world"[32]; "one's focus on, and/or reverence openness, and connectedness to, something of significance believed to be beyond one's full understanding and/or individual existence"[33]; "a feeling of being connected to something larger than oneself"[34]; "the experience of a sense of timelessness and spacelessness".[35] The authors approach spirituality as a predisposition, partially embedded in the human genetic endowment, for experiencing self-transcendence.

Spirituality is also related to personality, but contrary to religiosity it is more strongly associated with openness to novelty, fantasy, and universalism. Spiritual people seem to be similar to religious people in the latter's prosociality and conscientiousness, but not in their conservatism and authoritarianism, and their low inclination towards self-direction and hedonism. In this respect spiritual people

[27]Glaeser and Sacerdote (2008), Schieman et al. (2010). Johnson (1997), Baker (2008), Sherkat (2008, 2011).

[28]For instance, Verhage (1964), Poythress (1975), Wenegrat (1990, 88), Bell (2002), Lynn et al. (2009), Nyborg (2009), Kanazawa (2010).

[29]Lynn et al. (2009).

[30]Norris and Inglehart (2004).

[31]Zuckerman (2005), Halman and Draulans (2006), Paul (2009), Harris (2010).

[32]Armstrong (1993, 3).

[33]Krippner (2005, 81).

[34]Wilson (2002, 3).

[35]Saver and Rabin (1997, 507).

much more resemble non-religious people who highly value self-direction and universalism[36] and novelty.[37]

Unfortunately, little is known about the spirituality of non-religious people who cherish modern secular ideologies such as humanism, socialism, liberalism, feminism, ecologism.

3.2.5 Relations Between Religiosity, Spirituality and Morality

Although different, spirituality and religiosity are usually strongly interrelated.[38] It will be seen from heritability research that spirituality has a stronger genetic component in variability than religiousness, which is primarily transmitted through learning processes.

Although spirituality is often linked to religiosity, this need not necessarily be the case. Spiritual experiences can be independent of religiosity, beliefs in gods, or adherence to religions.[39] Spirituality can exist outside religion; it can be embedded in many other human spheres of interest such as philosophy, ethics, politics, environmentalism, art, and even science. Spiritually driven people also differ from religious people, for instance, in showing more compassionate and altruistic behaviour towards those who differ.[40]

Although interrelated, religion, religiosity and spirituality are apparently often confounded with each other. Religion is an institutional characteristic; religiosity and spirituality are personal characteristics. From heritability studies it appears that both religiosity and spirituality can be strongly influenced by environmental factors (see Sect. 3.6.1). Moreover, in comparison to spirituality, religiosity appears to be more susceptible to memetic than genetic factors.[41] This implies that people with strong predispositions to spirituality need not necessarily be religious. Indeed, it should not be forgotten that religion/religiosity results from an interaction of specific cultural, economic and ecological living conditions and the evolved human mind.[42]

As far as the relation between religion and morality[43] is concerned, religious people have often been made to believe that morality is contingent upon religiosity, that only religion can give meaning to life on the planet, that without the existence of a God there would be no basis for sustaining the moral order, and that non-believers and atheists lack values and are incapable of moral behaviour.[44]

[36]Barnea and Schwartz (1998), Caprara et al. (2006).
[37]McCrae (1996).
[38]Hamer (2005, 13), Dennett (2007, 7).
[39]For instance, Piedmont (1999, 988), Zinnbauer et al. (1999, 899), Hill et al. (2000), Anandarajah and Hight (2001, 83), Vannelli (2001, 223), Hill and Pargament (2003, 72), Miller and Thoresen (2003, 29), Zinnbauer and Pargament (2005, 27), Harris (2014).
[40]Saslow et al. (2013).
[41]Bouchard and Loehlin (2001), Koenig et al. (2005).
[42]Bulbulia (2004), Whitehouse (2008).
[43]For the definition of morality see Chap. 2 Sect. 2.2.
[44]See the discussions in Holloway (1999), Buckman (2000), Wielenberg (2005), Hauser and Singer (2006), Epstein (2010).

This view is embedded in the Holy Scriptures. According to Psalm 14 of the Bible, fools who say there is no God "are corrupt", "do abominable deeds", "have no knowledge", are incapable of doing any good, and "are in great terror, for God is with the generation of the righteous". This ancient position is still well spread today, not only in the minds of religious leaders[45] but also among ordinary religious people.[46]

Many survey investigations on people's views about the relation between religiosity and morality find a positive relationship between degree of religiosity and moral behaviour. Immoral behaviour is often intuitively judged as being more prevalent among nonreligious people.[47] However, most of those studies are strongly biased because they usually consider only one part of the ideological spectrum in a population. They distinguish several grades of moral involvement on the religious side of the spectrum, but ignore completely the degree of moral involvement on the non-religious side of the ideological spectrum. It is much more challenging to measure the degree of ideological involvement/activity/participation of the non-religious part of the population. The ideological activism of non-religious people may be based on a diversity and/or combination of secular ideologies, as it may build on values of humanism, ecologism, or feminism for example.

In fact, contrary to the widespread conviction that religious believers are more moral than non-religious people, there are some indications that exactly the opposite might be the case. For instance, a recent comparative international study in six countries (Canada, China, Jordan, Turkey, USA and South Africa) on altruistic behaviour of children from religious and non-religious households showed that religiousness was inversely predictive of children's altruism and positively correlated with their punitive tendencies: children from religious families were less likely to share with others than were children from non-religious families; and a religious upbringing was also associated with more punitive tendencies in response to anti-social behaviour.[48]

Although assimilating morality with religion is a quite general belief among religious people, there is much evidence that it is an erroneous conception.[49] The idea that morality is contingent upon religiosity and that moral norms originate from a religion is a typical in-group prejudice. The moral rules among hunter-gatherers,[50] the non-theistic ethics in ancient Greece and in Confucian China,[51] as well as the ethical basis and practice of many modern secular ideologies, show that morality can develop without any connection to the moral commandments of a supernatural power. Some moral standards have developed and are

[45]For instance, Pope Pius XI, 1930; D. Wuerl, Archbishop of the Roman Catholic diocese of Washington, DC (https://www.priestsforlife.org/magisterium/bishops/wuerl-2006-red-mass.htm).
[46]See, for instance, the survey by Shermer and Sulloway quoted in Shermer (2003).
[47]Pew Research Center (2007), Gervais (2014).
[48]Decety et al. (2015).
[49]Hauser (2006, 421).
[50]Wright (2009, 23–26).
[51]Martin (2008, 251).

being implemented despite official religious teaching, as is the case with gender equality, for example. Empirical confirmation of high moral standards is found in many strongly irreligious populations in societies which are characterised by excellent educational systems, prosperous economies, high health care and social protection systems, low rates of violent crime and corruption, and whose citizens, on average, score very high on happiness indices. This is the case in countries such as Denmark and Sweden.[52] As Daniel Dennett[53] stated:

> There is no reason at all why a disbelief in the immateriality or immortality of the soul should make a person less caring, less moral, less committed to the well-being of everybody on Earth than somebody who believes in "the spirit".

As will be argued below, there is a clear causal sequence starting from the basic needs of human nature (and its evolution), then the development of morality in more complex social groups, to the religious consecration and imposition in large agrarian societies.[54] Morality does not derive from religion: rather the opposite is the case.[55] Moral principles arose in human societies long before the major world religions developed. Hominin biological evolution towards the development of genetic predispositions to cooperative behaviour and moral sensibilities preceded by eons of time the emergence of the organised religions of the agrarian era.[56]

Major components of morality appear to be universal.[57] Moral principles, such as the Golden Rule, altruism, bravery, generosity, and prohibition of in-group killing, stealing and lying, evolved as a consequence of natural selection in our highly socialised species.[58] In their cross-cultural, socio-historical analysis of personality traits that are universally considered to be moral virtues, Christopher Peterson and Martin Seligman[59] identified the following: wisdom, courage, humanity, justice, temperance, and transcendence. Matt J. Rossano[60] correctly argues that religion may positively contribute to the development of such moral virtues. However, it is difficult to argue that non-religious people, and in particular atheistic humanists, are less supportive of those virtues. Even transcendence, which is believed by many to be inherently connected with religion, is not just applicable to supernatural experience.

[52]For instance, Zuckermann (2008), see also Epstein (2010), Shults (2015).
[53]Dennett (2007, 305), see also Aronson (2008), Norenzayan (2014).
[54]Teehan (2006, 748), Lahti (2009, 69).
[55]Broom (2006), Hauser (2006), Pyysiäinen (2006).
[56]See also Beit-Hallahmi (2010, 130).
[57]Stenger (2009, 150).
[58]Broom (2004; 2006).
[59]Peterson and Seligman (2004).
[60]Rossano (2010, 186ff).

3.3 Origin and Evolution of Religion/Religiosity/Spirituality

All students of religion are unanimous in asserting that the prevalence of religion and religiosity is a universal characteristic of the human species.[61] Some scholars have baptised the human species as *homo religiosus*[62] because of its nearly universal belief in supernatural agents.

This universality is a strong condition, which hypothesises that religion/religiosity has played an important role in the biological evolution of humankind. It also forms an argument for presuming that such a universally shared behavioural pattern partially has a genetic basis.[63]

Although the idea that biological (more particularly genetic and neurological) factors might be involved in forms of behaviour that lie at the basis of spirituality/religiosity and/or religious practice has strongly gained momentum in recent years,[64] this presumption is not at all new. Many anthropologists and other scientists acquainted with evolutionary science have since long fostered the idea that the human brain has been selected evolutionarily for the conceptualisation or at least the receptivity of religious beliefs and practices.[65] Its function evolved as a response to the survival challenges in a harsh natural and competitive social environment.[66] There is a view that the biological predisposition to religiosity and religious memes emerged and proliferated as a function of biological evolutionary needs related to survival and reproduction. This view has nowadays become commonplace among evolutionary scholars who address those issues.[67]

Stewart E. Guthrie[68] argues that religion finds its origin in animism, an interpretative perceptive strategy "that aims too high, attributing to things and events more organisation than they have." In the harsh and dangerous Pleistocene environment, this strategy must have had an evolutionarily advantageous pay-off, because in the case of a perceptive ambiguity—for instance, mistaking a boulder for

[61]For instance, Brown (1991), Armstrong (1993), James (1902; 1997), Atran (2002), Kardong (2010).

[62]Eliade (1961).

[63]Alper (2006, 62), Churchland (2011, 108).

[64]For instance, d'Aquili and Newberg (1999), Hamer (2005), McNamara (2006; 2009).

[65]For instance, Darwin (1871), James (1902), Harrison (1909), Cattell, (1938, 1972), Gallus et al. (1972), Stent (1976), Wilson (1978), Kieffer (1979), Reynolds and Tanner (1983), Baril (2006).

[66]Tremlin (2006, 141).

[67]For instance, Crippen and Machalek (1989), Boyer (2001), Atran (2002), McClenon (2002), Wilson (2002), Voland and Söling (2004), Dawkins (2006), McNamara (2006), Graffin and Provine (2007), King (2007), Bulbulia et al. (2008), Wolpert (2008), Ellsworth (2009), Feierman (2009), Voland and Schiefenhövel (2009), Wright (2009), Kardong (2010), Teehan (2010), Voland (2010).

[68]Guthrie (1993, 39–61), see also Barrett (2000, 31–32), Atran (2006, 189), Teehan (2010, 45).

a bear—the overestimation of a potential danger would have been a life-saving decision. Hence, the animistic interpretative strategy is a more general application of Pascal's famous wager.[69]

3.3.1 Earliest Signs of Religion/Religiosity/Spirituality

Some anthropologists have interpreted the occurrence of cannibalism and skull preservation amongst *Homo erectus pekinensis*, whose age is estimated to be 500,000 years, as the very first signs of a mystical belief system.[70] Also the nature of the Sima de los Huesos human remains (Homo heidelbergensis, estimated at 430,000 years old) was interpreted as the earliest signs of human burials.[71]

Ceremonial burials amongst *Homo sapiens neanderthalensis* are better established facts, indicating that this hominin devised a belief that death could be transcended.[72]

The earliest signs of a mortuary ritual amongst *Homo sapiens sapiens* have been discovered on the Herto skulls from Middle Awash in Ethiopia, radioisotopically dated to between 160,000 and 154,000 years ago.[73] For *Homo sapiens sapiens* many more paleo-anthropological or archaeological findings are known to be indicative of the ancient presence of a belief system.[74] One of the most remarkable discoveries appertain to the famous Chauvet cave in the Ardèche Gorge in France, where in addition to the amazing cave paintings that are estimated to be some 32,000 years old, remains of a cave bear worship cult have been found.[75]

Students of morality perceive in the history of humankind a gradual and hierarchical shift from a pre-moral sociality to a tribal morality and then towards a universal morality.[76] Even in the mid nineteenth century Charles Darwin[77] characterised the history of human morality by a gradual extension of our moral sensibility and awareness to ever larger and more inclusive social units. David C. Lahti[78] sees this extension of the circle of moral considerability in the shift from pre-religious animism to tribal religion and then towards universalising religion.

[69]Pascal's wager (Pascal 1670) suggests that in a bet on whether God exists or not, a rational person should live as if God exists and try to believe in God because of the benefits to be expected in case that God really exists and has the powers attributed to him. If God does not exist, losses will only be finite.

[70]Weidenreich (1943), Hayden (2003), Rossano (2007). See also Defleur et al. (1999) for cannibalism among the neanderthalers.

[71]Carbonell and Mosquera (2006).

[72]For instance, Louwe Kooijmans et al. (1989), Trinkhaus and Shipman (1993), Defleur et al. (1999).

[73]White et al. (2003).

[74]Hayden (2003).

[75]Chauvet et al. (1995), Bocherens et al. (2006).

[76]Singer (1981, 120), Lahti (2009, 70), Bellah (2011, 104).

[77]Darwin (1871), Chaps. 4 and 5.

[78]Lahti (2009, 70).

Based on A.F.C. Wallace's classification of cult institutions, Stephen Sanderson[79] distinguishes four major stages in the social evolution of religion: shamanic religions (containing only individualistic and shamanic cult institutions), communal religions (containing individualistic, shamanic, and communal cult institutions), Olympian religions (also containing ecclesiastical cult institutions and the worship of numerous gods), and monotheistic religions (identical to Olympian religions, but worshipping a single god). These four types can be regrouped into two major stages, namely kinship-based or tribal religions (including shamanic and communal religions) and organised world religions (including the large polytheistic and monotheistic religions).

3.3.2 Kinship-Based/Tribal Religions

Past and present hunter-gatherer populations are all characterised by the presence of kinship- or tribal-based religions, also sometimes called folk religions. They are animistic in orientation, meaning that spirits are not only present in humans but also in other components of nature. Animism attributes life to the lifeless, i.e. overvalues the real value of things and events.[80] Hunter-gatherer populations are often also characterised by the presence of belief in an afterlife, shamanism, and ancestor worship, but seldom adopt moralising high gods.[81] The earliest forms of religion were strongly characterised by the prevalence of rituals and sacred narratives that strengthened bonding among members of the group and promoted the community's survival. They had little to do with matters of theology. Tribal-based religions predate the present-day organised or so-called world religions in agrarian and modern societies.[82]

3.3.3 Organised or World Religions

Scholarly experts in the history of religions unanimously hold the view that the transition from animistic/kinship/tribal religions into organised or world religions occurred with the emergence of agrarian-pastoral stage in the socio-cultural and technological development of humankind, during which human populations established permanent settlements.[83] These religions developed in order to solve the social problems and conflicts that occurred as small societies based on kinship evolved into larger societies, composed of more numerous and less closely related people.[84] Hence, organised religions aim to promote broader forms of social

[79]Sanderson (2008, 3).
[80]Guthrie (1993, 6).
[81]Peoples et al. (2016). High God: is a "spiritual being who is believed to have created all reality and/or to be its ultimate governor" (Swanson 1960). High gods may "vary in their activity in human affairs and their concern with human morality" (Johnson 2005, 418).
[82]Dennett (2007), Steadman and Palmer (2008, x), Wright (2009).
[83]Armstrong (1993), Giovannoli (2000, 81), Dennett (2007).
[84]Teehan (2010, 66), see also MacIntyre (2004), Van Schaik and Michel (2016).

cooperation beyond narrow kin-based relations.[85] Indeed, several researchers have found relationships between demographic and socio-cultural characteristics such as population size and density, the practice of agriculture, animal husbandry, and political complexity, and the development of beliefs in moralising gods or high gods.[86] Interestingly, environmental duress, either in the form of resource scarcity and/or ecological instability or aridity, also appears to contribute to the development of such beliefs.[87]

However, organised or world religions are not only characterised by the belief in powerful, all-knowing, moralising, rewarding and sanctioning high god(s), but also by several additional features such as routinised rituals, 'fictive kinship' and in-group markers, elaborate value and norm ruling, mainly in the domains of self-control and sexual and reproductive behaviour. All of these are aimed at dealing with the challenges provoked by the new living conditions in order to promote the cultural, political, demographic and biological success of their communities.[88]

The emergence of the agrarian-pastoral stage in human history was characterised by a large number of technological inventions. There was increasing subsistence productivity, changing living conditions and social relations within human societies, strongly characterised by increasing inequalities between the sexes, social classes, and whole societies.[89] There was also worsening of the living conditions, health, and quality of life of large portions of the population (with the exception of the ruling classes) due to increasing susceptibility to infectious diseases as a result of increasing population size and crowding, animal husbandry, and the dependence upon nutritional monocultures.[90] The early agrarian societies were strongly stratified theocracies in which elites safeguarded their social and economic privileges and reproductive advantages by suppressing and exploiting their subordinates.[91]

In a multi-regression analysis of the evolution of religion, Stephen Sanderson and Wesley Roberts[92] identified two major predictors of the change of tribal to organised religion: the development of agriculture and the invention of writing.[93] The millennium BCE was characterised by the emergence of important new philosophical or religious innovations (Hinduism and Buddhism in India,

[85]Recently Norenzayan et al. (2016) labelled the organised religions as 'prosocial religions'. Although the organised religions are characterised by a number of cultural features that strongly enhance the social relations and social cohesion within their societies, identifying them as 'prosocial' does not distinguish them from the previous stage in the development of religions, because the animistic or folk religions also had the promotion of social cohesion in their smaller communities as one of their major attributes and benefits.

[86]Roes and Raymond (2003), Johnson (2005), Botero et al. (2014), Purzycki et al. (2016).

[87]Snarey (1996), Botero et al. (2014).

[88]Norenzayan et al. (2016, 13).

[89]Lenski (1984), Heilbroner (1995, 30).

[90]Veenhoven (2005), Steckel and Wallis (2009).

[91]Pinker (2011, 57), Peoples and Marlowe (2012).

[92]Sanderson and Roberts (2008).

[93]Farmer (2006) considers this transformation to be influenced by the expanded availability of lightweight reading materials.

Confucianism and Taoism in China, Israelite monotheism, philosophical rationalism in Greece) called the Axial Age[94] by Karl Jaspers.[95] One of the major religious characteristics of this change was the shift from polytheism to monotheism. Stephen K. Sanderson[96] explains this philosophical/religious transformation as the consequence of the massive increase in the scale of warfare, related to the development of iron weapons and to the rapid and large-scale urbanisation. Both of these were socially disruptive and psychologically anxiety-producing phenomena, which decreased people's attachments to close kin and other social intimates. Invoking Lee A. Kirkpatrick's[97] thesis that God or Gods are primarily substitute attachment figures for kin figures, Sanderson sees the Axial Age disruptions as the cause of the greater need for a substitute attachment figure in the person of a compassionate, all-powerful and loving God. Indeed, the technological (invention of agriculture, iron tools and writing) and socio-demographic (larger and more strongly stratified populations) changes allowed the formation of socially stratified, belligerent or war-torn states in which Axial sages tried to develop a just, equitable and compassionate ethic.[98] At the same time, rulers used the new religions as an instrument of social control and domination.[99]

In their extensive evolutionary analysis of the Bible, Carel Van Schaik and Kai Michel[100] argue pertinently that the narratives of this Holy Scripture are largely representing cultural strategies aimed at resolving or adapting to the challenges and crises that emanated from the largest behaviour change in the history of humankind —the establishment of permanent settlements in the agrarian revolution. Indeed, this change was accompanied by multiple and more intensive calamities such as strong population growth and density (with their breeding ground for infections and impersonal relations) and ownership of real estate, land, husbandry and … women (with its associated stronger social stratification and increased social inequalities and inequities, transference of pathogens from animals to humans, and increased intergroup violence and war).

The organised religions that emerged in agrarian cultures are usually classified into two major groups of religions: Eastern religions and Mediterranean religions. The Eastern religions consist of the religions and philosophical traditions originating in the Far East (India, China, Japan and Southeast Asia). The Mediterranean religions emerged in the Middle East and include Judaism, Christianity and Islam.[101]

Eastern and Mediterranean religions differ philosophically or theologically in a number of respects. Whereas the Mediterranean religions believe in a monotheistic

[94]Ara Norenzayan et al. (2016, 24) argue that the societal and religious changes of the 'Axial Age' long preceded and even followed that period, and moreover developed very gradually.
[95]Jaspers (1949, 15), see also Armstrong (2006).
[96]Sanderson (2008, 153).
[97]Kirkpatrick (2005).
[98]For instance, Armstrong (2006, 397).
[99]Wade (2009, 124).
[100]Van Schaik and Michel (2016).
[101]Pinxten (2010).

God that is strongly involved in the daily life and afterlife of its believers, the Eastern religions and philosophical traditions have a more comprehensive conception of the divine. They also show a larger heterogeneity—some being (poly) theist such as Hinduism, Sikhism and Taoism, others being non-theist such as Buddhism and Confucianism; Shintoism is an animistic folk religion.[102] Eastern religions/philosophies are either Dharmic (Hinduism, Buddhism) the goal of which is to liberate oneself from the suffering of the Earth, or Taoist (Taoism, Confucianism, Shintoism) that preaches harmony with the underlying natural order of the universe.[103] Whereas the religious-philosophical conceptualisation of the Eastern and Western traditions differ quite substantially, their ethical prescriptions or strivings show much more similarity, for instance, in commending moral virtues such as reciprocal altruism, enlightenment, austerity, familial duty, loyalty, humaneness, honesty, truthfulness, humility.

This discussion is largely limited to the three major organised religions—the so-called Abrahamic[104] religions (Judaism, Christianity and Islam) that emerged in the Middle East for two reasons. Firstly, the authors are somewhat more familiar with those religions, through the reading of their original scriptures—the Old and New Testaments of the Bible and the Qur'an—as well as through professional research and personal interactions and confrontations with their adherents and scholars. Secondly, and more importantly, those religions or at least two of them—Christianity (particularly Catholicism) and Islam—play currently an active role in slowing down some aspects of the modernisation process in many countries. Resistance to change is also seen at the intergovernmental level, for instance, at the occasion of the United Nations Conference on Population and Development in Cairo (1994), and the United Nations World Conference on Women in Beijing (1995) on issues such as family planning, sexuality and gender equality.[105]

3.3.3.1 Judaism

The Hebrew Bible, which closely corresponds to the contents of the Jewish Tanakh and the Protestant Old Testament, narrates the largely mythical history of the Jewish people in ancient times—from the biblical creation of the world,

[102]Nevertheless, ordinary people in these religions often believe in and pray to a series of gods and spirits that behave counterintuitively and unintelligibly to factual and logical reasoning (Atran 2006, 188).

[103]For instance, Coogan (2005).

[104]Abrahamic religions are the monotheistic faiths emphasising and tracing their common origin to Abraham, the mythical ancestor of several Middle East tribes, with whom, according to the Hebrew bible, God made a covenant about his worship, future descendants and land (Genesis, 17). Christians see Abraham as their spiritual and physical ancestor (Rom. 4:17). Muslims see Abraham as a prophet in the line from Noah to Muhammad, all to whom Allah sent revelation (Qur'an, 4:163).

[105]At the conferences of Cairo (1994) and Beijing (1995), the Holy See, some Catholic countries and some Islamic countries, after having endeavoured and partly succeeded in watering down the conference recommendations, expressed many reservations on the conference consensus that was reached (United Nations 1994, 1995).

approximately 4050 BCE,[106] to the times of Nehemiah, governor of the province of Judah in the Persian Empire, approximately 450 BCE. However, scientific evidence does not support the historical actuality of the events recorded in the Old Testament, in particular for the period preceding the Babylonian captivity in the sixth century BCE.[107]

From the Tanakh/Old Testament, it appears clearly that Judaism is an ethnocentric religion, inspired and dominated by a jealous God[108] who favours his chosen people[109] by establishing a covenant with its tribal patriarchs,[110] by giving land to them and all of their offspring,[111] and by blotting out their enemies.[112] Richard Dawkins[113] characterised Judaism as:

> originally a tribal cult of a single fiercely unpleasant God, morbidly obsessed with sexual restrictions, with the smell of charred flesh, with his own superiority over rival gods and with the exclusiveness of his chosen desert tribe.

In a longer term historical perspective the Judaist belief system with its strong in-group genealogical identity-oriented tenets and its rigorous adherence to formalistic behaviour in matters such as sexual, nutritional and vestimentary rules, contributed considerably to maintaining the survival of its population throughout history, despite the frequent persecutions to which it was subjected. Furthermore, notwithstanding that the confession of Judaism is largely limited to Jewish people, its ideological and historical significance is more important, because of its early move from polytheism to monotheism, but also through its influence on the much more widespread Christianity and Islam.[114]

3.3.3.2 Christianity

The message of Jesus, a charismatic Jewish rabbi in early first century Palestine, was exclusively addressed to Jews[115] and had little adherence.[116] It was Paul of Tarsos who broadened it to all of humankind, beyond national or ethnic borders.[117]

---

[106]Ussher (1650).

[107]Thompson (2000), Lazare (2002), Silberman and Finkelstein (2002).

[108]Exodus, 20:5: "You shall not bow down to them (i.e. *other gods*) or serve them, for I the Lord your God am a jealous God, …."

[109]Deuteronomy 7:6: "For you are a people holy to the Lord your God. The Lord your God has chosen you to be a people for his treasured possession, out of all the peoples who are on the face of the earth."

[110]Genesis 17:7: "And I will establish my covenant between me and you and your offspring after you throughout their generations for an everlasting covenant, to be God to you and to your offspring after you."

[111]Genesis 17:8: "And I will give to you and to your offspring after you the land of your sojournings, all the land of Canaan, for an everlasting possession, and I will be their God."

[112]Exodus 23:22: "But if you carefully obey his voice and do all that I say, then I will be an enemy to your enemies and an adversary to your adversaries."

[113]Dawkins (2006, 37).

[114]Wilson (2002, 133).

[115]Matthew 10:5-6; Vermes (2004, 414), Wright (2009, 267).

[116]Stark (1997).

[117]Galatians 6:10; Sim (1997, 192), Vermes (2004, 417), Dawkins (2006, 37), Wright (2009, 267).

It became a universal belief, not constrained to any ethnic group. The only requirement for belonging was belief in the Christian God.[118]

The theme of Jesus' teaching was strongly eschatological[119] in orientation.[120] He preached submissiveness, repentance, unconditional love, forgiveness of sin, grace, and supported morally those who were traditionally treated as the 'wretched of the earth' such as slaves, women, the poor, ethnic outsiders, children, prostitutes, the sick and prisoners. During the Roman Empire, Christianity partly thrived due to the misery and misfortune related to social inequality, oppression and slavery.[121]

In many respects, Christian ideology, as it appears from the Sermon on the Mount[122] and other sections in the New Testament,[123] is partly at odds with some of our inborn human drives; and it contrasts strongly with the evolutionary mechanisms as they functioned in pre-modern tribal living conditions.[124] For instance, Christianity extended morality to all persons regardless of kin or tribal relatedness, thus adapting to a more complex, multi-ethnic environment.[125] However, notwithstanding its broadened vision, Christianity retained the old-time in-group/out-group bipolarity, now based on a moral divide instead of non-shared ancestry.[126] Nevertheless, from an evolutionary perspective, Christianity was really a revolutionary movement in its time and place—"a kind of mutation on the bio-cultural scene" as Philip Heffner[127] puts it, only followed and further broadened almost two millennia later by the innovative ethical principles of the Western Enlightenment.

Some of the moral concepts found in the Sermon of the Mount, in particular the ethic of reciprocity, were not original or unique in the history of humankind, but were widespread among the peoples of the Middle East.[128] The Golden Rule,[129] for instance, is virtually a universal ethical precept.[130] However, it must be acknowledged that the love message of Christianity expressed this value in a very positive and explicit way.

From the present day perspective, Jesus' teachings also include many elements that are maladaptive from a biological evolutionary perspective,[131] or reprehensible from a modern evolved moral perspective.[132] The Bible contains, not only in the

[118]Matthew 25: 31-46; Keith (1946, 73), Teehan (2006, 768).
[119]Eschatology: is the branch of theology dealing with the final events of history, or the ultimate destiny of humanity.
[120]Vermes (2004, 343).
[121]Cattell (1972, 272).
[122]Matthew 5-7.
[123]For instance, John 1:9; Matthew 28:19; Gal. 3.28; Rom. 2:11; Cor. 12:13.
[124]Keith (1946, 69), Teehan 2009, 244; 2010), de Duve (2009; 2011).
[125]Lahti (2009, 85).
[126]Lahti (2004, 143), Teehan (2010, 129, 142).
[127]Hefner (1999, 495).
[128]For instance, Vermeersch (2016, 62–65).
[129]Golden Rule or ethic of reciprocity states that "one should treat others as one would like others to treat oneself".
[130]For instance, Blackburn (2001).
[131]For instance, Matthew 5:17; 5:28-30; 6:7-11; 6:25-34; 10:35.
[132]For instance, Matthew 5:32; 10:34.

Old Testament but also in the New Testament, moral directives that are unjustifiable from a modern moral perspective.[133] This is the case, for instance, concerning slavery,[134] non-believers[135] and women.[136]

Christians appeared to be only one of the numerous new religious sects in the late Roman Empire.[137] Many scholars have pondered on why it became such a successful movement. Various explanations have been suggested, which are probably not mutually exclusive. For instance, the Christian movement appealed to the many 'wretched of the Earth'. Pagan women in Roman times perceived the living conditions and expectations of Christian women better than their own (more faithful and less abusive husbands), and/or the fact that Christians developed communal and caretaking practices that protected them better against illnesses and epidemics,[138] allowing them to reproduce at higher rates.[139] Compared to the then existing belief systems in the Roman Empire, Christianity excelled due to its positive concepts of a commonsensical morality, salvation, and resurrection in a paradisiacal hereafter.[140] However, in the authors' view, there may have been an additional and even more fundamental reason for the initial Christian success: the Christian love message agreed very well, particularly in the socially disruptive environment in which it emerged, with human innate predispositions to the needs of empathy, altruism and reciprocity. The evolutionary background of this predisposition was only scientifically discovered during the Second Darwinian Revolution (see Chap. 1, Sect. 1.1.4). In their evolutionary analysis of the Bible, Carel Van Schaik and Kai Michel[141] also point to the importance of the intuitive moral component of early Christianity that wanted to go back to the biosocial and moral roots of the hunter-gatherer era with its stronger intimate and more equal social relations.

Of course, in the end, politics also played a crucial and even decisive role in the firm footing of Christianity: soon after Emperor Constantine I issued in 313 CE the Edict of Milan, legalising Christian worship, Emperor Theodosius declared in 380 CE Christianity to be the only legitimate religion in the Roman Empire. Although the imperial conversion to Christianity might have occurred out of sincere faith, this monotheistic religion with its hierarchical, centralised structure, preaching charity

[133]Williams (1893, 522).

[134]For instance, Ephesians, 6:5-8: "Slaves, obey your earthly masters with fear and trembling, with a sincere heart, as you would Christ, not by the way of eye-service, as people-pleasers, but as servants of Christ, doing the will of God from the heart, rendering service with a good will as to the Lord and not to man, knowing that whatever good anyone does, this he will receive back from the Lord, whether he is a slave or free."

[135]For instance, Mark 16:16: "Whoever believes and is baptized will be saved, but whoever does not believe will be condemned."

[136]In Judeo-Christian Scriptures, women are expected to be submissive to their husband and their primary role is childbearing. For instance: Timothy 2:11-15; see also Corinthians, 11:3 and 14:34.

[137]Jones and Reynolds (1995, 299).

[138]Kitcher (2007, 143), Richerson and Boyd (2005, 210).

[139]Wilson (2002, 148).

[140]Adams (1995).

[141]Van Schaik and Michel (2016, 334).

and docility to its submissive believers, fitted well with the worldly ambitions of the imperial Roman rulers.[142] In general, autocratic regimes choose to value religious beliefs as a socially binding agent and support for their government.

Christianity became a political power that served as an instrument of rulers' divine justification of their exercise of power, religious warfare, ethnic cleansing, Crusades, torture, witch hunting, persecution and burning of heretics, Inquisition, Jewish pogroms, and intolerance towards non-believers in general.[143] In his remarkable study of violence in human history, Steven Pinker[144] rightly concluded that Christendom in medieval times evolved to a culture of cruelty.[145]

In particular, it became an institution that, in opposition to the preaching of its founders, ideologically justified and politically supported the rulers of the earth (kings, emperors, and dictators of all kinds) and enriched itself at the cost of its followers.[146] Although at odds with some of evolutionary mechanisms in its original teaching, Christianity applied age-old evolutionary methods of in-group/out-group competition during its expansion.[147]

In modern times, even up to today, several branches of Christianity, and in particular Catholicism, issue and impose behavioural norms in the domain of sexuality and reproduction, inspired and justified by ancient principles that are maladapted to the novel living conditions of modern culture and society.[148] Typical examples are opposition to birth control and the imposition of a single acceptable model of the family.

In most present-day developed countries Christianity (or at least several of its major denominations) appears as a tolerant ideology, accepting ideological pluralism and involved in inter-ideological cooperation, also at the political level. However, this is a very recent development. It is a result of a long and hard confrontation and struggle with the secular ideologies and political movements that emerged in the footsteps of the Enlightenment and the development of the sciences to which Christianity eventually largely adapted. Whenever or wherever Christianity succeeded in preserving its ideological monopoly and political power, the positions of non-believers, non-Christians or even 'deviant' Christians were/are much more precarious, often characterised by societal ostracism or exclusion. This dominant behaviour is, of course, not specific to Christianity, but can be observed wherever a dogmatic or doctrinarian ideology succeeds in seizing total power—cf. the fascist and communist authoritarian regimes of the twentieth century or the current-day Islamic theocracies.

[142] O'Grady (2013, 347–352).
[143] For instance, Keith (1946, 73), Harris (2004), Hitchens (2009), Nicey (2017).
[144] Pinker (2011, 132).
[145] For a more general discussion of religious violence, see for instance Haught (2002) or Deschner (1986–2013).
[146] Lowell (1967), Sheils and Wood (1987).
[147] Keith (1946, 73).
[148] Cf. the Roman Catholic position on contraception (Pope Pius XI, 1930; Pope Paul VI 1968), abortion (Pope John Paul II, 1995), in vitro fertilisation (IVF) (Congregation for the Doctrine of the Faith, 1987), homosexuality (Catholic Church, 1993), eugenics (Pope Benedict XVI, 2009), and euthanasia (Sacred Congregation for the Doctrine of the Faith, 1980).

3.3.3.3 Islam

Muslims consider the Qur'an, revealed to Muhammad (570–632 CE) through the angel Gabriel as guidance for mankind,[149] as the literal word of God.[150]

The Qur'an's main message is to believe in the monotheistic, almighty and all-knowing, but unpredictable and capricious Allah in need of worship.[151] The Qur'an is largely influenced and referenced by the Judeo-Christian scriptures.

Its moral commandments largely reflect the male dominated social power relations in the ancient agrarian-pastoral, tribal-structured Arabian society—with the ideological endorsement of women's submissive position,[152] protection of male sexual prerogatives,[153] maintenance of polygyny,[154] acceptance of slavery,[155] exclusion or even extermination of people with deviant opinions,[156] incentivising 'jihad' (='striving in the way of Allah'),[157] and carrying out cruel punishments.[158]

---

[149]Qur'an 2:185; see also Qur'an 25:5-7.

[150]According to independent Islam-scientists, the Qur'an is, just as the Old and New Testaments of the Bible, the result of a long editorial process that took several centuries and in which several authors were involved (e.g. Warraq 1998; Mulder and Milo 2009; Ohlig and Puin 2009).

[151]See also Edis (2007, 153).

[152]Qur'an 4:34: "... if you fear high-handedness from your wives, remind them (of the teachings of God), then ignore them when you go to bed, then hit them."

[153]Qur'an 55:70-74: "There are good-natured, beautiful maidens... Dark-eyed, sheltered in pavilions... Untouched beforehand by man or jinn... Which then, of your Lord's blessings do you both deny?"

[154]Qur'an 4:3: "If you fear that you will not deal fairly with orphan girls, you may marry whichever (other) women seem good to you, two three or four. If you fear that you cannot be equitable (to them), then marry only one, or your slave(s): that is more likely to make you avoid bias."

[155]Qur'an 33:50: "Prophet, We have made lawful to you the wives to whom you have granted dowries and the slave girls whom God has given you as booty."

[156]Qur'an 5:33: "Those who wage war against God and His Messenger and strive to spread corruption in the land should be punished by death, crucifixion, the amputation of an alternative hand and foot, or banishment from the land: a disgrace for them in this world, and then a terrible punishment in the Hereafter, unless they repent before you overpower them—in that case bear in mind that God is forgiving and merciful." See also Qur'an 9:73.

[157]Qur'an 4:74: "Let those fight in the way of Allah who sell the life of this world for the other. Who so fights in the way of Allah, be he slain or be he victorious, on him We shall bestow a vast reward."

[158]Qur'an 24:2-3: "Both the male and female who are guilty of adultery or premarital sex are to be flogged with one hundred lashes. Absolutely no mercy is to be given. It is to be witnessed by a group of Muslims. The adulterers can only marry a person who has been found guilty of the same crime or an unbeliever in the religion or Islam."

Qur'an 5:38: "Men or women who steal must have their hands cut off as a reward for their deeds. This will be an example for others."

Qur'an 8:12-14: "The hearts of the infidels will be terrorized so Muslims should attack with courage and behead them and cut off all their fingers. Maiming your victims will show that opposing Allah and Mohammed results in severe punishment. They are going to Hell."

Qur'an 5:33-34: "Those who make war on Allah and Mohammed or strive to spread disorder in the land should be killed, crucified, have their hands and feet cut off on alternate sides or be exiled. They must be degraded in this world and doomed in the afterlife except those that repent before the Muslims capture them. In their case Allah will forgive them."

Just as Jesus Christ (or his followers) evoked divine intervention to promote his (their) teachings and religious-ethical standards, Muhammad[159] must have been a very charismatic, intelligent and shrewd man who understood very well how to appeal to his followers with his divinely inspired revelations. Historically, it appears that evolution towards family-transcending morality required stronger coercive imposition and punishment.

From a present day ethical point of view, it must be acknowledged that the Qur'an advocates some laudable, though not original moral principles,[160] and sounds even quite modern in its style of theological argumentation.[161] However, overall, it has more similarities with the Old Testament than with the quite innovative teachings of Christianity in the New Testament as expressed in the Sermon on the Mount.[162]

Most puzzling to non-religious scholars is the inconsistency between the idea of an all-mighty, all-knowing and all-compassionate God who does whatever he wants[163] and the alleged freedom of people to choose between good and bad and between belief and disbelief. It may be argued that Islam's theology contains a number of features and controversies that make it difficult for Islamic believers and their societies to fully contribute to and participate in the scientific and moral innovations of modern times.[164] Examples are the literal reading of the Qur'an and the infallible status of Muhammad, its divinely derived fatalism, its focus on the hereafter and its glorification of martyrdom, its custom to give individuals the power to enforce Islamic law by commanding the good and forbidding evil, its reliance on the ancient Shari'ah,[165] and its commandment to wage jihad.

Perhaps more important than the reflections about theological subtleties of Islam is the fact that the initiators of this religion succeeded in promoting the Arabian political and cultural identity and its expansion.[166] Historically, Islam spread over a large part of the Middle East, Africa and Asia through conquest or migration, easy

[159]Or his followers who drafted the Qur'anic texts in the two or three centuries after Muhammad's death (see, for instance, Rodinson 1996; Warraq 1998; Ohlig and Puin 2009; Hazleton 2014).

[160]Apparently, Islam's initiators (Muhammad and/or his followers) felt the need to react against the thriving but ruthless capitalism in the economically booming and successful Mecca of their time, causing them to preach that Muslims ought to develop a just and fair society in which the less fortunate and more vulnerable are treated decently (Armstrong 1993, 156, 167; see also Hazleton 2014).

[161]Wright (2009, 397).

[162]Matthew 5-7.

[163]For instance, Qur'an 2:272: "Not upon you, [O Muhammad], is [responsibility for] their guidance, but Allah guides whom He wills…"

[164]Hirsi Ali (2015, 34).

[165]Shari'ah is Islamic religious law derived from the Holy Qur'an and the Sunnah. It refers to the sayings and practices of the Prophet Muhammad as recorded in a hadith. It is based on divine authority, and embodies broad, general rules that are immutable.

[166]Armstrong (1993, 158), Kennedy (2007), Wade (2009, 622).

conversion (but difficult apostasy), strong ideological indoctrination of children[167] and adult believers,[168] and high fertility (primarily indirect, through Islamic views about the social role of women).[169] In recent decades, Islam has also gained a foothold in several European countries, through immigration from former colonies or recruitment of immigrant guest workers, family reunion policies, and high fertility.[170] In most recent years this is also occurring via refugees and asylum seekers from conflict or war areas, prompted by a variety of causes such as population pressure, climate change, oil interests, internal or regional political power struggles, and Western neo-colonialism.

Present-day countries where Islam is the predominant religion have not undergone their own Reformation or Enlightenment.[171]As a consequence the scientific and moral benefits of modernity are difficult to achieve, such as freedom of thought and speech, the acceptance of ideological and political pluralism and tolerance, the separation of religion and state, the creation and dissemination of scientific knowledge (e.g. concerning evolution science), the realisation of sexual equality, the right to individual self-realisation.[172]

An example is the difficulty to reconcile the full body of values articulated in the United Nations Universal Declaration on Human Rights by many Islamic states. In the 1990 *Cairo Declaration on Human Rights in Islam* (CDHRI), adopted by the Nineteenth Islamic Conference of Foreign Ministers of the Organisation of the Islamic Conference, Article 24 of the CDHRI states that it is "subject to the Islamic Shari'ah", and its Article 25 confirms that Shari'ah "is the only source of reference for the explanation or clarification of this Declaration."

The CDHRI is clearly at odds with the principles of freedom of worldviews and life stance of the UN human rights charter. The Cairo Declaration makes no mention of provisions for democratic principles, protection for religious freedom and change of religion, freedom of association and freedom of the press, as well as equality of rights and equal protection under the law.[173] In contrast, it includes a

[167]See for instance Stamos (2008, 183).
[168]According to David N. Stamos (2008, 183), the requirement of public prayer five times a day especially helps preserve the Islam meme complex in populations with low literacy levels.
[169]For instance, Coleman, in Jones and Reynolds (1995, 240).
[170]Haddad (2002), Bawer (2006), Nachmani (2009).
[171]Bruce (2002), Rushdie (2005), Pope Benedict XVI (2006), Harris (2007), Kaufmann (2011, 11), Hirsi Ali (2015), del Valle (2016).
[172]Du Pasquier (1992), Lewis (2003), Manji (2003), Edis (2007), Van Rooy and Van Rooy (2010), Hirsi Ali (2015).
[173]Littman (2003).

striking discrimination against non-Muslims,[174] women,[175] and non-married people.[176] It also restricts certain rights and freedoms of expression,[177] even to the extent that certain essential provisions are below the legal standard in force in a number of Islamic countries. Under the cover of the 'Islamic Shari'ah' it even legitimises corporal punishments that in some countries (e.g. Saudi Arabia, Pakistan, Iran, Somalia, Sudan and northern Nigeria) include floggings, stoning, and amputations.

The fundamentalist Islamist wave and its terrorist actions that have ravaged parts of the planet in recent years have several obvious geopolitical causes. This includes national and international political power struggles and economic inequalities, and in particular the hegemony of the United States, the developmental retardation of

[174]The CDHRI includes several articles which are exclusive for non-Muslims, for example Article 1:

(a) All human beings form one family whose members are united by their subordination to Allah and descent from Adam….

(b) All human beings are Allah's subjects, and the most loved by Him are those who are most beneficial to His subjects, and no one has superiority over another except on the basis of piety and good deeds."

Article 2:

(a) Life is a God-given gift…

Article 10:

Islam is the religion of true unspoiled nature. It is prohibited to exercise any form of pressure on man or to exploit his poverty or ignorance in order to force him to change his religion to another religion or to atheism.

Article 11:

(a) Human beings are born free, and no one has the right to enslave, humiliate, oppress or exploit them, and there can be no subjugation but to Allah the Almighty.

Article 23:

(b) Everyone shall have the right to participate, directly or indirectly in the administration of his country's public affairs. He shall also have the right to assume public office in accordance with the provisions of Shari'ah.

[175]Under Shari'ah law, Islamic marriages require that a Muslim woman may marry only a Muslim man and that her guardian must give permission for a virgin female to marry. Women's testimony is only worth half that of a man, but also their freedom of movement is strictly limited. Women are not even free to choose their work. Orthodox Islam forbids women from working outside the home (Warraq 2009).

Article 6 of the CDHRI stipulates the following about women:

(a) Woman is equal to man in human dignity, has her own rights to enjoy as well as duties to perform, has her own civil entity and financial independence, and the right to retain her name and lineage.

(b) The husband is responsible for the maintenance and welfare of the family.

[176]The CDHRI states in its Article 5 among others:" (a) the family is the foundation of society, and marriage is the basis of making a family. …".

[177]The CDHRI includes the following Article 22:

(a) Everyone shall have the right to express his opinion freely in such a manner as would not be contrary to the principles of the Shari'ah….

(b) Everyone shall have the right to advocate what is right, propagate what is good, and warn against what is wrong and evil according to the norms of Islamic Shari'ah.

(c) Information is a vital necessity to society. It may not be exploited or misused in such a way as may violate sanctities and the dignity of Prophets, undermine moral and ethical values or disintegrate, corrupt or harm society or weaken its faith.

the Islamic world, the Israeli-Palestinian crisis, and the power struggles for oil-control. However, it cannot be denied that the conceptual breeding ground of this movement is strongly fuelled by the controversies in the contents of the Islamic religious-political ideology.

In pluralistic environments—for instance at international conferences where people are present from different belief and ideological systems, religious and secular—Muslims often present their faith as one of peace and tolerance.[178] It is also striking that, in the aftermath of the terrorist attack on the French magazine '*Charlie Hebdo*' early in 2015, so many western politicians, representatives of Muslim communities, and media commentators publicly denied that Islam—"a religion of peace"—had anything to do with that evil deed. This denial may be intended to appease the internal and international Muslim community, but it is not at all helpful to prevent terrorist acts or promote Muslims' social inclusion into modernity. When one reads carefully the Islamic scriptures[179] or consults the national legislations of some Islamic countries, a totally different picture emerges. The Flemish philosopher Maarten Boudry[180] recently suggested to moderate Muslims not to use direct quotes from the Qur'an in disputes with Islamic fundamentalists. When they endeavour to evince the peaceful character of Islam and distance themselves from Islamic fundamentalists on the basis of the rare Qur'an verses expressing peacefulness or tolerance towards 'others', they expose themselves to quotes by fundamentalists who then highlight the uncountable number of hate verses scattered all over the Qur'an that call for jihad, or express intolerance, hate, damnation, or violence against unbelievers or apostates.[181] The authors consider that coherent reflections rather than direct quotes from the Holy Scriptures are more appropriate in debates about values and can address better challenges of the power struggles in modernity.

3.4 Some Anthropological Questions and Paradoxes About Religions as Sources of Morality

From a human-evolutionary perspective, a question inevitably arises why the divinely inspired prophets/messiahs such as Moses, Jesus and Muhammad, and their scriptures, appeared so late in human history. These religions do not explain why the souls of the many thousands of generations of *Homo sapiens sapiens* who predated classical antiquity were not worth saving. A conservative estimate is that *Homo sapiens sapiens* is 150,000–300,000 years old.[182] What about the souls of

[178]For instance, Rauf (2004).
[179]In addition to the Qur'an, see the Hadith collections of Muhammad Ibn Ismail al-Bukhari (810-870): *Sahih Bukhari*, and Muslim ibn al-Hajjaj al-Qushayri (821-875): *Sahih Muslim*.
[180]Boudry (2014; 2015, 187).
[181]http://www.answering-islam.org/Quran/Themes/jihad_passages.html.
[182]Vigilant et al. (1991), White et al. (2003), Trinkaus (2005), McDougall et al. (2005), Hublin et al. (2017), Richter et al. (2017).

the major hominins[183] who preceded *Homo sapiens sapiens,* namely *Homo erectus*, whose age may be estimated to be tenfold higher, namely 1.5–2.0 million of years?[184] Why did Jaweh/God/Allah's revelations only occur in what is from an evolutionary time perspective to be defined as the very recent past, and only in some semi-desert regions of the Middle East? Religious believers, even when they address this problem,[185] have never provided a convincing answer to this fundamental question. Whenever they try, as for instance Gary Emberger from the Department of Natural Sciences at Messiah College in Grantham, Pennsylvania,[186] the present-day scientific achievements and insights on the evolution of the cosmos, the planet, life on this planet, and in particular the hominisation process, appear to be largely misunderstood, ignored, or even distorted. To a great extent real insight is absent regarding the complexity, comprehensiveness, and cohesiveness of the present-day huge body of scientific knowledge of the broad variety of the natural, social and cultural sciences. Confronting the high age of humankind and the evolutionary continuity between modern humans and their hominin predecessors, it is understandable that theologians are desperately, and in vain, struggling to reconcile those facts with myths such as the biblical Genesis story.[187]

Equally incoherent is the assertion that God is almighty and omniscient, creator of everything, and at the same time morally perfect, immensely benevolent and loving. Philosophers have amply shown the logical incompatibility between the alleged multiple attributes of God and, in particular, the inconsistency between those attributes and the existence of evil.[188] Indeed, how to explain the ruthlessness of the evolutionary mechanism (mutation, natural selection) and its biological effects (genetic impairments and infectious diseases) or natural catastrophes? An omnipotent supernatural power that produces such human and animal suffering can only be labelled a sadist. Of course, there is the narrative of original sin[189] that might explain human-made calamities, such as wars, famines, environmental destruction, and a variety of criminal acts, but how to explain genetic impairments or natural disasters for which the human cannot be responsible? Furthermore, how to justify the punishments of innocent people for the assumed mischiefs and vices

[183]In present-day taxonomic terminology, *Hominina* is a sub-tribe (including modern humans and their extinct relatives) of the Family *Hominidae* (including the great apes and humans) (cf. Goodman et al. 1998).

[184]Grine et al. (2009), Fleagle et al. (2010).

[185]For instance, Miller (1999), Giberson (2008), Deane-Drummond (2012), Walton (2012).

[186]Emberger (1994).

[187]Moritz (2012).

[188]Russell (1957, 1997), LaCroix (1974), Stump and Murray (1999, 153), Everitt (2004, 228, 2006), Teehan (2013). See, in particular, the excellent collection of philosophical papers in the anthology of Michael Martin and Ricki Monnier (2003), especially Part 2 (59–124): Deductive evil disproves of the existence of God.

[189]The Augustinian notion of an angelic fall giving rise to the natural evils of our world and the corruption of an originally perfect creation (Campbell 1975; Hick 1966; Emberger 1994; Williams 2001; Haught 2004, 2010).

of their far-off ancestors? The fundamental question of *Le problème du mal*[190] has never been given a satisfactory religious explanation, despite the innumerable ingenious, intellectual acrobatics of apologists and theologians to explain that phenomenon.[191] Hence, it is not surprising that most philosophers consider the existence of evil the strongest argument against the existence of the type of deity worshipped in the Abrahamic religions.[192]

Moreover, why would an omnipotent, omniscient, benevolent, and merciful supernatural power only favour 'the chosen ones' and not the others (as in Judaism), or 'the true believers' and not the non-believers, the pagans, the agnostics, and the atheists, who apparently have not been granted God's grace[193] (as in Christianity and Islam)? In particular, why would a just God have created a human species in which there are such salient, non-self-induced inequalities and inequities, either of genetic, ontogenetic, or purely social origin? Why are there these discriminations amongst its creations?

Why would a just and benevolent God be so sexist by only favouring the male gender, and ideologically and socially discriminating the female gender, as can be witnessed from the prayer of male Jews "Thank you, God, for not making me a woman",[194] or from the Catholic tradition that only men can become priests, bishops, or popes,[195] or from Islamic customs that women, and not men, should wear a headscarf, a veil, or even a burqa, or in the hereafter serve as one of the 72 virgins for the male martyrs of Jihad?[196] The founding fathers of the three Abrahamic religions were awful sexists and/or driven by a psychopathological fear of femininity. In either case they faithfully reflected the mores and practices of the ancient patriarchal, male dominated type of society.

[190]Leibnitz (1710), Teilhard de Chardin (1956, 345), Barbour (1997, 300). See also the discussions, for instance, in Stump and Murray (1999, 153), Everitt (2004, 228), Kitcher (2007, 123), Ehrman (2008) or Vermeersch (2016, 35ff), about the logical inconsistency of evil with the existence of a being which is all-powerful, all-knowing, and perfectly good.

[191]For instance, Emberger (1994), Barbour (1997, 300ff), Crean (2007), McGrath (2004), McGrath and McGrath (2007), Berlinski (2008), Haught (2008, 2010), Southgate (2008), Fleck (2011), Coakley (2013).

[192]Martin and Monnier (2003, 59–124), Everitt (2004, 228), Stenger (2009, 107), Vermeersch (2016, 34ff).

[193]"For by grace you have been saved through faith: and this is not your own doing, it is the gift of God—not because of works, lest any man should boast" (Eph. 2: 8-9).

[194]The Talmud Menahoth 43b-44a says that a Jewish man should pray the following prayer each day: "Thank you God for not making me a gentile, a woman, or a slave".

[195]Corinthians 14:34-35: "the women should keep silent in the churches. For they are not permitted to speak, but should be in submission, as the Law also says. If there is anything they desire to learn, let them ask their husbands at home. For it is shameful for a woman to speak in church."

[196]"It was mentioned by Daraj Ibn Abi Hatim, that Abu al-Haytham 'Adullah Ibn Wahb narrated from Abu Sa'id al-Khudhri, who heard Muhammad saying, 'The smallest reward for the people of Heaven is an abode where there are eighty thousand servants and seventy-two houri, over which stands a dome decorated with pearls, aquamarine and ruby, as wide as the distance from al-Jabiyyah to San'a." (Imam at-Tirmidhi in his "Sunan", Volume IV, chapters on "The Features of Heaven as described by the Messenger of Allah", chapter 21: "About the Smallest Reward for the People of Heaven", hadith 2687).

Why has humankind known so many Gods, almost as many (and in polytheism even more) as there were tribes or civilisations? Why was there a Brahma (Hindu), a Shangdi (Chinese), an Anu (Babylon), a Zeus (Greek), a Jahweh (Judaism), a Jesus (Christianity), a Quetzalcoatl (Aztec), a Viracocha (Inca), an Al-Lah (Islam), out of the 217 gods listed in an overview of the world's major deities?[197] If God exists and wanted to reveal himself to humankind as his favoured creature, why has he not been revealed everywhere as the same and, moreover, unique creator and benefactor? The authors suggest some pathways for addressing these questions in an evolution science perspective.

3.5 Biological Determinants of Religiosity and Spirituality

3.5.1 Genetics of Religiosity and Spirituality

The idea that religiosity is influenced by genetic factors is not at all new, for example in the nineteenth century Francis Galton[198] stated in his *Hereditary Genius* "a pious disposition is decidedly hereditary". Nowadays, evidence for this view comes from two domains of genetic research: behavioural genetics and molecular genetics.

3.5.1.1 Heritability

Features such as spirituality and religiosity are typical characteristics that show a continuous variation within a population: there are few people with strong spiritual and religious drives and there are few people who score very weakly on these variables; most people take an intermediate position.[199]

On the hypothesis that such behavioural characteristics might be influenced both by genetic and environmental factors, some behavioural geneticists have applied standard behavioural genetic research methods[200] to estimate the degree to which

[197] ancienthistory.about.com/library/bl/bl_myth_gods_index.htm.

[198] Galton (1869, 274).

[199] For instance, Inglehart et al. (2004), Pickel and Müller (2009).

[200] These standard methods are based on the variance analysis of different groups of people, varying in their degree of genetic similarity and differences in the environment in which they have been raised. A classical method consists of cross-comparing identical and non-identical twins, raised in the same or in different families (e.g. Plomin et al. 2008). The fraction of the phenotypic differences between individuals that can be attributed to genetic differences is called heritability and the fraction that can be identified as environmental variance is called modificability. One of the most important subdivisions of environmental variance is the division between shared and non-shared environmental influences that can affect members of the same family. The notion 'shared environment' refers to between-family non-genetic differences that make siblings more similar than children reared in different families. Social class and parental differences in childrearing styles are examples of between-family variation. The concept of 'non-shared or unique environment' refers to within-family non-genetic variance that makes siblings in the same family different from one another. Within-family non-genetic differences include prenatal and biological conditions as well as psychosocial events that affect one sibling in a different way to another.

genetic and various environmental factors influence those differences between individuals in a population.[201]

Spirituality and religiosity has been found to have a moderate heritability of around 0.40–0.60.[202] This means that both culture and genes play a role in the variation of spirituality and religiosity. However, genetic factors are more important for variation in spirituality, whilst variation in religiosity depends more on environmental factors.[203] Whereas variation in church attendance appears to be influenced both by genetic and social inheritance,[204] variation in religious affiliation is primarily transmitted culturally.[205] All of the empirical data fit well in an evolutionary perspective: from this it can be assumed that variation in religion/religiosity is more strongly determined by culture than variation in spirituality, which is more linked to the basic personality traits which are more strongly influenced by genetics.[206]

Hence, a variation in individual religiosity should not be at all surprising. There may be some within-population variation due to genetic or ontogenetic factors responsible for the variation in spirituality, but within as well as between-population variation in religiosity must be largely the result of ecological, economic

In the coming years the methods to study the causes of within- and between-population variance in behaviour and measurement of the effects of genetic and environmental factors will be further refined, taking into account the new insights from molecular genetics showing that traits are not only influenced by genetic and environmental factors, but also that the effects of DNA are partly contingent on the environment. In the domain of social behaviour, some scholars already speak of an emerging new sociogenomics (e.g. Robinson et al. 2005, 2008; Roberts and Jackson 2008; Slavich and Cole 2013).

[201]For instance, Loehlin and Nichols (1976), Truett et al. (1994), Beer et al. (1998), D'Onofrio et al. (1999), Kirk et al. (1999), Koenig et al. (2005; 2008), Bradshaw and Ellison (2008), Button et al. (2011), Kandler and Riemann (2013).

[202]Bouchard and Loehlin (2001), Koenig et al. (2005). For instance, on the basis of the Minnesota twin study (Waller et al. 1990; Bouchard et al. 1999; Koenig et al. 2005) in which the heritability of intrinsic religiousness (a proxy for spirituality) was examined, heritability was estimated at 43%. Identical results were obtained on the basis of the Australian Twin registry (D'Onofrio et al. 1999; Eaves et al. 1999a, b; Kirk et al. 1999a, b) in which the heritability of spirituality was investigated by means of a self-transcendence questionnaire: the estimated heritability was 37% for men and 41% for women. Regarding the effects of environmental factors, it appeared that the non-shared environment accounted for 42–50% of the observed variance, whilst the effect of the shared environment was insignificant. In contrast, for religious service attendance the shared environment amounted to 43% of the variance, whereas the remaining variance was due to a mixture of the non-shared environment and a limited genetic factor. In the Virginia Commonwealth University twin study (Kendler et al. 1997; 2003; 2009) differences in religious affiliation are culturally determined, whilst variation in religious attitudes and behaviour are subject to varying degrees of genetic effects. Recent findings from the National Survey of Midlife Development in the United States (Beer et al. 1998; Bradshaw and Ellison 2008) confirm this by showing that genetic influences are relatively smaller on religious attendance (32%), somewhat larger for conservative beliefs (41–44%) and religious coping strategies (42%), and quite strong for religious transformation and commitments (65%).

[203]D'Onofrio et al. (1999), Bradshaw and Ellison (2008).

[204]Truett et al. (1994).

[205]Boomsma et al. (1999), D'Onofrio et al. (1999), Eaves et al. (1990, 1999, 2008).

[206]Saroglou and Muñoz-García (2008, 88).

and cultural factors. Both explanations—genetic and environmental—are mutually not exclusive or contradictory.[207] One can even imagine that people who have a strong genetic or ontogenetic endowment for spirituality, and would have become strong religious believers in pre-modern cultural stages, can end up as conscious atheists in modernity. This means that variability in religiosity and spirituality does not at all refute evolutionary theories of religion, contrary to what some scholars have argued[208].

3.5.1.2 Molecular Genetics

Although still in its infancy, molecular genetics has already found significant relations between personality traits indicative of self-transcendence and spirituality, and specific gene variants that play a role in the brain chemistry of neurotransmitters.

David E. Comings et al.[209] were the first to find a correlation between the self-transcendence subscore—and more particularly the spirituality subscore of the Temperament and Character Inventory (TCI)[210]—and the DRD4 gene which codes for a receptor that senses the presence of dopamine[211] in the brain. The researchers suggested that the DRD4 gene might be one of the genes that play a role in the personality trait of spiritual acceptance.

In his book *The God Gene*, Dean Hamer[212] reports no association between DRD4 and any aspect of transcendence, but a clear association between the VMAT2 A/C polymorphism and self-transcendence.[213] People with C/C and C/A genotypes had increased self-transcendence scores compared to people with an A/A genotype. However, the VMAT2 polymorphism explained only a minor fraction of the self-transcendence variance, meaning that many more genes must be involved in this complex behavioural trait. Calling the VMAT2 the God Gene is, of course, a gross exaggeration as Hamer admits himself, but apparently such language generates interest in the book.

Meanwhile, several other research teams have established relations between personality traits related to spiritual and self-transcendent experiences and specific genes related to brain chemistry.[214]

In conclusion: several studies have found significant relations between personality characteristics related to self-transcendence (and spirituality) and some genes

[207]Ellsworth (2009).

[208]For instance, Thagard (2005), Zuckerman (2005).

[209]Comings et al. (2000).

[210]Cloninger et al. (1993).

[211]Dopamine is a neuroendocrine transmitter that performs several functions in the body. In the brain it is a neurotransmitter, which is a chemical released by nerve cells sending signals to other nerve cells.

[212]Hamer (2005, 72).

[213]Vesicular monoamine transporter 2 (VMAT2) is a protein that in humans is encoded by the *SLC18A2* gene. VMAT2 is an integral membrane protein that transports monoamines, particularly neurotransmitters such as dopamine, norepinephrine, serotonin, and histamine.

[214]For instance, Borg et al. (2003), Ham et al. (2004), Beaver et al. (2009), Sasaki et al. (2011, 2013).

that play a role in the physiology of several neurotransmitters in the brain. Although those findings will have to be confirmed by studies replicating those results and be extended to other genes involved in spirituality-related brain chemistry, they are in accordance with the findings of behavioural genetic research as well as neurological research.

3.5.2 Neurological Basis of Religiosity and Spirituality

A first question is whether religious behaviour results from cognitive processes or from emotional predispositions. There can be little doubt that cognition plays an important role in the development of religious beliefs. This was particularly advantageous in the pre-scientific era, when people did not have exact knowledge about the physiology of life and death. Human cognition succeeded in creating ingenious narratives about supernatural forces through which natural, biological and social events were rationally explained. Nowadays, neurological and other scientific findings also increasingly point to the important role of emotions in the sensitivity to or experience of religiosity and spirituality.[215] In particular, several studies have found a link between measures of empathic concern and religious or spiritual belief. In contrast, non-believers may have personality profiles that are more conducive to analytic thinking, which is, in turn, more associated with a naturalistic/materialistic worldview.[216]

A variety of neurological and neuropharmacological studies have documented associations between neurological processes and religious-mystical experiences[217]: some aspects of religiosity have been found to be associated with cortical volume differences[218]; electroencephalographic studies have recorded increased electric activity in the frontal lobes during various types of meditation[219]; ambulatory blood pressure and stress-induced hypertension are reduced during contemplative meditation[220]; transcerebral magnetic stimulation may cause a sensed presence of God[221]; and neuropharmacological studies have revealed dopaminergic activation as the most important neurochemical factor related to religious activity.[222]

Certain pathological conditions (temporal lobe epilepsy, Parkinson's disease, schizophrenia, head injuries) can influence religious attitudes and behaviours.[223]

[215] Hinde (1999), Fuller (2006), Oviedo (2009).

[216] For instance, Jack et al. (2016).

[217] See overviews in Persinger (1987), Joseph (2000; 2001), Newberg et al. (2001), Newberg (2006), Previc (2006), McNamara (2006; 2009), Jeeves and Brown (2009), Chiesa and Serretti (2010), Tiger and McGuire (2010), Shermer (2011), Wlodarski and Pearce (2016).

[218] Kapogiannis et al. (2009), Miller et al. (2014).

[219] For instance, Banquet (1972), Benson et al. (1990), Inzlicht et al. (2009).

[220] For instance, Timio et al. (1988), Wenneberg and Schneider (1997).

[221] Persinger (2003).

[222] Previc (2006), Brugger (2007).

[223] For instance, Persinger (1987), Ramachandran and Blakeslee (1998), Siddle et al. (2002), Wuerfel et al. (2004), Rogers and Paloutzian (2006), McNamara et al. (2006), Schachter (2006), Harris and McNamara (2009), Johnstone et al. (2011).

Bipolar disorder (mania/hypomania), obsessive compulsive disorder (OCD), schizophrenia, and temporal lobe epilepsy (TLE) are all associated with frontal-temporal over-activity and elevated dopamine, and may show heightened religiosity, religious experiences, and/or religious practices.[224] Some of the world's spiritual prophets, founders and leaders of major religions, such as Paul, Muhammad, and Joseph Smith, might have been sufferers of temporal lobe epilepsy.[225] Some religious writings—a textbook example is *The Revelation to John* in the New Testament—are rife with fantasies, visions, and hallucinations typical of schizophrenic disorder or hallucinogenic drugs.[226]

Indeed, various environmental factors have also been found to induce heightened levels of religiosity—hallucinogenic drugs,[227] high mountains (associated with acute and chronic hypoxia),[228] and hot deserts.[229]

Novel neuroimaging techniques, such as positron-emission tomographic scanning (PET), single photon emission computed tomography (SPECT) and functional magnetic resonance imaging (fMRI), have shown that mental activities of a religious-spiritual nature are associated with changes in particular cortical activity levels, primarily involving the ventromedial temporal and frontal regions.[230] According to Patrick McNamara[231] the limbic system (in particular the amygdala), portions of the basal ganglia, the right temporal lobe (especially the anterior portion of the medial and superior temporal lobe), and the dorsomedial, orbitofrontal, and right dorsolateral prefrontal cortex are important brain centres involved in religiosity.

A remarkable fact is that so-called sacred values—whether of religious or secular nature or inspiration—have been found to be associated with increased activation of several brain regions, particularly those involved in semantic rule retrieval, suggesting that sacred values affect behaviour through processing rights and wrongs (deontological imperatives) and not through evaluating costs and benefits (utilitarian concerns).[232]

---

[224]Dewhurst and Beard (1970), Previc (2006), Dein (2011).

[225]McKinney (1994), Saver and Rabin (1997), Newberg et al. (2001), Gazzaniga (2005), Trimble (2007), Comings (2008). Regarding Paul's experience of being blinded by a bright light in the sky on the road to Damascus, William Hartmann (2015) suggested that this might have been produced by a fireball meteor, implying that Paul's Damascene conversion and subsequent important role in the development of Christianity might have been strongly influenced by a random space rock entering the Earth's atmosphere.

[226]See, for instance, Wilson (2012, 263).

[227]For instance, Pahnke (1967), Batson and Ventis (1982), Schultes et al. (2001), Goodman (2002), Nichols and Chemel (2006).

[228]Arzy et al. (2005).

[229]Bennion (2004).

[230]Azari et al. (2001), Azari (2006), Borg et al. (2003), Newberg and Lee (2006), Newberg and Waldman (2007, 2009), Beauregard and Paquette (2006).

[231]McNamara (2009, 105), see also Previc (2007, 527).

[232]Berns et al. (2012), Atran and Ginges (2015).

Relying on neuroimaging findings, some authors started to speak about neurotheology (the study of theology from a neuropsychological perspective).[233] However, some authors, such as Armin W. Geertz,[234] pertinently warn against introducing religious agendas into scientific work because of the risk of teleologically tainting such research.

Nevertheless, various indicators of neural functioning suggest that our brain has the capacity to experience or be receptive to spiritual or religious phenomena. Although most studies have been done on specific population samples,[235] some neurological investigations point to the existence of a general genetic variability in spirituality or religiosity and its susceptibility to environmental factors. This is not an uncommon phenomenon, and is even to be expected for a complex human behavioural trait such as spirituality/religiosity.[236]

The neurological findings about brain activities related to religious experiences have raised the question about the existence of a God module in the human brain:

> …cross-cultural and time immemorial beliefs in gods, soul, and after-life suggest that there is a God part in the brain, that God is an inherited characteristics hardwired in the neurophysiology of the brain.[237]

This idea is based on the theory of the modular architecture of the brain, meaning that the mind consists of a set of discrete and functionally specialised problem-solving modules, each one related to the management of specific adaptive problems, resulting from millions of years of natural selection.[238] Although the entire brain system is involved in mental processes related to beliefs about supernatural agents, two mental tools are thought to be of great importance in this respect: the Agency Detection Device (ADD), which detects the presence and activities of other beings around us,[239] and the Theory of Mind Mechanism (ToMM),[240] which consists of the ability to attribute mental states to others and to interpret their intentions.[241]

In conclusion, present cognitive science helps us to understand how and why beliefs in supernatural agents are so resilient. The capacity to develop such beliefs relies on powerful cognitive systems that developed in the course of hominin evolution and that contributed to survival and reproductive fitness.[242] As a result,

[233]For instance, McKinney (1994), d'Aquili and Newberg (1999), Peters (2001), Joseph (2002), Alston (2007).

[234]Geertz (2009, 324).

[235]For instance, Buddhist monks, nuns, experienced meditators.

[236]For instance, Vance et al. (2010).

[237]Albright (2000), Alper (2006, 151).

[238]Fodor (1983), Tooby and Cosmides (1992), Geary (1998), Gazzaniga (2005), Ellsworth (2009), Tremlin (2006).

[239]Guthrie (1999), Barrett (2000, 31), Boyer (2001).

[240]Premack and Woodruff (1978), Povinellia and Preuss (1995).

[241]Tremlin (2006, 75, 105).

[242]Tremlin (2006, 132).

god concepts are extremely easy to acquire and transmit. This also explains children's sensitivity to religious indoctrination, so that three to five year olds have a predisposition to believe in an omniscient God.[243]

Consequently, beliefs in gods seem to be quite 'natural'. Justin L. Barrett[244] argues pertinently that it would be preferable but more difficult to explain why people do not believe in god(s):

> Being an atheist is not easy. In many ways it just goes against the grain. As odd as it sounds, it isn't natural to reject all supernatural agents.

Some authors nevertheless continue to believe that physicalism[245] is not capable of explaining spiritual experiences by the material functioning of the brain alone. They believe that the soul[246] remains a necessary concept for understanding the originality of the human.[247] Sometimes, anomalies of behaviour and experience are also invoked to explain religious beliefs and spiritual practice, such as telepathy, clairvoyance, precognition, psychokinesis, near-death, out-of-body and past-life experiences, for which there currently appears to be no conclusive natural explanation.[248]

However, it is not because the neural machinery and/or environmental causes of some mental processes[249] have not yet been completely discovered or explained that the material mechanisms of natural phenomena should be rejected or would need to be complemented by extra-natural mechanisms that are influenced by covert religious agendas. Notwithstanding rare exceptions,[250] mainstream neurologists are of the view that there is no neurological argumentation for accepting such thing as a soul. Human mental life, including spiritual life, is a product of neurochemical and electrical activity of the brain. The neurophysiological basis of supernatural experiences is no longer a complete mystery.

---

[243]Evans (2000), Barrett (2001), Kelemen (2004).

[244]Barrett (2004, 108).

[245]Physicalism or materialism: this theory claims that reality consists entirely of physical matter which is the sole cause of every possible occurrence, including human thought, feeling, and action. (http://en.wikipedia.org/wiki/Materialism).

[246]'Soul' in Christian thought is a concept that has functioned in at least two ways: to designate the enduring facet of the self (that in some sense survives death), and as a label for the aspect of the human that is accountable for moral choices and capable of communication with God (Brown et al. 1998).

[247]For instance, Bulkeley (2005, 1), Jones (2005, 56).

[248]Krippner, in Bulkeley (2005, 68).

[249]For instance: meditation, hypnosis, and other fields of psychophysiology, and rare parapsychological phenomena, such as telepathy, clairvoyance, psychokinesis, near-death, out-of-body and past-life experiences.

[250]For instance, Beauregard and O'Leary (2008).

3.6 Evolutionary Advantages and Disadvantages of Religion

The universality of religious beliefs and the discovery of genetic and neurological factors being related to spirituality/religiosity means that, at the ultimate level, an evolutionary advantage must have been involved in the ancestral environment. It is highly unlikely that such a widespread and deep-seated phenomenon could have developed purely by chance. In other words, religiosity/religions must have favoured, through social cooperation and by means of natural and cultural selection, differential survival and reproduction.[251]

That insight has raised the question as to whether spirituality/religiosity/religion is an evolutionary adaptation of the human species[252] or merely a (multiple) by-product of adaptations selected for other domains of human life, such as the human evolved cognitive capacities for cultural creation and social interaction.[253]

Arguments in favour of the adaptationist position are that religion is a universal phenomenon, that its acquisition occurs relatively effortlessly, and that it has an associated biology with genetic, neurological and chemical components.[254] In the behavioural domain it generally increases individual health and well-being; and it promotes social solidarity and intra-group cooperation resulting in social selection[255] or group selection.[256] It often also successfully incites its adherents to produce more offspring than less religious or nonreligious people.[257]

The main arguments for the by-product view are that there are no specific genes for religious belief, that there are no evolved psychological mechanisms designed to produce religious beliefs, and that religiosity and religious adherence vary quite considerably between individuals and between cultures/societies. In this view, religiosity is a spandrel,[258] an exaptation-like[259] by-product of social-cognitive mechanisms that evolved for purposes other than religion itself.[260] For instance, Newberg et al.[261] argue that the neurological basis of spirituality may have arisen

[251]For instance, Grinde (1998), Saxton (2009), Voland and Schiefenhövel (2009), Daecke and Schakenberg (2000), Kardong (2010), Bellah (2011), Norenzayan et al. (2016).
[252]For instance, Irons (2001), Wilson (2002; 2003), Bulbulia (2004), Adams (2005), Alcorta and Sosis (2005; 2012), Dow (2008), Norenzayan and Shariff (2008), Purzycki and Sosis (2008), Sanderson (2008), McNamara (2009), Sosis (2009), Kardong (2010), Baril (2013).
[253]For instance, Darwin (1871), Sperber (1985), Guthrie (1993), Hinde (1999), Kirkpatrick (1999; 2006; 2008), Pyysiäinen (2001), Atran (2002), Bering (2006), Newberg et al. (2001), Boyer (2003), Atran and Norenzayan (2004), Pinker (2004), Dawkins (2006), Granqvist (2006), Hauser (2006), Pyysiäinen and Hauser (2010), Van Schaik and Michel (2016).
[254]McNamara (2009).
[255]Dow (2008).
[256]Wilson (2003, 2; 2005, 385).
[257]For instance, Cliquet and Maelstaf (1977), McQuillan (2004), Frejka and Westoff (2008), Sanderson (2008), Weeden et al. (2008, 2013), Blume (2009).
[258]Spandrel: a cognitive by-product of other adaptive systems (Gould and Lewontin 1979).
[259]Exaptation: refers to shifts in the function of a trait during evolution (Gould and Vrba 1982).
[260]Boyer (2003), Kirkpatrick (2008); see also the discussion in Boudry (2015, 198).
[261]Newberg et al. (2001, 125).

from neural processes that evolved to address more basic survival needs such as mating and sexual experience.

According to Ryan M. Ellsworth,[262] the adaptation view is mostly found among behavioural ecologists and evolutionary anthropologists, whilst the by-product view finds most of its advocates within evolutionary cognitive psychology. However, the discussion about whether religion is an evolutionary adaptation or an evolutionary by-product of the evolved specific human cognitive capabilities is, in the authors' view, quite irrelevant for its ultimate outcome. Religiosity/religion can, via its proximate effects, indirectly but ultimately influence the differential intergenerational transmission of particular genes and memes. Even if religiosity would appear not to have originated as an adaptation in the hominin evolutionary history, it may have been subsequently co-opted for adaptive purposes.[263] For instance, religious fundamentalism may be adaptive in some contexts without being an adaptation in an evolutionary sense. Moreover, it should not be forgotten that complex adaptation processes evolve from the building blocks from previous adaptation processes.[264]

Recently Ara Norenzayan and colleagues[265] tried to reconcile the by-product and adaptationist approaches by arguing that religious elements arose originally as evolutionary by-products of ordinary cognitive functions that subsequently appeared to have great adaptive advantages, mainly in intergroup competition.

The authors tend to share the view of Robert Wright[266] who wrote

> Religion arose out of a hodgepodge of genetically based mental mechanisms designed by natural selection for thoroughly mundane purposes.

Religious faith is a natural consequence of fundamental mental functions, and its ethical tenets are the cultural extension of cognitive capacities and emotional drives and needs.[267] From an evolutionary point of view, it is striking how strongly essential religious beliefs and practices are focused on sociobiological behavioural phenomena that result from evolutionary mechanisms such as selection in general, but also kin selection, reciprocity selection and group selection.[268]

3.6.1 Proximate Advantages of Religion in the Pre-scientific Era

The question now is which proximate advantages spirituality/religiosity and religious beliefs/religions produced favourable survival and reproductive effects in the

[262]Ellsworth (2009).
[263]Powell and Clarke (2012).
[264]Voland (2009, 11).
[265]Norenzayan et al. (2016, 5ff).
[266]Wright (2009, 482).
[267]Teehan (2006, 749–751).
[268]For instance, Reynolds and Tanner (1983), Wenegrat (1990), Jones and Reynolds (1995), Burkert (1996), Grinde (1998), Hinde (1999), Boyer (2001), Atran (2002), Wilson (2002), Steadman and Palmer (2008), Ellsworth (2009), Reiss (2009), Voland and Schiefenhövel (2009) Kardong (2010).

pre-scientific environment? In literature, a multitude of factors, not necessarily mutually incompatible but rather complementary, have been proposed: explaining life and death, coping with anxiety, stress, calamity and death, healing disease, mastering contingencies, controlling sexual and reproductive behaviour, promoting social cohesion, preventing cheating behaviour, safeguarding social dominance, defining behaviour towards strangers, justifying wars, etc.[269]

Trying to understand the emergence as well as the effects of religious beliefs, the authors think that the proximate advantages should be considered at two different levels: the individual and the societal level. Nevertheless, some basic moral principles of the major religions, such as the love messages of Buddhism, Hinduism and Christianity, are difficult to classify according to their individual or social effects, because they pertain both to individual well-being and social welfare.

At the individual level, two major categories can be distinguished: (1) understanding/explaining the facts of life and death; and (2) mastering the events of life and death. At the societal level, also two major categories can be distinguished: (1) promoting social cohesion and solidarity, mainly in view of defence against competing groups/populations, or conquering others' territories and resources; and (2) reinforcing the social dominance of the leadership.

3.6.1.1 Understanding and Explaining the Facts of Life and Death

The increasing brain capacity during hominin evolution was selected for exploring and mastering more thoroughly the properties of the environment (as well as self). These increasing capabilities also elicited interest, fascination, and wonder about the facts of nature, and the causes of life and death.[270]

It is not surprising that, in the pre-scientific stage of the cultural history of humankind in which there was still no profound knowledge of the real determinants of natural (and cultural) phenomena, humans turned to supernatural forces to explain the facts of life and death. Indeed, all religions devised ingenious stories and myths to help people understand the origin of life and death and to give meaning to life and death. This need to understand and explain is the direct consequence of the hominisation process, resulting in an increased intelligence and, above all, a stronger self-conscious awareness.[271] Matthew Alper[272] formulated this in a pertinent way:

> In summary, what I'm suggesting is that at some point in the last two million or so years, during the emergence of the later hominids, a cognitive adaptation emerged that enabled us to cope with our awareness of death, while at the same time allowing us to maintain self-conscious awareness.

[269]For instance, Diamond (2012, 367).
[270]Dawkins (2000), Atran (2002), Bulkeley (2004), Fuller (2006).
[271]See Newberg et al. (2001).
[272]Alper (2006, 129).

Many authors[273] consider the emergence, function, and persistence of religions/religious beliefs as a coping mechanism to relieve our existential fears and anxieties, to comfort us in a contingent, bewildering, threatening and dangerous world, to alleviate pains of disease and other forms of misery,[274] and to fight the fear of and terror about the absurdity of death, "probably the most intolerable of all absurdities",[275] "the ultimate indignity".[276] Religions provide a perspective of continuity of life, often in a heavenly hereafter as an antidote to the cold prospect of post-life nothingness.[277]

One can ask, what has the religious mysticism and myth-formation about life and death to do with evolutionary survival and reproduction? There is no direct causal relationship, but the solutions which religions devised to cope with matters of life and death (such as offering the perspective of a heavenly eternal life in an imaginary hereafter, and the existential advantages and satisfaction people experience thanks to their neural capacity for spirituality and religiosity), might have resulted in diminishing or eliminating all sorts of life anxieties, and might also have provided people with positive feelings such as optimism, altruism, action-oriented motivation, self-sacrifice. Such feelings are positively correlated with improved mental and physical health, which, in turn positively influences survival and reproduction.[278]

The same question has been raised about the evolutionary meaning of the development of spirituality and religious feelings in general. Some scholars are of the view that the evolutionary toolkit, and in particular natural selection, has no relevance to emotional predispositions for phenomena such as spirituality, mysticism, religiosity, aesthetics and art.[279] Those phenomena should be valued in themselves, beyond survival or reproductive processes. It is granted that the human brain is capable of developing many cognitively and emotionally driven forms of behaviour, which are experienced as ontogenetically highly satisfactory. It would be a mistake to completely dissociate those multifaceted phenomena, some of which are so important for social relations, such as spirituality and religiosity or aesthetics and art, from the indirect effects they may have on survival or trans-generational processes.

3.6.1.2 Mastering the Events of Life and Death

The human brain wants not only to understand life and death, and to buffer the anxieties they evoke, but also and even mainly to control those phenomena.

[273]For instance, Newberg et al. (2001), Spilka et al. (2003), Voland and Söling (2004), Ostow (2006).
[274]Boyer (2003, 121).
[275]Klarsfeld et al. (2003, 184).
[276]FM-2030, 1989, 199.
[277]McGuire and Tiger (2009, 132).
[278]For instance, Hamer (2005, 143), Inglehart and Welzel (2005), Alper (2006, 104), McGuire and Tiger (2009, 132).
[279]For instance, Thagard (2005), Oviedo (2009, 146–148).

Religious ruling in this respect pertains to four main domains—death, health, self-oriented innate drives, and especially sexual and reproductive behaviour.

Regarding death control, religions could not do very much in the pre-scientific era of human evolution and history. People had no realistic insight into the real causes of disease and deterioration, but devised a multitude of magical practices that were supposed to master or counteract contingency and life-threatening events. Above all, they invented comforting narratives about beliefs in an eternal life in a hypothetical hereafter, thus mentally extending the basic drive for survival into eternity. At the same time they were creating hope for a better life than the one they experienced on earth, with its threats of disease, famine, accidents, crime and war, and other calamities.[280] Victor Stenger[281] explains the popularity of Christianity and Islam "because of their appeal to our most selfish instinct with the promise of eternal life".

Improving health is another important action domain of religions,[282] not only practiced via hygienic habits[283] and health care but also by means of spiritual healing practices and rituals, food prohibitions, and above all miracles.[284] Many authors are of the view that the religious promotion of physical and mental health had important adaptive benefits.[285] Indeed, already and especially in ancestral times, fertility was positively correlated with good health.[286]

Regarding self-oriented innate drives, Donald T. Campbell[287] has argued pertinently that religions have traditionally played an important biosocial role in trying to master and counteract a multitude of innate ego-centred drives, such as selfishness, greed, aggression and spite, with the view to promoting social cohesion and cooperation. By imposing moral standards, among others, via the authority of a supernatural power (and the threat of supernatural punishment[288]), religions tried to deal with the important problem of countering free-riding and cheating by internalising social control.[289]

Last but not least, religions are particularly involved in controlling sexual and reproductive behaviour. From reading scriptures and other basic documents of the three Abrahamic religions, the authors perceive three concerns that dominate the sexual and reproductive rulings of those religions: (1) facilitate males, with their strong, regular ejaculation pressures, sexual accessibility to the precious source of

[280]Gailliot et al. (2008).

[281]Stenger (2009, 16).

[282]Chatters (2000), Koenig et al. (2002), Sloan and Bagiella (2002), Kardong (2010, 36ff).

[283]Cleaning practices evolving into religious rituals may, without a conscious knowledge about their hygienic effects, nevertheless have caused a positive relation between such religious customs and health promotion (Kardong 2010, 36; see also Van Schaik and Michel 2016, 160ff).

[284]Theologians define miracles as phenomena that do not obey laws of nature.

[285]McNamara et al. (2006), McClenon (2002), Newberg et al. (2001), Rossano (2010), Van Schaik and Michel (2016).

[286]For instance, Easterlin (1975), Porter (1999), Grundy and Tomassini (2005).

[287]Campbell (1975).

[288]Johnson (2005), Johnson and Kruger (2004), Johnson and Bering (2006), Schloss and Murray (2011), Laurin et al. (2012), Johnson (2013).

[289]For instance, Wilson (2002).

female bodies; (2) protect the transmission of male genes against those of potential competitors; and (3) promote the spreading of the faithful's genes and memes.

A broad diversity of religious commandments and taboos, such as women's veiling[290] and sequestration, sexually differential mating rules, but also anti-contraception and anti-abortion stands and campaigns must be understood in that way.[291] They are all the result of the interaction of some human-specific sexual dimorphic characteristics. They include such diverse features as the larger female role in reproduction, the presence of concealed ovulation, and male–male competition for young fecund mates. Furthermore, the social and ideological conditions in male-dominated patriarchal societies with their sexual power divisions were

[290]In recent years several Western countries have experienced a hot debate about whether and in what circumstances Islamic women are to be allowed or forbidden to wear a headscarf, or even a niqab or burqa. In the headscarf debate all sorts of arguments for and against are used which typically only cover superficial elements. Rarely, if ever, are the root causes of this behavioural phenomenon addressed, which some consider to be insignificant—'a futile detail'! Headscarves, hijabs, chadors, niqabs and burqas are, along with many other discriminatory behavioural phenomena towards women, ultimately a result of what biologists call 'sperm competition' or the somewhat less distant 'male-male competition'. In male-dominated patriarchal societies, as male property women were socially repressed, sexually segregated or monitored, possibly even veiled, and so protected (!) against the genetically cuckoo risks arising from possible contacts with male competitors. The headscarf is a cultural relic of the male cuckoo syndrome and the derived masculinist dominance urge that culminated in the agrarian-pastoral era in the biosocial and cultural oppression and exploitation of women. In the agrarian-pastoral cultural phase, the masculinist sexism was conveniently justified, confirmed and strengthened in the religions that developed in that era. Although female subordination and sequestration is a general feature of agrarian-pastoral religions, it has been maximised in Islam with its harem culture, veils, hijabs, chadors, niqabs and burqas. The headscarf—historically one of the symbols of female oppression and sexual chastity—is nowadays often represented as an expression of personal identity, comparable to the Christian crucifix, the humiliating origin of which also became the mark of exquisite identity. However, this is a fallacy! Why do only Muslim women have to wear headscarves, hijabs, chadors, niqabs or even burqas, while Muslim males are relieved from such forms of cover? Advocates in the West of the sexually differential vestimentary Muslim codes should visit fundamentalist Islamic countries to observe and experience how this difference is an expression of the fundamental sexual discrimination existing in those countries. Hence, the importation of sexually differential vestimentary codes, expressing sexual discrimination, should not be favoured, particularly for female compatriots with an immigrant background who need to integrate culturally and socially in modern host societies. Any form of cultural, religious, social, or political sexism should be discouraged, if not fought. It should not be forgotten that behind the claimed right to cultural identity, there often exists a hidden agenda to slow down the upward emancipation and social mobility of minority groups, and especially of the female members of those groups. Women in general are still struggling to fight the 'glass ceiling': this ceiling often lies much lower for immigrant women. All modern secular ideologies—liberalism, socialism, feminism, humanism, etc.—implicitly or explicitly reject inequity and inequality between the sexes and are also in favour of a clear separation of church and state. Whilst they have more or less succeeded in neutralising the social and political dominance of the Christian churches, they should avoid reversing this process with respect to the Islam or other religious belief systems that, via immigration, try to get foothold in modern societies. Accepting the headscarf or other behavioural manifestations that originated in the oppression and exploitation of women is—in the modern, secular society that resulted from the scientific revolution and the Enlightenment—an example of a regressive evolution.

[291]Batten (1994).

sanctified by the organised religions that emerged and evolved in those societies. Paradoxically, the religious doctrines aimed at protection against sexual infidelity may not only have been in the genetic interest of males by avoiding cuckoldry[292]: they may also have had advantages for females by the avoidance of the male partner investing his resources in other females and their children.[293]

The strong religious rulings about realising a high fertility—"be fruitful and multiply, and fill the earth and subdue it…"[294]—has a direct effect on the multiplication of the religious followers—their memes and genes. Hence, the memetic religious rulings about sexual and reproductive behaviour are ultimately strongly genetically oriented and explain the reproductive and evolutionary success of those belief systems. Even in modern times, practicing religious people usually realise higher fertility than non-religious people.[295] However, this is a recent phenomenon, because nearly everybody was religious in earlier times.[296]

All three major incentives for the religious control of sexual and reproductive behaviour—guaranteeing males accessibility to female bodies, protecting males against cuckoldry, and spreading the faithful's memes and genes—are the ultimate causes for women's profound social oppression in human evolution and history. Traditionally, both in matters of sex and reproduction, religions see women as the precious resource, as subjects for men's sexual gratification and as producers and carers of offspring.

3.6.1.3 Promoting Social Cohesion and Cooperation

Many scholars believe that religion is adaptive because it favours 'fictive kinship', fostering family-like cooperation between non-family members. It supports group living in larger communities,[297] and promotes altruistic and cooperative behaviour within the in-group.[298] Moreover, religious groups appear to cooperate more within themselves than nonreligious groups—religion is a typical in-group marker[299] and in-group reinforcer. This need for increased cooperation, solidarity and commitment, especially in larger, genetically less or non-closely related groups of individuals, resulted in the invention and imposition of costly signaling religious practices.[300] Cross-cultural and even experimental research suggests that beliefs in a

[292]Strassmann et al. (2012).
[293]Harris (2010, 147).
[294]Genesis 1:28; Deuteronomy, 7:13-14; see also the discussion in Betzig (2005).
[295]For instance, Cliquet and Maelstaf (1977), McQuillan (2004), Weeden et al. (2008), Frejka and Westoff (2008), Zhang (2008), Blume (2009), Vaas and Blume (2009), Rowthorn (2011), Weeden and Kurzban (2013).
[296]Vaas and Blume (2009, 220).
[297]Roes and Raymond (2003), Shariff and Norenzayan (2007), Norenzayan et al. (2016).
[298]For instance, Sosis (2000), Irons (2001), Wilson (2002), Sosis and Ruffle (2004), Voland and Söling (2004), Alcorta and Sosis (2005), Johnson (2005), Johnson and Bering (2006), McNamara (2006), Norenzayan and Shariff (2008), Steadman and Palmer (2008), Bulbulia et al. (2008), Soler (2008), Bulbulia (2012), Preston and Ritter (2013).
[299]Yamamoto et al. (2009, 225).
[300]Cronk (1994), Sosis (2003, 2006), Bulbulia (2004), Alcorta and Sosis (2005), Sosis and Alcorta (2008).

supernatural punishing agency and in an afterlife may have been important promoters of in-group social cooperation.[301]

In the past religious prosociality, supported by divine revelation and 'In-God-We-Trust' institutions, contributed considerably to the phenomenon of in-group identity and its distinction from belonging to out-groups.[302] In this respect it must also have been a powerful instrument in intergroup confrontation and competition that was a predominant phenomenon in human evolution and history.[303] The religious aversion or even enmity towards out-groups is a striking phenomenon that has initiated, justified and fuelled wars and genocides throughout human history.[304] Apparently, beliefs in the supernatural have a strong social uniting effect on individuals[305] even for the annihilation of 'others'.

3.6.1.4 Reinforcing the Social Dominance of the Leadership

Several authors have also stressed that religions, especially when they become the official societal or state ideology, not only favour in-group cohesion but also strengthen in particular the power and prestige of society's leadership—males, kings, emperors, popes, and kleptocracies in general—facilitating them to control or even to submit their subjects.[306] Religions took advantage of the ancient (primate) tendency of obedience and submission to dominant alpha-males in order to support ideologically and socially the power position of social leaders,[307] resulting in what Friedrich Nietzsche[308] once referred to as '*Sklavenmoral*' ('slave morality') for the populace. Indeed, as Stanley Milgram[309] showed in his bewildering experiment on obedience to authority, most people are intuitively very obedient to authority. However, the degree to which authority is blindly obeyed varies cross-culturally.

It must also be acknowledged that a religiously supported and justified leadership may also have had a positive organisational effect in society. It enabled protection against criminals and especially defence against out-group enemies and the conquering of others' territories and resources—particularly female mates.[310]

[301]Shariff and Norenzayan (2007), Atkinson and Bourrat (2011), Keltner et al. (2014).
[302]Voland (2009, 16).
[303]Divale (1972), Ember (1978), Alexander (1979, 1987), Diamond (1992), Van der Dennen (1995), Keeley (1996), LeBlanc (2003), Dunbar (2004), Gat (2006), Pinker (2011).
[304]Van der Dennen (1995), Juergensmeyer (2001), Haught (2002), Avalos (2005), Diamond (2012, 367).
[305]Gorelik et al. (2012).
[306]Marx (1867), Alexander (1987), Cronk (1994), Diamond (1997, 277), Stenger (2008, 246).
[307]Grinde (1998), Glass (2007).
[308]Nietzsche (1887, 267).
[309]Milgram (1974).
[310]Dennett (2006, 56).

3.6.2 Ultimate Advantages of Religion in the Pre-scientific Era

This book's classification of proximate advantages of religion in the pre-scientific era is the basis for an answer to Daniel Dennett's[311] pertinent question: "*cui bono*? Individuals, elites, or society as a whole?" Clearly, the advantages of religion concern all three.

Although based on fantasy, as Richard Dawkins[312] pertinently argues in his book *The God Delusion,* religiosity and religions must have had considerable advantages in the pre-scientific era of human evolution and history, even though they involved high material, emotional and cognitive costs for individuals.[313]

It helped individuals to understand incomprehensible facts of nature in a simple way: it was a support to cope with anxiety and accept death and hope for a better and eternal life in the mythical hereafter; it stimulated spiritual experiences, it incited people to ethical behaviour; and it facilitated better endurance of misfortunes such as the loss of beloved ones, hereditary impairments, infectious diseases and natural catastrophes.

Regarding elites, religions were particularly advantageous for individuals and groups who occupied dominant positions in society, because religious institutions usually gave them a divine justification for their leadership and related privileges.[314]

At the societal level, religions reinforced sociality, solidarity, and mutualism, and strengthened group cohesion in the combat with rival populations, and conquering new territories and resources.[315]

Furthermore, religious beliefs not only had advantages at the proximate level, but through those effects may also have had ultimate advantages. For many people the proximate effects may have had 'fitness'-enhancing effects.[316] The increase of the number of descendants over many generations may be ultimately the most important effect of religious behaviour.[317] Hence, Richard Alexander[318] and Bernard Crespi and Kyle Summers[319] rightly interpret the origin and maintenance of

[311]Dennett (2007, 90).
[312]Dawkins (2006, 165–166).
[313]Atran (2002); see also Vaas and Blume (2009, 115–126).
[314]For instance, Marx (1867), Alexander (1987).
[315]For instance, Batson (1983).
[316]For instance, Jones and Reynolds (1995), Alper (2006, 239), Dennett (2007, 69), Harris (2004, 15), Steadman and Palmer (2008, ix), Blume (2009, 119).
[317]The celibacy, and consequently the zero reproductive fitness of Catholic priests seems, at first sight, to contradict the proposition that religions are strongly focused on the reproduction of their adherents, but this is not necessarily so. If non-reproducing priests succeed in motivating their kin and non-kin adherents of their faith to strongly reproduce in order to spread their faith and the growth of their church—as everybody knows who has ever heard priests delivering sermons at the occasion of baptism ceremonies—one can witness the reproductive effects of a kind of kin selection by the indirect transmission of genes and memes of the celibatarians via their fellow believers (Livi-Bacci 1971; Crook and Crook 1988; Hill 1999; Deady et al. 2006).
[318]Alexander (2013).
[319]Crespi and Summers (2014).

religion and the concept of God as the function of maximising inclusive fitness "through serving the interests of one's circle of kin and one's larger-scale social and cultural groups."

From a biological evolutionary viewpoint, in the pre-scientific era, religious-magic systems had to be considered as vehicles for the development, justification, imposition and transmission of ethical systems of vital importance for developing and transmitting human life. From a psychological as well as a social point of view, they were a 'natural security system'.[320] In those living conditions they fulfilled the function of an exo-somatic developmental and survival mechanism of the same importance as biological organs.[321]

Considering all of the foregoing—at the proximate level: the genetic and neurological determinants of spirituality/religiosity, the individual and societal advantages of religious adherence, the on-going religious indoctrination in many societies; and, at the ultimate level, the evolutionary effects regarding the multiplication of genes and memes favouring religious adherence and obedience—it becomes understandable why religiosity and religious beliefs continue to endure. It is not difficult to understand *Why god won't go away* as Andrew Newberg, Eugene d'Aquili and Vince Rause[322] put it—as religions maintain their power positions in modern society, despite the fact that science has completely undermined their ideological narrative about origin and source of moral authority.[323]

3.6.3 Place of Religious Beliefs and Religions in Modernisation

Whilst religions may have had proximate (ontogenetic) as well as ultimate (phylogenetic) advantages in pre-scientific living conditions, a quite different picture emerges in modern culture. Notwithstanding that religious beliefs and religions still can and do have several important advantages for many individuals, overall, neither their foundations nor much of their rule-giving and practices, are well adapted to the novel environment that is being created by the modernisation process. In many respects, religious beliefs, Holy Scriptures or prophecies on which they build, and religious institutions are maladapted to modern culture. Belief in supernatural beings is a typical example of a phenomenon that had adaptive advantages in the environment in which our ancestors evolved, and the same can be said about organised religions in the pastoral-agrarian era, but this may no longer be the case in the novel environment of modernity.[324]

[320]For instance, Wiebe (2013).
[321]Cliquet and Thienpont (2002, 601).
[322]Newberg et al. (2001).
[323]For instance, Dawkins (2000; 2003; 2006), Edis (2002), Stenger (2003; 2008; 2011), Dubessy et al. (2004), Hitchens (2008), Isaacson (2012), Krauss (2012).
[324]Wenegrat (1990), Roele (1993), Reynolds (1995), Bulbulia (2004), Dennett (2007), Davis (2009), Coyne (2012), Wiebe (2013).

However, the 'maladaptationist hypothesis' does not imply that all of religious morality is retrograde or maladaptive. The reason theretofore is that much of the religious moral rulings are the result of evolutionary selective processes that have survival value, not only in pre-scientific living conditions but also in the novel environment of modernity.[325] As will be argued in Chap. 5 on the contents of an evolutionary based ethics, many religiously inspired or supported moral rules fit with the logic of evolutionary morality.

3.6.3.1 Advantageous Religious Spillovers in Modernity

Today, religions have some persisting advantages, at least for their adepts, at the individual level, the religious in-group level, and the societal level.

At the individual level: for many individuals who have genetically or neurologically strong predispositions to spirituality/religiosity, religious adherence and practice may still be very satisfactory for coping with the anxieties of misfortune and death.[326] Religious beliefs may give those people a meaning to life and death.[327] It may help people who have not acquired much education and knowledge to behave morally, and motivate them to be socially cooperative members of their society.[328] Deeply religious people may enjoy the considerable emotional and cognitive pleasurability of their religiosity/spirituality.[329] Increased religiosity is associated with increased agreeableness and benevolence.[330]

People with low personal control or insecure attachments may find relief in the belief in God.[331] Some research indicates that people who believe in a loving, caring, forgiving, and approving God report having or show signs of higher self-esteem,[332] positive mood,[333] better psychological well-being,[334] more life satisfaction and higher quality of life,[335] stronger self-improvement,[336] and show fewer symptoms of anxiety and depression, and other psychiatric symptoms.[337]

At the religious in-group level: many studies have found that religious involvement is correlated with decreased morbidity and mortality.[338] However, some critical appraisals of the relevant literature show that there is little empirical evidence for the proposition that religiosity has more beneficial health outcomes.[339]

[325]For instance, Teehan (2006, 769).
[326]For instance, Vail et al. (2010).
[327]For instance, Hogan (2004, 738.7).
[328]For instance, Newberg et al. (2001, 81).
[329]For instance, Grinde (1998), Purzycki and Sosis (2008), McGuire and Tiger (2009).
[330]Saroglou (2002), Saroglou et al. (2004).
[331]Kay et al. (2010), Granqvist et al. (2010).
[332]Benson and Spilka (1973), Francis et al. (2001).
[333]Levin (2001), Krause (2005).
[334]Bradshaw and Ellison (2008), Heinemann and Wörmann (2010), Flannelly and Galek (2010), Sasaki et al. (2011).
[335]Kirkpatrick and Shaver (1992), Pargament et al. (1998), Flannelly et al. (2010).
[336]Sedikides and Gebauer (2010).
[337]Phillips et al. (2004), Bradshaw and Ellison (2008), Flannelly et al. (2010), Levin (2010).
[338]Koenig et al. (2000), Miller and Thoresen (2003), Newberg and Lee (2006), Hall et al. (2008), Levin (2010), Waldron et al. (2011).
[339]For instance, Sloan and Bagiella (2002).

As is often the case in research into religious attitudes or behaviour—and this probably applies to all of the above mentioned findings—few investigations have been undertaken in which control groups of people with secular ideological convictions such as humanists were involved. The risks of social assortment bias in the composition of the researched population samples in most studies of religious attitudes and behaviour are not carefully taken into account. This is especially true for the United States where the majority of the above citied findings come from and where the overwhelming part of the population believes in God.

It is generally believed that religiosity fosters cooperative and altruistic behaviour.[340] Whoever has had intense contacts with religious officeholders—priests, monks and nuns—is, in most cases, struck by their kindness, humane concerns, and sociability in general. However, here again comparative investigations about the social behaviour of people with other ideological convictions are lacking. People committed to social justice or equity in education, embedded in values of humanism, may equally have highly humane concerns and be generous. In addition, like any other ideology, religious adherence and practice, even when not explicitly articulated, may have important social or financial benefits and facilitate the acquisition of wealth and power, either via social status/prestige, social identity[341] or via in-group nepotism.[342]

At the societal level: religions contribute substantially to justify and legitimise morality, which is an important factor promoting the viability of groups.[343] Religion also fosters social cohesion through its shared values, meaning systems, rituals and lifestyles.[344] Moreover, religious institutions are well known for organising social initiatives, which implement altruistic behaviour.[345] In general, religions have a positive and durable effect on the establishment and maintenance of social capital.[346] Finally, religions may strongly contribute to controlling socially disrupting forms of behaviour.[347] To sum up, potentiality for doing good is undeniably present, albeit predominantly directed towards believers.

3.6.3.2 Disadvantageous Effects of Religious Beliefs in Modernity

Notwithstanding persistent advantages, the maladaptation of religious beliefs and religious adherence in modernity pertains to many issues, not only concerning non-believers but also for the population and society as a whole.

Scientific knowledge has largely discredited the basic foundations of religious beliefs and ideologies embedded in sacred texts. Many religious beliefs about the origin and evolution of the cosmos, about the origin and evolution of life (and in

[340] Sosis (2000), Sosis and Ruffle (2004), Finkel et al. (2010).
[341] Ysseldyk et al. (2010).
[342] Bellow (2003).
[343] Atran and Norenzayan (2004).
[344] Hogan (2004).
[345] For instance, www.caritas.org.
[346] Putnam (2000, 65–79).
[347] For instance, Johnson (2005).

particular about human evolution), about the sense of life and death, about the source and content of some values and norms guiding behaviour, are incompatible with the findings of the natural and human sciences. In many respects religious beliefs even prevent people from acquiring knowledge and understanding of the facts of life and death.[348]

In matters of education, religions indoctrinate children[349] and adolescents[350] one-sidedly at an early age. Their sense of reality is distorted, introducing serious confusion into their minds between what has to do with religious narratives from the sacred texts and what has to do with science. In terms of pedagogics religious indoctrination is imprinting children and adolescents with feelings of shame, remorse, or guilt over their own development and growth that reverberate for many years in adulthood.[351] The paradoxes of modern times are present in human rights conventions issued as result of compromises between states having different worldviews. Indeed, one can go as far as to argue that the early religious indoctrination of children is in contradiction with Article 14.1 of the United Nations Convention on the Rights of the Child[352] which stipulates "States Parties shall respect the right of the child to freedom of thought, conscience and religion".[353] Freedom of thought and religion do not go hand in hand within religious communities and faith organised states. A specific educational problem concerns pressures from some religious denominations in countries such as the United States, Islamic states, or Christian and Islamist groups in Europe, to eliminate biology and other natural sciences from the school curricula. There are also pressures to counterbalance scientific education with narratives about creationism or intelligent design, which reject the scientific theory of evolution.[354]

In matters of mental and physical health, religions have largely lost their credibility for healing practices and have been replaced by the knowledge acquired by the modern biomedical sciences. The selective anti-interventionist ideology of religions in matters of life and death (e.g. anti-contraception; anti-abortion; anti-euthanasia; even general anti-medical intervention in some denominations, such as the Christian Scientists) is counter to the progress in medical life saving and caring achieved in modern culture. The right to die in dignity is not a life saving strategy. However, it may be considered as a caring strategy at times when modern

[348]For instance, Kitcher (1982, 2007), Godfrey (1983), Tiffin (1994), Larson and Witham (1998), Stenger (2003; 2008; 2011; 2012), Dawkins (1986; 2000; 2003; 2006), Russell (1997), Edis (2002), Harris (2004), Scott (2005), Skybreak (2006), Young and Edis (2006), Fuller (2007), Hitchens (2008), Coyne (2009), Isaacson (2012).

[349]Harris and McNamara (2008), Heimlich (2011).

[350]Alcorta and Sosis (2005).

[351]Dawkins (2006, 315), Council of Europe (2007), Dennett (2007, 56).

[352]United Nations (1989).

[353]However, it should be mentioned that, as is often the case in UN charters, the same Article 14 includes in its §3 a restrictive condition, worded as follows: "Freedom to manifest one's religion or beliefs may be subject only to such limitations as are prescribed by law and are necessary to protect public safety, order, health or morals, or the fundamental rights and freedoms of others."

[354]Scott (2005), Numbers (2006), Jalajel (2009), Hameed (2010), Riexinger (2010).

medical and pharmaceutical industry produces the means for prolonging the dying process without enhancing the potential for life (See also Chap. 6, Sect. 6.2.2.2).

The sexist tenets of the organised religions that emerged and flourished in the agrarian-pastoral cultural era were either based on masculine domination and devised as a function of what biologists call 'sperm competition',[355] or resulted from psychopathological (male) fears of the female gender[356]—these have become obsolete in modern culture. The biological knowledge about the sexes, the generalisation of education and access to information, technological advances, and ideological emancipation movements, have thoroughly, albeit not yet completely, changed the power relations between the sexes in modern societies.[357] Religious rulings about sexual and reproductive behaviour have induced several forms of sexual repression that are incompatible with knowledge acquired by science and/or by present-day secular ethical standards. Well-known examples of such repression are: the rejection of non-marital sex (in some regions or social environments leading to the practice of honour killings[358] or stoning[359]), the condemnation of masturbation,[360] the belief that sex serves only as a function of fertilisation, the condemnation of homosexual behaviour,[361] and the practice of ritual genital mutilation (male circumcision,[362] female genital cutting such as clitoridectomy or infibulation[363]).

At the societal level, the religious ideological support of social dominance by a hereditary ruling class is scientifically unjustified because of the scientific knowledge about segregation and recombination of genes in a sexually reproducing species such as *Homo sapiens sapiens*.[364] Due to Mendelian genetics, children do not necessarily show the same capabilities as their parents and cannot be expected to be able to assume the same responsibilities. Moreover, the maintenance of a hereditary ruling class contradicts all modern emancipatory ideologies and thus it is in sharp contrast to newly emerged moral standards.

[355]Baker and Bellis (1995), Shackelford and Pound (2006).
[356]Augustine of Hippo, 398; 426.
[357]Cliquet (1984).
[358]For instance, Meetoo and Mirza (2007).
[359]For instance, Terman and Women Living Under Muslim Laws (2007).
[360]For instance, Cornog (2003).
[361]For instance, Siker (2007).
[362]For instance, Denniston et al. (2010).

NB. In May 2012, in a historical and sensational verdict the district court of Cologne (Germany) ordered that boys who are circumcised for religious reasons is an offense because it is an irrevocable physical injury, arguing that the right of the parents nor the constitutionally enshrined freedom of religion justifies impeding the child's right to bodily integrity and self determination. The Jewish and Muslim communities condemned the ruling as antireligious, and by the end of 2012, not surprisingly, the German Parliament voted, in opposition to its own constitution that guarantees bodily integrity, a law protecting religious circumcision (http://dipbt.bundestag.de/dip21/btd/17/112/1711295.pdf). For a detailed discussion of this issue, see Aurenque and Wiesing (2015).

[363]For instance, Skaine (2005), Odeyemi (2008).
[364]Cliquet (2010, 406–416).

The ‘in-group’ orientation of religions (‘the chosen ones’, ‘the illuminated ones’) is resulting in the strengthening of intolerance towards ‘others’ and is easily supported by instruments of modern mass media, leading to ethnocentric, racist and xenophobic attitudes and practices, including social ostracism of apostates and non-believers, as well as justifying war and genocide. Religions claim universality but in fact their institutions are in-group oriented and concerned with winning over and enlarging the pool of faithful. In an increasingly globalising world, and with the presence of weapons of mass destruction, such an ideological divide between ‘true’ believers having the monopoly to the true word and ‘out-groups’ has not only become obsolete but is also a danger to our survival as a species. Hence, it may be affirmed that organised religions and religiosity based on beliefs and in-group morality transmitted through religious institutions are no longer instruments of human survival.[365] They ceased to be adaptive to human survival and have become maladaptive instruments.

In the history of humankind the organised religions have systematically been a source and cause of intergroup violence and war, resulting from their in-group oriented morality that, in turn, is rooted in our evolutionary heritage.[366] Religiously inspired authors evidently refute this finding.[367] However, in-group favouritism and out-group aversion or even out-group enmity of religions towards non-believers is a quite general phenomenon. It accompanied Christianity particularly during Crusades and Inquisition times and nowadays, it is particularly pronounced in jihad ideology and among Islamic fundamentalists who openly foster a culture of death, with salvation for its martyrs in a hypothetical hereafter.[368]

The reproductive ideologies of organised religions—“be fruitful and multiply, and fill the earth and subdue it, …”[369]—aimed at spreading their own beliefs are, in the presence of overpopulation and over-exploitation of available resources in many parts of the world, in direct conflict with the necessity to limit births, and to diminish the population size and density through birth control.[370] The current economic and ecological problems, which many countries and regions are struggling with, are in part the direct consequences of the norm systems of religions that did not succeed soon enough in changing their reproductive-growth ideologies into responsible birth control ideologies. The consciously sustained drive by some religious denominations to further expand their memes (and genes of their followers) through high(er) fertility remains highly visible in present times.

Several decades ago Julian Huxley[371] argued that theistic religions, with their divine revelations and dogmatic theologies, are not only an impediment to scientific

[365]For instance, Huntington (1996), Haught (2002), Harris (2004), Dennett (2007), Saxton (2009), Graham and Haidt (2010).
[366]Nelson-Pallmeyer (2003), Teehan (2010, 147).
[367]For instance, Armstrong (2014).
[368]See the argumentation in, for instance, Du Pasquier (1992), Jansen (1997), Lewis (2003), Manji (2003), Van Rooy and Van Rooy (2010).
[369]Genesis 1:28.
[370]For instance, Cohen (1996), Ehrlich and Ehrlich (2008), Ewing et al. (2010).
[371]Huxley (1964, 108); see also Stenger (2009, 47).

and social progress but are also obstacles to the emergence of new forms of belief systems that are compatible with our present knowledge and that are necessary for the future survival and progress of humanity.

In his recent book, *The Folly of Fools*, about the evolutionary logic of deceit and self-deception in human life, Robert Trivers[372] pertinently argues that certain features of religion provide a recipe for self-deception, disregarding any rational thinking. For instance, this is the case with the presence in many religions of a unified, privileged view of the universe for the own group (supporting the in-group/out-group syndrome), the belief in a series of interconnected phantasmagorical things (e.g. afterlife, God, miracles, immaculate conception, last judgment, resurrection of the dead), the deification of prophets, the theistic revelation of Holy Scriptures, the prevalence of faith over reason, and the conviction about the moral superiority of the own in-group.

3.6.4 Is God Redundant?

On the basis of the current knowledge in the natural sciences (in particular physics, geology and cosmology), the life sciences (in particular evolutionary biology, anthropology, sociobiology, evolutionary psychology, genetics and neurology), and some cultural sciences (in particular ethnology and archaeology), there can be no doubt that religions are man-made. They are products of human evolution and the human need, in the pre-scientific era of human history, to master the human fears of finiteness (death), to control human evolution (destiny), to deal with diseases and natural catastrophes, to strengthen social cohesion, to submit the masses, and to defeat the enemies.[373] The authors acknowledge that "absence of evidence is not necessarily evidence of absence".[374] They maintain that, to the best of their knowledge, there is no scientific evidence of the anthropomorphist, creationist, interventionist, omnipotent, omniscient, omnibenevolent, miracle-wreaking God, as worshipped by Jews, Christians, and Muslims. It may be affirmed beyond a reasonable doubt that such a supernatural agent does not exist.[375] Science has no need to resort to a creator to explain the origin and evolution of the cosmos, or the origin and evolution of life in general and of the hominins in particular. It also does not need to rely on supernatural forces to explain spirituality and justify morality.[376] As Matthew Alper[377] wrote:

> … humankind can no longer be viewed as a product of God but rather God must be viewed as a product of human cognition.

[372]Trivers (2011, 282).
[373]Tremlin (2006, 6).
[374]Mark Bekoff, quoted in Moritz (2012).
[375]Philipse (1995, 2012), Martin and Monnier (2003), Everitt (2004), Dawkins (2006), Stenger (2007, 11; 2009, 12; 2012, 78), Paulos (2008), Krauss (2012), Vermeersch (2016).
[376]Stenger (2012, 78).
[377]Alper (2006, 97).

Or as Michael A. Persinger[378] concluded:

> God is an artefact of the brain.

As Stephen Hawking and Leonard Mlodinow[379] conclude in their recent synthesis *The Grand Design*:

> But just as Darwin and Wallace explained how the apparently miraculous design of living forms could appear without intervention by a supreme being, the multiverse concept can explain the fine-tuning of physical law without the need for a benevolent creator who made the universe for our benefit.

Or as Lawrence M. Krauss[380] concluded in his impressive *Why there is Something rather than Nothing. A Universe from Nothing*:

> I find oddly satisfying the possibility that, in either scenario (a unique universe or a multiverse of universes), even a seemingly omnipotent God would have no freedom in the creation of our universe. No doubt because it further suggests that God is unnecessary – or at best redundant.

The writers of the Holy Scriptures, as well as their divinely inspired messiahs and prophets, had no idea of the real causes of human life, death and disease, or of the determinants of natural events and processes. The contents of the Tanakh, the Bible and the Qur'an on origins and causes of life, death and natural events are contradictory to the real knowledge sciences are acquiring at last.[381]

However, one can only express admiration for the ingenious ways in which faiths, in the absence of knowledge, have designed myths about the origin of life and humankind, about the reasons for man's finiteness, about the causes of diseases and disasters, and about the rewards and punishments in a promised hereafter.

Reading the basic scriptures of Judaism, Christianity and Islam is extremely revealing, not only from a cultural-historical point of view but also from an evolutionary-biological point of view. They provide a lot of reflection and divine justification of the cultural-technological stage of development in which they emerged and thrived, namely the patriarchal, male dominated agrarian-pastoral stage of cultural development. Nevertheless, the non-believer is overwhelmed by inconsistencies in their narratives, the awkwardness of their superstitions and taboos, and the lack of knowledge about the real origins of morality.

The evolutionary biologist is especially struck by the ambiguity of the Abrahamic religions regarding biological evolutionary matters. On the one hand, those religions are in some respects going against evolution in their conceptualisation of the creation of the cosmos and life.[382] On the other hand, many of the moral rules are in accordance with the necessities of evolutionary processes, and are compatible

[378]Persinger (1987), quoted in Murray (2008).
[379]Hawking and Mlodinow (2010, 165).
[380]Krauss (2012, 185).
[381]For instance, Godfrey (1983), Tiffin (1994), Wilson and Dolphin (1996), Pennock (1999, 2003), Moore (2002), Forrest and Gross (2004), Young and Edis (2004), Stenger (2008, 2011), Krauss (2012).
[382]See the Genesis discourse in the Bible, even when this is not taken literally.

with the biological needs of individuals and the functioning of societies in pre-modern times.[383] They applied evolutionary strategies to maintain their tribal in-group integrity through endogamic rules governing sexuality and reproduction (as in Judaism), and/or to spread their beliefs by conquering, converting or exterminating other populations (as in Christianity and Islam).[384]

3.7 Science and Religion

Some further reflection is needed about (in)compatibility between religions and science with respect to the origin of morality.

First of all, it must be acknowledged that the scientific community is divided in that respect[385]: there are compatibilists and incompatibilists.[386] Some are of the view that science and religion are intrinsically 'non-overlapping magisteria'.[387] Many others believe that there is no conflict between religion and science (and in particular evolution science)—as long as religion does not make quasi-scientific claims about the factual nature of the world[388]—and make considerable intellectual efforts to reconcile or integrate both domains.[389] However, others think that science and religion are totally incompatible.[390]

Regarding the question as to whether science and religion are non-overlapping magisteria, the authors share the position taken by Victor Stenger[391] who refutes, for instance, the standpoint of the US National Academy of Sciences[392] which states that:

> Science is a way of knowing about the natural world. It is limited to explaining the natural world through natural causes. Science can say nothing about the supernatural.

[383]Hinde (2002), Teehan (2006, 2010).
[384]Keith (1946).
[385]For instance, Barbour (1990), Kurtz (2003), Lüke et al. (2004), Graffin and Provine (2007), Viney (2008), Reiss (2009).
[386]Clayton and Schloss (2004, 320), Feierman (2009, xv).
[387]Non-overlapping magisteria: a concept proposed by Gould (1997, 1999), defined as different domains of inquiry, each one of which based on specific and non-overlapping domains of teaching authority. (See also, for instance, Anderson and Peacocke 1987; Barbour 1997; National Academy of Sciences 1998; Wilson 2002, 41; American Association for the Advancement of Science 2006; Ayala 2007; Reis 2009; Grassie 2010).
[388]Miller (1999, 169), Shermer (2004, 6), Ruse (2000; 2001; 2008; 2010), Rolston (1999), McGrath (2004), Collins (2006), Roughgarden (2006), Pope (1994; 2007), Lüke et al. (2004), Armstrong (2009).
[389]For instance, Teilhard de Chardin (1956), Sharpe (1991), Hefner (1993), Williams (1996, 2001), Haught (2000, 2010), Collins (2006), Pope (2007) Feierman (2009).
[390]For instance, Huxley (1894), Dewey (1922), Huxley (1927), Monod (1970), Russell (1997), Edis (2002), Dawkins (2006), Dennett (2007), Stenger (2008; 2012), Harris (2010), Philipse (2012), Valdecasas et al. (2013).
[391]Stenger (2007, 28; 2012, 290).
[392]National Academy of Sciences (1998, 58).

Victor Stenger correctly argues that the supernatural can be studied by science if it is alleged to be able to affect natural phenomena and play an important role in the functioning of the universe and human life. Religions as a rule make claims about the natural world. They devised creation myths and express views on the origin of life and the cosmos, and prescribe rules about human moral conduct. These all relate to matters that belong to the realm of science. Furthermore, religion is an aspect of human culture that can be observed and studied by scientific methods.[393] This means that rejecting this approach would imply the acceptance of a dogmatically naturalistic nature of science. Hence, Mark Perakh and Matt Young[394] state correctly:

> Science is neither based on methodological naturalism nor restrained by it: it is restrained by one and only one requirement: it requires evidence.

Also the assertion of the US National Academy of Sciences[395]

> Whether there is a purpose to the universe or a purpose for human existence are not questions for science

is, in the authors' view, totally wrong. It is understandable that the American National Academy of Sciences does not want to wage war on its population, 90% of which believes in a personal God. The view that the existence of purpose is not amenable to scientific study is quite widespread, especially in religious quarters.[396] However, since the question of purpose of (human) life is an essential aspect of (and might be of fundamental importance for) the ontogenetic development and phylogenetic evolution of humanity, it must be a fully legitimate research subject of science. It might be that science will discover that there is no purpose for human existence, but that is another problem for which the implications would also have to be investigated.

There may be several reasons why so many scientists prefer to consider science and religion to be two different ways of knowing and think that there is no inherent conflict but also no meeting point between science and religion.

Some are of the view that science is unable to deal with the challenges of human existence, because it is believed that science can only describe but not prescribe. It would only be able to study what is but not what ought to be. (This matter was discussed in more detail in Chap. 2, Sect. 2.2.1.1). In contrast, religion would be able, because of its divine revelations, to address individuals' and societies' moral needs forever. Robert N. McCauley[397] even argues that "no development in science will ever seriously threaten the persistence of religion" because religion would be a "natural" phenomenon while science would be "unnatural".

---

[393]Stenger (2012, 290); see also Rachels (1991, 99ff), Feierman (2009, xvi).
[394]Perakh and Young (2006, 191).
[395]National Academy of Sciences (1998, 58).
[396]For instance, Mix and Masel (2014, 2444); Delhez (2015, 35ff).
[397]McCauley (2000).

Some scientists may not have thoroughly reflected on the substantive differences and incompatibilities between scientific findings and religious teachings or are, in particular, not sufficiently informed about evolutionary science or natural sciences in general. Indeed, present-day science has become so voluminous and complex that it needed to be organised in many specialised disciplines: however, the deplorable result is that many practitioners are losing touch with what should be generalised knowledge. Others may not want to consider those issues, either because of their strong religious socialisation and heritage[398] or loyalty to family traditions, or fear of losing their nepotistic or social privileges (status, prestige, job, promotion, funding), or simply for political reasons or even physical dangers or bullying in religiously dominated societies or environments. Still others may foster religious beliefs, which have a substantially different content from the tenets of the organised religions with their creationist, interventionist, personal God. Some may believe that "the great masses of humanity are best kept sedated by pious delusions".[399] Finally, a few may be so overwhelmed by their spiritual drive—or simply because their belief is consoling[400]—that all other considerations fade away and they continue to cherish the old dualistic conception of human existence—"immaterial soul versus material body"—in their way of double thinking and living. They compartmentalise their scientific activity and religious beliefs, although one may argue that in fact they compromise their scientific principles.[401] This double thinking can even be perceived in the way some authors continue to contrast natural sciences to the social sciences and humanities, ignoring the unity of scientific methodology.[402] Indeed, it is not impossible that the evolutionarily based predispositions for religious beliefs even play tricks on some scientists whose religious beliefs continue to satisfy them emotionally and intellectually, or who perceive evolution science as antagonistic to the way in which they conceptualise their existence.[403] Indeed, it must be admitted that scientific approaches to sense of purpose and morality appear to be emotionally less attractive and satisfactory to many people than mysticism, myth, ritual, magic, or religion.[404] Thus, some authors argue that the ambiguous position of scientists who continue to cherish their theistic beliefs is a prime example of the manifestation of self-deception.[405] Indeed, in their efforts to reconcile religious doctrines with scientific facts or theories, religious scientists risk corroding or usurping scientific truths in unverifiable ways.

Let us take the example of Francis S. Collins,[406] the renowned head of the Human Genome Project, currently director of the US National Institutes of Health, and co-discoverer of the genetic misspellings that cause cystic fibrosis,

[398]Ecklund and Scheitle (2007).
[399]This issue is discussed in Harris (2010).
[400]Martin Gardner (1996), quoted in Shermer (1997, 133).
[401]For instance, Miller (1999), Collins (2006), Giberson (2008); see also Ashton (2001).
[402]See, for instance, several contributors to the edited book of Lüke et al. (2004).
[403]For a discussion of this issue, see Tremlin (2006).
[404]Shermer (1997, 277).
[405]Trivers (2011, 279).
[406]Collins (2006).

neurofibromatosis and Huntington's disease. Collins has recently written a successful popularising book: *The Language of God. A Scientist Presents Evidence for Belief*. Notwithstanding his excellent acquaintance with the natural sciences, Collins believes that there is "still the possibility of a richly satisfying harmony between the scientific and spiritual worldviews." He believes that "God cares for us and can intervene in human affairs—on rare occasions even miraculously." Originally an agnostic, or even an atheist, Collins became a believer after reading *Mere Christianity* by the Christian apologist Clive S. Lewis[407] in which this author discusses the universal morality or natural law commonly known throughout humanity. In his book Collins raises all the right questions, accepts major findings of the natural sciences—origins of the universe, evolution of life (and rejection of creationism and intelligent design), and molecular genetics—but fails to provide plausible, convincible religious answers to the philosophical and moral questions he raises. In his acceptance of the great monotheistic religions of the world, he is, among others, strongly influenced by the "powerful evidence provided by human altruism". Apparently, Collins did not consider adequately the achievements of the Second Darwinian Revolution in which the evolutionary causes and mechanisms of altruistic behaviour were more thoroughly searched and explained.

Another recent example is the edited book of Martin A. Nowak and Sara Coakley,[408] entitled *Evolution, Games and God* in which several contributors[409] try to reconcile or justify Christian belief components with the more recent sociobiological and evolutionary game theoretical findings about cooperative and altruistic behaviour. Although several of those contributors raise the right questions and seem to be acquainted with more recent achievements of evolutionary science, their religious preconceptions prevent them from interpreting the philosophical implications of the new knowledge in a scientifically correct and comprehensive way. Their theologically-inspired answers to pertinent questions, such as the origin of prehuman dimensions of creation, the arena of human freedom and creativity, and the problem of evil,[410] are, from a scientific point of view—with its toolkit of logic, considerable body of empirical observations, experiments and derived theories—remarkably and disappointingly poor, vague, or meaningless. For instance, there is no reason why some aspects of altruistic behaviour, particularly those some authors refer to as 'moral altruism',[411] 'sacrificial altruism',[412] or 'intentional altruism',[413] could not be explained by the current evolutionary toolkit. It is not evident why theological reasoning has to be involved, as a *deus ex machina,* for phenomena which are fully explainable within the current evolutionary framework. Finally, theologically inspired scholars easily disregard the overall picture that evolution

[407]Lewis (1952).
[408]Nowak and Coakley (2013).
[409]See in particular the contributions of Johnson (2013, 168–185), Schloss (2013, 201–219), Pruss (2013, 329–342), Clayton (2013, 343–361), Rota (2013, 362–374), and Coakley (2013, 375–386).
[410]Coakley (2013, 375).
[411]Pruss (2013, 332).
[412]Rota (2013, 364).
[413]Schloss (2013, 212).

science has acquired about the evolutionary mechanism and process. This is shown in their assertions that the scientific findings and theories about the evolution of cooperation and altruism would provide some evidence or arguments for God's existence.[414] Indeed, contrary to the tenet of some parts of this book that suggests a causal link from theology to evolution processes, in particular regarding cooperation and altruism, present evolution science suggests the opposite causal link: namely science provides explanations as to why hominin evolution incited religions to invent, in pre-scientific living conditions, moral rules and codes supporting the cooperative and altruistic needs of the evolving hominins.

Compared to religious belief indicators in the general population, the prevalence of religiosity among scientists is much lower, although not inexistent.[415] For instance, whilst more than 90% of Americans believe in God, a survey among a sample of eminent scientists, namely members of the US Academy of Sciences (NAS), resulted in quasi opposite figures: the highest percentage of belief in God was found among NAS mathematicians (14%), the lowest rate among biological scientists (5%).[416] However, most scientists who continue to believe in a supernatural agency appear to be methodologically secularists or naturalists.[417]

Science and religion are, of course, similar in some respects concerning questions they address. First of all, religious behaviour and abstract reasoning may have co-evolved, both being concerned with abstract concepts and comprehensive frameworks, for which the aim is to understand and influence reality.[418] Historically, science arose from religious and theological thinking.[419] Both try to understand/explain life and death, and both try to control life and death, but their methods (and sometimes also their ends) are completely different. Nevertheless, in many areas of basic morality they may arrive at identical conclusions and solutions. Since the divinely inspired moral rulings of religions are in fact man-made, many of them may be largely the result of rational thinking (and natural selection).

As F. March rightly pointed out:

> Religion and science are the products of reflective thought.[420]

However, whatever the origin of religious memes, in the same way as genes they are subject to Darwinian selection,[421] resulting in the maintenance and reproduction of moral prescriptions that make sense from an evolutionary point of view.[422] Indeed, many of the moral precepts that can be observed in (successful) religions are compatible with the biological needs of individuals and the social viability of societies.[423]

---

[414] Pruss (2013, 333), Rota (2013, 364), Clayton (2013, 347), Coakley (2013, 383).
[415] Ecklund and Scheitle (2007).
[416] Larson and Witham (1998, 313).
[417] Shults (2015, 736).
[418] Previc (2006, 525); see also Wilson (2002, 41).
[419] Van Schaik and Michel (2016, 392).
[420] March (2009, 16); see also Stanley (2014).
[421] Cziko (1995), Dawkins (1983), Corning (1997; 2005).
[422] Miller (1999), Clayton and Schloss (2004), Teehan (2006).
[423] Hinde (2002), Voland and Söling (2004, 53).

In that respect, religion and science may seem to be compatible, but this apparent compatibility is, from a conceptual and methodological point of view, an intellectual mirage. Conceptually and methodologically, religion and science are fundamentally different and irreconcilable: religions draw and justify their worldview and moral rulings from alleged, albeit unproven or ungrounded, divine revelations, whilst science builds its knowledge on empirical and measurable observation, corroboration, experimental testing, hypothesis and theory formulation, following the laws of logic. Moreover, science changes or refines its theories, explanations and predictions as soon as new evidence contradicts or complements earlier findings or views. Religions are constrained by their Holy Scripts.

Divinely revealed myths about the origin and evolution of the cosmos and life, and in particular human life, such as can be found in Holy Scriptures like the Bible and the Qur'an, or even the current-day more sophisticated theological interpretations about evolution,[424] are completely at odds to and incompatible with the present-day knowledge of the natural and human sciences, and in particular with biological evolution science.[425] Religious apologists defend their faith by arguing that the Holy Scriptures should not be literally interpreted: however, this is a very weak position that resembles the semantic relativistic approach of the Humpty Dumpty character in Lewis Carroll's famous 1871 novel *Through the Looking Glass*.[426]

The incompatibility between science and religion is particularly salient with regard to the origin and evolution of morality.[427] Theologies have not been able to explain, through empirically observable or experimentally controllable means, any phenomenon, be it natural or supernatural. In contrast, notwithstanding their short-lived existence, sciences have made huge and consistent progress in unravelling facts about the origin and evolution of life and the cosmos, and in recent decades have even developed plausible explanations for the origin and evolution of phenomena such as gods, spirituality, religiosity, religion, and morality—domains that were traditionally considered to be the exclusive territory of organised religions.

[424]For instance, Teilhard de Chardin, 1950; Pope John Paul II, 1996; Haught (2004; 2010).

[425]On the basis of the present-day scientific acquisitions, it is fully justified to speak about 'evolution science', and no longer about evolutionary theory. The innumerable empirical observations and experiments of the present-day natural sciences show that biological evolution is a fact, and not just a theory, let alone a hypothesis. Allegations that evolution is a myth or a religion, e.g. "Evolution is sometimes the key mythological element in a philosophy that functions as a virtual religio" (Harrison, 1974, 1007) result from an incredible lack of knowledge and insight into sciences such as cosmology, geology, genetics and bioanthropology, cultural anthropology and archaeology. Indeed, evolution science includes not only knowledge from the life sciences but also involves knowledge from many other scientific disciplines, natural sciences as well as social sciences and humanities.

[426]See, for instance, the pertinent critique of Paul Cliteur (2010, 248–254) of the liberal interpretation of the Holy Scriptures by Karen Armstrong (1993; 2007; 2009).

[427]Stenger (2012, 45).

Given the fact that so many people, even in developed countries and particularly in the United States,[428] do not accept the findings of evolutionary science, and in particular of anthropology, and continue to believe in fantastical stories about divine creationism and, more recently intelligent design, it is necessary to briefly address these issues. The reader will mainly be referred to the extensive relevant scientific literature about the present-day scientific achievements concerning those issues.

The development of science, with its many and complementary discoveries in various disciplines, has been a challenge for religious believers; this is because crucial elements of religious belief have been gradually and systematically refuted and replaced by scientific explanations. It started with the replacement of the geocentric cosmic model by the heliocentric one of Copernicus-Keller-Galilei in the Renaissance. It was followed in the nineteenth century by the replacement of the anthropocentric and creationist model (in which humankind is considered to be the central focus of the universe) by the evolutionist model of Darwin and Wallace, further supported by the nineteenth and twentieth century shift from an organism-centrist to a gene-centrist view of life,[429] and completed in the twentieth century with the progress of physics and cosmology which are able to explain the evolution of the cosmos without the need for anything beyond physical laws.[430]

Darwinism, or more generally evolution science, is a painful thorn in the side of religious believers. John Micklethwait and Adrian Wooldridge[431] are probably right in pointing out that for religious believers the controversy about Darwinism is not only or not so much natural descent but also the theorem of survival of the fittest. This is considered as amoral or even immoral, as the idea of the interaction between random variation and natural selection is impossible to reconcile with an all-knowing, benevolent God.

Nevertheless, every new scientific discovery, in domains as different as astronomy, physics, chemistry, geology, palaeontology, anthropology and archaeology, molecular genetics, neurology, sociobiology and evolutionary psychology, has confirmed or even strengthened the scientific approaches and refutes the religious belief systems.[432] The common feature of the findings in those scientific disciplines is that the supernatural, immaterial or spiritual approaches and explanations of religions have been refuted and are replaced by natural and material

[428]In an analysis of recent survey results on adults in 31 European countries, Japan, Turkey and the US, in which the question was asked whether it is true that "Human beings, as we know them, developed from earlier species of animals", it was found that 50–80% of the respondents in European countries answered positively (the highest percentages were obtained in Iceland, Denmark, Sweden, France, Japan, UK and Norway); in the ranking of the countries included in the study the US took the penultimate place (with 40%), only followed by Turkey with 25% (Miller et al. 2006). A Gallup Poll of 2005 revealed that 53% of US adults still believe that God created man exactly how the Bible describes it.

[429]Galton (1865; 1869; 1883; 1889), Weismann (1868; 1892; 1902), Dawkins (1976), see Tanghe (2013) for a general overview of the shift in the organismcentrist-genecentrist paradigm.

[430]Hawking and Mlodinow (2010), Stenger (2011; 2012), Krauss (2012).

[431]Micklethwait and Wooldridge (2009, 40); see also Edis (2007, 142).

[432]For instance, Dawkins (1996; 2006), Miller (1999), Stenger (2007).

approaches and explanations, even for behavioural phenomena such as altruism, reciprocity, sociality, love, spirituality, and religiosity.

Contrary to what, for instance, Alister E. McGrath[433] argues on the basis of the analysis of the writings of authors such as Richard Dawkins[434] and Daniel Dennett[435]about the inadequacy of the their atheist apologetic "Darwinian worldview", the refutation of religious belief systems is not only based on nineteenth century Darwinism, and not even on the whole of the twentieth century 'Modern Evolutionary Synthesis', but it is also founded on the whole body of scientific findings, including those of anthropology, archaeology, neurology, geology, physics and cosmology. Most interesting in this scientific development is the enormous coherence and complementarity of the findings of the different relevant scientific fields going from natural sciences, then biosocial sciences, to cultural sciences. Hence, it is incomprehensible why some scholars, for instance John Caiazza,[436] continue to speak about the evolutionary mystique.

The recent literature in defence of theistic worldviews by authors who simultaneously accept the findings of science,[437] notwithstanding its ingenious efforts, is fundamentally disappointing and unconvincing in its defence of the God concept. Furthermore, the arguments put forward concerning the alleged convergence or reconciliation between the findings of science and religious belief systems are weak, even when explicitly intended to be based on hard science approaches to transcendental grounds for existence.[438] In fact, they only illustrate the ever widening of the fundamental cleavage between the scientific and religious approaches. Admittedly, many of the current-day theological writings take into account modern achievements in the physical sciences (physics, geology, cosmology), and take Darwinism into consideration; but they often fail to fully acknowledge the current Modern Evolutionary Synthesis, and in particular the Second Darwinian Revolution (sociobiology and evolutionary psychology), especially with respect to their implications for morality and religiosity.

3.7.1 Creationism, Creation Science and Intelligent Design

As the theme of this book deals with evolutionary ethics, it is necessary to elaborate briefly on the controversy between evolution science and (neo)creationism.

[433]McGrath (2010).
[434]Dawkins (2000; 2003; 2006).
[435]Dennett (1995; 2007).
[436]Caiazza (2005, 105).
[437]For instance, McFague (1993), Johnson (1996), Barbour (1997), Van Till (1998), Peacocke (1998), Miller (1999), Haught (2000; 2004; 2010), Kaufman (2001), Schroeder (2001), Hunter (2003), McGrath (2004; 2007; 2010); Collins (2006), Cornwell (2007), Crean (2007), D'Souza (2007), Flew and Varghese (2007), McGrath and McGrath (2007), Holloway (2008), Armstrong (2009), Dowd (2009), Delhez (2015), Wilcox (2016).
[438]For instance, Davis (2016) in the 'Scientific God Journal' (http://scigod.com/index.php/sgj/index).

Creationist views are not at all new. They were also generally accepted among scientists before the Darwinist revolution. Creation narratives have been elaborated in almost all religions and cultures of the world.[439] The creationist views of the three Abrahamic religions are partly based on the Babylonian creation myth[440] and go back to the Hebrew Old Testament. God is reported to have created in six days "the heavens and the earth", "the living creatures", and "man in our image, in our likeness".[441] According to the Ussher chronology,[442] based on the analysis of the biblical genealogies, the biblical creation occurred 4004 BCE,[443] a viewpoint that is still being shared by one of the variants of the present-day creationist movement, namely the 'Young Earth Creationists' (YECs).[444]

However, it must be stressed that many mainline religions or denominations declared that they see no conflict with evolution,[445] although some of those statements have to be taken with a grain of salt as, for instance, can be seen from the position of some of the recent Roman Catholic popes.[446] The situation is much worse in the Muslim world where most people believe that the Qur'an is the direct word of God, implying belief in a creationist view of life.[447]

3.7.1.1 Creationism and Neo-creationism in the United States

Soon after Darwin's *On the Origin of Species* appeared, evolutionary theory became almost entirely generally accepted in the global scientific community. That success, and its perceived threats to a religious interpretation and guidance of human life (and death), probably caused religious fundamentalists to severely oppose evolutionary theory. A revival and popularising of creationist beliefs started early in the twentieth century, mainly by fundamentalist Protestants in the United States. Inspired by George McCready Price's[448] pseudoscientific book *New Geology* in which the Genesis flood was considered to be the central geological event in the history of the earth, Harold W. Clark[449] introduced the term creationism in 1929.

---

[439] Sproul (1979), Leeming (2009).

[440] Stenger (2009, 165).

[441] Genesis 1:1-31.

[442] Ussher 1650.

[443] According to the present stage of scientific knowledge, the Earth is approximately 4.6 billion years old (Prothero and Dott 2009) and, according to the evidence of earliest fossils, life appeared on Earth at least 3.8 billion years ago and evolved gradually (http://en.wikipedia.org/wiki/Timeline_of_evolution), with the appearance of the first hominins 7 million years ago, and the present *Homo sapiens sapiens* between 150,000 and 300,000 years ago (Stringer and Andrews 2005; McDougall et al. 2005; Hublin et al. 2017).

[444] Whitcomb and Morris (1961), Ham (1987), Ashton (2001), Morris (2007).

[445] Matsumura (1995).

[446] Pope John Paul II (1996), Pope Benedict XVI (2007).

[447] See, for instance, the discussion in Edis (2007, 115ff).

[448] George McCready Price (1870–1963): Seventh-day Adventist and amateur geologist; see McCready Price (1923).

[449] Harold W. Clark (1891–1986): prominent creationist in the middle of the twentieth century.

Several creationist organisations were founded such as The Religion and Science Association,[450] The Deluge Geology Society,[451] the American Scientific Affiliation,[452] the Geoscience Research Institute,[453] and the Creation Research Society (CRS).[454]

The American creationists succeeded temporarily to get the teaching of evolution banned from public schools in several states (mostly in the southern 'Bible Belt'). It was only in 1968 that the US Supreme Court decided that State statutes banning the teaching of evolution are unconstitutional because they violate the constitutional separation between Church and State.[455]

In response to this ruling, the creationist movement decided to implement another strategy and began to argue that creationism is a science, just like evolution science.[456] In 1970, the Creation-Science Research Center (later changed to Institute for Creation Research)[457] was established as the research division of Christian Heritage College in San Diego. Creation science attempts to provide scientific support for the Genesis story of the Bible and to refute the scientific evidence for evolution. Due to the instigation of the scientific creationists, Balanced Treatment

[450]The Religion and Science Association (RSA), founded in 1935, was the first antievolutionary organisation in America (Numbers 2006, 123).

[451]The Deluge Geology Society (1938–1948) was a creationist organisation promoting flood geology.

[452]The American Scientific Affiliation (www.asa3.org), created in 1941, is a Christian religious organisation the purpose of which is to investigate any area relating to Christian faith and science.

[453]The Geoscience Research Institute (1958) (http://www.grisda.org/) is an official institute of the Seventh-day Adventist Church established to address the scientific evidence concerning origins.

[454]The Creation Research Society (CRS) was founded in 1963. The statement of belief of the Creation Research Society (CRS) (http://www.creationresearch.org/) includes: "(1) The Bible is the written Word of God, and because it is inspired throughout, all its assertions are historically and scientifically true in the original autographs. To the student of nature this means that the account of origins in Genesis is a factual presentation of simple historical truths. (2) All basic types of living things, including man, were made by direct creative acts of God during the Creation Week described in Genesis. Whatever biological changes have occurred since Creation Week have accomplished only changes within the original created kinds. (3) The great flood described in Genesis, commonly referred to as the Noachian Flood, was an historic event worldwide in its extent and effect. (4) We are an organisation of Christian men and women of science who accept Jesus Christ as our Lord and Savior. The account of the special creation of Adam and Eve as one man and one woman and their subsequent fall into sin is the basis for our belief in the necessity of a Savior for all mankind. Therefore, salvation can come only through accepting Jesus Christ as our Savior."

[455]Supreme Court of The United States (1968), Epperson v. Arkansas. No. 7. 393 U.S. 97. Argued October 16, 1968. Decided November 12, 1968; see also Flank (2006).

[456]For instance, Whitcomb and Morris (1964), Morris (1985; 2007).

[457]The scientific creationist principles of the Institute for Creation Research (http://www.icr.org/) include among others: "The physical universe of space, time, matter, and energy has not always existed, but was supernaturally created by a transcendent personal Creator who alone has existed from eternity; The phenomenon of biological life did not develop by natural processes from inanimate systems but was specially and supernaturally created by the Creator; The first human beings did not evolve from an animal ancestry, but were specially created in fully human form from the start. Furthermore, the "spiritual" nature of man (self-image, moral consciousness, abstract reasoning, language, will, religious nature, etc.) is itself a supernaturally created entity distinct from mere biological life."

bills mandating equal classroom time for creation science and evolution science were passed in several southern states of the United States (Tennessee 1973; Arkansas 1981; Mississippi 1981; Louisiana 1987).. Other moves by the scientific creationists consisted of arguing that evolution is a 'religion of secular humanism', or requiring that all science textbooks contain a printed disclaimer stating that 'evolution is only a theory, not a fact' (Alabama 1995; Washington 1998). However, all of these moves were rejected by federal courts or the Supreme Court of the US, arguing that creation science is a religious issue whilst evolution science is a matter of science.[458]

Unfortunately, the US courts used the wrong argument to reject the teaching of creationist science or intelligent design theory: those theories should not be rejected because of their religious nature (which they are), but because they are bad science[459] or pseudoscience.[460]

Finally, following their legislative defeats, the creationist movement tried to adapt and further evolved in the early 1990s with a new and subtler variant of creationism, namely Intelligent Design[461]—the view that is historically the original explanation for the remarkable adaptations of living beings to their environment.[462] Intelligent Design organisations emerged such as the Discovery Institute[463] and the Intelligent Design and Evolution Awareness Center.[464]

The Intelligent Design movement is a neo-creationist endeavour that fights the modern evolutionary synthesis with pseudo-scientific arguments. It seeks to explain evolution as the result of the intervention of a so-called superior intelligence or intelligent designer, without explicit references to God or the Bible. In this way it strives to avoid the constitutional opposition which occurred in the past.

Intelligent Design advocates demand that their ideas be taught in school science curricula alongside, and as a scientific alternative, to evolution science. They are strongly supported by public opinion in America. A recent survey of the Pew Research Center[465] shows that 64% of Americans favour the teaching of intelligent design theory and 38% would support the total removal of the teaching of evolution in schools. Hence, public opinion in America is largely at odds with the scientific knowledge of its scientific community.

---

[458]For an overview of the judicial events procedures, see Flank (2006).
[459]See Stengers (2009); footnote 84.
[460]See Perakh and Young (2006, 195).
[461]Johnson (1991; 1997), Davis and Kenyon (1993), Behe (1996), Dembski (1998, 2003, 2004), Dembski and Witt (2010).
[462]Miller (1999, 99), Young and Edis (2006, 1).
[463]The Discovery Institute (http://www.discovery.org) wants "to promote, as a scientific theory, the idea that life was designed by an intelligence". Its "work includes a belief in God-given reason."
[464]The Intelligent Design and Evolution Awareness Center (http://www.ideacenter.org) "believes that life is not the result of purely natural processes, but that it was in some way designed by an "intelligence" and "that the identity of the designer is the God of the Bible".
[465]Masci (2009).

Creationist science theory and Intelligent Design theory are totally absent in the publication record of recognised and peer-reviewed scientific journals.[466] These theories are considered by the mainstream scientific community as bad science, anti-science, dead science,[467] or pseudoscience.[468] Evolution science is based on sound scientific evidence whilst the theory of intelligent design is not.

Morton M. Hunt[469] expressed very well the current mainstream scientific view about creationism and Intelligent Design:

> Creationists and intelligent designers are one of the 'New Know-Nothings' in modern culture.

Excellent and critical historical overviews of modern creationism can be found in Edward J. Larson, Robert T. Pennock, Eugenie C. Scott, and Ronald L. Numbers.[470] For the history of modern creationism from a creationist point of view, see Henry M. Morris.[471] For the ubiquitous creationist obscurantism in the Muslim world, see the excellent overview in Taner Edis'[472] *An Illusion of Harmony*.

In numerous scientific publications, scientists have properly refuted the creationist and/or intelligent design theses or have tried to popularise evolutionary science.[473]

An interesting publication in this perspective is Kenneth R. Miller's[474] *Finding Darwin's God* in which the author, biologist and Christian, criticises and scientifically refutes in a well documented and pertinent way the arguments of the so-called scientific creationists and intelligent design advocates, such as Henry M. Morris, Phillip E. Johnson and Michael J. Behe,[475] arguing that 'God' is neither a charlatan, a magician or a mechanic. However, Miller himself leaves the evolutionary path when he deals with issues of social and moral values, which he considers to be fundamentally different from evolution, not willing or able to see that spirituality, religiosity, and morality are—just as morphological and physiological features—subject to and the result of evolutionary processes and mechanisms. Miller raises the right questions, but does not apply his biological knowledge to produce the scientific answers; therefore he remains caught in the traditional dualistic thinking about material and spiritual matters.

Reputed scientific organisations, such as the National Academy of Sciences (1999) and the Association for the Advancement of Science (2006) in the US, and

[466] Young and Edis (2006, 1).
[467] Kitcher (2007, 8).
[468] Lecointre et al. (2004), Perakh and Young (2006, 185).
[469] Hunt (1998).
[470] Larson (1985), Pennock (1999), Scott (2005), Numbers (2006); See also Gonzalez (2009).
[471] Morris (1984).
[472] Edis (2007, 115–151).
[473] For instance, Alexander (1978), Godfrey (1983), Kitcher (1982, 2007), Dawkins (1986), McKown (1993), Wilson and Dolphin (1996), Pennock (1999), Moore (2002), Manson (2003), Perakh (2003), Forrest and Gross (2004), Shanks (2004), Skybreak (2006), Young and Edis (2006), Shermer (2007), Coyne (2009).
[474] Miller (1999).
[475] Morris (1974), Johnson (1991; 1997; 2002), Behe (1996).

the Royal Society (2006) in the UK, rejected the non-science of the creationists and Intelligent Design advocates. On 21st June 2006, the InterAcademy Panel on International Issues (IAP)[476] issued a statement, signed by the academies of sciences of 70 states, on the necessity to teach evidence-based facts about the origins and evolution of the Earth and of life, and to reject the concealment or denial of evolution science, or the creation of confusion with theories not testable by science.

However, given the perseverance (and financial resourcefulness) of the (neo) creationists,[477] and the persisting lack of access to scientific knowledge in large parts of the population, especially in the United States, the scientific and educational community as a whole has, so far, largely failed to respond adequately to the spreading of the (neo)creationist myths and the implications for educational politics.

It is striking that a country that has produced breakthrough scientific achievements perpetuates a dual educational system and large shares of its population lack access to good education.

How is it possible to explain the following paradox? In this country, which is at the vanguard of scientific innovation and progress, not the least in philosophically and ethically relevant scientific domains (such as bioanthropology, sociobiology, evolutionary psychology, molecular genetics and neurology), 92% of its population continues to believe in the existence of God or a universal spirit, 74% believes in the existence of a heaven, 63% believes that their faith's sacred texts are the word of God, 60% of adults believe in a personal God, 59% believes there is a hell, and only 48% agrees that evolution is the best explanation for the origins of human life on earth.[478]

Several explanations have been given for this remarkable phenomenon. Some scholars link the American 'exceptionalism', in one or another way, to the process of immigration. For instance, Richard Dawkins[479] hypotheses that immigrants might have embraced religion as a kind of kin-substitute for the loss of the stability and comfort of their extended family in the country of origin. Another possibility is that immigration into the US was selective with regard to strong religious beliefs, so that these beliefs were transmitted, culturally or even genetically, to subsequent generations.[480] Other scholars point to the fact that religiosity is strongly coupled to societal insecurity and societal dysfunction—a phenomenon that is particularly striking in the United States with its excessive economic inequalities, dual education systems, lack of a trustworthy social protection[481] at times of unemployment,

---

[476]InterAcademy Panel on International Issues (IAP) (2006).

[477]For instance, in 2013, public indignation arose about the decision of the Texas state Board of Education to appoint a review team consisting of a majority of scientifically unqualified creationists to review the science textbooks to be used in public schools for the next decade. (http://www.slate.com/articles/health_and_science/science/2013/09/texas_science_textbooks_creationists_try_to_remove_evolution_from_classrooms.2.html).

[478]Pew Forum on Religion and Public Life (2007).

[479]Dawkins (2006, 40).

[480]Alper (2006, 195), Lynn et al. (2009); see also Bruce (2002, 219).

[481]For instance, Mickelthwait and Wolldridge (2009, 150) refer in this respect to the internationally well-known inverse relationship between the generosity of the welfare state and the success of religion: the more generous the secular welfare state, the less important become religious-based charities and the demand for religion in general.

sudden or prolonged ill health, lack of general long-term care for elderly, high crime and other social pathologies.[482] According to Steve Bruce[483] the key factor regarding the exceptional religious situation of the USA is the federal and diffuse structure of its polity: this allows its people to create their own subcultures in which their faith strongly retains features of pre-modern times. Finally, one wonders to what degree the ideological-religious pillarisation in the United States with its strong 'in-group' ethnic-racial splits has also contributed to maintain and nourish the religious-ideological divide in the country.

3.7.1.2 Creationism Goes Global

In the recent expanded version of his classic historical account of American creationism, Ronald L. Numbers[484] included a chapter on the global spread of antievolutionism, showing that the American creationists are making serious efforts to go global and to spread their unscientific views abroad—to Asia, Africa, Europe and South America. They are doing this with some considerable success in some regions of the world that are lagging behind in quality education for all.

In Europe, people generally seem to be better informed on evolutionist matters or have been less subject to fundamentalist indoctrination by the creationist lobby.[485] Nevertheless, some American Creationist organisations, such as the Institute for Creation Research and Answers in Genesis,[486] are increasingly and not completely unsuccessfully making efforts to bring their non-science ideas to Europe.[487] It is even more disquieting that in several European countries ministers of education, belonging to certain Christian denominations, have supported creationist movements in their efforts to get creationism and/or intelligent design, along with evolutionary theory, to be taught in biology classes.[488]

Efforts to disseminate creationist beliefs and disavow evolution science in Europe are also coming from the Islamic immigrant community supported by Islamic countries. For example, in 2007 the prolific Turkish creationist Harun

[482]For instance, Paul and Zuckerman (2007), Stenger (2009, 231), Delamontagne (2010), Harris (2010), Coyne (2012).
[483]Bruce (2002, 219).
[484]Numbers (2006, 399).
[485]For instance, Anderson and Peacocke (1987), Miller et al. (2006), Blancke (2011), Blancke et al. (2013).
[486]Answers in Genesis (http://www.answersingenesis.org/) is the notorious creationist ministry, founded in 1993 by the Australian Ken Ham, that, as part of its aggressive creationist dissemination strategy, has set up a Creation Museum near Petersburg, Kentucky, which gives an overview of the origins of the universe, life, and mankind based on a literal interpretation of the chapter Genesis in the Bible. The exhibits show that the Earth and all its life forms were created 6000 years ago, over a period of six days and that humans and dinosaurs once coexisted!
[487]Examples of creationist organisations in European countries: UK: The Biblical Creation Society (http://www.biblicalcreation.org/); Netherlands: Mediagroep in Genesis (http://www.scheppingofevolutie.nl/); Belgium: Creabel (http://www.creabel.org/); Germany: Studiengemeinschaft Wort und Wissen (http://www.wort-und-wissen.de/); Italy: Centro Studi Creazionismo (http://www.creazionismo.org); Russia: Russian Creation Science Fellowship; Poland: Polish Creation Society (www.creationism.org.pl).
[488]See Blancke et al. (2013, 2014).

Yahya[489] sent copies of his pseudoscientific book *The Atlas of Creation,* which tries to refute the theory of evolution, to a large number of schools in several European countries.

It was probably such developments that incited the Parliamentary Assembly of the Council of Europe to adopt in 2007 Resolution 1580 on *The Dangers of Creationism in Education* in which it

> firmly opposes the teaching of creationism as a scientific discipline on an equal footing with the theory of evolution by natural selection and in general resist presentation of creationist theories in any discipline other than religion.

Valuable but complex concepts, such as cosmic and biological evolution, need to be guarded and duly protected, by formal, informal and non-formal education, against attacks by zealots who want to curb or reverse the development of the human species.

The missionary actions of the (American) creationists or creationist organisations[490] found fertile soil in the Islamic world where, traditionally, the Qur'anic tenets are interpreted literally and supernatural design is a common tenet in the Muslim world of thoughts.[491] Although Muslims are very much aware of the importance of science and technology, because of its considerable contribution to development and welfare, they adopt scientific knowledge very selectively. Parts considered acceptable mainly concentrate on applied science, and much less, if at all on basic natural or social and human sciences. In this way they try to avoid the inherent conflicts with their creationist Qur'anic beliefs.[492]

Wherever Muslim scholars deal with fundamental issues of origin and evolution of life, they take an inherently creationist position.[493] In fact, they are intelligent designers *avant la lettre*, because in their view everything is ultimately explained by divine providence. In this domain, whenever they borrow ideas from the West and in particular from the United States, it is not the advanced American knowledge about evolution science and related matters, but the non-science clutter from the Christian (neo)creationists.[494] Hence, it is not surprising that the Islamic scientific contributions in the fundamental domains of natural and social and human sciences, even in the oil-rich Arab countries, is virtual nil.[495]

[489] Yahya (2006–2007); see also Yahya (1999).

[490] According to Numbers (2006, 425), also the above mentioned Turkish prolific writer Harun Yahya and the Turkish Science Research Foundation (the Bilim Araştirma Vakfi, or BAV) are playing an active role on the international scene, particularly in the Islamic world, in propagating creationist beliefs and fighting evolution science.

[491] Edis (2006, 11; 2007, 115.)

[492] Edis (2007, 2009); Hameed (2010), Riexinger (2010).

[493] For instance, Nasr (1989, 1994), Bakar (1987, 2003), Shanavas (2005), Yahya (2006–2007), Ghafouri-Fard and Akrami (2011).

[494] For instance, Yahya (2006–2007).

[495] Edis (2007, 23).

3.7.1.3 Why Persistent and Even Expanding Creationism?

One wonders how it is possible that so many people, some of whom have had a scientific education, continue to believe in creationist and Intelligent Design narratives, which have so evidently been falsified and refuted by an overwhelming body and variety of scientific findings in virtually every relevant scientific discipline that has something to say about origins of life, Earth and the cosmos. To date, the natural sciences, and in particular evolution science, show that the idea of a grand design as a driving force in (human) evolution must be given up. Nevertheless, many people, in particular religious believers, even among social scientists, have difficulty accepting this fallacy. Several, mutually non-exclusive determinants probably contribute to the explanation of this denial.

First, it is possible that our homocentric vanity plays tricks on us and pushes us to assume that there is purpose behind all natural phenomena and that we are part of an ultimate design. Unfortunately, this teleological thinking is merely wishful thinking.[496]

Scholars such as Christopher P. Toumey and Robert T. Pennock[497] are probably right in pointing out that a key element for understanding creationist and Intelligent Design consists of the fundamentalist ethical concerns of strongly conservative citizens. They are worried about the path that modernisation is taking with its alleged or supposed increase in sexual laxity, marriage breakdown, abortion, homosexuality, euthanasia, pornography, etc.[498] It will not come as surprise that evolution science—'a satanic invention' according to Henry M. Morris, one of the prominent American creationists[499]—is often added to that hideous list of so-called degenerate phenomena. However, the scientific progress which has been made in recent decades in fields such as sociobiology and evolutionary psychology, cognitive science, molecular genetics, and even cosmology, increasingly enlarges the gap between science and the creationist and Intelligent Design prejudices about a divinely imposed, invariably natural morality.[500]

A more fundamental explanation for the strong urge to believe that we are part of a grand design—and that there is pre-determined purpose in our existence—follows from our cognitive capacity to observe and experience our strong innate drive for survival and development, as well as our evolved neurological receptivity for supernatural agencies. Moreover, with the development of science, people were able to observe and understand the evolution of life on our planet, the complexity and directionality of which is so tempting to be mistakenly explained as orthogenesis[501] or designed evolution.

[496]For a discussion of this issue, see Vannelli (2001, 1).
[497]Toumey (1993), Pennock (1999).
[498]See for instance, Johnson (1991; 1997), Wiker (2002); for the Muslim world, see the discussion in Edis (2007, 143).
[499]Morris (1984, 110).
[500]Edis (2007, 154).
[501]Orthogenesis is a nineteenth century theory that life evolves in a unidirectional fashion on the basis of an innate driving force (Haacke 1893; Eimer 1898).

Although it would be perfectly legitimate to scientifically investigate whether hypotheses such as creation or intelligent design can be supported by empirical and experimental findings, the problem with the creation and Intelligent Design 'research' institutes is that they formulate their 'research' aims not as hypotheses, but as (revealed) dogmas that they want to document by 'scientific' facts. However, so far, present-day reliable scientific knowledge does not support the revealed tenets: on the contrary, it refutes them.

3.7.2 Challenges for Replacing Religion by Science as the Source of Morality

Religious ideologies with their moral dogmas fulfilled an important survival function in the evolution and history of humankind in the pre-scientific era. However, in many respects these are no longer well adapted to the novel environment of modernity and its exigencies for further human and cultural development and evolution. Is it today possible, desirable or even necessary to replace religious dogmas about origin of morality by scientific knowledge about human source of morality?

The development of science has shattered all traditional worldviews.[502] It questions the logic and coherence of the assertion that God is almighty and omniscient, creator of everything, and at the same time immensely benevolent and loving.[503] The alleged revealed role of supernatural beings or forces as presented in religious ideologies cannot be confirmed by scientific observations: there is no empirical basis for the existence of the personal God of the Abrahamic religions. Last but not least, all observations about the origin and evolution of the cosmos, life in general and the hominins in particular, are explicable by natural mechanisms, without the necessity for intervention by divine powers. They contradict the basic tenets of the religious revelations.

There is no doubt that all successful religions and philosophical traditions share, together with many secular ideologies, a large number of ethical principles. Although usually strongly in-group oriented, they are important for human ontogenetic development, intergenerational continuity, and societal cohesion. Examples are the Golden Rule; in-group solidarity; reproductive continuity and parental care; control of sexual drives; and health protection. This is, in the authors' view, the reason why differences in their social engagement between religious and non-religious people are often less important than the divide between conservative and progressive (religious and non-religious) people.

As a result of the development of science and the emergence of modern secular ideologies, most religious orientations and philosophical traditions have adapted or

[502]Huxley (1927), Dawkins (1986), Russell (1997), Dennett (1995, 2007), Stenger (2008), Stewart (2008).

[503]In this respect Robert Wright (2000, 319) deserves to be quoted: "The kind of God that is hardest to find evidence of is the kind most people seem to believe in: a God that is infinitely powerful and infinitely good."

rather were forced to adapt (more or less) to the exigencies of the scientifically driven and ideologically pluralistic modernisation process. Still a major exception to this trend is Islamic countries that still need to separate religious belief from state organisation and governance.

The human brain was selected, either as an adaptation or as an exaptation, to be sensitive and receptive to spiritual and religious phenomena. In most societies powerful religious institutions continue to influence religious and moral thinking, and many people, even in the predominantly secular societies, continue to draw their motivation for individual and societal action from their religious beliefs. At the same time, it must be acknowledged that science has not been particularly successful in communicating and popularising its findings about the origin of morality to motivate people to adapt their worldviews: at first sight these findings appear difficult to understand and may be emotionally unappealing to so many people.[504]

Hence, when reflecting on a new, global ethic that would overcome the in-group morality in a further modernising and globalising world, it will be a challenge to examine how and to what degree religious beliefs and secular knowledge can contribute to the future development and evolution of the human species.

[504]See also Wilson (2002, 230).

Challenges of Major Secular Ideologies

4

Abstract

This chapter briefly discusses the major secular ideologies that developed in the wake of the scientific revolution and the Enlightenment: liberalism, socialism, feminism, nationalism, ecologism, and humanism. The main purpose is to look at and evaluate those secular ideologies from an evolutionary perspective and their significance for the development of evolutionary ethics. It is concluded that all of the major ideologies appear to include moral principles and practices that can be considered to be useful for evolution-based ethics. Albeit, they are only partial building stones for the design of a viable universal, evolutionarily grounded ethics in a further progressing modernisation. None of the secular ideologies have succeeded so far in elaborating a comprehensive worldview comparable to the major organised religious traditions. They excel as a result of their fragmented nature and, in most cases, short-term perspective.

4.1 Introduction

In the course of modernisation numerous secular ideologies emerged and contributed strongly to shaping the nineteenth and twentieth centuries.[1] Do they, or at least some of them, replace the out-dated ideologies of the pre-scientific era? Are they better adapted to the novel environment of modernity? Do they have better chances to guide the future evolution of humankind?

This chapter briefly deals with, and characterises from an evolutionary point of view, what is perceived to be the major secular ideologies in modern culture. Some of them, like humanism, are predominantly of a philosophical nature; most of them,

[1]Ebenstein et al. (1999).

R. Cliquet and D. Avramov, *Evolution Science and Ethics in the Third Millennium*,
https://doi.org/10.1007/978-3-319-73090-5_4

like liberalism, socialism, feminism, nationalism, and ecologism, are also or mainly political in nature. However, all of them have a fundamental ethical basis.

All of the secular ideologies that emerged and developed in modern times were or are endeavours which aimed to culturally adapt human societies to the challenges and opportunities of the scientifically driven and evolving modernisation process. The question is to what degree they can be considered as successful cultural adaptations. Unlike faith based aspirations and expectations that require belief but not necessarily evidence of outcomes as much is to occur in afterlife, secular knowledge-based ideologies must provide evidence of effectiveness which makes them inherently transient.

There are two ideological phenomena of a more general nature—secularisation and atheism—which are typical of modernising societies, but cannot be considered as ideological systems per se. However, both interact with many of the specific modern ideologies. They require some preliminary comments.

4.1.1 Secularisation

Religious values and norms predominated in the pre-scientific era, often justified and imposed by an alleged supernatural power; but with the advent of science and its manifold applications, human societies gradually started secularising. The sense of empowerment that humans are acquiring through the mediating role of science and technology is conducive to shared values and is having a transformative role in society. This sense of empowerment is also associated with a growing awareness that humans need to assume responsibilities regarding the species' future, instead of delegating choices to the supernatural.

Secularisation is the societal transformation from a situation in which religious beliefs, values, norms and institutions dominate within a context in which non-religious convictions, values and norms, mainly based on autonomous human reason, prevail and secular institutions rule.[2] The social process of secularisation must be distinguished from the ethical or political principle of secularism. Ethical or political secularism means that a society should not be based on the values and norms of a particular worldview, but on a morality that can be shared by all citizens of whatever theist, deist, agnostic, or atheist conviction.[3] Hence, the secular state is not necessarily an atheist one, but encapsulates and often even guarantees philosophical and political pluralism.[4]

Secularism can best be understood as a multi-dimensional phenomenon occurring at three levels: the societal level, the institutional level and the individual level.[5] At the societal level, secularism implies separation of church and state; at the institutional level, it covers the adaptation of ruling bodies to secularised society; at

[2]For instance, Berger and Luckmann (1966, 74), Bruce (2002, 3), Hunter (2015, 1).

[3]Cliteur (2010, 3–4).

[4]Halman and Draulans (2006), Beekman (2012).

[5]Dobbelaere (2002).

the individual level, it refers to the decline of private piety.[6] Hence, secularism does not imply that a society is homogeneously non-religious.

In culturally and technologically more advanced countries, such as the Scandinavian countries, large parts of the population became just formally religious or plainly non-religious—'apatheistic',[7] freethinking, agnostic or atheist. These various groups of non-religious people, but also many perhaps most religious believers, adopted societal or political ideologies that emanated from the development of science and its applications in technology and governance. They are embedded in movements or events such as the seventeenth-eighteenth century Enlightenment and liberalism, the French revolution (1789–1799), the United States Bill of Rights (1791), Marxism (1848), the feminist waves of the nineteenth and twentieth centuries, the United Nations Universal Declaration of Human Rights (1948), the Council of Europe Convention for the Protection of Human Rights and Fundamental Freedoms (1950), and the ecology movement that emerged at the end of the 1960s.

Representative surveys in modern(ising) countries all show a gradual decrease in religiosity over the course of the twentieth century.[8] Even in the United States, a modest decrease in religious belonging can be observed since WWII.[9] The secularisation process in developing countries is less well documented and has yet to take off in many cases.[10]

Some scholars believe that they perceive a renaissance or resurgence of religiosity or a desecularisation of the world, and even an upswing of religious fundamentalism,[11] because of people's disenchantment with science and modernity.[12] However, the resurgence of religiosity is only part of the picture and is often badly or incompletely explained.[13] Apparently, several factors may be at work in different parts of the world. Probably the most important factor, more particularly in modernising countries or regions still characterised by a cultural lag, is the effect of a conservative reflex against the modernisation process itself; this is mainly aimed at maintaining traditional sexist or classist power positions in the family or society. In many countries, especially those characterised by diversity in religions or other traits of group identity, religion may serve as a mark of cultural or national identity. Another phenomenon is the return to democratic rule in former authoritarian regimes who tried to constrain the power position of churches, resulting in an (temporary?) upsurge of religious and nationalistic feelings and aspirations. Furthermore, the asynchronic development of the demographic transition and cultural

[6]Kaufmann (2011, 5).
[7]'Apatheism', a contraction of 'apathy' and 'theism': indifference or lack of interest towards belief or disbelief in a supernatural being.
[8]For instance, Inglehart (1990), Clark and Schellenberg (2008).
[9]Gallup Polls (1948–2008), Bruce (2002, 204ff).
[10]Zuckerman (2005, 12–15).
[11]Fundamentalism: ideological convictions in which literal beliefs in ancient myths and legends, whether religious or not, predominate (e.g. Longman 2004, xi).
[12]For instance, Berger (1999), Almond et al. (2003), Micklethwait and Wooldridge (2009).
[13]Bruce (2002), Barna (2007), Rainer (2007), Paul and Zuckerman (2007), Paul (2010), Stonawski et al. (2015), Pew Research Center (2015).

emancipation in many modernising countries, as well as the migration flows to developed regions, result in the numerical growth of populations that are still under the spell of traditional religions. Withal, it is also a fact that in advanced modern societies, religious people have, on average, a higher number of children than non-believers. In some countries this fertility differential was neutralised and compensated for a long time by an increasing secularisation, but in recent years this balance may have been tilted in favour of the religious part of the population.[14] Finally, it is also not impossible that deep believers continue to respond to their neurologically-based sensitivities and receptibilities for spiritual/religious phenomena, especially when disenchanted with the materialistic and libertarian orientation of some components of the modernisation process and its excessive social inequalities.

Notwithstanding those diverse topical trends in the resurgence of religiosity, the twentieth century and beginning of the twenty-first century are mainly and overwhelmingly characterised by a further progressing gradual secularisation of modern(ising) societies[15]—statistically or qualitatively[16] much more important than the phenomena supporting the resurgence thesis.[17] Steve Bruce[18] sees no reason to expect secularisation trend to be reversed, given the presence of opportunities for ideological diversity and pluralism, effective social equity, economic prosperity, and societal progress.

4.1.2 Atheism

The atheist-theist divergence is one of the two[19] major groups of antagonisms over ideas that divide modern(ising) societies. Although some authors have assigned a number of values to an atheist life stance,[20] most scholars consider atheism not to be an ideology, i.e. a comprehensive normative system of ideas about how the world and life should be understood and society should be organised. Theism and atheism may be part of particular religious and secular ideologies respectively, but taken by themselves they are not ideologies. Atheism is simply absence of theism.[21] It is a way of conceiving the world and desiring to change the world on the basis of

[14]Kaufmann (2011), Meisenberg (2011).
[15]Bruce (2002, 241).
[16]An example of a qualitative shift in religiosity appears from US statistics: whereas 90% of the American population believes in a personal God, only 10% of the members of the US National Academy of Sciences—who are obviously at the vanguard of modern culture creation—does so (Larson and Witham 1998).
[17]The alleged religious resurgence is due to the increased visibility of religious activities through modern means of communication, the immigration flows from developing to developed countries, and the increased vocalisation of religious authorities against the spreading secularisation (FM-2030 1989, 174).
[18]Bruce (2002, 241).
[19]The second divide is between left wing and right wing socio-economic models.
[20]For instance, Stebbing (1941), Robinson (1964).
[21]For instance, Cline (2016).

autonomous reasoning, and making use of acquired scientific knowledge instead of relying on the authority of supernatural agents. As such, it forms a method that partly lies at the basis of many if not all of the secular ideologies. To the extent that this approach or method is based on reasoning and use of scientific knowledge, atheism is also an underlying principle in the development of evolutionary ethics. In this respect, F. Leron Shults[22] refers to the crucial role of atheism as a cultural adaptation in response to the crises of the Anthropocene,[23] which are produced by the impact of human activity on the Earth's environment mainly since the onset of the industrial era.

4.2 The Major Secular Ideologies of Modernity

4.2.1 Liberalism

This secular ideology is undoubtedly the first among the major modern non-religious ideologies that emerged in the age of the Enlightenment at the very beginning of the modernisation process, in the seventeenth and eighteenth centuries. It evolved in the course of modernisation to one of the most important political movements worldwide.

Liberalism—from the Latin *liberalis* 'of freedom'—is the belief in the importance of liberty, individual autonomy and rights, private property, freedom of choice, pluralism and tolerance, and democratic governance. It challenged the authority of absolutist rulers, who were grounded in divine right and supported by established religions. John Locke,[24] generally considered to be the father of liberalism, introduced the idea that people should be governed by the governing authorities, not by supernatural authorities.

In order to understand and evaluate liberalism as an ideology correctly, one has to distinguish its ideological tenets in relation to the socio-economic/capitalist realm and ideological convictions that pertain to other spheres of life, more particularly to various aspects of individual rights. This includes freedom of expression, sexual behaviour (e.g. premarital sex, consensual union, divorce, homosexuality), reproductive behaviour (e.g. contraception, abortion, IVF) and end of life decision making (euthanasia, palliative care). This is, for instance, important in order to understand the differences in meaning of the concept of liberalism in the United States from that in most European countries, more particularly since the late twentieth century. In the United States, liberalism is in the socio-economic realm more identified with welfare state policies comparable to the socio-democratic and socialist orientations in Europe, whereas in most European countries liberalism is still associated with its original historical goals, namely laissez-faire economics and

[22]Shults (2015, 726).
[23]Anthropocene is defined as the Earth's most recent geologic time period as being human-influenced (Crutzen and Stoermer 2000, 17; Steffen et al. 2011).
[24]Locke (1690).

limited state intervention. In non-economic domains, such as ethical questions related to sexual behaviour, reproductive behaviour and end of life decision making, American liberals are taking rather similar, i.e. progressive, positions as European liberals do.

In the course of its history, economic/capitalist liberalism has diversified quite strongly in the domain of views on socio-economic issues concerning the conflict between labour and capital. Broadly speaking, three major subdivisions can be distinguished: classical liberalism, social liberalism and neoliberalism.

Classical liberalism, the roots of which go back to the seventeenth century, stressed the importance of civil liberties, free markets and free trade; it supports a world that is relatively free from government intervention. Classical liberalism is in favour of a '*laissez-faire*' economics[25] and, consequently, a capitalist organisation of society.

However, in late nineteenth century, the failure of *laissez-faire* economics (and its associated raw capitalism) to produce social justice resulted in the emergence of a new conception of liberalism, called social or modern liberalism.[26] This variant of liberalism views the freedom of the individual as being in harmony with the good of the community; it assigns a greater role to the state for promoting equal opportunities and decent social living conditions. Social liberalism contributed strongly to the development of the welfare state in the twentieth century.

Economic developments in the late twentieth century—the capitalist crisis, with its shrinking profit rates, the collapse of the state planned economies, and the emergence of the developing world—led to the development of economic policies that are neoliberal in nature. Neoliberalism pushes for "a market-driven approach to economic and social policy, and seeks to maximise the role of the private sector in determining the political and economic priorities of the state." It advocates deregulation, privatisation, and withdrawal of the state.[27] Neoliberalism fosters a market fundamentalist capitalism, including monetarist "economic policies and a reduction in government provision of services", resulting in an increase in income variance once again, namely the rich grow richer and the poor grow poorer. Indeed, the main achievement of the recent neoliberal wave has not been an increase in worldwide growth, but in the restoration of power and wealth to the economic upper class.[28]

Evaluating liberalism from an evolutionary point of view is not an easy matter because liberalism can, as explained in the previous paragraphs, have such very different meanings: classical and neoliberalism on the one hand, social or embedded liberalism on the other hand, are almost the opposite of each other. Nevertheless, the various forms of liberalism do have some common aspirations: individual emancipation and rights, equality of opportunities, freedom of thought and enterprise.

---

[25]Smith (1776), Mill (1863), see also Gray (2002).
[26]Green (1884).
[27]Harvey (2007).
[28]Harvey (2007), Duménil and Lévy (2011), Steger and Roy (2010), McQuaig and Brooks (2013).

From an evolutionary point of view, these liberal characteristics must be evaluated positively because they accord with major ontogenetic developmental needs of human specificity. In addition, the social liberal variant of liberalism also pays attention to community development and has a more balanced approach to individual-societal relations and interactions.

On the contrary, classical liberalism (with its *laissez-faire* economics, driven by capitalistic profit acquisition) and more particularly neo-liberalism (with its one-sided market-fundamentalist driven approach), are neither very responsive to individual developmental needs nor to well-balanced individual-societal interactions.

In general, the liberal ideology is strongly focused on the individual (freedom, rights), pays less attention to societal phenomena, and keeps silent on phylogenetic matters and human-environmental interactions.

Therefore, the authors can follow neither the older so-called Social Darwinist adherents of liberalism, who presumed that their *laissez-faire* economics was supported by and founded upon Darwinian competition and selection theory,[29] nor more recent authors such as Larry Arnhart[30] who advocate a Darwinian liberalism:

> I have argued that Darwinian science is compatible with a classical liberal understanding of how moral order in a free society arises from natural desires, cultural traditions, and individual judgments... Evolution provides a purely naturalistic grounding for liberal thought, so that there is no necessity to appeal to the supernatural.

Arnhart largely misses the sociobiological findings on the important shift from competitive to cooperative behaviour in the evolution of the hominins, and the need for its implication in order to evolve social morality, so as to enhance hominisation in the context of a further progressing modernisation.[31]

From an evolutionary point of view, in the liberal ideologies and their associated variants of capitalism there are found only a very minimal, if any, response to the future needs of a further evolving humanity in a progressively modernising and globalising world.

4.2.1.1 Capitalism

In some of the paragraphs above the concept of capitalism was mentioned, but it requires some additional discussion, not only because it is the major economic system in the world but also because it is sometimes related to evolutionary biology.

According to some authors, capitalism is a highly complex economic system of social cooperation, competition and conflict on the one hand, and on the other hand it is an 'ideology' with a particular set of values and norms.[32] In the authors' view, capitalism is not an ideology comparable to modern societal ideologies such as

[29]Spencer (1851), Sumner (1883).
[30]Arnhart (2010), see also the response essays of Myers (2010), Tiger (2010), Gintis (2010).
[31]See also Gintis (2010), Dilley (2013).
[32]For instance, Corning (2010).

liberalism and socialism which aim to develop an ethical and political system that guides and stimulates human development and societal progress. Capitalism is an economic system that uses particular means to achieve the production and keep control of distribution of goods. Those means are owned or controlled by a small minority of people who produce and distribute those goods for a profit, in particular their own profit.

Capitalism is valued very differently from various ideological points of view: among mainstream liberals, the view predominates that capitalism is a method that allows entrepreneurs to innovate and produce goods that meet people's needs which resulted in the material prosperity of modern society[33]; on the contrary, among mainstream socialists one will rather find the view that the main capitalist drive for producing goods is to invent needs and sell goods for a profit.[34]

Indeed, both assertions are partly true. The capitalist drive to invent and produce goods which meet the needs of people cannot be adequately and efficiently performed when there is no prospect for making a profit. The capitalist motive for profit cannot function when, in the end, the goods produced do not meet at least some needs, or in some cases artificially-induced needs.

However, some liberal and socialist inspired discourses about capitalism both often forget that in the most advanced countries, where successful capitalist production coexists and interacts with high standards of welfare and well-being at the population level, this success story is the result of a mixed, socially controlled market economy, slowly evolving towards what some have called an inclusive capitalism.[35] Nevertheless, modern culture still has to go a long way in order to fine tune its capitalist economic system, so that it avoids or limits the persistent disadvantages of unbridled economic competition; for example extreme stress at work, disrupted family life (particularly work-parenthood incompatibility), overproduction, environmental disruption and pollution, unsustainable production, and international conflict over access to resources and markets. A particular cause for concern is the recent recurrence of a fundamentalist capitalism with its unrestrained drive towards profit, without much intervention by institutions or public policies aimed at the common good.[36]

Before reflecting on capitalism from an evolutionary point of view, it must be recalled that the interweaving of biology and capitalism has a long history, resulting in theories that aimed to 'naturalise' capitalism, so as to justify the capitalist social and economic order (and their numerous social evils).[37] This view believes that capitalism is rooted in biological-evolutionary processes, in particular inter-individual or intergroup competition resulting in 'natural' selection and survival of the fittest—the latter being the most successful economically. This theory,

[33]For instance, Thompson (1993).
[34]Marx (1867), Piketty (2013).
[35]See, for instance, Esping-Andersen (1990), Goodin et al. (1999), Prahalad and Hammond (2002).
[36]Piketty (2013), Izaka et al. (2015).
[37]Kortright (2008).

sometimes called 'Social Darwinism',[38] was used as a 'natural', perhaps divine, justification of '*laissez-faire*' economics with its ruthless business practices.[39]

The American steel magnate Andrew Carnegie[40] articulated this 'biological-evolutionary' ('Darwinist') theory eloquently:

> the law of competition, be it benign or not, is here; we cannot evade it; no substitutes for it have been found; and while the law may be sometimes hard for the individual, it is best for the race, because it ensures the survival of the fittest in every department.

Furthermore, the American oil magnate John D. Rockefeller[41] declared in a Sunday-school address:

> The growth of a large business is merely a survival of the fittest... The American Beauty rose can be produced in the splendor and fragrance which bring cheer to its beholder only by sacrificing the early buds which grow up around it. This is not an evil tendency in business. It is merely the working-out of a law of nature and a law of God.

The idea that capitalism is grounded in deep-seated genetically based drives and functions leading to economic mechanisms which are analogous to evolutionary processes, such as differentially selecting, retaining and multiplying genes, is still very much with us today, although current day adherents of 'bionomics' are much more nuanced in their discourse than the earlier 'Social Darwinists'. Nevertheless, many still hold the view that capitalism is a biological inevitability. For instance, Michael L. Rothschild[42] argues that capitalism is the natural state for an economic system, because biology and economics are both systems for organising and retaining knowledge. Nevertheless, Rothschild is not blind to some of the blatant weaknesses of the capitalist system, as appears in his valuation of 'corpocracies' that he describes as "parasites feeding on the energy of a system, but not contributing to the vitality of that system". Rothschild concludes: "bionomics is to find the economic incentives to encourage socially desirable behaviours (reducing poverty, eliminating homelessness, environmental responsibility, etc.) and then letting the self-organising free market determine the ways in which the business ecosystem will encourage those behaviours".

How can capitalism be evaluated from an evolutionary point of view?

First of all, it should be acknowledged that the capitalist doctrine, which seeks to base and justify its practice in evolutionary science (mainly Darwinism), has grossly and lopsidedly misinterpreted the evolutionary mechanism and process as being relentless individual and intergroup competition and 'survival-of-the-fittest' theory. In contrast, Charles Darwin[43] and many evolutionists after him, particularly from the recent sociobiological and evolutionary psychological schools, have always developed a comprehensive evolutionary theory in which individual and intergroup competition was well balanced with, but not overpowered by, mutually

[38]See e.g. Hofstadter (1955), Jones (1980), De Tarde (1984), Tort (1992).
[39]Josephson (1934), Bergman (2001).
[40]Carnegie (1920).
[41]Rockefeller, quoted in Hofstadter (1955, 45), Huber (1971, 66).
[42]Rothschild (1990).
[43]Darwin (1871).

beneficial cooperation based on altruism, reciprocity, and mutual aid. (See Chap. 2, Sect. 2.1.2). More particularly, the successful evolution of the hominins, especially of *Homo sapiens sapiens* in its most recent cultural stage—the present modernisation process—is largely due to ever-increasing levels and degrees of socialisation. As Peter A. Corning[44] wrote:

> We did not, most likely, evolve as isolated individuals pitted in relentless competition with one another.

Admittedly, the human biogram is embedded with several individual needs and drives—greed for resource and property acquisition, hierarchical social status seeking, nepotistic and in-group, envious zero-sum thinking—that easily turn into 'capitalist' forms of behaviour, particularly in some economic and ecological living conditions.[45] However, in the human species social cooperation and mutual aid are extremely important factors in meeting the basic human needs that guarantee survival and intergenerational continuity and evolution. Economic activity is undoubtedly an essential factor in satisfying human basic needs: however, this should not be done using ever increasing growth rates and the maximisation of profits, but in a socially well balanced and environmentally sustainable way.[46]

Advocates of the Social-Darwinist or bionomic discourse often misinterpret the essence of the evolutionary mechanism. Evolution is about the long-term differential reproduction of genes in an evolving but sustainable environment, and not necessarily about maximising financial profits by means of (over)production of goods in an environmentally unsustainable way. In other words, the millionaire is, contrary to what William Sumner[47] presumed, not necessarily the fittest survivor. (See Chap. 2, Sect. 2.2).

In conclusion, from an evolutionary point of view, capitalism alone is unfit to serve as a guiding principle for the future development and evolution of humankind. As Peter A. Corning[48] stated in his enlightening paper on the indispensability of an evolutionary ethics:

> ...capitalism has been an engine of economic growth and progress; it is a proven system. However, the megathreats and the severe economic challenges that almost certainly await us in the future require the development of a more enlightened capitalism.

The biological basis of human economic behaviour[49] easily leads to a capitalist organisation of society. However, in the context of modernity with its powerful means of influencing and harming the Earth's biosphere, that economic system may no longer be a well adapted system needed to guarantee long-term progressive development and evolution of humankind.

---

[44]Corning (2010, 5).
[45]Wilkinson (2005).
[46]Corning (2010).
[47]Sumner (1914, 90).
[48]Corning (2003, 16).
[49]Robson (2001).

4.2.2 Socialism

Modern socialism started as a response to *laissez-faire* liberalism and its devastating effects on the living conditions of labourers in the nineteenth century. It opposed the brutal and exploitative working conditions of the Industrial Revolution in the late eighteenth and early nineteenth centuries.[50] For example, Count Henri de Saint Simon (1760–1825) endorsed an economy that would restrain the immoderation of capitalism and would be based upon equal opportunities.

Karl Marx (1818–1883),[51] as the most important theorist of socialism, and Friedrich Engels (1820–1895),[52] his co-author and friend, are considered to have developed the scientific approach to socialism, explaining human history through the clash of classes. They saw the control over means of production by the proletariat as the key aim for achieving social justice.

Just as early Christianity stood for universalism as it broadened its religious concerns to all of humanity, Marxism is a secular ideology that built on the idea of universalism that transcends national borders and continents. With its slogan "*Proletarier aller Länder, vereinigt Euch!*"[53] Marxism was the initiator of a new aspiration for global social justice to emerge from the class struggle.

Marx and Engels[54] also made, on the basis of their general theory of society, a pertinent analysis of the role of religion, as a man-made instrument in the suppression of the masses, so well expressed in Marx's[55] famous metaphor "*Die Religion ist der Seufzer der bedrängten Kreatur, das Gemüt einer herzlosen Welt, wie sie der Geist geistloser Zustände ist. Sie ist das Opium des Volks.*"[56]

Socialism evolved into an ideology that aims to democratically control or even publicly own natural resources and means of production; and it strives for the distribution of wealth in the interest of the whole community.

In the course of its development, socialism was split up into several branches, the two most important being the revolutionary Marxism-Leninism and reformist socialism. Marxism-Leninism strived for the establishment of centrally planned economies, under the control of a single political party that owns the natural resources and means of production. Reformist socialism tried to gradually reform and control the capitalist economic system in order to make it more equitable and humane.

4.2.2.1 Marxism-Leninism

The first variant evolved into dictatorial communist regimes which, after initial remarkable progress in achieving more social equality and equity, evolved to

[50]Corning (2010).
[51]Marx (1867).
[52]Engels (1878).
[53]Marx and Engels (1848): "Proletarians of all countries, unite!"
[54]Marx and Engels (1958).
[55]Marx (1844).
[56]"Religion is the sigh of the oppressed creature, the heart of a heartless world, and the soul of soulless conditions. It is the opium of the people".

become societies in which an inefficient central planning economy failed to produce progress in sustained welfare. Furthermore, the original ideal of creating a classless society did not materialise. In their competition with the more successful capitalist economies, communist regimes disintegrated, metamorphosed, or disappeared from the world scene.

Although in their *Manifesto of the Communist Party* of 1848, Marx and Engels[57] contrast in a simplistic way 'proletarians' and 'bourgeois', it must be acknowledged that the crude capitalism of the industrialising nineteenth century produced horrendous living conditions for the working class—the proletarians who possessed nothing but their '*proles*'. Hence, it is not surprising that a revolutionary branch emerged in the bosom of the socialist movement, which wanted to use radical methods in order to provoke fundamental societal changes in social status and political power relations, leading to the replacement of the 'dictatorship of the bourgeoisie' by the 'dictatorship of the proletariat'.[58] Unfortunately, contrary to the idealistic vision of Rosa Luxemburg and notwithstanding the initial realisation of new life opportunities and more humane work conditions for the oppressed masses, the communist movement evolved in the communist countries, in the direction of the dictatorship of abusive, parasitical, bureaucratic, hypocritical and inefficient communist particracies, unable to guide their societies into sustained economic and social progress.

Just as Christianity undermined its social ideology through its later alliance with the power structure of the establishment, Marxism, when implemented at the level of nation-states, did not result in establishing a classless society and failed to establish global social justice transcending national borders.

4.2.2.2 Reformist Socialism

The second major socialist movement evolved into a variety of democratic socialist regimes advocating social reforms which were intended to alleviate the more harsh facets and intrinsic defects of the capitalist economic system. The aim was to establish partially controlled market economies. Modern democratic socialism aims to create a more humane society by reducing the inequalities and eliminating the injustices partly embedded in human nature, which had been reinforced and inherited from the feudal and early industrial eras. Despite its success in controlling or limiting the most disastrous side effects of an unbridled capitalism, democratic socialism, as well as its ideological cousins—social liberalism and Christian democracy—is still far away from having designed adequate universal policies. They are struggling to preserve the advantages of the capitalist economy for the national (or regional) general welfare enhancement through technological innovation thanks to free entrepreneurship. At the same time they are struggling with dependence upon international financial speculation, free-riding multinational enterprises, exploitation of workers, exorbitant competition, overproduction, and environmental pollution.

[57]Marx and Engels (1848).

[58]Marx (1875), Luxemburg (1918), Kautsky (1918).

Socialism is difficult to discuss from an evolutionary point of view, because of the wide range of different ideological and political agendas it embraces, the many different meanings it has acquired in the course of history, and the diversity in views and approaches by which it is still characterised today.

The early socialist writers—Karl Marx and Friedrich Engels—were delighted with the emerging biological evolutionary theory, mainly because it provided a materialistic, non-teleological explanation for change in the natural world. In a letter dated December 12, 1859, to Marx, Engels wrote:

> Darwin, whom I am just now reading, is splendid.[59]

In a letter to Engels on December 19, 1860, Marx wrote about Darwin's book on natural selection:

> Although it is developed in the crude English style, this is the book which contains the basis in natural history for our view.[60]

In addition, a month later, on January 16, 1861, he wrote to Lassalle:

> Darwin's book is very important and serves me as a basis in natural selection for the class struggle in history.[61]

However, Marx and Engels only partially supported Darwinism. They rejected population pressure as a selecting agent—Malthus' contribution to the theory of natural selection—and believed instead that adaptive modifications due to environmental effects were inherited.[62] They were total environmentalists and believed in Lamarckism—the inheritance of acquired characteristics—not realising that this would have implied that the economically less advanced peoples and classes would have become inferior in their heredity.[63] The environmentalist and Lamarckian beliefs of the Marxists became a dogma amongst their followers in the communist countries, in particular the Soviet Union. In the first half of the twentieth century, this led to the rejection of Mendelian genetics, the development of a disastrous Soviet agricultural policy, and the elimination and even physical liquidation of many Russian geneticists, the most famous of whom was Nikolai Vavilov. Genetic science was exchanged for quack genetics, promoted by the fraudulent Trofim Lysenko who remained in his post as director of the Institute of Genetics at the Academy of Sciences until 1964.[64] In 1964, physicist Andrei Sakharov[65] spoke out against Lysenko in the General Assembly of the Academy of Sciences:

> He is responsible for the shameful backwardness of Soviet biology and of genetics in particular, for the dissemination of pseudo-scientific views, for adventurism, for the degradation of learning, and for the defamation, firing, arrest, even death, of many genuine scientists.

[59]Engels, quoted in Zirkle (1959, 85).
[60]Marx, quoted in Zirkle (1959, 86).
[61]Marx, quoted in Zirkle (1959, 86).
[62]Zirkle (1959).
[63]Muller (1948), quoted in Zirkle (1959).
[64]Medvedev (1969), Joravsky (1970), Soyfer (1994).
[65]Sakharov (1964), quoted by Joravsky (1970).

Many Marxist theoreticians reject(ed) the view that genetic factors are involved in matters such as social stratification and social mobility in capitalist society. In their view, the explanation for social differentiation, as well as any cultural development, can be reduced to differences in economic production. The works of Karl Marx and Friedrich Engels do not include propositions about the role of natural selection in society.[66] Nevertheless, they declare themselves openly in favour of the view that environmental factors cause social differentiation. Even among many present-day enlightened socialists, the 'Blank Slate' view of human nature is a doctrinarian tenet and human nature is a taboo. Sociobiology, evolutionary psychology, behavioural genetics, and cognitive neuroscience are often assimilated into conservative, rightist political views.[67] As Edward Wilson[68] pertinently wrote:

> Marxism is sociobiology without biology.

Paradoxically, the refusal to recognise genetic variation as an important factor in social relations and economic production—simplistic egalitarianism/blank slate theory—is strikingly in contradiction with the principles advanced by early socialist writers, as for instance Saint Simon's position on equal opportunities and Louis Blanc's/Karl Marx's[69] well-known famous slogan

> From each according to his ability, to each according to his needs.

Peter A. Corning[70] has listed several other biosocial shortcomings of the present-day socialist doctrine: (1) socialist indifference to merit as well as the rewards for merit is linked to its neglect or denial of the importance of genetic variation in societal matters; in this regard socialists often confuse equity with equality; (2) socialists often disregard elements of 'reciprocity'—duties and responsibilities that need to balance benefits and rights—and hence allow or even favour abuses by cheaters and free riders[71] at both extremities of the social stratification, including the disproportionate, and in fact unmerited excessive salaries and bonuses of corporate CEOs and managers, or the profits of unproductive rentiers, as well as the unjustified milking of the social protection systems in welfare states; (3) socialists often underestimate the importance of capital and entrepreneurship; (4) finally, socialism usually takes a too simplistic negativistic attitude towards competition, the socially acceptable forms of which can play a vitally important

[66]Woltmann (1899).

[67]Pinker (2002, 279–284). The blank slate view of human nature is not only a view which predominates in socialist quarters: it was the dominant theoretical approach in much of the twentieth century sociology and cultural anthropology and even today it can be perceived among many social scientists who have not succeeded in becoming acquainted with the present-day state of the art of the biological sciences, particularly evolutionary science. (For a discussion of these issues see, for instance, Van den Berghe 1990; Ellis 1996; Lopreato and Crippen 1999; Pinker 2002; Niedenzu et al. 2008; Corning 2011).

[68]Wilson (1978, 191).

[69]De Saint Simon (1819), Blanc (1840), Marx (1875).

[70]Corning (2010).

[71]A free rider is someone who enjoys the benefits of cooperation without reciprocating for its costs.

positive role in our complex societies and economies, firstly by the ability to energise innovation; and secondly by safeguarding against the abuses from holding a monopoly.

Overall, socialist ideology is strongly focused on societal structural matters, has lopsided and obsolete views about causes of individual development and the role of genetic factors in social relations, and often lacks a comprehensive view of societal processes or is commendably but one-sidedly focused on the least advantaged in society. It is characterised by a striking absence of long-term genetic and ecological concerns. In contrast, it rightly stresses the overwhelming importance of the social dimension and solidarity factor for harmonious individual development as well as for adequate relations between individuals and groups of individuals.

From an evolutionary point of view, socialism lacks—just as its big opponent liberalism—a broad and holistic vision about the future needs of a further evolving humanity in a progressively modernising and globalising world. As Peter Singer[72] put it:

> The left needs a new paradigm.

4.2.3 Feminism

Feminism stands for a broad range of movements that, in general, aim to achieve equal political, economic and social rights and equal opportunities for women compared to men. Feminism is mainly focused on women's issues, but some feminists are of the view that feminism should also consider men's emancipation because many aspects of sexism involve interactions between the two sexes.[73]

Just as the liberal and socialist ideologies discussed above, the feminist movement also developed into different branches: major subdivisions are radical feminism,[74] equality feminism,[75] and difference feminism.[76] However, many more 'feminisms' are distinguished in the feminist literature, according to their ideological adherence (e.g. socialist feminism,[77] liberal feminism[78]) or their specific

[72]Singer (1999, 5).

[73]For instance, Daly (1978), Freedman (2003, 2007).

[74]Radical feminism considers the old-time and still persisting patriarchal domination and its related male supremacy as the major and universal cause of women's oppression (e.g. Koedt et al. 1973; Willis 1984; Radical Women 2001).

[75]Equality feminism emphasises, notwithstanding the biological differences, the strong similarity between the sexes. Human nature would be 'androgynous, neutral and equal' (e.g. Young 1999).

[76]Difference feminism stresses that men and women are 'ontologically' different versions of the human being (e.g. Zinn and Dill 1996).

[77]Socialist feminists see the cause of women's oppression to be in the capitalist system, and in order to win equality women workers must stand in solidarity with each other. Socialist feminism interprets women's oppression as the result of the class structure in society (e.g. Bebel 1879; Boxer and Quataert 1978; Radical Women 2001).

[78]Liberal feminism pursues gender equality through political and legal reform without altering the structure of society (e.g. Wollstonecraft 1792; Friedan 1963; Walker 1996).

domain of interest (e.g. ecofeminism,[79] lesbian feminism,[80] individualist feminism,[81] gender feminism,[82] new feminism[83]). There is even a conservative feminism[84]—which is obviously a *contradictio in terminis*.

The history of women's emancipation goes much further back in time than the two or three waves of the modern feminist movement suggest.[85] It was influenced by societal processes and events such as the emergence of capitalism and Renaissance in the fourteenth century, the Reformation in the sixteenth century,[86] the Enlightenment in the eighteenth century,[87] the French Revolution (1789), the Industrial Revolution, and, last but not least, the Women's Rights Movement, first in the US and later in Europe in the nineteenth century, which is now identified as the first major feminist wave.[88] This first feminist wave was principally concerned with equality. Participants in the movement devoted themselves to achieving equal legal rights, general suffrage, equal access to higher education, and equal entry into the professions in order to achieve self-determination and independence.

The second feminist wave in the second half of the twentieth century, the origin of which is said to be associated with Simone de Beauvoir's[89] *Le deuxième sexe* and Betty Friedan's[90] *The Feminine Mystique*, and has shown an accelerated development since the 1960s, shares the first wave's politics of legal, educational, and economic equal rights for women, but surpasses it in the breadth of its concerns and the depth of its critiques. This second wave, known as the women's liberation movement, concentrates on more subtle issues such as women's rights to bodily integrity and autonomy and reproductive rights, struggle against domestic violence,

[79]Ecofeminists consider men's control and destruction of the natural environment as the main cause of the oppression of women (e.g. Mies and Shiva 1993; Diamond and Orenstein 1990; Ruether 1993).

[80]Lesbian feminism focuses on the discrimination of lesbians and women in society. It refutes 'heteronormativity'—the assumption that everyone is 'straight' (e.g. Faderman 1998).

[81]Individualist feminism aims at protecting individual women by legal measures that eliminate male privileges. It wants women to take full responsibility for their own lives and bodies (e.g. McElroy 2002).

[82]Gender feminism: a 'gynocentric' and 'misandric' variant of feminism (e.g. Sommers 1994).

[83]New feminism: a form of difference feminism, mainly from Catholic inspiration, that emphasises the complementarity rather than the hierarchy of men and women. It acknowledges the biological specificity of both sexes, while recognising their equal worth and dignity. It stresses women's "obligation to give birth to and raise children" (e.g. Pope John Paul II 1995). Given its ideological origin, it is understandable that *bona fide* feminists are quite suspicious about 'feminist' variants such as 'new feminism' or conservative feminism, albeit those variants sometimes remind us rightly of the importance of some important biological facts to be taken into account in the design of a transgenerational morale.

[84]Conservative feminism rejects a feminism that "adopts a male model of careerism and public achievement as female goals, thereby denying women's need for intimacy, family, and children." It believes that promoting gender equality leads to the ruin of the family (e.g. Stacey 1983).

[85]Weisbord (2011).

[86]For instance, Agrippa (1529).

[87]For instance, Wollstonecraft (1792), von Hippel (1792).

[88]Jaggar (1983).

[89]De Beauvoir (1949).

[90]Friedan (1963).

sexual harassment, sexual assault, sexual stereotypes, sexism and patriarchy in all spheres of life, and on more general societal problems such as armed conflict, environmental pollution, and Third World development.

Some feminists also distinguish a third wave that would have developed since the early 1990s as a response to the perceived failures of the second wave. Third-wave feminism objects to the second wave's essentialist definitions of femininity, which would be too strongly focused on privileged women.[91] In addition, the emergence of a so-called 'post-feminism', which consists of a wide range of viewpoints that challenge previous feminist discourses, largely coincides with the third wave. Some post-feminists even claim that feminism has lost its relevance in present-day modern society.[92]

All the major modern political emancipatory ideologies—Marxism (with its socialist and communist variants), liberalism and Christian-democracy—eventually included principles and policies with a view of restoring—or perhaps better put, of establishing at last—social equity and equality between the two sexes. Wherever these societal ideologies gained political power, they usually contributed more or less to female emancipation, but they never succeeded in realising full gender equity and equality. Thus, it is not surprising that a specific women-oriented ideology—feminism—emerged to accelerate the gender emancipatory process.[93]

In the course of modernisation, particularly in the second half of the twentieth century, feminist ideology has had substantial successes as can be seen from the trends in the educational, social, economic, cultural and political participation of women. However, feminism still has a long way to go.

4.2.3.1 Biosocial Obstacles to Sex Emancipation

Due to the evolutionary heritage of the human sexual dimorphism, as well as the cultural survival of maladapted ideologies dating from the pre-modern eras, women and men suffer from a number of biosocial obstacles to a profound sex emancipation[94] with which the feminist movement did not cope very well.

In modern societies, men still have or take advantage of some of their secondary sexual characteristics which they have retained from their adaptation to the hunting-gathering phase of human evolution: namely their biologically selected physical strength and drives for risky, agonistic and competitive behaviour, allowing them to acquire dominance socially, also towards women.[95] By exacerbating socio-economic competition in many domains of modern social life, modernisation enhanced the opportunities for fully indulging those drives.

[91]For instance, Heywood and Drake (1997), Baumgardner and Richards (2000), Henry (2004), Krolokke and Sorensen (2005), Gillis et al. (2007).
[92]For instance, Cott (1987), Modleski (1991), Jones (1994), McRobbie (2004).
[93]Humm (1992).
[94]Cliquet (2010, 223–238).
[95]Miller and Hoffman (1995), Whitmeyer (1998), Forthun et al. (1999), Sherkat and Ellison (1999), Stark (2002), Miller and Stark (2002), Freese (2004), Gove (1985), Udry (1988, 2000), Julian and McKenry (1989), Dabbs and Morris (1990), Gottfredson and Hirschi (1990), Booth and Dabbs (1993), Collaer and Hines (1995).

Whereas men are undoubtedly the greatest obstacle to the emancipation of women, the evolutionary heritage of the latter has encumbered them with a number of biological features, mainly in the domain of reproduction and caring, that puts them in a socially weaker position in the competitive and mobile environment of modernity. Moreover, powerful conservative forces try to preserve the old male-dominated system with its prerogatives and advantages.[96] They are supported by traditional religions that ideologically codified the ancient biosocial dominance-subordination relations between the sexes and ascribed them to a divine origin and justification. However, the well-known finding that women are more religious than men, especially regarding the degree of personal piety, may not just be the result of religious indoctrination, but may also partially be the effect of evolved sex differences in brain functioning in general. Indeed, there are a number of female behavioural characteristics that can be seen as the direct effects of specific female brain features: women have higher abilities to empathise; are more intuitive; have a higher emotional intelligence; are more socially oriented, more friendly and more care taking; are better in emotional bonding; more easily recognise facial expressions and emotional overtones in others; have a more inward directed demeanour; display a more selective mating behaviour; have stronger parental impulses; display heightened social and personal concerns; experience more often anxiety; and show more interest in art, aesthetics, linguistics, and literature.[97] Evolutionarily, the sex differences in brain structure and functioning are the result of selective processes related to the sexual differences in mating and parenting strategies that developed in the ancient hominin environment of evolutionary adaptedness. The progressing hominisation—with its increasingly slowly maturing offspring and its increasing socialised hunting-gathering culture and intergroup competition—required sexually differentiated biosocial specialisations, mainly characterised by the evolution of male features adapted to hunting and war making, and female characteristics more strongly focused on reproduction, nurturing and socialising. Some scholars have expressed the view that women's greater existential vulnerability, related to her larger reproductive and parental investments, may have made her more sensitive than men to the social and psychic comfort of spirituality and religiosity.[98]

4.2.3.2 Biosocial Opportunities for Sex Emancipation in Modernity

Notwithstanding some sex emancipation-inhibiting features, such as its persisting and even increasing pressure for competition, modernisation entails a much larger number of positive emancipatory forces which create previously unseen new opportunities for achieving sex equality in rights and equity in opportunities.

[96]Sharma (1987).

[97]For instance, Geary (1998), Baron-Cohen (2003, 21ff), Jausovec and Jausovec (2005), Ellis (2011), Moir and Jessel (1992), LeVay (1994), Blum (1998), Schulte-Rüther et al. (2008).

[98]For instance, Blum (1998), Geary (1998), Mirola (1999), Baril (2006), Campbell (2008), see also DesAutels (2010).

In the first place modern sex emancipation is due to the new knowledge furnished by science in general, and the application of this new knowledge to many domains of social life. Ancient beliefs about innate female inferiority can no longer be successfully sustained. Biology simply swept away the traditional views on the nature of the sexes and destroyed the ideological foundations of sex inequality and inequity. A completely different picture has emerged from present-day (social) biology about women's biological nature. The development and application of science have not only altered the objective social position of women in society, but have also had a considerable impact on their subjective perceptions and experiences. Psychologically, women are moving into a completely different position of power and negotiating position in what some authors call 'the battle of the sexes'.[99]

Biomedical progress induced a revolutionary level of mortality control, the ultimate condition for women's new opportunities. Not only has it largely freed women from the risks of infant and maternal mortality and morbidity, but also the control of mortality allowed—and in the end even imposed—fertility control. This has liberated women from virtually permanent reproductive functions and allowed for the establishment of new balance between reproductive, productive, and recreational functions in modern society. The development and availability of safe and effective methods of birth control is of considerable importance in this respect, not only from a social but also from a psychological perspective.

The transition from agrarian-pastoral to industrial and post-industrial economy caused a shift from a family-based economy towards family-transcending types of economic production, increasing educational opportunities, female paid labour and financial independence. Finally, modern technology is increasingly eroding the traditional male physical advantage with respect to muscular strength and speed. Thanks to modern technological means of replacement or aid, women can now perform tasks for which men were, on average, better adapted in pre-industrial living conditions.

Lastly, the development of democratic societal organization with its pacification ethos and suppressive effect on internal and external violence[100]—ranging from domestic violence to group brutality or even war—is an important social-environmental condition that strongly contributes to enabling female emancipation movements leading to more substantial equality and equity between the sexes.

4.2.3.3 Biosocial Challenges to Feminism

One of the major challenges feminism faces is precisely how to integrate the biosocial dynamics of sex in various domains of social life such as the combination of productive, reproductive and recreational activities, the active participation of both sexes to political life and decision-making, and the disciplining and peaceful management of male drives for competition and dominance.

Many feminists have the tendency to minimise, if not ignore, the biological sex component, because of the classical misapprehension that biology is synonymous

[99]Van der Dennen (1995).
[100]Rummel (1994; 1995; 2002).

with unchangeability, determinism and reductionism. It is a shortcoming that is present both among equality feminists and among difference feminists.[101] The egalitarians completely miss the biological dimension in the sex controversy; they think that biological differences simply do not matter in social matters. However, many difference feminists overreach themselves. According to Maggie Humm,[102] for example, second wave feminism has focused on the specifications of women's differences from men and from each other: thus, it has turned to psychoanalytic and social theories about sex difference in order to explain such issues as the increase in sexual violence and to construct a fresh feminist ethics. It is not surprising then that this feminist theory has failed to completely succeed in developing appropriate strategies to change sex relations in modern culture, as it so blatantly neglects the fundamental background of the sex divide, namely the biosocial basis of sexual dimorphism.

Neglect of the biosocial factor in sex relations makes this feminist approach non-productive. From an analytical standpoint, making the transition from facts to norms, but also from norms to facts is a delicate matter.[103] Although some situations are considered undesirable, this does not mean that underlying facts should be ignored or underestimated. Undesirable social situations can only be changed when their real causes are adequately understood and addressed.

Moreover, some feminist ideological standpoints suffer from a certain contradiction. On the one hand it is asserted that men and women are equally suited to all jobs, but on the other hand it is argued that if women did particular jobs they would be done differently. For instance, compassionate values would predominate, more cooperation and less competition would be promoted, and there would be less war.[104]

From an evolutionary point of view, the feminist movement must be seen very positively, despite the fact that many—if not most—branches of feminism misinterpret or underestimate the relevance of biological evolutionary factors and their importance for the achievement of the feminist goals.[105] The feminist movement is contributing considerably to the development of the genetic potentialities present in 'the other half' of the human species.

Many feminist goals can be integrated into an evolutionary ethics for the future development and evolution of the human species in a further modernising world. However, feminism remains focused on one single aspect of human development and fails to provide a broader ethical framework for the long-term evolution of humankind.

[101]Gatens (1996).
[102]Humm (1992).
[103]Zeiss (1982).
[104]Ridley (1993).
[105]For instance, Cliquet (1984; 2010), Roede (1988), Segerstraele (1988), Gowaty (1997), Vandermassen (2005), Campbell (2006).

4.2.4 Nationalism

Nationalism is a political ideology in which a group of individuals, belonging to a particular genetic, ethnic, cultural or religious community, strongly identify with that community, and take actions to achieve or preserve cultural or political autonomy, self-determination, or sovereignty of that community.[106]

Although nationalism goes far back in history, the concept was only coined late in the eighteenth century by the German philosopher Johann G. Herder.[107] It was an important driving force in the emergence of many modern states in nineteenth century Europe, fuelled by the French and American revolutions, the liberal ideals of popular sovereignty and dated, deep-seated urges for in-group bonding.

In the twentieth century, many thought that nationalism would decrease in importance or disappear as a result of modernisation, industrialisation, individualism, and globalisation. However, the contrary came about to some extent: nationalism has increased in political importance in the world, either as emancipatory movements against former dominant or colonial states, or as a reaction against economic or political upscaling or capitalist globalisation. In the last case, some authors speak about the emergence of 'neo-nationalism'.[108]

History has shown that nationalism can take two major and antagonistic forms: it can be an emancipatory or even revolutionary movement, or it can be an imperialistic-protectionist movement.

4.2.4.1 Emancipatory Nationalism

As an emancipatory movement, nationalism means that an underdeveloped or suppressed community—a genetic, linguistic, religious, or otherwise identifiable population—strives to achieve self-determination, either in the form of cultural autonomy within a multicultural state, or in the form of political independence.

Nationalism is often associated with ethnocentrism—the care of the community's own ethnic, cultural or religious identity. In principle, there is nothing wrong with ethnocentrism, as long as it only fosters own healthy community development and is not accompanied by the contempt or suppression of 'others'.[109] It is, indeed, extremely important that the organic communities in which people live together, at the local, regional, national, continental and also finally global level, function properly, for the *bonum commune communitatis*—the common good of the community.

Hence, from an evolutionary perspective, nationalism is to be valued positively in its variant of emancipatory cultural community development. As such, nationalism as a liberation movement contributes to the development of the potentialities and talents present within a cultural entity. In the evolution and historical development from tribal-sized populations consisting of narrowly related people, to

[106]For instance, Smith (1991; 1995; 1999), Hutchinson and Smith (1995).
[107]Herder (1774–1787).
[108]Gingrich and Banks (2006).
[109]Salter (2003).

massive populations consisting of millions of unrelated people, the concern for and care of community identity is probably needed in order to foster the sense of group belonging, solidarity and security.

Clearly, despite the positive community developmental effects emancipatory nationalism may have, it is also only a very fragmented ideology. This falls short of the comprehensive evolutionary ethics necessary for the sustained future evolution of the human species as a whole in the context of a further progressing modernisation.

4.2.4.2 Imperialistic Nationalism

As an imperialist-protectionist movement, imperialist nationalism strives for domination over other populations or expansion into their territories, based on its own perception of cultural, religious or even biological superiority. This form of nationalism often leads to inter-ethnic or international conflict, war, and even genocide.

A particular form of imperialist nationalism, fascism,[110] is a form of totalitarian ideology that glorifies its own community above all others. It governs by the means of an authoritarian, dictatorial leadership, usually based on a single party system.[111] In the domain of economics, fascist regimes favour an autarkic, self-sufficient state planned economy, albeit via indirectly controlled private owners.[112]

Fascist regimes are ultra-conservative in many respects. They support the traditional, male-dominated family system and favour pronatalist policies. They fight signs of what they perceive as moral decay such as modern arts, individualism, and materialism.[113] They are usually notoriously xenophobic, and sometimes overtly racist, or imperialist expansionist.[114]

Nazism—the abbreviation for 'National Socialism'—was a unique variant of fascism in Germany in the 1930s and 1940s. It included a salient pseudo-scientific form of racism and anti-Semitism. The Nazis believed in the supremacy of an alleged Aryan[115] race and strived to extend its power, territory and resources mainly in the east of Europe at the cost of the local populations.[116]

---

[110]The term fascism derives from *fasces,* the Roman symbol of collectivism and power: a tied bundle of rods with a protruding axe.

[111]Lyons and Berlet (1996), Morgan (2002), Paxton (2005).

[112]Richman (2008).

[113]Boyanowski (2002), Britt (2003), Richman (2008).

[114]Britt (2003).

[115]The so-called Aryan race is not a biological subdivision within the human species. The Aryan concept is an English language loanword derived from the Sanskrit Arya ('Noble'), which refers to the Indo-European languages in general and their speakers. Hence, it is a cultural concept that has, in principle, nothing to do with genetic differentiation in the population. The Nazis usurped the term to identify their racial ideal, namely the 'Nordic variant of the European racial stock'.

[116]The holy scripture of the Nazis was of course Hitler's (1925–1926) '*Mein Kampf*' which is, apart from its fateful ideology, a monument of incoherent, distorted, pseudoscientific semi-truths, but it is very clear in its political intentions and predictions. It is still an enigma how the German population, with its sophisticated intellectual background and rich cultural heritage in general, could have been seduced by the superficial and irrational clutter of ideas that is bundled into that writing.

Nazism was not only responsible for one of the most odious genocides in history —the Holocaust on six million Jews[117]—but also carried out massive massacres in many Eastern European countries, not the least in the Soviet Union which suffered more than 20 million World War II casualties.[118]

Fascist regimes tend to have a relatively short life because they are usually soon overthrown by democratically inspired movements or neighbours. They are, at the population level, fundamentally adverse to basic human needs and aspirations. From an evolutionary point of view, fascism clearly has nothing valid and promising to contribute to the ethical system, since it fundamentally ignores or challenges the mental and social potentialities of the human species as a whole.

Imperialist variants of nationalism are—particularly in modern culture—anti-evolutionary obstacles to future international cooperative development. They build on only one component of our evolutionary past, namely intergroup competition, and leave no space for the second indispensible component of evolution which is cooperation. They are parasitic on others and suppress potentialities of other population groups.

4.2.5 Humanism

Historically, the word 'humanism' has been used with different meanings; for instance, the classical curriculum in education, the Renaissance movement to revive classical learning, the Enlightenment idea of a human-focused ethical alternative to the traditional religions, and finally the modern humanist movement.[119]

Here, this section will focus on humanism as the current-day secular ideological movement that fosters a humanist life stance. It is rooted in rational and free thinking, understanding our universe in scientifically based naturalistic rather than in revealed supernatural, superstitious and pseudoscientific terms. It provides a secular ethics grounded in human values such as individual rights and social responsibilities, social justice, human solidarity, tolerant pluralism and social, political and economic democracy, and cosmopolitanism.[120] A significant feature of the humanist movement is that it is focussed on the human species as a whole, meaning that its moral concerns transcend narrowly defined in-groups.[121]

Intellectually, humanism has roots in ancient China (Buddhism), India (Hinduism), Greece and Rome, early Christianity, fourteenth century Renaissance, eighteenth century European Enlightenment, the French Revolution, and nineteenth

[117]Gilbert (1986), Gutman (1990), Hilberg (2003).
[118]Russian Academy of Science (1995).
[119]Giustiniani (1985), Lamont (1996), Walter (1997), Norman (2004), Kurtz (2007), Pinxten (2007), Slembrouck (2010), Grayling (2014).
[120]Gasenbeek and Gogineni (2002).
[121]Teehan (2010, 218).

century evolutionary theory,[122] atheism,[123] liberalism[124] and Marxism.[125] However, humanist organisations are a recent phenomenon, the result of the search for an alternative to the out-dated religious traditional institutions.[126]

Nowadays, there are more than 100 humanist, rationalist, secularist, ethical, atheist and free-thought organisations in over 40 countries. They are internationally organised by the International Humanist and Ethical Union (IHEU) which was founded in 1952 in Amsterdam and is the umbrella organisation around the world.[127]

Bert Gasenbeek and Babu Gogineni[128] distinguish four successive waves of modern humanism, originating around 1850, 1890, 1918 and 1945 respectively. The oldest wave originated in the mid-nineteenth century in Western Europe and the United States: it consisted of atheists including freethinkers, rationalists and secularists, who explicitly reject all religion.[129] The second wave emerged in the final decades of the nineteenth century and consisted of free-religious or ethical culture groups.[130] The third wave emerged in the interbellum and consisted of the US humanists that considered themselves to be religious humanists[131] who founded the American Humanist Association. In 1933, a group of these humanists produced a first *Humanist Manifesto,*[132] later amended in second[133] and third versions.[134] Finally, in the aftermath of the Second World War the present wave of humanism arose and lead to the foundation of the current IHEU, a result of a synergic action from English-speaking associations (American Ethical Union, American Humanist Association, British Ethical Union), European continental associations (Vienna Ethical Society, Dutch Humanist League and Flemish Humanist League), and the Indian Radical Humanist Movement.

In its Amsterdam Declaration of 1952, updated in 2002, the IHEU affirms "the worth, dignity and autonomy of the individual and the right of every human being

---

[122]Darwin (1859).
[123]Feuerbach (1841).
[124]Smith (1776), Mill (1859), Green (1884).
[125]Marx (1867), Engels (1878).
[126]Gogineni (2006), Grayling (2014).
[127]www.IHEU.org.
[128]Gasenbeek and Gogineni (2002).
[129]World Union of Freethinkers, WUFT (1880).
[130]International Ethical Union, IEU (1896).
[131]It is difficult to call Humanism a religion, at least when it is conceived in its present common meaning as a life stance that is rooted in rational and free thinking, understanding our universe in scientifically based naturalistic terms rather than in revealed supernatural, superstitious and pseudoscientific terms. It is exactly the opposite to religion, the specificity of which consists of the belief in super- or extra-natural phenomena. Nevertheless, some people do consider humanism a religion, probably for a variety of reasons. Indeed, humanism as understood by modern humanist organisations fulfils social roles that are similar to those performed by traditional religions. However, whenever or wherever it is necessary to refer to ideological diversity in society, it would be more respectful towards non-religious citizens to refer to, for instance, 'religions and other philosophical convictions', as is done in some United Nations documents.
[132]Sellars and Bragg (1933).
[133]Kurtz and Wilson (1973).
[134]American Humanist Association (2003).

to the greatest possible freedom compatible with the rights of others". Humanism is stated to be democratic and aims "at the fullest possible development of every human being", but at the same time insists that "personal liberty must be combined with social responsibility". Humanists are said to "have a duty of care to all of humanity including future generations" and are expected to "use science creatively, not destructively, and to value artistic creativity and imagination". Humanism is considered to be "an alternative to dogmatic religions which claim to be based on revelation on the one hand, and totalitarian systems on the other".

Paul Kurtz[135] summarised the modern humanist paradigm admirably into six main characteristics: (1) a scientific method of inquiry; (2) a naturalistic cosmology; (3) a nontheistic orientation; (4) a commitment to naturalistic ethics; (5) a commitment to democratic forms of governance; and (6) a commitment to international cooperation.

Although the national humanist associations are usually small organisations with a limited membership, in many countries they have a much greater influence than their numerical strength would suggest, because their ideas are supported by a number of authoritative people, as well as by political movements such as liberalism and socialism. At the international level, the IHEU has consultative status with intergovernmental organisations such as the United Nations, the UN Human Rights Council, UNESCO, and the Council of Europe. As an NGO it has an important influence and sometimes succeeds in preventing the intergovernmental organisations from taking out-dated ethical positions that are inspired by religious or authoritarian ideologies.[136]

Looking at the humanist movement from an evolutionary point of view, this ideology most closely approaches an evolutionarily-based ethics for the future development and evolution of humankind. Humanism is based on a rational, scientific analysis of the human condition, and rejects the divinely revealed religious mythologies as a foundation for well-adapted ethics to modern living conditions. It advocates the fullest possible development of all human individuals, at the same time taking into account the necessity to promote social life and solidarity, within as well as between communities.

Overall, modern humanism stresses fewer long-term evolutionary goals, although some of its most prominent initiators explicitly developed and justified the humanistic ideology from a long-term evolutionary perspective. The most well known amongst them is undoubtedly the renowned evolutionary biologist Julian Huxley.[137] In his *Essays of a Humanist* he wrote:

> This new idea-system, whose birth we of the mid-twentieth century are witnessing, I shall simply call Humanism, because it can only be based on our understanding of man and his relations with the rest of his environment. … It must be organized round the facts and ideas of evolution, taking into account of the discovery that man is part of a comprehensive evolutionary process, and cannot avoid playing a decisive role in it.

[135]Kurtz (2007).
[136]For instance, Cherry and Brown (2009), Jeffrey (2011).
[137]Huxley (1942; 1957; 1964).

This view is more explicitly and extensively developed by a variant of modern humanism which is called transhumanism—a term also coined by Julian Huxley in 1957. Transhumanism aims to reach a posthuman stage in human evolution which would have characteristics such as higher-than-current intellectual heights; resistance to disease; increased longevity; unlimited youth and vigour; increased capacity for pleasure, love, artistic appreciation, and serenity; and experience novel states of consciousness. The transhuman is seen as an intermediate form between the present human and the posthuman. (See Chap. 5, Sect. 5.3.2.4).

A critique that is sometimes made against modern humanism is its too narrow focus on the human species, whilst neglecting the relations with other species and the planetary environment in general.[138] However, this critique is not fully justified. Indeed, many prominent modern humanists were/are ecologically very motivated and play(ed) an important role in raising and disseminating awareness about the environmental problems resulting from the world population explosion and unsustainable modern consumption patterns. For instance, one of the major founders of the Flemish Humanistic League in the 1950s, the biologist Lucien De Coninck,[139] was one of the earliest protagonists raising awareness about ecological issues; this influenced a whole generation of students and followers in their thinking and actions regarding environmental matters. Nowadays, several humanist associations pay a lot of attention to ecological matters, which can also been seen from the appearance of the term ecohumanism.[140]

Some authors[141] have expressed scepticism about the humanistic goals and strategies, arguing that too much attention is paid to deviant forms of behaviour, e.g. support for permissive education, permissive attitudes to crime and against punishment, non-acceptance of genetic and other individual and group differences, narcissistic demand for rights without duties, advocacy of free love and sexual satisfaction in or out of marriage, opposition to engendering guilt, and moral relativism in general. It is noted that these authors are not making preposterous religious imputations that secular humanists are "those who believe in no morals"[142] and that they are responsible for alleged social evils such as pornography, homosexuality, drug addiction and abortion.[143]

Although there is some truth in the observation that the humanist movement sometimes focuses one-sidedly on minority forms of behaviour—e.g. homosexuality instead of heterosexuality, birth control instead of fertility, divorce instead of partnership, criminals instead of victims—and that it is too deeply involved in fighting the traditional religions instead of developing its own humanistic principles and actions, there is an obvious and excusable explanation for those orientations. When looking at the sociological and legal positions of non-religious people in many countries, many inequalities and inequities between religious and

[138]Badmington (2000), Wolfe (2009).
[139]Decraemer et al. (2010).
[140]For instance, Tapp (2002).
[141]For instance, Cattell (1972).
[142]Whitehead and Conlan (1978).
[143]For instance, LaHaye (1980).

non-religious people and institutions continue to prevail. This occurs not only in the developing world—particularly in the Islamic world where religion and State are not separated—but also in advanced countries, where free thought and freedom of life stance are constitutionally guaranteed. Hence, it is understandable that the humanist associations have the priority to deal with abolishing existing inequalities and inequities, if not discriminations, in the domain of worldviews and life stances.[144]

Another weakness of the humanist movement is its relative silence about economic matters and societal policies in general. This is probably due to the fact that its membership largely consists of people who profess either a socialist or a liberal ideology and consequently may foster quite fundamentally different views on these matters.

Another critique that could be levelled against humanists and free-thinking citizens in general is that, compared to religiously motivated people, they may be less committed to and involved in personal actions and interventions in daily human and social relations and interactions. Of course, humanists and humanist organisations are not subject to pressures to deal with such matters by a clergy. In addition, they do not have the advantage of the long historical organisational tradition, and also often generous state support that organised religions can rely upon. However, in general, secular ideologies may have succeeded less in evoking the required drives for active personal involvement in dealing with the individual problems of 'others'.

A judgment that is sometimes verbalised is that, according to the humanists' vision of life, every person would himself determine his own values and norms, and there would be no body from outside or above who can or would be allowed to do this.[145] Although this seems to be a somewhat lopsided and too simplistic assessment, it is nevertheless true that the freethinking, atheist, humanist part of humanity has not yet achieved the rich and well developed formal institutional structures that can or want to display moral authority. Whereas, for instance, the Pope regularly expresses—*urbi et orbi*—on behalf of the Roman Catholic Church moral positions on major social, economic, and political problems the human species or some of its deprived groups are struggling with, the international humanist community remains largely silent. There are several explanations for this striking difference in expressing and disseminating moral stances. The Roman Catholic Church has the advantage of being a historically well-established institution, not only with ample financial and intellectual resources but above all a hierarchical structure and ideology which facilitates its fast and adequate positioning in crisis situations. In contrast, the international humanist community undoubtedly has the intellectual resources, it lacks the financial means and institutional structures to systematically make its voice heard. However, it is an even more important question as to whether

[144]Consult, for instance, the table of contents of the major periodic publication of the International Humanist and Ethical Union: International Humanist News.
[145]For instance, Demaerel (2015).

the humanist community, because of the present state of its ideology, would want or allow its organisation(s), such as the IHEU, to take up that role.

In conclusion, notwithstanding some weaknesses, modern humanistic principles constitute a most important ideological input into an evolutionarily inspired ethics for the future development and evolution of humankind.

4.2.6 Ecologism

Ecologism is a political ideology that resulted from the confluence of the scientific discipline of ecology and awareness about increasing environmental threats in the second half of the twentieth century. Ecology—a term originally coined by the German evolutionist Ernst Haeckel (1834–1919)—refers to the study of the interaction between living organisms and their environment. As a political ideology, ecologism suggests that the biosphere and physical environment on our planet, which are of fundamental importance for our existence, should be the subject of moral concern and, hence, should be taken into account in policies.[146]

The ecologist movement was virtually non-existent in the first half of the twentieth century. Environmental issues were of primary concern for very few people, even among the scientific community. However, in the second half of the twentieth century, due to the accelerating population explosion at the global level on the one hand, and the increasing environmental pollution due to the invention and massive use of ever more artificial and environmentally detrimental molecules, on the other hand, environmentalist concerns and actions gradually emerged. In the United States, for instance, Rachel Carson's[147] publication *Silent Spring,* in which she denounced environmental pollution induced by pesticides, was a milestone in the growing public environmental awareness.

In the last quarter of the twentieth century, the ecological movement deepened and radicalised its approach to the need to preserve the environment. Arne Naess,[148] the Norwegian founder of deep ecology philosophy, made the distinction between shallow ecology which he defined as just environmentalism, the aim of which is to protect the environment against inadequate human interventions, and genuine or deep ecology which concerns the biosphere as a whole. Deep or fundamentalist ecologism fosters a shift from anthropocentric environmental values towards biocentric or ecocentric values, and from short-term to long-term goals.[149]

Politically, the ecological/environmental movement led to the establishment of national Green parties and continental Green federations or networks.[150] This development started in 1972 with the creation of Green parties in Tasmania and New Zealand, followed by France and UK in 1973 and many other European

[146]For instance, Baxter (1999).
[147]Carson (1962).
[148]Naess (1973).
[149]Dobson (1990), Smith (1998), Baxter (1999).
[150]Blakers (2001).

countries in the 1980s. In the 1990s many countries in other continents followed. Currently there are Green parties in more than 100 countries worldwide.[151]

The political platforms of the Green parties are not only focussed on environmental issues with the view to achieving a sustainable ecological future, but also foster socially oriented policies, health protection, social cohesion, globalisation based on solidarity that is ecologically sustainable, changing lifestyles and patterns of consumption, energy efficiency, energy saving and the development of renewable energies, and democracy and diversity.[152]

Although the Green parties are still a small minority in most countries, they are having a positive influence on the ecological agenda of other political ideologies. Several modern ideologies now have an ecological offshoot—cf. ecosocialism,[153] ecofeminism,[154] ecohumanism[155]—or have even largely integrated the ecological agenda into their platforms for action. A weakness of the ecological movement is that its policy proposals seldom have a sufficient systems-theoretical view that would replace the traditional approaches to environmental matters.[156] So far, they have not elaborated more concrete policy proposals to change or replace the capitalist, liberal or socialist approaches to economic production without fundamentally disrupting the welfare and well-being of citizens in modern(ising) societies.

However, civil society with numerous national and international environmental and nature organisations,[157] has probably had even more influence on recent policy making than the Green parties themselves.

In this context, it is appropriate to refer in particular to the work being done by the United Nations Intergovernmental Panel on Climate Change.[158] It is not only remarkable in its important climate assessment reports (1990, 1995, 2001, 2007, 2013) but also the massive (voluntary) contribution and cooperation of the world's scientists in the endeavour to commonly investigate and advise policy makers on the anthropogenic climate effects and their consequences for the environment and the biosphere, including human well-being.

In contrast to its remarkably rapid take off that the ecological movement achieved in raising public awareness about the environmental deterioration in modern times, the movement has produced a very limited practical result until now. It is sufficient to look at the major indicators of environmental degradation: the unprecedented further increases in atmospheric CO_2 levels with its threat of climatically warming the planet; the still increasing species extermination (the sixth

[151] http://www.globalgreens.org/parties.
[152] For instance, European Greens (2006).
[153] For instance, Wall (2010).
[154] For instance, Hobgood-Oster (2005), Gaard (2011), Sturgeon (2016).
[155] For instance, Tapp (2002).
[156] Wijkman and Rockström (2011, 18).
[157] For instance, Spiro (2007), Tava (2013).
[158] IPCC (2013).

extinction)[159]; the on-going environmental pollution; and the continuing deforestation of the tropical rainforest.[160] This is not completely surprising, given the short period of time in which the ecological awareness emerged and evolved, the exponential population growth that many developing countries are still experiencing, the desire of those countries to further develop, the insufficient motivation of people to change their consumption patterns, the financial interests of well-established economic sectors such as the oil industry, and related to all this, the insufficient action taken by policy makers to change those trends.

Evaluating ecologism from an evolutionary point of view is much easier than assessing liberalism or socialism with their diverse subdivisions. Indeed, both major subdivisions of ecologism—shallow environmentalism and deep ecologism—have much more in common with each other in reality and moreover, they both connect much more closely, although partially, with an evolutionary ethical perspective.

Advocates of ecologism—shallow or deep—are undoubtedly right in stating that the human species is part of the biosphere on this planet. It is not possible to isolate ourselves from the rest of nature. We profoundly depend for our well-being, and even for our mere existence, on the planet's biosphere and physical environment.[161]

However, by imposing moral duties towards the earthly biosphere as a whole, an essential feature of biology should be taken into consideration: the interdependence between species is largely based upon competition between species. Whether we like it or not, the human species is part of this interdependent and competing system. Humans are dependent upon other species—microorganisms, plants and animals—but at the same time we are in competition with many of them, either because they are deathly menaces to us, or because they are life-sustainable resources for us as providers of oxygen, basic nutrients or symbiotic support. Consequently, contrasting a biocentric view of life with an anthropocentric one, as advocates of deep ecologism do, is quite an artificial or unrealistic approach. As a dependent and competing species, humans cannot afford to take an indiscriminate or absolutely altruistic position towards other life forms. We can value them and protect them, provided we can protect ourselves from disease and death, as long as biodiversity gratifies our existential, aesthetic, or other mental needs and to the extent we depend on them—for breathing, nutrition or symbiosis. Indeed, the intrinsic value (deep) ecologists attach to nature and biodiversity, independent of the practical usefulness for the human, probably responds to important mental pleasure functions of an existential, aesthetic, scientific, or other perceptual nature.

[159]Since the origin of life on Earth, life has experienced five mass extinction events, all caused by natural phenomena. The sixth extinction is the result of human interventions. It began some 100,000 years ago when the anatomical modern human emigrated from Africa and dispersed all over the planet; it accelerated about 10,000 years ago when humans developed agriculture, and has further intensified since the onset of modernisation 400–500 years ago. The biodiversity in the planet is currently being diminished at a rate that parallels the five natural extinctions of the past (see Leakey and Lewin 1995; Novacek 2001; Kolbert 2014).

[160]For instance, Chew (2001, 2008).

[161]Dobson (1990).

Hence, the bio- or ecocentrism of the deep ecologists is, in fact, an expression of an enlightened anthropocentrism.

Indeed, where does our admiration for nature come from? Why does nature appeal so strongly to our aesthetic feelings or other mental needs? How can the extension of our altruistic feelings and concerns to (some) other species be explained? How did ecologism emerge in present-day modern culture?

The authors' hypothesis would be that the psychological satisfaction and pleasure felt in the presence of a rich, diversified nature results from the evolved disposition and experience that such a nature guarantees the existence of rich and abundant (nutritional) resources. The aesthetic and altruist feelings towards (some) other species may also have developed in civilisations that acquired more knowledge about nature or reflected more thoroughly about the meaning and future of life on this planet. The ecological concerns in present-day modern societies are undoubtedly evoked by the observations and experiences about the environmental harm modern culture produces, and are facilitated by the nutritional security that modern societies could achieve: however, ecological concerns are, alas, much more difficult to cherish under conditions of undernourishment or starvation.

In conclusion, ecologism brings an extremely important dimension to an evolutionary approach to ethics. It emphasises not only the strong dependence of the human species on other living forms but also guards against the destruction of natural ecosystems, depletion of natural (renewable and non-renewable) resources, and the reduction or elimination of biodiversity. It fosters a long-range time perspective and helps to recognise the importance of intergenerational processes. Finally, it reminds us of the aesthetic or other mental pleasures a rich nature provides us with. Overall, ecologism is one of the essential building blocks of an evolutionary ethics that envisages a long-term progressive future for the human species and life on this planet in general, despite its utopism of wanting to replace an anthropocentric ethics by an eco- or biocentric one.

4.3 Constraints of Secular Ideologies as Sources of Universal Morality

Modernisation is characterised by the development of a number of secular ideologies that incorporate many moral norms that were also part of religious teachings, but on the whole they do not draw on the sacred texts. Due to their relatively recent origin, in general they are philosophically less well substantiated, often lack a holistic approach, and many are not yet embedded in well-structured secular institutions.

Almost all secular ideologies include or highlight partial ethical elements that are valuable components for an evolutionarily inspired, universal ethics that might guide the future of the human species.

However, compared to the religious traditions that consist of comprehensive ideological systems responding to all major questions and challenges of life and death, they show three major shortcomings: (1) their fragmented nature; (2) their short-term approach; (3) their macro-level approach.

4.3.1 The Fragmented Nature of Secular Ideologies

The most striking characteristic of all major secular ideologies is that they approach and deal with humanity's presence and future in a very fragmented way. Moreover, several of those ideologies often seem to be mutually incompatible. In contrast to the traditional religions that provide an all-encompassing, cohesive and overarching worldview designed by a superpower,[162] so far the modern man-made secular ideologies have been unable to design a holistic worldview. They do not add up to a set of guiding values and norms for the solution of humanity's present day challenges and do not offer a coherent inspiration for its long-term evolutionary future. Most are primarily focussed on a limited target group: liberalism emphasises the development of the individual and neglects collective impacts; socialism concentrates on societal issues often at the expense of individual self-realisation; feminism is mainly focused on gender issues; ecologism is environmentally and ecosystem oriented; nationalism is mainly, perhaps only, concerned about national in-group interests. The only modern ideology that has a somewhat broader and more comprehensive approach is humanism. From an evolutionary point of view, transhumanism is particularly inspiring. However, overall, modernity lacks a secular worldview that gives meaning and purpose to human existence and continuity. The fragmented modern secular ideologies lack a grand and holistic vision leading to an inclusive morality.

Modern secular ideologies, philosophically and scientifically of Western origin, mainly concentrate on material welfare and the individual's well-being—they often lack ethical depth and spiritual inspiration. Some argue that they are characterised by ethical decline and regression in values and norms compared to religious morality. However, notwithstanding their fragmented nature, as a whole those secular ideologies constitute the multi-faceted and mature components of an ethical system with the potential for the highest development of individual emancipation and societal organisation ever achieved. Nonetheless, the modern secular ideologies still need to acquire a holistic perspective, an overall integration, and a longer-term vision about intergenerational goals. If it becomes well integrated and with a longer-term view, the Western ethical system could form a superior system to any of the religious-civilisational ethical systems with their religiously dogmatic, classist, sexist, and in-group oriented values and norms. At first sight this conclusion may sound Western-centric biased and naively optimistic. It is in fact a call for assuming responsibility that is proportionate to the opportunities opened up by modernity. A comparison of the various ideological-ethical systems that the human

[162]See also, for instance, Saroglou and Muñoz-García (2008, 97), Stewart (2008).

species has developed so far, their merits and limitations and their implications at the population or species level, leaves us with no other choice than to shape the future of our species based on choices to the best of our knowledge.

A major problem associated with the fragmented nature of modern secular ideologies is their mutual conflicting nature. Although most secular ideologies share some common philosophical propositions, such as emancipatory and democratic objectives, a number of their specific characteristics also differentiate them and often bring them in mutual competition and conflict.

Hence, modern ideological controversies and conflicts not only result from the clash between traditional religious beliefs systems (and their institutions) and modern secular ideologies but also, if not even more, between some of the modern ideologies themselves. Examples of such conflicts are: the opposition between capitalism and the various forms of socially-oriented ideologies (socialism, communism, social liberalism, social-christianism); the opposition between various forms of in-group favouritism and globalism; the opposition between the economic interests and aspirations of the developed and the developing world; and the opposition between unbridled capitalist growth ideology and ecological sustainability ideology.

A factor which further complicates the life stance and worldview profile in modern societies is that the theist-atheist opposition and the modern secular ideological diversity partially intersect with each other. This has led to theist and atheist views permeating the nascent modern worldviews in various combinations.

4.3.2 The Short-Term Approach of Secular Ideologies

Modern secular ideologies are not only characterised by their fragmented nature but usually also by their short-term perspective. They mostly lack, perhaps with the exception of humanism and ecologism, a vision about the long-term future. They are quite reserved, if not totally silent, about intergenerational goals. From an evolutionary point of view, their short-term approach is an inherent shortcoming.

Life is essentially an intergenerational process that should be duly taken into consideration and shaped by morality. This is of particular importance in modernisation because of its possible harmful effects on future generations and a lasting detrimental influence on the biosphere and even the physical environment.

4.3.3 The Macro-level Approach of Secular Ideologies

Most modern secular ideologies strive to address morality at the macro-level in an effort to respond to the immensely increased population size. They devise principles and practices to deal with human, societal, or ecological challenges at structural levels. They deal with the innovative traits of the modern world, and form a justified replacement for the alms-like charitable approaches of the past. However, most or all secular ideologies have miserably failed or neglected to deal equally well with the modern challenges at the individual levels. They have wrongly assumed that efforts in devising and implementing well-intentioned systemic or structural solutions for societal challenges would automatically meet the human needs at the individual level.

Modern secular ideologies have neglected to strongly enhance educational and agogic efforts to apply 'The Golden Rule' or the need for a message about solidarity in the anonymous and anomic modern societies with their millions of citizens. Several modern secular ideologies—in particular humanism—have broadened the moral rules related to altruistic, reciprocal, and mutual and spontaneous helping behaviour to all of humanity, regardless of race, ethnicity, religion or ideology: but they have mainly applied them at an abstract global level and much less at the level of individuals. Modern ideologies have forgotten to respond efficiently to the old-time profound needs for interpersonal relations, assistance, care and love, to which the human was genetically adapted in Pleistocene times, when it lived in small communities. Humans are still not very effective in bonding in the global world.

Evolution-Based Universal Morality

5

Abstract

This chapter highlights first the rationale for evolution-based ethical choices in modernity and proposes the main arguments in favour of this position. The core of this chapter concerns the identification and justification of evolution-based general ethical goals for the future. Two evolutionary-based prerequisites for the further development and evolution of the human species are distinguished: ecological sustainability and cultural progression of the modernisation process. The main aim is defined as the phylogenetic enhancement of the hominisation process; from this main aim five major derived objectives are inferred: the ontogenetic development of human-specific potentialities, the promotion of quality of life, the promotion of equity, the shift from competitive toward cooperative efforts, and the promotion of universalism. The rationale for each one of those general ethical goals for the future development and evolution of humankind is explained and justified on the basis of the confrontation of the long-term hominisation process with a further progressing modernisation. As far as the future of the hominisation process is concerned, four alternative scenarios are discussed: extinction, regression, stabilisation and progression.

5.1 Need for a Universal and Inclusive Morality

In modernity the human no longer needs a traditional religion as a source of moral norms but does need a well adapted morality—progressive, inclusive and universal ethics. World challenges are global in nature and traditional moral norms, which are strongly in-group oriented, are slowing down or obstructing further progress of the species. Many of the traditional moral norms are exclusionary as they propose, for example, different standards for men and women, and for believers and non-believers. Indeed, numerous authors call for a new global ethic, a new global

R. Cliquet and D. Avramov, *Evolution Science and Ethics in the Third Millennium*,
https://doi.org/10.1007/978-3-319-73090-5_5

morality, a new Enlightenment, a cosmopolitan humanism, a universal secular morality, and a new evolutionary worldview, just to mention a few.[1]

This idea of the development of a universal or autonomous morality, independent of religious beliefs and faith based institutions, is not at all new. In fact, it has its roots in ancient Greece and became the central idea of the eighteenth century European Enlightenment.[2] Since then many authors have advocated the development of a universal morality. For instance, Paul Cliteur,[3] a Dutch jurist and philosopher, pertinently argues in his recent book *Moral Esperanto* in favour of the need for an autonomous ethics. He sees the separation of morality from religious divine command belief systems as the only way to make modern(ising) multicultural societies, consisting of a broad range of religious and secular ideologies, peacefully living and working together for the benefit and happiness of all, believers as well as non-believers.

In recent decades, considerable concrete impetus to such a new global ethics has been given by the United Nations. It is sufficient to mention the Universal Declaration of Human Rights (1948),[4] the many topic-specific charters, for instance on population (the World Population Plan of Action 1974[5]; the Cairo Action Plan, 1994[6]), women (the Beijing Declaration and Platform for Action, 1995),[7] climate (the United Nations Framework Convention on Climate Change, 1992),[8] and environment (the Rio Declaration on Environment and Development, 1992).[9]

The same applies to the many charters of the Council of Europe; in the first place its Convention for the Protection of Human Rights and Fundamental Freedoms (1950–2010),[10] but also its topic-specific documents such as the European Social Charter (1965), the European Convention on the Exercise of Children's Rights (2000),[11] and the Convention for the Protection of Human Rights and Dignity of the Human Being with regard to the application of Biology and Medicine: Convention on Human Rights and Biomedicine (1999).[12]

Modern international charters dealing with ethical and policy issues are a giant stride forwards in the development of a new global morality. However, they are often partially or even largely influenced or curbed by traditional religions and are strongly dominated by national, in-group interests. Often they do not sufficiently

[1]For instance, Huxley (1927; 1964, 84), Keith (1946, 10), Manenschijn (1987), Global Ethic Foundation (1993), Küng and Kuschel (1993), Katz (1999, 237), Loye (1999), Katz (2000), Peters (2003, 334), Dawkins (2006, 262), Kwame (2006), Young (2006, 44), Cliteur (2007, 219ff.), Stewart (2008), Hitchens (2009, 277), Haught (2010), Hinde (2011), Wilson (2012, 287).
[2]Cliteur (2007, 223ff.)
[3]Cliteur (2007).
[4]United Nations (1948); see also Meloni (2016, 137).
[5]United Nations (1975).
[6]United Nations (1994).
[7]United Nations (1995).
[8]United Nations (1992a).
[9]United Nations (1992b).
[10]Council of Europe (1950).
[11]Council of Europe (2000).
[12]Council of Europe (1999).

deal with the issues on a global level and in a long-term evolutionary perspective. They are frequently more focused on the preservation of the *status quo* rather than adjusting to new and laying grounds for future developments.[13] They generally build on the lowest common denominator of rights already proclaimed or achieved in the most powerful states. In that respect they preserve the status quo in these countries but set their worldviews as norms for other countries.

While acknowledging the tough negotiations and compromises made in order to pass intergovernmental charters, it can be observed that many—and in particular the Universal Declaration of Human Rights—are focused on individuals whilst individual-transcending issues such as collective rights are not addressed. They also mainly focus on rights whilst neglecting to address responsibilities. They fail to address many socio-economic rights while focusing primarily on the political ones.

The charters that resulted from the UN World Population Conferences of 1974, 1984 and 1994 referred to rights *and* responsibilities in several places[14]; but when policy makers or activists use those documents, they mostly focus on rights—sexual rights, reproductive rights, abortion rights, developmental rights, etc.—and forget the related responsibilities. For example, abortion rights also entail the responsibility to avoid unwanted pregnancy; reproductive rights need to be associated with responsibilities to provide effective socialisation and education for both boys and girls.

Those global charters are the result of intergovernmental negotiations and consensus, resulting in qualified compromises. Often the underlying issues at stake are not even considered, particularly those resulting from the clash between our evolutionary and cultural heritage and the challenges of the novel environment of modernity. The salient political weakness of those intergovernmental institutions, in particular the United Nations, is the lack of implementation of those instruments. This implies that moral principles in the fields of population control, environmental protection, economic inequalities, gender inequities, climatic regulations, and international conflicts for example, are believed to have a moral weight but are applied *à la carte* by states.[15]

Efforts by the intergovernmental bodies to promote universal values clearly lack the authority of the word of God and the fear of God. They can be relativised. Most importantly, over past half a century or more the community of nations has not set up enforcing mechanisms with a set of rewards for compliance and sanctions for breaches and free loaders.

Both the traditional religions and the nascent secular ideologies are partly deficient in dealing effectively with the ethical problems of humankind in modernity, and to safely guide the human species through new subsequent stages of biological evolution and cultural development. Broader reflections are needed to adapt and develop our values and norms for the future development and well-being of humans and humankind, as well as of the future of our planetary environment.

[13] See also Cattell (1972, 439).
[14] For example, in the ICPD Action programme of the Cairo conference (1994, 10).
[15] See also Baofu (2010, 229).

The thread of thought throughout this book is that the existential challenges facing the human species in a further modernising and globalising world can only be effectively engaged through a confrontation of the biological hominisation process and the opportunities and challenges of the scientifically driven modernisation. The authors want to give impetus to the idea that evolution science can provide, in the context of a further progressing modernisation, a framework for elaborating global ethics to guide humanity to higher levels in the hominisation process. The authors want to examine the degree to which the traditional and modern ideologies can contribute to and fit into this endeavour. Whereas it is not a new idea, it is a grand challenge around which people from very diverse ideological origins might want to unite.

The authors are well aware that such an endeavour is a daring enterprise. Indeed, when defining ethical goals for the future of humankind, there are fundamentally two possible strategies one can choose: looking for directions in which way humanity should go or identifying directions in which we should not go. Considering that science and technology are evolving at an extremely fast pace, the renowned evolutionary scientist Edward O. Wilson[16] believes that it is quite precarious to make predictions and precepts about the longer-term future course humanity should take. Hence, in his recent book *The Social Conquest of Earth*, he opts for the goal where *not* to go. By contrast, in this book, it is opted for the choice where to go, although it does not exclude the identification of future directions to be avoided. Moral choices, indeed, inevitably include positive as well as negative recommendations.

5.2 Rationale for Evolution-Based Ethical Choices

In previous chapters, some arguments have already been advanced to justify the derivation of values and norms from evolution science. However, it is necessary to address this issue more explicitly and in a more comprehensive and nuanced way. Why, indeed, evolutionary ethics? Why not simply ethics based on biology, since so many basic needs are so clearly of a biological nature? Or even on culture, as so many scientists, even some evolutionists, argue?[17]

The authors see three major reasons for the omnipresent importance of ethics in human life: (1) the incomplete genetic programming of human development, requiring complementary, exo-somatic, cultural intervention in order to achieve optimal biological functionality; (2) the social character of human nature, requiring cultural ruling to master inter-individual and inter-group relations; and (3) the cultural evolution of the human species, resulting in the modernisation process with its considerably increased capacity to master the environment and address

[16]Wilson (2012, 287).

[17]For instance, Ehrlich (2001), Ayala (2009).

traditional human adversities—ignorance, superstition, disease, hunger, war and environmental destruction.

The fundamental argument in favour of evolutionary ethics, based on an evolutionary consciousness,[18] is that life is an intergenerational phenomenon and, even more important, an evolving phenomenon. Knowledge about the evolutionary history of the hominins allows, by means of the comparative method of analysis, the identification of the biological originality and specificity of present-day humankind.[19] It also allows the detection of the dominating trend(s) in the hominin evolution and extrapolates those trend(s) into the future and possibly identifies desirable future goals.[20] Acknowledging the considerable biological variability within the hominin tribe, and the cultural variability within the present *Homo sapiens sapiens*, the evolutionary approach provides a standard against which human variability (and adaptability) can be measured and evaluated.[21] Due to its broad, intergenerational time dimension, an evolutionary biological account of human values and norms can be so much better identified and evaluated than a static, intragenerational biological account, or even a comparative cultural account, the time dimensions of which are so much more limited. The specificity of a species is the result of a very long-term process of phylogenetic evolution, resulting in the assemblage of a large number of inter-depending traits (morphological, physiological and behavioural) which define the originality and identity of a species and make it possible to capture the limits and range of the natural security of a species.[22]

Evolutionary consciousness may be "a romantic idea of the highest calibre" as Gregory Gorelik and colleagues[23] positively evaluate it. It might sound romantic at first glance, but the evolutionary approach is a much better basis for developing a universal morality than any other value-based account about God, reason, emotion, nature, individuals, or society[24] that humans have ever devised. It can be a holistic approach that takes into account genetic and environmental factors, individual and group elements, emotion and reason, biological and cultural factors, and above all, long-term intergenerational processes. It makes it possible, because of its holistic and long-term time dimension, to deal with all of the major philosophical issues ethicists raise, such as the question of intentionality ("how people ought to act toward one another"), the authorative question ("why ought one be moral?"), the distributive question ("whose interests ought to be promoted?"), and the substantive question ("which interests ought to be promoted?").[25]

The evolutionary paradigm also allows the identification of mismatches between our biological evolution and our cultural development. This mismatch is evident for two reasons: (1) partly because of biological maladaptations to the novel

[18]Gorelik et al. (2012).
[19]For instance, Chapais (2013, 53).
[20]For instance, Huxley (1957), Cattell (1972), Kaufman (1997); see also Kelly (2016).
[21]For instance, Cattell (1972), Richards (1986), Masters (1989).
[22]Sagarin and Taylor (2008, 261).
[23]Gorelik et al. (2012, 353).
[24]See, for instance, McShea and McShea (1999, 310).
[25]See, for instance, Gewirth (1993, 245–248).

environment of modernity; and (2) partly because of the fact that some moral codes and conventions which were developed and well adapted to life in the hunter-gatherer, and particularly the agrarian-pastoral era, are now no longer appropriate to resolve biosocial or biocultural challenges in modernity. Regarding biological maladaptations, it may well be that they not only apply to some physical traits but also to evolved emotional predispositions that generate intuitive moral judgments.[26]

The identification of evolutionary mismatches between biological and cultural developments which Julian Huxley[27] called "challenging monsters on our evolutionary path" and Paul R. Ehrlich[28] named "evolutionary hangovers"—"structures or behaviours that once were adaptive but whose positive influence on reproductive performance has declined or disappeared"—can help to develop, in an informed way, better adapted human values, social structures, and/or technological means to accommodate the mismatches. We are better placed to develop alternative, viable and long-term strategies to cope with the biological-cultural maladaptations that arose in the process of modernisation.[29]

The evolutionary approach to ethics in modernity can also contribute to resolving conflicting ethical choices, e.g. freedom of speech, right to privacy, individual versus societal interests, intragenerational versus intergenerational care, and help to rank values and norms which, at first sight, all seem to be equally important.[30]

The evolutionary perspective—focused on an intergenerational and long-term perspective—is also needed to evaluate and take advantage of new opportunities that scientific inventions have created. This is the case not only in the domain of the biomedical sciences but also in the domain of within- and between-group relations, and ecological relations, to consciously lead human biological development (ontogeny), human evolution (phylogeny), human society and ecology in the desired direction.

Finally, evolutionary ethics can replace the traditional religious ideologies which rely upon fantastical revelations that made sense in pre-modern eras but have become obsolete—if not downright dangerous—in a modern context. A typical example of a threat is the in-group ideology which is inherently built into religious thinking. An evolution-based ethics might also complement the fragmented modern ideologies that have failed to design holistic goals for future progress. Evolutionary ethics exceeds by far traditional ethics or modern fragmented ideologies because of its holistic approach, including not only individuals and populations but also past, present and future generations.[31] Evolutionary ethics is not just a competitor or complementary to traditional ethics, but it is its successor.[32] It is especially

[26]Hauser (2006, 418).
[27]Huxley (1964, 82); See also Hinde (2002).
[28]Ehrlich (2000, 34); See also Keith (1946), Burnham and Johnson (2005).
[29]Huxley (1957), Cattell (1972), Hinde (2002).
[30]Richards (1999), Corning (2003).
[31]See also Shermer (2004, 10).
[32]Bradie (1994, 7).

important in this perspective that one of the major achievements of evolutionary science, and hence one of the major tenets of evolutionary ethics, is that any notion of a supernaturally inspired or determined ultimate design and purpose in life and human history has to be abandoned.[33] Thanks to science and reason, it is now known that we humans have to define for ourselves the design and purpose we want to see in life and history. This new phase implies that humans can and ought to take up the responsibility for identifying needs and the instruments for the implementation of evolving values.

In the introductory chapter, it was stated that the goal of this book is to address the necessary value and norm changes, with the prospect of evolving to higher levels of hominisation, adapting humanity and its culture to the opportunities and challenges of the modernisation process, and in particular enlightening or maturing modernity. Hence, the intended evolutionary ethics is not just based on evolutionary science, but on the confrontation of the knowledge about the evolutionary process with the modernisation process. The ethical choices that will therefore be suggested and justified in the following sections and chapters are: (1) a further progressing hominisation; and (2) a further progressing enlightened modernisation.

5.3 Evolution-Based General Ethical Pursuits

In this chapter a number of basic ethical goals of a more general nature are addressed that apply to several, if not all, specific domains of human life.

Three types of more general ethical goals to be pursued are addressed: two prerequisites, the main aim and five derived goals. The two evolutionarily-based prerequisites for the further development and evolution of the human species are: (1) ecological sustainability; and (2) cultural progression of the modernisation process.

The main aim of an evolutionarily based ethics is the phylogenetic enhancement of the hominisation process. From this main aim five major derived objectives are inferred: (1) the ontogenetic development of human-specific potentialities; (2) the promotion of quality of life; (3) the promotion of equity; (4) the shift from predominantly competitive towards cooperative social relations; and (5) the promotion of universalism.

5.3.1 Ethical Prerequirements

5.3.1.1 Ecological Sustainability

The present variant of the hominins, *Homo sapiens sapiens*, emerged some 150,000–200,000 years ago in East Africa and initially included only a few thousand individuals. Since then the human species spread all over the planet, settled in

[33] See, for instance, Vannelli (2001, v).

practically all possible environments, and considerably increased its numbers, particularly during the twentieth century. The human species is nowadays seven billion people and will, probably before the end of the present century, reach a maximum of eleven billion before stabilising at or decreasing to a lower but still very large world population size.[34]

The population increase of the human species has, in all of its major cultural stages, provoked ecological catastrophes. It resulted in the eradication of many other species, as for instance in the Upper Palaeolithic, or turning arable land into (semi-) deserts in the agrarian-pastoral era. It is producing multiple ecological disasters in the current industrial phase.[35] The human species, with its apparently insatiable drive for possession and power, is not by nature a cautious and provident conservationist. Contrary to earlier romantic views, neither hunter-gatherers nor agriculturalists-pastoralists developed a substantial conservation ethic and achieved long-term conservation.[36] The abuse of the planet's resources and environment has become markedly worse in modern times. Humankind is now producing an anthropogenically caused 'sixth extinction'.[37] Many so-called advanced nations ravaged not only their own resources and environment but also usurped the resources of other populations through imperialist (neo)colonial domination and exploitation. In addition, they have now started to prey on their own children and grandchildren by making and accumulating debts to be covered by future generations.[38] Hence, the fully justified conclusion of Matt Ridley[39]:

> … there is no instinctive environmental ethic in our species - no innate tendency to develop and teach restrained practice. Environmental ethics are therefore to be taught in spite of human nature, not in concert with it. They do not come naturally.

It is only in the second half of the twentieth century that the disastrous effects of industrial culture on the environment elicited an ecologically grounded ethical and political awareness: however, the impact is still very limited.[40] As Paul R. Ehrlich[41] pertinently stated:

> Until recently, people have not paid much attention to the long-term environmental effects of their behaviour but rather have focused on the satisfaction of their immediate needs.

The current ecological track record of the human species is lamentable: due to its population growth and production and consumption patterns, humankind is decimating the planet's biological abundance and diversity at an accelerating pace. Humans are overusing natural resources, producing habitat loss and environmental degradation, depleting non-renewable resources, polluting sea, air and soil,

[34]United Nations Population Division (2015).
[35]For instance, Ehrlich and Ehrlich (1981), Goudie (2005), Wijkman and Rockström (2012).
[36]Low (1996), Wilson et al. (1998).
[37]Leakey and Lewin (1996), Novacek (2001), Kolbert (2014).
[38]For instance, General government gross debt in EU countries (2002–2013). http://epp.eurostat.ec.europa.eu/.
[39]Ridley (1996, 225).
[40]Vermeersch (1988), Penn and Mysterud (2007, 28).
[41]Ehrlich (2000, 320).

aesthetically defiling the environment, anthropogenically changing the climate, and damaging several other critical planetary ecosystems.[42] As Warren M. Hern[43] thoughtfully noted:

> The human species is a rapacious, predatory, omniecophagic species engaged in a global pattern of converting all available plant, animal, organic, and inorganic matter into either human biomass or into adaptive adjuncts of human biomass.…In this respect, the human species is an example of a malignant ecotumor, an uncontrolled proliferation of a single species that threatens the existence of other species in their habitats.

In 1980, William R. Catton[44] stated pertinently that the post-exuberance age in human history has been reached and that there is an urgent need to change our ecological paradigm from exuberance to sustainability. There is a gradual increase in public awareness of the long-term devastating effects of the dominant type of economy and especially its current variant—the neoliberal fundamentalist free market capitalism—on the planet's ecosystems and in particular its climate. Nevertheless, there is a risk that remedial measures will only deal very superficially with the fundamental causes of the ecological demise. As Naomi Klein[45] argues so well in her recent book *This Changes Everything. Capitalism vs. The Climate* that we will need a fundamental alternative worldview based on mutual dependency instead of hyper-individualism, reciprocity instead of dominance, and cooperation instead of hierarchy.

One may even argue that there is, behind the ecological concerns of some states, a hidden agenda to preserve the status quo in power and developmental opportunities. Therefore, the authors believe in the notion of ecological justice which requires more thorough reflection and the setting up of institutions which have powers of action.

The authors consider the commitment to ecological sustainability a *sine qua non* for the survival, the long-term development and the further evolution of the human species on this planet. There is a need to preserve crucial components of the physical environment such as inhabitable and arable land, water, and climate. On the one hand we are in competition with many other life forms; on the other hand the human species vitally depends on the existence of other life forms for breathing, nutrition, digestion, and even for more subtle psychological needs. The priority ranking of the ecological goal as one of the two ethical prerequisites for the future development and evolution of humankind is fully justified from an evolutionary perspective. We are in the critical stage of humans having a disruptive impact on life and the environment of this planet.

In technologically developed countries many of the basic biological needs of the vast majority of the population are largely met. Under these circumstances, the preservation of the diverse life forms and landscapes may have become an element in satisfying our cognitive/emotional/spiritual needs, appealing to our aesthetic and

[42]Rockström et al. (2009).
[43]Hern (1993, 16).
[44]Catton (1980, 58).
[45]Klein (2014, 513).

exploratory needs. Although the ugliness of many human settlements often takes unimaginable proportions and there are few individual and even group concerns for cleanness and a pristine environment, it is also true that people experience extraordinary pleasure from green spaces, natural landscapes and beautiful habitats,[46] as well as from leisure travel. Our modern environmental aesthetics may be the result of our affluence culture in which basic survival needs are covered and leave room for other needs. However, as mentioned in Chap. 4, it is not impossible that the sense of environmental beauty may also be partially related to remnants of ancient drives which developed for survival purposes in the hunter-gatherer era of human evolution: a spacious environment with a diverse and lush flora and fauna which guaranteed feeding and safety needs.[47]

5.3.1.2 The Enhancement of the Modernisation Process

The fundamental feature of modernisation is the development of science, which brought not only a more thorough knowledge and understanding of reality but also made possible more effective ways of inter-vening in life, society and the environment. This interventionist nature of science is embedded in modern culture and has fundamentally changed humankind's biosocial and biocultural relationships. It is not only offering new opportunities for the future but also bringing about new challenges.

On the one hand the achievements of science and technology increasingly allow humans to master their environment and their life; on the other hand modernisation has been, and is being, accompanied by a number of unfavourable and potentially catastrophic effects. The challenges and choices for the future are essentially of a moral nature. The authors argue that, in combination with the humanistic principles of the Enlightenment, modernisation may result in a sustained enhancement of quality of life, provided that maladaptive practices are countered by an evolutionary-based ethics.

The increased capacity for intervention in modern culture is characterised by the potential to gradually replace the major mechanisms of biological evolution by cultural mechanisms. The perceived inhumane mechanism of natural selection can be replaced by cultural forms of selection, and natural chance events can be replaced by consciously chosen acts.[48] With modernisation we witness a transition from trial and error to what John Stewart[49] calls 'intentional evolution'. This capacity for intervention is the basis of a new evolutionary worldview.

Given its important achievements in acquiring scientific knowledge about nature, its enormously enhanced opportunities to influence natural processes, and its tremendous improvement in the quality of life, the authors consider the modernisation process to be a positive achievement by humankind. They do acknowledge risks and challenges and negative side effects, but consider them transient and

[46]See, for instance, the 'Biophilia Hypothesis' of Wilson (1984) or the 'Savanna Hypothesis' of Orians and Heerwagen (1992, 557).
[47]Wilson (1984), Kellert (1996), Falcicchio and Barbiero (2015).
[48]For instance, Galton (1883), Huxley (1894), de Duve (2009).
[49]Stewart (2008); see also Harris (2007, 4).

amenable.[50] This is why it is argued that the furthering of the modernisation process, in its enlightened or matured form, is the second prerequisite of an evolutionarily based ethical approach.

This implies that evolutionary founded values and norms cannot simply be derived from a narrow Darwinian calculus in a vacuous space. They can only be elaborated on the basis of the interaction of our evolved dispositions with the essential characteristics of modern culture.[51]

In emphasising the modernisation goal, the authors do not imply the superiority of the Western culture. However, it is assumed that if people are given a real choice based on adequate knowledge about alternative living conditions, no one—apart from romantic unworldly dreamers or religious zealots—would prefer constant confrontation with risks of early death, disease, starvation, poverty, inter-group conflict, and the misery of ignorance and superstition.[52]

From discussions at world assemblies of the United Nations, such as the World Population Conferences,[53] it appears that all governments want to develop, particularly those from developing countries. They seem to want to join what Joseph Henrich and colleagues have labelled WEIRD—Western, Educated, Industrialised, Rich, and Democratic—societies.[54] However, pathways to development build on different values. Technological achievements seem to appeal to many conservative regimes who want also to preserve their traditional patriarchal, sexist, and oligarchic cultural and religious values. These states are committed to leaving their women in the pre-scientific era. Some authors see in this endeavour a desperate attempt to maintain acquired power positions in male-dominated families and societies at large.[55]

Praise for the modernisation process should not make us blind to its disastrous ecological effects: the loss of biodiversity, the destruction of natural ecosystems, the depletion of natural resources, pollution, and the aesthetic degradation of the natural environment.[56]

Equally disastrous is the ruthless economic exploitation of developing nations. Colonisation practices enriched 'advanced' nations at the expense of oppressed peoples, thereby setting the standard for developing nations' 'elites' who, in turn, often exploit(ed) their own populations to an even larger degree.[57] Modernisation was also accompanied by ideologies which conflict fundamentally with the ethical ideals of the Enlightenment (liberty, equality, and fraternity).

[50]See also Häggström (2016).
[51]Kitcher (1985), Levy (2004).
[52]See also Kurzweil (2005).
[53]United Nations (1975, 1984, 1994).
[54]Henrich et al. (2010).
[55]Cliquet and Thienpont (1995).
[56]Ehrlich and Ehrlich (1981, 2008), Goudie (2005), Carlson and Lintott (2008).
[57]For instance, Sklar (1979), Sachchidananda and Lal (1980), Bakre (2008).

There are also the exponentially increasing risks of a technogenic catastrophe and the self-annihilation of our species in advanced technologies[58] of the ABC weaponry development[59] and the GNR revolution.[60]

However, just as the crimes committed in the name of Christianity or Marxism are no ground for rejecting the positive messages these ideologies contain, so should the detrimental ecological and humanitarian effects of modernity not lead us to abandon the principles of modernisation, as done by some critics of modernity.[61]

Although it may not be possible to anticipate the effects of future inventions in a modernising world, the authors are of the view that humanity is only at the beginning of the modernisation process. Great scientific innovations in genetic engineering, robotics and nanotechnology will lift the quality of life to unprecedented heights. Those inventions will radically change our perceptions about the exhaustion of limited material resources to sustain human life.[62] Forecasters have generally failed dismally to foresee the drastic changes brought about by completely unpredictable discoveries.[63]

In addition to science and technology, modernisation embraces tolerance toward variation and even deviance from mainstream cultural characteristics.[64] This is a wise response toward the factual variation which exists in human populations with respect to drives, aspirations, and behaviours—provided they do not harm others (cf. The Golden Rule). However, just as is the case with genetic variation, it is a safety valve to allow adaptations in altered environmental conditions. Furthermore, it appears from research that culturally important inventions and innovations, whether of a technological or moral nature, have only been made in societies that tolerate deviance.[65]

Multiple Modernities/Multiculturalism. Some scholars have challenged the idea of a further progressing modernisation that would spread all over the world. They hypothesise that the future of modernisation will express itself in the form of multiple modernities that are based on fundamentally different cultural values.[66] It is a view that lies in the prolongation of Raymond Cattell's[67] idea of cooperative competition among culturally diverse civilisations and Samuel P. Huntington's[68] Clash of Civilizations. Given the importance people give to cultural—and more particularly religious values—it is not inconceivable that the modernisation process will, at least during some of the time, show some differentiation between major geographical areas of the globe. Indeed, differences may nowadays be observed between the more individualistically and religiously oriented United States and the more social(ist)- and

---

[58]Bostrom (2002), Verdoux (2009), Bostrom and Cirkovic (2011).
[59]ABC weaponry: atomic, biological an chemical weapons of mass destruction.
[60]GNR revolution: the explosion of knowledge of genetics, nanotechnology and robotics.
[61]For instance, Vaknin (2009).
[62]Kurzweil (2005).
[63]Rees (2003).
[64]Jones and Reynolds (1995).
[65]Goldstone (1987), Mokyr (1990).
[66]See, for instance, Eisenstadt (2000), Taylor (2007), Haynes (2013).
[67]Cattell (1972, 105).
[68]Huntington (1996).

secular-oriented European Union, or between the more individualistically oriented West and the more community-focused Far East. Another example of an even more fundamental cultural divide is the opposition between the West and the Islamic world where religion and state organisation have not yet been clearly separated. There are still very strong attempts by some religious communities to accentuate their uniqueness and monopoly on the divine truth.

In recent years the discussion about multiculturalism is linked to the idea of multiple modernities. It started to take a more prominent place in many European countries with the arrival of numerous immigrants from more distant countries with religious backgrounds typical of the agrarian-pastoral cultural stage in human history. These immigrants were more recognisable phenotypically or in clothing and decoration, manifested their cultural features more explicitly, and were given fewer opportunities or were less willing to integrate into the receiving society. In some cases the pressure on recent immigrants to maintain distinct cultural/religious features, while living in European countries, may be traced back to the policies and financial support of some mono-cultural and anti-democratic theocracies that want to support and further spread their own religion in the world.[69] Immigrants with such cultural and ideological backgrounds bring values and norms acquired during their socialisation in their native country which can, in various ways, be extremely different from and clash with the mainstream normative basis of the receiving country. Examples may include the discriminatory perception and treatment of women, arranged or forced marriages, physical punishments, honour crimes, genital mutilation, polygyny, fatwa's against persons who hold different opinions, and attempts to censor teachings about evolutionary biology. Such opinions and practices are considered incompatible with the level of emancipation reached by modern culture, which highly values science, rights related to sex/gender equality, freedom of expression, norms promoting individual development, and human rights—all of which are embedded in the legal systems of virtually all modern democracies.[70] Countries that were tempted into an improvident, if not naïve, multiculturalist policy are now reaping the bitter fruits of their sloppy endeavours—the presence of a demographically growing, socially and culturally insufficiently integrated, rancorous religious minority, worshipping values that run counter to the achievements of modern society and re-assurgency of right-wing aversion to diversity as a reaction by significant shares of natives. In some cases, modest but largely insufficient efforts are being made to reorient the earlier policies.[71]

However, the modernisation process is mainly driven by the progress of the modern sciences which, de facto, have advanced much more in a uniform globalisation than other aspects of human culture. Science transcendent religious, social, or national in-group drives, even in the most powerful groups and nations, tend to ripe benefits only for themselves. The authors believe that eventually the sciences will, like a cosmic black hole, swallow up the remaining maladaptive cultural

[69]Safa (1997), Van Rooy (2008).
[70]Avramov and Cliquet (2005, 201ff).
[71]For instance, Carle (2006).

differences, most of which are hangovers from earlier cultural stages in human evolution and history. With this point of view, once people are fully confronted with and aware of the achievements of sciences, is unlikely that they will be inclined to cling to pre-modern values and norms, and reject ideas such as freedom of expression, individual self-actualisation, social solidarity, tolerance towards diversity, sexual equality, global responsibility, and ecological sustainability. If a future multiplicity in the modernisation process develops, in the long run it will certainly not evolve in contradiction with the achievements of science or the universal values which evolved in the footsteps of the European Enlightenment.

5.3.2 The Main Aim: The Phylogenetic Enhancement of the Hominisation Process

The phylogenetic future of humankind can, in principle, evolve in four possible directions: (1) extinction of the human species; (2) regression of human-specific characteristics; (3) stagnation of evolution at the present level of development; and (4) the further progression of the hominisation process (Fig. 5.1).

In addition, one can imagine combinations between some of those alternative paths, as for instance suggested by Nick Bostrom[72] who also distinguishes a recurrent collapse path, a future fluctuating evolution with alternating stages of progression and regression. However, such a future scenario might fit more into a framework of alternative possible ontogenetic developments.

In this section the discourse is concentrated on the genetic changes, namely on the phylogenetic future of the present representative of the hominins, whilst matters related to the ontogenetic enhancement of human characteristics will be discussed below in Sect. 5.3.3.1.

5.3.2.1 Human Extinction: Possible but Avoidable Future

Evolutionary extinction is a common phenomenon. It has been estimated that 99.9% of all species that ever existed on Earth are already extinct.[73] Catastrophic events are menacing events that can extinguish a species. Examples include the impact of a planetoid 65 million years ago in the Chicxulub region (Yucatan Peninsula, Mexico)[74] that led, probably in combination with massive volcanism,[75] to the extinction of the dinosaurs, and the major eruption of the Toba volcano on Sumatra some 74,000 years ago that produced an important demographic bottleneck for *Homo sapiens sapiens* in the Upper Pleistocene era.[76]

[72]Bostrom (2009, 551).
[73]Hallam and Wignall (1997), Newitz (2013).
[74]Alvarez et al. (1980), Hildebrand et al. (1991).
[75]For instance, Keller (2014).
[76]Ambrose (1998).

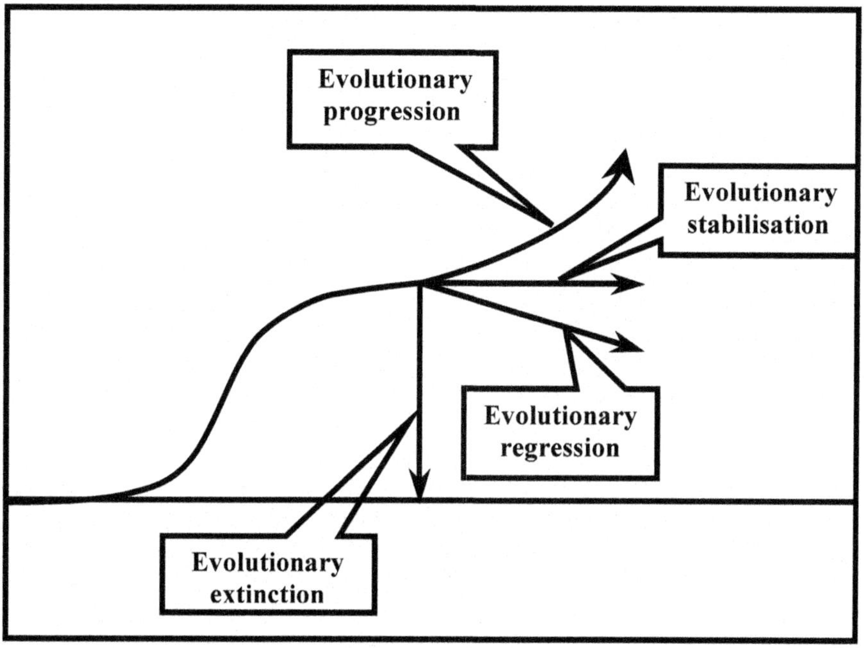

Fig. 5.1 The long-term genetic future (Cliquet 2010, 611)

Human extinction means the disappearance on this planet of the hominin phyletic line, namely the disappearance of the present *Homo sapiens sapiens* and any further hominin stage that might evolve in the hominisation process.[77] If humanity does become extinct the hominisation process would have to restart from one of the extant primate species: this path would, if it ever came so far, take many million years. We should be aware that the emergence of the hominin phyletic line on this planet was a singularity due to the unique sequence and convergence of a series of environmental conditions and biosocial preadaptations, which has an extremely low probability of being repeated.[78]

Several scholars have examined the probability and the possible causes of a sudden human extinction.[79] They unanimously reach the conclusion that there is a serious risk that humanity's future is under threat of a premature end. This is mainly

[77]In order to distinguish some scenarios in which the human species would go extinct because it evolves to a new stage in the hominin evolution from a scenario in which the hominin phyletic line disappears from this planet, Bostrom (2002) devised the 'existential risk' concept which he defined as "an adverse outcome that would either annihilate Earth-originating intelligent life or permanently and drastically curtail its potential".
[78]For instance, Wilson (2012, 45).
[79]For instance, Joy (2000), Posner (2005), MacKenzie (2008).

due to technical advances which in themselves render society more vulnerable to disruption, either by error or terror. For instance, Nick Bostrom[80] believes that the risk of extinction due to existential disaster lies above 25%. Also John Leslie[81] estimates that the humanity has a 30% chance of annihilation by war, disease, or some other catastrophe in the next 500 years. Sir Martin Rees[82] is even more pessimistic and fears that the chance of humanity surviving the twenty first century is only 50%. The latter author considers that it is a miracle that humanity survived the twentieth century without a major catastrophe for the species.

A broad variety of possible extinction causes have been suggested, ranging from natural disasters such as a comet, asteroid or planetoid impact, solar flares, supernovae, black hole explosions or mergers, gamma-ray bursts, galactic centre outbursts, or supervolcanoes, to an extreme Ice Age due to passage through an interstellar cloud. Extinction may result from a pandemic with high virulence and 100% mortality rate among infected individuals, or an essentially unpredictable breakdown of a complex system—annihilation by extraterrestrials—as investigated by chaos theory. It may occur due to man-made disasters such as a world-wide nuclear war, chemical or biological warfare, bioterrorism, destruction of the atmospheric ozone layer, the production of designer pathogens, a runaway global warming, poisoning by pollution, the development of high-energy particle accelerator experiments, destructive uses of advanced molecular nanotechnology, self-enhancing artificial intelligence, or badly programmed superintelligence.[83] Nick Bostrom[84] is of the view that, currently, the greatest existential risks are of an anthropogenic nature, namely those that derive from the application of present or anticipated modern technologies.

Many of those possible disasters might, as an existential risk, only lead to a considerable population decrease or a fundamental societal collapse. A worldwide nuclear war, a massive meteorite impact, a supervolcano eruption, or a deadly pandemic might decimate the human world population, but some remote populations might survive, as was the case after the Toba supervolcanic eruption 74,000 years ago.[85] However, in extreme conditions such global disasters might annihilate the total human species, leaving no survivors (and cultural knowledge) to rebuild modern culture. There would be no next hominin stage. Therefore, the authors agree with John Leslie[86] who concluded his well-known book on *The End of the World: The Science and Ethics of Human Extinction* with:

> …we have a strong duty not to risk the extinction of the human race, and above all not to risk it for utterly trivial benefits.

[80]Bostrom (2002).
[81]Leslie (1998, 146).
[82]Rees (2003; 2004).
[83]Leslie (1998), Bostrom (2002; 2014), Lorenc (2015), Stroeykens (2016).
[84]Bostrom (2002); see also Bostrom and Cirkovic (2008).
[85]Rampino and Self (1992), Ambrose (1998).
[86]Leslie (1998, 155).

Contrary to the somewhat otherworldly Voluntary Human Extinction Movement[87] which sees the voluntary extinction of *Homo sapiens sapiens* as an alternative to the extinction of millions of species of plants and animals, the authors consider avoiding or reducing the risk of human extinction as a moral obligation which should be pursued purposefully, deliberately and actively.[88] Although some scholars are of the view that the moral obligation to preserve the human species cannot be derived from purely biological premises,[89] the authors think, on the contrary, that our innate survival drive constitutes the natural basis for that moral choice. Admittedly, in a very distant future, when the physical living conditions on this planet will have become unbearable due to the evolution of our sun, and emigration to other planets or solar systems might prove to be impossible, our then living successor in the hominin tribe might indeed decide—or rather have to decide—otherwise.

5.3.2.2 Evolutionary Regression: Possible but Undesirable Future

Whereas human extinction is a matter that concerns the human species as a whole, human regression is a process that has to be considered at the individual level, as well as at the species (or population) level.

As far as the species (or population) level is concerned, many of the potential existential risks mentioned in the previous section on human extinction also apply—perhaps even more appropriately—to a possible biological and/or cultural regression. Indeed, many of the above-mentioned risks might considerably reduce human numbers, break down our complex global society, and regress human culture far below the most advanced levels that have been reached in modernity. Apart from unexpected disasters, such as a sudden new pandemic or the impact of a giant meteorite, the current anthropogenically caused ecological degradation is probably the most imminent threat due to overpopulation, overconsumption and pure mismanagement of the Earth's natural resources.[90] Moreover, some scholars fear that the degree of complexity acquired by modern society inherently implies the risk or danger of an inevitable collapse.[91]

Biological regressive processes may also manifest themselves at the individual level. The knowledge that we have about the evolutionary mechanism and the hominisation process suggests that, in the absence of conscious human intervention, in the future humanity will experience a number of regressive phenomena that are consistent with the loss-mutations experienced in the past for features in which a functional loss occurred. Humans retain features such as rudimentary ear muscles, the appendix, regressing wisdom teeth, and a reduced coat of hair. Continued evolution along these lines of loss of usefulness could lead to atrophied lower limbs, non-lactating mammary glands, weakened auditory and visual powers, and a

[87]http://www.vhemt.org/.
[88]See also Matheny (2007), Newitz (2013).
[89]For instance, James (2011, 141).
[90]For instance, Eckersley (2001).
[91]For instance, Tainter (1988), MacKenzie (2008).

further reduced and degenerated set of teeth. Humans could see an increase in all possible physical and mental disorders for which replacement therapies are developed or relaxation concerning selection is made possible.[92]

In modern culture humans have succeeded in efficiently intervening against contagious disease and early death, with the result that a considerable part of modern populations reaches a much higher age than in pre-scientific living conditions. The successful phenotypic care provided in modern culture leads to a relaxation of natural selection. Numerous less favourable alleles, which in pre-scientific living conditions were rapidly barred from the gene pool, can at present be preserved thanks to replacement therapies or other protecting factors; in many cases their carriers are also able to reproduce. Because of this, the frequency of such less favourable alleles is increasing.[93]

Furthermore, the relaxation of selection caused by morbidity and mortality control may be reinforced by an increase in reproductive fitness. Surviving individuals may find a partner or partners and produce children. This effect has already been shown for several impairments, such as diabetes,[94] schizophrenia,[95] and phenylketonuria,[96] for which replacement therapies or other types of medical treatment have been effectively developed. The reproductive fitness of such patients has been enhanced, so that an increase in the frequency of the alleles responsible for these conditions may be expected to increase. Although many congenital defects are known to result in lower marriage rates, in infertility or are associated with low fertility,[97] the effect of modern culture is that, through replacement therapies, mating and reproductive opportunities for those with genetic disorders are improving.[98]

Another contraselective effect of modern culture may result from differential reproduction with respect to cognitive abilities. The transition from chance to planned fertility has been accompanied by social differentials in fertility that can affect the genetic composition of the population. In the United States, Robert D. Retherford and William H. Sewell[99] calculated that the generational change in measured intelligence was a decline of 0.8 of an IQ point, resulting in a generational genotype decline of about one-third of an IQ point. On the basis of a negative correlation between IQ and fertility ($r = -0.73$) across nations, Richard Lynn and John Harvey[100] estimated that the world's genotypic IQ declined by 0.86 IQ points from the year 1950 to 2000 and they project a further decline of 1.28 IQ points in the world's genotypic IQ between the years 2000 and 2050.

[92]Glass (1966).
[93]Dobzhansky (1962), Thibault (1972).
[94]Aschner and Post (1956/57), Post (1971).
[95]For instance, Erlenmeyer-Kimling and Paradowski (1966), Lane et al. (1995), Avila et al. (2001).
[96]Howell and Stevenson (1971).
[97]Reed (1971), Slater et al. (1971).
[98]Teitelbaum (1972).
[99]Retherford and Sewell (1988).
[100]Lynn and Harvey (2008).

Overall, medical interventions and changes in reproductive behaviour in modern society seem to have a slight dysgenic effect. Some argue that the present course of human evolution is not going in a desirable direction. However, this is probably only a temporary consequence of a major shift in cultural change and its associated demographic regime. In the near future—namely in the course of this millennium—this dysgenic effect might be neutralised, if not reversed, by future improvements in genetic knowledge and genetic modification, and by the adaptation of behavioural norms to the new genetics and demographics.

5.3.2.3 Evolutionary Stabilisation: A Conservative Illusion

Evolutionary stabilisation means that the present human gene pool is maintained in future generations. In evolutionary terms this would mean stagnation. The ideal of maintaining the status quo is a position taken by bioconservatives and religious believers. They think that human nature should not be changed either because it has been created by a supernatural power and/or because change would corrode human dignity.[101] However, this position is based on misapprehension. It does not take into account the Red Queen[102] Effect, which means that life must constantly evolve in order to preserve the survival of a species. This race is determined by the biological competition between and within species.[103] It also has an effect on the intergenerational processes causing the accumulation of unfavourable mutations.

Indeed, evolutionary stabilisation would require the avoidance of the accumulation of unfavourable mutations that occur throughout generations. In the context of modernisation, the elimination of unfavourable mutations implies the application of eugenic engineering, either via manipulation of genes or via differential reproduction (see Chap. 8, Sect. 8.3.3.3). In order to avoid evolutionary regression, a commitment to achieve evolutionary stabilisation requires purposeful intervening in the production and multiplication of deleterious genetic variants.

The avoidance of evolutionary regression, namely the accumulation of mutations which are considered to be unfavourable, can in the first place be achieved by preventing culturally induced mutations. Mutagenic radiation or chemicals are to be avoided or applied in such a way that they cannot have a harmful effect on the human gene pool. Secondly, the reproductive behaviour of the carriers of genes that are regarded as unfavourable can be restricted or regulated in such a manner that unwanted genes are not transferred to future generations. Such strategies would maintain an evolutionary stabilisation of the human species. As di Lampedusa[104] wrote in '*Il Gattopardo*':

> Se tutto deve rimanere *com'è, è necessario che* tutto *cambi.*

Hence, it is an illusion that human stabilisation would mean non-intervention. On the contrary, stabilisation would require major human interventions. People

---

[101]For instance, Kass (2002), Sandel (2007), Tirosh-Samuelson (2011).
[102]Carroll (1872).
[103]Van Valen (1973), Dawkins and Krebs (1979).
[104]di Lampedusa (1958: "…*if everything is to remain the same, everything will have to change.*").

often think that maintaining the *status quo*, i.e. undertaking no action, is on the safe side of the action-inaction antagonism, but as John Harris[105] pertinently argues, such a belief is without foundation.

From an evolutionary perspective, humans are clearly in a too early stage in human evolution to strive towards stabilisation. Opportunities for further development and evolution of human cognitive and physical capacities and new forms of social relations that match human potentialities are still significant.

5.3.2.4 Evolutionary Progression: The Sensible Option

In principle, there are two major ways to enhance the hominisation process, namely (1) the enhancement of specific human characteristics through cultural influences upon ontogenetic development; and (2) change the genetic composition of the human species, eventually leading to a next stage in the evolution of the hominins. For the latter development, several names have been suggested for the successor of the present *Homo sapiens sapiens,* e.g. *Homo sapientior,*[106] *Homo superior,*[107] *Homo cyberneticus,*[108] *Homo provectus,*[109] *overhuman,*[110] *posthuman,*[111] or even *Homo deus.*[112]

Friedrich A. Hayek's observation in his remarkable book *The Political Order of a Free People* that "to pretend to know the desirable direction of progress seems to me to be the extreme of hubris"[113] is pertinent. Nevertheless, reflecting on the desirable future is an imperative. Indeed, the authors define the central aim of evolutionary ethics as the active advancement of the hominisation process for the long-term future of humanity. Further hominisation is not just a possible futuristic scenario but also an ethical choice. This implies that humanity should steer its own future evolution through the means of conscious genetic and ontogenetic interventions, namely by promoting human-specific features directed at a continuing hominisation.[114] The 'natural' Darwinian evolution mechanism, based on random genetic mutations and natural selection, would be replaced by what Simon Young[115] calls Designer Evolution. Indeed, actively and deliberately pursuing the furtherance of the hominisation process comes down to what Steve Fuller[116] calls the achievement of a "secular sense of intelligent design"—a formidable moral innovation and challenge. As Edward O Wilson[117] recently stated:

[105]Harris (2007, 80).
[106]Overhage (1977, 228).
[107]Stapledon (1935), Kirk (2002).
[108]Young (2006, 21) (*Cyberneticus:* from the Greek *kubernetes* = the steersman of a ship).
[109]Mulhall (2002, 85) (*Provectus:* from Latin = more advanced, more highly developed).
[110]Nietzsche (1883–1885), More (2010, 1).
[111]Nichols (1988), Braidotti (2013).
[112]Harari (2017).
[113]Hayek (1979, 169).
[114]Cliquet (1961, 59; 1996–1997; 2010, 524ff).
[115]Young (2006, 38); see also Harris (2007, 4), who speaks about "enhancement evolution".
[116]Fuller (2007, 153).
[117]Wilson (2014, 14).

> We are about to abandon natural selection, the process that created us, in order to direct our own evolution by volitional selection – the process of redesigning our biology and human nature as we wish them to be.

This idea of seeking to actively advance the hominisation process is not new. It can be found as a constant in the thoughts or writings of evolutionary ethicists, although formulated in somewhat different terms, for instance, by J.B.S. Haldane as science and the future,[118] by Julian Huxley as evolutionary humanism,[119] by Hermann Muller as guidance of human evolution,[120] and Raymond Cattell as Beyondism.[121]

For the authors of this book inspirational was the view of Sergius G. Kiriakoff, at Ghent University who in 1955 argued that the aim of ethics is the promotion of the hominid phyletic line, and C.H. Waddington's[122] *The Ethical Animal*:

> The biological function of ethics is to promote human evolution.

The human drive to improve oneself is a fundamental characteristic of our species. The desire to acquire new capacities is as ancient as our species,—"a natural outgrowth of our human intelligence, curiosity and drive".[123] Progressive aspirations are already present in the earliest recorded human cultures, from the Babylonian Epic of Gilgamesh (eighteenth century BCE), the Buddhist millennial mythos (fifth century BCE), to the Jewish and Christian traditions. Whereas in pre-modern times, religions mostly channelled those aspirations into a heavenly hereafter, the Enlightenment, to start with Frances Bacon's *New Atlantis*,[124] pushed these desires

---

[118]Haldane (1924).

[119]Huxley (1942, 576) saw evolutionary progress "as consisting in a raising of the upper level of biological efficiency, this being defined as increased control over and independence of the environment". Aware that it is not possible to discover a purpose but only a direction in evolution, Huxley proposed that the past human direction serves as a guide to formulating our purpose for the future: "The future of progressive evolution is the future of man. The future of man, if it is to be progress and not merely a standstill or a degeneration, must be guided by a deliberate purpose." (577).

In later publications Huxley (1964) called this new purpose "evolutionary humanism": "... because it can only be based on our understanding of man and his relations with the rest of his environment. ... It must be organized round the facts and ideas of evolution, taking into account of the discovery that man is part of a comprehensive evolutionary process, and cannot avoid playing a decisive role in it. ... Such an Evolutionary Humanism is necessarily unitary instead of dualistic, affirming the unity of mind and body; universal instead of particularistic, affirming the continuity of man with the rest of life, and of life with the rest of the universe; naturalistic instead of supernaturalistic, affirming the unity of the spiritual and the material; and global instead of divisive, affirming the unity of all mankind."

[120]Muller's (1960) 'Guidance of Human Evolution' is an extremely well-developed paper that deals with all of the essential issues related to the steering of humankind's future course.

[121]Cattell's (1972) "A New Morality from Science: Beyondism" pursues in great detail the development of a new moral value system derived from (evolutionary) science, calling for the cultural and genetic progress of man to be equal, and paying equal importance to individual self-realisation, societal development and intergenerational change.

[122]Waddington (1960, 69).

[123]See also Bostrom (2005), Naam (2005), Harris (2007, 16).

[124]Bacon (1626).

into scientific discoveries and their applications in technology and societal organisation for improving our earthly welfare and well-being.[125]

From Homo sapiens sapiens to Homo sapientior? The authors define human-specific potentialities that should be further developed as characteristics that are: (1) specific for the hominisation process; and (2) conducive to further progressing modernisation. They distinguish them from human potentialities in general, which cover all human capacities made possible by the human genome at the individual level, and the human gene pool at the population or species level. Not all human potentials that developed during the hominisation process merit further promotion. For the future, it is valuable to have the human-specific cognitive and emotional capacities and the human-specific potentialities toward high levels of sociability. Potentialities associated with strong drives for inter-group competition and conflicts in the context of the novel environment of modernity are maladaptive remnants of the evolutionary past that do not merit promotion.

Human-specific features find their highest expression in the singularity of the human brain, which is responsible for human's high cognitive performances, refined emotional life and strong sociability. The active advancement of the hominisation process may need to correct our current genetically or environmentally caused weaknesses. Intervention may need to be directed towards preventing the spread of evolutionary regressive phenomena, and to lead future human evolution in the direction of an improvement in human-specific features.[126] They encapsulate cognitive abilities (including biological instruments of communication, memory, information-processing, reasoning), emotional personality characteristics that facilitate sociability, altruism, reciprocity, mutualism and cooperation, and other desirable human attributes such as creativity, health, immunity to diseases, longevity, physical vigour (speed, strength, endurance), sexual arousal and orgasm, euphoria, and physical attractiveness.

In the realm of mental powers, Hermann Muller[127] specified more profound analytic abilities, multi-dimensional thinking, more creative imagination, and the development of new mental faculties such as telepathy. In the physical realm he added qualifications such as reduction in the need for sleep, better management of the effects of sedation and stimulation, and a general increase in physical tolerance and aptitudes.

From a historical point of view two major evolutionary enhancement movements can be distinguished, namely the eugenics movement and the transhumanist movement. In the writings of some of the earlier proponents of eugenics, a prefiguration of some of the transhumanist aspirations can be found.[128] Indeed, in 1939

[125]Hughes (2012, 757).

[126]See, for instance, Huxley (1957), Muller (1960), Cattell (1972), Ettinger (1972), FM-2030 (1973), Parens (1998), Thornhill (1998), Grammer et al. (2003), Bostrom (2004), Hughes (2004), Glannon (2007), Harris (2007; 2009), Furnham and Swami (2008), Jordan (2008), Savulescu et al; (2011), Buchanan (2011).

[127]Muller (1960).

[128]For instance, Haldane (1924), Muller (1935), Huxley (1957).

in the journal Nature the joint statement published by a group of prominent American and British biologists, called *Social Biology and Population Improvement*, widely referred to as *The Eugenic Manifesto*,[129] is in its objectives and wording rather modern and in many respects still valuable today.

The Eugenics Movement. The eugenics movement emerged in the wake of Darwin's evolutionary theory and builds on the explanation of the evolutionary mechanism. Its aims were usually articulated as the improvement of the physical, mental and social capacities of the human species.[130] It can be argued that the essence of the movement boils down to the furtherance of the hominisation process. As such, the eugenic movement is not fundamentally different from some aspects of the more recent transhumanist movement. The more specific aims and methods of eugenics are extensively discussed in Chap. 8, Sect. 8.3.3.3.

The Transhumanist Movement. Recent advances in various biomedical sciences (e.g. molecular genetics, medically assisted reproduction, (psycho)pharmacology, cosmetic surgery) have elicited an abundant—positive as well as negative—scientific literature on bioethics and biophilosophy.[131] They also address the biological future of humankind,[132] dealing with the prospects for influencing the future ontogenetic development of humans and steering the future evolution of *Homo sapiens sapiens*. In particular, the World Transhumanist Association[133] and its adherents have advocated the idea of enhancing human intellectual, physical and psychological capacities, more specifically by the means of modern technologies.[134] Meanwhile, in the last two decades Transhumanism has evolved into a number of subdivisions such as extropian libertarianism, liberal democratic World Transhumanist Association/Humanity+,

[129]Crew et al. (1939). [This document is reproduced as Appendix 1 in Glad (2006, 108–112)].

[130]See, for instance, Galton (1883), Pearson (1909), Crew et al. (1939), Osborn (1940), Sutter (1950), Blacker (1952), Huxley (1962), Bajema (1976), Lynn (2001).

[131]For instance, Ramsey (1970), Fletcher (1974), Kieffer (1979), Reiss and Straughan (1996), Walters and Palmer (1997), Mataré (1999), Skene and Thompson (2001), Kass (2002), Holland (2003), President's Council on Bioethics (2003), Engelhardt (2006), Zycinski (2006), Glannon (2007), Sandel (2007), Lindsay (2008).

[132]For instance, Roslansky (1966), Graham (1970), Ettinger (1972), Howard and Rifkin (1977), McFaul (1978), Diamond (1992), Leakey and Lewin (1995), Kitcher (1996), Silver (1999), Joy (2000), Glannon (2001), Ward (2001), Fukuyama (2002), Mulhall (2002), Stock (2002), Habermas (2003), Kilner and Mitchell (2003), Mehlman (2003), Baillie and Casey (2004), Bostrom (2004; 2009), Hughes (2004), Gerdes (2006), Glad (2006), Garreau (2006), Stewart (2008), Agar (2010), de Duve and Patterson (2010), Gordijn and Chadwick (2010), Savulescu et al. (2011), Palme (2012), Persson and Savulescu (2012), Hurlbut and Tirosh-Samuelson (2016).

[133]The World Transhumanist Association was founded in early 1998 by Nick Bostrom and David Pearce (Bostrom 2005; www.Transhumanism.org).

[134]Bostrom (2003; 2014), Hughes (2004), More and Vita-More (2013); see also the special issue of the Journal of Medicine and Philosophy (2010, 35, 6: 617–720) that provides an overview of the origin, the nature, and the aims of the transhumanist movement as well as a critical assessment.

Singularitarian millennialism, religious transhumanism,[135] and radical democratic transhumanism or technoprogressivism.[136]

Transhumanism, a term coined by Julian Huxley[137] in 1957 and re-introduced by Max More[138] in 1990, aims to improve existing human capacities or to create new capacities.[139] It endeavours to reach human stages characterised by higher-than-current intellectual heights; resistance to disease; increased longevity[140]; unlimited youth and vigour; increased capacity for pleasure, love,[141] artistic appreciation, and serenity; experience of novel states of consciousness; and heightened capacity for sociability,[142] etc. The prefix 'trans' in transhumanist implies that humans should transcend their present-day capabilities.[143] Transhumanists understand by enhancement: improvements in the current performances of the human organism or creation of new abilities or faculties as, for instance, suggested by Hermann Muller in the early 1960s.[144]

Hence, the ideas of the transhumanists relate both to the ontogenetic and phylogenetic enhancement of the human species. For the moment this section will only dwell upon their ideas and strategies about the (phylo)genetic improvement of humanity.

One of the merits of the Transhumanism Movement is that it distinguishes two stages in the future evolution of humankind: the transhuman stage and the posthuman stage. The enhancement of specific human qualities which transhumanists want to achieve does not just imply the promotion of variants that lie at the higher tail-end of their present frequency distributions—the traditional eugenic goal. The transhumanists want more. They also want to promote the physical, intellectual, emotional and social abilities to such an extent that they come to lie outside the upper margin of the present variation, and greatly exceed the mental,

[135] Although transhumanism is inherently secular, some authors (for instance, Tirosh-Samuelson 2012, 721) are of the opinion that this ideology shares several features with traditional religions such as "the pursuit of perfection and focus on human improvement; the concern for the betterment of society by eliminating social ills such as poverty, sickness, and suffering; the progressive understanding of human history that sees the future as necessarily better than the past; and the preoccupation with transcendence". This is undoubtedly true, and there is no reason why religious people would not be able to share the ideals of the transhumanist movement (see, for instance, Garner 2005). Nevertheless, this does not imply that transhumanism is a religion the essence of which is, in the authors' view, the belief in a supernatural being who created us, cares about us, imposes moral rules, and judges us.

[136] Hughes (2012, 757).

[137] Huxley (1957, 17).

[138] More (1990; 2010).

[139] Buchanan (2011, 23).

[140] The idea of pursuing substantial human enhancement and longevity is not an entirely new proposition. According to Stambler (2010), it can be traced back to several authors at the end of the nineteenth/early twentieth centuries.

[141] For instance, Earp et al. (2015), Giubilini (2015).

[142] Increasing sociability implies that some human potentialities, for instance, in the domains of violence and aggression, should be reduced instead of enhanced (e.g. Bostrom 2003).

[143] Wood (2006, 109), Naam (2005, 5), Harris (2007, 44).

[144] Bostrom and Roache (2008).

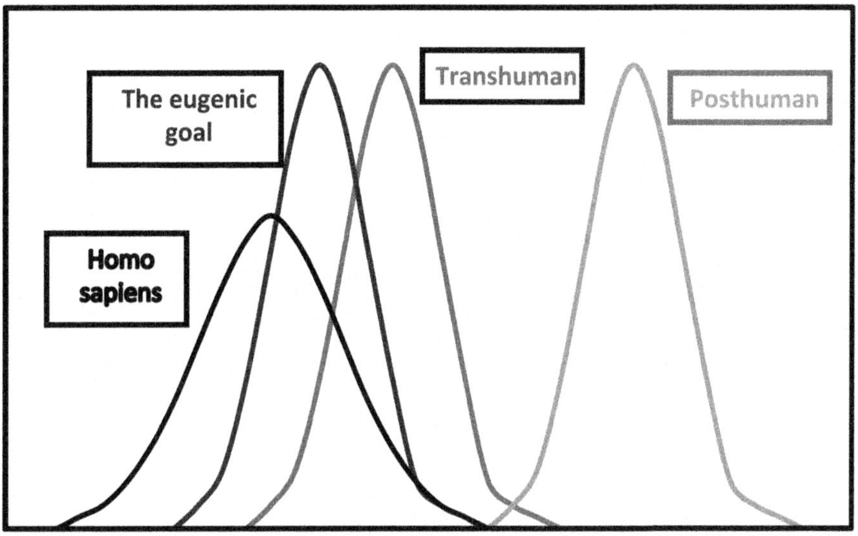

Fig. 5.2 Frequency distribution of a hypothetical human variable in the present, the transhumanist and the posthumanist stages

physical and social capacities of the present *Homo sapiens sapiens*[145] (Fig. 5.2). The latter goal ultimately aims to achieve a posthuman or overhuman[146] stage, the *Homo sapientior* as Paul Overhage[147] named it.

Homo sapientior or the posthuman stage could, in principle, be seen as the successor to the present *Homo sapiens sapiens,* just as *Homo sapiens sapiens* was the successor of *Homo erectus* in the hominin phyletic line. The transhuman may be seen as an intermediate form between the present human and the posthuman, and would form the evolutionary link to the coming era of posthumanity.[148]

On the basis of the exponential velocity of science, some futurologists are of the view that the transition from the present human to the transhuman and to the posthuman might be imminent. This idea is well expressed in the title of Ray Kurzweil's book *The Singularity is Near.*[149] He argues that contrary to the earlier hominin transitions (for instance, from *Homo erectus* to *Homo sapiens sapiens*) which took many thousands of generations, the future transition might take only a few generations.[150]

[145]Bostrom (2008).

[146]Present-day transhumanists have divergent views about the degree to which Nietzsche's (1883; 1901) concept of the 'overhuman' ('Übermensch') corresponds to the 'posthuman' ideal of the Transhumanists. For instance, Sorgner (2009) sees strong similarities between those two concepts, but Bostrom (2005) claims that Nietzsche cannot be seen as an originator of transhumanism.

[147]Overhage (1977, 228) coined the term, but as a theologian he was against pursuing a more advanced hominin.

[148]FM2030 (1973), Sorgner (2009).

[149]'Singularity'—a concept first used by the famous mathematician John von Neumann (1950)—is a period of such fast and profound technological change that it will irreversibly transform human life (Kurzweil 2005, 7); See also Vinge (1993), de Garis (2005).

[150]Kurzweil (2005), Rees (2003).

Transhumanism has rightly been viewed as an outgrowth of the Enlightenment and secular humanism.[151] For instance, James Hughes[152] calls the transhumanist movement a modern form of Enlightenment techno-utopianism. It seeks to improve the condition and performance of humans by differential reproduction, and cultural means, such as education and cultural refinement, but also by using high-tech biomedical interventions to transcend the human's natural attributes in health, life-span, and physical and mental performances.[153]

Transhumanists strongly promote individual rights and liberties in making decisions about whether to reproduce, how to reproduce, and which technological methods to use in reproduction. They condemn coercion and fiercely reject racialist and classist approaches. Transhumanists are of the view that human enhancement technologies should be within the reach of everyone, and that individuals should have, on the basis of informed consent,[154] the right and freedom to apply these technologies to themselves (morphological freedom)[155] and to use them for having children (reproductive freedom).[156] In the mind of many transhumanists, individual

[151]Tirosh-Samuelson (2012), Beland and Patenaude (2013).

[152]Hughes (2012, 757).

[153]Bostrom (2003; 2005; 2007), Hughes (2004, 156).

[154]Gunderson (2008, 86).

[155]Morphological freedom: the fundamental right of people to decide for themselves the constitution of their physical make-up (Sandberg 2001).

[156]The Transhumanist Declaration of the World Transhumanist Association (http://www.transhuamism.org, 1998–2009) was originally crafted in 1998. It been modified over the years. It was adopted by the Humanity+Board in March 2009 (http://humanityplus.org/learn/transhumanist-declaration/): "Humanity stands to be profoundly affected by science and technology in the future. We envision the possibility of broadening human potential by overcoming aging, cognitive shortcomings, involuntary suffering, and our confinement to planet Earth. We believe that humanity's potential is still mostly unrealized. There are possible scenarios that lead to wonderful and exceedingly worthwhile enhanced human conditions.

1. We recognize that humanity faces serious risks, especially from the misuse of new technologies. There are possible realistic scenarios that lead to the loss of most, or even all, of what we hold valuable. Some of these scenarios are drastic, others are subtle. Although all progress is change, not all change is progress.
2. Research effort needs to be invested into understanding these prospects. We need to carefully deliberate how best to reduce risks and expedite beneficial applications. We also need forums where people can constructively discuss what should be done and a social order where responsible decisions can be implemented.
3. Reduction of existential risks and development of means for the preservation of life and health, the alleviation of grave suffering, and the improvement of human foresight and wisdom should be pursued as urgent priorities, and heavily funded.
4. Policy making ought to be guided by responsible and inclusive moral vision, taking seriously both opportunities and risks, respecting autonomy and individual rights, and showing solidarity with and concern for the interests and dignity of all people around the globe. We must also consider our moral responsibilities towards generations that will exist in the future.
5. We advocate the well-being of all sentience, including humans, non-human animals, and any future artificial intellects, modified life forms, or other intelligences to which technological and scientific advance may give rise.
6. We favour allowing individuals wide personal choice over how they enable their lives. This includes use of techniques that may be developed to assist memory, concentration, and mental

freedom is considered to be of paramount importance for deciding about their own body and the characteristics of their children.[157]

Transhumanists are well aware that the future technological manipulation of genes should be undertaken with great care and gradually in order to avoid adverse side-effects, at the individual as well as at the population or societal level. They also realise that currently broad social support does not exist for such a venture. This has always been the case with innovative ideas, as the recent history of major social or moral movements such as shown by the pristine Christian movement, the early labour movement, the first feminist wave, the ecological movement, the family planning movement, and the right to euthanasia movement. Since humanity currently faces severe current challenges, such as overpopulation, socio-economic imbalances and inequalities, environmental pollution, depletion of natural resources, and reduction of biodiversity, it is understandable that longer-term concerns are not perceived as an urgent priority. Nevertheless, it is also important that current challenges are addressed from the perspective of a longer-term vision and objective.[158]

The narrative about the freedom of individual rights and choices concerning themselves and their offspring may be appealing at first sight. However, they impose a number of ethical concerns. From an evolutionary ethical point of view, which would aim at the furtherance of the hominisation process and progressing modernisation, there can be no absolute morphological and reproductive freedom without responsibilities.

A distinction should be made between the freedom of adults, able to take up responsibility for their choices about their own morphological freedom, and their freedom to make choices for their offspring. Able and informed adults can have morphological freedom with regard to their own features (= morphological freedom/rights), on the condition that their choices do not harm others or induce unfavourable effects on society (= morphological responsibility). Able and informed adults can have the choice to apply or not apply enhancing procedures for their children (= reproductive freedom/rights). However, choosing whatever feature they want to impose upon their (future) children needs to entail reproductive responsibility. Here, their choices need be embedded in new moral standards that are supported by socially responsible institutions.

Stefan L. Sorgner,[159] in his discussion of a deaf couple's wish to have a deaf child, argues that this couple makes use of their right to procreative freedom, that a deaf person can lead a good life, and that a culturally accepted subculture shares the concept in question. According to Sorgner there is nothing wrong with such a wish

energy; life extension therapies; reproductive choice technologies; cryonics procedures; and many other possible human modification and enhancement technologies."

[157]For instance, Naam (2005, 7), Young (2006, 67).

[158]Muller (1935), Elgin (1993), Muhlhall (2002), Stock (2002), Hughes (2004), Glad (2006), Dvorsky (2008), Stewart (2008), Savulescu and Bostrom (2009), Persson and Savulescu (2012).

[159]Sorgner (2010); see also Sandel (2007, 1), Glannon (2007, 60) or Scully (2008, 60) about a lesbian deaf couple who wanted (and got) two children through artificial insemination from a donor who also had a heritable form of deafness.

and no one is harmed. On the other hand, one can argue that hearing is a useful faculty of the human species that was not only of crucial importance in the EEA but is still very useful for social life and for functioning effectively in modern society. Human culture, and in particular modernity, has created a multitude of rich cultural and technological products—music, theatre, opera, cinema, TV, radio and several other listening devices—which can only fully be enjoyed by means of a hearing capacity. Consciously choosing to deprive future children from this potentiality could be considered unethical and even cruel.[160]

In designing enhancement strategies/policies, the rights of children to optimal conditions and perspectives for developing human-specific potentialities should be embedded in the new ethics and be given priority over the wishes of self-centred parents.[161]

As Francis Crick[162] correctly emphasised:

> If we can get across to people that the idea that their children are not entirely their own business and that it is not a private matter, it would be an enormous step forward.

In principle, there are two major strategies through which the genetic composition of the human population could be enhanced in the direction of a progressing hominisation: eugenic engineering and behavioural changes in (differential) reproduction, although in some interventions both approaches will be involved (see Chap. 8, Sect. 8.3.3.3). However, transhumanists seem to be more interested in the (future) technological manipulation of genes (eugenic engineering—the new eugenics) than in traditional behavioural differentiation of reproduction (sometimes called the old eugenics). This is comprehensible in view of their goal to transcend the genetic capabilities of the present *Homo sapiens sapiens* beyond the present natural human range. Changes in reproductive behaviour can only affect the distribution of genetic variants within the present genetic range, whilst biotechnological innovations might exceed that range.

Indeed, the transhumanists' idea is to change, by means of biotechnological interventions, the genes of our future children and to create so-called designer babies. Thus, future generations would savour genetic dispositions which allow them to reach higher than present levels of health, physical attractiveness, libido, artistic experience, sociability, longevity, and physical and intellectual performance in general.[163]

In this respect, Julian Savulescu[164] defends the principle of procreative beneficence, by which parents-to-be would have the moral obligation to select that child who can be expected to have the best life, not only by avoiding the reproduction of disease genes but also by choosing genes that enhance the child's prospects for

[160]See also relevant counterarguments against consciously imposed deprivations in Hughes (2004, 140) and Singer (2009, 278).
[161]Feinberg (1992, 77).
[162]Crick, quoted in Cattell (1972, 349).
[163]Bostrom (2003, 2008), Steinbock (2008).
[164]Savulescu (2001); see also Glannon (2001, 76).

normal (= non-disease linked) life opportunities and performances in matters of health, intelligence, emotional satisfaction and sociability.[165]

From the available transhumanist literature, the impression is gained that many transhumanists are enthusiastic about the progress to be expected from future technological innovations, and want to avoid any association with the dysgenism of the twentieth century authoritarian ideologies, in particular Nazism.[166] However, they neglect the possible eugenic measures and procedures to improve the genetic composition and structure of the present human population(s) by means of differential reproduction, i.e. a differential by which carriers of genes considered as favourable would produce a more numerous offspring than others.

This traditional eugenic strategy, as may be seen in Fig. 5.2, would not exceed the upper bound of the present genetic variation. Nevertheless, it would move the distribution of desirable traits in the domains of health, cognitive abilities and sociability: the population average would be increased and the genetic variance would be reduced, particularly at the lower end of the existing variation. Of course, this strategy raises the ethical questions of reproductive rights and responsibilities of individuals and societies. However, the preoccupation with the perspective of future technological inventions and innovations should not make us blind to delicate and difficult decision-making about present-day opportunities and necessities. (This matter is discussed more extensively in Chap. 8, Sect. 8.3.3.4).

Some authors are of the view that the individual freedom in reproductive choices should be restricted in case of possible unfavourable societal effects.[167] The question of possible unfavourable effects of individual reproductive choices at the population or societal level has to be carefully evaluated. Insufficiently thought considerations or short-term perspectives might easily lead to restrictive, conservative measures that are contrary to the longer-term progressive hominisation goals.

A particularly interesting, but highly disputed question is whether the transhumanist enhancement should also be applicable to moral predispositions or behaviours.[168] The specificity of the transhumanist approach is that, in addition to the traditional educational and training efforts to develop moral behaviour, modern biomedical means would also be used to influence or change the underlying biology or genetics of moral behaviour, not only in the case of extreme socio- or psychopathologies but also in general.[169] Some of the transhumanist advocates of moral enhancement are even of the view that this type of enhancement should be a

[165]Harris (2007, 19) went even a step further in claiming that 'enhancement is a moral duty'. See also in this respect to the discussion in Brassington (2010).

[166]See Lynn (2001), Cliquet (2010, 535–537).

[167]For instance, Buchanan et al. (2001, 183), Harris (2007, 74), Gyngell and Douglas (2016, 249).

[168]Hopkins (2008, 4), Persson and Savulescu (2010), Savulescu et al. (2011), Persson (2012), Tennison (2012), Eberl (2014), Rakić (2014), de Melo-Martin and Salles (2015), Douglas (2015), Hughes (2015), Murphy (2015), Tonkens (2015), Harris (2016). See also the discussion forum in the 2010 volume of Politics and the Life Sciences (Bucy 2010) about Walker's (2009) proposal for developing a 'Genetic Virtue Program (GVP)', and the open peer commentaries on Shook's (2012) target article in AJOB Neuroscience.

[169]For instance, Douglas (2008), Persson and Savulescu (2008; 2012), Kabasenche (2012), Shook (2012), Gyngell and Easteal (2015).

precondition for applying enhancements in other domains, particularly in the domain of cognitive capacities.[170]

Moral enhancement advocates, such as Mark Walker,[171] are fully aware of the conceptual, scientific (and moral) difficulties of endeavours to induce genetic changes and improvements on personality characteristics which may change (im)moral behaviour, as formulated by the forum discussants in the 2010 volume of Politics and the Life Sciences.[172] Advocates of moral enhancement usually emphasise the need for a general, population-focused approach,[173] a holistic, multi-virtues approach,[174] or a problem-focused cost-benefit-analytic approach.[175]

Many authors oppose the idea of biotechnologically induced moral enhancement, either because of the insufficiently thought out implications, or because of philosophical or theological objections.[176] It is interesting that many opponents of genetic enhancements of moral behaviour do not object to moral improvements through socialisation.

There are some important differences between (genetic) enhancements targeted at physical and mental health characteristics which require further reflection in the context of ethical challenges of modernity and transhumanist aspirations. Individuals choosing to enhance their athletic ability or increased intelligence may do so primarily for their own benefit. Choices to enhance personality characteristics which are related to (im)moral behaviour affect relations with others more profoundly.

Bioconservative Counter-Arguments about Human Enhancement. Biocon-servatives[177] oppose the use of technology to modify human nature, and are particularly against enhancements that involve germline interventions[178] in human beings.[179] However, strong opposition not only exists regarding genetic enhancements but moral indignation also often arises when ontogenetic interventions are suggested, not for therapeutic reasons but to reach higher levels of 'normal' physical, emotional or cognitive capacities.[180] Hence, bioconservative opposition to enhancement concerns genetic features as well as (some) phenotypic characteristics.

---

170 Persson and Savulescu (2008; 2012), Rakić (2015).

171 Walker (2009, 2010); see also Hughes (2015), Murphy (2015).

172 Andreadis (2010), Agar (2010), Arnhart (2010), Blackford (2010), Bronstein (2010), Sprinkle (2010).

173 For instance, Murphy (2014).

174 For instance, Hughes (2015).

175 For instance, Carter and Gordon (2015).

176 For instance, Jotterand (2014), Jebari (2012), Beck (2015), Carter and Gordon (2015), Douglas (2014), Tonkens (2015), Wiseman (2016).

177 Bioconservatism is a modern movement which opposes specific or general technological development and emerging technologies (e.g. Hughes 2004). Opposing bioconservative positions to transhumanist goals is a somewhat too simplistic classification. In fact three main positions can be distinguished: permissive, restrictive and conservative positions (see, for instance, Giubilini and Sanyal 2015).

178 Germline interventions: biotechnologically induced changes in the molecular structure of genes in reproductive cells, so that the change can be transmitted intergenerationally.

179 For instance, Habermas (2001), Annas et al. (2002), Fukuyama (2002), Kass (2002).

180 Sandel (2007, 8, 12).

Bioconservatives are found among religious apologists and many religious believers[181] who think that God is the engineer of humanity and that it is preposterous to try to replace him. Bioconservativism is also supported by some secular allies.[182] Some leftists[183] fear that technological enhancements will make society more unequal. Fundamentalist feminists[184] are suspicious of the patriarchal appropriation of the female body. Deep environmentalists[185] think that human beings are not smart or wise enough to be able to improve nature.[186] Several authors[187] have extensively refuted the often superficial and badly substantiated bioconservative objections. This book will briefly discuss the major bioconservative counter-arguments from the pertinent literature and propose some ways forward.

Many objections have been formulated against transhumanism or some of its propositions or methods. Some objections pertain to particular aspects of the transhumanist goals or the specific proposals of some transhumanist proponents. This is not surprising given the diversity in transhumanist approaches that already exists, or the large number of its productive and sometimes imaginative adherents. It is beyond the scope of this book to deal with all of the criticisms that ever have been advanced about human enhancement. The discussion will concentrate on what the authors perceive to be the major general objections to the transhumanist paradigm, namely human enhancement technologies that: (1) undermine human dignity and dehumanise humanity; (2) challenge the wisdom of nature and pervert the God-given natural order; (3) damage the human species; (4) increase inequalities; and (5) provoke conflicts between humans and posthumans.

Human Dignity/Human Nature. Regarding the bioconservative objection[188] about the subversion of human dignity,[189] there is lack of robust arguments to demonstrate how the enhancement of specific human abilities could undermine human dignity or have a dehumanising effect. Apart from the fact that the use of the concept of dignity—also in international charters on bioethics[190]—is very vague and weak, the authors think that biotechnical interventions to ameliorate human potentials mean exactly the opposite—at least that is the purpose of enhancement goals. In contrast to what the anti-meliorists fear, improving human-specific features will, by definition, also enhance human dignity and increase the humane

[181]For instance, Council for Biotechnology Policy (2002), Kilner and Mitchell (2003), Colson and Cameron (2004).
[182]For instance, Fukuyama (2002), McKibben (2003).
[183]For instance, Rifkin (1983), Winner (1989).
[184]For instance, O'Brien (1981), Corea (1986), Raymond (1994).
[185]For instance, Kimbrell (1998).
[186]See the extensive discussion in Hughes (2004, 107–153).
[187]See, for instance, Hughes (2004), Harris (2007), Jackson (2008), Bostrom and Savulescu (2009), Buchanan (2011), Temkin (2013), Murphy (2014).
[188]For instance, Lewis (1943, 2001), Giesen (2004), de Rus (2006).
[189]Dignity, etymologically derived from the Latin term 'dignitas' (= worthiness, elevation, excellence, distinction) is a concept that is usually understood to be that a human being has an innate right to respect and ethical treatment. For a modern 'naturalistic' view of human dignity see, for instance, McClelland (2011).
[190]For instance, the Universal Declaration on Bioethics and Human Rights by UNESCO (2005) http://unesdoc.unesco.org/images/0014/001461/146180e.pdf.

nature of our species.[191] Also, the bioconservative reproach that transhumanist endeavours are an expression of human vanity[192] is ill-placed. Attempts to enhance one's aptitudes in domains such as health, physical and mental performances, libido, and longevity have little to do with vanity or *hubris*, but are directed towards better functioning, well-being and happiness.[193] Even strivings for enhancing features such as physical attractiveness and libido are, from an evolutionary point of view, instrumental in mating behaviour and ultimately in reproduction.[194]

Bioconservatives often cherish an idealistic biological folk view of human nature.[195] As a rule it is based on a lopsided intuitionism or theological doctrine. The concerns that bioconservatives raise about human dignity mainly originate from religious beliefs about the divine creation of humanity and are reflections of a conceited view of ourselves. A more modest attitude would be appropriate, considering the trail of humane and ecological disasters that the present human species has marked in its history, as well as the accelerated destructive way in which it is currently managing this planet. Overpopulation, overconsumption, environmental pollution, depletion of natural resources, and decimation of the planet's biodiversity and volume have yet to be addressed by bioconservatives. The present human species can pride itself on many merits, but is still far from a résumé that guarantees a long-term survival and further evolvement toward higher stages of biological evolution and cultural development as anticipated by Duane Elgin[196] in his book *Awakening Earth: Exploring the Evolution of Human Culture and Consciousness.*

There are indeed risks of producing unfavourable side effects when technically intervening and changing essential features of humanity which are part of evolved complex interdependencies between large numbers of features. Enhanced features of any nature will require adaptations for many other morphological, physiological and psychological characteristics. For example, it is known how our present cognitive capabilities are related to brain size, energy expenditure, body height, pelvis size, pregnancy duration, emotional personality characteristics, sociability, etc. Such complex interdependencies will have to be cautiously taken into consideration in enhancement efforts, which will probably have to evolve along paths of gradualness in order to avoid undesired side-effects or outcomes.[197] Thus, some aspects related to enhancement are predominantly of a technical nature.

Advocates of human enhancement are usually very well aware of the risks and dangers of (onto)genetic interventions such as accidents, imbalance of complex interdependencies, decrease of genetic variability, social misuse or political abuse, and commercial exploitation.[198] Hence, it is incorrect when the meliorist

[191]See also Bostrom (2005), Murphy (2014), Chan (2015).
[192]For instance, Habermas (2003), Sandel (2004, 2007).
[193]For instance, Caplan (2006, 35).
[194]Grammer et al. (2003).
[195]Mameli and Bateson (2006), Linquist et al. (2011).
[196]Elgin (1993).
[197]See also the argumentation of Glannon (2001, 39), Agar (2004; 2010) and Baylis and Robert (2004) against radical enhancement and in favour of moderate enhancements and the need to proceed with caution in this endeavour.
[198]For instance, Glover (1984), Hughes (2004), Buchanan (2011).

opponents[199] evoke the doomsday scenario of Aldous Huxley's[200] dystopic *Brave New World* to use it in arguments against the transhumanists' enhancement goals. What the transhumanists want to achieve is exactly the opposite to the *Brave New World.* Rebecca Roache and Steve Clarke[201] document that many transhumanists are in fact biomoderates rather than bioliberals.[202]

The objections of the bioconservatives against technical interventions on human nature, resulting in alterations which would repair or enhance the humane character of the human species,[203] are also rather superficially thought out. The choice of non-interference with natural processes, such as mutation and natural selection, would mean that we would have to abandon the achievements of biomedicine and even many environmental and social interventions which have resulted in heightening the quality of life and happiness.[204] We would need to return to 'natural' levels of morbidity and mortality. However, it must be admitted that most bioconservatives do not object to therapeutic interventions, but are only opposed to enhancement goals. However, it is not always clear how they distinguish between the two practices and objectives.

A pertinent question advanced by bioconservatives is whether the future transhuman, and particularly the future posthuman, will still be a human. From an evolutionary point of view, it can be argued that the posthuman will be a hominin, but a higher evolved hominin than the present one. He will represent a more advanced stage in the hominisation process, with characteristics that surpass the ones seen in our cognitively, socially, artistically or physically most distinguished current conspecifics.

Bioconservatives often also evoke objections to enhancements because of the fear of loss of freedom and autonomy.[205] Assuming that the present non-enhanced human has an unlimited freedom and autonomy in moral choice, they choose to ignore that the present human condition, with its current limited degree of freedom and autonomy, is the result of our phylogenetic heritage which has been moulded by natural selection in the EEA. Here, the religious secular divide acquires a central position in ethical choices regarding the future of the human species. Religious believers humbly glorify the God-given blessings of life. However, for non-believers freedom and autonomy are not a priori being threatened by enhancement efforts. They argue that religious-based refusal of human enhancement stems from a romanticised, but unrealistic, view of the 'unbidden giftedness' of human nature, which ignores the resulting hazardous genetic gambling in the natural lottery. Non-believers argue that genetic enhancements will contribute to

[199]For instance, Kass (2002, 5), Fukuyama (2002, 3ff), Mehlman (2003).

[200]Huxley (1932).

[201]Roache and Clarke (2009, 2).

[202]Roache and Clarke define bioliberals as people "who don't view enhancement as unusually risky, and believe that enhancement should generally be permitted", whereas biomoderates would be "more aware of the potentially undesirable consequences of particular enhancements as reason to restrict their use".

[203]For instance, Kass (2002, 22).

[204]For instance, Glover (1984).

[205]Mehlman (2003), Sandel (2007), Blackford (2010).

amplify choices, enrich lives, and further liberate the human from the armour of natural constraints. In fact, transhumanists aim for genetic enhancements which are the extension of the social, educational and current medical efforts that modern societies make in order to improve the living conditions and well-being of their citizens.

Genetic enhancement is also often objected to because children would not have the freedom/autonomy/self-determination to decide upon their enhanced features.[206] This is, of course, true. However, do children now have the right to choose their 'natural' predispositions, which are largely the result of chance combinations of genes present in the reproductive community from which they emerge? It is clear that they do not. Children do not have the possibility to consent to their genetic endowment, which shapes their physical characteristics and hereditary health issues and may include physical and mental disabilities. If children did have a choice it is not difficult to imagine that they would choose to be born in good health.

Wisdom of Nature/Playing God. The possibilities for enhancing human potentials, particularly by replacing the forces of mutation and natural selection by more humane procedures in order to avoid the reproduction or multiplication of unfavourable characteristics or improve human-specific traits, are certainly challenging the ethics that builds on the wisdom of nature. They are, no doubt, shaking up the notion of God-given natural order.[207]

Religious believers consider the human species to be the result of a divine creation. They believe that life is sacred. The principle of the sanctity of life implies that human life starts at conception and does not end with death. Consequently, interventions such as contraception, induced abortion, eugenics, and euthanasia are usually rejected. Intervening in human life, and more particularly in the creation of life, is considered a form of *autopoiesis* (self-creation). Playing God is considered to be the supreme form of *hubris* (haughtiness, arrogance) and a sign of disrespect for our supposed creator.[208] Efforts of self-creation and self-expression are considered as non-natural rights.[209]

Scientifically based interventions to push humanity to higher levels of consciousness and performance are, indeed, tantamount to humanity Playing God.[210] Such interventions are based on the principle of quality of life, which defines human life in terms of psychological, cultural, intellectual, moral, and relational indicators of personality.[211] Here, the choice to intervene in life is not assessed on the basis of the sense of purpose given at conception and life after death, but rather on the basis of the extent to which intervention promotes quality of life.

Evolutionary scientists are usually impressed by the miraculous effects of the evolutionary mechanisms (mutation, natural selection) on the evolution of life and in particular the hominisation. However, they also acknowledge the roughness,

[206]For instance, Habermas (2001), Mehlman (2003, 81).
[207]Peters (2002), Council for Biotechnology Policy (2002), Colson and Cameron (2004).
[208]For instance, Overhage (1977), Knapp (1989), Peters (2002), Zycinski (2006).
[209]Kass (2002, 226).
[210]Campbell and Walker (2005), Harris (2007, 35).
[211]For instance, McFaul (1978), Holland (2003).

cruelty and defectiveness of the evolutionary system, which is based on chance events. It is proceeding at a very slow tempo, using and wasting enormous amounts of energy, and resulting in far-from-perfect adaptations, senescent degeneration, morbidity and mortality.[212] If life was planned and designed by a supernatural power, it is a bad and wasteful planner and designer. Allan Buchanan[213] picturesquely suggests that the evolutionary mechanism is the tool of a tinkerer, not of a Master Engineer.

When rethinking about the natural evolutionary mechanism there is reason to suppose that the human enhancement aimed at intentional genetic modification may be much more efficient, time- and energy-saving, and especially more humane in procedure and result. It could be more effective at promoting reproductive fitness and human well-being and quality of life up to an old age. As Christian de Duve[214] rightly noted:

> Natural selection has not privileged the foresight and wisdom needed for sacrificing immediate benefits for the sake of the future.

Some authors argue that the religiously inspired opposition to the enhancement aspirations of the eugenics and transhumanism movements is quite amazing, if not outright odd. Transcendent religions traditionally tried to surpass sickness and death, and even often devised ingenious narratives about the eternal prolongation of an ideal life after death, freed from the earthly miseries, or some other form of current-life transcendence. In several respects, transcendent religions and transhumanist goals are quite identical in aspirations, but strongly different in methodologies and the take up of responsibility.[215] In fact, as Steve Fuller[216] pertinently points out, an enhanced humanity would bring us closer to the divine aspirations of many religions.

The genetic enhancements goals of the eugenic and transhumanist movements lie not only in the extension of many efforts in non-genetic domains, but also constitute more consciously pursued influences in the domain of human genetics, which humans have exerted since time immemorial.[217]

Incompetence and Maleficence of Humankind. There are complementary beliefs or fears about human's incompetence and maleficence which are related to the bioconservative objections concerning the wisdom of nature or the magnificence of its supernatural creator. The historical record on human-caused social and ecological disasters provide the bioconservatives with strong arguments against the development of innovative technologies, such as genetic engineering, robotics and nanotechnology, which are considered to be too invasive and dangerous.[218]

[212]See e.g. Darwin (1871), Huxley (1894), Galton (1883), de Duve (2009).
[213]Buchanan (2011, 184).
[214]De Duve (2011, 148).
[215]See also Hopkins (2005), Jordan (2008).
[216]Fuller (2011, 209).
[217]Murphy (2014, 338).
[218]Joy (2000), McKibben (2003).

Advocates of enhancement technologies are well aware of the dangers and entrapments of scientific inventions and applications that can be used for good or evil.[219] Indeed, some of the newest technologies, for instance, in the field of genetic engineering, might create deadly synthetic pathogens against which no defence is available or could be developed.[220] In the domain of nanotechnology, the erroneous production of out-of-control irreversibly self-replicating nanobots leading to an end-of-the-world 'grey goo' scenario is also mentioned.[221] Some go as far as to suggest the relinquishment of research areas such as biotechnology and nanotechnology.[222] The authors, together with many researchers, argue that it would be much wiser, as precautionary measures, to develop and enforce strict ethical guidelines for the future development of such fields.[223]

Should planes have been abolished after the 9/11 event in the United States, when terrorists made use of modern technology to promote their religiously inspired ideals? A similar comment can be made about bad governance. The governments of US President George W. Bush and UK Prime Minister Tony Blair mislead their population and the world community by applying sophisticated technologies for military purposes in order to preserve or enlarge their national economic and political power positions.[224] Is that a technological or an ethical challenge? Humanity will have to make considerably enhanced efforts in order to avoid the misuse of future advanced technologies for selfish or malicious religious, nationalistic, economic or political goals. In the authors' view, the choices to be made are not technical/technological but inherently ethical.

Increasing Inequality. The inequality argument is a tricky one: theoretically, the differential application of enhancement efforts could, indeed, increase inequalities between people, especially if the enhancement enterprise is market instead of socially driven. Although Maxwull J. Mehlman's[225] fear about the emergence of a genobility, or a genetic overclass, a caste of enhanced humans who would rule over the rest of us, the unenhanced underclass[226] is useful basis for reflection, this fear falls within the domain of science fiction. It is in complete contradiction with the Hardy-Weinberg law. However, there is of course a possibility, albeit unlikely, that the enhanced people would succeed in establishing a separate reproductive community that would aim at dominating the unenhanced.

It is more likely that richer or better informed people might, in the absence of socially distributive ethics and policies, initially have better opportunities to take more or earlier advantage of the new enhancement techniques.[227] In fact, innovations that improve the quality of life—whether in the domain of education, health,

[219]Rees (2003, 75), Kurzweil (2005, 396).
[220]For instance, Tucker and Zilinskas (2006).
[221]Drexler (1986), Drexler et al. (1991).
[222]McKibben (2003).
[223]For instance, Jacobstein (1999), Mehlman (2003, 155ff), AusBiotech (2005).
[224]Rich (2006); http://en.wikipedia.org/wiki/Iraq_and_weapons_of_mass_destruction.
[225]Mehlman (2003); see also Fukuyama (2002, 157).
[226]Wolbring (2006, 125–128).
[227]Baylis and Robert (2004, 11), Mehlman (2003, 108ff), Sandel (2007, 15), Fukuyama (2002, 16, 157).

well-being, or comfort in general—have initially always advantaged privileged individuals and groups in society before socially-driven policies resulted in reaching the population(s) as a whole. Hence, as John Harris[228] argued, this fact should not be used as an argument to ban enhancement innovations before they become available to everyone because they might not become available at all. However, *bona fide* enhancement adherents have never made propositions to limit enhancing innovations to the privileged alone. On the contrary, enhancement advocates always propose that the enhancement enterprise should be socially-driven, democratically controlled and regulated,[229] and that the first efforts to be made would consist of: (1) the decrease of well-being-harming genetic impairments; and (2) the increase in the mental, social or physical capabilities of those at the lower end of the current normal range, instead of interventions aimed at raising the upper bound of the normal distribution of desirable characteristics first.[230] Such efforts would diminish the genetic variance and increase the population average in the variables aimed at, moving the Gauss distribution of biological characteristics to the higher parts of the current range (Fig. 5.2). This means that the problem is not one of eugenic engineering, but of social morality. The genetically impaired or less able would have equal opportunities to take advantage of the new enhancement technologies.[231] This is quite unlikely in the present, but probably temporary, climate of intensified capitalist competition and the decline of a more socially oriented economy. However, it is incorrect and highly unfair to link the transhumanist endeavours with capitalist practices. That is what Katherine Hayles[232] does by stating that all versions of transhumanism "carry into the new millennium some of the most questionable aspects of capitalist ideology."

The transition toward a more evolved stage of hominisation can only be achieved when accompanied by a more highly developed sense of social morality. In fact, the egalitarian and equal opportunity problems raised by (genetic or ontogenetic) enhancement technologies are not different from the societal challenges in those domains. Modern human societies are currently facing challenges of excessive inequalities with varying success. It should not be forgotten that a societal commitment to high quality and equity in education produces more knowledgeable adolescents, as the OECD Programme for International Student Assessment (PISA) of 15-year-old school pupils' scholastic performance on mathematics, science, and reading shows year in and year out. There is a relatively high degree of consensus that social engineering which produces better educated citizens is socially desirable.

In his *Case against perfection* Michael J. Sandel[233] formulates in this respect some social concerns that should be seriously taken into consideration. First of all, this author fears that a genetic enhancement policy would weaken efforts for social engineering. This is possible, but not inevitable. On the contrary, genetic

[228]Harris (2007, 31).
[229]For instance, Hughes (2004, 233).
[230]For instance, Buchanan (2011, 198).
[231]Hughes (2004, 233), Singer (1999, 60), Lindsay (2008).
[232]Hayles (2011).
[233]Sandel (2007, 52, 97); see also Mehlman (2003), Zanc et al. (2010).

enhancement efforts could reinforce social amelioration policies, provided that the transhumanism or new eugenics is socially, and not individualistically, driven. Second, Sandel also draws attention to the increasing competition in various domains of modern life, already resulting in several unhealthy practices of hyper parenting.[234] Indeed, the currently increasing educational overuse practices might invade the genetic domain—again if future policies in this domain are merely individually and not socially oriented. Hence, let us recall wise advice by earlier, prominent scholars, such as Hermann J. Muller, Julian Huxley, and Frederick Osborn,[235] that genetic enhancement policies can only be societally successful if pursued within the framework of a perspective that envisages the *bonum commune communitatis* (common good of the community). A further evolved humanity is only possible when accompanied by higher levels of social morality.

Bona fide eugenic and transhumanist goals have always included the strengthening of the social fabric of our societies—in addition to enhancements of individual characteristics such as intelligence and mental and physical health. This implies that future humans should be enhanced to be genetically more sensitive and adaptable to social life, and to cooperation with larger numbers of individuals and groups of individuals. Hence, the promotion of sociability, ontogenetically as well as genetically, is a *conditio sine qua non* for reaching higher levels of hominisation and humanisation.

Another concern of the anti-meliorists is that the eugenist or transhumanist enhancement goals and procedures would worsen the moral and social status of disabled people and lead to their discrimination.[236] This is really turning the argument upside-down. The efforts of humanitarian and egalitarian oriented societies to provide care and assistance for disabled people, and to promote solidarity between the more and less gifted, should not be an argument against efforts to prevent the formation or development of disability as much as possible. Denying parents the possibility to prevent disabilities before conception or birth of a child, or deny people the freedom to have healthy children does not appear to be ethical.[237] The hope is that the currently limited biomedical toolbox—contraception, artificial insemina-tion by donor (AID), ovum donation, in vitro fertilisation (IVF), embryo transplant, embryo selection, and selective abortion—will soon be extended to higher performing eugenic engineering of gametes or embryos, namely germinal gene therapy. This is the technique in which sex cells are treated with recombinant DNA in order to genetically alter germ-line cells in order to replace unfavourable genes with wanted genes.[238]

However, there can be no doubt that the goal of the furtherance of the hominisation process includes, in a longer-term perspective, not only the prevention of genetic impairments but also enhancement efforts to create genotypes which transcend the upper side of the present variability of desirable features. In time, such a

[234]Rosenfeld and Wise (2001).
[235]Muller (1934), Huxley (1936), Osborn (1940).
[236]For instance, Kass (2002, 130).
[237]See also Harris (2007, 95ff).
[238]Wheale and McNally (1988), Friedmann (1998), Stock (2002).

shift might result in the evolvement of the present human, over the transhumant, to the posthuman or *Homo sapientior*.

In the modernisation process it is not only the eugenic and transhumanist enhancement movements that are driven by a particular normative view about the future of humankind. The dynamics of the modernisation process itself, with its fast evolving and increasingly complex attributes, forms a strong incentive for highly valuing and socially rewarding particular abilities. As a consequence an increasing proportion of 'normal' people risk becoming less well adapted or less competitive and start considering themselves—and are seen by others—as disadvantaged, if not socially disabled.[239] The biological future of humankind cannot be contemplated in a cultural or ecological vacuum, but must be considered in the context of the modernisation process we are experiencing and the future that we are aspiring for.

Conflicts between Humans, Transhumans and Posthumans versus Hominin Advancement. Risks of increased inequalities due to enhancements evoke the danger of possible outbreaks of conflicts between co-existing humans, transhumans and posthumans. One scenario is that posthumans, as a superior (sub)species, could use its enhanced abilities to suppress and exploit the unenhanced, the Left Behind[240] and thus perpetuate or even increase human insecurity.[241] An alternative scenario would be that a further progressing hominisation would be accompanied by increasing levels of tolerance of diversity, thus avoiding the threat of human-posthuman conflicts.[242] However, given the past hominin practice of provoking or at least contributing to the extermination of earlier, less evolved hominin variants, the former scenario is not an impossible doom scenario for present-day humanity and its moral standards which strongly favour competition. The *Homo erectus* might not have been so happy with the more advanced *Homo sapiens*.

Given our knowledge of hominin evolution, we should be aware about our own limitations and potential for failures. Knowledge of history reinforces the need for further hominin advancement which is not only desirable but might be necessary for the long-term survival of the hominin phyletic line.[243] Hence, we should embrace humility and accept, promote and welcome a future replacement of our own species by a more advanced one. It would honour present-day humanity to have contributed to the evolvement of our descendants to higher levels of hominisation, namely higher levels of physical, mental, social and moral capabilities. As Ramez Naam[244] anticipates, we may be the prospective parents of "new and unimaginable creatures, … initiators of a new genesis".

A more imminent conflict-related concern is the military or counter-intelligence development and application of biotechnological enhancement procedures[245] in the in-group/out-group conflicts. American scholars write openly about that aspect and

[239] Fuller (2011, 110, 157).
[240] For instance, Closson (2006), Agar (2010), Oderberg (2014).
[241] For instance, Fukuyama (2002), McIntosh (2008).
[242] Hughes (2004, 179).
[243] See, for instance, Harris (2007).
[244] Naam (2005, 234).
[245] For instance, Armstrong (2010), Royal Society (2012), Tennison and Moreno (2012).

refer to US governmental institutional and financial support of research in that domain for military purposes. It would be highly surprising if other countries did not have the same intentions.

The Rationale for a Progressing Hominisation. The rationale for a progressing hominisation as an ethical objective is that the past direction in human evolution is characterised by a process of increasing evolutionary complexity; this has resulted in an increased potential to understand the world, to adapt better to environmental diversity and challenges, to master our biology and environment, to satisfy our needs and desires, and to reach higher levels of quality of life and happiness. The authors are of the opinion that the continuation of this process could further increase the human-specific potentials and further welfare and well-being of humans. This goal is completely in line with the strivings and aspirations that have characterised the human species all along the course of its evolution and history.[246] As Gordon D. Kaufman[247] wrote:

> In the hope that our biohistorical trajectory may move creatively toward a more humane and ecologically well-ordered world, we can be motivated to give ourselves in strong commitment to its continuing growth and development.

Although the evolutionary process is not necessarily a synonym for progress,[248] in the case of the hominisation process, overall a progression toward increasing abilities to control natural processes can be observed, resulting in increasing survivability and well-being. The hominisation process is a teleonomic process[249] characterised by directionality in its evolution, albeit not a teleologic process resulting from a preplanned intelligent design.[250]

Thanks to the acquired insight into the evolution of life, we can now consciously choose to push the hominisation process in the direction of ontogenetic and phylogenetic developments beyond the limits of the present human condition. Such a goal may lead us to a better understanding of the universe, further improve our quality of life and happiness, and even help us, as Simon Young[251] states, to "discover a new sense of purpose, direction, and meaning to life". Hence, whereas we are, so far, not the result of an intelligent design, we might decide to become our own intelligent designer.[252]

Some authors go even further and believe that the next step of human progress will be to inhabit and eventually transform the universe, and ultimately even save the universe from the Big Chill or the Big Crunch.[253] Whereas it is true that the emergence of human consciousness and intelligence is—at least on our planet—a novel and

[246]See also van Niekerk (2012).
[247]Kaufman (1997, 187).
[248]Huxley (1942, 535); see also Ayala (1988), Ruse (1996), Bregman (2013).
[249]Teleonomy: the apparent purposefulness of living organisms that results from their evolutionary adaptation; it stands in contrast with teleology which applies to processes that are purposefully designed by an agent.
[250]See, for instance, Gould (2008, 336).
[251]Young (2006, 19).
[252]Young (2006, 32); see also Fuller (2007, 153).
[253]See for instance Zey (2001; 2006), Stewart (2008).

unique evolutionary development, the belief that the human might also evolve to imbue the whole cosmos is an interesting—but quite a bold, if not over-ambitious—idea that considerably transcends the capabilities of the present human species.

Nevertheless, now that the human species has acquired insight into the evolutionary process and mechanism, it is fully understandable that it contemplates and speculates about the cosmic significance of its existence and future. Reflections are no longer in terms of a hereafter, as was done in the pre-Darwinian era, but in terms of its possible far future evolution.[254] We have not yet been able to observe whether intelligent life exists elsewhere in the universe.[255] The scientific community is strongly divided about the question as to whether intelligent life is a unique phenomenon limited to our planet[256] or if it might be a quite common phenomenon in the universe.[257] However, the questions of our uniqueness and our far future destiny are a fascinating challenge.[258]

One can obviously question the meaningfulness of the furtherance of the hominisation goal when considering the extremely long time perspective in which the known universe emerged, evolved and will, in the end, disappear. The creation of our present universe 13.82 billion years ago[259]—the Big Bang—is either a unique cosmic event or is the latest stage in a cosmic oscillation: big bang followed by big crunch, in turn followed by another big bang, etc.[260] In both cases, the evolution of organic life in our present universe is, as far as can be ascertained today, doomed to end in the far future.[261]

It is estimated that our sun will collapse within 7 billion years, after having evolved to a red giant (in about 5 billion years) and ending as a white dwarf (in about 7 billion years), and a black dwarf, trillions of years later. It is expected to absorb our planet, after it has entered the red giant phase and expanded beyond the Earth's current orbit.[262]

However, life on our planet will probably end much earlier because the gradually brightening sun will cause the decrease of carbon dioxide levels in the atmosphere, thereby rendering it insufficient to support photosynthesis in about 600 million years,[263] and the evaporation of the Earth's oceans into space by about 1.1 billion years from the present.[264] On the basis of another method of calculation, Andrew J. Rushby and colleagues recently estimated that our planet would have an additional habitable zone lifetime of 1.75 billion years.[265]

[254]For instance, Rees (2003; 2004).
[255]Stenger (2007, 144).
[256]For instance, Ward and Brownlee (2000).
[257]For instance, Darling (2002).
[258]See also Elgin (1993), Chaisson (2001).
[259]http://www.esa.int/Our_Activities/Space_Science/Planck/Planck_reveals_an_almost_perfect_Universe
[260]For instance, Schröder and Smith (2008).
[261]For instance, Ward (2001).
[262]For instance, Sackmann et al. (1993), Schröder and Smith (2008).
[263]Caldeira and Kasting (1992).
[264]Kasting (1988).
[265]Rushby et al. (2013).

In the course of the gradual heating of the sun other celestial bodies, such as Mars, and later Jupiter's moon Europa[266] or Saturn's moon Titan,[267] may become temporarily habitable (or may be made habitable) for the human species or its successor. However, before our sun collapses into a white dwarf, our descendants will have to migrate to habitable planets of other solar systems or galaxies. Would such migration ever become possible? Our present knowledge of the laws of physics does not support such a hypothesis and some consider it to be a delusion.[268]

The distant time horizon of the inhabitability of our present planet (let us say, 600–1.100 million years), compared to the past life span of the human species (say 200,000 years), and the extremely short time lapse of the present scientific era (say 400–500 years) leaves our descendants ample time to think and prepare for interplanetary, intersolar, intergalactic, or even inter-universe migrations. The human species is facing short-term—let us say this millennium—challenges and risks, such as a nuclear disaster, a deadly pandemic, environmental pollution and ecological destruction, climatic change, unsustainable population growth and size, persistent, perhaps increasing inequality levels, and dysgenic trends. These are serious enough to absorb our attention and effort, in order to avoid extinction or regression long before our present planet becomes uninhabitable because of cosmic events.[269] This millennial perspective gives ample justification for effort to sustain progressive hominisation.

Even mid-term (e.g. 100,000 years) risks, such as the next Ice Age, the possible impact of another giant asteroid, the inability of the human species to reach higher levels of social integration and cultural creation, should already be subject to reflection and preparatory action.[270] It sense to consider such threats, not only because they may endanger the far future trajectory of humanity or its hominin successor but also because some of these catastrophes could occur in the near future.[271]

In any case, the assumption is that a further-evolved hominin than the present *Homo sapiens* might be in a better position to cope with the cosmic, biological and socio-cultural challenges with which it will be confronted in the future. We have to be aware not only of the present limitations of our ability to understand the universe[272] but also of our appalling inability to resolve our present problems of population control, ecological management, inequality reduction, and in-group/out-group conflicts. The wise words of Julian Huxley should be remembered[273]:

[266]For instance, Mendez (2009).
[267]For instance, Lorenz et al. (1997).
[268]For instance, Wilson (2012, 296).
[269]Bostrom (2002), Matheny (2007).
[270]See, for instance, Elgin (1993).
[271]Baum (2015).
[272]Glover (1984, 179).
[273]Huxley (1964, 254).

(The human) is not merely exceedingly young; he is also exceedingly imperfect, an unfinished and often botched product of evolutionary improvisation.... The process of hominisation is very far from complete.

5.3.3 General Ethical Derivations from a Progressing Hominisation

5.3.3.1 The Ontogenetic Development of Human-Specific Potentialities

A first general derived objective from the hominisation goal concerns the ontogenetic development of human individuals.

Knowledge about the hominisation process enables us to identify the human-specific features and potentialities, and also to understand why the ontogenetic development in the human is only very partially genetically programmed. It will also help us understand why environmental factors, including cultural values and norms, are needed to complement the genetic endowment to fully develop the human-specific potentialities.

Indeed, as explained in Chap. 2, Sect. 2.2.3, the hominisation process was characterised on the one hand by a major shift in the genetic programming of ontogenetic development, from a predominantly instinctive behavioural control toward a more conscious control of behaviour through the develop-ment of the large brain hemi-spheres; on the other hand it is marked by the relatively short human preg-nancy duration which caused women to give birth prematurely, before the baby's brain had fully matured. These changes in the evolved ontogenetic development of humans necessitate the development of culturally developed and transmitted values and norms, co-steering the ontogenetic development of the individual. The helplessness of the human infant requires many years of intensive social care and culturally determined learning and socialisation. Even the motivation for social care, cultural learning and socialisation must be stimulated by means of value and norm systems. In many respects, the human no longer knows instinctively what and how to teach his offspring.

The progressive hominisation goal in an evolutionarily based ethics inherently implies the development of cultural values and norms supporting the development of human-specific ontogenetic potentialities in the life stage that Joseph Henrich referred to as the Environment of Ontogenetic Adaptiveness (EOA).[274]

The ontogenetic development of the human individual needs not only environmental and in particular socio-cultural intervention to complement the genetically determined human-specific biogram, but it also needs also the fixation of the degree of ontogenetic development. Indeed, human-specific features can be developed at varying levels: the development of physical, mental and social potentialities of humans can vary along a broad range, from a minimal to a maximal

[274]Henrich (2008).

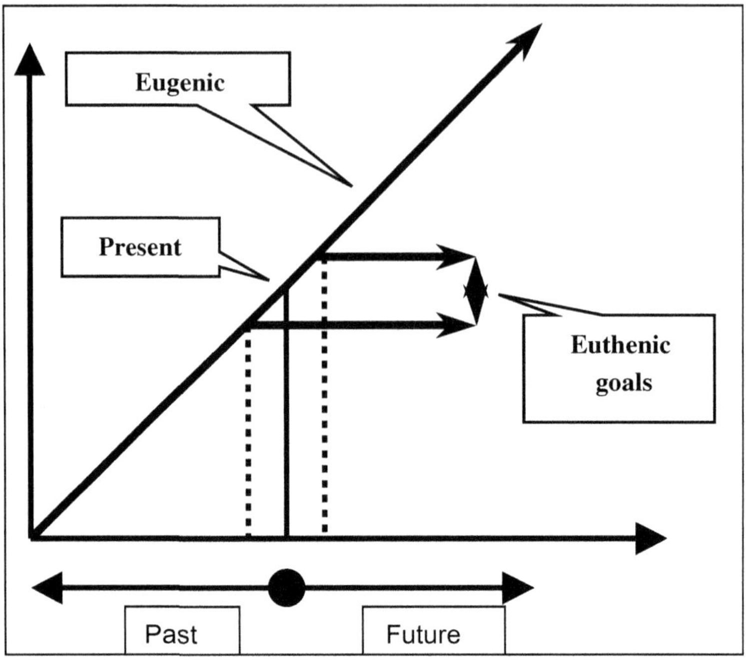

Fig. 5.3 Euthenic and eugenic goals for future ontogenetic and phylogenetic development (Cliquet 2010, 524)

level (Fig. 5.3). In other words, the ontogenetic development of the human species can be realised at different levels of quality of life.

The authors are well aware of the difficulty in defining the concept of quality of life, since it includes not only objective, material elements of the development of human potentialities, but it also embraces subjective elements of what people consider and value as quality (of life). Nevertheless, the evolutionary approach can help us to identify major characteristics of what is to be understood by quality of life.

Indeed, the central goal of evolutionary ethics—the furthering of the hominisation process—implies not only that the ontogenetic development of human-specific potentialities is to be achieved at its highest possible level (the euthenic goal in Fig. 5.3) but also that the human-specific ontogenetic potentialities are to be further enhanced beyond the current biological range.

Transhumanists advocate that the ontogenetic enhancement of human-specific features—bodily capacities, health characteristics, cognitive abilities, emotional personality characteristics and sociability—are to be pursued not only by means of the currently known educational and biomedical procedures but also by the development of new technological means.[275] It is believed that such interventions

[275]Bostrom and Sandberg (2007), Bostrom and Roache (2008), Bostrom and Savulescu (2009).

could improve us more profoundly than the traditional procedures produced by beliefs, habits, culture and education.[276] Moreover, we should be aware that many biotechnological forms of human enhancement are already currently applied in various domains of human physical and mental performances,[277] such as muscle enhancers (e.g. anabolic androgenic steroids) which influence the structure and function of skeletal muscle, sexual enhancers (e.g. Viagra and Cialis) which enhance erectile function, and cognitive enhancers (e.g. Ritalin and Adderall) which influence cognitive functions including short-term memory, concentration, comprehension and alertness.[278] Many therapeutic and preventive therapeutic interventions are even more important: they succeed not only in healing or preventing disease but also prolong longevity and maintain or even improve the physical and mental functionality of many citizens up to an old age.

An important aspect for enabling a sound ontogenetic development concerns the avoidance of existential risks, particularly of an anthropogenic nature.[279] Some of those risks are already well known—population explosion, natural resource depletion, environmental pollution, human-induced climatic change, global nuclear war—but the existential risks of other technological innovations—e.g. designer pathogens, experiments with high-energy particle accelerators, advanced molecular nanotechnological inventions, self-enhancing artificial intelligence—will also have to be considered in a relatively nearby future.[280] The occurrence of an existential disaster might only cause a temporary regression, with time allowing for a resurrection and regeneration, but it would, in addition to the reappearance of dated hardships, considerably delay further progression.

5.3.3.2 The Promotion of Quality of Life

The discussion about the ontogenetic objectives in an evolutionary founded ethics, led us to the conclusion that quality of life should not only be fully maximised but should, in a longer-time perspective, also be further enhanced, even above currently achievable levels.

This brings us to the delicate question of the relationship between quality and quantity. Indeed, the promotion of quality of life should be weighed not only against the holiness of life as argued above but also against the quantity of life. Due to the scarcity of resources that humans have almost constantly had to cope with in the course of our evolution, the pursuit of high quantity was often a guarantee for promoting quality. This quantity-quality relationship has run like a thread throughout human history: high fertility was a protection against high mortality, provided a large number of workers in family businesses, and more security for

[276]Bostrom (2007).
[277]For instance, Rothman and Rothman (2003), Carter (2016).
[278]https://humanenhancementdrugs.com/.
[279]Bostrom (2002).
[280]For instance, Bostrom (2002; 2010), Mulhall (2002), Diamond (2005).

parents in their old age; a bountiful harvest preserved people from dreaded starvation in winter; a large number of soldiers were a barrier to possible conquerors or formed an instrument for the conquest of new territories.[281]

The development of modern culture tends to turn the former positive association between quantity and quality into a negative one. High fertility leads to a demographic explosion under conditions of efficient mortality control. Ultimately this could lead the population to exceed the carrying capacity of the environment and create intolerable conditions for humans. Moreover, in modern living conditions the human predisposition to maximise the inclusive fitness or the related greed can easily lead to various forms of overconsumption with pernicious consequences for quality of life. For example, high production/sale/use of all kinds of weaponry leads to high rates of murder and/or war. Overconsumption creates epidemics of obesity, abuse of stimulants, high rates of traffic accidents, environmental pollution, and resource depletion.

Assuming that modern culture will keep on developing, it can be argued that the relationship between quantity and quality will have to be reconsidered and, especially, be redefined in the light of concern for sustainable growth. Given the finiteness of the planet and its limited capacity to sustain life, the exploitation of the Earth's resources will eventually reach a point at which the further improvement of quality of life will become inversely proportional to the growth in population size. After all, quality and quantity both require escalating use of the restricted amounts of available raw materials, space, and energy, which imposes a growing burden on the environment and Earth's ecosystems. This relationship is stated in Ehrlich and Holdren's well-known formula: $I = P \times A \times T$.[282]

Such a vision unequivocally promotes the improvement of the human quality of life, if necessary at the expense of the decreased pace of population growth in numbers. In modern culture, we cannot continue indefinitely to "…be fruitful and multiply and fill the earth…".[283] We may need to take up the responsibility for controlling growth and achieving better sharing of resources. This matter will be discussed in more detail in Chap. 8, Sect. 8.2.3.

5.3.3.3 The Promotion of Equity

Genetic variation is, from an evolutionary-biological point of view, an extremely significant phenomenon. At the population level, and in a long-term perspective, it is a relative safety valve for adaptation to changing living conditions; at the individual level, and in the short-term, exceptional biological capabilities are the engine for cultural achievements and innovations.

However, there is another side of the coin: unfavourable genetic mutations, ontogenetic accidents, infectious diseases, natural disasters, adverse life course events, and social inequities can all produce psychophysical differences in individual potentialities, competences and performances. For many individuals such

[281]Parsons (1999).
[282]Ehrlich and Holdren (1971): I = impact; P = population; A = affluence; T = technology.
[283]Genesis 1:28.

traumatic events lower their quality of life, their life satisfaction, and their chances for happiness. From an egalitarian ethical point of view, biological variation is, in many respects, a nightmare.

Most value systems, particularly in modern times, try to reconcile the facts of biological diversity with the ideals of equality and equity by promoting these values without completely excluding diversity. In modern democratic societies the concepts of equality—likeness or sameness in quality, status, or degree—and equity—encompassing ideals of justice and fairness under conditions of heterogeneity—are usually understood as equality of opportunity. Given individual differences in abilities and work effort, it is generally assumed that equality of opportunity provides each person, regardless of ascribed characteristics such as family background, worldview, ethnicity, race, or gender, the same chance of acquiring a favourable cultural or socio-economic position.[284]

However, the establishment of equal opportunity does not necessarily imply that people will end up culturally, socially or economically equal, since differences in abilities or work effort normally result in differences in performance and are usually differentially valued and rewarded. Some people are in a more vulnerable situation, either because of genetic heritage or due to life course events: for instance people who are mentally challenged, physically less able, long-term diseased, less skilled or unemployed. In order to safeguard such people from social exclusion or misery, modern advanced democracies have developed social protection systems.[285]

As a consequence of increasing knowledge and its dissemination via education and modern means of communication, it is no longer possible to justify extreme forms of social inequity and inequality or other forms of social exclusion and exploitation.[286] Wherever such social forms of inequity/inequality still exist, eventually they are or will be vehemently challenged, and it can be expected that they will attenuate due to social pressure, especially as societies modernise. Nevertheless, even in the future human societies will have to deal with the discrepancy between biological diversity, due to genetic or environmental causes, and the necessity to create equal opportunities for all and avoid the social exclusion and indigence of people with weaker abilities or competencies. In this respect Friedrich A. Hayek's view should be taken seriously into consideration[287]:

> It is just not true that humans are born equal;… if we treat them equally, the result must be inequality in their actual position; …thus, the only way to place them in an equal position would be to treat them differently.

Clearly, the ideal would be that modern societies would develop to the fullest the potential of individual students with different backgrounds and talents. Torsten Husén[288] referred to this vision of equal educational opportunity in these terms:

[284]Schaar (1967), Rawls (1971), Parelius and Parelius (1987), European Commission (2011).
[285]For instance, Deleeck (1992).
[286]For instance, Avramov (2003).
[287]Hayek (1960, 76–77).
[288]Husén (1972, 26).

...every student should have an equal opportunity to be treated unequally.

In conclusion, an evolutionarily-based ethic will, in the living conditions and opportunities of modernity, simultaneously have to positively value genetic diversity, favour—depending on the ecological, economic or cultural circumstances—particular genetic variants, enhance equity and equality of opportunities, and, in order to facilitate the social inclusion of disadvantaged population categories,[289] develop adequate social protection systems. These matters will be discussed in greater detail in Chap. 8, Sect. 8.2.3.

5.3.3.4 The Shift from Competitive Towards Cooperative Efforts

Biological evolution proceeds mainly via one of its major basic mechanisms—natural selection—and it functions on the basis of competition between individuals.[290] However, one of the most salient trends in the hominisation process is the increasing importance of social cooperation.

As explained in Chap. 1, Sect. 1.1.3, in the second half of the twentieth century sociobiologists discovered specific biological mechanisms, namely kin selection, various forms of reciprocity selection and group selection, that explain the presence and evolution of altruistic behaviour in social animal species and humans; these form the biological basis of social bonds that transcend parental-offspring relations.

In the same period, anthropological causal explanations of the hominisation process shifted from ecological theories[291] and technological theories[292] to biosocial theories.[293] The biosocial theory of the increase in (social) intelligence during the evolution of the primates, and in particular of the hominins, states that (social) intelligence is selected not only on the basis of the group size but mainly as a function of the complexity of the social interactions between the group members with whom one lives. Our brain has been selected to cope with the challenges raised by increasingly complex social interactions.[294]

Due to the recent development of modernity (in evolutionary terms) with its incredibly increased number of anonymous interactions, within and between large populations, our genetic predispositions (selected and adapted to deal with social interactions in and between small populations) are not in equilibrium with the cooperative requirements in our novel modern environment.[295]

Humans remain endowed with drives toward both competition and cooperation. Human social life originally evolved to support the survival of the individual and it still continues to serve that purpose. However, human societies gradually

[289]For instance, Avramov (2003).
[290]Williams (1966), Dawkins (1976).
[291]For instance, Clutton-Brock and Harvey (1980).
[292]For instance, Oakley (1959), Washburn (1960).
[293]For instance, Humphrey (1976), Byrne and Whiten (1988; 1997), Dunbar (1992; 1998; 2003), Emery et al. (2007).
[294]For instance, in this respect McNamara (2006, 195) points to the role of the frontal lobes in the development of cooperation: "Frontal lobes are the neural systems most consistently activated in association with decision-making around cooperation dilemmas."
[295]Burnham and Johnson (2005).

developed, via cultural creation and accumulation, as complex autonomous entities in which cooperation increasingly became a crucial instrument for the survival of individuals *and* society. This process accelerated strongly from the hunter-gatherer stage, through the agrarian era, to the present modern context of demographic and cultural development.

Hence, it is not surprising that a large part of ethics relates to the regulation of social life. The societal functions of value and norm systems need little justification. The egocentric drives of the individual must be moderated and mastered for successful social life, so that sociality, and also the individual emancipation of all group members, can be optimised. A more difficult issue concerns the desirable value and norm systems regulating the relations between groups of individuals, and in particular between large societies. In-group/out-group relations belong to the most difficult issues in ethics.

It is to be expected that the future development of modern culture will further intensify the tension between competition and cooperation, within as well as between societies. Therefore increasing efforts will have to be made to find a viable balance between both, thus avoiding hyper-individualism and absolutist groupism. All this entails an ever-increasing need for a further shift from competitive toward cooperative efforts. Indeed, as Fereidoun Esfandiary[296] noted

> …competition is antifuture.

5.3.3.5 The Promotion of Universalism

The specific human genome emerged and evolved in Pleistocene times, in the EEA, when people lived in small tribes and interacted with other small tribes.[297] Cooperation within groups had survival and reproductive advantages. Increasing their inclusive fitness via reinforcement of the in-group solidarity and cohesion facilitated defensive or offensive actions against competing out-groups.[298] Even in the agrarian and early industrial phases of humanity's cultural history, the more evolved forms of ethnocentric tribalism—patriotism and nationalism—had adaptive advantages and explain much of the exploratory and conquering success of expanding nation-states. Group conflict appears to have been one of the striking universals in human existence.[299]

Evolutionary science ultimately explains humanity's strong in-group reflexes and their offshoots, such as nepotism, tribalism, ethnocentrism, xenophobia and racism, as a result of the drive to protect and spread one's own genes. Together with the drives toward greed and demographic expansion, the in-group syndrome is responsible for much of the pronounced between-group enmity that characterises human beings.

[296]FM-2030 (1989, 66).
[297]Aiello and Dunbar (1993), Dunbar and Spoors (1995).
[298]Ember (1978), Alexander (1979, 1987), Keeley (1996), Diamond (1992; 2005).
[299]For instance, Brown (1991, 130).

However, the novel environment that emerged from modern culture is characterised by a number of features and trends that causes the in-group syndrome to become too dangerous. The development of ABC weaponry is so life-threatening that the in-group syndrome has largely lost its adaptive advantage. Furthermore, the exponential increase in scientific knowledge, as well as the improvement of between-group communication by means of ICT (information and communication technologies), international educational and commercial exchange, travel and tourism, all together fundamentally undermines the in-group syndrome by breaking through group isolation and eroding two of its most basic breeding grounds—ignorance and prejudice. Previously separated human populations and nations are now growing toward a single world community, the components of which show an increasing mutual dependency. Modern culture is becoming a globalised phenomenon and has implications at the planetary level. A typical example of the transitional stage, in which humanity finds oneself on a trajectory to higher levels of future stages of culture, is the present unaccomplished extension of within-group amity to the whole of the human species as a single community with a universal morality.[300]

Although universalism contravenes the strong in-group focused human instincts,[301] human evolution and history is characterised by a gradual expansion of our circle of moral considerability from the family, to the tribe and the nation, religious confraternity, then to the entire human species.[302] The scientific knowledge and technological instruments of modern culture facilitate—and even compel—people and societies to think and act in a more global perspective.

The discoveries of molecular genetics show that people of different racial groups probably have about 99.9% identical DNA.[303] The molecular-genetic reinforcement of this factual datum might facilitate and support, philosophically, the idea of universalism in ethical matters. Another fundamental argument in favour of a universal human ethics relies on the fact that neurological research increasingly points to the idea that our brain functions on the basis of a universal set of biological responses to moral dilemmas.[304]

With the prospect of safeguarding the survival, as well as furthering the well-being, of the human species, many evolutionary scientists[305] are of the view that an evolutionary ethics in modernity should be of a universal or global nature. This implies that our moral norms regarding rights and responsibilities should be extended to all human beings. This moral universalism requires the in-group/out-group relations to be replaced by alternatives to the traditionally hostile and aggressive patterns. In other words, modernisation requires a shift from in-group toward out-group relations. Scientific knowledge, and in particular

[300]Alexander (1993, 180).
[301]Wells (1905), quoted in Keith (1942, 51).
[302]For instance, Singer (1981; 2002), Lahti (2009).
[303]Jorde and Wooding (2004).
[304]Gazzaniga (2005, xix).
[305]For instance, Darwin (1871), Keith (1947), Katz (1999), Singer (1981; 2002).

evolutionary science, can help to reflect on and work towards a new global, if not cosmic, ethic.[306]

The plea for a shift towards an increasing globalism does not mean that community development at local, regional and national levels are not important. People live and function in such sizeable groupings on which they depend for their identity and daily living conditions.[307] Individual emancipation, as well as societal progress, strongly depends on community coherence. However, in modernity they need to be steered beyond the old-time narrow in-group mindedness. Community development and globalisation no longer need to be incompatible.

5.4 Evolutionary-Based Specific Ethical Challenges Related to Sources of Biological Variability

In earlier work on biosocial interactions in modernisation,[308] the domains of cultural intervention were classified according to the source of biosocial variability: individual variability, age variability, sexual variability, various forms of group variability and intergenerational variability. In this book, which is focussed on human variation and relations that need to be subject to ethical consideration, largely the same logic is followed and the following domains are distinguished, hierarchically classified according to their degree of complexity: (1) At the individual level: individual variability regarding biological traits in general, and age and sex in particular; (2) At the group level: variability of and relations with groups of people, according to kin relatedness, social class, race/ethnicity/ideology, and statehood; (3) At the generational level: intergenerational replacement.

From an evolutionary point of view, each of these domains requires attention for specific ethical considerations and reflection that will be discussed in the following chapters.

Contrary to what is sometimes believed, human morality applies to two major domains of life: development of 'self' and relations with 'others', implying obligations toward oneself as well as toward related and non-related others.[309]

Development of 'self' is concerned with ontogenetics (the development from conception to death) which is necessary for individual survival and fulfilment; this development requires moral guidance because the individual developmental processes in the human are no longer completely genetically programmed and environmental factors can steer the individual development in various directions and levels. These issues are dealt with in Chap. 6.

[306] Loye (1999), Chaisson (2001), Stewart (2008).
[307] Salter (2003).
[308] Cliquet (2010).
[309] Rottschaefer (1998; 2000, 239), Hinde (2002, 178).

Variation of and relations with 'others' is a much vaster domain which includes several important fields for moral action: relations with kin and family, as well as relations with larger groups of individuals—social classes, ideological groups, ethnicities, races, states. These issues are dealt with in Chap. 7.

A specific domain of variation of and relation with others concerns intergenerational change. Issues of intergenerational replacement are addressed in Chap. 8.

The justification for moral action in all of these domains is also concerned with the incomplete genetic programming of the human biogram and the 'Janus'—double sided—character of human nature regarding moral issues. Indeed, the human species has several genetic and neurological predispositions (partly innate, partly learned) that elicit and support moral behaviour toward 'self' and/or 'others', but is also equipped with drives that suppress or even oppose morality.[310] In the first category the authors would list predispositions which lead to capabilities for self-actualisation, caring, empathy, sympathy, altruism (generosity, sharing, solidarity, mutualism, reciprocity), sociability (cooperation, loyalty, tolerance, reconciliation, consolation, conflict intervention, and mediation), and honesty (truthfulness). The second group consists of predispositions toward selfishness, aggressivity, greed, spite, dishonesty, free-riding, cheating, thieving, cruelty, anti-social behaviour, dominance, and violence. An evolutionarily-based universal ethic in the novel environment of modernity can and should contribute to bolstering the former to the detriment of the latter.

[310]For instance, Eibl-Eibesfeldt (1971), Masters (1989), Ehrlich (2000).

6 Evolution-Based Ethical Challenges Related to Individual Variability

Abstract

This chapter deals with age variability, sex variability and individual variability in general. In addition to age and sex relations, interpersonal relations in general are addressed. For each of those issues the salient aspects regarding their evolutionary background, recent trends in modernity, and ethical reflections about the future are discussed. Regarding age variability, the discussion considers the accelerated biological growth process and the prolonged social maturation period in the life course, the increased life expectancy at all life stages, and the prolonging of the terminal phase of life. Regarding sex variability, particular attention is given to the specificity of human sexuality, trends in sexual behavioural in modernity, and ethical reflections about relations between the sexes in the future. Regarding individual variability in general, special attention is paid to the importance of the maintenance of variation, the control of maladaptive traits and attributes, and the containment of individualism. Regarding interpersonal relations in general, the dynamics of individual competition and cooperation, and the causes of individual maladapted behaviour are addressed.

6.1 Introduction

Individual variability appertains to what the previous chapter referred to as 'ontogeny'—the development from conception to death of individuals. However, ontogenetic variability applies not only to differences within individuals during their life course but also to differences between individuals which emerge and get established during the life course.

R. Cliquet and D. Avramov, *Evolution Science and Ethics in the Third Millennium*,
https://doi.org/10.1007/978-3-319-73090-5_6

Due to their specific ethical implications, three main aspects are discussed: (1) variability of age-specific characteristics; (2) variability of sex-specific characteristics; and (3) variability of individual characteristics in general.

For each of those issues the evolutionary background is recalled, trends in modernity are highlighted, and some crucial ethical implications for the future are reflected upon.

6.2 Age Variability

In the human life course three major stages can be distinguished: childhood, adulthood and old age. All three of these life course stages are associated with age-specific challenges which largely relate to dependency and intergenerational solidarity. As a result of their stronger implications in the evolutionary context, due to changes in intergenerational dependency the discussion will be focused on relations between adults and children and between adults and older persons.

6.2.1 Evolutionary Background of Growth and Senescence

Two major processes characterise the individual life course: growth and senescence, the latter ending in the dying process.

6.2.1.1 Growth

Growth consists of early processes that enhance the functional capacities of the individual. During hominisation the pace of the human growth process evolved, so that postnatal growth acceleration is followed by a period of interruption of growth acceleration in the time before adulthood. This postponement of maturation until puberty, controlled by the hypothalamus, relates to the long period that the human brain needs to become fully functional. During this period of pre-puberty maturation and socialisation the growing individual is still quite docile. From puberty onwards, an individual comes into sexual competition with other adults. Ascending the primate phylogeny, this pre-pubertal period becomes ever longer. In the human it is the longest.[1]

6.2.1.2 Senescence

Senescence refers to the age-related changes that lead to the gradual and generalised regression of mental and physical functions that end in death.[2] The evolutionary explanation of the occurrence of senescence is that ageing is caused by a decrease in the force of natural selection with increasing age.[3] Selection against genes that

[1]Tanner (1962).
[2]For instance, Comfort (1956), Finch (1990).
[3]Medawar (1952).

manifest themselves early in the life course affect a larger number of individuals than selection against genes that reveal themselves at advanced age, when the number of survivors and their reproductive capacity are smaller. Genes with late detrimental consequences can accumulate and result in senescence among individuals who live sufficiently long enough. Senescence is the inevitable result of the fact that selection has a greater impact on genes that only affect survival or fertility early in life than genes whose effects are only manifest late in life.

The human lifespan[4] has substantially increased over the last few million years of evolution. Evolutionary theory explains this increase as an adaptation related to the increase in brain size. Larger brains allow a better control of the environment and result in a reduction of mortality; larger brains require a longer maturation time; in turn, a longer maturation requires a larger birth interval, as well as a shift from a multiparous towards a monoparous gestation, both of which lead to a lower age-specific fertility. All of these features require a longer lifespan: on the one hand this allows for the more intensive parental care of long-term needy youngsters, and on the other hand it allows for an overall fertility rate that is sufficient for generational replacement. A longer lifespan requires a larger investment in somatic maintenance and repair, in the end this results in a postponement of senescent processes. Over the last few million years the increase in brain size and its co-evolving cultural development in the course of hominid evolution have contributed significantly to the increase in human longevity.[5]

6.2.2 Developments in Modernity Regarding Age Variability

6.2.2.1 Growth

On the subject of human growth, the contemporary industrial cultural phase is characterised by a temporal growth acceleration. This phenomenon occurs in all countries or regions where modern culture develops: this has been true for some 150 years in Western countries, but also recently in modernising developing countries.[6] This temporal growth acceleration includes a moving forward of the beginning and the end of the maturation and growth processes, and the achievement of a larger end result. The temporal growth acceleration has been observed for body height, body weight, skeletal age, dental eruption, menarche/first ejaculation and other sexual maturation characteristics, several serological characteristics, and measured intelligence (the so-called Flynn-effect).[7] Whereas growth acceleration is

[4]Lifespan: refers to the typical length of time that an organism can be expected to live.
[5]Carey (2003), Baltes et al. (2006), Gurven and Kaplan (2007).
[6]Meredith (1974), Bodzsar and Susanne (1998), Krawczynski et al. (2003), Zhen-Wang and Cheng-Ye (2005).
[7]Tanner (1962), Flynn (1987).

taking off in many modernising developing countries, it seems to have weakened or even stopped in the upper social strata of the most advanced industrial countries where it has, at this time, apparently reached its limits.[8]

In modern culture, the evolved postponement of biological maturation is considerably prolonged by an additional period of social maturation. Indeed, there is an increasing gap between biological maturation and social maturity. An adequate socialisation and learning period requires much more time than the biologically programmed period, which corresponds *grosso modo* to the end of primary school at 12 years of age. In order to be able to function well in modern culture the average citizen in complex modern society needs a much longer learning period than the duration of primary school. The social maturation time needed transcends the biological (sexual) maturation period by many years. Moreover, the time gap between biological and societal maturity is constantly increasing. On the one hand biological maturity now comes earlier in the life course due to the temporal growth acceleration; on the other hand societal maturity takes more time due to increasing needs for schooling and training. This expanding gap between biological maturation and societal maturity is a typical example of the asynchronic biosocial development in modern culture, as biological-evolutionary adaptation is not fast enough to respond appropriately to cultural change over time. This asynchrony is a major cause of the generational conflicts between parents and their adolescent children in affluent modern societies. Numerous cultural, economic and technological innovations have vastly empowered and raised the awareness of adolescents but at the same time educational needs have prolonged their dependency.[9]

6.2.2.2 Senescence

The Extension of Life Expectancy. Modern societies are characterised by a revolutionary extension of life expectancy, whereby the causes of death have largely shifted from external (mainly infectious diseases) to internal factors (senescent deterioration). Most people spend the largest part of their old age in good health as modern health and welfare care succeed in mitigating or even considerably compensating senescent deterioration.[10] More and more people are protected against or successfully treated for infectious diseases in old age and are thus exposed to senescence as a gradual and generalised regression that results in death. In fact, senescence has become the major cause of illness, disability, dementia, and finally death.[11]

In modern culture, the average life expectancy at birth[12] has more than doubled over the past two centuries and it has roughly tripled over the course of human

[8]Sinclair and Dangerfield (1998), Tanner (1978), Vercauteren and Susanne (1985), Hauspie et al. (1996), Krawczynski et al. (2003), Zellner et al. (2004).
[9]Noom (1999), Arnett (2001).
[10]For instance, Doogle et al. (1988), Avramov and Maskova (2003), Jacobs et al. (2004).
[11]For instance, Bostrom (2005, 100).
[12]Life expectancy: the average number of years of life remaining at a given age, i.e. the average expected lifespan of an individual.

history.[13] Life expectancy among hunter-gatherers varied between 21 and 37 years.[14] In the most advanced regions of the world it amounts nowadays to above 80 for men and above 85 for women.[15] Currently it continues to increase by approximately three months every year. Also the maximum reported age at death has systematically been increasing,[16] currently lying at around 120 years.[17]

Modernisation has succeeded in enabling more people to reach the biological potential lifespan. However, not much progress has been made so far in life-extending technologies, which could move the current species-specific lifespan to a higher age. However, it is possible that in the near future biomedical inventions will enable the extension of longevity beyond the present species-specific lifespan, in addition to further increasing the number of people who survive up to the maximum lifespan. Since the human species-specific lifespan is not very well known, it is extremely difficult to predict the ultimate reachable limit of old age.[18]

Increasing life expectancy clearly influences the age composition of the population. However, population ageing can be the result of population greying as well as of population dejuvenation. The first is caused by an increase in the proportion of the older age groups—a swelling of the age groups at the top of the age pyramid; the second results from a decrease in the proportion of younger age groups in the population, for instance as a consequence of decreasing fertility, which shrinks the age groups at the bottom of the age pyramid. Population greying can be reinforced when larger birth cohorts reach the age categories that are conventionally considered to be elderly or seniors, as is currently the case with the post-World War II baby boom generation reaching seniority.

The likelihood of substantial increases in average lifespan and the potential for extreme longevity raises a host of societal issues: huge increases in the number of old and very old persons; massive growth in health expenditures for the elderly; labour shortages among the prime working age groups; depletion of social security and pension contributors; intensifying intergenerational tensions; and potential for decreasing quality of life for all. Hence, it is not surprising that population ageing is a very important concern for many policy makers today, but it is an even greater ethical challenge for the third millennium as it puts a strain on the traditional intergenerational models of care and transfer of resources. Indeed, ageing costs. It costs in terms of public pension schemes, meeting the new care needs of older people, and expanding health services so that they can meet the growing demand arising from the ageing of the population. It is remarkable that, from the ethics perspective, the achievement of high life expectancy is considered to be a "great triumph of civilisation",[19] for which people have strived for so long and so hard, yet

[13]Wilmoth (2000).
[14]Galor and Moav (2005), Gurven and Kaplan (2007).
[15]http://epp.eurostat.ec.europa.eu/statistics_explained/index.php/Mortality_and_life_expectancy_statistics.
[16]Wilmoth and Lundström (1996).
[17]Robine and Vaupel (2002); see also Dong et al. (2016).
[18]Finch (1997, 245).
[19]Notestein (1954).

we are struggling to develop ethical standards for intergenerational equity to manage this triumph. Beyond the issue of financial costs, opportunity costs for the elderly and their families are also at stake. Societies struggle to find socially useful and individually meaningful activities for its growing elderly population.[20]

Ageism. It is not surprising that ageism[21]—meaning marginalisation or discrimination based on age, and especially prejudice against the elderly—is so widespread in modern societies.

Surveys amongst the elderly on attitudes or behaviour towards people of higher ages have systematically found that the experience of ageism is widespread, frequent and multiple.[22] Seniors often experience disrespect, or are considered to be less able or productive. Older people as a group experience negative treatment in terms of poor access to transport and housing, low incomes, and inadequate nursing home care. There is evidence of extensive de facto health care rationing or overtreatment by providers on the basis of age.[23] While few have experienced overt or brutal ageism, interaction in everyday life involves some negative treatment, and only occasional positive sageism,[24] i.e. acknowledgement of and respect for the wisdom of older people.

In industrial societies older workers, say 50–60, often experience discrimination in the labour market, have fewer opportunities at job solicitations, and are more often fired or forced into retirement. There is higher than average unemployment among older workers. Employers invest less in elderly workers to keep them competitive in the regular labour market. In the most developed countries elderly workers are often being made redundant, given casual jobs or being pushed into early retirement by their employers.[25]

In the twentieth century the biology of ageing—resulting in longer and healthier old age—and the societal perception of ageing—as lacking a role in society—clearly evolved in opposite directions.[26]

The Prolongation of the Dying Process. The use of advanced medical technologies not only contributes to increasing longevity but also, in a growing number of cases, to prolonging the dying process. This issue is receiving increasingly prominent attention in the scientific and ethical literature.[27] Death control, either in the form of palliative care or euthanasia, includes practices that deal with the unintended and undesired effects of partial medical successes whereby people are kept alive in conditions of severe terminal suffering or degrading regression. Death control practices—dignified end-of-life decision-making and humane care for dying patients—clearly relate to delicate and difficult issues on which fundamentally

[20]Avramov and Cliquet (2005).
[21]Butler (1969).
[22]Avramov and Maskova (2003), Avramov and Cliquet (2005).
[23]Kapp (2001).
[24]Minichiello (2000), Palmore (2001).
[25]Andrews (1999).
[26]Avramov and Cliquet (2006).
[27]For instance, Logue (1993), Nuland (1994), Gorsuch (2006), Yount (2007), Griffiths et al. (2008).

differing philosophical and religious views exist in all societies.[28] In the light of the religious versus secular and conservative versus progressive divide on how to view the prolongation of the dying process, it is necessary to reiterate that this generalised prolongation is not a natural or supernatural creation, but a technical/technological possibility created by science.

6.2.3 Ethical Reflections Regarding Age Variability in Modernity

6.2.3.1 Bridging the Gap Between Biological Maturation and Social Maturity

Emphasising the importance of caring for children would only be kicking open doors. The weak genetic programming of human behaviour and the long maturation process of the human child are the basic roots and form the evolutionary explanation for this ethical principle. The increasing gap between biological maturation and social maturity in modernity only reinforces the survival value of the norm that adults must provide care for children.

The specificity of modernity requires a high level of ontogenetic development aimed at the full actualisation of the human-specific potentialities and abilities, allowing an adequate harmonious phenotypic adaptability to this type of culture. It necessitates, moreover, a rather fundamental change in the contents and methodology of formal, non-formal and informal education; this includes among others a thorough acquaintance with the essential features of modernity and a moral education adapted to diversity and dynamics.

It is reasonable to assume that in the third millennium the value of education—as the systematic instruction provided in schools—and non-formal and informal learning—which enhance knowledge, skills, competences and understanding—should increase in the direction of science-based rather than faith-based rationale.

6.2.3.2 Overcoming Ageism/Promoting Active Ageing

Individual and population ageing are not only the achievements of modernisation but they are also the result of consciously pursued goals. It should be acknowledged that humans have, as all living beings, a strong drive for survival and a desire for long (healthy) life. Hence, it is understandable that modern humans take ample advantage of the extraordinarily increased opportunities offered by science and its applications in medicine and welfare care, which increase life expectancy and fight the unfavourable side effects accompanying senescence. Reaching old age at the individual level—that great triumph of civilisation—should consequently be highly welcomed.[29] Ageing at the population level, at least in its component of population

[28]For instance, Brock and MacLean (1993), Dowbiggin (2005), Paterson (2008), Wilcockson (2008).
[29]See also Bostrom (2005).

greying, is a consequence of the progress in increasing individual life expectancy and should equally be accepted as a phenomenon to which modern societies have to adapt.

As the prospect of surviving to a very old age affects the overwhelming majority of the population, ageism is a delusion of youth and excessively competitive working conditions. As people reach more advanced stages in their life course, they should have the opportunity and responsibility to continue being active and fully included in societal processes, especially now that their educational and health levels are improving so much, at least of the younger seniors. In this regard, some current policy measures or behavioural patterns are obsolete such as early retirement schemes and relatively low mandatory retirement age. It is obvious that in modernity people cannot continue being gainfully employed until death but opportunities for flexibility and variability in individual ageing and health conditions need to be built into the currently rather rigid age categories that govern intergenerational solidarity. The traditional categorisation of paid work and measurements of productivity need to be reconsidered—taking into consideration society as a whole and not only the enterprise/industry level.[30]

People are living longer and healthier lives and are spending ever more years in passive leisure. On the one hand reflections are necessary about the place of elderly in society and their responsibilities regarding the take up of health and welfare provisions over a very long period; on the other hand working conditions are maladapted to the ageing of the work force. These issues cannot be addressed at the enterprise level and need to be supported by the new ethics and built into legislative institutions.

Active ageing is promoted in national and international documents but has yet to become the norm in a further modernising humanity.[31] Ageism is incompatible with the ethical and scientific-technological achievements of modern culture, which has produced a revolutionary increase in life expectancy and vitality at higher ages. In modern culture, ageism associated with the sidelining of elderly persons has become a maladaptive practice as it is a striking example of waste of human creativity and productivity.

Too often Alfred Sauvy's notorious image is seen:

de vieilles gens qui ruminent de vieilles idées dans de vieilles chaumières[32]

Ageism should get the same negative connotation as racism, sexism, or inequity and needs to be addressed via equal opportunity ethos.

6.2.3.3 Reaching the Species-Specific Lifespan in Good Health

From an evolutionary point of view, the resources that modern culture invests, or should invest, in the extension of life expectancy at older ages seem to be contrary

[30]For instance, Avramov and Cliquet (2003).

[31]Commission of the European Communities (2002), World Health Organisation (2002), United Nations (2002), Avramov and Maskova (2003).

[32]"old people who ruminate old ideas in old dwellings" (authors' translation).

to the principle of maximising inclusive fitness. It may absorb resources necessary to assist the reproduction, education, health care and socio-cultural and socio-economic development of younger generations.

However, a closer look reveals that this apparent contradiction needs to be qualified. Avoiding morbidity and mortality at a younger age and allowing people to reach old age gives people the opportunity to fully develop their own genetic potentialities, and also creates the conditions for their parental and grandparental investment. It also prolongs the potential for the overall contribution to society by each individual.

The further increase of life expectancy is to be evaluated as a positive strategy as long as the results do not decrease quality of life. Not only does it enhance the opportunity to valorise the genetic potentials present in the gene pool of a population; it also promotes the emancipation and happiness of individuals, allowing them to complete every stage of their ontogenetic development. This strategy forms the basis for people to make an optimal contribution to society.

The obvious, and consequently recommended, method of further increasing life expectancy up to the species-specific potential is compressed morbidity,[33] resulting in a retardation of senescent processes and the maintenance of a healthy life course in old age up to or close to the end of life.[34] In this way, lifespan in old age would be characterised by a reduction of frailspan in favour of healthspan,[35] "adding life to years, not years to life".[36]

Investing resources in increasing life expectancy when senescent deterioration has gravely progressed, or when the dying process has irrevocably set in, needs to be carefully evaluated both from individual, family and societal points of view. This issue evokes the delicate and difficult problem of death control.

6.2.3.4 Managing the Dying Process

Sir Francis Bacon, an English philosopher and statesman, devised the concept of euthanasia early in the seventeenth century.[37] Derived from the Greek words *eu* (good) and *thanatos* (death), euthanasia means a good or easy death. Today euthanasia has come to mean an act or practice of painlessly putting to death persons suffering from incurable conditions or diseases—euthanasia *sensu largo*. In currently existing legislation, as well as in most attitudinal surveys about euthanasia, the concept is defined in a more restricted sense. Euthanasia defined *sensu* stricto is the prescription or administration of drugs by a physician with the

[33]Other anti-ageing strategies such as 'decelerating ageing' (=senescence processes are delayed and average life expectancy and maximum life span are increased) and 'arresting ageing' (=senescence processes are reversed in adults, restoring vitality and function) (de Grey 2000) which are primarily aimed at extending human lifespan, would also have a substantially favourable effect on health at a greater age.

[34]Fries (1980).

[35]Bostrom (2005), Bostrom and Roache (2008).

[36]Post and Binstock (2004).

[37]Yount (2000).

intention of ending the suffering of an incurably ill patient at his/her explicit request. It entails the capacity of the patient to give informed consent.

Euthanasia—in its narrow as well in its broader sense—is obviously a subject on which people of different worldviews foster very different opinions.[38] Based on the ideologies of the great religious traditions, euthanasia is still illegal in most countries of the world. Muslims believe that only Allah has the right to end life; both Hindus and Buddhists teach respect for life and the belief that euthanasia is an interruption of karma; Jews and Christians base their objections on the Biblical commandment, "Thou shalt not kill".[39] Consequently, in many countries another form of death control is being advocated and developed, namely palliative care.[40]

Euthanasia and palliative care are considered to be fundamentally different practices, mainly due to ideological reasons, but in the view of the authors this opposition is in many respects quite artificial. The two approaches seem to be rooted in different ideological values. Euthanasia is ending the incurable suffering or decay by facilitating death; palliative care is end-of-life care helping patients to bear the suffering leading to death. However, the outcome of procedures for stopping treatment of the underlying disease and/or providing relief from symptoms in palliative care and euthanasia is quite similar in the sense that both relate to the ending of life, only with some difference in timing. Palliative sedation may be considered to be slow euthanasia.[41]

The religious opposition to euthanasia is based on the sanctity of life principle, because it is believed that all life comes from God and only God has the right to decide about its termination.[42] There are also objections against euthanasia which are based on secular arguments, for instance, that human life is a basic good[43] and there are concerns about possible abuses.

The proponents of euthanasia rely on the combination of the principles of quality of life and personal autonomy or self-determination. They argue that terminally ill individuals should themselves have the right to make end-of-life decisions, in case they consider that their quality of life is degrading to such an extent that it is not worth being further artificially prolonged.[44] Oddly enough, both advocates and opponents to euthanasia often invoke the principle of human dignity to justify their positions.[45]

[38]For instance, Torr (2000), Haley (2003), Allen et al. (2006), Lindsay (2008).
[39]Allen et al. (2006, 6).
[40]Although often considered a form of compassionate medical care for the terminal ill, palliative care (from the Latin palliare, to cloak) is a much broader medical specialisation which is aimed at reducing the severity of disease symptoms, rather than providing a cure, in order to prevent and relieve suffering and to improve quality of life for people facing serious, complex illness. Ventafridda (2006) defines it as a multidisciplinary approach towards patients and their families during the progression of incurable illness, the advanced stages of disease and the last hours of life.
[41]Distelmans (2012, 23).
[42]For instance, Pope John Paul II (1995, 102).
[43]For instance, Boyle (1989), Gorsuch (2006).
[44]For instance, Kohl (1992), Kurtz (1992), Lindsay (2008, 119).
[45]Behuniak (2011).

The ethical dimension of the management of death is accentuated today and it may be expected to be even more prominent in the future, due to technological advances that allow for artificial prolongation of some vital functions. Artificial prolongation is in fact enhancement of the duration of the dying process, without improvement of the potential for life or the quality of life.

So far, only a few political jurisdictions have started to legalise euthanasia *sensu* stricto and have set up a number of rigorous requirements so that both patients and the medical profession are protected.[46] These include three European countries—The Netherlands (in 2001), Belgium (in 2002), and Luxemburg (in 2008). Colombia legalised euthanasia in 1997 but the lack of a clear regulatory framework has limited its practice. In Switzerland euthanasia *sensu* stricto is not prosecuted when it is done without self-interest.[47] Assisted suicide[48] is legal in Germany, Japan, Albania, and in the US states of Washington, Oregon, Vermont, New Mexico, Montana and California.[49]

In many countries euthanasia is more and more being applied tacitly (usually in the form of strong palliative sedation), because the legislation is lagging behind societal and technological developments.[50] Legalisation would bring existing practices of (slow) euthanasia out of the grey area, and make them more open for the control of legal liability and protection of legal rights and obligations of both patients and medical care personnel.

Euthanasia *sensu largo*, meaning the terminating the life of patients suffering from incurable conditions or diseases irrespective of their own will, clearly raises another major ethical issue in the euthanasia debate, as it includes patients both with and without informed consent or an explicit request. Euthanasia *sensu largo* can, moreover, apply both to mentally competent as well as to mentally incompetent terminally ill patients.

The euthanasia legislation presently existing in the few countries that have taken this route is still very limited. For instance, in Belgium euthanasia is only allowed in its restricted sense (ending the suffering of an incurably ill patient at his/her explicit request) and is not applicable in the case of acquired incompetence (dementia). Furthermore, elderly people fall outside the law if they are not suffering unbearably from an incurable disease, but want to step out of life because they are suffering from many geriatric ills, lost their partner and have no meaningful future life perspective. There are continuous debates in Belgium about the extension of the right to euthanasia, for example in case of poly-pathologies being considered unbearable.

In 2014 the Belgian euthanasia law was extended to minors. The conditions are that the minor is 'judgment proficient' and that the parents or legal guardians agree

[46]Deliens and Van der Wal (2003), Nys (2003), Griffiths et al. (2008).

[47]Giroud et al. (1999), Bosshard et al. (2002).

[48]Physician-assisted 'suicide' implies that a physician provides a lethal substance to a patient who can apply it at his/her own convenience. It stands in contrast to voluntary active euthanasia in which a physician performs the intervention requested by a patient.

[49]Wikipedia: legality of euthanasia.

[50]Gastmans et al. (2006).

to the request. The law provides for the possibility of euthanasia for minors only in the case of unbearable and hopeless suffering physically.

Regarding the delicate problem of active ending of life on infants, a remarkable and responsible initiative was developed at the Department of Paediatrics of the University Medical Centre Groningen (UMCG) in the Netherlands in the early 2000s, known as 'The Groningen Protocol' which is now applied all over the Netherlands.[51] The Groningen Protocol deals with cases in which a decision is to be made to actively end the life of an infant (1) with no chance of survival, (2) with a very poor prognosis and dependence on intensive care without hope of recovery, and (3) with a hopeless prognosis, who experiences what parents and medical experts deem to be unbearable suffering. The procedure is subject to very strict conditions such as full agreement of the parents on the basis of a thorough explanation of the condition and prognosis and the agreement of a team of physicians, including at least one that is not directly involved in the care of the patient. After the intervention an outside legal body assesses the justification for the decision.

Two of the general ethical goals developed earlier—ontogenetic development of human-specific potentialities and the promotion of quality of life—lead to the logical conclusion that favours managing death. In the present phase of modernisation medicine is not yet able to prevent extreme human suffering, and it even produces unintended prolongation of severe terminal suffering or degrading regression. In the light of the use of life-supporting technology, it seems appropriate that people should have the right to refuse artificial prolongation of the dying process and/or the end to suffering that inevitably leads to death. In pre-modern living conditions, natural selection automatically and rapidly—although often painfully—eliminated serious forms of cognitive and physical deterioration and degeneration. This was nature's way of avoiding the survival of human life that had lost its specific human nature. Euthanasia is a new form of cultural management of the dying process in which a person concerned has the last word.

6.2.3.5 Increasing the Human Species-Specific Lifespan Potential?

Whereas very few people will reject the goal of reaching the species-specific lifespan in good health, strong differences of opinion exist in the bioethical profession and literature on the question regarding whether to also pursue efforts to increase the human species-specific lifespan potential.

Bioconservatives usually fiercely reject the idea of striving towards extending human longevity beyond the current human species-specific potential. Some of them substantiate their claim on the basis of religious beliefs which accept, value, and submit to the way in which the professed God(s) created human nature, including the limited human lifespan and its embouchement in death. Leon Kass[52] even makes the case for the virtues of mortality on the basis of a strange gallimaufry of moral benefits such as "interest and engagement", "seriousness and aspiration", "beauty

[51]Verhagen and Sauer (2005).
[52]Kass (2002, 264).

and love", and finally "virtue and moral excellence". Other opponents of lifespan extending strategies refer to the fact that death is an inherent part of life and is needed for the continuation of the human species.[53] Some others invoke dangers of increasing inequalities and injustices in the availability or appropriation of (expensive) lifespan extending techniques, or fear that life-extending strategies would strengthen the age-graded hierarchies or considerably aggravate the currently already serious demographic imbalances regarding diminishing shares of young people.[54]

At present, it is not yet possible to prolong the species-specific human lifespan for adults.[55] Nevertheless many biogerontologists are convinced that the future progress of the biomedical sciences will ultimately and perhaps even quite soon—possibly in this millennium—succeed in discovering the genetic mechanisms which result in senescent degeneration and death. By knowing the mechanisms, it may become possible to circumvent senescence, either via genetic engineering or by the means of technological interventions in ontogenetic developmental processes.[56] Contrary to what some bioconservatives moot,[57] no biogerontologist in their right mind ever suggested prolonging the human lifespan with its present load of degenerative phenomena in very old age. They only advocate prolongevity, namely a significant extension of the human lifespan, free from senescent diseases and disabilities.[58] Some visionaries even prophesy that future biomedical progress will not only prolong the natural human lifespan potential by a few tens of years but by many hundreds, if not thousands of years.[59]

The perspective of extending the species-specific human lifespan also raises the thorny question of immortality.[60] Extending lifespan how far? Ten years, one hundred years, one thousand years? Why not infinitely?

How can lifespan extending aspirations and immortality ambitions be evaluated from an evolutionary point of view?

From an individual ontogenetic perspective there can be no doubt that, due to the individual drive for self-realisation and self-preservation, people would want to live a long (and healthy) life; and why not an eternal life? The comforting belief in an eternal hereafter in many religions (and the success of the many books on life after death)[61] is an eloquent expression of this desire to perpetuate oneself in the (far) future. However, it can be expected that life extending aspirations in the modern knowledge-based world are no longer concentrated on the hereafter. Today science is advancing toward this goal through gene manipulation. At the same time advances in scientific knowledge render the probability of an existence of life after

[53]For instance, Hayflick (1996).
[54]For instance, Lewis (2001), Fukuyama (2002, 57ff), Chapman (2004).
[55]For instance, Olshansky et al. (2002).
[56]Bostrom (2005).
[57]For instance, Fukuyama (2002, 68).
[58]Gruman (2003).
[59]For instance, Bacon (1627), Metchnikoff (1907), Stapledon (1930), de Grey (2007).
[60]For instance, Ettinger (1965), Ettinger et al. (2005), Harrington (1969), Adams (2004), Zey (2006).
[61]For instance, Chopra (2006), D'Souza (2009), Kübler-Ross (2012).

death ever smaller. Whenever or wherever anti-ageing procedures are discovered, not only resulting in decreasing or eliminating senescent degeneration but also leading to a longer lifespan, there can be little doubt that many, if not most, people would want to take advantage of those discoveries. We see today the exponential growth of beauty- and health-improving cosmetic and medical industries that help to preserve youthful looks.

The desire to extend human longevity beyond its current natural boundary will have to be carefully matched against the risks of poor quality of life at very old ages and the prolongation of the dying process in pain and discomfort.

Furthermore, its meaningfulness will have to be compared to the length of the other stages of the life course.[62] Indeed, from an evolutionary point of view, the potential lifespan of the human species is a biological feature that has been selected on the basis of the long maturation time of its offspring and relatively low human fecundity. As long as the duration of human biological maturation does not increase or other reproductive traits do not require a longer reproductive life phase, natural selection will not push toward a longer lifespan. From that viewpoint an additional prolongation of the potential lifespan seems to be a redundant, useless investment.[63] However, modernity has fundamentally changed the natural rules of the evolutionary game: the human species no longer wants to be the plaything of natural biological processes. It has embedded in its culture a strong desire to direct its own future evolution. The question is then, should efforts be made to prolong the human natural lifespan, should we strive for immortality?

Further extending the human-specific lifespan makes sense, not only from an ontogenetic point of view (because of our desire to live long and healthily) but also evolutionarily, with the perspective of a progressing hominisation which, by definition, will be characterised by a further extension of brain capacity. Based on the past hominisation trend, which was characterised by a positive correlation between brain expansion and lifespan increase,[64] it is quite likely that a further increase in brain capacity would also need a further extension of longevity. This is not only because of a possibly longer maturation time but also because of the longer time a more highly performing brain might (have to) be creative and functional.

From an individual transcending evolutionary point of view, lifespan extension will somewhat slow down the evolutionary process over time—as generation length is one of the determinants of the possible rate of intergenerational genetic change. However, this should not be a major evolutionary obstacle, given the very long time perspective during which life can be envisaged to further evolve on our present planet, namely some 500 million to one billion years. More important for the survival, or at least the well-being, of the species is that a longer lifespan will have an effect on the potential future size of the human species, the numerical stabilisation of which at a stationary or even reduced level might have to be compensated by even stronger birth controls than without further increasing longevity. Also, the

[62]See also Temkin (2008).
[63]Hayflick (2000), Perls and Fretts (2001).
[64]Sacher (1959), Judge and Carey (2000).

concerns about the risks of increasing social inequalities and injustices in the availability or appropriation of (expensive) lifespan extending techniques have to be seriously taken into consideration. However, these social concerns, as well as the demographic effects related to further extending lifespan longevity, are primarily of an ontogenetic nature and may only indirectly be of phylogenetic importance, namely influencing the genetic future of the human species or its successor(s).

Enhancing individual longevity—within certain limits—does not seem to fundamentally endanger the future evolution of the human species or its successors, at least if ontogenetically it were to be properly managed. By contrast, the pursuance of individual immortality would considerably challenge further evolution of humankind. Immortality combined with the continuation of reproductive capacity could imply the end of generational replacement and continuous population growth. Immortality could fundamentally challenge strivings for further hominisation, which requires intergenerational genetic change. Indeed, we should be well aware of the fact that individuals cannot evolve: only populations (species) evolve. Hence, achieving individual immortality might, in theory, mean the end of hominin evolution. Some argue that preventing individual immortality is a necessary condition for achieving species immortality in the sense of the appearance and evolution of an endless series of successive future hominin stages.[65]

Prevention may not be necessary as the practical realisation of immortality might prove to be less achievable than its theoretical premise. Indeed, John Harris[66] reminded us of the high probability that immortals, although genetically predisposed to live eternally, might not be completely invulnerable to life course accidents which might, on average, reduce their theoretical immortality to an admittedly high, but nevertheless time-limited longevity.

6.3 Sex Variability

6.3.1 Evolutionary Background of Sex Variability

6.3.1.1 Specificity of Human Sexuality

As already discussed in Chap. 2, Sect. 2.1.2.4, sexual selection shaped the specific human sexual dimorphic features in the course of the hominisation. The specificity of human sexuality is related to the decrease of the difference in parental investment between the two sexes and to the development of cooperative breeding.[67] This resulted in the establishment of more enduring relations between the sex partners. The ultimate explanation for the human sexual specificity is to be found in the singularity of the human offspring, which requires enduring and intensive care, due

[65]For instance, Gyngell (2015) advances evolutionary arguments against radical increases in life expectancy in general.
[66]Harris (2007, 68).
[67]Isler and van Schaik (2012).

to its long-term neediness. The need for sustained care of children, not just their procreation, lies at the basis of the specificity of human sexuality.[68]

The need for longer lasting sexual relations, increase in paternal involvement, and investment in ever longer maturing offspring in the course of the hominisation, necessitated a feminisation of the hominin male which resulted in a less aggressive, more cooperative, caring, providing and protecting partner, as well as a decrease in male-male competition.[69] Also, the hominin female developed specific sexual characteristics—such as concealed ovulation, large breasts, multiple erogenous zones, face to face sex interaction, shift from a cyclical toward a non-cyclical sexual readiness, capacity for orgasmic sex—all favouring the enduring interest of the male partner.[70]

An evolved aspect of the specific human sexuality is the strong sexual drive, resulting in coital frequencies that surpass by far the amount needed for generational replacement. The strong sexual drive in the human is often misunderstood. It is not a relatively recent uneconomic maladaptation, as Raymond Cattell[71] wrongly hypothesised, but one of the instruments produced by natural selection in order to facilitate the establishment and maintenance of enduring partner relations.

In the course of the hominisation process, dimorphism in secondary sexual characteristics, such as body size, musculature, assertiveness and energetic activity, competitive and aggressive behaviour, has not yet disappeared completely. The hominin transition from a scavenger-gatherer economy to a hunter-gatherer economy was accompanied not only by increased paternal investment but it also resulted in, and even necessitated, a sex-specific task and role division. Women continued to specialise—very successfully—in caring for and socialising ever slower maturing children and in more sedentary food gathering.[72] In contrast, men concentrated on strongly mobile (group) hunting, initially on small game, and later also on larger prey, and on the defence and conquest of women and territories.[73] This task and role division engrafted itself onto the existing sexual dimorphism in robustness of the early hominins. However, robustness has decreased due to the increasing paternal investment and relaxation of natural selection due to the development of technology.[74] It is not impossible that still other factors, such as protection against predators,[75] the effect of larger body size,[76] or the remnants of ancestral genetic

[68]The care of the big-brained, long maturing hominin infant influenced not only the evolution of the cooperative breeding of the parents but also involved support from other family members such as juveniles (e.g. Kramer and Otárola-Castillo 2015) and grandmothers (e.g. Hill and Hurtado 1991).

[69]Symons (1979), Cliquet (1984), Rancour-Laferriere (1985), Lancaster et al. (1987), Ridley (1993), Geary (1998), Miller (2000), Cieri et al. (2014).

[70]Lancaster (1985), Rancour-Laferriere (1985).

[71]Cattell (1972, 245).

[72]Dahlberg (1981), Slocum (1980).

[73]Chagnon (1990).

[74]Brace and Ryan (1980), Frayer (1980), Frayer and Wolpoff (1985).

[75]DeVore and Washburn (1963).

[76]Leutenegger and Cheverud (1982).

make-up,[77] have also played a role in the reduction of sexual dimorphism in *Homo sapiens sapiens*. Very probably our current sexual dimorphism is multifactorial in origin.[78]

The moderate sexual dimorphism of *Homo sapiens sapiens* is a good example of an evolutionary compromise, whereby selective pressures operated in different directions. It resulted in the preservation of (reduced) male robustness and aggressiveness; at the same time it adapted to the needs of group hunting and territorial defence and conquest and to the enhanced requirements of cooperation and sociability with respect to relational and parental investments in large brained and long maturing offspring.[79]

The still existing sexual dimorphism between males and females, not only in body build, genital sexuality and reproduction but also in mind (mental aspects of sexual differentiation),[80] has an important consequence that males are somewhat more oriented toward impersonal sexuality (sex dissociated from love) whereas females are more focused on personal sex (sex with carefully selected partner).[81] This helps to explain phenomena such as the higher prevalence of polygyny over polyandry, the stronger desire or actual behaviour of many men in having sexual intercourse more quickly, more often and with more partners. Another consequence of the human sexual dimorphism is that, in strongly hierarchical societies, men not only compete with other men but also control and exert power and dominance over women.[82] This is, for instance, manifested in the association of patriarchy and non-democratic societies, with the higher prevalence of male sexual intimidation, harassment, abuse, and even rape in some circumstances. The sexually generously equipped but reproductively poorly armed human males strived to control women by all means including segregation and imposing submissiveness due to the fear of the cuckoldry syndrome.[83]

There is strong evidence that the hominisation process was characterised by a shift from promiscuous sexuality, over polygamy and toward monogamy.[84] As a long-term mating strategy, monogamy had numerous important evolutionary advantages. Children in a monogamous family have a higher coefficient of relationship than in any other social unit;[85] women can garner far more resources for their children through a single spouse than through several temporary sex partners.[86] Furthermore, monogamy increases paternal certainty; and children's survival and later reproductive success is more likely as a result of higher paternal investment.[87] Monogamy may be said to be the expected outcome of K-selection

[77]Cheverud et al. (1985).
[78]Plavcan (2001).
[79]Buss and Malamuth (Eds.) (1996).
[80]Cliquet (2010, 199–210).
[81]Malamuth (1996, 275).
[82]Buss (1996, 306), Wrangham and Peterson (1996).
[83]Hiatt (1989), Taylor (2002), Geary (2006).
[84]For instance, Marlowe (2003), Chapais (2013).
[85]Melotti (1980).
[86]Buss (1999).
[87]Buss (1999).

processes in which parental investment is equalised between the sexes.[88] Monogamy decreases sexual competition and increases social cooperation and bonding between males.[89] In the transition from hunter-gatherer to agrarian economy, with its increasing residential population and higher risk of sexually transmitted infections, socially imposed monogamy groups may have reduced the exposure to such diseases.[90] The adoption of intensive agriculture favoured the establishment of monogamous unions because ownership of land became critical to productive and reproductive success.[91] Monogamy contributes to the establishment of democracy and the harmonious functioning of larger societies in which sexual competition between males is further reduced and their likelihood of cooperativeness increased.[92] Normatively imposed monogamy reduces the number of unmarried men, also resulting in a decrease in crime, rape and murder, thereby also heightening success in competition between populations.[93]

Polygamy, in particular polygyny, is obviously in the interest of dominant powerful males, producing salient male-male inequalities.[94] It is usually prevalent in societies with strong differences in wealth between males, and with authoritarian governments needing to suppress the sexually have-nots.[95] It is a form of partnership that is inadequate in modern, large size societies where male-male relations have strongly shifted from competition to cooperation and to more egalitarianism.

However, in socially monogamous societies extra-pair mating is a not uncommon practice. For males, extra-pair mating is evolutionarily explained as a secondary mating strategy aimed at spreading genes over a larger number of partners. For females, the evolutionary advantage is thought to rely on obtaining an increase in quality of genetic offspring.[96] Recent genetic research findings confirm the genetic underpinnings of variation in extra-pair mating in humans and the ancient biological evolutionary background of this mating strategy.[97]

An important factor for understanding the sexually differential mating strategy in the human is the strongly age-linked sexual differences in the capacity to produce additional viable offspring. In the human species, the male's reproductive capacity is huge and is, from puberty onward, relatively independent of age. In contrast, women's fecundity is low and strongly time dependent.[98] Women's reproductive capacity is strongly associated with age-varying physical and behavioural features such as general body build, in particular waist-hip ratio,[99] youthful facial traits,

[88]MacDonald (1995).
[89]Chapais (2013).
[90]Low (1990), Bauch and McElreath (2016).
[91]Fortunato and Archetti (2010).
[92]Alexander (1971; 1979), Betzig (1986), Holcomb (1993), Ridley (1993), Voland (1993), MacDonald (1995), Van Schaik and Michel (2016, 136).
[93]Henrich et al. (2012).
[94]Kanazawa and Still (1999), Marlowe (2000), de la Croix and Mariani (2015).
[95]Betzig (1995), Arnhart (1998), Geary (1998, 156), Rubin (2002, 118).
[96]Buss and Schmitt (1993).
[97]Garcia et al. (2010), Zietsch et al. (2015).
[98]Eaton and Mayer (1953), Charbonneau (1979), Nieschlag (1986).
[99]Singh (1993), Hughes and Gallup (2003), Streeter and McBurney (2003).

breast form, skin and muscle tone, fat distribution, and energy level.[100] Physically attractive and behaviourally dynamic features peak at younger ages. Therefore, youth and physically attractive features were in the past considered to be strong indicators of high reproductive value[101] and are still highly valued by men today.[102] Indeed, women tend to maximise their youthfulness and display their youthful physical appearance.[103] Women systematically show a stronger preference for resourceful, somewhat older, caring men who are willing to invest time, energy and emotion. Women also have a preference for men with bodily and facial features that express strength, social dominance, but also sociability and confidence. Height is a typical and well-known example of such a feature: it is a perceived as a sign of social status, dominance and protection.[104]

6.3.1.2 The Origin and Evolution of Love

Love is one of the fundamental behavioural characteristics of the human species. It is largely biologically determined.[105] In the expert literature, two major aspects of love are distinguished: compassionate love (sexual desire) and companionate love (romantic love). Although both are usually related, they are considered to have a different evolutionary origin and function—the first being an expression of the reproduction related mating system, the second an expression of the pair-bonding system that evolved as a function of raising slowly maturing offspring. Both components are controlled by distinct patterns of brain activation.[106]

The origin of love between heterosexual adults can be traced back to the earliest hominin ancestors, because it formed the psychological foundation of the social bond that made, via increasing bi-parental care, an essential contribution to the survival of children.[107] The experience of love can be detected in virtually every human population: strong evidence of its existence is found not only in all literate civilisations[108] but also in most preliterate societies.[109]

The presence of biological predispositions for love does not mean that they will automatically and universally manifest themselves. These predispositions need to be socially and culturally fostered. Love may get competition from other drives or its expression may be suppressed in particular living conditions. The experience of love may depend upon the degree of enduring compatibility between partners. Last but not least, there may be individual variation in the (genetic and ontogenetic) ability to develop long-lasting affectionate feelings. Consequently, one may expect

[100]Barber (1995).
[101]Fisher (1930), Williams (1975), Howell (1979).
[102]Buss (1989), Thiessen et al. (1993), Bereczkei and Csanaky (1996), Wiederman and Kendall (1999), Lee et al. (2014), Antfolk et al. (2015).
[103]Kachigan (1990), Moir and Jessel (1992).
[104]Ridley (1993).
[105]For instance, Fisher (1992), Bartels and Zeki (2000), Emanuele et al. (2007), Savulescu and Sandberg (2008).
[106]See, for instance, Diamond (2004), Fisher et al. (2005), Gonzaga et al. (2006).
[107]Mellen (1981), Gangestad (2011, 125).
[108]For instance, Mellen (1981), Buss (1994).
[109]For instance, Westermarck (1922), Jankowiak and Fischer (1992), Jankowiak (1995).

a substantial variation, within as well as between populations, in the prevalence and the degree of development of durable affectionate relations. Most men and women continue to opt for durable attachments also in situations where people have a relatively broad opportunity to choose the type of partnership and living arrangement.[110] Even most of those who end a cohabitation or marriage often seek a new durable partnership.

The evolutionary origin of love, and particularly of its companionate component, is grounded in the cooperation between the sexes which is aimed at securing parental investment in offspring.[111] The view that human love is an evolved feature, selected for its function in meeting the needs of slowly maturing offspring, is perhaps less easy to accept than the conclusion that relationships between parental affection and child development have an evolutionary basis. Indeed, in many cultures, the drives for protection, survival and economic security in partnerships seem to be preponderant over affectionate needs, for instance, as can be seen in arranged marriages. Moreover, cross-cultural studies of marital patterns,[112] as well as studies of sexual behaviour in modern culture,[113] have documented the mildly polygamic nature of the human—at first sight this feature might seem to be at odds with the need for enduring love. However, the presence of and competition between several drives, such as the need for love, the desire for several sexual partners, and the urge for resource acquisition, does not repudiate their existence and functionality. Clearly, in some socio-ecological conditions one or the other of these drives may become predominant. It is quite understandable that economic drives may have prevailed over affectionate needs in ancestral conditions of material hardship where mere survival was at stake. It is also quite understandable that, according to Louis Roussel,[114] from the end of the eighteenth century the idea of happiness became the preponderant foundation of marital and family life in the West. In conditions where vital basic needs (essential for individual or group survival) are satisfied, emotional needs, and also derived and acquired needs, may take precedence.[115] Thus, the process of modernisation is very probably characterised by a major shift from survival to love as the basis of family life.[116] However, it would be a mistake to think that romantic love is a Western invention. There is sufficient evidence that shows the universal nature of love.[117]

6.3.1.3 Evolution of Sexual Deviations

The human species is characterised by the presence of several forms of sexuality that deviate from the majority pattern of human heterosexuality and reproduction.

[110]Laumann et al. (1994), Corijn and Klijzing (2001).
[111]For instance, Buss (2007, 379, 380).
[112]Westermarck (1922), Murdock (1961), Mellen (1981).
[113]Kinsey et al. (1948; 1953), Spira et al. (1993), Laumann et al. (1994), Wellings et al. (1994).
[114]Roussel (1989).
[115]Montagu (1955), Maslow (1972).
[116]Farrell (1993), Gray and Garcia (2013).
[117]Jankowiak and Fischer (1992), Buss (1994, 2007), Fisher (2004).

Examples are celibacy, homosexuality, philandering, incest, prostitution, sadomasochism, fetishism, rape, paedophilia, exhibitionism, voyeurism and transvestism.[118]

From an evolutionary perspective, sexual deviations are quite specific behavioural phenomena, several of which are difficult to explain. With the exception of evolutionary studies about incest,[119] homosexuality,[120] rape[121] and occasionally pornography[122] and prostitution,[123] very few evolutionary analyses have been made of the evolutionary origin of sexual deviations.[124]

John A. Hewitt's[125] thesis is that some sexual deviations may be by-products of particular sexually related forms of behaviour that had evolutionarily meaningful functions such as male-male bonding (leading to homosexuality), sexual dominance/submissiveness (leading to sadomasochism, rape, and even coercive forms of marriage arrangements), the need for long-term child care or even the preference for youthful reproductive traits (leading to paedophilia), and the need for division of labour (leading to fetishism). If such a by-product is reproductively maladaptive, the evolutionary trend that produces the sexual deviation will be counteracted by a lower reproductive fitness, resulting in a stable evolutionary state with the deviation reduced to a demographic rarity.

Sexual deviations occur in both sexes but most, if not all, are much more prevalent in men than in women.[126] There are probably several proximate factors causing this striking difference, some of which may also be deviation specific. Hewitt has given a plausible evolutionary explanation: since men can fertilise several women, on average they are reproductively more disposable than women and can acquire traits that fulfil other functions. Women's reproductive investment in each offspring is much higher, costly and makes her more vulnerable and selective in her reproductive strategies.

Another way of viewing some sexual deviations consists of the possibility that sexual deviations are by-products of the evolution of the hominin brain which

[118]In the sexuological literature, the definition of sexual deviation shows some variation. Some authors limit it to activities involving a non-human object, a non-consenting partner such as a child, or pain or humiliation of oneself or one's partner (http://www.thefreedictionary.com/Sexual+deviations); others include many more variants such as homosexuality, exhibitionism and voyeurism (e.g. Hewitt 2002). Here, the authors also include forms of behaviour such as celibacy and promiscuity, which deviate from the 'normal' pattern of human heterosexuality and reproduction.

[119]Westermarck (1906; 1922), Van den Berghe (1980), Shepher (1983), Wolf (1995), Wolf and Durham (2004), Turner and Maryanski (2005).

[120]Muscarella et al. (2001), Wilson and Rahman (2008), Vasey and VanderLaan (2014).

[121]Buss and Malamuth (1996), Thornhill and Thornhill (1983), Thornhill and Palmer (2000), Brown Travis (2003).

[122]For instance, Seto et al. (2001).

[123]For instance, Burley and Symanski (1981), Buss (1994), Schmitt et al. (2001), McGuire and Gruter (2003).

[124]For instance, Goodman (1998), Hewitt (2002), Shepher and Reisman (1985).

[125]Hewitt (2002).

[126]McAnulty and Burnette (2006).

allows it, in combination with some other biological features or drives, to develop behavioural patterns which are modifications of the basic biological drive.

For instance, the origin of human homosexuality, in the sense of a lasting erotic and emotional preference for members of the same sex as a persistent form of behaviour during the life course, has been considered as a by-product of some facets of the hominisation process. Notwithstanding its lower reproductive success,[127] i.e. its largely maladapted nature,[128] homosexuality is maintained in the population at a relatively high frequency[129] which cannot only to be accounted for by recurrent chance mutations. Hence, evolutionary scientists have been looking for compensating reproductive advantages that this type of behaviour might have had in human evolution and its genetic or ontogenetic mechanisms.

Several explanations have been suggested: homosexuality has some selective advantage due to the presence of genetic variants with pleiotropic effects (sexually antagonistic selection promoting higher fecundity in females and homosexuality in their male offspring)[130] particularly in socially stratified societies,[131] or genetic variants that interact with other alleles in particular genotypes resulting in heterosis.[132] Another group of explanations relate to interactions between individuals, including theories based on kin selection,[133] reciprocity,[134] parental manipulation[135] and homosociality.[136] A third group of explanations suggest that homosexuality is a by-product of the feminisation process during human evolution.[137] Sydney Mellen[138] sees homosexuality as a side effect of a galloping hominisation that advantaged less aggressive, more social, sensitive and communicative males, resulting in an excessive feminisation of some male individuals. An alternative possibility is that it is a consequence of neotenic changes, which caused certain terminal stages of male behavioural differentiation to arrest at a late premature stage in some individuals—a development that went a little bit too far in its trend. This feminisation side effect theory would also explain why homosexuality is much less prevalent among women who had, in the EEA, no reason to become more mas-

[127]For instance, Bell and Weinberg (1978), Schwartz et al. (2010).
[128]Ellis and Symonds (1897).
[129]For instance, Spira et al. (1993), Laumann et al. (1994), Wellings et al. (1994).
[130]Hamer and Copeland (1995), Camperio-Ciani et al. (2004; 2012), Zietsch et al. (2008), Bonduriansky and Chenoweth (2009), Schwartz et al. (2010), Blanchard (2011), Camperio-Ciani et al. (2015); see also Chaladze (2016).
[131]Barthes et al. (2013).
[132]Hutchinson (1959), Kirsch and Rodman (1982), MacIntyre and Estep (1993), Miller (2000).
[133]Wilson (1975), Weinrich (1978), Blanchard (1997), Apostolou (2013), VanderLaan et al. (2013).
[134]Trivers (1971).
[135]Trivers (1974); see discussion in McKnight (1997, 145–159), and in Kirkpatrick (2000).
[136]Kirkpatrick (2000), Ross and Wells (2000), Fleischman et al. (2014).
[137]Mellen (1981), Miller (2000), Rahman and Wilson (2003).
[138]Mellen (1981).

culine.[139] Also, Geoffrey Miller[140] has suggested that male homosexuality is a by-product of variable brain feminisation, produced by a polygenetic system in which single alleles for greater sensitivity, empathy, tenderness, and kindness make heterosexual carriers of those genes better fathers and more attractive mates. Rahman and Wilson[141] proposed that variations in genotypes produced hominin males who were more feminine in behavioural traits and bisexual in sexual preferences. Females were attracted to such males because they were associated with decreased aggression and infanticide, and superior parenting behaviour. Over time, this choice led to the evolution of alleles associated with exclusive homosexual interest.

6.3.2 Changes of Sex Relations in Modernity

Scientific knowledge, in interaction with modern ideologies—the ideals of the Enlightenment and the French Revolution, liberalism, socialism, humanism, and in particular feminism—is profoundly changing traditional sexual attitudes and behaviour in modernity.

6.3.2.1 Gradual Shift from Women's Subordination to Women's Empowerment

Hunter-gatherer societies are, on average, characterised by relatively equal sex relations. By contrast, due to the emergent availability of energy- and nutrient-rich staple food or livestock, the agrarian-pastoral cultural phase made use of the sexual dimorphic heritage of the hunter-gatherer era to establish or strengthen the patriarchal and socially dominant position of men in a societal context of increasing social inequalities of all kinds. Hence, social and sexual suppression of women increased considerably in agrarian-pastoral societies; this was characterised by an accumulation of material property in which women were, just like staple food or livestock, considered to be property.[142] Sexist[143] attitudes developed in all its aspects—exclusion of women from economic production and control, resulting in low social status; prohibition from owning property, involvement in politics, and receiving education; seclusion of women and the restriction of many of their

[139]However, the greater sexual fluidity in women might have been an adaptation to the need for cooperative breeding which was a strategy through which ancestral women obtained additional investment for their offspring in order to compensate for common crises such as rape, paternal desertion or death (Kuhle 2013; see also Kanazawa 2016). Another explanation for the divergence of exclusive heterosexual orientation in women is that, in pre-modern living conditions, selection pressures against alleles for such orientations were weak because of the strong parental and partner control of female mating behaviour (see Apostolou 2016).

[140]Miller (2000).

[141]Rahman and Wilson (2003).

[142]Wood and Eagly (2007, 387); see also Konner (2015).

[143]The term 'sexism' has been coined to define ideological and social systems in which sexual variation is used as a primary criterion to assign normatively differentially valued roles and tasks in society (Duberman and Azumi 1975).

activities and rights; confinement to what is called in German language the notorious three Ks: "*Kirche, Küche und Kinder*"[144]; close surveillance over female sexuality, premarital virginity and extramarital sex; development of double standards in sexual matters; and last but not least, the religious-ideological foundation, endorsement and enforcement of the biological inferiority and social subordination of women.[145]

Sexist attitudes and behaviour continued to prevail in early modernisation, and even today they are not completely eradicated. There may be four main reasons for this phenomenon: (1) the religious values and norms of the agrarian-pastoral era, with their ideological affirmation of women's biological inferiority and social subordination, are still with us and continue, albeit with decreasing intensity, to exert their influence on sexual relations;[146] (2) the social structural features of the agrarian era largely continue to survive in modern society, which is structurally and functionally still strongly tailored to men; (3) the biological specificity of men allows, or even forces, them in the context of the competitive nature of modern culture, to make use of their larger male body build and greater muscular strength and speed, to satisfy their strong ejaculatory needs, and to exert their drives toward energetic activity, assertiveness and risk-taking behaviour—all features which, in particular circumstances, can easily turn into aggressive and violent behaviour (toward women); and (4) last but not least, the human male remains confronted with a human female whose specific biological characteristics—such as being the preferred object for sexual intercourse and satisfaction, characterised by hidden ovulations, and being the bearer of children—not only make him compete with other males for mating and reproductive behaviour, but make him—in the modern emancipatory, egalitarian, and contraceptive culture vis-à-vis females—a requesting and, to make matters worse, depending party.

The interaction of the traditional ideological indoctrination on sex differences, the social structural and functional features of modern society favouring men's social status, and the biological specificity of human males, continue to make many men believe in their biological superiority and, hence, their social prerogatives.[147]

Although the authors share the view of Ross Honeywill[148] that modernity was—initially—largely a masculine enterprise, it is thanks to the development of modernity and, in particular (biological) science, that knowledge about the sexes and their traditional power relations started to change fundamentally. New knowledge about the differences in the nature and the abilities of both sexes makes it possible to have the necessary changes in conceptions, attitudes, values and

[144]"*Kirche, Küche und Kinder*": church, kitchen and children.
[145]Karimi-Boosherhi and Rasouli-Nia (1988), Jogan (1989).
[146]Due to immigration from developing countries, in particular with an Islamic religious or cultural background in which women are considered male property, Western societies are increasingly confronted with an upsurge of traditional macho or sexist attitudes and forms of behaviour; they are even experiencing behavioural phenomena such as veiling, honour killings and female genital cutting which had belonged for a long time in the past or they were never part of Western cultural traditions (Manji 2003; Bawer 2006; Van Rooy and Van Rooy 2010).
[147]For instance, Goldberg (1973, 1993), Farrell (1993), Connell (1995), Bly (2004).
[148]Honeywill (2016, 17).

norms. Biology simply swept away the traditional views on the nature of the sexes and destroyed the ideological foundations of sex inequality and inequity.

However, the application of science in the fields of medicine, economy, and technology is at least equally important. In the first place, biomedical knowledge has induced a revolutionary level of mortality control, the ultimate condition more particularly relevant to women's new opportunities. Modern medicine largely freed women from the high risks of infant and maternal mortality and morbidity, whilst control of infant mortality allowed—and in the end favoured—fertility control. This liberated women from virtually permanent pregnancies and allowed the establishment of a balance between reproductive, productive, and recreational functions in modern society. Of considerable importance in this respect is the development and availability of safe and effective methods of birth control. Modern contraception has had a considerable impact on the subjective perceptions and experiences of both sexes. Women are moving into a completely different power position.[149] Men have lost their position of control over women's sexual and reproductive behaviour and now need to negotiate out of a much more humble and equal status position than in the pre-contraceptive era.

6.3.2.2 Progressive Shift to More Sexual Equality

Overall, modernisation is characterised by a gradually decreasing inequality in a broad range of traditionally different behavioural patterns between the sexes, regarding sexuality, education, occupation, economic activity and political activation.

Several characteristics of modern societies—monogamous mating, humanistic and egalitarian ethos, democratic decision making, aggression prohibiting or controlling polity, complexity of cooperative interactions between citizens—have a suppressing effect on the traditional physical male-male competition. They are producing a shift in this competition away from physical confrontations toward various domains of social competition such as educational, occupational and economic success.[150] At the same time, the modernity features, and their effect on the changing male-male competition and decreasing male dominance over female behaviour, allow a fuller expression of female choice in various domains of social life, not the least in mating behaviour but also in domains such as education, occupation and leisure.

Some authors have advanced the thesis that modernisation is also influencing the physical condition of males, with important consequences for sex relations. The modern cultural environment could accelerate the male feminisation process that was phylogenetically concomitant to the hominisation process. Modern living conditions are indeed characterised by a decrease in the need for the typical old-time masculine attributes such as muscular strength, physical endurance and aggressiveness.[151] Modern technological developments in economic production,

[149]Van der Dennen (1995), Batten (1994), Cronin (2006).
[150]For instance, Geary (1998, 142).
[151]For instance, McAllister (2009); see also Farrell (1993), Tiger (1999), Sykes (2003).

but also in warfare, as well as the humanisation of interpersonal and intergroup relations, make the old-time masculinity features less necessary or desirable. Moreover, due to the increasing female emancipation, economic independence through work, social protection measures and modern methods of birth control, men have largely lost their double social control system—providing and impregnation—over women.[152]

6.3.2.3 Steady Shift from Sexual Repression to Sexual Emancipation

Modernity is also fundamentally changing the attitudes and behaviours regarding the repression of sexuality, more particularly—albeit not only—for women. The traditional suppression of any sign of a healthy sexual behaviour and pleasure is gradually being replaced by the acceptance of sexuality as an important and humane source of energy and happiness. In modernity, at last we are witnessing the first signs of development of an emerging sexual education, a real sexual emancipation and a true sexual culture.[153] However, there is still a long way to go.

For instance, sex education was only introduced in most advanced countries in the course of the second half of the twentieth century.[154] However, in some countries where sex education is offered, such as the United States, parents have the right to refuse to let their children take part in the lessons.[155] Often sex education is limited or diluted to physical or health and reproductive aspects and does not sufficiently deal with the specificity of human sexuality, namely affectionate relations. Also, in many countries delicate aspects such as contraception and abortion are not dealt with adequately. In the United States, many schools only deal with the so-called abstinence-only approach of sex education that promotes abstinence from sex but does not teach about contraception or condom use and avoids discussion about abortion.[156] Research reveals no evidence that abstinence-only sex programmes would delay sexual initiation, decrease the frequency of sexual intercourse or number of sexual partners, or reduce sexual transmitted infections (STIs) or pregnancy.[157] By contrast, more comprehensive sex education programmes appear to produce these effects.

6.3.2.4 Moderate Shift from Sexual Abuse to Sexual Safety

The gender emancipation in modernity has had an effect that traditional forms of sexual coercion, such as forced or arranged marriages, or hidden forms of sexual abuse such as incest, rape, forced prostitution or paedophilia, are no longer tolerated or come into the open and can be more effectively fought.

[152]For instance, Tiger (1999).
[153]Abramson and Pinkerton (1995), Kontula and Haavio-Mannila (1995), Comfort (2003).
[154]Bruess and Greenberg (2008), Sauerteig and Davidson (2008), Schroeder and Kuriansky (2009).
[155]http://www.ncsl.org/research/health/state-policies-on-sex-education-in-schools.aspx.
[156]For instance, Santelli et al. (2006), Ott and Santelli (2007), Kohler and Lafferty (2008).
[157]For instance, Collins et al. (2002).

The humanistic ideals of the modern democratic ideologies reject and condemn all forms of sexual coercion, also those that were accepted as normal or even the norm in traditional ideologies. In particular, this refers to forced or arranged marriages which (often) implied that the resulting sexual relations were not the outcome of personal choices or love of the partners, but were forced upon them because of ideological, social, cultural, economic or political interests of the parental family or broader community.

Although strong forms of incest are rejected in most cultures (incest taboo), the repression of sexuality in traditional ideologies had as a—presumably unintentional—consequence that incestuous behaviour was often hidden, unreported and, consequently, unpunished. The more open sexual climate in modernity facilitates incestuous abuses to become (more) public and to be sanctioned against. The changing climate also raises the moral awareness about incestuous abuses and may have a preventing effect. The increasing number of reported cases of incestuous abuse has raised the question as to whether the less repressive sexual climate is not the cause of this increase. Expert opinion seems to tilt toward the view that the increasing recorded frequency of forced incest is indeed the result of the greater openness to deal with this social pathology which remained hidden in earlier times, rather than a real increase due to a relaxation of the sexual mores in modernity.[158] However, it is not impossible that incest between consenting adult relatives might somewhat increase, now that efficient contraception or selective abortion can prevent the unfavourable genetic effects of inbreeding. Although the human species is, just as many other organisms, endowed with an incest avoiding predisposition, this aptitude is relatively weak and variable, and therefore it needed to be reinforced by a culturally induced incest taboo.

Forced prostitution and rape are practices that are rejected and punished in all cultures—except perhaps in conditions of war.[159] Although in pre-modern times rape may have been an evolved alternative reproductive strategy for males who were unable to successfully attract desirable females,[160] in modern civilisation it is considered as a severe form of sexual abuse. In the past, and often even today, due

[158]Feldman et al. (1991).

[159]Gottschall (2004), Kivlahan and Ewigman (2010).

[160]In contrast to the feminist theory (e.g. Brownmiller 1975; Mardorossian 2004) and the social learning theory (e.g. Malamuth 1980) of rape, the evolutionary theory of rape endeavours to look at the more distant (ultimate) causes of this form of sexual deviance, also taking into account proximate factors of social and biological nature. From an evolutionary perspective, rape is seen as a secondary reproductive strategy of individuals, mainly males, who have no ability or occasion, either due to female rejection or parental objection, to establish a long-lasting sexual relationship or investment in offspring. The predominance of rape among the male sex is explained by the fact that human males are still characterised by a smaller parental investment in offspring, and that they may increase their reproductive fitness by inseminating several females. Although there are strong selective forces against rapist behaviour (from females as well as from 'other' competing males) natural selection has apparently succeeded in preserving this violent form of sexual behaviour as a minority phenomenon (Denno 1998–1999; Thornhill and Palmer 2000; Brown Travis 2003; Zeedyk 2007; Apostolou 2013). It can be expected that in modernity, where contraception and induced abortion are broadly available, this form of behaviour will be more strongly selected against (see also Blum 1998, 251).

to overt or covert forms of sexism and misogyny, as well as the sexually repressive climate of traditional, male dominated society, rape resulted in laxity in prosecution, or even generation of guilt feelings among the victims. In many cases, due to the persistent sexist climate—cf. the widespread rape myths[161]—victims do not even report this crime because of the shame and social stigma associated with this humiliating and damaging sexual assault.[162] The raising standards on quality of life and social justice in modernity gradually result in an increased reporting of rape and sexual assault and conviction of rapists. Nevertheless, research indicates that many sexual assaults are still never reported to police.[163] Modern(ising) societies still have a long way to go in order to adequately deal with this unacceptable scourge.[164]

Paedophilia—sexual activity by an adult with a prepubescent child—is a practice that is not accepted in modern society where the social, spiritual and moral well-being and physical and mental health of children are highly valued. Children should have the right and opportunity to develop, at their own pace, their own sexuality and not be abused by dominant and egocentric adults who may be genetically and/or environmentally predisposed to such behaviour. Art. 19.1 of the United Nations Convention on the Rights of the Child[165] explicitly states that the child should be protected "from all forms of physical or mental violence, injury or abuse, neglect or negligent treatment, maltreatment or exploitation, including sexual abuse, while in the care of parent(s), legal guardian(s) or any other person who has the care of the child." The recent scandals in the Roman Catholic Church, concerning sexual abuse of minors by clergy, which have become public in many countries, are a typical example of the increasing public awareness and rising moral standards in the domain of sexual relations involving adults and children.[166]

6.3.2.5 Demographic Shift from Life-Long to Serial Monogamy

The life-long monogamy ideal in traditional Western societies has never existed in practice, either due to high mortality of women at childbirth, or to various forms of hidden or more or less tolerated extramarital sexual relations. Partnership in modernity is evolving towards a larger variety in partner relations and the number of partners.

The number of sexual partners over the life course has been increasing, due to the increasing occurrence of premarital sex in adolescence, the increase in successive relations as a result of the breaking-up of unions (separation or divorce), reconstitution of new partnerships (remarriage, successive cohabitations), and the increase in multiple partnerships, either simultaneous or subsequently.

[161]For instance, Suarez and Gadalla (2010).
[162]For instance, Kilpatrick et al. (1992); Kilpatrick (2000), Suarez and Gadalla (2010).
[163]Kilpatrick et al. (1992), Jones (1999), Kilpatrick (2000), Ellis (1989, 3).
[164]Westmarland and Gangoli (2012).
[165]United Nations (1989).
[166]Pilgrim (2011).

However, from recent representative sex surveys[167] it appears that the large majority of people want and live in long-lasting monogamous relations, or reconstitute such type of relations after a break-up. Successive monogamy has again become a statistically important phenomenon, albeit because of separation or divorce instead of the earlier mortality of one of the spouses, in particular of women at childbirth. Extra-pair relations, multiple partnerships or even philandering appear to be all forms of minority behaviour.

6.3.2.6 Modest Shift from Homophobia to Homophilia

The traditional homophobic attitudes derive from Judeo-Christian-Islamic ideology that strongly condemns, punishes and persecutes same-sex relations, which were considered as a socio-pathological behaviour.

Judd Marmor[168] sees fears and anxieties behind homophobia, which is fostered by one or more of three major factors: deep-seated insecurity concerning one's own sexuality and gender-identity, strong religious indoctrination, and simple ignorance about homosexuality. From an evolutionary point of view, a more basic cause for homophobia may be attributed to the fact that in order to maximise their inclusive fitness, people do have an interest in the sexual orientation of their offspring.[169] Just as is the case for parents whose offspring remains childless for other reasons, such as celibacy, sterility or infertility, parents of homosexuals faced the prospect of having no grandchildren and not seeing their genes perpetuated in the future.

In recent decades, homosexual behaviour has become a subject of greater interest in modern society. Various factors may have contributed to this rising interest: the increasing ideological pluralism in advanced democratic societies, the progress of egalitarianism in general, the advance in scientific knowledge about homosexuality, the shift from a belief-based towards a knowledge-based ethics, and the impact of the holebi social movement (homosexuals, lesbians and bisexuals) that has succeeded in putting their case on the public agenda. The attention given to homosexuality in recent years is linked with deep-rooted discrimination and not with the demographic prevalence. It appears from probability sample surveys in different countries that the prevalence of same-sex couples is statistically very limited, ranging from three to five percent among men and one to three percent among women.[170]

In the immediate future, as homosexual relationships become less stigmatised, their prevalence may be expected to further increase, or at least become more visible. Since homosexuality is largely determined by biological (genetic and/or ontogenetic) factors, and with the shift from a belief-based towards a knowledge-based ethics and egalitarian pluralism, it may be anticipated that there will be increased social acceptance of homosexual households and families as a minority variant. In ethically advanced countries same-sex couples can now register

[167]Spira et al. (1993), Laumann et al. (1994), Wellings et al. (1994).
[168]Marmor (1980).
[169]Gallup and Suarez (1983).
[170]Diamond (1993), Spira et al. (1993), Laumann et al. (1994), Wellings et al. (1994).

their partnership and in some countries[171] they can marry and have the same rights and obligations as heterosexual couples.

A possible longer-term consequence of the lifting of the taboo on homosexual relationships might be that the genes for same-sex preference may decrease in the gene pool, since the transmission of genes for homosexuality via (forced) heterosexual relations will be reduced. The increasing number of homosexual unions might decrease pairings of unknowing heterosexuals with covert homosexuals. In other words, the selection intensity against alleles inducing homosexual orientation may increase.[172] Social selection against gay genes may also be reinforced through prenatal intervention, when such genes might become detectable prenatally because of parental preferences for heterosexual orientation of their offspring.[173] Furthermore, in the modern contraceptive society, where high parity is increasingly avoided, the increased risk of adult homosexuality due to the fraternal brotherhood effect[174] may be expected to become less prevalent. Finally, environmentally induced homosexual behaviour, for instance, related to early life experiences such as overprotective motherhood and authoritarian fatherhood,[175] might decrease thanks to progress in psychological and pedagogical insights and their social dissemination.

However, if (male) homosexuality is, indeed, evolutionarily linked to gracilisation, in particular the feminisation of the human male in the hominisation process, and these gracilisation and feminisation trends continue to progress, the reducing trends sketched above might be somewhat counteracted.

6.3.3 Ethical Reflections Regarding Sex Variability in Modernity

Modernity is characterised by a salient pluralism in viewpoints about sexual values and norms.[176] A broad multiplicity of elements underpins this variability in attitudes, in particular the persistent confrontation of the traditional ideological values of the Abrahamic religions in the West and Islamic countries with the scientific knowledge and the secular ideological movements of modernity. None of the modern ideologies have been able to produce, let alone impose, a generally accepted or acceptable norm system guiding and ruling sexual behaviour in

[171]Belgium, Brazil, Canada, Denmark, France, Luxembourg, the Netherlands, New Zealand, Norway, Portugal, South Africa, Spain, Sweden, the United Kingdom, and in some regions of Mexico and the United States.

[172]See also Apostolou (2016).

[173]Ehrlich (2003).

[174]Fraternal brotherhood effect: the probability of displaying a homosexual preference increases with the number of older brothers (Blanchard 1997; Blanchard and Klassen 1997; Barthes et al. 2013).

[175]However, solid evidence in support of the social environment as a causal factor in the development of sexual orientation does not yet exist. In general, children raised in homosexual environments have a heterosexual orientation (Swaab 2004).

[176]For instance, Cozic (1995), Williams (2000).

modernity. Therefore, an evolutionarily based approach to sex relations in the context of a further progressing modernisation, allowing for a broader time frame and conceptual distance, might relativise some of the current controversies or ambiguities and help to solve them.

Moral norms have always been strongly directed towards regulating sexual relations. The explanation for this concern is not always well understood. At the proximate level, sex is one of the strongest competitive (and often also aggressive) drives in human behaviour and consequently it has to be ordered socially. At the ultimate level, sex is the instrument for reproductive behaviour and consequently for the (differential) transmission of genes—the pre-eminently most important factor at the intergenerational level.

In small communities (of the hunter-gatherer type) and in larger, belligerent societies, sexual relations were strongly under societal control because the reproductive behaviour of its members was of crucial importance for the survival or the expansion of the community. Modern, and in particular peaceful, societies with their millions of citizens are much less vulnerable from a reproductive point of view and, hence, can be much more relaxed in matters of sexual relations. Societal continuity no longer depends on the reproductive performance of virtually every individual or couple. Therefore, a broader variety of sexual relations can be tolerated.

This greater tolerance towards sexual diversity, however, does not imply that the ancient biosocial needs for well-defined values and norms can be replaced by moral relativism. Contrary to what some unworldly advocates of total sexual freedom preach, not everything is possible. Furthermore, modern societies will need to continue to value evolution-based essential biosocial needs, such as long-lasting partner relations, child care and fertility levels, that guarantee individuals' and couples' quality of life as well as society's intergenerational continuity. Tolerance toward some sexual variants, such as homosexuality and childlessness, can be permitted but cannot be expected to become the norm.

A particularly important general rule in matters of sexual relations which are so strongly apt to competition, dominance, aggression and abuse, is the 'Golden Rule or ethic of reciprocity'. In the Confucian version it states that one should not do unto others what you do not want them to do to you.[177] However, despite its importance, this rule is insufficient as the sole foundation for a modern sexual ethics.

6.3.3.1 Furthering Sexual Emancipation

Just as many other human-specific characteristics, human sexuality has to be culturally supported because its biological foundation is only partially genetically programmed. The development of human sexuality can, indeed, be steered in different directions—with an accent on sheer physiological processes and functions, a merely ego-centred genitality, a dominance focused need satisfaction, etc.; or it can

[177]Gensler (2012); see also the discussion of the Golden rule in its negative and positive versions in Churchland (2011, 171).

be oriented towards the development of what is specific in human sexual relations, namely the use of the sexual drive as an energy source for the development of enduring, affective relations.

In such a perspective there is no place for the repression of a healthy, normal sexual behaviour in which partners can make use of the broad range of seducing and satisfying sexual practices. Further developing the human-specific sexuality requires a comprehensive sexual education, including all relevant aspects related to sexual and reproductive processes. Obviously, a comprehensive sexual education should not only include information on physical (genitality, reproduction, birth control) and health (avoidance of STI) aspects but also and in particular include education on psychological and relational aspects—the specificity of human sexuality.

6.3.3.2 Establishing Sex Equality and Enhancing Women's Empowerment

A future sexual ethics in a further modernising context may be expected to strongly be equality- and equity-oriented, with equal investment into girls and boys.

Sexism of whatever nature or degree may need to be further combated. We should, indeed, be aware that even the most advanced modern societies are still plagued with unconscious and hidden sex-related biases and so-called glass ceilings and walls.

Contrary to what some strictly equality-oriented ideologues preach, for instance, as can be witnessed in 'equality feminism', sex equality and equity does not imply that women and men should perform, in equal numbers, exactly the same functions in society or should be equally present in all occupations. The biological heritage from our ancient EEA will inevitably continue to produce some sex differences in occupational abilities and preferences. In contrast, it is important that sex-linked differences in occupational activities should be equally valuated (and remunerated). However, the authors are of the opinion that there is one domain where sex equal numerical representation should be guaranteed and imposed at all levels, namely ethical and policy decision-making.

Overall, the sex equality- and equity-oriented morality should concentrate on creating equal opportunities. This principle implies the establishment of sex-specific selective measures in many domains of social life, but especially in the domains of educational, occupational and retirement policies, taking into account the inevitably larger maternal investment in offspring. Despite much well-meant lip service being paid to this matter, the real challenges in this domain are still largely underestimated. It might be useful in this respect to call up the forgotten knowledge about the positive experiences of women-friendly measures in former communist countries.[178]

A cultural, regional or country specific problem is prenatal sex selection—the practice of diagnosing and aborting female foetuses in countries, such as China and

[178] Avramov and Cliquet (2005).

India, where pre-modern preferences for male children still prevail.[179] Although the evolutionary and social background factors of this pre-modern preference for male children are well understood, in modernity it is a maladapted practice. Not only does it offend against the equality principle but the seriously distorted sex ratios, as observed in China and India, could have a number of unfavourable societal side effects, such as a shortage of eligible female marriage partners, and increased risks of several forms of sexual deviant behaviour such as prostitution, socially induced homosexuality, and rape. In contrast, in cultures where prejudice against female offspring has vanished, it is difficult to find rational arguments against pre-conception sex selection aimed at satisfying couples' 'king wish' to have offspring of both sexes.

Overall, notwithstanding the considerable progress that has been made in matters of sexual equality/equity, at least in modern(ising) societies, there is still a long way to go and considerable efforts to be made.[180]

6.3.3.3 Securing Sexual Safety

The larger male body build, stronger musculature, masculinised brain physiology, higher levels of androgen production, and regular ejaculation urge can, in conflict or socio-pathological situations, easily lead to aggressiveness and dominance against women.

In modernisation, sexual violence of any nature has become an unacceptable form of behaviour—battering women, threats and harassments, abuse, and rape. Proper normative education and judicial punishment—zero tolerance appropriately introduced here—should guarantee sexual safety, in particular for women and children.

Forced or arranged marriages, female seclusion and mate guarding are considered as reprehensible forms of sexual coercion in modern societies. In many pre-modern societies and ideologies dominant males succeeded in morally and socially imposing such forms of sexual coercion on women (and men) and suppressed their freedom of choice. In modern societies, where personal freedom and self-actualisation are highly valued and belong to the prime human rights,[181] forced or arranged marriages and other forms of behaviour intended to limit the sexual autonomy of women are unacceptable—regardless of whether they are considered as normal or even the norm in traditional societies or ideologies.

Modernity is sometimes trapped in its appreciation of diversity and practices such as genital mutilations[182] that are not completely banned. Particularly detrimental is female genital cutting—aimed at controlling the sexual and reproductive life of women and ultimately decreasing their capacity for sexual pleasure. Except for medical reasons, male circumcision is an equally obsolete custom, a remnant of ancient mythical beliefs.[183]

[179] Hesketh and Xing (2006), Lai-wan et al. (2006), Prabhat et al. (2006).
[180] For instance, Baumgardner and Richards (2000).
[181] United Nations (1948), Council of Europe (1950; 2010).
[182] Denniston et al. (1999; 2010).
[183] Doyle (2005).

Sexual safety also includes the prevention of sexually transmitted infections, as well as the social safety of sexual minorities such as homosexuals.

A topic which is usually also dealt with under the heading of sexual safety is preventing unwanted or untimed pregnancy, but this issue will be discussed later in Chap. 8, Sects. 8.3.2.1 and 8.3.3.1.

6.3.3.4 Promoting Quality Sexual Relations

As explained above, the specificity of human sexuality as well as the neuro-hormonal equipment that facilitates the development and maintenance of enduring and affective partner relations are, ultimately, a function of long-maturing and dependent offspring. During several millions of years, hominins have been selected for this type of sexuality and neurology.[184]

This human-specificity will not only be with us for an additional long time but will also need to be furthered in a progressing modernising culture: evolving standards of quality of life, both for adults and children, will constantly further increase; economic living conditions will no longer be an obstacle to the fulfilment of emotional and affective (more advanced psychological) needs; and the raising and education of offspring may be expected to last ever longer.

Hence, the development of values and norms targeted at the promotion and support of a stable and affectively based partnership may remain a priority goal and preferential role model of modern sexual morality.[185]

This does not mean that only the old-time ideal of a lifelong monogamous partnership should be promoted. On the contrary, in cases where a partnership does not succeed in achieving or maintaining a high standard of affective relationship, separation/divorce should be easily achievable so that the establishment of another partnership, if so desired, would be possible.

6.4 Individual Variability in General

Chapter 2, Sect. 2.1 explained how several evolutionary mechanisms and processes result in the production of genetic differences between individuals. It also discussed why our evolutionary heritage makes environmental factors, including moral rules (values and norms), so important for the harmonious development of our human-specific characteristics.

In this section attention will be drawn to the evolutionary significance of individual variation; and it will discuss the adaptive and maladaptive value of self-oriented drives in the original environment in which individual differences emerged.

[184]Mellen (1981), Lampert (1997), Fisher (2004), Pedersen (2004).
[185]Avramov and Cliquet (2005).

Individual variability can be due to differences between individuals in one or another single trait; but it can also be due to the way in which the variability of several characteristics are combined within individuals, resulting in a much larger variability between individuals.

6.4.1 Evolutionary Background of Individual Variability

6.4.1.1 Evolutionary Importance of Individual Variability

Individual genetic variability is, within certain limits, important from an evolutionary point of view. Variation between individuals in their characteristics, as well as in the combination of their characteristics, is a safety valve with respect to evolving environmental conditions leading to changes in the intensity and direction of selective pressures. Indeed, evolutionists generally reject the idea of creating a uniform humankind because genetic uniformity might seriously hamper adaptability to future changing environmental living conditions.[186]

However, the initial variation of single biological characteristics at several life course stages—conception, birth, adolescence, adulthood, seniority—may not be optimally adaptive in some cultural or ecological conditions. Indeed, the initial genetic or congenital variation of most biological characteristics, amplified by environmental causes of variability during the life course, is usually quite broad; but as the life course progresses it narrows due to selective pressures. For instance, at birth some newborns with very low birth weight make them susceptible to high rates of neonatal mortality, resulting in the elimination of all their genes from the gene pool of the population.[187] The result is that the variation in body weight at later stages in life is much smaller than at birth, though still not negligible.

Variation in the combination of various characteristics in the individual—for instance, high potentialities for strenuous physical performance combined with lower cognitive abilities, or vice versa; exceptional artistic creativity combined with weaker health characteristics; high intelligence combined with low fertility—may have various effects on the overall individual phenotypical adaptability on the one hand, and on the intergenerational transmission—namely the genotypic adaptability —of the various variables on the other hand. The combination of high and low phenotypes of various types of characteristics may nevertheless have positive phenotypical adaptability effects on the overall performance of individuals. Excelling in particular features or behaviours may contribute to the maintenance and intergenerational transmission of less favourable genetic variants of co-varying characteristics.

[186]For instance, Dobzhansky (1962), Glover (1984; 2003).
[187]For instance, Stewart et al. (1981).

6.4.1.2 Adaptive and Maladaptive Value of Self-oriented Drives in the Environment of Evolutionary Adaptedness

The human species emerged and evolved in Pleistocene times and was equipped with features and potentialities that it inherited from earlier evolutionary periods or that emanated in the EEA itself. Most of its features and potentialities, including its strong self-oriented drives, allowed it to survive and reproduce in that harsh ancient environment[188] and thus had huge adaptive value.

However, the success of the hominin lineage in the EEA was largely due to the increasing importance of the extent, intensity, and complexity of its emerging social life for survival, reproduction and intergroup competition. Stronger biological predispositions for behavioural phenomena developed, such as altruism and reciprocity, and also cultural mores evolved that reinforced sociability and restrained ego-centred drives. Unrestrained self-oriented drives became partly maladaptive in the EEA and were thus under biological and cultural pressure.

6.4.2 Developments in Modernity Regarding Individual Variability

6.4.2.1 General Individual Variability in Modernity

In Chap. 1, particularly in Sect. 1.3, the enormously enhanced capability of modern culture was highlighted, mainly through improvements in environmental living conditions, to stimulate the development of characteristics or processes to previously unprecedented heights such as growth, longevity, health, physical and mental performances, and well-being in general.

Here, it is important to emphasise that the concept of well-being associated with improvements in living conditions not only includes the state of the organism itself but also other keys of life satisfaction, and in particular relations with others—family, work environment, overall community, the human species as a whole, and last but not least the planetary environment. Altruistic behaviour, even in its pure forms, produces neurological responses which results in positive emotional feelings, the so-called warm-glow effects.[189]

Achievements in phenotypical enhancements resulting from the modernisation process still exhibit significant inequalities. Large parts of the populations living in developing countries are still struggling with the traditional scourges of infectious disease, malnutrition and conflict which depress the development of their potentialities. Also, the much better off developed countries have still not reached a stage in which all of the traditional environmental causes of phenotypic variation are under control, particularly those that are linked to social class hierarchies.[190]

It is important to comprehend that modern culture gradually succeeds not only in combating or offsetting the old-time natural causes of environmental deprivation or

[188]For instance, Volk and Atkinson (2013).
[189]Andreoni (1989), Harbaugh et al. (2007); see also Manner and Gowdy (2010).
[190]For instance, Berkman and Kawachi (2000), Bauer et al. (2008).

threat, such as infectious diseases, starvation and war, but also increasingly provides opportunities for artificial enhancement of specific human performances. This implies that the composition of the population changes. The transhumanist goals of enhancing phenotypically human performances by means of technological interventions are already being applied in many domains of life. In this regard, Antonio Sandu[191] rightly argues that the current historical moment can already be considered as the beginning of the transhuman civilisation. Many therapeutic and preventive therapeutic interventions, as well as all kinds of enhancement drugs, already succeed in improving physical, intellectual, creative, emotional, athletic, sexual, energetic and even moral performances.[192] As the biomedical sciences further progress, increasingly effective phenotypic interventions will be possible, pushing the population averages upward, narrowing the population variability and transcending the upper bounds of present capabilities.

6.4.2.2 Well-Adapted and Maladapted Features in Modernity

Modern culture has undoubtedly increased the survival chances of individuals. However, it may not have always increased their phenotypical adaptability. In some cases it has increased the physical and/or mental performance and, hence, the adaptability of those individuals; whereas in other cases it may have failed to improve their functionality and, on the contrary, increased their dependency burden. Both paediatrics and gerontology—in fact all medical fields—struggle, from a medical technological as well as from an ethical point of view, with delicate questions about the medical and ethical meaningfulness of sophisticated medical interventions in cases of considerably diminished, or even absent, human-specificity or unbearably increased physical or mental suffering of patients (and their relatives).

Thanks to improving environmental living conditions in general, and biomedical care in particular, the effects of 'natural' selection are being changed. In many cases this resulted in a broader variability, but in some cases in the reduction of the initial variation. The improving phenotypic care in modernity results in contraselective effects of medical replacement therapies for deficient or deleterious genes, increasing the viability of genotypes and phenotypes which would have been eliminated at much earlier life stages in pre-modern living conditions. This relaxation of natural selection has an effect that alleles are not only preserved, thanks to replacement therapies or other protecting factors, but are in many cases also multiplied because their carriers are able to reproduce. The frequency of such 'weak' alleles is, in other words, increasing via two channels—decreased mortality and increased fertility.[193] Contraselection is, in fact, an issue that is more directly related to reproductive behaviour. Hence, it will be more extensively discussed in Chap. 8, Sect. 8.3.2.2.

[191]Sandu (2015, 3); see also Forlini and Hall (2015).
[192]For instance, Rothman and Rothman (2003), Carter (2016).
[193]Dobzhansky (1962), Thibault (1972).

Modern culture, however, faces another challenge related to the extant genetic variability of many characteristics that are remnants from adaptive processes for survival in EEA hunter-gatherer ancestral living conditions and that are still present in our current gene pool. Indeed, some of the genetically determined potentialities and drives which had a high adaptive value in EEA living conditions may have become less optimal, if not outright harmful, in modernity with its new opportunities, requirements and exigencies. Most of those drives concern relations with others, for instance, nepotism—the tendency to favour family members—or xenophobia—the tendency to foster negative feelings against strangers—or out-group enmity—the tendency to foster hostile feelings toward other groups or populations.

Some self-oriented drives are also maladaptive. A salient example of a harmful self-oriented drive in modernity is related to the strong human craving for resource acquisition. In pre-modern living conditions characterised by high risks of resource scarcity or unpredictable fluctuations in resource availability in domains such as food, energy and mates, the craving was the basis for the development of risk management.

By the time of the agrarian era, with its enlarged surpluses in subsistence means, all major religions included norms and rules to control excesses in major self-oriented cravings. In the West, the 'seven sins' of Christian theology—gluttony, sloth, avarice, wrath, envy, pride, lust—are a well-known example. Terry Burnham and Jay Phelan[194] relate gluttony to unknown contingencies for the future, lust to excesses in mate acquisition, and pride, envy, sloth, and rage to drives for dominance hierarchy.

In modernity, with its strongly increased affluence of resources, the innate drive for resource acquisition easily turns into overconsumption, with its multiple negative effects on health, social life and the environment. Everyday examples of over-consumption are: excessive intake of food (overweight) and stimulants (addiction), too intense social status competition (excessive stress), philanderous mating behaviour (sexually transmitted diseases, couple dissolution), and excessive pressure on the environment (ecological damage). Affluence easily leads to 'affluenza' and 'luxury fever'[195]—all-consuming epidemic.[196]

[194]Burnham and Phelan (2000, 120); see also Chapman (2004, 103), Krebs (2011, 94).
[195]Frank (2010).
[196]Dodds (2008, 121); see also Krebs (2011, 87).

6.4.2.3 Individuality Versus Individualism

The individual has a genetically unique identity,[197] during the life course it is enlarged by environmental living conditions and experiences. This became known in German as '*individualität*'—individuality, the notion of individual uniqueness, originality, and self-realisation.

In the English-speaking world, the notion of self-determination, self-reliance and the full development of the individual, and pride in personal freedom became popularly known under the concept of individualism. This content differs quite strongly from its historical French meaning that had a pejorative connotation, being largely equated with egoism and selfishness.[198] In many European continental countries, the notion of individualism still has this French meaning.

In literature, the view seems to predominate that modern culture is characterised by a gradual increase in individualistic attitudes and behaviour.[199] Countless survey investigations, such as the World Values Survey and the European Values Study,[200] indeed appear to confirm that, in recent decades, attitudinal and behavioural changes in different domains of life show an increasing tendency for individuals to concentrate on themselves, to withdraw from social groups, institutions, and anything outside themselves.[201] This is revealed in a particularly striking way in the research about changes in family values, family relations and structures.[202]

Alexandra Maryanski and Jonathan Turner[203] have argued that our individualistic drive is a typical feature that results from our hominoid heritage, but it was suppressed in the early stages—hunter-gatherer and agricultural-pastoral stages—of our cultural development. This suppressed drive acquired a new impetus in the present scientific-industrial stage. Those authors developed their original thesis in a well documented study. However, there might be another explanation, namely that the reblooming individualism in the scientific-industrial culture is not so much linked to our hominoid heritage, but is due to the increasing cognitive abilities hominins acquired during the hominisation process. These cognitive abilities allow, in the novel environment of modernity, for a stronger development of individuality.

Undoubtedly, there are many factors in the development of modern societies that allow for or even encourage an increase in individualism: technological innovation,

[197]Due to the combination of the large number of genes in the genome, and the processes of meiosis and fertilisation, an endless number of genetically different individuals can be formed. Current estimates indicate that all humans are approximately 99.6–99.8% identical at the nucleotide sequence level. Within the remaining 0.2–0.4% genetic material, approximately 10 million DNA variants can potentially occur in different combinations. This represents a very small fraction of the total genome, but it is vastly more than enough variation to ensure individual uniqueness at the DNA level (Tishkoff and Kidd 2004). With the exception of monozygotic (identical) twins, where the segregation-recombination—mechanism is bypassed, no two individuals have the same genome.

[198]Lukes (1971).

[199]For instance, Schmid (1984).

[200]Halman et al. (2007).

[201]Glenn (1987).

[202]Lesthaeghe (2002), Hofferth (2003), Karraker and Grochowski (2005).

[203]Maryanski and Turner (1992).

in particular ICT, geographical and social mobility, affluence, social security, independency of jobs, exposure to media, small family size, rapid social change, open borders, and last but not least, the resurgence of an aggressive fundamentalist capitalism with its increased economic competition and globalisation.[204]

Thanks to its increased socialisation and technological innovation, modern culture has considerably enhanced the opportunities for individual emancipation and self-actualisation. This has resulted not only in improved physical and mental performance but may have also induced amongst many individuals the subjective feeling of heightened independence and freedom, which hence lay the grounds for an amplified individualism. The growing opportunities for individual emancipation in modern culture are only possible due to higher levels of socialisation and cooperation, and particularly thanks to the actions or influences of ever larger numbers of anonymous and unrelated people. Socialisation and cooperation allowed for transgressing the boundaries of one's own kinship group or even nation. However, some aspects of individual emancipation and basic features of societal progress associated with the interdependence of individuals have evolved in opposite and conflicting directions. The pursuit of more individuality with the growth of human potential, which was made possible by cooperation, contrasts with the quest for individualism, which is associated with lower levels of cooperation. This trend often penetrates national or international ethical or political discussions and their resultant charters in which individual rights are advocated, but individual responsibilities are neglected.

6.4.3 Ethical Reflections Regarding Individual Variability

The authors have already identified three principles which are of high relevance to individual development: the ontogenetic development of human-specific potentialities, the promotion of quality of life, and the promotion of equal opportunities.

Ontogenic development of human-specific potentialities, promotion of quality of life and the promotion of equal opportunities and equity are, in the authors' view, the key ethical objectives. However, some more specific issues should also be stressed.

6.4.3.1 Maintenance of Genetic Variability

Genetic variation ought to be maintained, yet within certain limits—variation between various characteristics, but also variation within characteristics. The evolutionary justification for this principle is that the preservation of a variety of characteristics, as well as the continuing maintenance of a variation within individual characteristics, enables genetic adaptability opportunities in changing environmental living conditions.

By way of example, although a high evaluation of cognitive abilities in the present cultural evolution is comprehensible, maintaining and cultivating attributes

[204]Elliot and Lemert (2005).

such as physical resilience, strength, endurance and speed should also not be derided. In the event of whatever catastrophe that might disrupt the delicate nature of modern culture, which relies so strongly on the use of technical aids and non-human sources of energy, human physical abilities might be a precious source for survival and save us from regression if not extinction. Hence, a diversified gene pool is a protection against the nasty effects of unforeseeable natural catastrophes.

Also, the preservation of variability per characteristic might be important, for example body height. It is well known that modernisation has been accompanied by a gradual increase in body height and that tallness is a positively valued feature.[205] The modern increase in body height is largely due to improved environmental living conditions, not least a higher and richer food intake, but selective processes may also be involved because a richer nutrition and other favourable environmental living conditions relax selective pressures against large body size, in particular at birth.[206] However, the fluctuations of body size in history show[207] that in poorer living conditions smaller body height has a higher survival value.[208]

6.4.3.2 Phenotypic Enhancement of Human-Specific Potentialities

Evolutionarily desirable human potentialities in a further progressing modernisation are generally identified as physical and mental health (absence of disease or impairment), immunity (resistance to disease), energism, beauty, longevity, cognitive and emotional intelligence, creativity and above all, sociability.[209] This list should also include healthy self-oriented drives that are necessary for individual survival and fulfilment, which are to be distinguished from extreme selfishness, narcissism and solipsism.

Many scholars addressing the issue of developing human-specific potentialities do this in terms of increasing the satisfaction of basic needs,[210] happiness[211]—some speak even about Darwinian happiness[212]—or quality of life, e.g. Julian Huxley:[213]

> Man's dominant aim must be increase in quality – quality of human personality, of achievement, of works of art and craftsmanship, of inner experience, of quality of life and living in general.

Quality of life is one of the major ethical goals that has been derived from an evolutionarily founded ethics focused on the future development of humanity in a further modernising context.

[205] For instance, Meredith (1974).
[206] Thomson (1959), Nettle (2002).
[207] For instance, Ruff (2002).
[208] For instance, it is well-known that tall inmates of the Nazi concentration camps were much more vulnerable to the meagre rations they received (Baker et al. 2010; Wachsmann and Caplan 2010).
[209] For instance, Muller (1958), Huxley (1964), Cattell (1972), Bostrom and Savulescu (2009).
[210] For instance, Ehrlich (2000).
[211] For instance, Cattell (1972).
[212] Grinde (1996).
[213] Huxley (1964, 246).

Many socially important biological characteristics, in particular those determined by polygenes, can be influenced by environmental factors in their phenotypical expression during ontogenetic development.[214]

In modern culture with its strongly increased exigencies and opportunities for intellectual, artistic and technological innovation, human societies have, more than ever before in history, an interest in the human-specific potentialities of as many citizens as possible being maximised, or at least optimised. This is important not only because all the talents available are needed to further the modernisation process, and to push the scientific, cultural, social, and technological developments up to still higher levels, but also because citizens are needed who can absorb and adequately function in the rich but also demanding novel environment that is being created.[215]

As argued above, the current practice of technological enhancement of ontogenetic processes and desirable characteristics, which are aimed at optimising or maximising human-specific potentialities, should be further pursued as the transhumanist school of thought advocates.[216]

6.4.3.3 Prevention, Suppression, or Transformation of Maladaptive Traits

It must be acknowledged that the human-specific potentialities not only include individually and socially valuable traits but also several drives leading to self-destructive behaviour that decreases or impairs the above-mentioned positive physical, mental and behavioural characteristics.[217] Furthermore, modern technology has refined and multiplied many drugs, and tools such as guns and fast-running vehicles, which can increase the risks of lowering or even annihilating valuable human characteristics or attributes. Moreover, modern society shows a number of more general features, such as its larger size, its more competitive nature and its larger anonymity, which enhance the risk of maladaptive behaviour by several groups of more vulnerable citizens.[218]

An evolutionarily based set of ethics, aimed at the enhancement of the hominisation and modernisation processes, needs to prevent or suppress phenotypically and, in the end, also genetically maladaptive individual drives such as (self-induced) mutilations, substance abuse (resulting in dependence), gluttony (resulting in obesity), sloth (resulting in apathy), various kinds of excessive risk-taking behaviour including violent sports which degrade or basically harm, physically or psychologically, the practitioners.

[214]Mather and Jinks (1971), Lynch and Walsh (1998), Kearsey and Pooni (1998), Plomin, et al. (2008).
[215]Bajema (1971).
[216]FM-2030 (1970; 1973; 1989), More (1990), Bostrom (2003; 2005), Hughes (2004), Kurzweil (2005), Young (2006), Savulescu and Bostrom (2009).
[217]Masters (1989).
[218]See also Mealey (1995, 166).

6.4.3.4 Fostering Equal Opportunities

Chapter 5, Sect. 5.3.3.3 dealt with the question of the promotion of equal opportunities under conditions of diversity, a matter which is of essential importance for cohesiveness in society.[219] This principle is also of the highest relevance for the topic of individual development.[220] Whereas the ideal of equality may, given the individual variation be unreachable, the promotion of equal opportunities is a reasonable and fair proxy.

The principle of equal opportunities is well defined in the empirically grounded ideological paradigm that Peter A. Corning refers to as the biosocial contract.[221] It consists of three complementary normative precepts, namely (1) equality in the distribution of basic goods and services within a society; (2) equity in the distribution of surpluses beyond the provisioning of our basic needs according to merit; and (3) reciprocity in the contribution to the collective survival enterprise in accordance with our ability and performance.

Notwithstanding the considerable progress that has been achieved in modern culture in implementing the equal opportunity principle, especially with respect to education and work, there is still a long way to go. Deep-seated drives such as nepotism, in-group favouritism, sexist, classist, ethnocentric or racialist prejudice still have to be mastered.

A particularly difficult and delicate problem raised by a continuously progressing modernisation, with its ever increasing complexity and enlarging exigencies and responsibilities, is the social inclusion of individuals who are phenotypically—either due to genetic endowment, environmental accidents or life course events—less well adapted to the novel environment of modern culture. The less healthy, the less physically able, the less cognitively able, the less educated, the weaker emotional personalities, that is to say persons with special needs, are part of the human variability.

Pending future more highly performing biotechnical interventions that would succeed in enhancing the genetic and ontogenetic composition of modernising populations, as the transhumanists anticipate, modern societies are meanwhile striving to develop innovative forms of life-long education and provide social support in a broad range of living and working conditions. This support builds on the ethical principle related to the quest for individual realisation of potentialities but also to the fact that no well-functioning society can allow itself to leave a significant part of its population behind, as lack of support to the socially vulnerable undermines the very basis of social cooperation.

6.4.3.5 Promotion of Individuality, not Individualism

The importance assigned to the person and the increase in opportunities for individual development and emancipation in modernisation is extremely valuable. In modern culture increased opportunities are considered necessary conditions for

[219] Tancredi (2005, 192); see also Brosnan and de Waal (2003).
[220] Schaar (1967).
[221] Corning (2010).

well-being and happiness, and also because historically they have been proven to enhance cultural creativity and social progress.[222]

However, giving an absolutistic priority to individualistic endeavours, without taking into account the various between-individual and individual-societal interdependencies, is increasingly a maladaptive practice. As Frans de Waal[223] stated:

> A morality exclusively concerned with individual rights tends to ignore the ties, needs, and interdependencies that have marked our existence from the very beginning.

The social protection systems based on mutual solidarity, which gradually developed and matured in the course of the twentieth century in modern welfare states, do involve the risk of some individualistic excesses. Friedrich Hayek[224] even speaks of "the re-emergence of suppressed primordial instincts". Free-riding in any social domain—whether it be social welfare abuse, financial speculation, delinquency, environmental pollution, machismo, aggression, to name only a few—is incompatible with the subtle interrelationships and interdependence between the needs of the individual and community. Modern society requires a vision of humankind that, as Willy Wielemans[225] formulates it, goes '*voorbij het individu*' ('beyond the individual'); it also needs to promote structures that foster cooperation rather than competition and attempts to channel competitive drives to socially desirable ends.[226]

Whereas the understanding of individuality as an expression of uniqueness of a human being is valuable for the preservation of life, individualism as freeloading or abusive behaviour towards the other and/or collectivity is maladaptive.

Our ethical norms should foster 'individuality', not individualism since we no longer live in the small communities of the EEA in which human genomes and the current human gene pool emerged. Instead, we live in hugely populous and highly complex societies with considerably enhanced opportunities for individual emancipation and self-actualisation, as well as many maladaptive characteristics with respect to individual-societal inter-dependency. Individuality does not conflict with a global ethic, whilst individualism is narrowly self-centred behaviour.

An evolutionarily founded ethics must, in the context of a further progressing modernisation, be strongly focused on the maximisation of the self-actualisation of valuable human-specific characteristics, in other words, the promotion of individuality. Not only does this need to include self-oriented features but also personality traits that accommodate relations with others—family, community, the human species, even the planetary environment as a whole. In this respect, the innate predispositions to spirituality, which fulfilled an important role in our evolutionary past, could be further enhanced culturally towards a greater sense of belonging to the global world, and could be an evolutionary substitute for the drive towards the runaway individualisation that we are experiencing now.

[222]Beck and Beck-Gernsheim (2002).
[223]De Waal (1996, 167).
[224]Hayek (1979, 165).
[225]Wielemans (1993).
[226]For instance, Singer (1999, 44).

Hence, individualistic attitudes and behaviour, in the sense of its original French meaning, are to be rejected or discouraged. They are a maladaptive, although comprehensible side effect of the modernisation process which has, unfortunately, so far been insufficiently accompanied by educational efforts to make individuals understand and adapt to the novel environment of modernity. The complex modern society and culture depends mainly on the interdependence, cooperation, interaction and mutualism of ever increasing numbers of individuals that hugely transcend the kin- or tribe-related in-group, a topic which will be pursued in more detail in Chap. 7.

6.5 Interpersonal Relations

In addition to the specific age and sex relations, interpersonal relations based on evolutionary predispositions also have to be dealt with. These types of relations became more important as human populations gradually grew in size, but are of particular significance in modernity where nations include millions of people and where interpersonal relations can even go far beyond their own community or national borders, due to progressing internationalisation and globalisation of human activities.

6.5.1 Evolutionary Background of Interpersonal Relations

6.5.1.1 Competition Theory

Human beings, just as other living organisms, are not only equipped with extremely strong drives towards self-actualisation and survival but also with drives to compete with conspecifics for acquiring desirable resources—territory, nutrition, wealth, prestige, and above all sex and reproduction. Darwin's theory of natural selection provides an evolutionary explanation for the competitive drive, but many ideologies, some of a very different nature, e.g. classical or neoliberalism and Marxism, have tried to use (and abuse) his theory to give a scientific foundation and justification to their ideological viewpoints. Even many pre-Darwinian ideologies often based, evidently unknowingly, their beliefs and practices on Darwinian selection principles.[227]

The competitive drive can express itself through various forms of behaviour, e.g. selfishness, assertiveness, hard work, social dominance, aggressiveness. Each of them is subject to variation—genetically, ontogenetically and/or environmentally/culturally determined. Extreme variants of those drives can, in particular social or ecological conditions, result in antisocial behaviour.

[227]See Keith (1946, 64), Teehan (2006; 2009).

Aggression[228] is one example. Although most traditional and modern ideologies, in general, reject and condemn aggressive behaviour, individual aggressiveness is not only a ubiquitous phenomenon in the human species[229] but in mild forms it appears to have evolutionarily adaptive advantages. Namely it may enhance, via the achievement of increased social dominance, reproductive success.[230]

Whereas mild forms of individual aggressiveness are tolerated in most societies, even positively valued and institutionally regulated, extreme forms of individual violent behaviour are usually rejected and combated, except in intergroup conflict or war. The existence of extreme forms of aggressiveness can be due to normal variation of the aggression drive or to the genetic or environmental damage of the aggression inhibition system.[231] However, extreme aggressiveness is usually the outcome of interactions between specific genetic predispositions and environmental deprivation or assault.[232] Notwithstanding the social opposition and control of extreme forms of aggressive behaviour, such behaviour may have had some reproductive success, for instance via social dominance, coercion, deception, or rape, and allowed the intergenerational transmission of the genes involved.

Politically or religiously induced values and norms can exacerbate competitive forms of behaviour, usually resulting in a broad range of unfavourable social effects, in particularly enhanced social or sexual inequalities and violence. In the economic sphere such norms result in intensified class struggles and social aggressiveness and violence. In the sexual sphere cultural norms allowing or promoting polygamy, which leaves many young men without a sexual partner or strongly restricted sexual intercourse among youngsters and young adults, result in an increased incidence of social unrest or forms of sexual violence. Some studies show that the proportion of unmarried young men is a good predictor of social or sexual violence in society.[233]

6.5.1.2 Cooperation Theory

Before embarking on the discussion of the evolutionary background of cooperative behaviour, it is necessary to dwell for a moment on the relation between the concepts of altruism and cooperation. *Mutatis mutandis*, this also applies to the relation between selfishness and competition. Indeed, it is often implied that altruistic behaviour and cooperative behaviour are linked, just as it is assumed that selfish behaviour and competitive behaviour are equivalent. However, as Peter Corning argues pertinently in his book *The Synergism Hypothesis*, they are not.[234] Altruism (sacrificing oneself for others) and cooperation (acting/working together with others) are different forms of behaviour, but altruistic behaviour is clearly a strong incentive or condition for cooperative behaviour. This is even truer for the

[228]Ferguson and Beaver (2009).
[229]McCall and Shields (2008).
[230]Hawley and Vaughn (2003), Smith (2007), Ferguson (2008).
[231]Beaver et al. (2009), Rietveld et al. (2003), Wright et al. (2008), Nettle (2006).
[232]For instance, Caspi et al. (2002).
[233]Hauser (2006, 132).
[234]Corning (1983, 84).

relation between reciprocity/mutualism (which may be altruistic or selfish in nature) and cooperation. Hence, the evolutionary mechanisms responsible for the genetic transmission of predispositions to altruistic, reciprocal and mutualistic behaviour are to be considered as strong vehicles for promoting various forms of cooperative behaviour. The sociobiological revolution of the 1960s and 1970s—labelled in Chap. 1 as the Second Darwinian Revolution—pertinently corrected the ideological misinterpretation of Darwin's evolutionary theory by discovering evolutionary mechanisms that explained the genetic transmission of behavioural characteristics, such as altruism and reciprocity or mutuality, that provide a biological-evolutionary basis for cooperative behaviour—the basic ideas that were, in fact, already anticipated in Darwin's[235] *The Descent of Man and Selection in Relation to Sex.*[236]

As already explained in Chap. 2, Sect. 2.1, recent sociobiological developments have resulted in the establishment of evolutionary mechanisms through which predispositions facilitating or inducing cooperative behaviour can be transmitted and evolve beyond narrow kinship relations. Theories on various forms of reciprocity selection, coercive selection, group selective processes including cultural group selection, and gene-culture coevolutionary interactions have been elaborated into an overall evolutionary theory of cooperation.[237] Randolph M. Nesse[238] has reminded us that, in addition to kin selection and reciprocity selection, there are several complementary evolutionary mechanisms—sexual selection,[239] social selection,[240] social commitment,[241] manipulation[242] and coercion[243]—that have the potential to shape altruism and morality because they provide individuals or social groups, or even society as a whole, with a fitness advantage. The fact that evolution has the ability to generate cooperation in an overall context of competition incited Martin Nowak[244] to consider natural cooperation as a third fundamental principle, beside natural selection and mutation. Although cooperation cannot be considered as an evolutionary mechanism comparable to mutation and selection, the strong biological predispositions for cooperation are, as an evolutionary result of those mechanisms, to be considered as one of the most specific

[235]Darwin (1871).
[236]See also Weiss and Buchanan (2009).
[237]Recent overviews on evolution of cooperation can be found in: Wilson and Sober (1994), Axelrod (2001), Sanderson (2001), Barash (2003), Hammerstein (2003), Kappeler and van Schaik (2006), Weiss and Buchanan (2009), Bowles and Gintis (2011), Krebs (2011), Nowak and Highfield (2011), Sussman and Cloninger (2011), Voland (2013).
[238]Nesse (2000, 229, 230).
[239]Miller (1998); see also Chap. 2, Sect. 2.1.2.4.
[240]Alexander (1987), Frank (1998), Nesse (2009), Boehm (2014); see also Chap. 2, Sect. 2.2.2.3.
[241]Schelling (1960), Hirshleifer (1978), Frank (1988), Nesse (2001).
[242]Trivers (1974).
[243]Van den Berghe (1979), see also Chap. 2, Sect. 2.2.2.7.
[244]Nowak (2006, 1563).

features of the human species. Without those predispositions, it would have been impossible for *Homo sapiens sapiens* to develop complex forms of societal life and civilisation.[245]

In recent years, the biological origin and evolution of human cooperation continues to fascinate the scientific community as can be seen from the lasting flow of publications on cooperation theories,[246] the comparative analysis of cooperation in the primate and hominin lineage,[247] ontogenetic studies,[248] evolutionary game experimental studies,[249] and empirical studies on a broad range of specific topics such as the effects of various behavioural predispositions to cooperation,[250] the importance of basic emotions—love, gratitude, deference, righteousness, pride, but also anger, envy, shame, guilt, contempt, and moral outrage—for cooperation,[251] the role of cultural factors or gene-culture coevolution,[252] or the effects of demography and ecology.[253]

6.5.1.3 Competition/Cooperation Balance

First, it must be reiterated that the evolutionary theories of kin selection, reciprocity selection and group selection have resolved a fundamental paradox in evolutionary theory: natural selection which promotes self-advancement at the expense of others, namely competition, is complemented by various mechanisms which explain the genetic transmission of predispositions and practices leading to altruistic or mutualistic behaviour, namely they promote cooperation.

Hence, contrary to some ideological views in which people are seen as either intrinsically selfish or fundamentally altruistic beings, evolutionary science has well established that social species, and in particular humans, are endowed with drives both for competing and cooperating behaviour.[254] Pro- and antisocial forms of behaviour may compete and interact with each other. In virtually all viable or successful societies, social rules or legislations have been struggling to devise a just

[245]For instance, Gorelik et al. (2012).

[246]For instance, Nowak et al. (2010), West et al. (2011), Tomasello et al. (2012), Krasnow et al. (2013), Rand and Nowak (2013), Smaldino et al. (2013), Zaki and Mitchell (2013), Keltner et al. (2014).

[247]For instance, Brosnan (2010), Melis and Semmann (2010), Langergraber et al. (2011), Silk and House (2011), Barrett et al. (2012), Grueter et al. (2012), Burkart et al. (2014).

[248]For instance, Tomasello (2009), Hamlin et al. (2011), House et al. (2012), Sebastian et al. (2013), Kuhlmeier et al. (2014), Jensen et al. (2014).

[249]For instance, Rosas (2010), Fehl et al. (2011), Marlowe et al. (2011), Xia et al. (2011), Eriksson and Strimling (2012), Garcia and Traulsen (2012), Hwang and Bowles (2012), Zhuang et al. (2012), Capraro et al. (2014), Stewart and Plotkin (2014), Hoffman et al. (2015).

[250]For instance, Smith (2010), Marlowe et al. (2011), Xia et al. (2011), Garcia and Traulsen (2012), Hwang and Bowles (2012), Nowak and Highfield (2012), Tomasello et al. (2012), Rand and Nowak (2013), Wang (2013), Capraro et al. (2014), Smaldino (2014).

[251]Fessler and Haley (2003).

[252]For instance, Atkinson and Bourrat (2011), Chudek and Henrich (2011), Dijker (2011), Ihara (2011), Bogin et al. (2014), Phillips et al. (2014).

[253]For instance, Lamba and Mace (2011), Powers and Lehmann (2013), Krasnow et al. (2013), Tan et al. (2013).

[254]Masters (1989, 1), May et al. (1989), Hinde (2002, 178), Kümmerli et al. (2010).

balance between competitive and cooperative behaviour. However, from an evolutionary point of view—a long-term perspective—increasing complexity and progression results from a process in which cooperation overtakes competition in importance.[255] Samuel Bowles and Herbert Gintis[256] list three reasons why altruistic behaviour, supporting cooperation, outcompeted selfish, competitive behaviour: (1) it protects altruistic members from exploitation by the self-interested (shunning, ostracism and execution of free-riders); (2) it favoured systems of socialisation that led individuals to internalise the norms that induce cooperation; and (3) it was an advantage in between-group competition for resources and survival that was, and remains, a decisive force in human evolutionary dynamics.

The sociobiological discoveries about the biological-evolutionary mechanisms underlying altruistic and reciprocal or mutualistic behaviour were received with initial enthusiasm in some quarters. It was believed that they scientifically confirmed the value and truthfulness of ancient ideological viewpoints or appealed well to some of our cherished universal moral tenets such as the Golden Rule or the love message of Jesus Christ.[257] However, initial enthusiasm has been fundamentally shattered by the discovery that altruistic and reciprocal or mutualistic behaviour is in fact, at least in its ultimate effects (namely the differential transmission of alleles), to be equated with genetically selfish behaviour. In other words, social cooperation, without excluding moderate forms of individual competition, fulfils the same functions as competition. We are not only genetically predisposed to altruistic behaviour but also culturally conditioned to such behaviour, both by ancient (e.g. "…thou shalt love thy neighbour as thyself"[258]) and modern (e.g. '*fraternité, liberté, égalité*'[259]) ideologies, so that we have difficulties in digesting this "bitter pill of our fundamental selfishness" as Randolph M. Nesse[260] put it. In biological-evolutionary terms, true or genuine altruism—or ascetic altruism, as Joseph Lopreato[261] calls it—can only be defined as helping behaviour that decreases the inclusive fitness of the altruist and increases that of the beneficiary:

> Ascetic altruism is behaviour, conscious or unconscious which, guided by innate predispositions, potentially reduces the inclusive fitness of the dispensers and potentially increases the fitness of the recipients.

How is group cooperation to be reconciled with the competitive self-interest of individuals? Paradoxically, in particular conditions of in-group or out-group threats and competition, the development of altruistic behaviour favours cooperation which, in turn, is advantageous for individual survival and hence the transmission of genes. As already mentioned in Chap. 2, Sect. 2.1.2.8, human evolutionary history is, contrary to intuition or moral conviction, in all probability characterised

[255]For instance, Nowak and Highfield (2011, 267ff).
[256]Bowles and Gintis (2011, 4).
[257]For instance, Pope (1994; 2007).
[258]Leviticus 19:18.
[259]National motto of France, originating in the French revolution (Latham 1906).
[260]Nesse (2000, 228).
[261]Lopreato (1981, 117).

by a co-evolution of group-beneficial altruistic/cooperative behaviour and violent intergroup competition and warfare.[262]

The altruistic/mutualistic drives of the human must have been a very early hominid adaptation. Human grouping tendencies developed as early as the Pleistocene, in the EEA. It was an adaptation for living in small groups in which people were genetically closely related and where people also had the opportunity to get acquainted with all of the group members.[263]

6.5.1.4 Evolutionary Causes of Antisocial Behaviour

The biological-evolutionary theories about the predisposition and transmission of altruistic and cooperative behaviour do nevertheless raise the question as to why some people continue to behave in a largely or even exclusive selfish or antisocial[264] way? Why are there free-riders and cheaters in our species? How can evolutionary theory explain the widespread occurrence of genetically determined or influenced antisocial behaviour that, in principle, must be seen as a maladaptation in such a highly socially developed species as *Homo sapiens sapiens*?

Just as the evolutionary background of the prevalence of forms of maladapted moral behaviour has been explained (see Chap. 2, Sect. 2.2.6), different elements also have to be considered regarding the evolutionary background of antisocial behaviour. In general, both genetic and environmental factors may, in complex interactive processes, contribute to the development of antisocial behaviour.[265]

Predispositions to selfish and/or altruistic behaviour may be subject, just as any other biological trait, to genetic variation resulting from mutations or particular gene recombinations at conception. As is the case with all complex forms of behaviour, those predispositions are only partially determined by genetic factors. Consequently they can also vary in their ontogenetic development due to environmental influences, and not the least result from socio-cultural determinants. Some extreme genetic variants of selfish behaviour, leading to strong expressions of dominant or aggressive drives, are usually subject to negative selection due to their higher risks. However, in particular circumstances they may have had some reproductive advantage and success and consequently have been intergenerationally transmitted. Most importantly, since most people have the ability to develop or display selfish as well as altruistic forms of behaviour, the combination of these predispositions—as well as the social or environmental challenges people have to cope with—may tilt the adaptive strategy they choose in one direction or the other. Consequently, on an evolutionary scale, these elements may result in the simultaneous co-existence of different behavioural strategies that compete with each

[262]Keeley (1996), Gat (2006), Choi and Bowles (2007), Lehmann and Feldman (2008), Bowles (2009; 2012), Ginges and Atran (2011), Halevy et al. (2012), Gneezy and Fessler (2012), Gorelik et al. (2012), Konrad and Morath (2012), Rusch (2014), Puurtinen et al. (2015).

[263]Wilson and Sober (1994).

[264]Antisocial behaviour is any behaviour that causes damage to other persons or even to society as a whole. It is to be distinguished from asocial behaviour which refers to the absence of interaction with other people.

[265]Moffitt (2005), Rutter et al. (2006).

other.[266] Evolutionary game theory also helps to understand how alternative competitive strategies may result in evolutionarily stable or slightly fluctuating population equilibria, with the frequencies of different traits or behavioural patterns dependent on their reproductive payoffs in varying living conditions.[267]

Obviously, in most cases, or on a long-term scale, antisocial behaviour remains a minority phenomenon, simply because reproductively it is a less advantageous form of behaviour than cooperative behaviour in a social species and it is subject to negative selection. Moreover, in a species as highly socialised as the human, antisocial forms of behaviour are fiercely combated by enforced community rules.[268]

Nevertheless, sociobiology and evolutionary psychology have developed several specific evolutionary theories of antisocial behaviour with the goal of explaining its relatively high frequency,[269] particularly among males.[270] Four major theories can be distinguished, which partially overlap in several respects.[271] Indeed, all of these theories have the common assumption that antisocial forms of behaviour produce, especially in particular living conditions, reproductive advantages for individuals who victimise others, allowing their own genetic predisposition to be transmitted to future generations, or prevent the reproduction of others. These four theories are the cheater (or 'cad vs. dad') theory of antisocial behaviour, the r/K theory of antisocial behaviour, the coincidental status-striving theory, and the cuckoldry-fear theory.

The cheater (or cad vs. dad) theory of antisocial behaviour[272] is the expression of genetic programmes for an evolved male strategy that is aimed at adopting a cad reproductive strategy rather than a dad strategy. In other words, this theory describes males who focus on obtaining numerous mating opportunities by means of highly deceptive and/or forceful strategies, rather than assisting their partner in caring for the offspring they sire. Several forms of sexual harassment and assault, forced copulation, and of course especially rape[273] are explained by this theory.

The r/K theory of antisocial behaviour[274] assumes that people who have the tendency to harm others may be manifesting a more r-oriented reproductive strategy by producing more offspring than law-abiding citizens.

The coincidental status-striving theory[275] states that competition for status and resources favours males who attract mates and sire a disproportionate share of offspring. Males who are the most extreme in their overtly competitive and status-striving activities are more likely to violate the within-group order. Many

---

[266] Thomas (1984), Barr and Quinsey (2004).
[267] For instance, Colman and Wilson (1997).
[268] For instance, Masters (1989).
[269] Rowe (2002), Mealey (1995), Walsh and Ellis (2003), Gottschalk and Ellis (2009).
[270] Wrangham and Peterson (1996), Ghiglieri (1999).
[271] Ellis (1998).
[272] Macmillan and Kofoed (1984), Harpending and Draper (1988), Gottschalk and Ellis (2009).
[273] Thornhill and Thornhill (1983), Thornhill and Palmer (2000).
[274] Ellis (1987).
[275] Alexander (1979), Ellis (1990).

cases of homicide, as forms of elimination of male competitors, can be explained in this respect.[276]

The cuckoldry-fear theory[277] explains some forms of antisocial behaviour based on the idea that males, due to the uncertainty of their paternity, are more susceptible to the risk of being cuckolded in cases of their partner's infidelity. Thus, males would be under additional natural selection pressure to use violent tactics in order to curtail sexual infidelity. Various forms of male partnership tactics—mate guarding, partner sequestration, female genital mutilation, spouse abuse, assault and murder[278]—as well as some forms of child neglect and abuse and infanticide, particularly in cases of non-relatedness, come under this heading.[279]

The two major components of antisocial behaviour—violence and cheating—have a clear evolutionary basis: they formed as an adaptation to the EEA, albeit as occasional or secondary strategies. However, it is important to keep in mind that behaviours and emotions which evolved as reactions against antisocial violence and cheating are also part of our evolutionary heritage.[280]

Obviously, not all antisocial behaviour is ultimately related, consciously or unconsciously, to reproductive drives as proposed by the above-mentioned evolutionary theories. A substantial part of antisocial behaviour appears to be life-course persistent, influenced by variations in cognitive and emotional personality, as well as neurological and hormonal functioning. These are caused by unfavourable genetic or developmental factors, or the combination of the presence of a specific biological endowment and the perceived or actual social inequalities in the accessibility or availability of desired goods and services in society.

6.5.2 Interpersonal Relations in Modernity

As explained extensively, human populations evolved from very small numbers in the hunter-gatherer phase of cultural development to many millions in modern societies, a novelty to which *Homo sapiens* has not yet adapted genetically.[281] Therefore, modern million-member societies face many biosocial constraints and conflicts resulting from the fact that the human mind, with its specific evolved psychological mechanisms and design as an adaptation to ancient living circumstances, is not yet adapted to the novel environment we have created.[282] Modern societies consequently struggle with many problems related to the necessity to induce cooperative behaviour amongst huge numbers of people with whom individuals have no close genetic relatedness, and whom they mostly do not know very

[276]Daly and Wilson (1988), Duntley and Buss (2005).
[277]Hiatt (1989), Geary (2006).
[278]Buss (1994; 2002).
[279]Lightcap et al. (1982), Burgess and Garbarino (1983), Ellis (1998), Daly and Wilson (1988).
[280]Raine (1993).
[281]Campbell (1975), Masters (1989, 158).
[282]Bowlby (1969), Irons (1998).

well[283] or with whose values they do not identify well. Finding the right balance between cooperation and competition is one of the main endeavours of modern societies.[284] Human nature has partly stalled, and therefore partly maladapted, in its evolutionary transition from individual to group,[285] and from small tribal societies to large million-member societies: this is well illustrated by the historically transient nature of very large societies in which competitive and cooperative forces so often could not be kept in balance.[286]

6.5.2.1 Competition and Cooperation in Modernity

The development of nation-states, with their well-organised and well-trained police forces and other bodies of law enforcement, has undoubtedly succeeded in humanising society and diminishing or controlling excessive competition such as aggression and violence on the one hand. The modal citizen no longer needs a personal escort or bodyguard team to travel from one city to another as in medieval times. Although this does perhaps not apply fully to children, young women, elderly people, or even a solitary walker in some quarters of many large modern cities, even in advanced countries assaults, thefts, and rape attempts still appear to be regular events.[287] On the other hand the development of modern culture, with its drives towards scientific, technological and cultural innovation, and ever-increasing (economic) growth and wealth, has raised the stakes of meritocracy in education and work, but it has also considerably increased the interpersonal competition in order to achieve excellence and access abundance. No domain of societal life escapes this trend of competition: education, science, economics, sport, literature and other forms of art.

Democratic and respectable policy decision makers have made increasing competition their primary political goal. In 2000 the EU Lisbon European Council set as a new strategic goal for the following decade:

> to become the most competitive and dynamic knowledge-based economy in the world capable of sustainable economic growth with more and better jobs and greater social cohesion.

In 2009, US President Barack Obama, speaking about education at the Hispanic Chamber of Commerce, said that the U.S. must improve education to stay competitive in the world economy and concluded his speech with the advice:

> … start to prepare every child, anywhere in America, to outcompete any worker anywhere in the world.

[283] Bowlby (1969), Tooby and Cosmides (1990), Wilson (1975; 1978), Burnham and Johnson (2005).
[284] Ridley (1993).
[285] Stearns (2007).
[286] Masters (1989, 21).
[287] For instance, Shichor et al. (1979), Kneebone and Raphael (2011).

Modern societies apparently tend to confuse competition with performance and choose to forget or ignore their own scientific findings about the negative effects of unbridled competition, which increases stress up to unbearable heights and disrupts family life, kills empathy, friendship and cooperation, increases anxiety and aggression, and in the end lowers performance, productivity and creativity.[288] The psychophysical potential for dealing with continuous and intensive competitive challenges has its limits. Overexerting human potential in competition may impair well-being and creative performance.[289]

On the one hand the intensity and amount of cooperation between large(r) numbers of people has been achieved via macro-social management in practically all domains of social life—education, labour, mobility, health and welfare care, social security, and police and military protection. Reciprocal altruism has been considerably enhanced via imposed taxation, redistribution of community resources and the development of social protection.[290] On the other hand at the interpersonal behavioural level far less progress has been made, such as when observing the indifference towards others or the lack of personal involvement and aid in many daily life situations that require a personal commitment towards unrelated citizens. Particularly in big cities, there is the often amazing indifference and non-assistance from bystanders at accidents or incidents,[291] and a lack of courtesy, even towards elderly people, pregnant women or people carrying babies. In many countries, at first sight trivial, the traditional custom of ceding his seat in a bus, tram or train has dwindled away.[292] One may even wonder whether several inherent features of modernisation—such as the increasing wealth of individuals and families, the market type of economy,[293] and the development of social protection measures at macro-levels as rights that carry little individual responsibility—are a partial cause of the lack or even decrease of altruism, reciprocity and mutualism at personal level.

Stronger interpersonal forms of cooperation appear to only exist in smaller size communities where everyone knows everyone, in societies where altruistic values are particularly strongly emphasised, or in ideological, ethnic or racial in-groups that are in competition with or threatened/discriminated by out-groups.

6.5.2.2 Antisocial Behaviour in Modernity

The situation regarding antisocial behaviour in modernity has profoundly changed in several respects compared to the context of earlier stages in human evolution and history. There have been important changes in the causes of antisocial behaviour and in what is considered to be antisocial behaviour. Also, the types and degrees of

[288]Kohn (1987).
[289]For instance, Cordes (2008).
[290]In this respect Richard Dawkins (1976, 105) characterised the welfare state as: "perhaps the greatest altruistic system the animal kingdom has ever known. But any altruistic system is inherently unstable, because it is open to abuse by selfish individuals, ready to exploit it."
[291]Latané and Draley (1970).
[292]For instance, Burton (2008).
[293]See, for instance, Kasper and Borgerhoff Mulder (2015).

control of social behaviour have changed and, above all, the number, aggregation and mobility of people within and between nations has increased tremendously.

Improving living conditions, educational efforts and efficient policing practices are considerably reducing earlier widespread forms of antisocial behaviour such as assault, theft and homicide. Living standards have improved in many respects, thereby eliminating extreme causes of shortages in livelihood and decreasing the need for antisocial actions in order to obtain elementary resources for survival and well-being—for instance, the well-known increases in many forms of social strive and criminality in situations when famine strikes.[294]

Concerning the content of what is considered to be antisociality, several domains of social interactions are characterised by increasing rigour (e.g. all forms of sexual harassment, child abuse, ecological mismanagement), but other domains continue to be approached with great laxity or incompetence (e.g. financial fraud, Internet abuse, international crime).

Furthermore, regarding moral control mechanisms of antisocial behaviour, moves in opposite directions can be discerned: on the one hand there is a decrease of socialisation and conditioning by traditional ideologies due to the weakening of their ideological foundation, but insufficient take-over by modern secular ideologies; there is also a decrease in family and nearby community control; on the other hand there is the increase of the broader, albeit insufficiently compensating, societal control mechanisms (police, ICT).

The increase of population size and of mobility of persons within and between states caused a considerable increase in the number and superficiality of interactions between genetically non-closely related citizens,[295] allowing free-riders, cheaters and downright antisocial individuals to take ample advantage of the resulting increased anomy. Modernity is probably also changing the population composition with respect to predispositions to antisocial behaviour through the conservation, or even promotion, of less favourable genetic mutations or behavioural patterns which were under heavy selective pressures in pre-modern living conditions, but now have sufficient survival value in the modern culturally, economically, or biomedically protected environment or are even fostered by such environments.

Criminality. Antisocial behaviour manifests often itself as criminal behaviour, although it is a broader concept than criminal behaviour, the latter being limited to acts that are defined by the criminal code of a country. Nevertheless, a person may exhibit antisocial behaviour without being a criminal, and a person can exhibit criminal behaviour without being antisocial. Notwithstanding their differences, they often coincide.

[294]Keys and Brozek (1950).
[295]Newson and Richerson (2009).

In recent decades, biocriminology has made enormous progress in identifying genetic and neuro-hormonal factors and their interaction with environmental and socio-cultural living conditions in the development of antisocial, and in particular criminal, behaviour.[296]

Present-day biosocial criminology is primarily based on recent developments in three large biological fields of study: genetics, neurosciences and evolutionary biology. The first includes molecular genetics,[297] but behavioural genetics is even more important.[298] The neurosciences include neurophysiology, psychophysiology and neuro-chemistry.[299] The third domain includes several fields of biological anthropology, in particular sociobiology and evolutionary psychology.[300]

It is, in particular, the progress in the evolutionary sciences that considerably contributed to the understanding of why modernity is still plagued by a substantial amount of antisocial behaviour and, in particular, criminal behaviour. The human species is still genetically and neuro-hormonally equipped with drives—and a variation in those drives resulting in the development of different coping and reproductive strategies—which were functional and adaptive in the ancient environment in which our specific gene pool took shape.[301]

In modernity, biological predispositions or life course events resulting in antisocial behaviour, and *a fortiori* criminal behaviour, have become socially maladaptive. However, the protecting environment of modernity might be increasing the reproductive adaptedness of such predispositions. For instance, Shuyang Yao and colleagues[302] found, on the basis of a nationwide study, that criminal offenders in Sweden had more reproductive partners and more children (from different partners) than non-offenders. Those authors consider criminality to be adaptive in contemporary industrial society, the result of an alternative adaptive reproductive strategy.

Modern societies are, notwithstanding the much improved living conditions including elimination of starvation and other extreme forms of shortages in livelihood, excellent breeding grounds for antisocial and criminal behaviour, due to (1) their huge population sizes with their numerous social interactions between genetically non-related or socially anonymous citizens; (2) the opening of national

---

[296]Mednick and Christiansen (1977), Buikhuisen (1979), Taylor (1984), Wilson and Herrnstein (1985), Mednick et al. (1987), Eysenck and Gudjonsson (1989), Denno (1990), Ellis and Hoffman (1991), Raine (1993), Moir and Jessel (1995), CIBA Foundation Symposium (1996), Rowe (2002), Walsh (2002), Walsh and Ellis (2003), Thienpont (2005), Wright et al. (2008), Walby and Carrier (2010), Rocque et al. (2012), Raine (2013).

[297]Rowe (2002).

[298]Mednick and Christiansen (1964), Rhee and Waldman (2002).

[299]Raine (1993).

[300]For instance, Mealey (1995), Pitchford (2001), Quinsey (2002), Walsh and Ellis (2003), Thienpont (2005).

[301]For instance, Mealey (1995), Thienpont (2005), Ferguson (2008; 2010), Duntley and Shackelford (2008).

[302]Yao et al. (2014).

borders to international traffic via migration, travel, commerce and ICT; (3) the strongly competitively oriented social fabric;[303] and (4) the still considerable social inequalities in opportunities which provoke feelings of relative deprivation. Moreover, modern culture has a protective environment in which the basic needs for security, nutrition, housing, education, labour opportunities, health care, and leisure are largely ensured; it also frees the individual from the need to respond to the challenges, exigencies and dangers of living in natural circumstances—such as the stresses of hunting and gathering, natural disaster, and war. Consequently, modern culture attracts some people, in particular young males, to sensational, adventurous, and occasionally antisocial, actions.[304]

Thanks to the considerable progress in the ethical, social, psychological, and biomedical sciences, the manifest and deeper lying causes of antisocial and criminal behaviour are now much better understood. Furthermore, increasingly better methods have become (or will soon become) available to prevent or cure undesirable deviant forms of behaviour. Indeed, one of the most important findings of recent decades is that increasingly the risk of antisocial or criminal behaviour can be predicted at a young age, on the basis of the combination of familial, social,

[303]See Callahan (2004).

[304]Recently, many commentators in Western countries have expressed surprise about Western-raised and educated young men of Muslim creed joining the ranks of the Jihad warriors in the so-called Caliphate ISIS that Islamic fundamentalists are trying to erect in parts of Syria and Iraq or undertaking terrorist attacks in Western countries. This is not surprising at all. First of all, the numbers of European jihadi in proportion to the total Muslim population in Europe is extremely small (<0.02%). Even when only younger age groups (20–40) are taken into account, the figures remain very low (<0.2%). Indeed, it is rather surprising that the pull is not much higher, given the Western policies towards the Middle East on the one hand, and the poor integration of many second- and third-generation migrant descendants on the other hand. It is a textbook example of a convergence of all necessary biosocial elements to produce such a phenomenon: (1) young men, (2) sexually highly aroused but frustrated, (3) many with relatively low cognitive abilities making them easily susceptible to simplistic and fulfilling instigations about aggressing and destroying out-groups, (4) and/or emotional personality characteristics, predisposing them to adventurous heroic, asocial or criminal behaviour (Victoroff 2005; Weenink 2015), in contrast with their often dull or banal preoccupations or even non-occupations, (5) experiencing feelings of injustice toward their own in-group and/or feelings of personal deprivation of societal benefits, (6) strongly in need of finding or affirming a socially high status identity, (7) preyed on by fundamentalist religious zealots who find ample incentives in their holy scriptures to arouse self-deception among their adherents (Fink and Trivers 2014) and to incite them to punish or kill the out-group of despicable infidels, amply taking advantage of the inborn drive to defend and promote the in-group and assault out-group(s), and (8) last but not least, originating from a religious group in which fundamentalist beliefs about the construction of an idealised world, especially in the hereafter, are largely prevailing (Koopmans 2013). In particular the last factor may be an important cause for the astonishing fact that some well educated youngsters, originating from privileged wealthy families, also appear to be attracted to the jihad calling.

An even more puzzling question is why Western-educated Muslim women migrate to ISIS, although here the numbers are even lower than for the male jihadists. From a preliminary analysis of the motives of such women (Hoyle et al. 2015), it appears that partly similar factors as for male ISIS adepts play a role in their decision, namely disappointment with Western society and policy, but also, based on their indoctrinated belief system, desire to be part of and contribute to an ideal Muslim society in the traditional, dutiful, submissive, domestic female role.

physiological and behavioural indicators. Neural defects, low arousal levels, low verbal intelligence, hormonal abnormalities, low serotonin levels, lack of ability to concentrate, irritability, impulsivity and aggressivity, experience of deprivation or abuse, insufficient moral training at a young age, and especially the combination of several of these indicators, form risk predictors for antisocial behaviour and criminality in adolescence and adulthood. Progress in the aetiology of antisocial and criminal behaviour, together with expanding possibilities for ethical, social, psychological and biomedical interventions, offer new opportunities for early, preventive treatments for children with multiple risks of criminal behaviour and may allow a shift from a juridical punitive approach toward a controlled psychotherapeutic and biomedical approach.[305]

Psychopathy/Sociopathy. The multidisciplinary study of antisocial and criminal behaviour has contributed to the identification of antisocial personality disorder (APD or ASP),[306] either in the form of psychopathy or sociopathy. The distinction between psychopathy and sociopathy is not always very clear, but the authors would opt for David T. Lykken's[307] distinction whereby psychopathy is considered to be the result of being born with abnormal emotional personality traits of congenital origin, whereas sociopaths are primarily or mainly the result of a failure to receive adequate socialisation due to weak or irresponsible parenthood or social deprivation in general. This distinction corresponds to the one Linda Mealey[308] makes between primary and secondary sociopaths.

The prevalence of psychopathy is estimated at about one per cent.[309] Its prevalence is much higher in prison populations, particularly among chronic criminal offenders.[310] Psychopathy is mostly a male phenomenon; it is almost four times as high in men than in women.[311]

Psychopathy is considered to be the lower extreme of the normal distribution of the capacity for empathy.[312] Psychopaths have a lack of empathy, remorse, guilt and shame. In the professional literature they are described as people without conscience, as being callous, selfish, dishonest, arrogant, aggressive, impulsive, irresponsible, hedonistic, promiscuous, as well as egocentric, glib and grandiose but superficial. They manifest a lack of concern for the suffering of others, are at ease with their exploitation of others and show extreme reluctance to be responsible for others.[313]

[305]Moir and Jessel (1995), Rocque et al. (2012).
[306]For instance, Black and Lindon (2013).
[307]Lykken (1995).
[308]Mealey (1995; 1997, 162).
[309]Hare (1999).
[310]Mealey (1995).
[311]Sigvardsson et al. (1982), Mulder et al. (1994), Cottler et al. (1995).
[312]See, for instance, Baron-Cohen (1997; 2011). Some authors consider psychopaths to be a subspecies of humans (Harris 1995; Kopenhaver 2010), or a special taxon (Harris et al. 1994), erroneously applying taxonomic concepts to phenomena which are expressions of normal variation within the human species.
[313]Cleckley (1984), Hare (1999), Churchland (2002), Millon et al. (2002), Blair et al. (2005), Blair, (2006), Farrow (2007).

They start sexual activity at an early age and often in an aggressive way. They tend not to form strong attachments or long-term relationships, and are characterised by marital histories of desertion, non-support and abandonment.[314]

They often display antisocial behaviour, which frequently results in a criminal record.[315] They lack the violence inhibition mechanism (VIM) in the presence of submission.[316] A specific characteristic of psychopathy is that extremely violent and antisocial behaviour appears at a very early age, characterised by a broad variety of antisocial forms of behaviour such as lying, theft, killing animals, truancy and disruptive actions at school, and delinquent behaviour.[317]

However, not all psychopaths are violent people or even criminals.[318] They are often perceived as attractive and charming; they are often sexually hyperactive and can be highly skilled at deception and manipulation. There is only a weak association between psychopathy and high IQ.[319]

Psychopathy has a genetic predisposition.[320] Its heritability has been estimated at 50 per cent.[321] Several differences in brain structures have been identified showing signs of abnormal neurophysiology,[322] including reduced activity in areas involved in emotional processing. There is no convincing evidence that psychopathy is the direct result of early social or environmental factors.[323]

Several characteristics of psychopathy—among others its relative high prevalence, its aggressive and sexual predatory nature—have provoked evolutionary scientists to wonder whether psychopathy is not just a pathology resulting from accumulated unfavourable mutations,[324] but might be a minority alternative life-history strategy that is evolutionarily adaptive.[325] According to this view, psychopathy is a strategy of individuals who are unsuccessful in the normal competition for resources or mates, therefore they use deception and cheating to get them without reciprocating. Psychopathy is a frequency-dependent strategy,

[314]Hare (1999), Fulton et al. (2010; 2014).

[315]Eysenck (1977), Raine (1993), Ellis and Walsh (2000), Noziglia and Siegel (2006).

[316]Blair (1997, 87), Blair et al. (2005).

[317]For instance, Harpending and Sobus (1987), Hare (1999).

[318]Dutton (2012).

[319]Cleckley (1984), Hare and Neumann (2008).

[320]Hare (1995), Plomin and McGuffin (2003), Jang (2005), Viding et al. (2005), Larsson et al. (2006).

[321]For instance, Mason and Frick (1994), Viding et al. (2005), Larsson et al. (2006), Ferguson (2010).

[322]Kandel and Freed (1989), Damasio et al. (1990), Smith et al. (1992), LaPierre et al. (1995), Raine et al. (2000), Tiihonen et al. (2000), Blair, et al. (2001), Laakso et al. (2001), Lacasse et al. (2003), Hauser (2006, 237), Kiehl et al. (2001; 2006), Rilling et al. (2007), de Oliveira-Souzaa et al. (2008), Weber et al. (2008), Wahlund and Kristiansson (2009), Yang et al. (2009), Rijsdijsk et al. (2010), Remmel and Glenn (2015).

[323]Pitchford (2001).

[324]Barr and Quinsey (2004), Glenn et al. (2011).

[325]For instance, MacMillian and Kofoed (1984), Harpending and Sobus (1987), Harris et al. (1994), Harris (1995), Mealey (1995), Baron-Cohen (1997), Colman and Wilson (1997), Pitchford (2001), Crawford and Salmon (2002), Wiebe (2004), Ferguson (2008), Verplaetse et al. (2009), Cartwright (2010, 427), Kopenhaver (2010), Glenn et al. (2011), Glover (2011).

meaning that the force of natural selection against it varies according its relative frequency in the population.[326] Although they are under risk of being socially combated or ostracised in one or another way, their promiscuous sexual and mating pattern in combination with their cheating and deceiving behaviour, make them succeed to a certain degree in transferring their genes intergenerationally.[327]

An alternative to the sexual/mating hypothesis is the warrior strategy:[328] between-group competition might have favoured the presence of warriors with personality traits that facilitated them to ruthlessly and recklessly fight and eliminate others, in the end contributing to group survival.

Some scholars have argued that modernity is not only an excellent breeding ground for criminal behaviour, due to its lack of small-scale social control,[329] but that the extremely competitive and cheating nature of the modern economy and corporate culture is an ideal soil in which psychopaths can thrive, more particularly in the higher ranks of corporate, military and governmental hierarchies.[330] Clive R. Boddy, Rick Ladyshewsky and Peter Galvin[331] identified greater amounts of psychopathy at more senior levels of corporations than at more junior levels; they argue that the ruthless, selfish and conscience-free approach to life of corporate psychopaths raises not only problems for corporations themselves but also for society as a whole, due to the global spread of a too strongly individualistically and supercompetitively oriented corporate economy. In a recent paper Clive R. Boddy[332] even argues that psychopaths working in financial corporations may have had a major part in causing the recent global financial crisis.

It leaves no doubt that the present competitive corporate culture, in which psychopaths thrive so well, is unsustainable in the long-term. However, it might take quite some time before modern societies are able to take adequate measures to change and contain this development, since it will need to be politically mastered at the international, if not the global, level. Hence, the economic success of psychopathic personalities cannot be expected to have a long fate. However, there is another reason why psychopathy will regress in modernity: just as is the case for machos in general, the strongly promiscuous psychopaths will no longer have the reproductive success they had in the past, because in modern societies women now possess efficient means of birth control through which they can avoid undesirable conceptions or pregnancies with partners who are not prepared to share long lasting parental responsibilities with them.

[326]Baily (1995), Mealey (1995), Colman and Wilson (1997).
[327]See also Hare (1995), Wiebe (2004).
[328]Baily (1995), Book and Quinsey (2004), Kopenhaver (2010).
[329]Boehm (2012, 338).
[330]Callahan (2004), Lobaczewski and Knight-Jadczyk (2007), Kopenhaver (2010).
[331]Boddy et al. (2010).
[332]Boddy (2011).

6.5.3 Ethical Reflections Regarding Interpersonal Relations

6.5.3.1 Containing Competition

From the preceding sections, it appears that modernity struggles with two fundamental problems regarding competitive behaviour. The first one concerns the individual drive for outcompeting others in order to obtain or secure resources for self-actualisation and reproduction. The second relates to the specific economic and technological features of modernity that incite individuals, or groups of individuals, to express their urge for resource acquisition far beyond their individual or family needs: they develop a culture of greed and grabbing that must inevitably lead to societal stress and strife, as well as ecological overcharge. Both the individual drive for resource acquisition and the culture of greed, which developed in agrarian and especially in industrial society, ultimately result from a genetics that emerged in ancient times of within- and between-group competition for scarce resources.

As argued above, modernity needs a substantial further shift from competitive toward cooperative efforts. Realists will argue that it will be impossible to completely eliminate the drive towards competition from the human genome or gene pool. Perhaps not, but in the modern culture of the future the ethical goal should be focused on the drive for individual self-actualisation and interpersonal cooperation instead of interpersonal competition. If this principle could be introduced throughout our educational system and political culture, and become better rewarded socially, it might have unexpected multiplicative dispersal effects in many other domains of social life.

The culture of greed, transcending the need for satisfaction at the individual or family level, which developed in agrarian but especially in industrial society, and not the least in its present global neo-liberal variant, has to be contained considerably in order to avoid its unfavourable social (i.e. growing social inequalities) and ecological (i.e. increasing environmental deterioration) consequences.[333]

6.5.3.2 Promoting Cooperation

The human species is a social species, but its genetic predispositions for sociality have been largely selected as a function of social relations with few (and mostly closely related) people, and ultimately with the view to individual self-interest. Moreover, intragroup cooperativeness was evolutionarily largely designed as a function of intergroup competitiveness.

As human societies grew in size and complexity, cultural systems (moral, religious, educational, legal) had to be developed to deal with interpersonal relations between ever larger numbers of people. However, so far humanity has failed to develop a moral system that would allow the human species to function as one single unit.[334] Indeed, we are genetically not so well prepared to deal with altruism at a distance. Our gene pool developed largely in the ancient EEA, when group sizes were small and our physical and mental horizon for mutual help was strongly

[333]For instance, Motesharrei et al. (2014).
[334]Alexander (1993, 180).

restricted.[335] The position that we have reached today is to develop ethics of altruism that generally ends at the border of sovereign states.

Modernity is characterised by two counter movements which impede the further development of a well-adapted social morality: (1) as argued above in Sect. 6.4.2.3, the development of welfare state institutions induced the subjective feeling of heightened independence and boundlessness amongst many individuals, and hence lay the grounds for an amplified individualism; and (2) the weakening of the traditional religiously based moral code systems, and their insufficient replacement by secular moral indoctrinating systems, also hinder progress in cooperative morality.

Hence, modern society should make considerable innovative efforts to develop a strengthened social morality via its ideological pillars, its educational system, its mass media and its policymaking. As Sam Harris[336] wrote recently in *The Moral Landscape*:

> There may be nothing more important than human co-operation… few subjects have greater bearing upon the question of human well-being.

Prosocial behaviour requires a subtle biocultural interaction: internally a sufficient neuro-physiological and endocrinological basis for social arousal needs to be present, but externally the culture must be sufficiently stimulating in order to satisfy the exploratory and danger defying neuro-hormonal constitution of the human, particularly the potentially aggressive drives of the male adolescent and young adult. We will have to foster our social intelligence—the capacity to understand the intentions, motivations and desires of other people[337]—and to organise mutual cooperation among strangers on a massive scale.[338]

6.5.3.3 Mastering Antisocial Behaviour

In modern society violent and cheating behaviour has become evolutionarily maladaptive: socially because the harsh living conditions (including intergroup conflict) of the original EEA no longer exist, at least in peacetime; and reproductively because women can now protect themselves efficiently against involuntary conception by means of modern birth control practices through which the reproduction of genes of sexual offenders and abusers can be avoided.

All forms of antisocial behaviour, in their violent or cheating variants, disturb normal social life and community welfare and well-being in way one or another.

[335]Hauser (2006, 10).
[336]Harris (2010, 55).
[337]Gardner (1983).
[338]Bowles and Gintis (2011, 199).

There is no place for antisocial behaviour, let alone criminality, in the modern welfare state that aims to offer care for all of its citizens, create equal opportunities for everybody, and keep inequalities within reasonable limits.

Whereas the social causes of antisocial behaviour can easily be eradicated through fair social policies, it is much more difficult to deal with any kind of biological causes that predispose individuals to antisocial or criminal behaviour. Predispositions or urges toward such behaviour should be canalised into socially positive activities. Where this proves to be impossible, such as may be the case for psychopathy, paedophilia or sexual assault, it can be hoped that further progress in biomedical sciences will soon allow complementary measures, or even the replacement of traditional juridical punitive measures, by psychotherapeutic and medical interventions in order to prevent such undesired forms of behaviour. Personal freedom in these matters absolutely needs to be subordinate to community welfare and well-being. As Sam Harris[339] pointed out pertinently:

> Evolution may have selected for territorial violence, rape, and other patently unethical behaviours as strategies to propagate one's genes—but our collective well-being clearly depends on our opposing such natural tendencies.

[339]Harris (2010, 101).

Evolution-Based Ethical Challenges Related to Group Relations

7

Abstract

In this chapter four major categories of group relations are addressed, which are based on: (1) kin and family; (2) social status; (3) race, ethnicity, worldview, or political conviction; and (4) statehood. The evolutionary background, developments in modernity and ethical reflections for the future are discussed for each of them. Family relations are discussed in the broadest possible meaning, in order to include all variants appearing in modernity. The evolutionary origin of the family is explained, the changes that families experience in modernity are sketched, and ethical reflections for the future are suggested. The evolutionary heritage and modern changes of our drives for status achievement, egalitarianism, and distaste for relative deprivation are described. Major ethical forward looking aspects in the evolutionary perspective are presented. Competition with respect to social disparities based on race, ethnicity, worldview or political conviction are highlighted in their common evolutionary background grounded in the in-group/out-group syndrome. Regarding relations between states, the evolutionary heritage of intergroup conflict leading to present day governance systems is recalled. Ethical aspects of international relations that represent challenges for the future are identified.

7.1 Introduction

Four types of group relations are distinguished because each one of them is characterised by a specific type of feature that is at the root of antagonistic behavioural patterns towards other groups. The first concerns kinship and family relations that are usually associated with nepotism. The second concerns distinctions which are characterised by differences in social status, resulting in a social class hierarchy,

R. Cliquet and D. Avramov, *Evolution Science and Ethics in the Third Millennium*,
https://doi.org/10.1007/978-3-319-73090-5_7

which is often subject to classism.[1] The third are distinctions characterised by racial, ethnic, or ideological features and are often subject to various forms or degrees of racism, ethnocentrism, or ideological intolerance. The fourth one concerns between state relations based on governance organisation, often associated with several forms or degrees of xenophobia.

7.2 Kinship and Family

Kinship and family relations are a primary form of interpersonal relations. There are several definitions of the term kinship. Here it is used in its biological meaning, namely people related by descent (genetic relatedness). The term family is a broader concept than kin, as it refers to a group of people affiliated by descent (= kinship, mostly a genetic relationship, but it can also include adopted or foster children), sexual affinity (marriage, consensual union, living-apart-together (LAT) relation and commuting relation), or coresidence (other household types).

The development of kinship and family relations in modernity requires the consideration of those relations in a broad context, including relations resulting from lineage (parent-children relations) and ascendance (adults-parents relations), relations based on affinity which may be of a sexual nature (marriage, consensual union, LAT-relation, commuting relation) or a social nature (non-sexual coresident adults). This approach allows the inclusion of biological as well as social family relations, and all types of family and household forms, not only traditional nuclear and extended families but also monoparental families, adoptive families, reconstituted families, gay and lesbian unions, etc.

Nuclear family relations include two major components: partnership relations and parent-children relations—the latter descending as well as ascending. Both fundamentally relate to care and well-being, be it of a sexual, psychological, social, educational, financial or health care nature.

7.2.1 Evolutionary Background of Family Variability

Some social scientists[2] believe that the family is not a natural phenomenon, but instead it is a social construct based on social and psychological factors. In contrast, evolutionarily inspired scholars maintain that the origin of the human family is of a biological nature and that, even today, especially in modern culture, the only functions that keep families together remain biosocial in nature.[3] The family is a typical sociobiological group phenomenon. The social-constructionist view of the family is an example of short-term, proximate and non-evolutionary thinking.

[1]Classism: prejudice or discrimination based on social class.

[2]For instance, Ditch et al. (1995), Zonabend (1996).

[3]Gough (1971), Van den Berghe (1979), Mellen (1981), Emlen (1995), Chapais (2008), Gorelik et al. (2010).

A long-term, evolutionary and cross-cultural approach, in contrast, shows that the ultimate and fundamental *raison d'être* of the human family is of a biological nature:

> The human family is, very simply, the solution our hominid ancestor evolved over three to five million years to raise brainy, slow maturing, neotenic, highly dependent, and therefore, very costly (in terms of parental investment) babies.[4]

7.2.1.1 Long-Term Childcare and Enduring Partnership

The origin, universal existence and future continuity of the (nuclear) family are the result of the hominisation process that produced the human-specific brain that requires, in turn, a long postnatal maturation and socialisation process.[5] Families are the social extension of uterine life, based upon the needs of slowly maturing human children—several of which must be produced in order to guarantee inter-generational continuity.[6] The predisposition to intense and long childcare is the result of selective processes in response (during hominisation) to the increasing ontogenetic dependence of the human infant and adolescent.

As argued in Chap. 6, Sect. 6.3.1, the sexual specificity of the human species, facilitating the establishment of enduring love relations between the partners, has been selected as a function of the increasing dependent long maturing offspring.

Evolutionary science provides us with four theories from which specific hypotheses about mate preference and choice in an enduring partnership can be derived: 'good genes' theory, parental investment theory, reproductive value theory, and paternity confidence theory.

The 'good genes' model refers to the preference for and choice of mates who possess features displaying viability, parasite resistance, immunocompetence and developmental stability.[7] Features expressing physical fitness, the population average, and bilateral symmetry are clearly preferred in mate choice, by both men as well as women.[8]

The concept of parental investment[9] implies, for a species such as the human where males contribute significantly to the upbringing of their offspring, which females seek to mate with males who have the ability and willingness to provide resources that will benefit their children.[10] This implies that females will value and seek out male personality characteristics such as intelligence, ambition, and industriousness; social features such as high educational level and high social status (which are associated with or result in increased earning capacity); and physical

[4]Van den Berghe (1988, 43).
[5]Gough (1971), Van den Berghe (1979), Mellen (1981), Emlen (1995), Bellah (2011, 104).
[6]Van den Berghe (1979), Filsinger (1988), Booth et al. (2000), Salmon and Shackelford (2007).
[7]Westermarck (1921), Barber (1995).
[8]Gangestad and Buss (1993), Singh (1993), Buss (1994), Perrett et al. (1999), Shackelford and Larsen (1999), Thornhill and Grammer (1999), Honekopp et al. (2004), Roberts and Little (2008), Craig and Little (2008).
[9]Trivers (1972).
[10]Alexander and Noonan (1979).

features such as height and strength.[11] Other desirable attributes might include: dominance, maturity, emotional stability, reliability, fidelity, willingness to provide for resources, sociability, and love.[12]

Reproductive value[13] is the degree to which individuals of a given age and sex have the capacity to produce additional viable offspring and, hence, to transmit their genes to future generations. In the human species, the male's reproductive capacity is huge and from puberty onward it is relatively independent of age.[14] In contrast, women's fecundity is low and strongly time dependent. Women's age-specific natural fertility varies quite substantially: it increases in the first years after puberty, peaks in the mid twenties, and thereafter it gradually and increasingly declines with age, especially after 35 years of age.[15] At the same time, women's future reproductive capacity is strongly associated with age varying physical and behavioural features such as general body build, in particular waist-hip ratio,[16] youthful facial traits, breast form, skin and muscle tone, fat distribution, and energy level. Physically attractive and behaviourally dynamic features peak at younger ages. Therefore, youth and physically attractive features are considered strong indicators of high reproductive value and are consequently highly valued by men since women's fecundity is a limited and, hence, precious resource.[17]

Paternity certainty[18] is a sensitive issue in a species where the male contributes quite substantially to parental investment. Whereas maternity can never be doubted, paternity can. Consequently, parentally investing males have a genetic interest in securing paternity confidence in order to avoid cuckoldry (the investment in offspring that is not genetically theirs).[19] Accordingly, paternal investment will lead to a relatively stronger experience of male sexual jealousy and to a stronger male preference for female chastity and fidelity. This evolutionary prediction finds confirmation in many traditional sexually asymmetrical sociocultural practices and double standards: female sequestration, veiling, genital mutilation, virginity protection, foot binding, mate guarding, chastity belts, spousal homicide, legal restrictions on female sexual behaviour, wife beating, penalties for adultery, conjugal dissolution, daughter guarding, etc.[20]

7.2.1.2 Elderly Care

Whereas there is a strong basis in evolutionary theory for explaining partnership behaviour and parental care for offspring, the question has been raised as to whether

[11] Barber (1995).
[12] Buss (1994, 2007).
[13] Fisher (1930), Williams (1975).
[14] Nieschlag (1986).
[15] Eaton and Mayer (1953), Charbonneau (1979).
[16] Singh (1993), Hughes and Gallup (2003), Streeter and McBurney (2003).
[17] Howell (1979), Buss (1989), Wiederman and Allgeier (1992), Ridley (1993), Thiessen et al. (1993), Bereczkei and Csanaky (1996), Wiederman and Kendall (1999).
[18] Daly and Wilson (1978).
[19] Platek and Shackelford (2006).
[20] Dickemann (1979), Daly et al. (1982), Betzig (1989), Batten (1992), Buss (2002), Perilloux (2008).

there are also evolutionary explanations for the care of people in their post-reproductive stage of life. Views differ on this matter. On the one hand some scholars state that "evolution does not care about elderly" because, as has been seen in Chap. 6, Sect. 6.2.1.2, genes protecting survival at post-reproductive age are less selected for[21]: the force of natural selection weakens with increasing age.[22] On the other hand some scholars argue that, since the elderly can increase the reproductive fitness of their adult children—cf. the grandmother hypothesis[23]—the latter have a reproductive interest in transferring resources to their parents.[24] Although the net flow of transfers within families goes universally from parents to children and grandchildren, there is often also a partial flow from adult children to elderly parents.[25] Moreover, the altruistic care given to an ageing parent could also be explained on the basis of kin selection theory[26] and reciprocity theory.[27] Finally, there are several theories—attachment theory, equity theory, and family obligation theory—which explain elderly care more on the basis of ontogenetic processes to which adult children have been acculturated.[28]

7.2.1.3 Nepotism

There is another important evolutionarily based family characteristic that is responsible for a particular form of family related attitudes and behaviours, also persisting in modern living conditions. It concerns the predisposition to favour people according to their degree of genetic relatedness, conventionally called nepotism.[29] Nepotism is defined as favouritism shown to relatives.[30] Behaving nepotistically, namely by favouring one's own kin, increases the probability of enhancing one's inclusive fitness by favouring the reproduction of the genes one shares with the recipients of one's altruism. Hence, nepotistic behaviour is a consequence of kin selection and is explained by inclusive fitness theory.[31]

7.2.2 Family Variability in Modernity

In their kin influence hypothesis, Lesley Newson and Peter J. Richerson[32] suggest that modernity is characterised by a considerable shift in people's social networks going from relatives to non-related persons.

[21]For instance, Bonneux et al. (1998).
[22]Medawar (1952).
[23]Williams (1957), Hill and Hurtado (1991).
[24]Coall and Hertwig (2010).
[25]For instance, Caldwell (1982), Lee (1997).
[26]Hamilton, 1964.
[27]Trivers, 1971.
[28]For instance, Cicirelli (1991).
[29]The word nepotism derives from the Italian '*nipoti*' which refers to any family descendent; in Latin '*nepos*' stood for grandson or nephew.
[30]Van den Berghe (1978), Alexander (1979), Bellow (2003).
[31]Hamilton (1963; 1964), Maynard Smith (1964).
[32]Newson and Richerson (2009).

Indeed, the twentieth century has witnessed remarkable changes in family structures and dynamics in Western Europe, North America and other overseas mainly English speaking countries, as well as in non-Western countries where modernisation is taking off[33]: smaller household sizes; continuing shift from extended to nuclear families; advance of the initiation of sexual partnership[34]; decrease in nuptiality, although eventually a large majority of couples marry[35]; increase in separation or divorce[36]; appearance of new forms of unions such as unmarried cohabitation[37] and living-apart-together[38]; increase in single-person households[39]; rise in the number of one parent families (mostly headed by women) either from separation, by accident or choice[40]; increasing number of remarriages[41]; changing sex relations among others due to increasing outside employment of women[42]; shift from incidental to planned parenthood[43]; changing relations between adults and their parents[44]; and, last but not least, a substantial decrease in fertility, often to below replacement levels.[45]

In the 1960s, several of these family changing phenomena started to accelerate because they spread over larger sections of Western or westernised populations, due to the combined effects of a number of interrelated and mutually reinforcing economic, technological and cultural factors which speeded up in that period.[46] This caused Ron Lesthaeghe and Dirk van de Kaa to define these changes as the Second Demographic Transition.[47]

The important changes in family structures observed in recent decades are partly the result, but also partly the cause, of changes in relational contents, dynamics and processes.[48] Both partner relations and parent-child relations have been affected by several changes in values, power and decision-making and in the emotional content of relationships. Partnership is shifting from complementarity toward egalitarianism, from normative behaviour toward individual choice and from a commanding toward a negotiating housekeeping. Similarly, parent-child relations have undergone changes including a shift from paternal power to parental authority, from

[33]Cliquet (2003), Avramov and Cliquet (2005).
[34]Bozon and Kontula (1997, 1998).
[35]Kiernan (1993), Sardon (2002).
[36]Deven and Cliquet (1986), Sardon (2002).
[37]Trost (1979), Corijn and Klijzing (2001).
[38]Rindfuss and Stephen (1990), Trost (1998), Fisher (1992).
[39]Cherlin et al. (1997), Corijn and Klijzing (2001).
[40]Deven and Cliquet (1986), Miller (1992), Barber (2005), Van Delft et al. (1988), Burghes (1993).
[41]For instance, Norton and Miller (1992).
[42]Gornick and Meyers (2003), Van Dongen (2009).
[43]Cliquet and Schoenmaeckers (1976).
[44]For instance, Rossi and Rossi (1990), Grundy (2008).
[45]For instance, Cliquet (1987), Chesnais (1998), Teitelbaum (1999), Frejka (2008).
[46]Hoffman-Nowotny (1987), Van de Kaa (1987), Cliquet (1991; 2003).
[47]Lesthaeghe and van de Kaa (1986), Lesthaeghe (1995; 2010).
[48]Deven (1996).

submission to self-development, from obedience toward exploration and from a unilateral toward a bilateral transmission of values and knowledge.

The many changes in family structures and functions observed in modernity undoubtedly have many effects. Overall, most changes correspond to basic needs and desires which can, in the more liberal moral climate and richer economic context of modernity, be achieved to a greater extent, and consequently they must be positively valued. This is particularly the case for women whose social and psychological position in family relations has considerably improved. Innovative features of modernity such as the scientific knowledge about sex differences, the emancipatory secular ideologies, the rise in educational levels of both women and men, the independent occupational opportunities for women, and last but not least, the effects of lower infant and maternal mortality and greater fertility control have contributed to the emancipation of women.[49]

However, some of the modern family changes also have unfavourable side effects to which societies have not yet adapted well. A few examples may illustrate this:

- The (partial) shift from the earlier social control of family structures and functions to personal choices and decision-making make both partnership and parent-child relations subject to much higher cognitive and emotional requirements. Consequently, although family relations have become more satisfying, they are also more vulnerable and less stable[50];
- The increase of outside employment for women has several advantages, for instance for the economic/financial situation of the family as whole, and especially for the empowerment and independence of women themselves. However, working conditions remain tailored to men and, due to insufficient societal adaptations to the new family situation, the dual career of parents is not easy to manage. In many cases there is simply not enough parental time left for adequate parenting. The increasingly competitive nature of the market economy, pushing people to ever higher levels of production, goes in the opposite direction to what is needed in terms of quality parenting[51];
- There is evidence that the modernisation process is characterised by a major shift from the authoritarian toward authoritative, permissive, uninvolved or nurturing styles of family relations.[52] Whereas research seems to suggest that the authoritative and nurturing styles are the most recommendable, there is evidence that these ideal styles are still far from being generally realised;

[49]Cliquet (1984), Aarssen (2007).
[50]Cliquet (2003).
[51]For instance, Gauthier and Smeeding (2004).
[52]Behavioural scientists distinguish several parenting styles and practices in early child development. Parenting (or child rearing) is the process of promoting and supporting the physical, emotional, social and intellectual development of a child from infancy to adulthood. For instance, Diana Baumrind (1971) distinguished four major parental types: authoritarian, authoritative, permissive, and uninvolved; Dacey and Packer (1992) also distinguished the nurturing parent model.

- Couple dissolution and divorce have undoubtedly contributed to resolving union or marital conflicts. However, they also result in a substantial number of cases of increasing poverty, more particularly of female headed single parent households[53];
- The effects of couple dissolution or divorce on children is a complex issue, which depends on a broad variety of separation/divorce contexts such as the age of the children, the nature of the separation process (peaceful or warlike), the future parental roles of both parents, and in the case of remarriage by one or both parents, the quality of the reconstituted family climate. No doubt, separation/divorce is for most children (and parents) a painful and stressful process, the development and consequences of which are not always properly dealt with[54];
- Growing numbers of elderly people, and especially women, are living alone in old age. The majority of elderly people living alone or in couples seem to be satisfied with their independence.[55] Nevertheless, due to the increasing life expectancy and the shift in the causes of morbidity and mortality from exterior to internal factors, an increasing proportion of elderly need permanent assistance or care in old age. This care, due to a variety of reasons—geographical mobility, desire of adult children as well as the elderly persons or couples to remain independent, outside employment of the traditional female carers of older people —can in many cases no longer be provided by adult children[56];
- There is also the phenomenon of below replacement fertility that will be discussed in Chap. 8, Sect. 8.2.2.2.

7.2.3 Ethical Reflections About Family Variability in Modernity

7.2.3.1 Promoting Quality Family Relations

Different views have been expressed about the (desirable) future of the family.[57] Observing the many changes that the family is undergoing in modernity, some have advanced doom-mongering views about the disappearance of the family[58] while others long for a return to the traditional family: these are views that are obviously widespread in ideologically (religiously and/or politically) conservative quarters, if not in extreme right wing circles.[59] Among futurists, some more eccentric visions blossom such as the idea that "social institutions like the family become obsolete in

[53]For instance, Avramov (2002).
[54]For instance, Deven and Cliquet (1986), Fomby and Cherlin (2007), Van Peer (2007).
[55]Lee (1997), Dooghe et al. (1988), Jacobs (2004).
[56]Vanden Boer (1999).
[57]For instance, Cornish (1979), Roussel (1989), Duvold (1995), Moynihan et al. (2005).
[58]See discussion in Berger and Berger (1983), Wright and Jagger (1999).
[59]Abbott and Wallace (1992), Gilbert (1999).

scientifically and technologically advanced societies",[60] or that there is "no necessary reason why partnership has to be limited to a dual unit, but could consist of three, four, or even more consenting adults who wish to get married and have children".[61]

On the one hand such visions greatly undervalue existing biosocial knowledge about the evolutionary origin of family bonding, regarding partnership as well as parent-children relations; and on the other hand they undervalue modern experiments regarding alternative family models. Taking those insights into account, there can be little doubt that we will, for a very long time, be genetically predisposed to drives towards emotionally high quality partnership and to continue self-rearing of one's own children. Evolved partnering love, as well as parent-child love, has been thoroughly built into our genetic heritage and it is not likely to change for a long time.

Representative surveys about sexual behaviour or family relations, including the opportunity to express preferences for a wide variety of sexual and partner relations, result in massive preferences for enduring love relationships.[62] However, less conventional forms of sexual relations, some of which may represent alternative forms of reproductive strategies, seem to attract more attention or succeed better in getting into the public spotlight, thereby seriously distorting the representation of reality.

Research on children who have grown up in institutional environments—even with perfect material living conditions—show that the absence of individual emotional bonding leads to behavioural disturbances (aggressivity, delinquency, antisociality), intellectual retardedness (lower IQ), and even physical retardation (stunted growth, illness, and increased mortality).[63] Whenever attempts have been tried to collectivise child rearing outside the nuclear family—see the experiments in the Oneida community,[64] the early Soviet Union,[65] and the Israeli Kibbutzim[66]—they soon had to be abandoned, due to the considerable economic costs, the mediocrity of the results, or the dissatisfaction of the parents about their reduced parental role towards their own children. The suppression of the bond between parent, especially mother, and children is felt to be emotionally intolerable in the large majority of cases.[67]

The principal challenge regarding family life in modernity and a further progressing modernisation consists of providing overall support to families to deal with the negative side effects of modernisation such as the higher emotional and intellectual demands on family relations, the work life/family life incompatibility and the more complex relations between descendants and ascendants. In other words,

[60]For instance, FM-2030 (1973), Bainbridge (2010).
[61]For instance, Sorgner (2010).
[62]Spira et al. (1993), Laumann et al. (1994), Wellings et al. (1994), Corijn and Klijzing (2001).
[63]Bowlby (1951), Montagu (1957), Rutter (1972).
[64]For instance, Foster (1984), Coşgel (2000).
[65]For instance, Prigent (1955).
[66]For instance, Sharabany et al. (2001).
[67]Arnhart (1998).

the modernisation process requires ethics that enhance the overall quality of partner relations and parent-children relationships.[68]

7.2.3.2 Accepting Diversity in Family Structures

As already argued in Chap. 6, Sect. 6.3.3, modern societies with their millions of citizens are reproductively much less vulnerable and, hence, can be much more receptive to behavioural variation in the domain of sexual, family and reproductive patterns. Moreover, a diversity in family structures will better harmonise with the broad diversity of social and psychological needs and philosophical convictions present in modern populations. A further modernisation can embrace a variety of family forms, in terms of legal status (married, unmarried, divorced), living arrangements (cohabiting, living-apart-together, commuting), sexual orientation (heterosexual, homosexual), parenting (biparenting, monoparenting, coshared parenting), and primary or successive relationships.

Regarding successive relationships, it must be recalled that in the past enduring marriages or unions which lasted 50 or 60 years were exceptional because of high mortality. Nowadays, separation or divorce has largely replaced mortality as the precedent for the establishment of another union. From the perspective of a further evolving hominisation and progressing modernisation, a succession of relations during the life course might become a still more common phenomenon, due to the expectation that longevity will be further prolonged.

7.2.3.3 Revisiting Mutualism Between Descendants and Ascendants in Families

Due to the nuclearisation of the family, increasing female outside employment, geographical mobility and individualisation in general, in many cases grandparenting is losing its important supporting parenting role in modernity.[69] There is perhaps no immediate solution for situations in which grandparenting is impossible, but the enduring importance of grandparenthood can only be underlined, not just for grandparenting itself, but in many cases as support for family expansion.[70]

The same family and societal developments—nuclearisation of the family, female outside employment, geographical mobility and individualisation in general —make it increasingly difficult to obtain family support for older adults in need of assistance or care. Various alternative, but undoubtedly more costly, solutions are being or will have to be set up on mass scale such as community welfare care, domiciliary care, residential care, sheltered housing and institutional care.[71]

In view of the increased longevity and the future expected further increase in life expectancy, the ethical basis of the family solidarity that goes predominantly from descendants to ascendants needs to be revisited.[72]

[68]Cliquet and Avramov (1998), Avramov and Cliquet (2005).
[69]For instance, Silverstein and Long (1998).
[70]For instance, Hočevar and Černič Istenič (2010).
[71]For instance, Vanden Boer (1999).
[72]Avramov and Cliquet (2003).

7.2.3.4 Addressing Nepotism in Modernity

Favouring one's own kin is a deeply rooted human drive that, in all probability, cannot be disposed of easily. At first sight there is nothing wrong with helping kin, particularly offspring. However, modern social processes and interactions between individuals have become so complex and diverse that traditional kin nepotism has to become subordinate to broader moral rules, such as guaranteeing equal opportunities, without which broader social interactions cannot function properly. This matter will be addressed in greater detail in Sect. 7.3.3.1.

7.3 Social Status Hierarchies

All known human societies show relatively strong differences in the distribution of wealth, power, and prestige. Sociologists have amply documented the phenomenon of social stratification in which people are ranked into a number of hierarchically differentiated layers.[73] Societies are, however, not only structurally stratified: the different positions and functions of its members are also differentially evaluated and rewarded.[74]

Depending on the stage of a society's cultural development—gathering-hunting, agrarian-pastoral, industrial, post-industrial—the allocation of differentially rewarded social status positions are made on the basis of different criteria, such as descent, personal wealth, military, political or religious status, economic and financial status, education, and personal qualities,[75] or these criteria are differently weighted or valued.

7.3.1 Evolutionary Background of Social Status Hierarchies

The existence of social status inequalities in human societies is consistent with dominance/competence hierarchies that are generally found in social animal species,[76] and especially in the closely related African great apes.[77]

During the hominisation process, social status differences evolved in a fluctuating manner: these have ranged from strong despotism, as present in the great apes, relatively egalitarian relationships among nomadic hunter-gatherers, strong social hierarchies in the agricultural-pastoral and early modern eras, to gradually decreasing social inequalities and inequities in the more advanced modern societies.[78] Overall, humans are characterised by a universal drive to social

[73]For instance, Sorokin (1927).
[74]For instance, Barber (1957), Ellis (1993).
[75]Barber (1957), Grusky (1994).
[76]Hinde (1974), Wilson (1975), Omark et al. (1980), Trivers (1985), Wilkinson (2005).
[77]De Waal (1982), Boehm (1999, 64), Mazur (2004), Chapais (2015, 163).
[78]Knauft (1991, 391), Boehm (2012), Gavrilets (2012), Harvey (2014).

dominance.[79] What appears to change over time is the social validation of the source and the acceptable degree of power and influence underpinning hierarchies.

At the individual (or family) level, social hierarchies are the result of within-group competition for scarce resources such as territory, food and sexual mates. The drives toward high status and priority access to resources are related to the principle of the maximisation of inclusive fitness.[80] Status and prestige are the markers of social success. Social success is the route to power which, in turn, is the key to the universal human desire of resource acquisition.[81] High status is associated with greater access to resources, wealth and mates, which can be converted into the production of a larger number of offspring: eventually this means a higher likelihood of intergenerational transmission of genes.

At the group level, social dominance relations (leadership-followership relations) may have had several other advantages during hominisation such as maintenance of group stability,[82] facilitation of the transmission of communication,[83] and coordination and collective action in intergroup conflict or competition for resources.[84] Essentially, groups organised in hierarchies appear to be more efficient at intergroup combat than groups that are organised in other ways. D. Cassill and A. Watkins[85] even argue that cooperative hierarchies provided a greater probability of survival to a greater number of members than egalitarian cooperatives. Leadership, either despotic or democratic, clearly becomes more important as group size and social complexity of societies increase.[86]

Hence, individual survival drive as well social efficacy may be reasons why predispositions for deference and submissiveness toward authority were embedded in the human gene pool that so strongly influenced social status hierarchies and the moral rules that regulate them.[87]

The evolutionary background of social status differences is ultimately explained with differential reproductive fitness, in other words, with Darwinian selection. Much has been written to explain why people spend so much time and energy on status related behaviours such as flaunting outward appearances. Present-day people are obsessed with a youthful appearance: both women and men use various forms of enhancement, for instance, to appear taller or acquire excessively big breasts in the case of women, and drive huge or powerful cars in the case of men. Several explanations have been given for this remarkable phenomenon. Amotz and Ashivag Zahavi[88] explain this by what they call the handicap principle, a phenomenon that acts as a signal of high status and availability of abundant resources which is

[79]Boehm (1999, 147).
[80]Hamilton (1964), Williams (1966), Alexander (1979).
[81]Betzig (1986).
[82]de Waal (1996, 128).
[83]Omark et al. (1980).
[84]Alexander (1979, 1993), Van der Dennen (1995), Sidanius and Pratto (1999), Flinn et al. (2005), King et al. (2009), Gavrilets and Fortunato (2014).
[85]Cassill and Watkins (2010).
[86]King et al. (2009).
[87]Boone (1998), Krebs (2011, 75ff).
[88]Zahavi and Zahavi (1997).

ultimately important for survival and reproduction. Appearance is a tool used to acquire greater admiration and greater social status. In order to acquire a gracious gait and appear taller, many women wear stiletto shoes. Advantages obtained from wearing this uncomfortable footwear may be compared to peacock tails or deer antlers.

Since the space for high social status in society is by default scarce, achieving such a status is the source of intense competition and conflict. Differences in social status often involve social inequalities associated with dominance styles varying from tolerant to despotic.[89] Strong unequal social status relations usually elicit movements to restrain the power and privileges of more dominant strata and to achieve a relatively egalitarian distribution of resources.[90] Indeed, notwithstanding his strong predisposition to seek high social status, *Homo sapiens sapiens* has, in the course of his evolution, also been endowed with a longing for egalitarianism[91] and distaste for relative deprivation.[92] Several factors and processes have been suggested for the evolution of the egalitarian drive: counter-dominant coalitionary behaviour, envy, social selection for altruism and self-restraint, moralistic punishment, and intergroup conflict.[93] The ethnological and sociological literature as well as results from economic game experiments motivated Herbert Gintis[94] to introduce the epithet *Homo egualis,*[95] the human who tries to keep inequalities within narrow margins, albeit with some variation in degree—"a weak urge to reduce inequality when on top, and a strong urge to reduce inequality when on the bottom".[96] Gerald Gaus[97] even speaks about the egalitarian species. From an evolutionary perspective, it may be argued that people will further develop a dislike of excessive social disparities that give groups of people power over others.

7.3.2 Developments in Modernity Regarding Social Status Allocation

Modernisation is characterised by a clear shift from authoritarian toward democratic organisation in demographically large and socially complex societies, which have a generalised statutory basic education and large proportions of highly educated populations. Nevertheless, modern societies still struggle to find the right balance between human predispositions to hierarchically structured social status differences and human urges for egalitarianism.[98] Democratisation is, in other words, not only

[89]de Waal (1996, 125).
[90]Marx (1867).
[91]Erdal and Whiten (1994), Boehm (1999).
[92]Fehr and Schmidt (1999).
[93]Boehm (1993; 1997; 1999; 2012), Erdal and Whiten (1994), Dawes et al. (2007), Gavrilets (2012).
[94]Gintis (2000, 252).
[95]*Homo egualis*, from Italian '*eguale*', Latin '*aequalis*'.
[96]Gintis (2000, 258); see also Guinote et al. (2015).
[97]Gaus (2015).
[98]Somit and Peterson (1997, 3).

impeded by specific social characteristics of complex modern societies but also by human evolutionary heritage.

In modern culture, the hierarchy of functionally necessary social activities increasingly needs to be determined by the degree of people's personal capacities and competences to acquire knowledge and take up responsibility.[99] These factors require a particular biological (physical as well as mental) endowment and equipment of the individual as well as the necessary educational (and moral) training to valorise those biological potentialities.

Due to Mendelian segregation and recombination of genes, as well as other life course events, biologically partially assorted social classes become more heterogeneous intergenerationally. To a certain extent social stratification needs to be reassorted in each generation. Modern societies need to enhance social mobility in order to remain functional and assign jobs and social status on the basis of personal abilities and performances instead of family bonds.[100]

Modernity has made huge progress in creating democratically more equal opportunities for all its citizens by enlarging the opportunities for education and assigning jobs according to abilities and performances, and by achieving movement towards meritocracy. However, due to various deeply embedded ancient drives, such as nepotism, in-group bias, tribalism, ethnocentrism, racism and xenophobia, modern societies have not completely succeeded in building up their social stratification through a shift from descent to capacity, performance and production.[101]

The experience of planned economies has shown that complex societies must maintain a reasonable gradation in labour remuneration in order to motivate citizens for their study and labour efforts.[102] Communism kept the income variance for workers low and failed to stimulate the more creative and productive people. On the other hand, in recent decades in many Western market economies the income differential has increased substantially between the average worker and the average top manager. The accumulation of wealth in the hands of very few has further peaked, exacerbating social differences and inequity.[103] Mainly in the US and other English speaking countries, income inequality has exploded since the 1980s; this is not only on the basis of an unprecedented increase in wage inequality but also due to a growing inequality of capital income.[104] In this context, it is appropriate to refer to the recent research on the increase in social inequalities and use of resources by Safa Motesharrei and colleagues[105] in which they show that not only overexploitation of natural resources but also exuberant social inequalities can independently result in a general societal collapse.

[99]Barber (1957), Chapais (2015).
[100]Sorokin (1927), Burt (1961), Cliquet (1968), Salter (2008).
[101]Savage and Egerton (1997), Rubin (2002).
[102]For instance, Rawls (1971, 100).
[103]For instance, Hacker and Pierson (2010), McQuaig and Brooks (2013), Piketty (2013).
[104]Piketty (2014, 300).
[105]Motesharrei et al. (2014).

Contrary to what one would expect from our egalitarian heritage that took shape in the hunter-gatherer era[106] and the social emancipatory and egalitarian oriented ideologies of the agrarian-pastoral era[107] and class struggles in the industrial[108] era, the longer-term hominisation may require greater ethical weight to be given to equity.

7.3.3 Ethical Reflections Regarding Social Status Acquisition

7.3.3.1 Meritocracy and Equal Opportunities

The very nature of modern society with its large population that requires complex organisation, and the still present human predisposition to social dominance, make the establishment of a well functioning democratic organisation a difficult endeavour. Nevertheless, the long-term ethical goals of development of human-specific potentialities, promotion of quality of life, promotion of equal opportunities and moving from competitive toward cooperative efforts can only be achieved in a well functioning system in which all citizens benefit from the abundance and are part of the decision-making processes.

From an evolutionary perspective, social hierarchy was necessary and it was embedded in deep-seated human drives for self-actualisation and advancement, interindividual competition, resource acquisition, and ultimately reproductive success. It is necessary today because of the size and complexity of modern culture, whose functioning is based on the diversification of tasks and responsibilities.[109] Some authors argue that control is in the existential interest of both higher and lower status groups.[110] However, sources and degrees of hierarchisation require social control by means of values, norms and institutions without which they easily degrade to social inefficiency and injustice.

From a reproductive fitness point of view, social hierarchies are, in the modern contraceptive culture, perhaps less important than they were originally in the EEA where they ultimately fulfilled an important role in the maximisation of inclusive fitness. Nevertheless, to the extent that modern social hierarchies are, even more than in the past, associated with and caused by differentials in biopsychic potentialities, their positive association with reproductive performances remains essential for the future evolution of humankind.

The complexity of modern societies requires the diverse talents present in the population to be well developed and adequately allocated to specific tasks and

[106] Boehm (2012), Gavrilets (2012).

[107] See, for instance, the 'Sermon on the Mount' in the Bible, Matthew, 5: 1–12.

[108] Marx (1867), Green (1884), Leo PP XIII (1891).

[109] Masters (1989), Sober and Wilson (1998), Wilson (2002, 36).

[110] Social control is in the interest of higher social status groups because the enduring welfare and success of the privileged also depends on the harmonious functioning of society as a whole (Diamond 2005, 513). Social control is also in the interest of the lower social status groups because leadership is an important condition for the proper functioning of complex societies (De Waal 1996, 128).

responsibilities. Modernity can only exist, be sustained and further evolve when it is based on the principle of meritocracy and equal opportunities for people to develop their abilities.

Our sexually reproducing system implies that with each generational change our genotypes are segregated and our genes are recombined. As a result children do not always completely resemble their parents. This genetic recombination and segregation system requires that, in a complex, diverse society, people in every generation are to be biosocially reassorted over the various social strata. Modern societies must be hierarchically open societies in which social mobility is easily possible. Meritocracy is a precondition for achieving greater intergenerational equality of opportunities. Survey research also finds a massive preference for meritocratic distributions.[111]

In addition to the Mendelian segregation and recombination of genes, ontogenetic developmental processes occurring during the pre- and/or postnatal life course may also affect the physical or mental performance potentialities of individuals and require changes in occupational activities.

Hence, from an intergenerational point of view, rigid class or caste structures are incompatible with the requirements of modern societies, as they need the right people on the right jobs in order to maintain their dynamic functionality and creativity. Modern societies need strong democratisation policies to reassort, across the different social strata in each generation, the talents and performance capabilities of individuals in the population.

The Mendelian transmission system, as well as the environmentally sensitive and impressionable long maturation process, requires a broad recruiting and assorting system based on the principle of providing equal opportunities for all, especially via education.

A clear distinction must be made between difference and inequality. Most of the time differences are products of nature (or culture). For example, women and men are different but given equal opportunities for education and status acquisition they do not have to be socially unequal.

7.3.3.2 Elites

Social hierarchies have been and are associated with the privileged status of elites. Elites—understood as groups of people who fulfil an eminent intellectual, social, cultural, economic or political role in society—are of considerable importance in modernity as drivers of innovation from which all people benefit.[112] This includes, inter alia, scientists, engineers and biomedical researchers who made modern scientific-industrial culture, but also the philosophical, ethical, legal, and political thinkers and activists who contributed to shaping the democratic and humane social order of modern civilisation.

The notion of elite has acquired a pejorative connotation in some quarters because some high status actors in modern economics and politics have not been

[111]Marshall et al. (1999), Baumard et al. (2013).
[112]Cattell (1972, 339), Cela Conde (1987, 148).

able or willing to master some of the ancient drives, such as selfishness, greed and cheating, and have enriched themselves disproportionately and immorally at the expense of their own and other populations. Some researchers argue that people of high social status behave more unethically[113] and show less communal and prosocial behaviour[114] than lower class individuals. Raymond Cattell[115] rightly advocated that our educational system should increase the correlation between intelligence and character values because:

> A society which allows its most highly intelligent individuals to manifest ill-balanced, emotionally perverted and essentially dishonest character traits is in grave danger indeed.

Roger D. Masters[116] also emphasised the crucial importance of the moral responsibility of modern elites:

> From the perspective of evolutionary biology, the survival of a mature civilization depends in good part on the virtue of its members. … This risk of overexploiting the benefits of civilization can arise either from the behaviour of elites, who seek to gain benefits beyond those available from social surpluses, or from the population at large, unable to justify restraints in private consumption.

A further progression of human evolution and culture requires considerable enhancement of the social morality of humanity's elites. It also needs considerable knowledge and skills in the general population so as to control the most advantaged. From the evolutionary perspective, particularly endowed people should be fully aware of the fact that they are the privileged outcomes of genetic recombinations in the human gene pool, often reinforced with favourable social or environmental events in human culture or society. Due to this privilege they need to assume higher social responsibilities and obligations. High moral virtues are required in order to take up a biosocially rooted endowment that allows for successful leadership. One should not lose sight of the evolutionary understanding of the genetic recombination in the human gene pool via sexual reproduction. This basic mechanism implies that there is the need for intergenerational replacement of elites based on the merits of each succeeding generation.

7.3.3.3 Equity and Social Inclusion

As the modernisation process proceeds, a substantial proportion of the populations will, due to their limited capabilities of genetic or ontogenetic endowment, or as a result of life course accidents, have difficulties in adapting to or being fully functional in this novel environment. As society becomes more complex, the increasing societal exigencies put people with limited biological abilities or lower social capital in a socially less favourable position.[117]

[113]For instance, Piff et al. (2012).
[114]For instance, Guinote et al. (2015).
[115]Cattell (1972, 377).
[116]Masters (1989, 185).
[117]Gottfredson (1997, 2004), Bostrom and Sandberg (2007).

It is well known that people differ in their cognitive or other abilities and that in advanced modern societies, in which such a high value is placed on intellectual abilities and qualifications, an increasing proportion of the population may not be able to meet the requirements of a competitive, creative, and fast evolving economy. There are historical examples of societies failing to resolve the conflicts between within-generation and intergenerational sources of biological variation and social stratification inequalities. The communist regimes in Eastern Europe had full employment policies but were unsuccessful in motivating their more endowed workers. Today, many Western European countries with generous welfare policies suffer from high and non-random rates of unemployment.

The challenge of integrating less adaptable and vulnerable members of society, and employing less capable members of society may be expected to accompany societies in the decades to come. It is clear that considerable efforts will have to be made to improve educational qualifications, but educational efforts alone will not resolve the problem of different abilities—educational systems are by default meritocratic. Modern societies will have to continue providing jobs or create meaningful activities which allow less capable and qualified people to remain functional and socially included in the novel environment of modernity.[118]

7.4 Race, Ethnicity, Worldview, and Political Conviction

Populations within current nation states have a miscellaneous composition with respect to various characteristics stemming from group differences such as: (1) population genetic differences, often commonly referred to as racial differences[119]; (2) ethnic differences, characterised by differences in cultural traits such as language or customs; (3) differences in worldview ('Weltanschauung', philosophy of life), including religious and various forms of non-religious beliefs such as atheism, agnosticism and freethinking; and (4) differences in political conviction or adherence.

Several of those group differences often coincide, not because of genetic causes but due to historical processes or migration histories, or socio-political movements. These group differences may be partly correlated with social status differences.

[118] Avramov (2003).

[119] In the biological sciences the concept of 'race' has a very specific meaning. In population genetic terms it is defined as a population that is statistically significantly distinguished in its allele frequencies from other populations, is distributed within a more or less localised territory, and may interbreed with neighbouring populations in areas of geographical overlap. However, the number of allele pairs to consider in identifying a race is an arbitrary matter. The genetic categorisation of races is a probabilistic matter. Hence, many racial classifications are possible. Usually, populations that differ in only a few of their allele frequencies are not characterised as races. As a rule, the concept is reserved for important biological subdivisions of a species which are distinguishable by a substantial combination of genetic characteristics resulting from their evolutionary past.

The reason for dealing with these four groups of differences under the same heading is that they evoke remarkably identical, although differently labelled, between-group patterns of behaviour. They are embedded in the in-group/out-group syndrome.[120] Indeed, racism,[121] ethnocentrism,[122] xenophobia,[123] and various forms of philosophical, religious or political intolerance show identical sociological and psychological traits. They can all be traced back to the same evolutionary background.

7.4.1 Evolutionary Background of In-Group/Out-Group Relations

In-group/out-group distinctions, whether rooted in genetic, cultural, linguistic, or ideological (religious or political) identification, but also occupational, or even recreational (mainly competitive sports) identification, are often associated with attitudes and various forms of behaviour that express stronger degrees of amity, attachment, empathy, trust, altruism, or favouritism towards the own group. They are associated with indifference, avoidance, rejection, disapproval, discrimination, or even straightforward enmity toward out-groups. A wealth of empirical investigations,[124] together with many laboratory experiments in recent years[125] as well as

[120]The concept of 'in-group/out-group syndrome' bundles together all possible forms of social behaviour for situations in which social entities are opposed to each other; it is characterised by a variety of attitudes or feelings of alienation but can also be associated with attitudes and forms of behaviour that involve feelings of superiority versus inferiority and can even be a source of latent or open animosity. The in-group can be defined as a couple, a nuclear or extended family, a circle of friends, a sports club, a clan, a tribe, a social class, a religious or philosophical group, a linguistic group, a cultural community, a nation, a race, or a species. The opposite of the in-group is the antagonistic out-group, the strangers, the 'others' (Thienpont and Cliquet 1999).

[121]The term 'racism' describes the belief that genetic differences between human populations, which determine particular socially or culturally relevant biological and psychological qualities, justify and legitimate making a discriminating distinction between people belonging to or descending from those populations, and thus treating them differently. However, this term is often used in an inappropriate way, namely referring to ethnic or religious traits that have nothing to do with genetic differences.

[122]Ethnocentrism refers to feelings of loyalty toward one's own cultural community, usually coupled with negative attitudes toward other, different communities. Ethnocentrism is a broader concept than racism because different types of qualities can characterise culturally identifiable groups: language, values, norms and customs, religion, etc.

[123]Xenophobia (<Greek: xenos = strange, foreign; phobos = fear) concerns feelings of fear or aversion of, if not hatred for, foreigners. Xenophobia is simply the flip side of the same coin as ethnocentrism, although it does not inevitably derive from ethnocentrism. One can be ethnocentric without detesting others. Ethnocentric feelings can turn into hostile feelings as a consequence of negative experiences with neighbours (Van den Berghe 1999).

[124]See for instance, Worchel and Austin (1986), Reynolds et al. (1987), Brewer (1999), Thienpont and Cliquet (1999), Salter (2004), Giles, et al. (2010), Hruschka and Henrich (2013).

[125]See, for instance, Bernhard et al. (2006), Efferson (2008), Yamagishi and Mifune (2008), Mifune et al. (2010), Halevy et al. (2012), Masuda (2012), Mussweiler and Ockenfels (2013).

mathematically modelling studies,[126] have extensively documented in-group bias and negative attitudes and behaviour toward out-groups.

Many explanations have been given for phenomena such as racism, ethnocentrism, xenophobia, and religious or political intolerance.[127] However, why are in-group/out-group relations, and particularly in-group amity and out-group enmity, so strongly linked to similarities between the members of the in-groups and dissimilarities with members of the out-groups? Furthermore, why do the similarities and dissimilarities apply to phenotypic as well as to social and cultural characteristics?

The formation of effective group alliances had many obvious proximate advantages, especially in situations of between-group competition,[128] and in-group alliances can easily lead to in-group egoism.[129] However, ultimately, the fundamental explanation for the in-group/out-group forms of behaviour relies in the genetic selfishness grounded in kin selection theory,[130] reciprocity theory,[131] and similarity theory.[132] These three theories—kin selection, reciprocity, and similarity theory—converge in inducing behavioural patterns that favour the intergenerational transmission of an individual's own genes, either via kin or via more remotely related consorts. The selfish gene theory[133] is the ultimate common denominator of these theories.

However, empirical evidence shows that the in-group/out-group syndrome is not only applicable to genetically different groups. It is also suitable for other forms of groupism ranging from homogamous partner choice, sex- and age-related bonds, friendships, occupational units, to ideological (religious and political) groups, and obviously ethnicity.[134] Apparently, kin-related phenotypic markers of genetic similarity are extended to ethnic or other sociocultural group characteristics in larger communities; these are defined by the degree of perceived relationship, which can go far beyond any genetic affinity.[135] The nepotistic drive is simply extended from kin to other forms of group identity that involves similarity in social, cultural, linguistic, religious, or political characteristics. Peter Richerson and Robert Boyd define them as symbolically marked subgroups[136]—groups to which people belong and identify themselves on the basis of beliefs. It is interesting that the phenotypically weakest markers of similarity (ideological groups) make the strongest use of kin-related nomenclature categories—fatherhood, motherhood,

[126]For instance, Fu et al. (2012), Nakamura and Masuda (2012).
[127]For instance, Banton (1987).
[128]Silverman and Case (1998).
[129]Tullberg and Tullberg (1997).
[130]Hamilton (1964).
[131]Trivers (1972).
[132]Rushton et al. (1984), Salter and Harpending (2013).
[133]Williams (1966), Dawkins (1976).
[134]Cliquet (1965).
[135]Van den Berghe (1978; 1999), Eibl-Eibesfeldt (1998).
[136]Richerson and Boyd (2005, 234).

brotherhood, sisterhood. In this context Ara Norenzayan and colleagues[137] even use the concept of 'fictive kinship' relations.

Several scholars, beginning with Charles Darwin[138] himself, consider the in-group/out-group syndrome to be a major determinant in the development of morality, either because moral principles appear to be mainly focused on one's own group,[139] or because the function of moral systems is to enable the own group to compete successfully with other human groups.[140] Indeed, as Richard Alexander and several other scholars have argued that[141] rather than predators other human groups, as well as disease or scarcity of resources, may have been an important selective factor in the hominisation process, resulting in the promotion of biological traits and moral rules that reinforced in-group solidarity and cooperation. Roger D. Masters[142] considers that both intragroup cooperation and intergroup competition were particularly instrumental in offsetting the selective disadvantages of the enlargement of human groups beyond the narrow borders of kin-related group members.

Peter Richerson and Robert Boyd[143] argue even that the tribal level of social organisation in the EAA resulted, through cultural group selection and mainly as a consequence of intergroup conflicts, in the establishment of dispositions which they label as tribal instincts.

Herbert Gintis[144] devised the epithet *Homo parochius* for this human tendency to divide human groups into insiders and outsiders at a net material cost to oneself. In recent publications, some authors now use the term parochialism for this phenomenon.[145] However, this is a less fortunate concept for designating in-group/out-group antagonisms because it incorrectly suggests that this syndrome would be mainly related to religious divisions.[146]

The in-group/out-group syndrome is probably one of the strongest and multi-dimensional innate drives that characterises the human biogram. In recent years, more investigations have begun to reveal some genetic determinants[147] and neuroendocrinological bases[148] of in-group bias and out-group derogation and intergroup conflict. In-group favouritism is already present in young children.[149] Nevertheless, it has also been documented that the intergroup variation in

[137]Noranzayan et al. (2016, 13).
[138]Darwin (1871).
[139]de Waal (1996, 30).
[140]Alexander (1987, 142).
[141]Alexander (1975; 1990), Wrangham (1999), Flinn and Coe (2007, 341).
[142]Masters (1989, 173).
[143]Richerson and Boyd (2005, 214ff).
[144]Gintis (2000, 252).
[145]For instance, Bernhard et al. (2006), Van Dijk and Feith (2010), Bowles and Gintis (2011, 133ff), Hruschka and Henrich (2013).
[146]The concept 'parochialism' is a derivation of the concept parish (from Latin 'parochia' and Greek 'paroikia') which is a small unit of believers in Christian denominations.
[147]For instance, Guo (2006), Lewis and Bates (2010), Hatemi et al. (2010), Lewis et al. (2014), Kandler et al. (2015).
[148]For instance, De Dreu et al. (2011).
[149]Jordan et al. (2014).

in-group/out-group attitudes and behaviour is related to a broad variety of environmental cues—for instance, moral learning, infectious disease risk, thermic stress, economic insecurity, experience of war, exposure to globalisation—and, in some cases, even to short- or long-time scale processes of gene-environmental interactions.[150]

7.4.2 Developments in Modernity of In-Group/Out-Group Relations

The evolutionary analysis of the in-group/out-group syndrome clearly shows that this was a necessary and successful strategy in the EEA. Small population units had survival and reproductive advantages, thus increasing their inclusive fitness via reinforcement of the in-group solidarity and cohesion, thereby facilitating defensive or offensive actions against competing out-groups. Even in the agrarian phase of humanity's cultural history, the more evolved forms of ethnocentric tribalism—patriotism and nationalism—had adaptive advantages.

However, the historic in-group/out-group syndrome has largely lost its adaptive advantage in the novel environment that emerged from modern culture. This results from its increased needs for cooperation and interaction between ever increasing numbers of people, and its growing mobility and communication opportunities, leading to a globalisation of science and technology, commerce, tourism, culture and politics.[151] On the contrary, in several respects it may have become a maladaptive strategy. Nevertheless, we are still equipped with this deep-rooted urge to form groups on the basis of biological, cultural or political similarities and by distancing ourselves against the 'others'. Although positive contacts between in-groups and out-groups reduce prejudice,[152] many norms and institutions in modernity are still based on tribal social instincts.[153]

7.4.3 Ethical Reflections About In-Group/Out-Group Relations

7.4.3.1 Ethnocentrism and Racism

It is self-evident that efforts at education and socialisation will have to be substantially increased in order to counteract the ancient innate drives toward nepotism, tribalism, ethnocentrism, racism, and xenophobia, as well as the traditional religious belief systems and modern political ideologies that support or reinforce those drives. This action is necessary because the speed of modern cultural development, with study, research, industry, trade, tourism, policymaking, high rates and degrees

[150] See for instance, Hruschka and Henrich (2013), Krosch and Amodio (2014).
[151] Avramov and Cliquet (1999), Avramov and Cliquet (2007), Avramov (2008).
[152] For instance, Christ et al. (2014).
[153] Richerson and Boyd (2005, 229).

of mobility, migration, and international relations, has considerably outpaced the rate of genetic change.[154]

Combating the various forms of prejudice and discrimination of individuals or groups who are perceived as different does not contradict the desirability of promoting community identity and cohesion, whatever the community characteristics may be—ancestry, ethnicity, worldview, etc.[155] Promoting community identity is a necessity for a harmonious individual and societal emancipation, but should not be a reason or cause of demeaning or discriminating 'otherness', because tribalism has become a maladaptive practice.

7.4.3.2 Ideological Pluralism

The populations of modernised countries are characterised by ideological pluralism, namely with respect to worldviews and political conviction. The first manifests itself in a plurality of religious adherence and practice and in various forms of non-religious convictions such as freethinking, agnosticism and atheism; the second in the plurality of political movements and parties and the possibility of having democratic free elections.

Whereas political pluralism is the rule in all modern democracies, only a limited number of countries constitutionally recognise and/or institutionalise pluralism in worldviews. Religious institutions are still very powerful members of the governance systems, even in modern states. Let us take the example of Belgium where, in addition to the ethnic diversity (Dutch, French and German speaking communities), two major ideological dividing lines run through the country: the first concerns the plurality in worldviews, mainly characterised by a dividing line between religious believers (mainly Roman Catholics) and freethinking people; the second concerns the political pluralism, mainly characterised by the opposition between left ($\approx$ socialist) and right ($\approx$ liberal) political convictions.[156] Regarding the pluralism in worldviews, Belgium is one of the very few countries in which its public primary and secondary educational systems include, in addition to religious education (mainly Catholicism, but also Protestantism, Orthodoxy, Anglicanism, Judaism, and Islam), a course in non-confessional ethics.[157] Children and their parents can choose which pillar to follow. Public and non-Catholic free universities have departments for training teachers in non-confessional ethics. However, in recent years, the meaningfulness of the pillarised, expensive and complicated structure of the ideological education in Belgian public schools (in Catholic schools obviously only Catholic religious education is provided) is being questioned. Different sides are proposing that the specific ideological education (choice of denominational or

[154]For instance, Reynolds et al. (1987), Thienpont and Cliquet (1999).
[155]Salter (2004).
[156]Huyse (1970; 1987).
[157]In addition, Belgium has a cultural pact that obliges public authorities to involve the various ideological and philosophical orientations in the preparation and implementation of cultural policy (Belgisch Staatsblad 1973). The pact also includes guidelines for a balanced ideological composition of governing bodies of institutions, infrastructures and services that were established by the government, or are governed under a public authority.

freethinking ethics) should be replaced by a general philosophical-ethical education on citizenship. In this way young people would be educated about all major philosophical-ethical and religious orientations and approaches. This would not only make them philosophically-ethically more literate but would also allow them to understand how to handle philosophical-ethical and religious differences in a globalising world.[158] However, these proposals have collided with fierce opposition, not only from the Catholic side and other denominational groups but also from mainstream freethinking quarters. It will probably take quite some time before the ideological and professional vested interests of different groups are overcome.[159]

In his *A Theory of Justice*, John Rawls[160] raises the difficult question of the tolerance of the intolerant. Departing from his justice principles, he pertinently argues that righteous citizens should strive to scrupulously preserve the fragile right to free expression for themselves and also the intolerant, as long as freedom itself is not jeopardised. In modernity, the intolerant might have to be educated and obliged to respect the freedom of others. Indeed, it must be above any doubt that historic achievements in the area of universal rights cannot be waived, for example freedom of expression, freedom of belief (religious or atheistic), self-actualisation of the individual in the field of education, sexuality, family formation, reproductive and sex/gender equality. Modern democracies should closely monitor their science and Enlightenment based moral achievements in the field of freedom of expression and freedom of thought (religious or otherwise). However, at the same time they must confront authoritarian/fundamentalist ideologies which exploit these freedoms and abuse them with the aim of abolishing diversity and the cultural and moral progress humanity has reached, and to bend them in a cultural regression in which intolerance, ignorance and irrationality again prevail.

7.5 Competition and Cooperation Between States

7.5.1 Evolutionary Background of Relations Between States

The discourse on in-group/out-group relations also applies to between population relations where the biological and/or cultural differences may be even more outspoken and between which competition for territorial, energy and other resources may be more strenuous. Hence, it can be expected that the drives (discussed earlier)

[158] See, for instance, the Flemish association '*LEF (Levenbeschouwingen, Ethiek, Filosofie)*' (= Worldviews, Ethics, Philosophy) (http://www.levensbeschouwingen.be/).

[159] The opponents of a general, pluralistic philosophical/religious/ethical course do have a strong didactic argument in their conviction that it is practically impossible for teachers to objectively treat such sensitive, delicate and engaged matters as the different and often opposite world views in a non-prejudiced way. In their mind education in worldviews—whether of a religious or non-religious nature—requires a personal engagement in the specific ideology being dealt with.

[160] Rawls (1971, 242).

for nepotistic favouritism and in-group or tribal preferential treatment and out-group enmity play a greater role.

The archaeological and historical record of between population relations, strongly demonstrates in all three major cultural stages—gathering-hunting, agrarian-pastoral and industrial—that violent conflict played an important role in the evolution and history of humankind,[161] even to the extent that some scholars argue, in line with Heraclitus's notorious saying "war is the father of all things", that the progress of human civilisation and culture is a direct function of warfare.[162] As soon as the hominisation process, with its increasing cognitive abilities, allowed hominins to become ecologically dominant, it strengthened between group conflict in order to gain control over ecological resources.[163]

The present scientific consensus (particularly in scientific disciplines with an evolutionary background) about the cause of human violence, both at the individual and group level, considers that it usually results from the combination of biological predispositions towards competitive and expansionist drives, and scarce or unequally distributed resources needed for survival, social power magnification, biological reproduction and demographic multiplication.[164] The biological predispositions to individual and collective violence and out-group enmity have been under strong selective pressures in the course of the hominisation process. Some scholars consider them to be evolutionary adaptations, others as evolutionary by-products.[165] This complex history is often sidelined in some policy circles as may be seen in the well intentioned but lopsided *Seville Statement on Violence 1986*, subsequently adopted by UNESCO in 1989,[166] which largely denies the role of biological (genetic) factors in violence.

The evolutionary background of warfare is thought to go back to the common ancestor of chimpanzees and hominins, *P. prior*, who consisted of patrilocal, multi-male, polygamist groups that competed for females.[167] In the course of the hominisation process, this intergroup competition for females resulted, by means of sexually selective[168] and group selective processes,[169] in the establishment of genetic predispositions to male bonding, in-group cooperation and favouritism, and

---

[161]Harris (1974), Eibl-Eibesfeldt (1979), Ferguson (1984), Shaw and Wong (1989), Crook (1994), Van Der Dennen (1995), Keeley (1996), Wrangham and Peterson (1996), Guilaine and Zammit (2005), Kelly (2005), McCall and Shields (2008), Bowles (2009), Gat (2009), Baofu (2010), Pitman (2011), Gorelik et al. (2012).
[162]For instance, Mataré (1999, 31), Turchin (2011), Turchin et al. (2013).
[163]Geary (2007, 306); see also Bowles and Gintis (2011, 133).
[164]For violence at the individual level, see Ghiglieri (1999), Kanazawa and Still (2000), Moir and Jessel (1995), Pitchford (2001), Raine (1993), Rowe (2002), Walsh and Ellis (2003). For violence at the group level, see Shaw and Wong (1989), Low (1993), Crook (1994), Van Der Dennen (1995), Keeley (1996), Eibl-Eibesfeldt and Salter (1998), McCall and Shields (2008), Gat (2009), Teehan (2010).
[165]Durrant (2011).
[166]http://portal.unesco.org/education/en/ev.php-URL_ID=3247&URL_DO=DO_TOPIC&URL_SECTION=201.html.
[167]Wrangham and Peterson (1996).
[168]Rusch et al. (2015).
[169]Egas et al. (2013).

out-group enmity: all of these were instrumental to intergroup conflict and warfare. In the further course of the hominin evolution, with its gradually increasing cultural complexity and numerical size of human populations, intergroup competition for females expanded to the increasing need for a greater variety and amount of other natural resources.[170] With the emergence of the agricultural revolution, and later the early industrial revolution, not only did the need for territorial acquisition and protection surge but also the drive to socially dominate and exploit others thrived, as shown in the emergence or intensification of social phenomena such as social or ideological hierarchism, slavery and colonialism.

An intriguing point, already referred to in Chap. 2, Sect. 2.1.2.8, is that during the hominin evolution the spreading and intensification of within-group altruism and cooperation may have been strongly driven by between-group conflict and warfare,[171] or at least by group defence.[172]

In the gathering-hunting stage of human evolution there was a positive association between warfare and reproductive success, because mating and fitness benefits were worth individual lethal risks.[173] This was probably no longer the case in the agrarian-pastoral and industrial eras where warfare, fought by ever larger armies equipped with increasingly technological killing weapons, mainly had proximate benefits such as territorial expansion, economic power and resources acquisition. This gives less reproductive benefits, except perhaps for the successful warlords and other upper class members who take advantage of the warfare but are not directly involved in the war operations.[174]

The genetic effects of warfare has divided the scientific community for a long time[175]; but for modern societies, where armies are recruited on the basis of physical and mental health features, knowledge-based opinion seems to tilt towards contraselective effects. War is dysgenic in modernity.[176]

The current wars, especially proxy wars, are a confirmation of the deep-seated drive towards plundering the out-groups. However, evolutionary science shows that the human species is—in addition to selfish nepotistic and tribal drives—endowed with strong potentials and capacities for altruistic, reciprocal and cooperative behaviour that may, in different future living conditions, be extended across tribal or national borders, and even be applicable to the whole of the human species.[177]

[170] Pitman (2011).

[171] For instance, Boehm (1999), Smirnov et al. (2007), Choi and Bowles (2007), Lehmann and Feldman (2008), Bowles (2009), Ginges and Atran (2011), Halevy et al. (2012), Konrad and Morath (2012), Gavrilets and Fortunato (2014).

[172] Rusch (2014, 359).

[173] Rusch et al. (2015).

[174] Low (1993).

[175] Crook (1994).

[176] Cattell (1972, 202).

[177] See, for instance, Fehr and Fischbacher (2003), Manner and Gowdy (2010).

7.5.2 Developments of Relations Between States in Modernity

Modernity continues to be plagued by the presence of traditional warfare, increasingly through the use of modern technologies, on the one side[178]; on the other side, the relations between states are characterised by slowly increasing international cooperation, both at the regional[179] and global levels,[180] and by very modest degrees of international developmental assistance.[181]

Most, if not all, of the intergovernmental bodies are still in a very early stage of development, severely hampered in their functioning by traditional national interests, including some of their most powerful member states.[182] Another limiting factor in the development of efficient intergovernmental bodies is the presence in world gatherings of a large number of non-democratically elected governments who defend the interests of authoritarian oligarchies, and governments that are politically inspired by ideologies that emerged and flourished in pre-modern cultural eras.[183]

Despite omnipresent evidence of permanent warfare worldwide, modernisation is, overall, characterised by a decrease in violence.[184] As Steven Pinker[185] wrote:

> ...it is perhaps the most important and most underappreciated trend in the human saga: Violence has been in decline over long stretches of history, and today we are probably living in the most peaceful moment of our species' time on earth.

The increasing cooperation between states is embedded in four major features of modernity: (1) development of the ethics of democracy; (2) potential for generalised advantages of technological progress; (3) expansion of tools such as IT for dissemination of knowledge and ideologies; and (4) balance of terror due to the development of weapons of mass destruction.

According to Rudolph J. Rummel,[186] democracies are less warlike than their preceding political regimes. The more democratic nations are, the less likely they will be disrupted by domestic violence and the less likely they will provoke international conflict or be involved in war. Rummel's conclusion is straightforward: "democracy is a method of non-violence". This is particularly the case where

[178] For instance, World War I and World War II.

[179] For instance, European Union, EU; Shanghai Cooperation Organization, SCO; Association of Southeast Asian Nations, ASEAN; Arab League; African Union, AU; Organization of American States, OAS; Union of South American Nations, UNASUR; Pacific Islands Forum.

[180] United Nations.

[181] For instance, United Nations Development Programme (UNDP); International Monetary Fund (IMF); World Bank Group; United Nations Population Fund (UNFPA).

[182] Cf. the vetoing behaviour of the US, Russian Federation, and China in the Security Council of the UN.

[183] Cf. the opposition by fundamentalist Islamic regimes (and from the Holy See) to many of the draft articles on women's empowerment, sexual and reproductive rights and health of adolescents, abortion, and the family of what became in the end the Cairo Programme of Action of the third UN World Population Conference in 1994.

[184] Keeley (1996), Pinker (2011).

[185] Pinker (2007); see also Gómez et al. (2016).

[186] Rummel (1994, 1995, 2002).

democracy goes hand in hand with limitation of excessive inequalities. This is probably largely due to the fact that, with the exception of some psychopaths, sociopaths and religious or political zealots, the large majority of people in modern democracy—with the right to speak in political decision-making, and experiencing relative wealth and freedom from mass scale hunger and disease—have too much to lose from the waste and horror of war.[187] However, proxy wars are flourishing today, whereby technologically more advanced states can avoid own casualties by utilising out-groups.

For both individuals and societies, the spread of modern technology in all areas of life (education, research, industry, trade, transport, housing, health, nutrition, leisure and comfort in general) has benefits for both producers and consumers: the first have a financial interest in selling their products, while the latter also want to acquire access to skills for social mobility and to enjoy the pleasures of modern comfort.

Dissemination via ICT has become an extremely powerful means to spread modern knowledge and ideologies to deprived population groups and countries which are governed as oligarchies. The inverse is true also.

The development of weapons of mass destruction, namely nuclear, biological, or chemical weapons, includes the risk of annihilating all human life on the planet or at least reducing it to more primitive levels of development and civilisation. It may be considered as a miracle that the huge, but now slightly decreasing although still considerable, stockpiles of nuclear weapons in the world have not yet, by chance or on purpose, led to a nuclear Armageddon.[188] The balance of terror whereby no state could be spared has so far been reasonably effective in preventing cataclysmic warfare.

7.5.3 Ethical Reflections about Relations between States

All specific features of modernity—the present world population size, the eradication of mass hunger and disease in prosperous economies, the development of democratic decision-making, the risks of using weapons of mass destruction, the dysgenic and other miserable effects of modern warfare—point to the necessity to stop using warfare as a means of resolving international conflicts. A much better approach, supported by an effective world police force, should be able to prevent or to contain emerging international conflicts[189] than the currently ill designed,[190] manipulated[191] and prejudiced[192] planetary machinery for controlling and avoiding international warfare.[193]

[187] Cattell (1972, 198).
[188] For instance, Croddy et al. (2004).
[189] See also Cattell (1972).
[190] Cf. the veto rights of large powers, some of which are dictatorial regimes, whereas some others still pursue an imperialist policy.
[191] Cf. the US pressures on judges of the International Criminal Court.
[192] Cr. the highly selective character of recent warlords being brought or not brought to court, for example some of the small fry from former Yugoslavia, but not the Western war criminals responsible for the 2003 Iraq invasion.
[193] The UN Security Council, the International Court of Justice and the International Criminal Court.

As Samuel Bowles and Herbert Gintis concluded:

> Altruism and parochialism may have a common evolutionary origin, but the two are not inseparable… parochialism need not be our destiny.[194]

The solution for improving the relations between states is very simple, although probably very difficult to realise. It is simple because there is only one solution, namely to increase international and intergovernmental cooperation; it is most difficult due to the tenacity of the in-group/out-group syndrome and the vested interests linked to it.

All evolutionists who have dwelled on this question advocate the same precept: universalism; a world brotherhood of mankind[195]; expanding the kind of morality that we now express within groups[196]; broadening of the target of our altruism[197]; expanding the circle of our concern and involvement.[198] Or, as the great visionary scientist Julian Huxley[199] stated:

> To have any success in fulfilling his destiny as the controller or agent of future evolution on earth, he must become one single inter-thinking group, with one general framework of ideas: otherwise his mental energies will be dissipated in ideological conflict.

Some scholars think that the forthcoming global society will be characterised by a unity in diversity.[200] Earlier reference has been made to Raymond Cattell's[201] remarkable *Beyondism* in which he proposed, as a goal of evolutionary progress, the development of an ethos or atmosphere that he defined as 'cooperative competition' among groups, in contrast to a culturally homogeneous universalism. This is one of the very few issues in which the present authors fundamentally disagree with Cattell because his cooperative competition is in essence contradictory to the spirit and trend of globalising and universalising modernity. Although it must be acknowledged that Cattell's model is the one that is currently emerging[202] as an intermediate stage between the former nationalist stage and the future universalist stage, it is nonetheless opposite to the spirit of and trend towards a universal modernity. Since the human species is culturally in a major transitional stage of its development, we see and we will, for quite some time, continue to see differences in degree and tempo of the shift towards the future universalist and sustainable

---

[194]Bowles and Gintis (2011, 147).
[195]Keith (1948, 58).
[196]Alexander (1987, 190), Boehm (2012, 343).
[197]Dawkins (1976, 10).
[198]Singer (1981, 2002, 111).
[199]Huxley (1964, 84).
[200]See, for instance, John Stewart's (2008) 'Evolutionary Manifesto'.
[201]Cattell (1972, 105).
[202]Consider some of the major present-day competitive cooperators on the world scene: the US, the European Union and China. They clearly represent three different societal models, but all are unsustainable in a long-term perspective: the US, dominated by its private capitalist enterprise culture resulting in an unbearable social climate of excessively large variance in income range and lack of social protection for huge segments of its population; the European Union, with its admirable social protection model, but insufficient political integration and policy decision-making; China with its stunning economic growth and development, but its lack of democratic decision-making policy.

modernity; however, seen from a long-term perspective this is probably only a temporary stage and not the ultimate goal.[203]

Although selective processes will inevitably further promote the degree and intensity of international cooperation in the cultural stage of modernity in which the human species finds itself, the processes of learning by trial and error might still take a lot of time whereas some features of modernity, such as the availability and diffusion of weaponry of mass destruction, or the exhaustion, destruction or pollution of indispensable precious resources, leave no time to let things take their course. Hence, in 1984 Robert Axelrod,[204] in his prescient book *The Evolution of Co-operation*, implored us to use our foresight by speeding up international cooperation.

In essence, globalisation is the outcome of the modernisation process whereby science and technology, commerce, culture and ideology are gradually, but at an increasing speed, disseminated all over the planet. Globalization has also opened up opportunities for more global social justice by lifting out of extreme poverty people in least developed countries who otherwise would have no access whatsoever to paid work. Obviously, humanisation of working conditions in industry in developing countries, and prevention of massive unemployment or impoverishment of (expensive) workers in developed countries is a major challenge ahead. Globalisation is characterised by an increasing connectivity and interdependence of the world's scientific, economic, and political processes, the apogee of which would consist of a world political federation.

In recent decades, the concept of globalisation has acquired a bad, or at least a mixed, reputation mainly due to the greedy excesses of the current runaway economic globalisation, which is characterised by an insufficient world community control and the harsh neo-liberal deregulation of commodity, capital and labour. World political authorities are either running behind the economic developments or are the playthings of the corporate interests of the very few.[205] However, it must also be admitted that part of the reluctance about globalisation is due to the persisting influence of our tribal instincts and our lack of insight into our fast changing existential condition.[206] As Paul L. Ehrlich[207] noted:

> … we are still primarily small-group animals, evolved genetically and culturally to operate in limited, homogeneous, largely closed societies.

Notwithstanding the current critical and inhumane economic and financial world developments, so pitifully illustrated by the recent banking crisis, the ultimate goal remains the achievement of a mature globalisation: present-day inequalities between states will have been erased and the human species will have reached a planetary moral, economic, cultural and political community, a unified, inclusive

[203]See Elgin (1993).
[204]Axelrod (1984, 191).
[205]Stiglitz (2003), Baylis et al. (2011).
[206]Rubin (2002).
[207]Ehrlich (2000, 301).

and highly evolvable global society.[208] Through globalisation, humanity can transcend its evolutionary past.[209] Hence, the need for developing a new, universal morality for all humanity.[210]

Notwithstanding its innate narrow-minded in-group drive, the human brain possesses neural capacities, such as emotional sharing, empathic concern and affective feeling,[211] which may achieve such a goal in the presence of, or under pressure from scientific enlightenment, proper educational/agogic action and creative political governance.

[208]For instance, Wells (1905), Keith (1946), Huxley (1964), Axelrod (1984), Elgin (1993), Stewart (2008).
[209]See also Modelski et al. (2008).
[210]Penn and Mysterud (2007, 290).
[211]Decety and Cowell (2015, 293).

Evolution-Based Ethical Challenges Related to Intergenerational Replacement

8

Abstract

In this chapter the evolutionary background, developments in modernity and ethical reflections for quantitative and qualitative aspects of reproductive behaviour are addressed. First, the evolutionary and historical background and the causes of the human species' numerical growth are described. Next, major specific developments in modernity are discussed: the demographic transition, the paradox between the maximisation of inclusive fitness theory and below replacement fertility, and the relations between world population size and ecological sustainability. Finally, ethical reflections on the future demographics of the human species are outlined. Regarding the qualitative aspects of human reproduction, phenomic and genomic aspects are distinguished. The main features of reproductive behaviour in modernity are elucidated and their effects on demography, sexual relations and genetics are examined. Ethical reflections regarding qualitative aspects of human reproduction include euthenic as well as eugenic aspects. Advantages and disadvantages of different eugenic methods and decision-making processes are assessed.

8.1 Introduction

From an evolutionary point of view, reproduction is undoubtedly the most important biosocial phenomenon. Reproduction secures the intergenerational perpetuation of life. Genetically linked differential reproduction allows for changes in the population-genetic composition from one generation to another, potentially resulting in adaptation to new environmental challenges. Differential genetic reproduction is the ultimate instrument for biological evolution.

R. Cliquet and D. Avramov, *Evolution Science and Ethics in the Third Millennium*,
https://doi.org/10.1007/978-3-319-73090-5_8

8.2 Quantitative Issues Regarding Reproductive Behaviour

Population growth is an extremely important characteristic of the twentieth century because of the historically unique, strong increase in the world population, which is expected to expand further during this century. According to the medium variant of the UN Population Division population prospects[1] the world population would, in the course of the twenty-first century, increase from six billion in 2000 to *circa* eleven billion by 2100 and then slowly decrease or fluctuate at a somewhat lower level (Fig. 8.1).

8.2.1 Evolutionary Background of Population Growth

8.2.1.1 Fecundity, Fertility, Mortality, and Population Growth in Pre-modern Times

Humans are characterised by a relatively high fecundity[2]: this feature has been selected as a function of the high mortality levels that prevailed in the ancient EEA. High mortality required a high realised fertility[3] in order to compensate for death of infants and children.[4]

Studies on the reproductive behaviour of hunter-gatherer and agrarian-pastoral populations show high fertility levels: the observed completed fertility often averaged five to six live births.[5] However, there was also a high between-population

[1]United Nations Population Division (2015).

[2]Fecundity or potential fertility refers to the number of live-born children that could be realised without the limiting effects of sexual taboos, late marriage, lactational amenorrhea, pathological sterility, contraception and induced abortion. On the basis of a number of empirical observations of populations in different stages of cultural development and by the means of different methodologies, the average total potential fertility in human populations has been estimated to be 15.3 live-born children, with a between-population range of 13–17. The within-population (interindividual) variability in fecundity in populations with an average of 15.3 and a standard deviation of 5.09 has been calculated to lie between 0 and 26 live births (Bongaarts 1978; Bongaarts and Potter 1983), see also Léridon (1973, 1987), Léridon and Menken (1979), Langhoff (2007).

[3]Fertility refers to the number of live born children actually born to a woman. Completed fertility or completed family size is the fertility a woman has realised by the end of her fecund life phase. Proximately, fertility is the result of the combined effects of sexual intercourse, contraception, abortion and lactation on fecundity. Ultimately, fertility is influenced by the level of desired family size—which is itself subject to influences from biological drives, cultural values and socio-economic living conditions.

[4]For instance, Volk and Atkinson (2013).

[5]For instance, Henry (1961), Léridon (1973, 1987), Xie (1990), Hewlett (1991), Wood (1994).

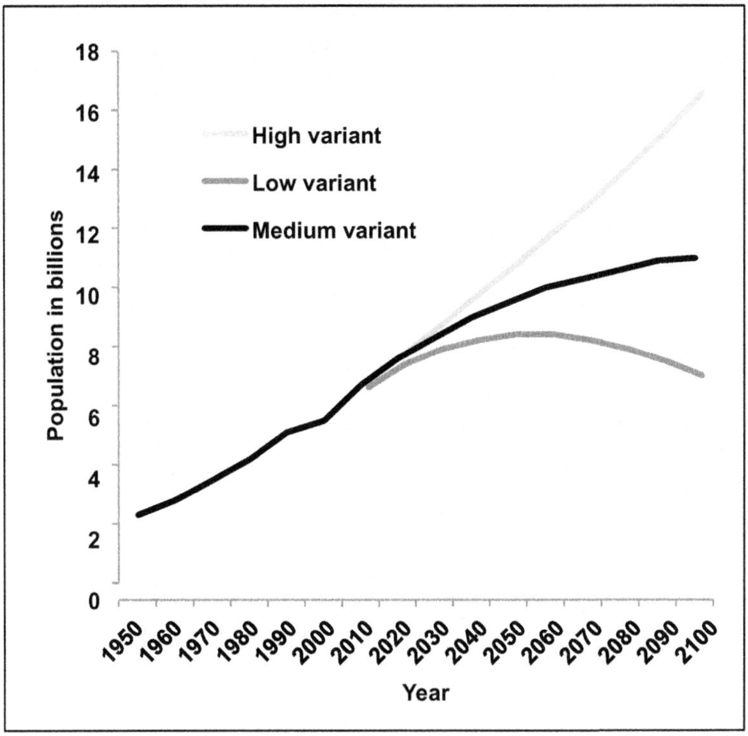

Fig. 8.1 World population prospects: the 2015 revision (United Nations 2015)

variance in pre-modern fertility.[6] Furthermore, the individual variation within those populations was considerable.[7]

Due to high mortality, intergroup conflict and periodical catastrophes, population growth in the hunter-gatherer era was very low and population size was almost stationary.[8] While fertility was five to six children on average, infant and child mortality reduced it to less than half; other calamities often reduced it further, resulting in zero growth.[9] The average intrinsic growth rate[10] was below 0.04 per

[6]For instance, Hewlett (1991) found among 23 hunter-gatherer societies TFRs (total fertility ratios) ranging between 2.6 and 7.8 live births, with a mean of 5.4; among 19 farming-herding societies the range is between 2.4 and 8.4 live births, with a mean of 6.1. Wood (1994) found among 70 natural fertility populations an average TFR = 6.1 with a between-population variance of 1.38.

[7]For instance, in the seventeenth-eighteenth centuries in rural Tourouvre-au-Perche, France, the completed fertility was 6.0 with a variance of 13.1 (live births ranging from 0 to 19); among the present-day Amish in Ohio the average number of live births is 6.3 with a variance of 11.1 (live births ranging from 0 to 15) (Charbonneau 1979).

[8]Hassan (1975), Polgar (1975).

[9]Birdsell (1968).

[10]The intrinsic rate of population increase is the instantaneous per capita birth rate, minus the instantaneous per capita death rate (Pianka 2000).

cent per year.[11] The size of the world population has been estimated as 50 million people in the period before the start of the agrarian-pastoral revolution.

Fertility increased in the agrarian-pastoral era[12] and, although mortality also increased,[13] it resulted in a higher intrinsic growth rate which allowed for a slow and gradual population increase.[14] This led to a world population of one billion before the start of the Industrial Revolution in the nineteenth century.

8.2.1.2 Maximisation of Inclusive Fitness

An important concept in the evolutionary study of reproduction is the theory of the maximisation of inclusive fitness.[15] This theory asserts that humans, like other organisms, developed through natural selection evolved behavioural tendencies in order to maximise their genetic representation in future generations in the context of constraints set by the environment and their phylogenetic past. Such optimal reproductive success, achieved through the production and survival of descendants and other blood relatives, results in evolutionary adaptiveness.

Both at the individual/family level and at the group level, and particularly in intergroup competition, demographic expansion was often an instrument of competition and conflict.[16]

8.2.2 Demographic Developments in Modernity

8.2.2.1 Demographic Transition

One of the important characteristics of modernisation is the demographic transition—the shift from high to low mortality and fertility levels.[17] This transition began in the eighteenth century in North Western Europe and the United States with a gradual decrease in mortality which in most cases was followed, approximately one century later, by an incremental but rapid decrease in fertility by using parity-specific birth control interventions.

Whereas the pre-transition phase was characterised by high mortality, high fertility and moderate population growth, the beginning of the demographic transition was characterised by mortality control and continued high fertility, resulting in explosive population growth (Fig. 8.2).

The theory of the demographic transition built on the premise that, following the mortality decrease, fertility levels would decrease and population growth would fluctuate around the population replacement level. The expected result of the

[11]Polgar (1975), Hassan (1975), Winterhalder et al. (1988), Gignoux et al. (2011), Zahid et al. (2016).
[12]In the agrarian era, fertility amounted, on average, to 6–7 live born children per women (for instance, Bentley et al. 1993, 274).
[13]For instance, Armelagos et al. (1991).
[14]Polgar (1975), Bocquet-Appel (2011), Gignoux et al. (2011).
[15]Hamilton (1964).
[16]MacDonald (1999), Parsons (1999).
[17]Landry (1934), Notestein (1945), Chesnais (1986), Lee (2003).

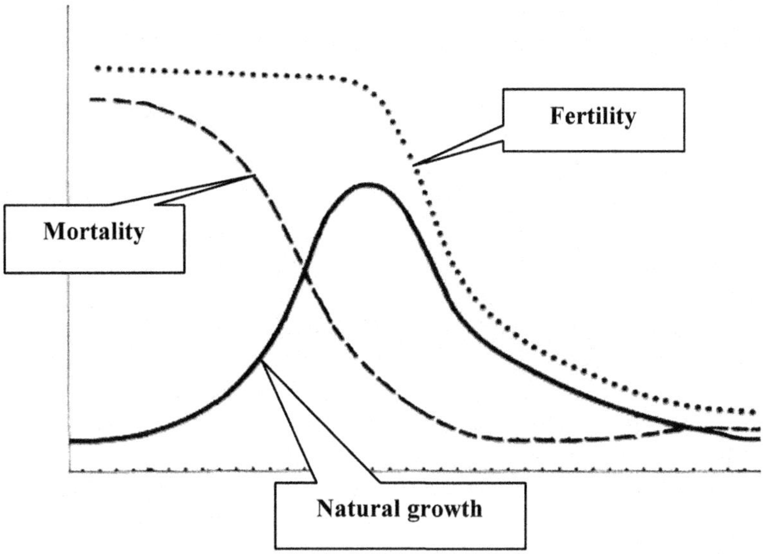

Fig. 8.2 Mortality and fertility decrease and population growth during the demographic transition (Based on Chesnais 1986)

demographic transition theory was a stationary population. However, in contrast to the theory, modern countries have not yet reached the expected stage of the demographic transition where balance would occur. Most developed countries appear to have leaped into a new phase, that of persistent below-replacement fertility.[18]

In the old industrial countries, the demographic transition gradually and slowly evolved in parallel with the Industrial Revolution in the eighteenth and nineteenth centuries. In the developing countries, mortality control and gradual fertility decline generally gained momentum in the course of the twentieth century. Some scholars distinguish two demographic transitions in the demographic changes that have accompanied the modernisation process, each characterised by a different pace of change: the first demographic transition, characteristic of the old industrial countries, and the present demographic transition,[19] typical of the developing world.[20] Indeed, the present demographic transition in the developing world is not only characterised by a later and faster decrease of mortality and fertility rates but also started at much higher levels than the first demographic transition in the old industrial countries.

[18]Teitelbaum (1999), Frejka and Ross (2001), Goldstein et al. (2003), Wilson (2004), Hoorens et al. (2011).

[19]Not to be confused with the above mentioned 'Second Demographic Transition' of Lesthaeghe and van de Kaa (1986, 1987, 1995, 2010) who refer to the recent accelerations in relational and reproductive behaviour in developed countries.

[20]For instance, Reher (2004), University of Michigan (2006).

Meanwhile, the move to below-replacement fertility is also spreading in fast emerging economies outside the old Western world.[21] Today, about 42% of the world population lives in countries with below-replacement fertility[22] and it is expected that it will spread further in the coming decades.[23] It is estimated that 75% of all countries will be reproducing at below-replacement levels by 2050.[24]

Below-replacement fertility has raised the question of how this phenomenon is to be reconciled with the biological imperative of fitness maximisation. How can this paradox be explained? In recent decades many hypotheses have been suggested to explain, from an evolutionary perspective, the reduced fertility in modern industrialised populations.[25] Many of these explanations are not mutually incompatible: this only goes to show how many factors are involved in the reproductive decision-making process and how modern culture has a multifactorial impact on it. Indeed, the new fertility regime is a deviation of the traditional adaptive strategy: it is a new regime on which selection has had little impact so far; the family context has changed profoundly; attitudes towards investment into and expectations from children have changed tremendously; the innate drive regarding resource acquisition has taken advantage of the increased modern economic benefits, etc.[26]

Some authors have argued that, from an evolutionary perspective, low fertility, or even (a temporary) below-replacement fertility, would be a maladaptive strategy.[27] This might be true from a short-term point of view, but when one considers low fertility simultaneously (1) in a long-term perspective, (2) from a population or global point of view, and (3) in a novel context of modernity with its high consumption patterns and high standards of quality of life (including low mortality), low fertility appears as a good adaptive strategy which, in the long run, will not only benefit the human species as a whole but also its individual members.[28] Maximising inclusive fitness in a context of population growth and resource

[21]For instance, Kinfu (2000), Caldwell and Caldwell (2003, June 2005), Jones (2007), Atoh (2008), Sutton (2009), Westley et al. (2010).
[22]Espenshade et al. (2003), UN Population Division (2015).
[23]Morgan (2003), Sobotka (2009).
[24]Lutz et al. (2001).
[25]Maladaptive strategy hypothesis (Hill 1984; Davis 1986; Vining 1986; Pérusse 1993; Borgerhoff Mulder 1998; Boyd and Richerson 2005); shortage of time hypothesis (Irons 1979); fertility control hypothesis (Turke 1989); breakdown of kinship network hypothesis (Turke 1989; 1990; Newson et al. 2007; Newson 2009; Newson and Richerson 2009); relaxed fertility selection hypothesis (Aarssen 2005); evolved two-child family hypothesis (Lopreato and Yu 1988); cultural evolutionary hypothesis (Boyd and Richerson 1985); transmission competition hypothesis (Aarssen and Altman 2006); quality of children hypothesis (Harpending and Rogers 1990; Rogers 1990; Kaplan 1996; Voland 1998); social status/competition hypothesis (Boone and Kessler 1999; Mace 2000; Johansson 1987; Mueller and Short 1983; Pérusse 1993, 1994; Borgerhoff Mulder 1998; Low et al. 2002); economic benefit hypothesis (Voland 1998; Kaplan and Lancaster 2000; Shenk et al. 2013); changed relations between ontogenetic fitness and reproductive fitness hypothesis (Caldwell 1982; Cain 1982; Handwerker 1986; Cliquet 2010), extrasomatic wealth hypothesis (Kaplan et al. 2002).
[26]Low (1994), Borgerhoff-Mulder (1998), see also Cliquet (2010, 319–324).
[27]Hill (1984), Vining (1986), Pérusse (1993), Borgerhoff Mulder (1998), Boyd and Richerson (2005).
[28]See also Krebs (2011, 186).

exploitation that already has an ecological overshoot of 1.5 Earths, is maladaptive (see below in Sect. 8.2.2.2). This overshoot can be expected to further increase when developing nations reach consumption levels of Europe or North America. H. Kern Reeve and Paul W. Sherman[29] suggest that the demographic transition is not a human maladaptation but rather a well-adapted life history decision-making, taking into account not only the number of offspring but also offspring quality and intergenerational resource generation and transmission.

8.2.2.2 Population Growth and Size During Modernisation

The onset and the intensity of the difference between the decrease in mortality and fertility during the demographic transition caused an exponential growth of the human species, which evolved from about one billion people around 1800 to two billion around 1930 and four billion around 1975. As already mentioned, the world's population now amounts to seven billion people and it is expected that, by the end of this century, it might peak at around eleven billion[30] and then start to decline (Fig. 8.1).

While the annual population growth rate in the period before the industrial cultural phase was very low and very slowly increased from 0.04 to 0.1%, during the short span of modernising it very rapidly and strongly rose to just above two per cent in the period 1965–1970. Since then, this figure has decreased again; by the end of this century it is expected to be back at the very low values it had during most of human prehistory. The present and past ages thus represent a unique and nonrecurrent moment in the demographic history of humankind.

The major part of the population growth since the middle of the last century took place and still does occur in developing countries: in many of these countries it forms a strong constraint on socio-economic development and preservation of ecological equilibria.[31]

Regarding the developed world, in the period before World War II fertility evolved to below-replacement levels in more than half of the European countries.[32] After a temporary post-World War II baby boom in the middle of the twentieth century, fertility resumed its decline—in some countries in the 1960s and 1970s, and in others as late as the 1980s. By the end of the century, fertility reached unprecedentedly low levels, notably in southern and particularly in eastern European countries where fertility took a steep plunge after the collapse of communism,[33] as well as in Japan[34] and some overseas English speaking countries.[35] By the turn of the century, the total fertility rate seems to have stabilised at more or less

[29]Reeve and Sherman (2007, 93).
[30]United Nations Population Division (2010).
[31]Ehrlich and Ehrlich (2008).
[32]For instance, Frejka and Ross (2001), Wilson (2004), Van Bavel (2008).
[33]Kohler et al. (2002), Goldstein et al. (2003).
[34]Japan: TFR 2000: 1.3 (Atoh 2008; UN Population Division 2010).
[35]Canada: TFR 2000: 1.52; Australia: TFR 2000: 1.75 (UN Population Division 2010).

below-replacement levels[36] and since then recovered even a little bit.[37] As explained above, the move to below-replacement fertility has also spread to fast emerging economies outside the old Western world.[38]

For a long time demographers have shown that a long-lasting subreplacement fertility would eventually lead to a virtual disappearance of the population.[39] This will not happen either because policies will try to reverse the low fertility trends or populations themselves will change their behaviour, as a result of experiencing the negative effects of a too long-lasting below replacement fertility, or even because some minority populations, who continue to realise high fertility levels, instigate resumed population growth and become the demographic majority.

Indeed, within more developed countries some ethnic or religious groups, as well as some immigrant groups from developing countries, maintain a (much) higher fertility than the rest of the population. Well-known examples are the Anabaptists (Amish, Hutterites, and Mennonites) in North America who continue to reproduce at natural fertility levels (on average about 7 children per woman),[40] and the Islamic immigrants in many European countries who have fertility rates which are double or triple that of the host population. However, the fertility rates of second and third generation descendants of immigrants tend to be much lower than those of the original immigrants.[41] In many developed countries religious people, in particular Catholics, have a higher fertility than secular people. However, in recent decades the fertility differentials of various ideological groups in developed countries has tended to become smaller. In many cases the higher fertility of particular religious groups results from religious positions that are remnants from pre-modern value- and norm systems. In some cases the higher fertility is deliberately maintained or stimulated in order to enlarge the quantitative representation of the in-group in the population. Eventually, this must inevitably contribute to aggravating the ethnic, racial or ideological in-group/out-group relations in the country.[42] From a demographical point of view, even relatively small fertility differentials within populations can, when persistent in the long run, reverse population growth trends.

8.2.2.3 Carrying Capacity of the Earth

The developmental or environmental effects of population size (and growth) cannot be estimated and evaluated as such, i.e. independently of the degree of quality of life and the kind and degree of consumption that we want to achieve or the amount of resources (energy and space) we (want to) use. Hence, we face an almost

[36]Bongaarts (2002).

[37]For instance, Lutz (2006), Sánchez-Barricarte and Fernández-Carro (2007), Hoorens et al. (2011), Goldstein et al. (2009).

[38]For instance, Caldwell and Caldwell (2003, June, 2005), Jones (2007), Atoh (2008), Westley et al. (2010).

[39]For instance, Bourgeois-Pichat (1988), Frejka et al. (2008), Reher (2007), Billari and Kohler (2004).

[40]Nonaka et al. (1994), Greksa (2002), Hurd (2006).

[41]Coleman (2006), Westoff and Frejka (2007), Caldwell (2009).

[42]Bialasiewicza (2006), Richerson et al. (2010), Kaufmann (2011).

insurmountable difficulty in assessing the relationship between the numerical development of the human species and the carrying capacity of our planet. This difficulty appears so clearly from the persistent diversity in attitudes, beliefs and policies on this matter. For instance:

> Population growth should be halted and a slow decline begun to a population size that, in a couple of centuries, might be environmentally sustainable[43]

> Overpopulation does not threaten the environment or humanity.[44]

There is considerable variation in the estimates of the number of people that the Earth can carry because of the different hypotheses that have been proposed. Many authors have tried to define the carrying capacity of the Earth, starting with Antonie Van Leeuwenhoek (1679) who estimated that the maximum number of people the Earth could support is 13.4 billion. Many more estimates followed of how many people Earth could support, ranging from less than 1 billion to more than 1000 billion. For 65 studies giving a range, the lower and upper bounds amount to 7.7 and 12 billion.[45]

A recent interesting approach has been the calculation of the Ecological Footprint by the Global Footprint Network.[46] Global Footprint Network's core research calculates both the Biocapacity (BC)[47] and the Ecological Footprint (EF)[48] for more than 200 countries, using over 5000 data points for each country per year, derived from internationally recognised sources; these have been used to determine the area required to produce the biological resources a country uses and to absorb its wastes and to compare this with the area available.[49] The ratio EF/BC is the estimated ecological overshoot. For 2007, the ecological overshoot (EF/BC) is 50% above unity; meaning, humanity used the equivalent of 1.5 Earths to support its

---

[43]Ehrlich (2000, 322); see also Grant (1992; 1996), Hern (1990, 1993), Hardin (1993), Pimentel and Pimentel (1997; 2005), Smail (1997), Margulis (1998), Grant (2000), Short and Potts (2009).

[44]Bailey (2006); see also Simon (1981; 1998), Connelly (2008), Angus and Butler (2011), Ellis (2013).

[45]Cohen (1995, 402–418).

[46]Wackernagel and Rees (1998), Ewing et al. (2010); http://www.footprintnetwork.org/.

[47]Biocapacity (BC) = area × bioproductivity (Ewing et al. 2010). The biocapacity is measured by calculating the amount of biologically productive land and sea area available to provide for the resources a population consumes and to absorb its wastes, given current technology and management practices.

[48]Ecological footprint (EF) = population × consumption × resource and waste intensity (Ewing et al. 2010). When the BC > EF, there is an ecological reserve; when the BC < EF, there is an ecological deficit. The ratio EF/BC is the estimated ecological overshoot. In their Ecological Footprint Atlas 2010 edition, the Global Footprint Network estimated for 2007 the world's biocapacity at 11.9 billion global hectares (gha) and the ecological footprint at 18.0 billion global hectares (gha) for a world population of 6.7 billion people. This gives an average biocapacity per person of 1.8 global hectares (gha) and an average footprint per person of 2.7 global hectares (gha), giving an ecological overshoot (EF/BC) of 1.5.

[49]The 'ecological footprint' is perhaps not a completely satisfactory instrument to measure the total ecological impact of humanity—for instance, it does not include the impact of the use of chemicals or the effects on biodiversity (see Wijkman and Rockström 2011, 150)—but it is an impressive and most elaborated proxy for measuring the human impact on the environment.

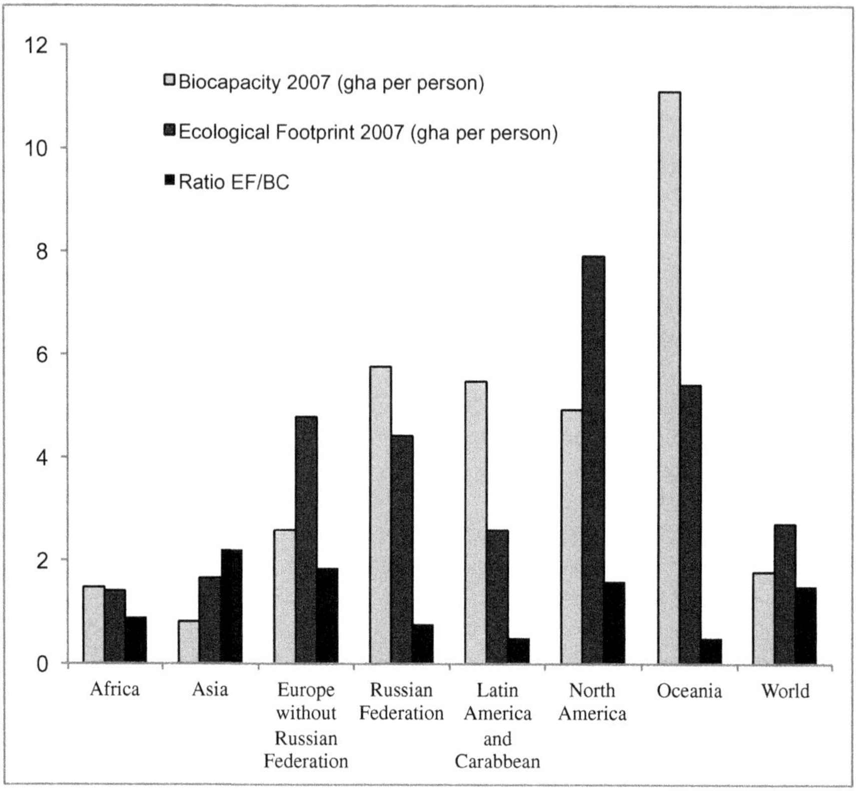

Fig. 8.3 Biocapacity, Ecological Footprint and EF/BC ratio in 2007, per continent and world (authors' calculations based on Ewing et al. 2010)

consumption. From Fig. 8.3 it can be seen that there are substantial differences in the ecological overshoot between continents.

In addition, it has to be observed that the current ecological overshoot of 1.5 Earths relates to a world population in which only about 20% is estimated to enjoy a standard of living typical of the developed world.[50]

In the hypothesis that the whole world acquires by 2050 the level of prosperity of Europe with its current consumption patterns, it can be estimated that humanity would need almost four Earths (Fig. 8.4).[51] If the European consumption further increased linearly between 2007 and 2050, as it did between 1991 and 2007, and this level of consumption is applied to the whole world population in 2050, humanity would need nine Earths. It is self-evident that this is impossible. Even with the hypothesis of the development of new technologies, it seems inevitable that a further increase in the quality of life on a planetary scale can only take place

[50]Smail (2002, 27).

[51]See also Wackernagel and Rees (1996), Smail (2002, 28).

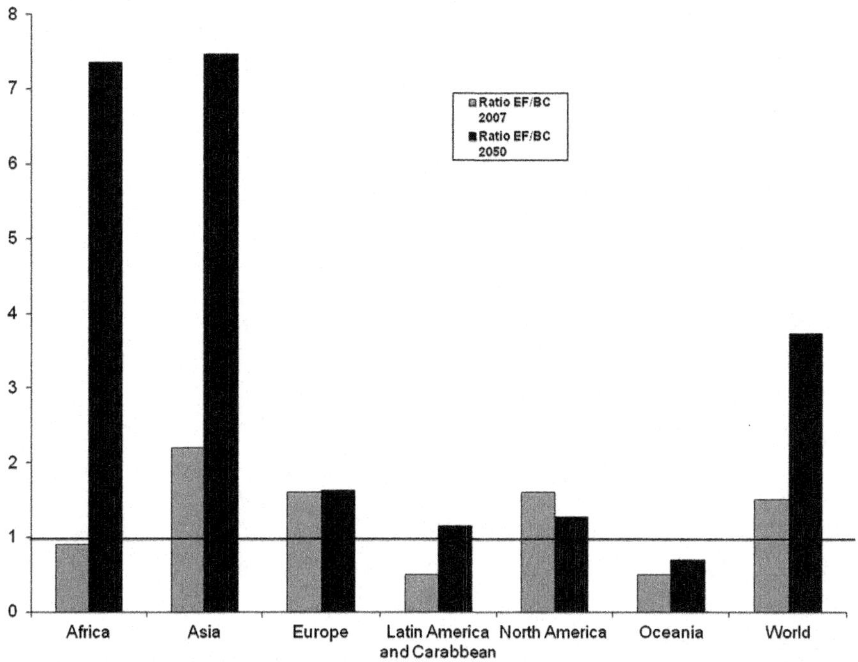

Fig. 8.4 Ecological deficit/reserve in 2007 and 2050, based on the hypothesis that the whole world acquires the level of prosperity and welfare of Europe with its current consumption patterns (authors' calculations on the basis of data from Ewing et al. 2010)

at the expense of a reduction in population size and change in consumption patterns.[52]

Identical reasoning can be made for many specific types of resources. Let us take the example of the needs, uses and distribution of fresh water, which is clearly an indispensible but finite resource. The increasing use of this natural resource is not only related to the world population growth but also, and in particular, to the advance of the modernisation process (Fig. 8.5).

Walter K. Dodds[53] estimated that the availability of freshwater on the Earth amounts to 9000 km^3, approximately half of which is currently being used by the human species, namely 678 m^3 per person. In the United States 2700 m^3 is used per person. If the US usage was extended to the total world population, more than twice the naturally available water resources would already be needed. Knowing that the world population will, in all probability, further increase by three to four billion people, it is without doubt that the American water consumption levels cannot be generalised at the global level. Even if the current excessively high US use and abuse of fresh water was substantially reduced, the expected further world

[52]See also Catton (1980), Pimentel and Pimentel (1991; 1997), Diamond (1992; 2005), Myers (1997), Ferguson (1999), Pimentel et al. (1999), Costanza (2000), Smail (2002), Rees (2003).
[53]Dodds (2008, 21); see also United Nations (2011, 99ff).

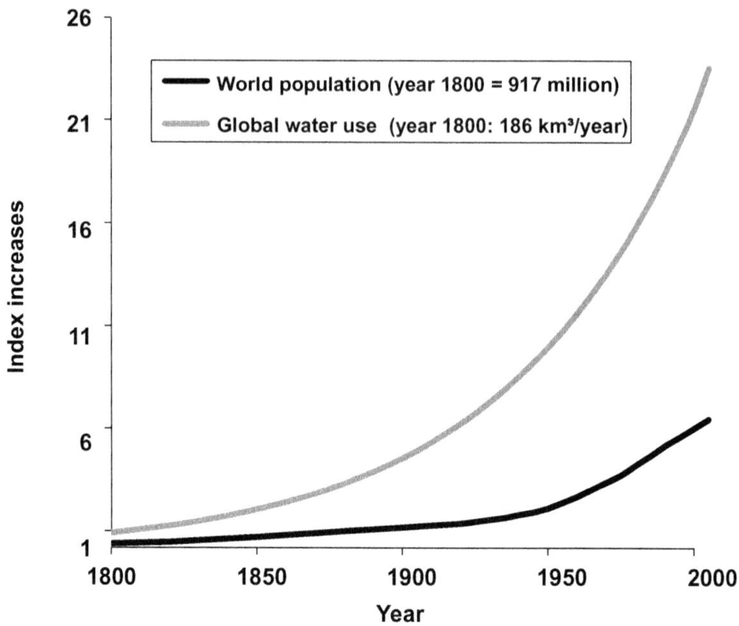

Fig. 8.5 World population growth and global water use between 1800 and 2000 (Dodds 2008, 215–216)

population increase up to eleven billion is incompatible with the goal of applying modern water consumption patterns.

If the current US use of fresh water (2700 m^3/per person) was generalised at the global level, the Earth, with its 9000 km^3 availability of fresh water, could support 3.3 billion people. If the data provided in Meadows, Randers, and Meadows[54] are used, roughly identical results are arrived at, namely 3.7 billion people.

Figure 8.6 represents the relationship between the Human Development Index (HDI) (2012) and the Ecological Footprint (2007) for 153 out of 192 countries in the world. It also illustrates well the double effort that would have to be made to reach a future sustainable human development—namely, moving the countries with a low or medium HDI, as well as the countries with a high or very high HDI, into the 'sustainable human development' quadrant of the graph.

8.2.3 Ethical Reflections Regarding Population Growth

Given the evolutionarily crucial importance of reproductive behaviour, it will not come as a surprise that individuals are equipped with a series of behavioural

[54]Meadows et al. (2004, 67): total availability of freshwater = 5.620 km^3, US use per person = 1500 m^3.

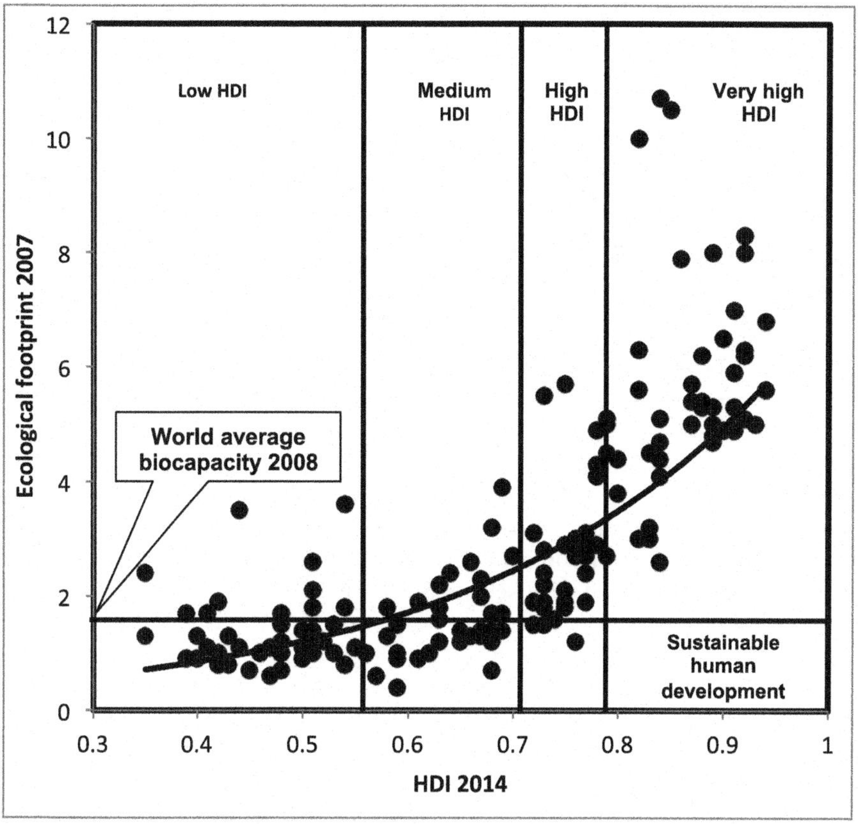

Fig. 8.6 Relationship between Human Development Index (2014) and ecological footprint (2007) for 155 out of 188 countries in the world (after UNDP 2013, 35), http://www.footprintnetwork.org/en/index.php/GFN/blog/human_development_and_the_ecological_footprint

predispositions, such as the drive for mate acquisition, greed for resource acquisition, territorial expansion, status seeking behaviour and love of children, that are—consciously or unconsciously—targeted at transmitting one's genes intergenerationally. Nor is it surprising that traditional ideologies, which played such an important role in the adaptability of societies in pre-modern times, attach paramount importance to reproductive values and norms. Consequently, ethical issues related to reproductive behaviour are a very sensitive matter.

Human reproduction needs to be seen in the context of ethical goals for a further evolution of the hominisation process: (1) the preservation of ecological sustainability; and (2) the cultural furthering of the modernisation process.

Given the knowledge we have about the past and present damage that the human species has caused to the planet's natural environment, together with the threat that current trends in ecological mismanagement may continue, unrecoverable damage to the available natural resources will occur with detrimental consequences for the human species itself. Hence, for pragmatic and moral reasons, we should protect the existing biodiversity, the natural ecological systems and the available natural resources that our planet possesses.[55]

The authors argue for a matured and further progressing modernity—i.e. a cultural stage based on the application of modern science and technology which allows the optimal development of human-specific potentialities, but without the current-day disastrous effects on biodiversity, depletion of natural resources, overpopulation, underdevelopment, use of environmentally harmful technologies, overconsumption, environmental pollution, ABC-weaponry threats for preserving consumption patterns and dominance of powerful nations and neo-colonial exploitation.

There is plenty of evidence concerning the desire of people in advanced countries to further progress on the path of modernisation, as well as the desire of people in second and third world countries to develop, in order to acquire levels of quality of life and well-being similar to those of the 'first' world countries.

It can be expected that further scientific and technological development through future inventions and interventions might resolve some of the shortages of vital resources that have been identified today. We are only at the very beginning of the modernisation process and great scientific innovations, for instance in the fields of genetic engineering, robotics and nanotechnology, which will further uplift the quality of life to unprecedented heights: this may radically change our present perceptions about overpopulation and the exhaustion of limited material resources to sustain human life.[56] Forecasters have generally failed dismally to foresee the drastic changes brought about by completely unpredictable discoveries.[57]

Apropos reproductive behaviour, we should always think about its possible impact on future generations. Indeed, in reproductive matters we need, in addition to a short-term perspective, a long-term, and even a very long-term approach.

Regarding the short-term—generational change in the narrow sense of the word—it is not difficult to demonstrate the indispensable role of value and norm systems. Just as in the case of ontogenetic development, the reproduction of a new generation is not completely genetically programmed. Value and norm systems also influence the number of children that women bear and people raise.

Regarding the long-term, ethical goals concerning generational replacement are more difficult to justify on the basis of objective scientific arguments. The acceptance of this type of goal depends on the way in which life is conceived. From the point of view of the individual's perception, ethical values do not extend beyond five or six generations—grandparents, parents, self, children, grandchildren, great-grandchildren. However, when the levels of personal perception and experience are transcended and the advances of modern science are taken into consideration—particularly with respect to knowledge about the evolution of life—ethical systems may take on a different dimension. From an evolutionary perspective,

[55]Ehrlich et al. (1977), Wilson (1992), Hardin (1993), Chew (2001).
[56]Kurzweil (2005, 13).
[57]Rees (2003, 14).

human life has a temporal dimension that extends over a period of several million years and is characterised by a process of hominisation, the essential feature of which consists of increasing encephalisation, resulting in a growing capacity for controlling life and the environment.

8.2.3.1 Achieving Stationary Population

Assuming the desirability of preserving the Earth's natural ecological systems and resources, as well as a further progressing modernisation, it is clear that lower mortality and fertility rates need to be achieved. As a pragmatic endeavour, a decrease in mortality is in general supported by strong ethical norms. Nobody wants to regress to living conditions where the life course of people was constantly threatened by infectious diseases, malnutrition, homicide and natural disasters. However, since mortality no longer needs to be compensated by high fertility in order to maintain population stationarity, we are at a crossroads.

Consider the following facts (1) the present world population already has an ecological overshoot of 1.5 Earths, (2) the developing regions in the world rightly want to acquire well-being levels comparable to those of the most advanced countries or regions, and (3) the highest population growth rates exist in the least developed countries or regions. Thus, a global aim should be set to achieve, as soon as possible, a zero world population growth.

Due to demographic inertia,[58] it will be necessary to temporarily reduce fertility in many regions of the world to below-replacement levels, in order to reach zero growth as soon as possible.

It will not be easy to set the achievement of a stationary population goal as an ethical objective. First of all, there are the explicit proponents of population growth, usually targeted to in-group expansion, so well expressed in the allocution of the late Algerian President Houari Boumédienne at the General Assembly of the United Nations in 1974: "*Nous vous vaincrons par le ventre de nos femmes*" or the recent statement of the present Turkish president Recep Tayyip Erdoğan who in 2016, urged Turkish women to have at least three children,[59] and called recently on Turkey's citizens in Europe to step up their rates of procreation and have five children each in order to increase their power and influence in Europe.[60] Moreover, there are still many, often silent, growth advocates even in the scientific community. There are even the advocates of explicit population growth such as the late Julian Simons, Ben Wattenberg, Nicolas Eberstadt, Ronald Bailey and Matthew Connelly.[61] The views of the silent advocates of further population growth are more

[58]Demographic inertia refers to population growth effects resulting from changes in the structure of the population age pyramid (Koons et al. 2007).

[59]Many reports in current or recent European mass media, for instance, http://www.dailymail.co.uk/news/article-3627087/.

[60]Many reports in current or recent European mass media, for instance, http://www.telegraph.co.uk/news/2017/03/17/.

[61]Simon (1981, 1998), Simon and Kahn (1984), Wattenberg (1987), Watson and Wattenberg (1989), Eberstadt (1997), Bailey (2006), Connelly (2008), Angus and Butler (2011).

often expressed as views against birth control or prohibiting abortion rather than explicitly favouring the present levels of population growth.

Silent population growth advocates are often found among economists and politicians in general, because of the idea that population increase (via more numerous younger age groups) will contribute to the increase in economic production or enhance political power, within or between populations. In the short-term, and as an isolated phenomenon, at first sight this idea seems to be correct. However, although this was true in earlier cultural stages of human evolution and history, the premise cannot pass the critical test of a sound scientific analysis in a modern context.[62] In particular, in a long-term perspective and from a more global approach, also taking into account broad ecological and resource concerns, avoidance of further population growth in an already overcrowded planet should be a concern.

Most individuals or couples, wherever in the world, are not motivated by world population growth issues in deciding how many children they wish to have and what family size they actually realise. However, enlightened ethical, economic and political decision makers, in discussions about the big world problems of today—being of an ecological, financial, economic or physical nature—seldom address the population dimension in an appropriate way.

It is a striking fact that the preponderant role of the population factor is so often concealed or at least unaddressed in all kinds of societal strife situations—starvation, water shortage, unemployment, in-group/out-group conflicts—or natural disasters—climatic changes, earthquakes, floods, tornados, droughts.[63] For instance, in the United Nations' *Millennium Development Goals Report 2011* the world population crisis is not addressed and the need to reduce fertility in demographically expanding regions does not figure among the eight—otherwise very laudable—goals proposed.[64] Even in the new set of 17 measurable *Sustainable Development Goals (SDGs)*, formally accepted by the UN General Assembly in 2015, the issue of the world population growth was totally out of sight.[65] Also, at the recent United Nations Conference on Climate Change in Paris,[66] population size or growth were not considered. It seems that human development and climate are dissociated from populations.

In addition to those who do not want, for expansionistic, nationalistic or economic reasons, to consider the negative effects of a further population growth, there

[62]See, for instance, Coleman and Rowthorn (2011).

[63]Smail (1997).

[64]United Nations (2011). The eight 'Millennium Development Goals' are: eradicate extreme poverty and hunger; achieve universal primary education; promote gender equality and empower women; reduce child mortality; improve maternal health; combat HIV/AIDS, malaria and other diseases; ensure environmental sustainability; global partnership for development.

[65]Among the 17 identified Sustainable Development Goals, only Goal 3.7 includes an indirect hint regarding fertility control: "By 2030, ensure universal access to sexual and reproductive health-care services, including for family planning, information and education, and the integration of reproductive health into national strategies and programmes." (https://sustainabledevelopment.un.org/sdgs).

[66]United Nations (2015).

are also the proponents of other ethical principles, such as individual freedom, individual reproductive rights, or the sanctity of life, who completely ignore the demographic, societal, genetic, or ecological dimensions of unrestrained population growth.[67] Individual rights activists and patriarchally oriented religious institutions often form surreal coalitions in these matters, as population and ecological experts were able to observe during the Prepcom negotiations for the 1994 United Nations International Conference on Population and Development.[68]

8.2.3.2 Decreasing the World Population Size

Considering the facts that the human species already transcends the carrying capacity of the planet, reduces the natural biodiversity, unbalances many natural ecosystems, depletes natural resources through deforestation and overfishing, depletes non-renewable energy sources, pollutes the environment (air, soil, rivers, lakes, seas and oceans), and is, in all probability, responsible for accelerating climatic changes through high carbon dioxide, methane and other chemical emissions,[69] and considering further that the quality of life of populations in developing regions should not only become identical to that of developed countries, but that the quality of life of all should be further enhanced albeit by replacing quantity by quality consumption, it is necessary to decrease the world population growth to reach size which would be ecologically and globally sustainable in the long-term perspective.[70] As Julian Huxley[71] stated several decades ago:

> The world has to achieve the difficult task of reversing the direction of its thought about population. It has to begin thinking that our aim should be not increase but decrease – immediate decrease in the rate of population-growth, and in the long run, decrease in the absolute number of people in the world, including our own countries.

Therefore, a temporary, slightly below-replacement fertility, as it manifests itself in many developed countries and newly emerging economies, should be considered highly welcome and should extend as soon as possible to developing countries, especially those experiencing high population growth or density.

Regarding the situation in Europe, Dirk Van de Kaa[72] noted rightly:

> Thus, developments go at least in the right direction and Europe may well benefit from them. Conceivably it might increase material wealth, help protect the environment, and increase the educational and other investments in the children who are born. Europe should again set an example for other continents to consider.

A generalised (obviously temporary) below-replacement fertility would lead to a decrease of the world population in the coming centuries as anticipated, for

[67] For instance, Connelly (2008, xii, 382).
[68] See, for instance, Cliquet and Thienpont (1995).
[69] For instance, Wijkman and Rockström (2011), Meadows et al. (2004), Dodds (2008), Cafaro and Crist (2012), IPCC (2013).
[70] For instance, Van de Kaa (1978), Cohen (1996), Hardin (1992), Pimentel and Giampietro (1994), Ehrlich (2000), Grant (2000), Smail (2002).
[71] Huxley (1964, 86).
[72] Van de Kaa (2010).

instance, by the low variant population prospects of the UN Population Division[73] (Fig. 8.1).

Taking the Earth's finite physical and ecological limitations into account,[74] what population size would the Earth be able to support with the hypothesis that the whole world population would reach the developmental and consumption levels of Europe? In the hypothesis that Europe's consumption pattern would further increase between 2007 and 2050, as it did between 1991 and 2006, a sustainable world population would amount to 1.03 billion people.[75] In the hypothesis that the whole world would reach the developmental and consumption levels Europe had in 2007, the planet could support 2.5 billion people.[76] These figures (1.3–2.5 billion) correspond quite well with other estimations, based on more sophisticated calculations available in the literature, which all range between one and three billion.[77] Assertions that the Earth might be able to support a population of 10, 15 or even 20 billion with a standard of living at the level of the most advanced nations of today are, as J. Kenneth Smail[78] writes, "not only cruelly misleading but almost certainly false".

Estimations of a sustainable population size with respect to the renewable fresh water requirements per person for irrigated agriculture and other requirements (industrial, hygienic, leisure) that are common in present-day advanced countries[79] also lie in the range of the low population figures quoted above. However, in the domain of water requirements much stronger regional differentiations in carrying capacity can already be observed and may be expected to increase considerably—although in many cases this problem could, admittedly, be resolved through the construction of water pipeline networks, just as we now have oil pipeline networks. However, the Earth's capacity for providing fresh water is limited, as explained above, unless a very cheap energy source became available that would allow massive desalination of seawater.

A generalised below-replacement fertility in the coming centuries would lead to a decrease in the world population such as appears from long-term, low-variant population scenarios.[80] Indeed, such long-term scenarios show that very small differences in average fertility can make a huge difference to the final size of the

[73]UN Population Division (2004).
[74]Wijkman and Rockström (2011, 4).
[75]This figure is obtained as follows: 9.306 (world population prospects in 2050, in billions)/3.8 (number of planets to support the world population at the developmental level of Europe in 2007)/ 2.38 (the extrapolation of Europe's consumption pattern between 2007 and 2050, as it did between 1991 and 2006) = 1.03 billion inhabitants.
[76]This figure is obtained as follows: 9.306 (world population prospects in 2050, in billions)/3.8 (number of planets to support the world population at the developmental level of Europe in 2007) = 2.5 billion inhabitants.
[77]Ehrlich and Ehrlich (1990, 1996), Grant (1992; 1996, 75; 2000), Hardin (1993), Pimentel and Pimentel (1991, 1997), Giampetro and Pimentel (1993), Wackernagel and Rees (1996), Pimentel et al. (1999, 33–34), Costanza (2000), Wackernagel and Yount (2000), Smail (2002), Pimentel and Pimentel (2005).
[78]Smail (2002, 27).
[79]For instance, Cohen (1995, 297–328); Dodds (2008, 21–26).
[80]UN Population Division (2004), Basten et al. (2013).

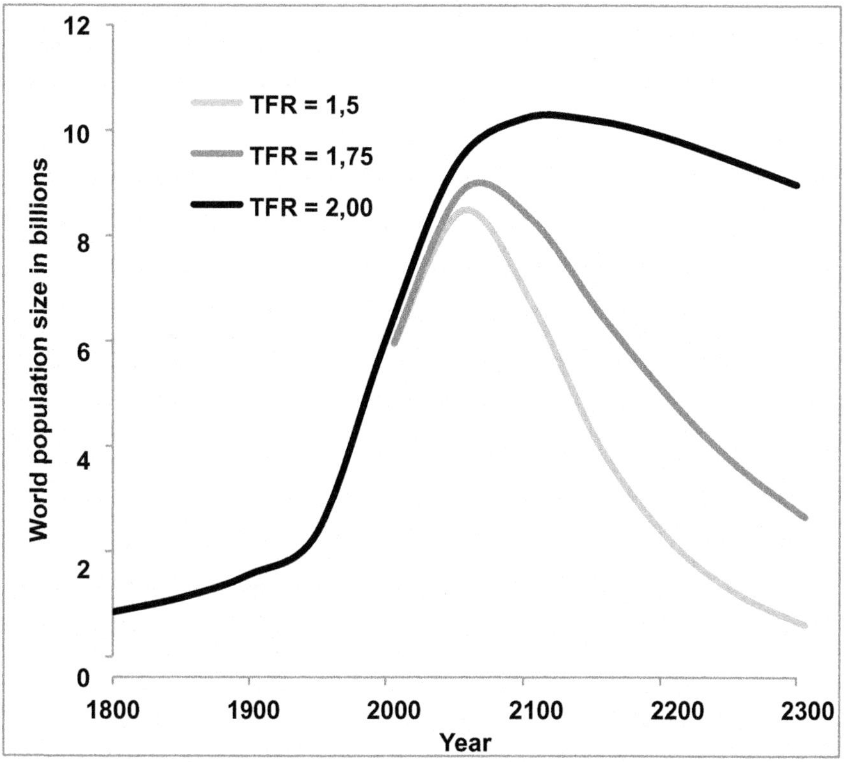

Fig. 8.7 World population increase of *Homo sapiens sapiens* since the beginning of the nineteenth century, and hypothesising that the future growth would evolve according to low and medium variants of the total fertility rate (TFR:1.50; 1.75; 2.00) and an average life expectancy of 90 years (based on Basten et al. 2013)

total world population. For example, with a sustained Total Fertility Rate (TFR) of 1.75 instead of 2.00 children per woman, the population on Earth by 2300 would only be three instead of eleven billion, i.e. eight billion less. If worldwide fertility decreased to the current average for Europe, the world population in 2300 would amount to approximately one billion (Fig. 8.7).

Given the quasi-certainty that the world population will further increase, in all probability up to eleven billion, temporary solutions will have to be found for those eleven billion people.[81] However, it is difficult to see how it would be possible to increase substantially, in the few coming decades that separate us from 2050 or 2100, the living standards of the present underdeveloped regions in the world up to the existing levels of the most advanced countries; it is also hard to envisage a substantial decrease in the production and consumption patterns and material living standards of the developed countries.

[81]Wijkman and Rockström (2011, 179).

A well-known and often heard alternative solution to the current ecological challenges is the reduction or change in the excessive production and consumption patterns in the developed world.[82] Consumption volumes and production patterns of the developed countries should take into account their unfavourable effects on the planetary environment, and they need to stop abusing resources from other regions or indebting their future generations. One must acknowledge that the present 'World Order' is all about preserving access to abundance to in-groups and silently or overtly blocking access to valuable resources to out-groups. Ideologies that sustain and encourage high fertility are playing an important role in keeping out-groups poor and powerless.

Indeed, the necessity to change the production and consumption volumes and patterns should not divert our attention from the equally important—and ultimately most important—issue of demographic numbers as, for instance, Ian Angus and Simon Butler[83] do from the ideological perspective of ecosocialism. Those authors rightly stress the disastrous ecological effects of the current consumption and particularly the production patterns, especially in the developed world, and rightly identify their root causes in the present form of capitalist societal organisation with its rampant profit-growth ideology, its considerable energy and resource wastage, and its huge within- and between-country social inequalities. However, they not only grossly underestimate the effects of population numbers but also misinterpret, because of their lopsided ideological framework, population concerns as expressions of socially conservative and reactionary attitudes aimed at maintaining a socially inequitable status quo. Even the consensus that was reached at the UN International Conference on Population and Development in Cairo 1994[84] is alleged to have given "new credibility to an agenda that has long been used to block social change".

Undoubtedly a lot has to be profoundly changed in matters of production and consumption, but it is illusory to think that populations in advanced countries or emerging economies will be prepared to substantially reduce their standard of living and quality of life—their achieved sense of well-being. It is also unimaginable that the populations in developing regions would stop aspiring to achieve the quality of life levels of the more advanced nations. Moreover, the further progression of the modernisation process will increasingly require energy and other resources. Hence, although considerable ecology friendly improvements in consumption and production patterns can and should be made, the most important ultimate trade-off will be between enhancing quality of life and decreasing population size worldwide.

The efforts to decrease population size should not necessarily be envisioned by country or be proportional to the present population size of each country. It would probably be wiser to consider larger geographical unities, such as continents or subcontinents, or regions with specific geographic capabilities or constraints, such as regions susceptible to natural catastrophes, mountain regions, (semi-)desert regions.

[82]For instance, United Nations (1992), de Geus (2003), Angus and Butler (2011).
[83]Angus and Butler (2011).
[84]United Nations (1994).

For example, looking at the current and expected ecological overshoot per continent, as represented in Figs. 8.3 and 8.4, it is quite clear that all continents, with the exception of Oceania, will have to reduce their population in order to maintain or increase their quality of life. Africa and Asia may have to make the biggest efforts. In this respect, the Chinese one-child policy has succeeded in avoiding an additional increase of 400 million people.[85] One might have reservations about the method of implementing that policy. However, the outcome at the population level is significant.

It is quite clear that the goal of decreasing the world population, to one third or even one tenth of what it is currently projected, will require a profound reassessment and change of humanity's moral, demographic, and economic goals and policies. It will also necessitate a major change in the traditional way of thinking by the religious institutions and secular political movements in the world: their support will be needed to reverse the current attitudes, behaviours and policies regarding demographic growth.[86]

It will be equally necessary to overcome our inborn drives and culturally reinforced desires oriented toward unlimited resource acquisition as can, unfortunately, be observed on a daily basis in national, continental or world politics. These biological drives and cultural norms will have to be changed when pursuing long-term sustainability. Neither the traditional religions nor the most modern secular ideologies, except the ecological movement and perhaps also the humanist movement, are intrinsically focussed on ecological sustainability.[87] On the contrary, traditional authoritarian regimes, as well as modern democratic regimes, are (for different reasons) all oriented towards economic (and often also demographic) growth.

8.3 Qualitative Issues Regarding Reproductive Behaviour

The numerical development of the human species is relatively easy to define and measure. The notion of the quality of parental investment is a much more difficult issue to define: it includes not only objective, material elements of the development of human potentialities but also subjective elements of what people consider and value as quality of life.

Quality of life fundamentally includes two partially interrelated aspects—the nature of genes and gene combinations we inherit from our parents, and the way in which our life course develops under the influence of our genome, i.e. the environmental/social influences to which we are subjected—and a third aspect, the environment with its material and immaterial components that we create for ourselves and our environment. All three groups of determinants of our quality of life

[85]Greenhalgh (2003).
[86]See also Smail (2002, 41).
[87]Dodds (2008, 153–163).

—the genetic, the external environmental/social, and the self-induced/created environmental determinants—can be achieved at very different levels.

In Chap. 5 it was argued that rationally grounded choices could be made in the following ways: (1) ontogenetically, the maximal development of the human-specific potentialities; and (2) phylogenetically, the progression of the hominisation process. What are the implications for ethical choices in the domain of reproductive behaviour?

8.3.1 Evolutionary Background of Qualitative Intergenerational Changes

This discourse can be very brief since the evolutionary processes resulting in the establishment or intergenerational change of qualitative traits in the population—whether genetic or cultural in nature—have already been extensively discussed in Chap. 2. The evolutionary background of quality of life issues essentially relates to the evolutionary mechanisms through which genes and memes are created and are all, or are not differentially, transmitted to subsequent generations. Genetic and cultural mutations provide qualitative variation; drift, migration and above all the various forms of selection to which genetic or cultural traits are subjected, result in their elimination, maintenance or increase.

Intergenerationally, the evolutionary toolkit results in the reproductive fitness of individuals or groups—the degree to which their genes and memes are intergenerationally transmitted and eventually result in the genetic and/or cultural adaptation of the organism/species to its environment.

8.3.2 Qualitative Reproductive Developments in Modernity

The modernisation process has a huge impact on reproductive behaviour. The following will be discussed: modern influences on reproductive behaviour which have a phenotypic effect, in other words influence aspects of quality of life which do not necessarily have transgenerational hereditary effects; and influences on reproductive behaviour that have a genetic effect.

8.3.2.1 Phenotypic Reproductive Developments in Modernity

Birth Control. In modernity two types of cultural-technological interventions in reproductive behaviour have to be distinguished, namely interventions to limit births and procedures to enhance fertility. Among the first group, contraception and induced abortion have to be considered[88]; regarding efforts to enhance

[88]Infanticide was a rather common practice in pre-modern eras, but it has become a rare phenomenon in modernity thanks to the general humanisation of modern culture and in particular due to the spread of contraceptive and abortive methods (Willianson 1978; Harris 1977; Hoffer and Hull 1981; Birdsell 1986; Herman-Giddens et al. 1999; Milner 2000; Putkonen et al. 2009).

fecundity/fertility, two types of methods are available for people who have difficulties in getting pregnant or completing a pregnancy with a life birth: biomedical interventions to enhance fecundity/fertility and adoption.

Contraception. Although birth control practices are not exclusively limited to modern culture (hunter-gatherers and agrarian-pastoral populations usually kept their fertility substantially below the biological potential through a variety of procedures),[89] modern societies underwent a revolutionary transition in fertility control, namely the general dissemination of parity-specific birth control practices through contraception and induced abortion.

The demographic transition was not caused by a major technological innovation in fertility limitation; although appliance methods, such as the condom, gradually became available they were not widely used initially. The major methods applied were *coitus interruptus* as well as several other traditionally known methods such as abstinence, extended nursing and late marriage. In the later stages of the demographic transition (the first decades of the twentieth century) induced abortion and —at least in some countries—mechanical barrier methods (such as the condom, pessaries, douche and spermacides) increased in importance.[90]

In the second half of the twentieth century, modernised countries experienced—with significant differences in pattern and pace—a second contraceptive transition, which was characterised by the development and dissemination of technological innovations: hormonal contraceptives, intrauterine devices and (somewhat later) more advanced medical techniques for female and male sterilisation (Fig. 8.8).[91] The new contraceptive technology was, moreover, paralleled by new surgical and pharmacochemical abortion procedures.[92]

The contraceptive profile in advanced countries continues to modernise.[93] In many overseas predominantly English speaking countries and several Western European countries, the most recent changes concern a notable increase in rates of female and male sterilisation. In most of those countries sterilisation has already become the most common method after people have reached their desired family size.[94]

A remarkable feature of the first contraceptive transition is that it occurred not only without much support from public or religious authorities but also against the opposition of various powerful groups of the establishment such as governments, legal authorities, churches, political parties and the medical profession. Despite its revolutionary character, the dissemination of birth control at the turn of the twentieth century took place in silence, if not in secrecy. Neo-Malthusian groups were

[89]Himes (1936), McLaren (1990), Riddle (1992).
[90]Dawson et al. (1980).
[91]Westoff and Ryder (1977), Cliquet and Lodewijckx (1986), Léridon et al. (1987), Benagiano et al. (2007).
[92]Cliquet and Thiery (1972), Baird et al. (1995).
[93]Frejka (2008).
[94]Ross (1992), Lodewijckx (2000).

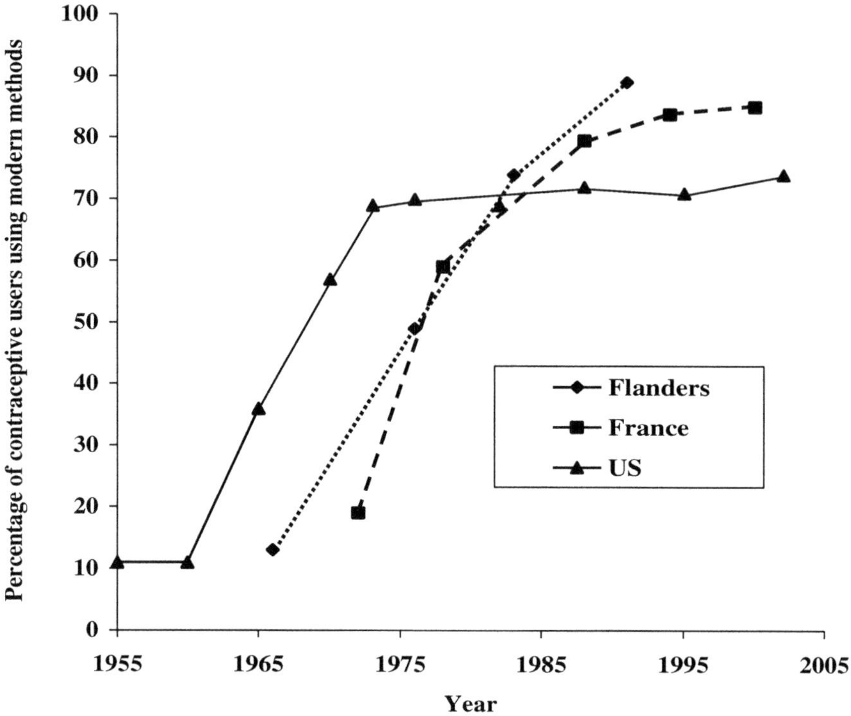

Fig. 8.8 The second contraceptive transition in Flanders, France, and the United States (Freedman et al. 1959; Whelpton et al. 1966; Westoff and Ryder 1977; Ford 1979; Mosher and Westoff 1982; Cliquet and Lodewijckx 1986; Mosher and Bachrach 1988; Cliquet and Callens 1993; De Guibert-Lantoine and Léridon 1998; Bensyl et al. 2005; Trussell and Wynn 2008; Fertility and Family Survey in the ECE Region; www.ined.fr/2009)

fought or boycotted from all sides.[95] The second contraceptive transition initially encountered identical, though less generalised resistances.

It is no surprise then that both the first and the second contraceptive transitions were characterised by substantial social and ideological differences in the use and effectiveness of contraception—less educated, poor people, as well as people with low cognitive skills, applied less contraception.[96] Practicing churchgoers used less effective contraception compared to freethinking and non-religious people.[97]

Although contraceptive differentials have decreased considerably, vulnerable groups remain in populations where sexual and reproductive education is still insufficiently embedded in the educational systems and mass media, or where medical and welfare care services do not yet fulfil their tasks properly. Present-day population subgroups with unmet family planning needs are those that have

[95]Stengers (1971), Van Praag (1979).
[96]For instance, Cliquet and Balcaen (1983).
[97]For instance, Cliquet and Schoenmaeckers (1976), Cliquet and Maelstaf (1977).

inadequate knowledge or are too poor to purchase modern contraceptive tools in order to control their fertility such as adolescents and people with weak cognitive abilities, and people who are still under strong in-group cultural pressures such as ultraconservative religious denominations,[98] socially insufficiently integrated immigrants from developing countries[99] and people living in poverty.

Abortion. In modernity, induced abortion is still quite frequent, albeit with strongly differentiating statistics according to the stage of the demographic transition, the contraceptive profile of the population, and prevailing attitudes, values, norms and legislation.[100] In general, it can be said that modernisation is characterised by an increasing acceptance and legalisation of induced abortion, a shift from non-medical (criminalised abortion) to medical forms of induced abortion (*abortus arte provocatus*), and a decrease in both non-medical and medical pregnancy interruptions—the latter being subject to an efficient contraceptive policy.[101]

From a statistical perspective, the indications for induced abortion are mainly of a social nature: age (too young, too old), marital status (unmarried, divorced), family size (excess fertility), financial constraints, and failed contraception. Medical indications are less common: maternal indications have become an exceptional phenomenon whilst relatively rare foetal indications can result in selective abortion.[102]

Biomedical interventions to enhance fecundity/fertility. Medically assisted fertility is a phenomenon that has developed gradually with modernisation; it involves a broad range of medical techniques ranging from general medical and genetic counselling to specific technical interventions that facilitate or replace natural conception. Such techniques may include donor insemination, ovum donation, in vitro fertilisation, gamete and zygote intrafallopian transfer, and embryo transplantation, as well as methods aimed at maintaining the pregnancy and inducing or facilitating delivery. In recent decades, the concept of medically assisted fertility has been increasingly understood in the narrow sense of methods that facilitate or replace natural conception.[103]

Whereas medical interventions related to the whole process of childbearing are quite widespread, conception related interventions are, at the population level, still quite rare. Nevertheless, it can be observed that increasing numbers of couples who experience difficulties in getting pregnant are turning to these techniques in order to fulfil their family-building desires.[104] In recent years, concern has grown about the possible long-term effects of proceptive interventions on foetal birth weight and

[98]For example, Eaton and Mayer (1953), Nonaka et al. (1994), Greksa (2002), Hurd (2006, 2011), Kaufmann (2011), Paltiel (2013).

[99]Jones et al. (1986), Bajos et al. (2002), Guldi (2008).

[100]Tietze (1981), Ketting and Van Praag (1983), Henshaw et al. (1999), Alan Guttmacher Institute (2000).

[101]Cliquet and Thiery (1972, 1985), Ketting and Van Praag (1983), Faúndes and Barzelatto (2006).

[102]Alan Guttmacher Institute (2000), Bajos et al. (2002).

[103]Bentley and Mascie-Taylor (2001), de Jonge and Barratt (2002).

[104]Delmotte and Cliquet (1983), Lodewijckx and Schoenmaeckers (1994), Schieve et al. (2002), Sobotka et al. (2008).

malformations.[105] Prudence is also called for in the application of proceptive medical techniques at ever higher ages for proceptive mothers due to increased foetal morbidity risks at those ages.[106]

Adoption. From an evolutionary point of view, adoption has been interpreted as a 'paradox', a 'misfire', even a 'double whammy', because adoptive parents not only fail to increase their own reproductive fitness but they also improve someone else's.[107] Adoption is one of those topics that anti-evolutionists like to use to refute evolutionary theory,[108] but this is an example of profound misunderstanding of the complexity of the evolutionary process and mechanism. First of all, in the past adoptive parents were often related to the natural parents. However, most importantly humans are also equipped with a natural desire to care for children. This caring predisposition can easily lead to adoption when people cannot have children of their own.[109] Adoption is a minority phenomenon: it is a secondary strategy in case natural parenthood is impossible.

Phenotypic Effects of Birth Control. In the first place birth control has quantitative effects. Below, demographic effects in general and effects on the differential growth of various ideological groups are briefly elucidated as examples. However, birth control can also have qualitative effects. It can influence many aspects of people's life experiences such as out of the house employment of women, partner relations, investment in children, and sexual satisfaction. Two examples are briefly discussed: effects on sexual behaviour in general and effects on male machismo behaviour.

Demographic Effects in General. The general spread of parity-specific birth control practices, starting at the end of the nineteenth and the beginning of the twentieth centuries, obviously had a strong effect on fertility: it formed the proximate cause of the fertility decline in the demographic transition. Some have raised the question as to whether modern contraceptives (and medical abortion methods) in the second half of the twentieth century also caused the resumption of the fertility decline that has been observed since the mid 1960s, after the baby boom of the post-war decades. There is considerable misinterpretation of the supposed causative relationship between the spread of efficient contraceptives—the pill, intrauterine device (IUD) and sterilisation—and the onset of the recent fertility decline that, indeed, remarkably occurred at the same time. It is often overlooked that in earlier periods—in the 1930s and even earlier—some populations succeeded in reducing their fertility to below replacement levels by using less effective methods (mainly withdrawal, sexual abstention and induced abortion). Successive fertility surveys—before and after the second contraceptive transition of the 1960s—have shown that the desired family size has decreased only slightly in recent decades. Moreover, desired family size appeared—and still appears to be—largely independent of the number of children actually produced: before the second contraceptive transition,

[105]Cetin et al. (2003), Dulitzki et al. (1998), Kozinszky et al. (2003), Belva et al. (2016).
[106]For instance, Salihu et al. (2003).
[107]Dawkins (1976, 100), Eisenberg (2001).
[108]For instance, Clayton (2002).
[109]Silk (1990), Arnhart (1998).

families did not differ substantially in their (original) desired number of children irrespective of completed family size (large and small). However, when the pill and other modern methods became available the unplanned or undesired large families (largely) disappeared. Subsequent fertility surveys that paid sufficient attention to the phenomenon of excess fertility[110] have shown that untimed and undesired births are increasingly averted by means of effective contraceptive methods. Hence, there can be no doubt that efficient birth control methods help women or couples achieve a better and more conscious control of the family building process. The avoidance of unplanned and especially unwanted pregnancies does have an influence on completed fertility.[111] Less apparent, though perhaps more important, is the indirect attitudinal effect of the availability of modern birth control methods on planned parenthood. The existence of such methods, and the knowledge that fertility can be mastered rationally and efficiently, may influence individual attitudes and motivations with respect to conscious decision-making about parenthood.

Effects on the Differential Growth of Various Ideological Groups. It is a well-known and well-established fact that the demographic transition progressed at a different tempo according to worldviews of people. Quite universally, religious people evolved more slowly in the reduction of their fertility than secularised people.[112] There have been several explanations for this quite general phenomenon, the most evident being the stronger individualistic values for self-fulfilment among secularised people and the stronger reproduction oriented teaching and indoctrination among religious people. Although, in general, there has been a gradual levelling down of the religious/non-religious differentials in fertility, even today religious people have more children than secular people.[113]

In addition to the generally positive fertility gradient related to the degree of religiosity, there are two specific forms of religious demography which amplify the fertility differentials according to worldview, namely (1) endogenous fundamentalist religiosity; and (2) immigration from less developed countries.

Whereas fertility differences between religions such as Protestantism, Catholicism, Judaism and Islam are decreasing, the differences between 'modernists' and 'fundamentalists' within each of those religions are persisting or even growing, which is changing the numerical proportions between religious moderates and fundamentalists. Hence, the classical secular-religious differential in fertility is exacerbated by fundamentalist religiosity, resulting in a fertility gradient going from the lowest values among secular people, to higher values among modernist religious believers, then the highest values among fundamentalist believers.[114] Extreme examples of endogenous fundamentalist growth are Haredi (ultraorthodox) Jews, Finnish Laestadian Lutherans, American Old Order Anabaptists such as the Amish and Hutterites, and Dutch Orthodox Calvinists. Some of these groups show

[110]For instance, Cliquet and Balcaen (1983), Lodewijckx et al. (1988).

[111]Calot (1990).

[112]For instance, Immerman and Mackey (2003), Kertzer (2006), Derosas and van Poppel (2006).

[113]For instance, Cliquet and Maelstaf (1977), Lutz (1987), Schoenmaeckers and Lodewijckx (1999), Philipov and Berghammer (2007), Frejka and Westoff (2008).

[114]Hout et al. (2001), Kaufmann (2011, 100).

fertility rates several times higher than the modernist believers, resulting in considerable demographic expansion.[115] In many cases fundamentalist movements or institutions are actively using their knowledge about the numerical (and ideological/political) effects of higher reproductive rates, inciting their followers to produce large numbers of children in order to demographically strengthen their position and outnumber the 'others'.[116]

Also, immigrants from developing to developed countries often have considerably higher, though gradually decreasing, fertility than the host populations[117] and are substantially increasing their representation in the population, at least in the agglomerations where they concentrate. For instance, in a few decades Brussels—the capital of the European Union—will have a majority population of Muslims, unless the secularisation process is accelerated.[118]

The fertility differentials between different ideological groups can change the numerical representation in countries; however, in some cases secularisation can be a powerful counterforce, offsetting the religious fertility advantage.[119] If the fertility differentials along the fundamentalist-secular gradient persist intergenerationally and are insufficiently compensated by secularisation, one may expect political and cultural implications. The greatest effect consists of a regressive cultural evolution regarding the achievements of modernity in domains such as scientific research, freedom of expression, individual self-actualisation, family planning, recreation, and rights of whoever fundamentalists consider to be undesirable or inferior such as atheists, women, homosexuals and apostates.[120]

Inspired by recent studies showing that variation in spirituality/religiosity is partly determined by genetic factors, Robert Rowthorn[121] raised the question as to whether the religious/non-religious fertility differentials might also have effects on the genetic composition of the population and the values it cherishes. Taking into account the findings of Laura B. Koenig and Thomas J. Bouchard,[122] who suggest that a genetic predisposition towards religion is associated with obedience to authority and conservatism, Rowthorn concludes that the higher fertility of religious people will not only diffuse religiosity genes in the population but also the values to which they are predisposed.

However, the future might not look so gloomy for free-thinking people and so rosy for religious fundamentalists. Modernity possesses increasingly powerful educational and technological means—obligatory schooling followed by ever prolonging education, TV, Internet, ICT in general—that disseminate the equally strongly increasing scientific knowledge base about the origin of the universe, of

[115]For instance, Eaton and Mayer (1953), Nonaka et al. (1994), Greksa (2002), Hurd (2006), Kaufmann (2011).
[116]For instance, Pride (1985).
[117]Lesthaeghe et al. (1988), Schoenmaeckers and Callens (1992), Coleman (1994, 2006), Lesthaeghe and Surkyn (1997), Westoff and Frejka (2007).
[118]Schoenmaeckers and Callens (1992, 144).
[119]See also Kaufmann (2011, 259).
[120]See also Kaufmann (2011, xxi), Longman (2004, 33).
[121]Rowthorn (2011).
[122]Koenig and Bouchard (2006).

life and of humanity, as well as the causes and mechanisms of biological evolution, social progress and cultural innovation. All these things influence people in their thinking and behaviour in a progressing modernising world. Increasing secularisation, or at least liberation from the traditional ideological straitjacket, seems to be inevitable; this will also result in more and more people with a strong predisposition towards spirituality looking for other social, cultural and political solutions than the ones offered or imposed by fundamentalist branches of traditional religious institutions. Nevertheless, increasing efforts will have to be made to fight obsolete ideologies that may possess considerable financial means for propagating their outmoded ideas. Examples are the financial power of some oil-rich Islamic theocracies that support the spreading of fundamentalist Islam outside the Near East,[123] or conservative American Maecenas who support the spreading of creationist or intelligent design ideologies as so-called 'scientific theories', even resulting in the erection of museums that illustrate 'scientifically' the truthfulness of the biblical genesis story.[124]

Effects on Sexual Behaviour. A major effect of modern, efficient contraceptive methods concerns their favourable influences on people's sexual life and satisfaction. The success of these modern methods is not only due to their efficacy in preventing conception but also their smaller or non-interference in coital behaviour. Those who have experienced the transition from the use of the traditional methods, such as withdrawal, periodical abstinence (and even mechanical methods), to the use of the modern, highly effective methods, such as oral contraception, IUD and sterilisation, testify how liberating those methods are. This is due to the reduction of the fear of unexpected pregnancy and in non-interference with lovemaking. This has been difficult to measure precisely because the users of modern effective methods have so much higher sexual expectations and standards than the users of traditional methods.[125]

Familiarity with efficient control over fertility permitted or favoured other sexuality related processes: premarital sex became less risky; marriage could be postponed or temporarily replaced by other types of union formation.

It is difficult to evaluate with precision whether modern contraception has substantially contributed to the emergence or advancement of the so-called sexual revolution, but its effect is probably largely overestimated.[126] Effective contraceptives may have furthered premarital sex, they may have contributed to the spreading of consensual unions, they may have facilitated extramarital relations and the formation of new partnering, but sexuological research shows that the arrival of modern contraception was not associated with a considerable increase of sexual promiscuity.[127]

[123]See for instance Ludlow (1998), Menelik (2010).
[124]See for instance, http://www.answersingenesis.org/; Numbers (2006).
[125]Van den Bogaert (1976).
[126]Antibiotic drugs may have had a more important influence in this respect (Sipe 1990).
[127]For example, Spira et al. (1993), Laumann et al. (1994), Wellings et al. (1994).

Effects on Male Machismo Behaviour. One of the longer-term evolutionary effects of modern contraceptives might be that their use will gradually change the sexual behaviour and perhaps even the genetic composition of the male population.

Women can now efficiently control their fertility through the use of efficient contraceptives: this has freed them from the psychological pressures and physical burden of unwanted or untimed pregnancy, resulting in decreasing numbers of unwanted adolescent births, premature marriages, forced marriages, and excess fertility. Perhaps efficient contraception has also facilitated for women the choice of multiple partnership and extramarital relations. However, for promiscuous males, modern contraceptive methods are an 'evolutionary dead end': the gene-spreading effect of their behavioural pattern is strongly reduced, if not completely eliminated. In the long run, modern contraception will erase the genetic effect of the machismo of philandering males since extramarital affairs will no longer have reproductive effects. Indeed, modern birth control practices are completely changing the rules of the game. Whereas in earlier times sexual promiscuity and extramarital sex might have increased the reproductive fitness of philanderers, in the modern contraceptive society women can effectively protect themselves against the unexpected or undesired side effects of adultery, namely against births which would not be fathered by their chosen partner(s). The same holds true for aggressive males who in earlier times, particularly in wartime, succeeded in increasing their inclusive fitness by means of rape or slavery, but in modern society they can achieve no such advantages. When parenthood tends to concentrate amongst couples that establish enduring relations, this personality type will have the advantage, both genetically as well as culturally. The long-term Darwinian effect of differential reproduction should not be underestimated.[128] If men want their genes to be reproduced, they will have to adapt their sexual behaviour much more than in the past to the desires and needs of women.

8.3.2.2 Genetic Developments Related to Reproductive Behaviour in Modernity

So far, reproductive behaviour in modernity has had two types of effects on the genetic composition of modern populations: unintended dysgenic changes as a consequence of contraselective effects and intended genetic changes as a consequence of eugenic interventions.

Dysgenic Effects of Modernisation. Up until now modern culture has had two types of dysgenic effects on the human gene pool: (1) contraselective effects due to a reversal of social status linked differential reproduction, particularly with respect to cognitive skills; and (2) selection relaxation effects of medical replacement therapies for deficient or deleterious genes.

Reversal of Social Status Linked Differential Fertility. It is generally believed that the pre-modern era in human evolution and history was characterised by a slightly positive relationship between social status and reproductive success, either because higher status males succeeded in acquiring more mates, and/or because

[128]Cliquet and Avramov (1998).

wealthier families possessed more material resources which slightly reduced mortality, in particular infant and child mortality.[129] According to Larry S. Milner,[130] the social class related differential fertility in medieval times was strongly related to the differential practice of infanticide. Hence, natural selection operated in pre-modern living conditions not so much via direct differences in fertility, but indirectly via mate selection or differential mortality.[131]

As mentioned in Chap. 5, Sect. 5.3.2.2, the slightly positive relation between social status and reproduction, which existed in pre-modern times, reversed during modernisation through the socially differential spread of birth control. The most developed and prosperous strata of the population generally came to control their fertility earlier and more effectively than the others.[132] The fairly significant correlation between cognitive abilities, education and socioeconomic status, as well as direct observations on the negative relation between cognitive abilities and fertility[133] (Fig. 8.9), gave rise to the theory of contraselection. In this theory the view has been put forward that modernisation would be attended by a decline in cognitive ability.[134]

Meanwhile, the negative association between fertility and socio-economic status (SES) indicators in industrial societies decreased quite substantially because birth control also became a common practice among lower educational and social status strata.[135] Some countries[136] even show the first signs of a possible return to a slightly positive relationship[137]; this supports Frederick Osborn's[138] 'eugenics hypothesis' which states that when individuals have freedom to make fertility related choices, more children will be born in the most favourable home environments. This implies that the end of the demographic transition might be characterised by a positive association between reproductive fitness and socially valuable biological traits.

Nevertheless, many industrial societies still show a moderate negative relationship between fertility and a variety of SES indicators (Fig. 8.10). However, some studies show that men and women have a different relationship between SES

[129]Retherford (1993), Chagnon (1997), Richerson and Boyd (2005), Gurven and von Rueden (2006), Nettle and Pollet (2008).
[130]Milner (2000).
[131]Cattell (1972, 287).
[132]Cochrane (1979), Vining (1986), Skirbekk (2007).
[133]Cattell (1937, 1950), Vining (1986), Glad (2003), Shatz (2008), Retherford and Sewell (1988).
[134]For instance, Graham (1970), Lynn (2001).
[135]Kirk (1969), Weeden et al. (2006).
[136]Kravdal (1992), Hoem (1993), Fieder and Huber (2007), Goodman and Koupil (2009; 2010).
[137]In several countries there appears to be a sex difference in this reversing trend: whereas fertility is slightly positively related to men's educational level, it is still negatively related to women's educational level (cf. Hopcroft 2006; Weeden et al. 2006; Fieder and Huber 2007; Keizer et al. 2007; Nettle and Pollet 2008), or intelligence level (Retherford and Sewell 1988). This sex difference in reproductive outcome is probably due to the difficulties which career women experience when trying to combine motherhood with occupational aspirations (as Muller had predicted in the 1960s), and perhaps also because of the postponement of childbirth (Hewlett 2002; Kemkes-Grottenthaler 2003).
[138]Osborn (1952, 1968), Osborn and Bajema (1972).

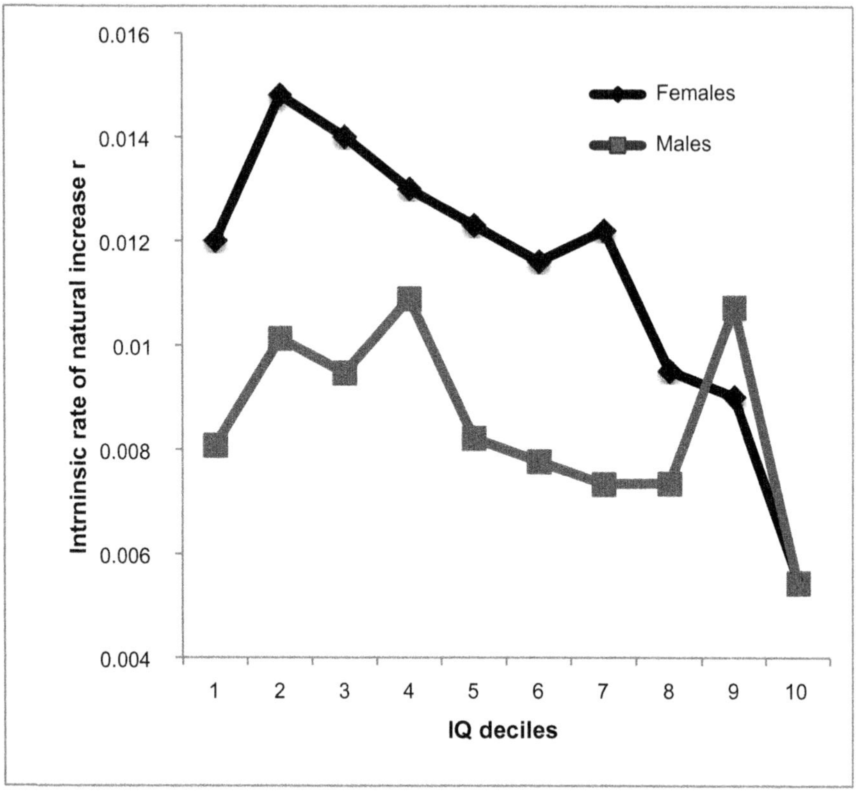

Fig. 8.9 The intrinsic rate of natural increase r by IQ deciles and sex, derived from the Wisconsin Longitudinal Study (based on Retherford and Sewell 1988)

indicators, such as educational level and income, and completed fertility or lifetime reproductive success. Whereas a flat or slightly positive relation is found for men, a more explicit negative association continues to exist for women.[139] In many countries considerable proportions of highly educated or professional women remain childless.[140] For instance, in Germany, a country characterised by a high rate of childlessness in general (about 25%),[141] 40% of women with an academic degree remain childless.[142] In the United States, between one third and one half of all high-achieving women have no children.[143] Apart from the long-term dysgenic effects that may be expected if this trend continues, there must also be unfavourable short-term social effects resulting from such huge proportions of voluntary

[139]For instance, Hopcroft (2006, 2015), Fieder and Huber (2007), Nettle and Pollet (2008), Huber and Bookstein (2010), Barthold et al. (2012).
[140]Kiernan (1989), Kirk (2001), Rowland (2007), Konietzka and Kreyenfeld (2007).
[141]Dorbritz and Schwarz (1996), Dorbritz (2008).
[142]Weiss (2002), Duschek and Wirth (2005).
[143]Hewlett (2002), Hopcroft (2006), Fieder and Huber (2007).

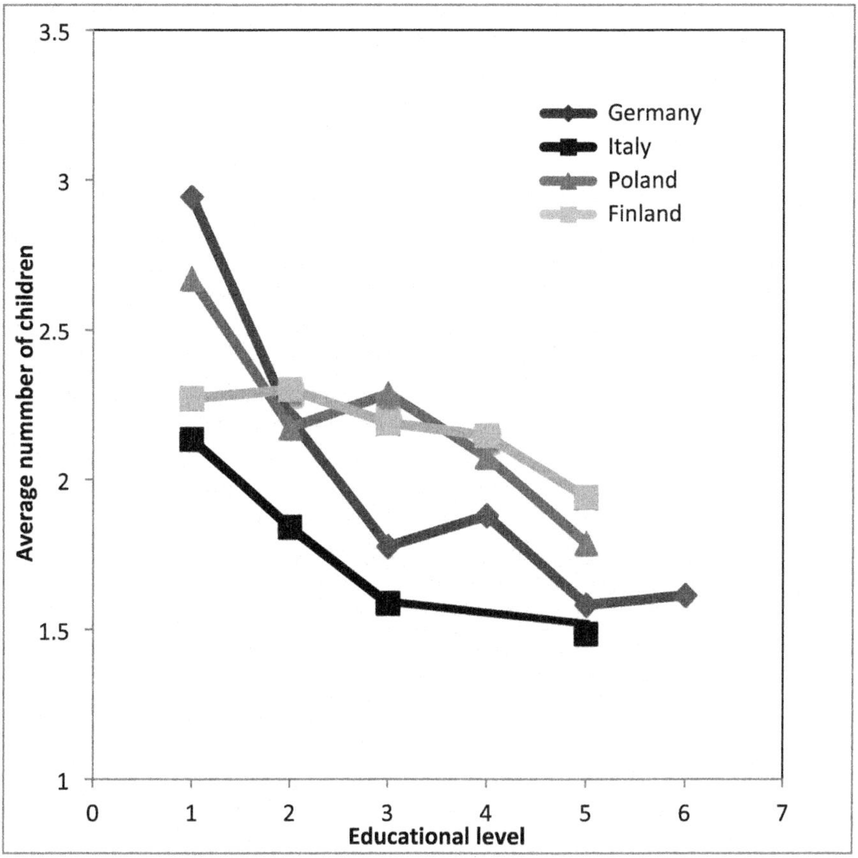

Fig. 8.10 Differential fertility by education of 40–65 year old women in selected European countries (Avramov and Cliquet 2008)

childlessness amongst the best educated and creative people who, one might suppose, are the able to provide a favourable social environment for raising children.

Overall, research indicates that the link between IQ and fertility in modern society seems to have had a slight dysgenic effect upon intelligence in the course of the twentieth century, even though birth control became a fairly generalised practice.[144] However, this is probably only a temporary consequence of a major shift in cultural development and its associated demographic regime. In the near future—namely in the course of this millennium—this dysgenic effect might be neutralised, if not reversed, by future improvements in genetic knowledge and genetic engineering, and the adaptation of norms to the new genetics and demographics. In particular, the low (or deficit) fertility of highly educated working women might be avoided if appropriate measures are taken in order to better reconcile work and

[144]Retherford and Sewell (1988), Lynn and Van Court (2004), Lynn and Harvey (2008).

family life for women, and to better balance childcare and household chores between men and women.

Selection Relaxation. In modern culture humans have succeeded in efficiently intervening against disease and death, with the result that a considerable part of modern populations reaches a much higher age than in pre-scientific living conditions. As mentioned in Chap. 5, Sect. 5.3.2.2, the successful phenotypic care provided in modern culture leads to a relaxation of natural selection.[145] Alleles, which in pre-scientific living conditions were rapidly barred from the gene pool, are currently preserved thanks to replacement therapies or other protecting factors and in many cases their carriers are also able to reproduce. This results in the increasing frequency of such 'weak' alleles.[146] Furthermore, the selection relaxation caused by mortality control, just as with morbidity control, may be reinforced by an increase in reproductive fitness. Surviving individuals may find a partner or partners and produce children. Although many congenital defects are known to result in lower marriage rates, in infertility or are associated with low fertility,[147] the effect of modern culture is that, through replacement therapies, mating and reproductive opportunities for those with genetic disorders are improving.[148]

The quantitative effect of selection relaxation depends on several factors: the relationship between mutation pressure and the degree of relaxation, the effect of replacement therapy on reproductive fitness, and the mode of inheritance (dominant, recessive or polygenetic).[149] Computer simulations show that for all types of inheritance the increase in allele frequency is very slow, especially for recessive alleles or polygenetic conditions, for which the most deleterious alleles are hidden in heterozygous combinations.

Eugenic Effects of Modernisation. There are two major types of eugenic effects of modernisation. The first one pertains to the general dissemination of modern birth control practices; the second is related to the development of genetic counselling.

Eugenic Effects of the General Dissemination of Birth Control Practices. As explained above, the demographic transition was initially characterised by strong differentials in birth control practices according to various indicators of social status (educational level, income, prestige, etc.). The increasingly positive association between social status and cognitive abilities in modern democratising societies, and the positive relation between social status indicators and birth control, leading to negative relations between those indicators and fertility, resulted in the theory of the contraselective (= dysgenic) effects of modernity. However, as modern birth control practices spread throughout the population, particularly with the development of effective contraceptives in the second half of the twentieth century (= the second contraceptive transition), the initially strong fertility differentials considerably

[145] Crow (1966), Cavalli-Sforza and Bodmer (2013).
[146] Dobzhansky (1962), Monod (1970, 179), Thibault (1972).
[147] Reed (1971), Slater et al. (1971).
[148] Teitelbaum (1972).
[149] Cavalli-Sforza and Bodmer (2013).

decreased and can even be expected to level out or even slightly reverse, as can be observed in the most advanced modern countries.[150] Hence, to the extent that a sound modern contraceptive (and educational) policy succeeds in reaching all people in the population, fertility differentials will no longer depend on the lack of knowledge or availability of means to control one's fertility, particularly among the less able, less educated, or underprivileged strata of the population.

Eugenic Effects of Genetic Counselling. Thanks to the explosive development of human genetics and biotechnology, present-day genetic counselling dispenses an increasing diagnostic and therapeutic toolkit, going from traditional genetic pedigree analysis, through biotechnological identification methods, to contraception, genetic-selective abortion and medically assisted fertility[151]: this increasingly allows the conception or birth of offspring with genetic impairments to be prevented through parental choice, and in more and more cases it allows couples to choose to have disability-free offspring. The rapid progress in molecular genetics will soon broaden medical genetics with the revolutionary tool of germinal therapy; this will make it possible to treat sex cells with recombinant DNA in order to genetically alter germline cells, so as to replace unfavourable genes with desired genes.[152] Medical genetics has a eugenic effect at the individual and family level and, if sufficiently available in society, also at the population level.[153] For instance, it is estimated that the new national screening policy for Down's syndrome in Denmark will make the country Down's free by 2030 as a result of parental choice.[154]

8.3.3 Ethical Reflections Regarding Qualitative Reproductive Issues

As could be expected from the preceding discussion, ethical dimensions of qualitative aspects of reproductive behaviour need to include two complementary domains of action: phenotypic (= euthenics) and genetic (= eugenics).[155]

The authors use the classical concepts of euthenics and eugenics to label those domains of action. Euthenics[156] refers to actions that improve people's functioning and well-being by altering their environment and living conditions. A subdomain of euthenics is euphenics, which aims to directly influence the phenotype rather than the genotype.[157] Although the concept of eugenics was originally defined by Francis Galton[158] as "the science which deals with all the influences that improve

[150]Kravdal (1992), Hoem (1993), Fieder and Huber (2007).
[151]Mashiach et al. (1990), Robertson (1994), Gardner et al. (2004).
[152]Wheale and McNally (1988), Friedmann (1998), Stock and Campbell (2000), Stock (2002).
[153]For instance, Holloway and Smith (1975), Devore (1992), Bernhardt et al. (2000), Castellani et al. (2009).
[154]Ekelund et al. (2008); http://www.ncregister.com/daily-news/the-fight-to-eliminate-down-syndrome-eugenics-at-work/#ixzz3cB9OyiXB.
[155]Huxley (1964), Cattell (1972), Calman (2004).
[156]Euthenics = from Greek: *euthenein*: thrive, flourish. See for instance, Richards (2011).
[157]Lederberg (1960).
[158]Galton (1883, 1905).

and develop the inborn qualities of a race to the utmost advantage"—implying that it includes both genetic and environmental actions to improve human-specific traits —it is usually interpreted as actions that aim to transmit favourable genetic characteristics—at the individual, family or population level—to future generations. Most people in scientific and lay circles choose to forget that eugenics implies also environmental actions.

Many people hesitate or even refuse to use the concept of eugenics due to excesses of the early, so-called 'Mainline Eugenic Movement',[159] and particularly because of the political and criminal abuse of the concept and practice by the Nazi regime in Germany during the 1930s–1940s.[160] Often eugenics is (wrongly) interpreted as attempts by the state to improve, whether or not via coercive measures, the genetic composition of the population.[161] In order to distinguish individual from social efforts to influence or change the genetic transmission of particular genetic characteristics, Julian Savulescu[162] devised the concept of procreative beneficence which aims to produce the best child, of the possible children, a couple could have. Although the authors understand the concern of Savulescu to create a distance from a concept that has been utterly abused, changing words does not alter the contents of a premise. Likewise, the authors do not reject all the ethical concepts of Christianity and socialism because they have also been utterly abused in some phases of human history.

[159]See Kevles (1985).

[160]See for instance Saller (1961), Klee (1983), Müller-Hill (1984, 1988), Kaiser et al. (1992), Glad (2006, 65–74). Contrary to what many scholars and commentators argue (for instance, MacKellar and Bechtel 2014, 19–21), Nazism had nothing to do with a genuine eugenic policy. The Nazi rhetoric on eugenics was absolute hypocrisy, a cover for other politically motivated actions that had no eugenic effect at all, and in some cases had precisely the opposite, namely a dysgenic, effect. The indiscriminate sterilisation of people with genetic impairments was scientifically unfounded (cf. Dahlberg 1948). The euthanasia of handicapped people had no eugenic repercussions. Most of the victims were seriously ill, were institutionalised and had no opportunity to transmit their genes intergenerationally. The aims of this policy were apparently more of an economic than a biological nature. With respect to the promotion of the so-called Aryan race, the Nazi policies were likewise deceitful because this so-called Aryan race does not exist as a biological entity. In so far as the German population was associated with the Nordic variant of the Caucasian race, there was not a single scientific argument or proof for the alleged biological superiority of this population-genetic variant. Also, the extermination policy toward the Jews had nothing to do with eugenics. On the contrary, by driving the more intelligent Jews and other valuable intelligentsia away as emigrants to other lands, or by eliminating them in concentration and death camps, Nazism in fact had a dysgenic effect, comparable to the persecutions of prominent thinkers by the Catholic Inquisition in the late medieval and early Renaissance eras. The intellectual superiority of the Jewish population was well known to the Nazi geneticists and anthropologists (cf. Lenz, in: Bauer et al. 1936). The Jewish achievements and intellectual capability were apparently a thorn in the flesh of the Nazis. The Shoah (or 'holocaust') was in fact a 'final solution' to the problem of competition with a socially creative and successful population group that was a traditional scapegoat in Christian Europe in times of crisis. Unfortunately, many people associating or confounding eugenics with Nazism do not read German, do not know the relevant literature, choose to ignore it, or do not critically think about the facts.

[161]For instance, Young (2006).

[162]Savulescu (2001, 2009).

In an excellent reassessment of eugenics, Richard Lynn[163] rightly states that it is quite unusual to reject a theory that is essentially correct. He sees the increasing precedence accorded to individual rights over social rights as the fundamental cause of the decline of eugenics in the later decades of the twentieth century—a phenomenon which he perceives as causing the rejection of eugenics and increasing the acceptance of individual rights in many other domains of life. Lynn's explanation about the shift from social to individual rights is absolutely correct, but this shift is probably also linked to two other twentieth century phenomena, namely the social abuses of authoritarian regimes (fascism, Nazism, communism) and the lopsided environmentalism that dominated the social sciences.

The recent strong development of molecular genetics and biotechnology in general has evoked an explosion of scientific and ethical-philosophical-religious publications, pro and contra, on what is now called the 'new eugenics',[164] or even 'neugenics'.[165]

8.3.3.1 Euthenic Reproductive Goals

In a modern(ising) context, with its strong mortality control, the promotion of birth control is a necessity. The degree to which births should be controlled—limited or enhanced—depends on the well-being of the children to be born, of the families that raise them, and of the population that supports them. The degree to which births are controlled also depends on the quality and sustainability of their environment in which they will grow up.

Promoting Contraception. From a modern perspective emphasising the need to promote the practice of birth control in the sense of reducing fecundability to zero in some life stages looks like kicking down an open door. However, despite the considerable progress that has been made in recent decades, there is still a substantial unmet need, particularly in developing countries. In the modernised countries there are still unmet needs among vulnerable population groups, such as adolescents, less able or less educated people, immigrants coming from developing countries and poor people, concerning both knowledge and the availability of modern, effective contraceptives. In many developing countries the unmet need still reaches levels that were typical of the advanced countries in their early demographic transitional phases at the beginning of the twentieth century.[166] We still face the fact that important ideological institutions, in particular the Roman Catholic Church, oppose the use of modern, effective contraceptives. The recommendations of the United Nations World Population Conferences of Bucharest (1974), Mexico City (1984) and Cairo (1994)[167] have not yet been put into practice all over the world or for all population categories.

[163]Lynn (2001); see also Paul (2014).

[164]Paul (1998), Habermas (2001), Lynn (2001), Duster (2003), Agar (2004), Glad (2006), Sorgner et al. (2006), Pope Benedict XVI (2009), Turda (2013), MacKellar and Bechtel (2014).

[165]Selgelid (2014, 6).

[166]Westoff (2006), The World Bank (2010), United Nations (2011), Darroch (2013), Darroch and Singh (2013).

[167]"All couples and individuals have the basic right to decide freely and responsibly the number and spacing of their children and to have the information, education and means to do so." (United Nations 1975; 1984; 1994).

Legalising Induced Abortion within the Framework of an Effective Contraceptive Policy. Several types of intervention—sexual abstention, contraception, induced abortion and infanticide—have traditionally been used to achieve birth control. There is a very simple biological reason why fertility limitation should primarily be achieved by using contraception: whenever there is a choice, intervention in biological processes should always be given preference to the minimalistic intervention.[168] Hence, contraception is preferable to induced abortion. Infanticide, as a practice of pre-modernity, has virtually disappeared as a means for limiting the family size.

As a means for controlling fertility, induced abortion is clearly a sign of failed contraception. Nevertheless, induced abortion has been legalised in modernised societies within the framework of a birth control policy that aims to reduce that type of intervention to a residual minimum. To be effective it needs to be set up in an educational framework of pre- and post-pregnancy counselling about abortion and contraception in which, however, the woman is free to decide by herself.[169] Empirical analyses about induced abortion practice and prevalence show that legalisation of abortion in the framework of an effective contraceptive policy succeeds in reducing abortion practice to a minimum.[170] In fact, legalising abortion reduces this practice much more than prohibitive measures and criminal prosecution. However, principled opponents of abortion care more about their principles/prejudices than about the actual positive outcome of legislative regulations.

Selective abortion, in cases of foetal or maternal health indications, is discussed below as an aspect of eugenics.

Enhancing Fertility? A minority of people—individuals or couples—cannot have children naturally, for instance, subfecund couples, single people and homosexual unions. Modern methods of medically assisted reproduction, as well as adoption, can increasingly help such people to satisfy their desire to have and raise children. There can be no doubt that such individuals or couples may be excellent parents.

However, in each of those cases there are arguments for applying due diligence measures in which the well-being of the children should be of primary concern. For instance, in the case of subfecundity, care is necessary in order to avoid medically assisted fertility interventions causing unfavourable genetic or ontogenetic effects on the offspring.

In a modern, technologically strongly, developed and million sized population context, society can deal with diversity in parenthood models including single parents and homosexual couples. However, since we are still a bisexual species in which children have an interest in having a father and a mother, the lone

[168]Opponents of contraception will argue that the minimalistic rule in fertility control would imply that preference should be given to sexual abstention over contraception. However, in the authors' view, sexual abstention is a much more profoundly intervening and disturbing intervention in an important biological process—sexuality—than modern methods of contraception.

[169]Cliquet and Thiery (1972), Staatscommissie voor de Ethische Problemen (1977).

[170]Ketting en Van Praag (1983), Faúndes and Barzelatto (2006).

mother/father or homosexual parental couple may be expected to remain a minority phenomenon.

Avoiding Excess Fertility and Excessively High Fertility. Due to inadequate contraceptive policies, many societies are still confronted with more or less high levels of excess fertility—pregnancies or births that are not wanted.[171] Obviously, a child may not be wanted at conception but it does not mean that it is not loved later on. However, excess fertility increases chances that all children may not receive sufficient care and support. Given the high developmental and educational standards that modern societies pursue, it goes without saying that excess fertility, resulting in parents having more children than they want, because they have no access to effective contraception, should be avoided.

Modern(ising) societies are, despite the progressing demographic transition towards low fertility, also confronted with an important residue of very large families—say five or more children. The (very) large family system is largely a remnant of the transition from a pre-modern to a modern demographic regime. In pre-modern living conditions people begot many children, but many of them did not survive the infant or juvenile stage. In modern living conditions, due to infant mortality control and social protection, large families can flourish thanks to the greatest altruistic system—the welfare state—that human societies have ever devised. As Richard Dawkins[172] argued:

> What has happened in modern civilized man is that family sizes are no longer limited by the finite resources which the individual parents can provide. If a husband and wife have more children than they can feed, the state, which means the rest of the population, simply steps in and keeps the surplus children alive and healthy....we have abolished the family as a unit of economic self-sufficiency, and substituted the state.

Nevertheless, the phenotypic features of children from (very) large families often contrast less favourably with children from small or moderate family sizes.[173] One cause is the less favourable living conditions in large families, particularly in families with low income, where there are simply not enough material means as well as parenting time available to meet the increased educational standards of modernity. Another cause is that people with weak personality characteristics, low cognitive skills, or low education do not succeed in adequately mastering their fertility. Both phenomena—unfavourable living conditions and biosocial assortment—often coincide.

A specific problem is population groups that continue to achieve high fertility because of their (usually religious fundamentalist) ideology that rejects so-called unnatural contraceptive interventions in their reproductive potentials—conveniently forgetting that the welfare state, which finances their high fertility, is equally unnatural. Some groups also want to expand their influence and power through demographic expansion to the detriment of the well-being of women and often

[171] For instance, Lodewijckx (1988), Tsui et al. (2010), Finer and Zolna (2011).
[172] Dawkins (1976, 125).
[173] Terhune (1974), Bossard and Boll (1975), Cliquet and Balcaen (1983), Desai (1995), Redmond (2000), Rodgers et al. (2000), Bradshaw et al. (2006).

children as well. This is a particularly thorny problem, because the proposition that such groups should lower their fertility to the average level of the general population might be interpreted as interfering with the principle of personal freedom. There is, however, the much more important issue that they compete in a domain that is no longer adaptive. The real competition in meritocracies is in cultural, social or economic domains typical of the novel environment of modernity. The seemingly moral principle professed by some ideological groups for achieving the biological maximisation of inclusive fitness principle in a modern(ising) environment is actually a form of maladaptive behaviour—it decreases chances of children to succeed. Some even call it immoral, as it may be regarded as denial of the opportunity to control fertility and non-assistance of socially vulnerable persons through the denial of knowledge and opportunity.

The demographic expansionism of religious fundamentalists in modern societies is largely possible due to the ideological and materialistic acquisitions of modernity (ideological pluralism and tolerance, infant and maternal mortality control, control of contagious diseases, generous social transfers), the very essence of which is in fact challenged by unsustainable fertility levels maintained through moral in-group pressure or denial of access to knowledge and means.[174]

8.3.3.2 Eugenic/Transhumanist Goals

In the science based eugenic literature,[175] as well in the more recent transhumanism literature,[176] three major human qualities can be identified that are subject to eugenic action: cognitive abilities, emotional personality characteristics that facilitate altruism and sociability, and mental and physical health. Ultimately, this comes down to what has been defined here as the enhancement of the hominisation process.[177] (See also Chap. 5, Sect. 5.3.2.4.) Indeed, both the eugenics and transhumanist proponents express considerable concern about the genetics of future generations: firstly, to avoid harm to our descendants; secondly, to improve their capabilities to adapt to future potential cultural developments and to push those developments to higher levels of performance. Eugenics/Transhumanism favours a long-term evolutionary perspective[178] in which, as Jonathan Glover states "equal weight is given to people's interests, whichever generation they belong to".[179]

Indeed, eugenics/transhumanism aims to transfer to the genetic and intergenerational domain what we already do in advanced societies in the domains of biomedical care and educational training: promoting the development of human-specific features, such as intelligence, morality (sociability, altruism,

[174]See also Kaufmann (2011, 259).

[175]Galton (1883), Muller (1935), Crew et al. (1939), Osborn (1940; 1968), Sutter (1950), Blacker (1952), Bajema (1976), Lynn (2001), Sorgner et al. (2006).

[176]For instance, Mehlman (2003), Agar (2004), Bostrom (2004), Miller and Wilson (2006), Savulescu (2009), Savulescu and Bostrom (2009), Knoepffler and Savulescu (2009), Buchanan (2011).

[177]Cliquet (1961; 1997–1998, 2010).

[178]See also Cattell (1972), Elgin (1993), Rees (2003), Glover (2006); http://longnow.org.

[179]Glover (1984, 140).

honesty, etc.), physical attractiveness, physical performance, sexual satisfaction, longevity, and health in general, thereby extending our concerns from the intra-generational to the intergenerational domain, from phenomic care to genomic care. So far, conscious human intervention has mainly been limited to phenotypic care and improvement, for instance through intragenerational intervention. Intergenerational intervention, particularly by using genetic manipulation, requires a generation transcending ethics and policy. Such an ethical orientation and policy programme would allow the human species to control not only its demographic growth and phenotypic development but also its genetic composition—in other words, its future evolution. This proposition may, at the beginning of this century, sound quite daring, if not scary, just as at the beginning of the twentieth century quantitative birth control was feared and vigorously opposed. However, from a longer-term perspective it is difficult to imagine that the human, once the technical know-how has been mastered, will not try to avoid genetically determined diseases and promote genetic features that are considered socially desirable or advantageous such as cognitive ability, sociability, physical attractiveness, and mental and physical health in general.

In the domain of human reproduction, the twentieth century was characterised by the massive shift from chance to choice in matters of quantitative birth control, whereas the twenty-first century might see the rapid shift from chance to choice in matters of qualitative birth control. Eugenics/Transhumanism is essentially aiming to replace the rather crude natural selection by a more humane cultural selection.[180]

Limiting Dysgenic Features or Trends. The human gene pool is continuously undergoing mutations. Most of these mutants have detrimental effects on the organism.[181]

In modern culture therapeutic practices produce, through selection relaxation, contraselective effects which maintain, or even reproductively increase, alleles in the gene pool that are less favourable for the physical or mental health or for the

[180]It is perhaps useful to remind the reader the sense in which the concept of selection is used: it is Darwinian or reproductive selection, i.e. the differential choice of particular characteristics (or their alleles, genotypes, or genomes) over other variants for transmission to future generations.

[181]According to recent studies (e.g. Giannelli et al. 1999; Nachman and Crowell 2000) the mutation rate in a human zygote is estimated to range between 128 and 175 mutations per diploid genome per generation; one out of three is estimated to be deleterious. Each individual carries five to seven lethal recessive genes (Muller 1950; Cavalli-sforza and Bodmer 2013; Larson 2002). Humans are estimated to have approximately 21,000 genes which can lead to abnormal phenotypes when mutating (OMIM 2017). Some of the genetic mutations are relatively minor or are amenable to treatment; others result in death or serious disability. This genetic load manifests itself partly in each generation by genetic impairments, most of which are eliminated in the very early stages of embryonic development as miscarriages or stillbirths. Some five per cent of newborns are born with a visible congenital impairment. Some other genetic diseases, such as Huntington's chorea, develop at a later stage in the life course. Most congenital impairments are genetic in origin, namely they are caused by some defect in the DNA of the carrier. In pre-modern living conditions, the mutational load was kept constant over succeeding generations by natural selection through the immediate or delayed elimination of deleterious alleles. Rare favourable mutations, in contrast, were preserved and spread through the increased reproductive fitness of their carriers and resulted in increased adaptiveness, which contributed to further hominisation.

social integration of their carriers. These 'saved' genes are added, in each generation, to newly arising mutations: slowly but systematically the population's mutational load therefore increases.[182] Whilst modern culture increases (through therapeutic interventions) the phenotypic adaptability of previously deleterious or harmful alleles,[183] a primary aim of eugenics is still to avoid genetic deterioration through an increase in less favourable genetic variants and to preserve the more favourable genetic variants in the gene pool.[184]

The arguments that eugenic action is inevitable[185] are as follows. In the long run, the accumulation of culturally (medically) preserved and reproduced deleterious genes together with the newly arising natural mutations in each generation would lead to a situation in which most, if not all, individuals in the population are endowed with innumerable hidden and conspicuous genetic defects; these would require such a scale and variety of medical treatment and social care that it would consume a lot of society's resources, leaving no surplus for other social or cultural needs.[186] Eventually, euphenic correction of the (increasing) genetic load would not be sufficient to guarantee the maintenance of mental, social and physical health at the population level, therefore eugenic intervention would have to be undertaken.

Modern culture also manufactures products that contain or emit ionising radiation and some molecules that have mutagenic effects which increase the mutational load in the human gene pool.[187] It is clear that eugenics aims to prevent the increase of the mutational load due to the use of such harmful products.

Finally, eugenics endeavours to change the trend toward dysgenic differential reproductive practices through which less desirable genetic variants of socially important continuous characteristics, such as low cognitive ability and some negative emotional personality characteristics, succeed in increasing their representation in subsequent generations.[188]

All procedures to limit dysgenic features aim to alleviate unnecessary human suffering and in James Neel's[189] words:

> Protect the gene pool against damage.

An argument that is sometimes invoked for preventing or limiting dysgenic developments is the fact that, in addition to the multiple psychological and physical suffering genetic impairments provoke for affected individuals, such impairments involve considerable costs to families and society as a whole.[190] In the authors' view, cost of are in terms of money is a relatively weak argument, considering the

[182]Muller (1960).
[183]Dobzhansky (1962; 1967).
[184]Huxley (1964).
[185]For instance, Carlson (1973).
[186]Muller (1960).
[187]Obodovskiy (2015).
[188]For instance, Buchanan et al. (2000), Lynn (2001), Savulescu (2001).
[189]Neel (1970, 820).
[190]See also the discussion in MacKellar and Bechtel (2014, 126–128).

huge financial resources modern societies have (and choose to spend and gamble away on very objectionable purposes such as military expenses).

Promoting Favourable Genetic Features. Modern societies must adapt to a rapidly changing environment, in which a high premium is being placed on high intelligence, creativity and energy. On the one hand such societies require individuals with high cognitive abilities and creativity in order to adapt to the rapidly changing social and technological environment; on the other hand individuals need high intelligence and creativity to be able to cope with the challenges of modern culture and to take advantage of the new opportunities for self-fulfilment.[191]

The continued development of culture toward levels and depths so far unseen, but conceivable, could be advanced by increasing key human faculties—cognitive abilities, sociability, and mental and physical health and longevity−to still higher levels which are beyond the present variation. This would allow the human species to have a deeper insight and understanding of itself and of nature in general; to further improve its mastery of the biological, physical and social environment; and to reach higher levels of well-being, both at the individual and the societal level, and increase the survival chances of the species. Thus, eugenics might contribute to this drive for mastery in the development of future stages of culture as, for instance, envisaged by Duane Elgin[192] or by the transhumanist movement with their visions of the future transhuman and posthuman.[193]

Once the direction of human evolution is set culturally and the social and medical techniques for acting upon that evolution have been developed, a number of individuals or even entire societies might attempt to genetically programme their offspring as favourably as possible. As we succeed more and more in treating and eliminating diseases and disabilities, it may also be expected that we will increasingly be inclined to take a preventive approach to ill health. Moreover, it is entirely possible that genetic engineering will not remain restricted to avoiding pathological situations, but will be broadened to improve 'normal' characteristics such as cognitive performance, memory, emotional life, sociability, physical attractiveness, sexual arousal and performance, and other desired human characteristics.

However, it is not impossible that some people will try to promote other, socially less valuable features such as, for instance, aggressivity. This raises the question of the relationship between the individual and the population. New biotechnologies will enable individuals and families to apply a kind of 'home-made eugenics'[194] in which they can decide on the type of children that they want to have. The dangers with this ought to be recognised, as already argued above in Chap. 5, Sect. 5.2.3.4. Society must ensure that parental choices do not harm the future well-being of their descendants or of society as a whole. In this respect Thomas Douglas and Katrien

[191] Bajema (1971).
[192] Elgin (1993).
[193] Bostrom (2004), Miller and Wilson (2006), Harris (2007), Savulescu (2009), Savulescu and Bostrom (2009), Knoepffler and Savulescu (2009), Buchanan (2011) (see also Chap. 6, Sect. 6.2.3.4).
[194] Kevles and Hood (1992).

Devolder[195] pertinently argued that procreative selection should not only focus on the well-being of the own offspring but should also take into account the effects on the well-being of others. Hence, the principle of 'procreative beneficience'[196] should be complemented or accompanied by what they call the principle of 'procreative altruism'—a principle that was already implicitly embedded in the 'old eugenics' through its emphasis on promoting genetic predispositions to sociability.

A specific problem related to human variation in a further progressing modernisation concerns the perspective of the ability to produce, in a not too remote future, probably already in this century, identical genetic copies of individuals by cloning. This technique, already applied in an experimental stage on animals, consists of producing offspring from a somatic cell. This asexual reproduction results in an individual which is genetically identical to the provider of the somatic cell. There already exists an extensive literature on the advantages and disadvantages, and more particularly on the ethical implications of cloning.[197] It is no surprise that bioconservatives are fiercely opposed to cloning.[198] In the context of the present discourse, there is only one aspect that is relevant to the discussion: to what degree would cloning affect human variability and the future hominisation process? The answer clearly depends on the degree to which cloning would be practiced and replace sexual reproduction. Whereas the cloning of exceptionally talented individuals might have positive effects on the short-term biological and cultural future of humanity, this practice must be discommended at the population level because, when massively practiced, it would reduce biological variability and counteract the advantages of genetic recombination linked to sexual reproduction.[199]

Objections to Eugenic Goals. The authors perceive four major objections to the pursuit of eugenic goals[200]: (1) objections related to recent historical events and developments; (2) objections of a religious or philosophical nature; (3) fears about increases in inequality within populations; and (4) fears about the position of people with disabilities. They are partly similar to the objections to the transhumanist goals described in Chap. 5, Sect. 5.3.2.4.2.

Objections Related to Recent Historical Events and Developments. These objections pertain to phenomena such as the coercive nature of earlier eugenically intended interventions, the class and race biases of some of the earlier eugenic movements, and especially the political and criminal abuse of the eugenic ideology in Nazism which, in reality, pursued practices that had nothing to do with eugenics or were even dysgenic in nature.[201] However, none of the science-based eugenic or

[195] Douglas and Devolder (2013).
[196] Savulescu and Kahane (2009).
[197] Silver (1998), MacKinnon (2001), Mackintosh (2012), Craig (2012), Philip and Cherian (2013).
[198] For instance, Cole-Turner (1998), Kimbrell (1998), Annas et al. (2002), President's Council on Bioethics (2003); see also article 11 of the UNESCO Universal Declaration on Bioethics and Human Rights (2005).
[199] See also Glover (1984, 2003, 2006), Glannon (2001, 132).
[200] For a more detailed overview of arguments supporting and opposing eugenics see, for instance, MacKellar and Bechtel (2014, 120–183).
[201] See footnote 162, p. 379.

enhancement treatises consider between-population genetic differences as one of the goals to be pursued. Eugenics and enhancement proponents are only interested in specific individual characteristics and their prevalence and intergenerational transmission within populations, which are independent of between-population-genetic differences. Nevertheless, some authors continue to suspect or warn against a possible hidden racialism in eugenic endeavours.[202] Although it is, indeed, quite likely that people will express population-genetic and even cultural 'in-group' concerns or preferences for normal traits in their choices of partners or germ donors for disease replacing or health promoting eugenic interventions,[203] in-group drives need to be taken into account—but they are not drivers of eugenics and, hence, should not be used as an argument against eugenic endeavours or practices.

Objections of a Religious or Philosophical Nature. The religious objection relates to the belief that the human species is the result of a divine creation of which humans are not entitled to usurp this supernatural power (see also Chap. 5, Sect. 5.3.2.4). However, considerations that are not purely religious, for instance, secular philosophical conceptions about human nature or human dignity, may also evoke concern or even opposition to any intergenerational genetic enhancement or betterment of the human species. Such religious and/or philosophical convictions probably lie at the basis of statements as can be found in the Council of Europe *Convention for the Protection of Human Rights and Dignity of the Human Being with Regard to the Application of Biology and Medicine*,[204] in which article 13 stipulates that "an intervention seeking to modify the human genome may only be undertaken for preventive, diagnostic or therapeutic purposes and only if its aim is not to introduce any modification in the genome of any descendants". This position is a typical example, not only of an extremely conservative point of view but also a complete lack of a broad, intergenerational vision of life as a dynamic, evolving phenomenon.

The Equality Question Revisited. As argued earlier, the equality question (see also Chap. 5, Sect. 5.3.2.4) is often raised by many opponents of human enhancement or eugenic policies because of the fear that they will widen the gap between haves and have-nots.[205]

However, it cannot be stressed enough that eugenic endeavours will first of all favour people with higher risks of genetic impairments, less abilities or capabilities to master their fertility and raise children in the best possible conditions—at least if eugenic policies were developed at the population level—thereby achieving improved health and well-being for all. Thus, the first major effect of a eugenic policy is its levelling effect on inequalities by the elimination of heavy genetic impairments that are an important cause of inequality and inequity.[206]

[202]For instance, Roberts (1997), Russell (2010), Phelan et al. (2013).
[203]For instance, Hanson (2001, 294).
[204]Council of Europe (1999).
[205]Sandel (2009, 74), Mehlman (2003), Wolbring (2006).
[206]Bostrom (2003).

Moreover, the importance of between-individual differences should be reiterated, and in particular the existence of higher abilities in some individuals which have occurred throughout history, and will more than ever be of importance for the further progression of human culture and well-being. The well-being of the biologically less endowed or less able depend largely on the presence and opportunities of the genetically privileged. Consequently, as Raymond Cattell wrote in his *Beyondism* several decades ago:

> We should accept individual difference with appreciation and admiration rather than envy.[207]

Last but not least, it should be recalled that eugenics is not only concerned with the genetic constitution of individuals but also the population, and even the human species as a whole. The purpose of eugenics is to push human-specific characteristics towards the higher tail end of their frequency distribution.

The Disability Question Revisited. The disability question is another one of the major social objections by bioconservatives against human enhancement. It is necessary to reiterate the fallacy of the arguments of the disability rights campaigners[208] who reject what Julian Savulescu and Guy Kahane[209] call the Selection Against Disability (SAD) View (see also Chap. 5, Sect. 5.3.2.4). Because of the ramifications of the positions taken by disability activists and bioconservatives for the ethics of intergenerational reproduction, the authors dwell somewhat more in detail on the pending controvercies.

When addressing the disability question, we should first of all try to understand the origin of this phenomenon. In each generation the genes in a reproductive community are recombined in genotype combinations some of which appear to produce phenotypes that show more or less serious disabilities, due to new unfavourable mutations or hidden deleterious alleles from former generations. In addition to genetically caused malformations, accidents during the early ontogenetic development (or even during later phases in life) can further increase the prevalence of disabilities. Most serious forms of disabilities are eliminated by natural selection in early (pre- or postnatal) stages of ontogenetic development[210]—and also in many pre-modern cultures by direct or indirect infanticide.[211] However, some impairments succeed, particularly in the medically and socially highly protective environment of modernity, in surviving the conception or pregnancy stage. As a branch of preventive medicine, eugenics simply wants to correct the imperfect and inhumane action of natural selection by using a more efficient and more humane cultural selection, by choosing that part of the initial biological variation that has optimal chances to live a healthy life; this is selected from the biological

[207]Cattell (1972, 379).
[208]For instance, Kass (2002), Wolbring (2003), Sandel (2007), Garland-Thomson (2012).
[209]Savulescu and Kahane (2009).
[210]Taking into account all prenatal losses (early losses, clinically recognisable spontaneous abortions and stillbirths), it is estimated that more than two thirds of conceptions end in intrauterine mortality (Hertig 1967; Léridon 1973; Macklon et al. 2002).
[211]For instance, Williamson (1978), Birdsell (1986).

(genetic and ontogenetic) variability that becomes visible at reproduction in each new generation. Since we have to severely limit the number of births anyway, at least in a modern context of successful mortality control, we might want to choose the better genetic genomes to constitute the next generation.

In other words, in the knowledge-based and humane oriented era of modernity it would be immoral to take an indifferent attitude towards the genetic and phenotypic diversity that results from our phylogenetic heritage. In its present early stage of development, modernity is clearly still in a transitional phase. Whereas in pre-modern living conditions disability problems were largely resolved in a 'natural' way, namely through spontaneous early mortality and/or non-intervention, now, thanks to the increasing resources for medical and social assistance and knowledge about phenotypic care, modern culture succeeds in substantially prolonging the life expectancy of heavily disabled people; however, this is not always done by redressing their impairment to levels of higher quality of life. In the future, as biomedical sciences progress further, the perspective exists that it will be possible to prevent the development of genetic impairments, and even to largely correct impairments arising during the life course. The development of a disability ethics[212] needs take biological evolutionary and cultural historical developments into account.

Advocates for the rights of the disabled oppose eugenic prenatal testing and the choice of selective abortion because they think that disability is part of the identity of the disabled person and that life with severe disability is worthwhile, or they believe that all diseases are part of the diversity of the human experience.[213] Feminist disability theorist Rosemarie Garland-Thompson[214] even considers a disability as a potentially generative resource rather than an unequivocally restrictive liability. Many scholars perceive here a fundamental difference in valuing life: whereas disability activists give priority to equal and inherent dignity and worth of all human life, eugenists give priority to quality of life for all human beings.[215] However, disability activists confuse moral equality with functional equality.

Disability activists not only completely lack insight into the evolutionary origin of congenital causes of disability, they also make no distinction between caring for people who were born with or acquired serious physical or mental impairments and correcting prenatally such impairments. Efforts to care for people with significant limitations, giving them equal opportunities and integrating them into society, should not be a reason for opposing endeavours to prevent the development of genetically or ontogenetically caused severe impairments: unfortunately, this distinction is not always made in disability rights discussions.[216] In fact, it can be

[212]Glannon (2007).
[213]Rock (1996), Wolbring (2000; 2003).
[214]Garland-Thompson (2002; 2012).
[215]For instance, Adorno (2010, 140).
[216]For instance, Amundson and Tresky (2008).

observed that people, when able to choose, a large majority prefer the avoidance of disability.[217] Most people, indeed, are very eugenic minded and behave accordingly, whenever they have the opportunity to do so and as far as they have not been indoctrinated otherwise.

As argued earlier, wherever modern science develops, disabled people are the first to take advantage of the scientific progress in reducing the suffering linked to disabilities. Fortunately modern science also increasingly succeeds in reducing—by prevention—the prevalence of disabilities.[218]

Disabled people and disability rights advocates should acknowledge that modern science is progressing in its endeavours. Not only are suffering and disability being relieved somewhat but also there is success in preventing more disabilities: disabled people know so very well, on the basis of their unfortunate experience and suffering, how much such a condition can poison one's life and prevent one from experiencing the normal joys of a disability-free existence. The old-time 'Golden Rule' should also be applied here and ethically guide people and especially the disability rights campaigners.

The conclusion of Calum MacKellar and Christopher Bechtel in their book *The Ethics of the New Eugenics* state that "…a compassionate civilised society should learn to accept all possible future children in an environment that reflects its unconditional and equal acceptance of the suffering as well as the happy child". While accepting the inevitable and ensuring care for the misfortunate is laudable, the conclusion of these authors follows from a number of salient *lakunae* in their overall scientific evaluation of the problem, as well as from a number of propositions resulting from traditional philosophical or theological tenets which are no longer well adapted to the novel environment of a further progressing modernity. The authors' main objections to those authors rejecting prevention of disability are as follows:

- The total lack of consideration of present scientific knowledge about causes and consequences of evolutionary mechanisms and processes;
- The lack of any reflection about the confrontation of those evolutionary mechanisms and processes with the novel environment of modernity, which increasingly allows people to control life processes and to avoid the miseries of the genetic and other life event lotteries to which they can be exposed;
- The lopsided and limited consideration of concerns of an individualistic nature, with total neglect of population (societal) or intergenerational dimensions;
- The oversimplified conceptualisation of life characteristics and potentialities by opposing, as mutually exclusive alternatives, equal and inherent dignity and worth of all human life versus quality of life and reduction of suffering, both of which should be thoughtfully taken into consideration. (Opposing 'equality/dignity versus quality of life' as mutually exclusive alternatives is a

[217]For instance, Evers Kiebooms (1987), Haddow and Palomaki (1996), Baruch et al. (2008).
[218]Hughes (2004, 12).

typical example of dualistic thinking, probably a remnant of traditional religious ways of conceiving matters of life and death);

- The conviction about the superiority or the one and only nature of the own worldview (in casu, favouring compassionate equality/dignity principles at the cost of enhancing health and quality of life of each human being), as if an alternative or divergent worldview (in casu, prioritising quality of life) could not equally be a sign of a civilised compassionate society. (Identifying equality/dignity values and norms as signs of 'civilised' societies is not only a very lopsided ideological position but also shows a salient lack of humility. It is a typical expression of cultural in-group thinking);
- The disregard of the fact that that earlier abuses of eugenics had nothing to do with a scientifically based eugenics. (These abusers' views were largely inspired by 'in-group'—nation, race, religion—considerations, whereas a scientifically based eugenics is concerned with the future of the human species as a whole);
- The lack of a broad time perspective in which the dynamics of changing viewpoints and experiences in the fast changing environment of modernity are duly taken into consideration (For instance, in the early decades of the twentieth century, quantitative birth control was initially rejected by virtually all of the renowned and respected political, military, and religious establishment; but in the course of that century, opinions and behaviours evolved towards a general acceptance and application of quantitative birth control. In the present century, the same change might occur regarding qualitative birth control);
- The denial of the reality that people are different, are being valued differently, are being paid differently, are being hierarchically classified according to their societal functions and performances, and that modern societies are organised and functioning on the basis of sorting and selecting procedures all the time and in every societal domain—in the educational system, in the labour system, for sport performances, and scientific and cultural output. (Evoking opposition between the equality/dignity principle and the quality principle in the domain of reproductive behaviour is a typical example of equivocal thinking, whereas this is not done in all other domains of social life);
- The view that the current 'civilised' opinion about the equality/dignity principle would have been largely inspired by traditional (i.e. Christian) values and norms associated with altruism and compassion for others, thereby benevolently ignoring the possibility that this principle might mainly be the result of individualistic selfishness resulting from concerns about the application of sorting and selective procedures to ourselves.

8.3.3.3 Eugenic Methods

There are three dimensions to be addressed in the choice of eugenic methods to be used: (1) level of intervention: individual and/or population; (2) type of eugenic approach: negative or positive eugenic interventions; and (3) type of technique to be used: reproductive behavioural strategies and/or eugenic engineering.

Individual and/or Population Eugenic Strategies. Eugenic action can be limited to targeting individuals or families, identified on the basis of pedigree analysis, who have a larger than a population risk of transmitting undesirable genetic traits. This is what is currently practiced in genetic counselling centres at universities or hospitals where parents with suspected hereditary defects are advised and possibly medically assisted on parenthood. Eugenic action can, however, be broadened by applying population level screening methods in order to identify at the population level the presence of undesirable genetic variants and apply reproductive or biotechnological procedures to prevent the intergenerational transmission of those variants (e.g. screening for sickle cell anemia).

Whereas in the domain of genetic counselling individual (or family) eugenic counselling is quite generally accepted and applied, population approaches are often met with more reluctance.[219] Part of this opposition might be due to the inherent practical difficulties (cost, organisation) of this approach, but some opposition might also be of an ideological nature—resistance against risk of abuse of social engineering policy approaches.

From an evolutionary point of view that is, by definition, focused on the human species, eugenic strategies should be targeted both at the individual and the population level of organisation. Furthermore, several of the major evolution derived ethical principles—the ontogenetic developmental principle, the equality principle, and the quality principle (see Chap. 5, Sect. 5.3.3.1–5.3.3.3)—point in the same direction.

Negative and/or Positive Eugenic Interventions. Negative eugenic methods consist of interventions that are aimed at preventing the transmission of unfavourable genetic traits; positive eugenics wants to enhance the frequency of favourable genetic variants.

Negative eugenics can make use of choice/change of partner, sexual abstention, contraception, genetic-selective abortion, and prevention of industrial practices (use of ionising radiation or of particular mutagenic molecules) that increase the mutational load in the population. Positive eugenics seeks to favour the transmission of desirable characteristics, either through the stronger reproduction of carriers of alleles for such characteristics, or through the use of biotechnological procedures such as medically assisted fertility interventions or, in the future, germinal engineering.

It should be stressed that negative and positive eugenic interventions are not to be seen as alternatives, but as complementary strategies. Hence, it should be negative *and* positive measures, instead of negative *or* positive ones. Although radical opponents to eugenics usually reject both strategies, many geneticists and ethicists who take a more moderate position towards eugenics more often favour negative eugenic measures than positive ones.

Reproductive Behavioural Strategies and/or Eugenic Engineering. The distribution of the genetic variants in a population can be changed through differential reproduction of the carriers of those variants—but only within the existing

[219]For instance, de Jong and de Wert (2015, 49).

range of variation—and/or by methods of eugenic engineering that might in the future also transcend the upper border of the presently existing variation. The first strategy is sometimes called 'the old eugenics', whereas the second one is referred to as the 'new eugenics'.

Reproductive Behavioural Strategies. In modern, democratic society with its increasing social assortment and intergenerational social mobility of available talents in the population, via educational selection systems and job recruiting procedures based on the principle of excellence, the genetic background of those talents —in particular intelligence, sociability and health characteristics—are differentially assorted over the different functional social strata (see Chap. 7, Sect. 7.3.2). In most modern societies the social and educational stratification is still negatively related to reproductive behaviour, although in the course of the progressing demographic transition (and the second contraceptive transition) the reproductive differentials have tended to decrease and level out.

From a eugenic as well as a cultural point of view, the ideal would be to have the existing negative relation between social/educational stratification and reproduction changed into a positive one.[220] Culturally, of course, it is necessary to enhance equal opportunity policies to buffer the negative effects of the genetic lottery.

People with genetic variants contributing to higher cognitive abilities, sociability and health characteristics should not have to encounter barriers in the work place which prevent them from having children. In the short run, facilitating the combination of parenthood with high creativity and social achievements in science and society would have positive cultural effects; in the long run it could also have favourable effects on the genetic composition of the population, despite the inhibitory effects of deleterious mutations, intergenerational recombination of genetic variants, and non-homogamous partner choice. A positive association between social/educational stratification and reproduction would also have a euthenic effect: more children in homes with better environmental opportunities.[221]

There are multiple means to achieve such a reversal in reproductive behaviour, but the most important are probably the dominant value and norm systems that should be changed through educational efforts and modernisation of ethics in the work place. Fertility research shows that highly educated women often want more children than they actually have, but are confronted with deficit fertility due to their long studies, their high professional responsibilities, and their geographical mobility leading to the absence of grandparental, other family and social support.[222]

Eugenic Engineering. A completely different, but in no way opposite, strategy consists of biotechnological interventions. This includes the current compensation for or correction of individual genetic or ontogenetic impairments and also

[220]Whereas Cattell (1972, 392) is in favour of fitting the birth rate to earnings in order to ensure a generally positive eugenic trend, the present authors would rather favour a positive relationship between educational level and reproduction. Earning differences may be partially an outcome of educational differences, but individual wealth is often determined by factors other than personal qualifications or performances.

[221]Cattell (1972, 348).

[222]Avramov and Cliquet (2003, 2005, 2008), Hočevar and Černič Istenič (2010).

improving, at the population level, the (genetic or ontogenetic) capabilities of 'normal' characteristics such as intelligence, sociability, physical attractiveness, sexual performance, physical abilities, longevity, and health characteristics in general—the essential aim of the transhumanist movement.

Eugenic engineering, now sometimes called reprogenetics,[223] includes micro-engineering medical techniques that influence the transmission of genes to the next generation or future generations via procedures such as medically assisted fertility or germinal gene therapy.

Medically assisted fertility is already comprised of a number of techniques which positively intervene in the generation of an embryo: artificial insemination by donor (AID), ovum donation, in vitro fertilisation (IVF), embryo transplant, embryo selection, and sex choice.[224] Whilst some of these methods may be intended for countering infertility—a disorder that is, however, not always genetic—all of them can be used to replace unfavourable characteristics and have genetic effects.

The discovery of the structure of the organic molecules responsible for genetic transmission (DNA) by Watson and Crick in 1953, together with the emergence of molecular genetics with its applications in biotechnology, undoubtedly provides the means for the future development of germinal gene therapy. This is the technique in which sex cells are treated with recombinant DNA in order to genetically alter germline cells to replace unfavourable genes with wanted genes.[225] Whilst changes to somatic cells affect only specific cells within a given organ system and are not passed on to future generations, germline changes affect every cell in the body and are passed on to future generations.

The genomic era of medicine began in 2003, fifty years after the first description of the structure of DNA, when the Human Genome Project completed the sequencing of the human genome.[226] Knowledge about the normal and pathologic variation for all human genes will allow the diagnosis of genetic diseases and in the future it will lead to effective treatments for them, particularly at the level of the genes themselves. It is generally acknowledged that, at the present stage of development of applied human genetics, human germline therapy is still too risky.[227] However, the recently developed powerful genome editing technique—known by the acronym CRISPR-Cas9—opens up promising perspectives for fast advances in the field of human gene therapy.[228] Nevertheless, researchers have also expressed the necessity to progress this with prudence so that genomic engineering and germline gene modification can be performed safely and ethically.[229]

[223]Silver (2000, 375).
[224]Mashiach et al. (1990), Robertson (1994), Gardner et al. (2004).
[225]Wheale and McNally (1988), Friedmann (1998), Stock and Campbell (2000), Stock (2002).
[226]Guttmacher and Collins (2003).
[227]Reiss and Straughan (1996), Stock and Campbell (2000), Stock (2002).
[228]Doudna and Charpentier (2014).
[229]Baltimore et al. (2015), International Summit on Human Gene Editing (2015), Doudna and Sternberg (2017).

As already explained in Chap. 5, Sect. 5.2.3.4.2, in their anti-enhancement crusade, bioconservatives are particularly opposed to germinal gene therapy[230] but their arguments have been amply refuted by enhancement proponents.[231] Contrary to what is sometimes envisioned, eugenic engineering will not harm future diversity and evolvability of the human species, but will "ensure rather than undermine its health and longevity".[232]

8.3.3.4 Eugenics: Who Decides?

A preliminary remark concerns the general social climate in which a eugenic policy can be developed. Sometimes, eugenics is labelled as a right wing, reactionary, conservative ideology that seeks to preserve the prerogatives of the ruling upper classes. It is true that a large part of the early twentieth century pseudo-eugenicist movement was class and race biased. However, the later science-based eugenics, as developed in what some historians now call 'reform eugenics' was promoted by socially minded scientists such as Havelock Ellis, J.B.S. Haldane, Julian Huxley, Hermann Muller and Frederick Osborn—all of whom advocated the improvement of the stock of the human species through selective breeding.[233] Even Francis Galton's principal successor in eugenics, Karl Pearson, was a 'Darwinian' socialist.[234]

Prominent eugenic thinkers, such as Hermann Muller, Julian Huxley, Frederick Osborn, Jean Sutter, Raymond Cattell and Carl J. Bajema,[235] were even of the view that it would be impossible to achieve eugenic goals in societies that were based on *laissez-faire* economics or authoritarian regimes. Democracy, guaranteeing personal freedom and a humanistic individual development by having a generous social protection system, was considered a necessary condition for sound and humble eugenics.[236] Socially motivated eugenicists also strongly favour female emancipation[237] and the widespread availability of birth control.[238] As has been seen in Chap. 5, Sect. 5.2.3.4, the transhumanists also attach great importance to democratic and generous social policies as a basic condition for achieving their eugenic goals.

Coercion versus free choice in the application of eugenic measures is one of the oldest debated ethical issues in the history of eugenics. This problem is not specific to eugenics because it applies to many other areas of social life, especially medical practice, but in the case of eugenics it always entails intergenerational implications.

---

[230]For instance, Kass (2002), McKibben (2003), Mehlman (2003), President's Council on Bioethics (2003), Sandel (2007), Sutton (2012).
[231]For instance, Fletcher (1974), Lyon and Corner (1995), Walters and Palmer (1997), Stock (2002), Baylis and Robert (2004), Glover (2006), Allhoff (2008), Lindsay (2008), Potter (2010).
[232]See, for instance, the discussion in Powell (2012).
[233]Paul (1998).
[234]Kevles (1985).
[235]Muller (1934), Huxley (1936, 1962), Osborn (1940), Sutter (1950), Cattell (1972), Bajema (1972).
[236]Osborn (1940).
[237]For instance, Muller (1960).
[238]For instance, Osborn (1940).

Individual decision-making—sometimes called liberal eugenics[239]—is often both opposed to and favoured over state regulated eugenics,[240] although both are identical in intentions in that they decide about the enhancement of other people.[241] Both individuals and states may make mistakes, although it must be admitted that the mistakes of governments can have much more serious consequences than those of individual citizens.[242]

An ethical dichotomy—individual free choice versus social (state-regulated) compulsory application of eugenic practices—is too simplistic a view of the ethical approach towards eugenics. In fact, it is possible to see a broad spectrum of gradually differentiated positions between these two extreme positions. At the one extreme—free choice—individuals may receive non-directive genetic counselling and make autonomously informed decisions about their future reproductive behaviour. The role of the genetic counsellor is defined as that of a neutral, non-judgemental information provider rather than a decision maker—a definition that would, according to Alison Pilnick,[243] enjoy widespread agreement amongst genetic counsellors in the United States.[244] At the other extreme—compulsory eugenics—individuals (identified as having a genetic risk) are required to participate in genetic counselling or screening and are, in cases of a diagnosed individual or family risk, obliged to follow legal provisions such as sterilisation or abortion.

A variety of other, more nuanced positions are possible in between these two extremes. For instance, whilst leaving the ultimate choice, in principle, to the patient, genetic counsellors could exert 'soft coercion'[245] or 'persuasion'[246] in convincing the patient to follow the counsellors' advice or decision. This viewpoint means that counsellors could positively interfere in the decision-making process in the interest of the patients, their families, and society overall.[247]

Compulsory rather than individual free choice might, but should not necessarily, be more common where the approach to eugenics is more socially oriented than individually or family oriented. In principle, however, all possible ethical positions regarding eugenic practices could be applied both in individually and socially oriented eugenics. It appears that when people can choose, a large majority select the avoidance of disability in their offspring.[248]

The best solution would probably be to (1) favour an educational/agogic approach—informing and motivating future parents to consider possible eugenic effects in their reproductive decisions; (2) promote or impose obligatory genetic

[239]For instance, Habermas (2001), Agar (2004), Sparrow (2011).
[240]For instance, Agar (1998), Savulescu (2001), Bostrom (2003; 2005), Glover (2006).
[241]Sorgner (2009).
[242]Bostrom (2005, 206), Glover (1984, 48).
[243]Pilnick (2002).
[244]See also the position of the Board of Directors of the American Society of Human Genetics which affirms its commitment to the fundamental principle of reproductive freedom and unequivocally declares its opposition to coercion based on genetic information (Reilly 1999).
[245]Bajema (1971).
[246]Savulescu (2001).
[247]Staatscommissie voor de Ethische Problemen (1976).
[248]For instance, Evers-Kiebooms (1987), Haddow and Palomaki (1996).

counselling/genetic screening; (3) keep in principle the ultimate decision as a right but also as a responsibility of individuals; and (4) set societal limits to genetic choices which could harm the children's well-being.

A particularly difficult issue arises in cases where patient(s) are (develop)mentally or socially incompetent to make a free and responsible decision about having children when there are high genetic risks. However, this question applies not only to cases where there are genetic risks but also to many other domains of social life (e.g. inheritance, parenthood in general, medical interventions, euthanasia). In the current stage of modern cultural development, the effects of the ideology of individualism on education policies and medical practice, as well as attitudes toward eugenics, cause disputes amongst professionals and the public at large.

Another disputed issue is the question as to whether parents have the right to choose the sex of their offspring—a question that is, strictly speaking, not a eugenic issue (see also Chap. 6, Sect. 6.3.3.2). Sex, indeed, is not a disease, but why should parents who already have one or more children of the same sex not be allowed to satisfy their 'king's/queen's wish' and have a child of the opposite sex? At the social level, this issue is clearly more a question related to sex relations and demographics: in some societies, where discriminatory attitudes toward girls are still prevalent, this may result in serious distortions of the sex ratio, as is, for instance, currently experienced in China and India.[249] It is understandable that many feminists fulminate against the practice of sex selection instead of focusing on social discrimination of girls and women.

Another series of ethical problems in eugenics concerns the issue of informing the patient about his/her risks of developing a genetic disorder[250] or his/her risks of transmitting the disorder to offspring. An even more tricky issue concerns the question as to whether relatives should be warned, possibly against the patient's wishes.

Related to the issue of the right to genetic privacy is the question of informing third parties, for instance with regard to job recruitment. Employers, both in the public and private sectors, have an interest in recruiting healthy employees who have a low risk of suffering from a long-term disease. Medical examinations are now a common part of the normal selection procedure for some long-term jobs. With the increasing possibilities of identifying the presence of genes that, in time, could cause unfavourable phenotypic effects, the question arises as to whether genetic screening procedures should be included in recruitment medical examinations.[251] This problem also emerges in the case possibly using genetic test results in insurance, with the risk of insurance clients being treated unfavourably on the basis of their genetics.[252] This question already exists for health conditions in general, since many insurance contracts are contingent on the absence of known phenotypic health risks.

[249]Johansson and Nygren (1991), Arnold et al. (2002).

[250]For example, with regard to Huntington's chorea, see Elger and Harding (2003).

[251]For instance, Draper (1991), Rothstein (1995).

[252]For instance, Joly et al. (2003), Zick et al. (2005), Wilkinson (2010).

It is understandable that germinal engineering often meets with more opposition and requires a more cautious approach than somatic engineering. Eugenics not only plainly contradicts the traditional conception of the creation of human life but it also has implications for the genetic composition of future generations.[253]

Implementing demographic selection poses several delicate problems, mainly related to the type of demographic method used to achieve the aims—such as partner choice, fertility, generation length and mortality. Whilst expanding personal freedom and opportunities in partner choice could have positive genetic effects,[254] the increasing claims for sexual and reproductive rights for all—including people with severe handicaps[255]—might have adverse effects if these rights are not combined with responsibilities and obligations for kin and society to assist these parents and their offspring.

Genetic selection by the means of mortality involves the delicate problem of induced abortion. Although a majority of countries in the world have already adopted laws liberalising pregnancy interruption for therapeutic reasons, including foetal indications, abortion will remain a controversial issue for some time, at least in those countries that have not yet modernised their legislation[256] or in countries where pro-life lobbyists actively try to reverse the existing pro-abortion legislation.

Achieving a differential fertility raises the problem of the methods to be used: free choice, education or coercion. In a value and norm system that prizes the enhancement of individual emancipation and autonomy, only free choice and education can be advocated. However, what happens if free choice and education cannot help? What are the social and intergenerational implications when the most educated and best trained citizens prefer to practise art or science, or to pursue other lifestyles, rather than raise children? What if persons with low cognitive skills cannot learn to realise that fertility limitation contributes to their own well-being, the well-being of offspring and the well-being of society? Some have expressed the opinion that in a number of cases soft coercion cannot be excluded[257] or that active intervention by educational authorities, welfare workers and public health workers is necessary.[258] In the authors' view much discussion, reflection and ethical consideration is still pending on these issues.

Now that opportunities for the qualitative control of births and deaths are increasing with the development of biotechnology, a societal debate on the limits to the use of these powers is inevitable and absolutely necessary. It is to be expected that the initial discussion of this will include a profound ideological debate, confrontation and political struggle. Although biotechnology is still in its early stages, it has already given an enormous impetus to the development of bioethics. Whereas there can be differences in views about the methods to achieve the eugenic goals, there should be no dispute as to the goals themselves.

[253]Stock (2002).
[254]Epstein and Guttman (1984).
[255]United Nations (2006).
[256]United Nations (1995).
[257]Bajema (1971).
[258]Staatscommissie voor de Ethische Problemen (1975).

From an evolutionary-based ethical point of view, it is essential that in matters of genetics, not only the interests and well-being of the individual making choices but also the health and welfare at the population level are considered; but above all, attention should be given to the well-being of future generations.

8.4 Complementarity of Quantitative and Qualitative Reproductive Goals

At first sight, the reader might be confused by the apparently opposite recommendations concerning quantitative and qualitative reproductive behaviour in a further progressing modernisation.

Indeed, on the one hand the authors argue in favour of quantitative birth control, bringing fertility levels (temporarily) below replacement level in countries or regions that are still confronted with high population growth or density, which prevents them from reaching the developmental levels of the most advanced nations in the world. On the other hand the authors advocate creating conditions and opportunities for creative and productive people to have children. This may seem to be a salient paradox.

However, on the one side fertility research has extensively documented that a high proportion of births are unwanted among people who do not know how or have no access to modern contraception; on the other side research has documented that high proportions of highly educated childless women had never decided not to have children. They have encountered obstacles in combining their careers with having children. Both outcomes, excess fertility and childlessness, raise the question of equality of opportunity.

Furthermore, it is not contradictory to simultaneously argue in favour of limiting fertility to replacement levels, or even decreasing temporarily fertility to below-replacement levels, as well as arguing for the maintenance of a variance in fertility within a certain range. From an ethical point of view, well adapted to the living conditions of a further progressing modernisation, there is no contradiction in promoting a quantitative birth limitation policy and a qualitatively oriented fertility policy—just as there is no contradiction between a policy favouring enduring partnership (via marriage or consensual union) and the legislation facilitating divorce, or between an educational policy avoiding abortion and the legalisation of abortion.

Changing the genetic composition of a population (the gene pool of a population) is a matter of differential reproduction, which is independent from the observed or desired demographic growth model. A qualitatively focused differential reproduction can be achieved in a demographically shrinking, stationary, or expanding population.

What are the chances that the quantitative birth control of the twentieth century will be perpetuated and generalised at a world scale, leading to a decrease in the world population (in view of developmental and ecological goals), and be complemented by a qualitative control (in view of enhancing the human-specific potentialities)?

There can be no doubt that those two population goals will be strongly opposed in some circles.

With respect to quantitative birth control, many politicians and economists still cherish (explicitly or implicitly) the old-time population growth ideology: they are not able to adapt to the novelty of modernity, or to see political, economic and ecological phenomena interacting at a planetary level. Fierce opposition to a (temporary) population decrease can also be expected to come from ideological groups, and particularly from religious groups, who still want their in-group to multiply and "fill the earth", conveniently forgetting or denying that the Earth is already overcrowded and plagued by excessive inequalities.

Opposition to the qualitative control of reproduction may be expected to come from some of the major traditional religious ideologies who continue, very selectively, to oppose cultural interventions on 'God's creation'. However, opposition can also be expected from some modern political ideologies such as liberalism that is predominantly focused on the individual; or socialism that cannot free itself from its nineteenth century (then justified) focus only on the deprivations of the least privileged classes; or some feminist fractions that perceive a menace from male dominance in every innovative intervention in the domain of reproduction.

Nevertheless, the shift to a qualitative control of births is already underway, pursued by individuals and couples who want to avoid the transmission of genetic impairments. In the future, when germinal engineering becomes available, many prescient people will certainly also endeavour to enhance 'normal' features of their offspring—just as individuals and couples started in the twentieth century controlling the number of their conceptions or births before the established ethical-religious and political rulers approved it.

It could well be that some cultures or regions in the modernising world will progress faster on this path than others. For instance, it would not come as a surprise that the modernising nations that are under the influence of the Eastern philosophies would progress much faster in adopting qualitative population policies than the countries which developed under the aegis of the Mediterranean religions. This might occur despite the fact that, as a consequence of the Enlightenment and the development of science, these countries initiated the eugenic, transhumanist and other qualitatively oriented ideologies. In the end, and as an intermediate stage, Raymond Cattell[259] might be partially or temporarily right with his idea of 'co-operative competition' as a grand experiment of deliberately setting up bioculturally diversified groups.

Hence, 'evolutionaries'[260] will have to make considerable efforts to disseminate their ideas and convince ethicists, religious institutions and especially policy makers about the utility of the two population goals to be pursued in the domain of reproduction: decreasing numbers and increasing capabilities of people.

[259]Cattell (1972, 106).
[260]Stewart (2008).

Conclusions and Final Reflections 9

Abstract
In this last chapter the major conclusions from the individual chapters in this book are summarised and integrated. The chapter closes with a section about the need to reconcile traditional and modern ideologies. It is argued that the evolutionary approach is a good way to bring together religious and secular ideologies in order to reflect upon and develop a new global ethics focussed on a long-term evolutionary perspective. It systematises the questions that need to be addressed regarding the values and norms to be adapted and shaped for the future successful enhancement of human potentialities conducive to further hominisation.

9.1 Ideological Conflicts in the Modern World and the Need for a Universal Ethic

In this book it is argued that, at the start of the third millennium, the populations living in the modernising environment are confronted with ideological clashes between traditional religious supernaturally revealed moral codes, modern secular ideologies that assign to humans the responsibility for developing ethical standards, and findings from science that are posing new challenges both for religious and secular worldviews. Furthermore, ideological conflicts are observed within as well as between states.

Both the traditional religions and the current secular ideologies are considered by the authors to be partly inadequate to deal with the ethical problems of humankind in modernity and to safely guide the human species through new subsequent stages of biological evolution and cultural development. The traditional religious

R. Cliquet and D. Avramov, *Evolution Science and Ethics in the Third Millennium*,
https://doi.org/10.1007/978-3-319-73090-5_9

ideologies of the pre-scientific era, which fulfilled an important survival function, are in many respects no longer well adapted to the novel environment of modernity and its exigencies for further human biological evolution and cultural development. The modern secular ideologies approach moral dilemmas and challenges in modernity in a strongly fragmented and often conflicting ways promoting only vested interests of specific groups. They usually lack a vision about the long-term future and focus on single issues or specific population groups. They underachieve in applying or even strengthening moral principles and practices at the interpersonal levels.

It is argued that dealing with the ethical challenges of humankind in modernity, and considering future opportunities opened up by science and technology, necessitates a universal morality that transcends the in-group foundations of religions and blends the fragmented secular ideologies. Religions claim universality but in fact they are in-group institutions concerned with winning over and enlarging the pool of the faithful. The universal morality of the third millennium needs to transcend in-group institutions, be it religious or secular. Evolution science can, in the context of a further progressing modernisation, provide the framework for reflecting upon and developing such a comprehensive ethic. The emergence and development of evolution science, with its consecutive Darwinian Revolution, Modern Evolutionary Synthesis, Second Darwinian Revolution and molecular revolution, provides a solid, naturalistic basis for embedding values and norms in order to secure a safe and progressive future ontogenetic development and phylogenetic evolution of humankind. Knowledge about the hominisation process provides an evolutionary indicator that makes it possible to distinguish future alternative directions in which humanity could evolve and make rational choices about the future trajectory to opt for. Likewise, the characteristics of the modernisation process make it possible to differentiate between various possible future orientations and make knowledge-based choices about the desirable directions to pursue.

Reflections about the value and norm changes needed, from a progressing evolutionary perspective, to adapt to the opportunities and challenges of a further evolving modernisation process, will need to be elaborated on the basis of the confrontation of the hominisation process with features of the modernisation process.

9.2 Evolution-Based General Ethical Goals for the Future

For any future of humankind the general evolution derived prerequisites are the preservation of ecological sustainability and the cultural furtherance of the modernisation process.

The central goal that the authors argue should be set is the phylogenetic enhancement of the hominisation process.

General derived goals are: the ontogenetic development of human-specific potentialities; the promotion of quality of life; the promotion of equal opportunities; the shift from strongly competitive toward more cooperative relations; and the promotion of universalism.

9.2.1 Preservation of Ecological Sustainability

Humanity is in a critical stage due to the disruptive human impact on life and the environment of this planet. It is facing serious, anthropogenically caused, ecological challenges such as overpopulation, overconsumption, abuse or depletion of natural resources, environmental pollution and climate change, and decimation of biodiversity.

The commitment to a long-term ecological sustainability is considered a *sine qua non* for the survival, the long-term development and the further evolution of the human species on this planet. Crucial components of the physical environment need to be preserved such as inhabitable and arable land, water, and climate. On the one hand we are in competition with many other life forms; on the other hand the human species vitally depends on the existence of other life forms for breathing, nutrition, digestion, and even for more subtle psychological needs. The priority ranking of the ecological goal as one of the two ethical prerequisites for the future development and evolution of humankind is, from an evolutionary perspective, fully justified.

9.2.2 The Cultural Furtherance of the Modernisation Process

An essential component of a future evolution-based morality concerns the fact that the modernisation process is a positive achievement of humankind, despite some of its temporarily unfavourable but corrigible side effects. The idea of a further progressing cultural furtherance, which would spread all over the world, has been challenged by the hypothesis of some scholars who argue in favour of the development of multiple modernities. However, the authors are of the opinion that the latter can only be a temporary or intermediate stage in the future cultural development of the human species, because the modernisation process is mainly driven by the progress of sciences which are the most powerful unifying element and universal mediator in modern culture.

The justification for the choice of a progressing modernisation is based on the fact that modernity is not only the current apogee of the cultural trajectory the hominins have achieved over time but also that it succeeded, better than any of the previous cultural stages, in mastering extrasomatic threats, satisfying human needs and achieving unprecedented high levels of welfare and well-being. Hence, the furthering of the modernisation process, in its enlightened or matured form, is considered as a priority goal of an evolutionarily founded ethical approach.

9.2.3 Phylogenetic Enhancement of the Hominisation Process

The central goal of the evolution-based ethics proposed here consists of the phylogenetic furtherance of the hominisation process for the long-term future of humanity, ultimately resulting in the evolution of posthuman stages. It is an old idea that, in recent years, has been actively revived by the transhumanist movement.

The rationale for choosing a further progressing hominisation goal is that the past direction in human evolution is characterised by a process of increasing evolutionary complexity that has resulted in an enlarged potential to understand the world, to adapt better to environmental diversity and challenges, to master our own biology and our environment, to increase our survivability and longevity, to satisfy our needs and desires, and to reach higher levels of quality of life, welfare and well-being. The extrapolation of this process will enable a further increase in the human-specific potentials and, through them, further enhance the well-being of humans. It is assumed that an evolved hominin stage beyond the present *Homo sapiens sapiens* stage might put future humankind into a better position to cope with the cosmic, biological and sociocultural challenges which will confront it in the future.

9.2.4 The Ontogenetic Development of Human-Specific Potentialities

Human ontogeny not only depends on its phylogenetic heritage but also on the physical, biological, social and cultural environment in which it develops. It can be subject to considerable variation, going from a minimal to a maximal development.

The evolutionarily founded central goal—the enhancement of the hominisation process—implies that not only can the ontogenetic development of human-specific potentialities be achieved at its highest possible level, but also that the further ontogenetic enhancement of human-specific features can be pursued at that highest level for more if not all people.

9.2.5 The Promotion of Quality of Life

Just as the nature of the phylogenetic goal in an evolutionarily embedded ethic—the enhancement of the hominisation process—defines the nature of the ontogenetic goal —the development of human-specific potentialities at the highest possible level—this ontogenetic goal, in turn, results in the pursuit of the maximisation of quality of life.

In the pre-modern stages of human evolution and history, the pursuit of high quality was often supported by achieving high quantity and the survival of the fittest. However, modernisation has already boosted world population size and consumption patterns to high levels, the combination of which now surpasses the ecological carrying capacity of the planet. In a further progressing modernisation, the traditional positive relationship between quality and quantity may have to be reversed to a negative one. Namely the further pursuit of ever higher levels of quality of life for the whole of humanity will have to be achieved at the expense of quantitative goals, in particular regarding population size.

9.2.6 The Promotion of Equity

Genetic variation is, from an evolutionary point of view, an extremely important phenomenon because it creates opportunities for adaptation in changing environmental living conditions. At the same time, genetic and other causes of biological variation do produce inequalities in potentialities that may be a source of social inequality, social friction, societal malfunctioning and social injustice.

Modernisation is characterised by strong efforts to reconcile biological diversity with the ideals of equity by creating equal opportunities. An evolutionary-based ethics will, in the context of a further progressing modernisation, simultaneously have to positively value some aspects of genetic diversity and value social institutions that support diversity but which also enhance equal opportunities.

9.2.7 The Necessary Shift from Competitive Towards Cooperative Behaviour

The hominisation process is characterised by a shift from competitive towards cooperative behaviour, although humans remain endowed with drives for both competition and cooperation.

The modernisation process has accelerated the trend towards increased cooperation. However, the genetic predispositions selected and adapted to deal with social interactions in and between small populations are not in equilibrium with human cooperative requirements in the novel modern environment. A further progressing modernisation will require an ever increasing need for cooperation at the world level.

9.2.8 The Promotion of Universalism

The specific human genome emerged and evolved in Pleistocene times, when people lived in small tribes and interacted with other small tribes, resulting in the development of strong predispositions towards in-group oriented behaviour. However, human evolution and history is marked by a gradual expansion of our circle of moral considerability that substantially transcends the narrow ‘in-group’ oriented human inner drives: this goes from the tribal level, to the nation level, and then to the level of the human species in its entirety.

An evolutionary-based ethics, which must operate in a further progressing modernisation, should be of a universal nature: this means that moral rights and responsibilities should relate to the whole human species.

9.3 Evolution-Based Specific Ethical Challenges

In this book, evolution-based specific ethical challenges have been discussed according to the source of biosocial variation to which they pertain, hierarchically classified according to their degree of complexity: (1) individual variability; (2) group relations; and (3) intergenerational replacement. For each of those aspects, the evolutionary background, developments in modernity and ethical reflections for the future have been discussed.

The evolutionary approach—within the framework of a further progressing modernisation—is an adequate method to identify crucial ethical challenges and designate the direction in which solutions could be pursued or identify which gaps in the moral codes need to filled in order to address the challenges of the third millennium. This applies to the major features of individual variability—sex, age and individual differences in general—as well as the various forms of group relations—kinship and family relations; social status hierarchies; racial, ethnic and ideological differences; and international relations. The long-term evolutionary approach is particularly relevant for intergenerational replacement and intergenerational solidarity.

9.4 Reconciling Religious and Secular Ideologies and Evolutionary Goals

The ethical foundations, and some of the goals, of the religious ideologies are largely maladapted to modernity. The secular ideologies are fragmentary and currently incapable of providing a coherent response to the future evolution and development of humanity. A huge number of current and expected challenges are not addressed by the current moral norms, which are lagging behind and fail to cope with new situations stemming from scientific knowledge and technology. Moreover, both the traditional religious and the modern secular ideologies seem either to have lost or not yet acquired substantive moral authority. All this implies, at first sight, that the future of the human species (and our planet) looks pretty bleak. For some demographic and ecological challenges there is not much time left to address urgent problems in order to avoid further damage.

Nevertheless, it can be expected that both the traditional religious belief systems and many modern secular ideologies will still be with us for quite some time. The persistence of religious traditions will continue to direct or influence the behaviour of believers and play an important role in the lives of many people; this is either because of their neurological predisposition to think and feel in such a way and/or because of the power or action of powerful religiously controlled societal institutions. The diversity of modern fragmented secular ideologies, resulting from individual or (in-)group interests or the incapacity to devise a grand vision for the future, will also continue to keep busy and divide the minds and actions of many agnostics and atheists and will perpetuate their mutual ideological conflicts.

However, when looking at the relatively short historical time span that the modernisation process has been embedded in science, there is room for moderate optimism. Modernisation has, in its short history, enabled many fundamental changes with profound ethical implications. They include abolition of slavery, regulation of labour conditions, removal of many barriers to upward social mobility, and achievement of a higher standard of living and quality of life for large segments of the population. Changes have taken place hand in hand with the growth in opportunities for self-actualisation, guaranteed freedom of individual expression, greater autonomy in the choice of personal relations, promotion of female emancipation, family planning, and equal gender rights. Societies have achieved the advancement of ideological pluralism and tolerance, establishment of democratic governance, enlargement of environmental awareness and development of intergovernmental cooperation.

In the course of modernisation, many traditional ideologies have in several respects adapted to the findings and opportunities of the sciences, achievements of technology, and social movements that strived towards more social justice and equity.

Adaptation is often labelled in such a way that it reconciles traditional and scientific approaches. By way of example, most Christian denominations use medical science to lower the suffering of their believers in the terminal stages of life. They have developed palliative care practices in cases of incurable disease, which is, in fact, a 'soft' and prolonged form of euthanasia. Furthermore, many religious people and organisations are very progressive with respect to promoting equal opportunities for all, and cooperation between individuals and societies, also at the global level.

Regarding the views of the Catholic Church towards evolutionary science, some change can be discerned. In the papal encyclical *Humani generis* of 1950 "concerning some false opinions threatening to undermine the foundations of Catholic Doctrine", Pope Pius XII speaks about the "doctrine of evolution… as if the origin of the human body from pre-existing and living matter were already completely certain and proved by the facts" and rejects polygenism (understood as "Adam represents a certain number of first parents"). In 1996 Pope John Paul II, addressing the Pontifical Academy of Sciences, stated: "Today, more than a half-century after the appearance of that encyclical, some new findings lead us toward the recognition of evolution as more than a hypothesis." Furthermore, Pope Benedict XVI stated at a meeting with clergy in 2007: "… there are so many scientific proofs in favour of evolution which appears to be a reality we can see and which enriches our knowledge of life and being as such." Nevertheless, all three popes continued to make a dualistic distinction between the evolution of the human body and the creation of the human soul, not willing or unable to accept that the human mind/spirit/soul is irrevocably linked to and dependent upon the material basis and evolution of the human brain and body.

A fact of considerable importance is the diversity in the adaptation process of religious tenets to the findings and insights of modern science. Some denominations (for instance mainline Protestants) seem to have no major problem in assimilating and adapting to scientific progress; others (for instance the Catholic Church) are clearly struggling more with the novelty of modernity; still others (for instance Christian Evangelicalism in America and fundamentalists in Islamic countries) still adhere to the literal interpretation of their Holy Scriptures. Even within many religious groups, diversity exists in theological interpretation of the Holy Scriptures.

It may be expected that some countries, particularly in the Islamic sphere of influence, will have more difficulty in adapting to modernisation and to adopt fully modern science together with its societal and moral implications. Many of these countries—where the Holy Scriptures are literally interpreted—accept technology but reject science. Indeed, Islamic countries struggle with three obstacles on their way to modernisation—philosophical, social and political. The philosophical obstacle relates to the very nature of Islam with its specific content which is in so many conceptual respects fundamentally challenged by the achievements of science; the social obstacle concerns the fact that science and modern secular ideologies were not developed indigenously[1]; finally, the political obstacle is that modernisation/Westernisation was/is introduced, if not imposed, by ideologically different and technologically superior foreign invaders, colonisers, and exploiters, which stimulated a strong in-group reflex.[2]

However, an important phenomenon is the historically well-documented change in attitudes and behaviours of many religious believers during the modernisation process, which was contrary to the doctrinarian positions of their churches or belief systems. Let us take for example, the contraceptive transition (the shift in use of inefficient to efficient contraceptive methods) among Catholics, as revealed through several fertility surveys in the second half of the twentieth century. Although the Catholic Church only allows the use of so-called 'natural' methods of family planning, the large majority of Catholic practitioners, albeit somewhat later than non-believers, switched their contraceptive behaviour from traditional to modern contraceptive methods.

In fact, there is no reason why religions, which believe that a supernatural power has created humankind and equipped it with mental powers to organise and develop its life and future, would not be able to find rational arguments in theology to adhere to the central goal of an evolutionary ethics, i.e. the enhancement of the hominisation process. In the pre-scientific era, the main biosocial function of most major religions consisted of furthering the survival of human societies.[3]

---

[1]See Edis (2007, 219).
[2]Hashemi (2010).
[3]See, for instance, Cole-Turner (2011), Green (2013), Haught (2013).

Furthermore, as modern secular ideologies progress in using findings and insights from science, and more comprehensively broaden their ideological horizon to demographic, genetic, ecological and intergenerational levels, there is even more of a reason to adhere to a long-term conceptualised evolutionary ethics rather than the traditional in-group ideologies.

Convincing religious and secular ideologies about including or integrating an evolutionary ethical approach in their basic philosophies is just one step towards enhancing the potential for universal ethics. Another step is to encourage them to move together in acting in an evolutionarily ethical direction. The same applies to several of the secular ideologies whose goals are often diametrically opposite. Indeed, our modern societies are characterised by an ideological pluralism, composed of several religious faith systems as well as different secular convictions. Religions or religious denominations compete, often fiercely, among themselves about the right God and his prophets; moreover, they also fight secular ideologies, in particular those that do not believe in an almighty and benevolent supernatural power. Secular ideologies, in turn, often exhibit a strong aversion to the religious doctrines and the religions' intrusion in a person's life, wherever those religions succeed in maintaining their political power positions. This repulsion is particularly salient in what is now called the 'New Atheism'[4] that dissects and discloses the inconsistency and maladaptiveness of the traditional (revealed) religions in modernity.[5] In turn, several secular ideologies fight each other fiercely—for instance, liberalism versus socialism, ecologism versus capitalism, egalitarianism versus equal opportunities/meritocracy—and lack a solid foundation for common action.

However, it appears that dividing lines in ethical and political decision-making often do not run between religious and non-religious people, but between conservatives and progressives. Among the progressives one finds freethinkers/atheists as well as religious believers who, confronted with present-day problems and taking into account present-day scientific knowledge, appear to find their motivation for innovative global moral positioning or political action in their own religion or ideology. The concrete moral or political outcome of their choices is often very similar. To the extent that both freethinking people and religious believers rely on, or take into account, scientific knowledge in their moral or political decision-making, their ethical viewpoints tend to converge, thereby considerably facilitating the establishment of common ground for action.

Hence, instead of ideologically confronting the traditional religious and modern secular ideological approaches and institutions, it might be more rewarding to take a pragmatic attitude and examine how to mobilise the religiously or secularly inspired drives for moral agency for societal reforms; this would be based on an evolutionarily founded analysis for pushing human development and evolution towards higher levels of quality of life, welfare and well-being. It would be a giant step

[4]Philipse (1995), Harris (2004, 2010), Dawkins (2006), Dennett (2007), Hitchens (2008, 2009), Stenger (2008, 2009), van den Berg (2009), Verhofstadt (2013).
[5]See also Teehan (2010, 204), Zuckerman (2010, 13).

forward if the major, ethically-driven, progressive secular and religious institutions in the world took such a position. Recently, May Evelyn Tucker[6] pleaded in the same way for making progress in the domain of an ecologically sustainable future.

In parallel, given the often rigid philosophical or theological doctrinarian nature of most ideologies, and in particular their institutions as vehicles of power, there is a strong need for: (1) interreligious dialogue between the different religious belief systems; (2) consultations between various mutually opposed secular ideologies; and (3) cooperation between religious and secular ideologies.

Several initiatives—including the global level—already exist in the domain of interreligious dialogue. For instance, in 1993 the Council for a Parliament of the World's Religions[7] produced a declaration "Towards a Global Ethic: An Initial Declaration"[8] in which a common set of values found in the teachings of the

[6]Tucker (2015, 401).

[7]www.parliamentofreligions.org/.

[8]Global Ethical Foundation (1993), Declaration Toward a Global Ethic: "The world is in agony. The agony is so pervasive and urgent that we are compelled to name its manifestations so that the depth of this pain may be made clear. Peace eludes us—the planet is being destroyed—neighbors live in fear—women and men are estranged from each other—children die! This is abhorrent. We condemn the abuses of Earth's ecosystems. We condemn the poverty that stifles life's potential; the hunger that weakens the human body, the economic disparities that threaten so many families with ruin. We condemn the social disarray of the nations; the disregard for justice which pushes citizens to the margin; the anarchy overtaking our communities; and the insane death of children from violence. In particular we condemn aggression and hatred in the name of religion. But this agony need not be. It need not be because the basis for an ethic already exists. This ethic offers the possibility of a better individual and global order, and leads individuals away from despair and societies away from chaos. We are women and men who have embraced the precepts and practices of the world's religions: We affirm that a common set of core values is found in the teachings of the religions, and that these form the basis of a global ethic. We affirm that this truth is already known, but yet to be lived in heart and action. We affirm that there is an irrevocable, unconditional norm for all areas of life, for families and communities, for races, nations, and religions. There already exist ancient guidelines for human behavior which are found in the teachings of the religions of the world and which are the condition for a sustainable world order. We declare: We are interdependent. Each of us depends on the well-being of the whole, and so we have respect for the community of living beings, for people, animals, and plants, and for the preservation of Earth, the air, water and soil. Parliament of the World's Religions Declaration Toward a Global Ethic, page 2 We take individual responsibility for all we do. All our decisions, actions, and failures to act have consequences. We must treat others as we wish others to treat us. We make a commitment to respect life and dignity, individuality and diversity, so that every person is treated humanely, without exception. We must have patience and acceptance. We must be able to forgive, learning from the past but never allowing ourselves to be enslaved by memories of hate. Opening our hearts to one another, we must sink our narrow differences for the cause of the world community, practicing a culture of solidarity and relatedness. We consider humankind our family. We must strive to be kind and generous. We must not live for ourselves alone, but should also serve others, never forgetting the children, the aged, the poor, the suffering, the disabled, the refugees, and the lonely. No person should ever be considered or treated as a second-class citizen, or be exploited in any way whatsoever. There should be equal partnership between men and women. We must not commit any kind of sexual immorality. We must put behind us all forms of domination or abuse. We commit ourselves to a culture of non-violence, respect, justice, and peace. We shall not oppress, injure, torture, or kill other human beings, forsaking violence as a means of settling differences. We must strive for a just social and economic order, in which everyone has an equal

religions is identified as the basis for a global ethic. Although religiously inspired and initiated, the declaration addresses all people, both religious and non-religious.

A somewhat broader initiative is the United Nations World Interfaith Harmony Week, originally proposed by King Abdullah II of Jordan at the 65th UN General Assembly in 2010[9] in which it is proposed that the first week of February in every year would be celebrated as a time when people of all faiths, and those of no faith, work together to promote and celebrate religious and cultural understanding and cooperation, under the aegis of the motto "Love of God and Love of the Neighbour or Love of the Good and Love of the Neighbour." Although strongly religiously inspired, the United Nations World Interfaith Harmony Week addresses both religious and non-religious believers, eloquently expressed by the alternative motto "Love of God …or Love of the Good".

There is a great need for interreligious dialogue, given the history of interreligious in-group/out-group competition, conflict and war. However, a significant philosophical divide lies between religious and secular ideologies.[10] It is here that dialogue efforts should also be concentrated. Unfortunately, this type of ideological dialogue—certainly at the international or global level—is much less developed, if at all. Humanist or atheist organisations are usually excluded from interfaith dialogues. Neither of the two sides takes effective initiatives.

Equally, a dialogue is needed between various modern ideologies whose fragmented and short-term partial approaches to current and future human challenges prevents them from pursuing comprehensive solutions, but whose fundamental goals, when considered in a more holistic and longer-term perspective, might converge their visions more closely.

Global dialogue among religions, among secular ideologies, and between religious and secular worldviews is often too ambitious or still too high level and is permeated with compromises, which results in agreements at the level of the lowest common denominator—these do not give sense and weight to effective action. A more promising pathway in the near future may be to intensify dialogue between religious and secular progressives who both embrace the challenges of science and technology and are committed to pursuing ways to fill the gaps in the ethical systems.

chance to reach full potential as a human being. We must speak and act truthfully and with compassion, dealing fairly with all, and avoiding prejudice and hatred. We must not steal. We must move beyond the dominance of greed for power, prestige, money, and consumption to make a just and peaceful world. Earth cannot be changed for the better unless the consciousness of individuals is changed first. We pledge to increase our awareness by disciplining our minds, by meditation, by prayer, or by positive thinking. Without risk and a readiness to sacrifice there can be no fundamental change in our situation. Therefore we commit ourselves to this global ethic, to understanding one another, and to socially beneficial, peace-fostering, and nature-friendly ways of life. We invite all people, whether religious or not, to do the same."

[9]United Nations Resolution A/65/PV.34.

[10]For instance, Aref Ali Nayed (2006) commented in a reaction to Pope Benedict XVI's (2006) lecture at Regensburg University: "Islam can actually be Christianity's best ally against the arrogant pretensions of scientistic positivism, and for a deeper and more spiritual Reason." (http://www.masud.co.uk/ISLAM/misc/commentary_on_benedict.php).

The pluralist approach might remain a necessary and viable solution in cases where, for reasons of cultural heritage or personal considerations, diversity in moral approaches and solutions is an adequate response to the cultural or behavioural diversity in the population. Alternatively, it might merely be an intermediary stage in the cultural and ethical changes that are inevitably linked to the transition from traditional to modern society and culture, and the evolution of a universal ethic. Moreover, just as the maintenance of individual biological variability is an adaptive safety valve against unexpected or unforeseeable environmental changes, there might be evolutionary (adaptive) advantages in the existence or emergence of a variation in different cultures that would peacefully coexist and include creative specificities for the long-term survival and further evolution of humankind—cf. Raymond Cattell's[11] hypothesis of cooperative competition among culturally diverse civilisations. A question to be addressed is whether the multiple modernities hypothesis can be sustained if some cultures deny knowledge and propagate prejudice.

Indeed, a future diversity of worldviews and cultures could not be viable without a universal core ethic, built on the knowledge of the evolutionary sources of human variability and specificity. A universal core ethics needs to be shared worldwide and include rights and responsibilities of cultures regarding the future course of the hominin's biosocial and biocultural development and evolution.

The idea of a move in which spiritual religion and secular humanism could converge is not at all new. It has been advanced and promoted by many people—scientists, philosophers, theologians and politicians—ever since the religious-secular divide took shape. Let us take the example of the American philosopher John Dewey[12] who wrote a book in 1934 entitled *A Common Faith*. An example of a different nature is the renowned '*Nederlands Gesprek Centrum*'[13] in the Netherlands, whose aim is to promote, by means of 'conversations' and publications, the mutual understanding and appreciation between different ideological and social 'pillars' about important, but difficult and sensitive societal issues. Overall, in many developed countries there are numerous examples of pluralistic dialogue initiatives between religious and freethinking individuals or groups trying to find common ethical grounds to bridge ideological ravines in society.[14] However, these are also examples of how some relevant movements remain limited in their impact to a single country or at best to a particular (economic interest) group of countries. Indeed, nation states (and many national institutions) may be seen today as the strongest obstacles to the implementation of principles of an authentic universal ethics. Strive to preserve own prerogatives is part of the human endowment. Indeed, kin-group ethics challenged in the past the broadening of the circle to include non-kin.

[11]Cattell (1972, 105).
[12]Dewey (1934), quoted in Kitcher (2007, 161).
[13]www.nederlandsgesprekcentrum.nl/ (in English: 'The Dutch Centre for Dialogue').
[14]For example, Duyndam et al. (2005).

Nevertheless, in many modern, advanced societies the confrontation between traditional religious and modern secular ideologies, as well as between various opposite secular ideologies, has resulted in the development of ethical compromises, the establishment of a common minimal ethic,[15] or the establishment of pluralism in personal forms of behaviour. This is the case with moral education (availability of courses in various religions and non-confessional ethics), choice of marriage rituals (religious or civil procedures), choice of type of spiritual assistance in disease or end of life situations, choice of type of partnership (marriage or consensual union), means of birth control (contraception, abortion, medically assisted reproduction), and death control (palliative care, regulated euthanasia). Moral progress does not have to imply absolute normative uniformity. Respect for fundamentally different moral systems already gives a universal status to a particular moral view.[16]

A scientific approach, and particularly an evolutionary approach, to ethical questions might be a good way to bring together religious and secular ideologies in order to contribute to the development of a new global ethic. Not only could this deal with the crucial and urgent current challenges that humankind is facing but also focus on a longer-term future, since it is only in such a perspective that a safe future of the hominin phyletic line can be secured.

[15] Küng (1996, 2).
[16] Tao and Yan (2006, 176).

Bibliography

Aarssen LW (2005) Why is human fertility lower in wealthier countries? The role of relaxed fertility selection. Popul Dev Rev 31(1):113–126
Aarssen LW (2007) Some bold evolutionary predictions for the future of mating in humans. Oikos 116(10):1768–1778
Aarssen LW, Altman ST (2006) Explaining below-replacement fertility and increasing childlessness in wealthy countries: legacy drive and the "Transmission Competition" hypothesis. Evol Psychol 4(1):290–302
Abbott E (2001) History of Celibacy: experiments through the ages. Da Capo Press, Cambridge
Abbott P, Wallace C (1999) The family and the new right. Pluto, London
Abe N, Greene JD (2014) Response to anticipated reward in the nucleus accumbens predicts behavior in an independent test of honesty. J Neurosci 34(32):10564–10572
Abouheif E, Favé MJ, Ibarrarán-Viniegra AS et al (2014) Eco-Evo-Devo: the time has come. In: Landry CR, Aubin-Horth N (eds) Ecological genomics: ecology and the evolution of genes and genomes. Springer, Dordrecht, pp 107–125
Abrams DM, Yaple HA, Wiener RJ (2011) A mathematical model of social group competition with application to the growth of religious non-affiliation. arXiv:1012.1375v2
Abramson PR, Pinkerton SD (1995) With pleasure: thoughts on the nature of human sexuality. Oxford University Press, New York
Acerbi A, Mesoudi A (2015) If we are all cultural Darwinians what's the fuss about? Clarifying recent disagreements in the field of cultural evolution. Biol Philos 30(4):481–503
Acharya S (1999) The Christ conspiracy: the greatest story ever sold. Adventures Unlimited Press, Kempton
Adams C (1995) Why did Christianity become sopopular? http://www.straightdope.com/columns/read/1167/why-did-christianity-become-so-popular
Adams MB (2004) The quest for immortality: visions and presentiments in science and literature. In: Post S, Binstock R (eds) The fountain of youth. Oxford University Press, Oxford, pp 38–71
Adams WY (2005) Religion and adaptation. CSLI Publications, Stanford CT
Adorno R (2010) Fondements philosophiques et culturels de l'eugénisme sélectif [Philosophical and cultural basis of selective eugenics]. In: Laffitte J, Carrasco de Paula I (eds) La génétique, au risque de l'eugénisme? Edifa-Mame, Paris, pp 129–141
Agar N (2004) Liberal eugenics: in defence of human enhancement. Blackwell, London
Agar N (2010) Enhancing genetic virtue? Polit Life Sci 29(1):73–75
Agar N (2010) Humanity's end: why we should reject radical enhancement (Life and mind: philosophical issues in biology and psychology). The MIT Press, Cambridge
Agar N (2010) Thoughts about our species' future: themes from humanity's end: why we should reject radical enhancement. J Evol Technol 21(2):23–31
Agrippa C (1529) Declamatio de nobilitate et praecellentia foeminei sexus [Declamation on the nobility and preeminence of the female sex]. Modern edition: University of Chicago Press, Chicago, 1996

R. Cliquet and D. Avramov, *Evolution Science and Ethics in the Third Millennium*,
https://doi.org/10.1007/978-3-319-73090-5

Aiello LC, Dunbar RI (1993) Neocortex size, group size, and the evolution of language. Curr Anthropol 34(2):184–193
Aknin LB, Barrington-Leigh CP, Dunn EW et al (2013) Prosocial spending and wellbeing: cross-cultural evidence for a psychological universal. J Pers Soc Psychol 104(4):635–652
Aktipis CA (2004) Know when to walk away: contingent movement and the evolution of cooperation. J Theor Biol 231(2):249–260
Alan Guttmacher Institute (2000) Readings on induced abortion: politics and policies. Alan Guttmacher Institute, New York
Albright CR (2000) The 'God Module' and the complexifying brain. Zygon 35(4):735–745
Alcorta C, Sosis R (2005) Ritual, emotion, and sacred symbols: the evolution of religion as an adaptive complex. Hum Nat 16(4):323–359
Alcorta CS, Sosis R (2012) Ritual, religion, and violence: an evolutionary perspective. In: Juergensmeyer M, Kitts M, Jerryson M (eds) Handbook of religion and violence. Oxford University Press, New York, pp 571–596
Alexander M, Christia F (2011) Context modularity of human altruism. Science 334(6061):1392–1394
Alexander RD (1971) The Search for an evolutionary philosophy of man. Proc R Soc Victoria 84 (1):99–119
Alexander RD (1974) The evolution of social behavior. Ann Rev Ecol Syst 5:352–383
Alexander RD (1975) The search for a general theory of behavior. Behav Sci 20:77–100
Alexander RD (1978) Evolution, creation, and biology teaching. Am Biol Teac 40(2):91–104
Alexander RD (1979) Darwinism and human affairs. University of Washington Press, Seattle
Alexander RD (1987) The biology of moral systems. Hawthorne, Aldine
Alexander RD (1989) Evolution of the human psyche. In Mellars P, Stringer C (eds) The human revolution. Edinburgh University Press, Edinburgh, pp 455–513
Alexander RD (1990) How did humans evolve? Reflexions on the uniquely unique species. The University of Michigan, Ann Arbor
Alexander RD (1993) Biological considerations in the analysis of morality. In: Nitecki MH, Nitecki DV (eds) Evolutionary ethics. State University of New York Press, Albany, pp 163–196
Alexander RD (2005) A biological interpretation of moral systems. Zygon 20(1):3–20
Alexander RD (2013) Religion, evolution and the quest for global harmony. In: Summers K, Crespi B (eds) Human social evolution: the foundational works of Richard D. Alexander. Oxford University Press, Oxford, pp 384–425
Alexander RD, Borgia G (1978) Group selection, altruism and the levels of organization of life. Ann Rev Ecol Syst 9:449–474
Alexander RD, Noonan KM (1979) Concealment of ovulation, parental care, and human social evolution. In: Chagnon NA, Irons WG (eds) Evolutionary biology and human social behaviour: an anthropological perspective. Duxbury Press, North Scituate, pp 436–453
Alford JR, Hibbing JR (2004) The origin of politics: an evolutionary theory of political behavior. Perspect Polit 2(4):707–723
Allchin D (2009) The evolution of morality. Evol Educ Outreach 2(4):590–601
Allee WC (1951) Cooperation among animals with human implications. Henry Schuman, New York
Allen J, Chavez S, DeSimone S et al (2006) Americans' attitudes toward euthanasia and physician-assisted suicide, 1936-2002. J Sociol Soc Welf 33(2):5–24
Allen JS, Damasio H, Grabowski TJ et al (2003) Sexual dimorphism and asymmetries in the gray-white composition of the human cerebrum. NeuroImage 18(4):880–894
Allen M, d'Alessio D, Brezgel K (1995) A meta-analysis summarizing the effects of pornography II: aggression after exposure. Hum Commun Res 22(2):258–283
Allen M, Emmers T, Gebhardt L et al (1995) Exposure to pornography and acceptance of rape myths. J Commun 45(1):5–26

Allen B, Nowak MA, Wilson EO (2013) Limitations of inclusive fitness. PNAS 110(50):20135–20139
Allhoff F (2008) Germ-line genetic enhancement and Rawlsian primary goods. J Evol Technol 18 (1):10–26
Allis CD, Caparros ML, Jenuwein T, Reinberg D (2015) Epigenetics. Cold Spring Harbor Laboratory Press, Cold Spring
Almond GA, Scott Appleby R, Sivan E (2003) Strong religion: the rise of fundamentalism around the world. University of Chicago Press, Chicago
Almroth L, Elmusharaf S, El Hadi N et al (2005) Primary infertility after genital mutilation in girlhood in Sudan: a case-control study. Lancet 366(9483):385–391
Alper M (2006; 2009) The 'God' part of the brain: a scientific interpretation of human spirituality and God. Sourcebooks, Naperville
Alston BC (2007) What is neurotheology?. BookSurge Publishing, Charleston
Altemeyer B, Hunsberger B (2005) Fundamentalism and authoritarianism. In: Paloutzian RF, Park CL (eds) Handbook of the psychology of religion and spirituality. Guilford Press, New York, pp 378–393
Alvarez LW, Alvarez W, Asaro F et al (1980) Extraterrestrial cause for the cretaceous-tertiary extinction: experimental results and theoretical interpretation. Science 208(4448):1095–1108
Alvaro-Gonzalez LC (2014) Neuroethics (I): moral pathways in normal brain. Rev Neurol 58 (5):225–233
Ambrose SH (1998) Late Pleistocene human population bottlenecks, volcanic winter, and differentiation of modern humans. J Hum Evol 34(4):623–651
American Association for the Advancement of Science (2006) Statement on the teaching of evolution by thé board of directors of the American Association for the Advancement of Science. American Association for the Advancement of Science, Washington
American Humanist Association (2003) Humanist manifesto III. American Humanist Association, Washington
Amundson R, Tresky S (2007) On a bio-ethical challenge to disability rights. J Med Philos 32 (6):541–542
Anālayo Bhikkhu (2014) Karma and female birth. J Buddhist Ethics 21:107–151
Anandarajah G, Hight E (2001) Spirituality and medical practice: using the HOPE questions as a practical tool for spiritual assessment. Am Fam Physician 63(1):81–89
Andaya BW (2002) Localising the universal: women, motherhood and the appeal of early Theravāda Buddhism. J SE Asian Stud 33(1):1–30
Andersson M (1994) Sexual selection. Princeton University Press, Princeton
Anderson SW, Bechara A, Damasio H et al (1999) Impairment of social and moral behavior related to early damage in human prefrontal cortex. Nat Neurosci 2(11):1032–1037
Anderson S, Peacocke A (eds) (1987) Evolution and creation: a European perspective. Aarhus University, Aarhus
André J-B (2010) The evolution of reciprocity: social types or social incentives? Am Nat 175 (2):197–210
André J-B, Morin O (2011) Questioning the cultural evolution of altruism. J Evol Biol 24 (12):2531–2542
Andreadis A (2010) The tempting illusion of genetic virtue. Polit Life Sci 29(1):76–78
Andreoni J (1989) Giving with impure altruism: applications to charity and ricardian equivalence. J Polit Econ 97(6):1447–1458
Andreoni J (1990) Impure altruism and donations to public goods: a theory of warm-glow giving. Econ J 100(401):464–477
Andrews M (1999) The seductiveness of agelessness. Ageing Soc 19(3):301–318
Angus I, Butler S (2011) Too many people? Population, immigration, and the environmental crisis. Haymarket Books, Chicago
Annas G, Andrews L, Isasi R (2002) Protecting the endangered human: toward an international treaty prohibiting cloning and inheritable alterations. Am J Law Med 28(2–3):151–178

Antfolk J, Salo B, Alanko K et al (2015) Women's and men's sexual preferences and activities with respect to the partner's age: evidence for female choice. Evol Hum Behav 36(1):73–79

Antón SC, Snodgrass JJ (2012) Origins and evolution of genus Homo: new perspectives. Curr Anthropol 53(S6):S479–S496

Appleton N (2011) In the footsteps of the Buddha: women and the Bodhisatta Path in Theravāda Buddhism. J Feminist Stud Relig 27(1):33–51

Apostolou M (2013) Interfamily conflict, reproductive success, and the evolution of male homosexuality. Rev Gen Psychol 17(3):288–296

Apostolou M (2013) The evolution of rape: the fitness benefits and costs of a forced-sex mating strategy in an evolutionary context. Aggression Violent Behav 18(5):SI 484–490

Apostolou M (2016) The evolution of female same-sex attractions: the weak selection pressures hypothesis. Evol Behav Sci. https://doi.org/10.1037/ebs0000072

Armelagos GJ, Goodman AH, Jacobs KH (1991) The origins of agriculture: population growth during a period of declining health. Popul Environ 13(1):9–22

Armelagos GJ, Harper KN (2005) Genomics at the origins of agriculture, part one. Evol Anthropol 14(2):68–77

Armelagos GJ, Van Gerven DP (1980) Sexual dimorphism and human evolution: an overview. J Hum Evol 9(5):437–446

Armstrong K (1993) A history of God. William Heinemann, London

Armstrong K (2006) The great transformation. The beginning of our religious traditions. A.A. Knopf, New York

Armstrong K (2007) The Bible: the biography. Atlantic Books, London

Armstrong K (2009) The case for god: what religion really means. Bodley Head, London

Armstrong RE (2010) Bio-inspired innovation and national security. National Defence University, Center for Technology and Natural Security Policy, Washington

Armstrong K (2014) Fields of blood. Bodley Head, London

Arnett JJ (2001) Readings on adolescence and emerging adulthood. Prentice Hall, Upper Saddle River

Arnhart L (1998) Darwinian natural right: the biological ethics of human nature. SUNY Press, Albany

Arnhart L (2010) Can virtue be genetically engineered? Polit Life Sci 29(1):79–81

Arnhart L (2010) Darwinian liberalism. http://www.cato-unbound.org/archives/july-2010-darwin-and-politics/

Arnold F, Kishor S, Roy TK (2002) Sex-selective abortions in India. Popul Dev Rev 28(4):759–785

Aronson R (2008) Living without god: new directions for atheists, agnostics, secularists, and the undecided. Counterpoint, Berkeley

Arzy S, Idel M, Landis T et al (2005) Why revelations have occurred on mountains? Linking mystical experiences and cognitive neuroscience. Med Hypotheses 65(5):841–845

Aschner BM, Post RH (1956/57) Modern therapy and hereditary diseases. In: Kemp T, Hauge M, Harvald B (eds) First international congress of human genetics. S. Karger, Basel, pp 362–369

Ashton JF (2001) In six days: why fifty scientists choose to believe in creation. Master Books, Green Forest

Association for the Advancement of Science (2006) Statement on the teaching of evolution. http://www.aaas.org/news/releases/2006/pdf/0219boardstatement.pdf

Atkinson QD, Bourrat P (2011) Beliefs about God, the afterlife and morality support the role of supernatural policing in human cooperation. Evol Hum Behav 32(1):41–49

Atlan M, Droit R-P (2012) Humain. Une enquête philosophique sur les evolutions qui changent nos vies. (Human. A philosophical inquiry into the evolutions which change our lives). Flammarion, Paris

Atoh M (2008) Family changes in the context of lowest-low fertility: the case of Japan. Int J Jpn Sociol 17(1):14–29

Atran S (2002) In god we trust. The evolutionary landscape of religion. Oxford University Press, New York

Atran S (2006) The cognitive and evolutionary roots of religion. In: McNamara P (ed) Where god and science meet: how brain and evolutionary studies alter our understanding of religion. Praeger, Westport, pp 181–207

Atran S, Ginges J (2015) Devoted actors and the moral foundations of intractable intergroup conflict. In: Decety J, Wheatley T (eds) The moral brain. A multidisciplinary perspective. The MIT Press, Cambridge, pp 69–85

Atran S, Norenzayan A (2004) Religion's evolutionary landscape: counterintuition, commitment, compassion, communion. Behav Brain Sci 27(6):713–730

Audi R (2004) The good in the right: a theory of intuition and intrinsic value. Princeton University Press, Princeton

Augustine of Hippo (426) (1952) The city of god. Shambhala, Boston

Augustine of Hippo (398) (2009) The confessions of Saint Augustine. CreateSpace, Seattle

Aurenque D (2015) Genetic diversity as a value: imposing fairness. Am J Bioeth 15(6):18–20

Aurenque D, Wiesing U (2015) German law on circumcision and its debate: how an ethical and legal issue turned political. Bioethics 29(3):203–210

AusBiotech (2005) Biotechnology industry code of conduct. Australia's Biotechology Organisation, Malvern

Avalos H (2005) Fighting words: the origins of religious violence. Prometheus Books, Amherst

Avalos H (2010) Health care and the rise of Christianity. Baker, Grand Rapids

Avila M, Thaker G, Adami H (2001) Genetic epidemiology and schizophrenia: a study of reproductive fitness. Schizophr Res 47(2–3):233–241

Avilés L (2002) Solving the freeloaders paradox: genetic associations and frequency-dependent selection in the evolution of cooperation among nonrelatives. PNAS 99(22):14268–14273

Avramov D (2002) People, demography and social exclusion. Population studies 37. Council of Europe Publishing, Strasbourg

Avramov D (ed) (2008) Acceptance of immigrants in Europe? Pro Business Verlag book-on-demand, Berlin

Avramov D, Cliquet R (2003) Economy of time and population policy: rethinking the 20th century life course paradigm in the light of below-replacement fertility. Z für Bevölkerungswissenschaft 28(2–4):905–938

Avramov D, Cliquet R (1999) The flaring-up of the ingroup-outgroup syndrome in Eastern Europe: the example of former Yugoslavia. In: Thienpont K, Cliquet R (eds) Ingroup-outgroup behaviour in modern societies. An evolutionary perspective. NIDI/CBGS Publications, vol 34. Vlaamse Gemeenschap, Brussel, pp 139–164

Avramov D, Cliquet R (2005) Integrated policies on gender relations, ageing and migration in Europe: lessons from the Network for Integrated European Population Studies (NIEPS). CBGS Publications. Garant, Leuven-Apeldoorn

Avramov D, Cliquet R (2006) Preferences and expectations about generational replacement and age at retirement in Europe. Results from the DIALOG project. Finnish Yearbook of Population Research XLII. The Population Research Institute, Helsinki, pp 13–26

Avramov D, Cliquet R (2007) Xenophobia and integration of immigrants. Z für Bevölkerungswissenschaft 32(3–4):533–560

Avramov D, Cliquet R (2008) The need to adapt and reform social policy: setting the stage for effective population-friendly policies. In: Höhn C, Avramov D, Kotowska I (eds) People, population change and policies. Lessons from the population policy acceptance study, vol 2: Demographic knowledge—Gender—Ageing. European Studies of Population, vol 16/2. Springer, Dordrecht, pp 267–286

Avramov D, Maskova M (2003) Active ageing in Europe. Population studies 41. Council of Europe Publishing, Strasbourg

Axelrod R (1984) The evolution of cooperation. Basic Books, New York

Axelrod R (1986) An evolutionary approach to norms. Am Polit Sci Rev 80(4):1095–1111

Axelrod R (2001) The complexity of cooperation: agent-based models of competition and collaboration. Princeton University Press, Princeton

Axelrod R, Hamilton WD (1981) The evolution of cooperation. Science 211(4489):1390–1396

Ayala F (1987) The biological roots of morality. Biol Philos 2(3):235–252

Ayala F (1988) Can "progress" be defined as a biological concept? In: Nitecki M (ed) Evolutionary progress. University of Chicago Press, Chicago, pp 75–91

Ayala F (1995) The difference of being human: ethical behavior as an evolutionary byproduct. In: Rolston H III (ed) Biology, ethics, and the origins of life. Jones and Bartlett, Boston, pp 113–136

Ayala F (2007) Darwin's gift to science and religion. Joseph Henry Press, Washington

Ayala F (2009) What the biological sciences can and cannot contribute to ethics. In: Ayala F, Arp R (eds) Contemporary debates in philosophy of biology. Wiley-Blackwell, Oxford

Ayala F (2010) The difference of being human: morality. PNAS 107(S2):9015–9022

Azari NP (2006) Neuroimaging studies of religious experience: a critical review. In: McNamara P (ed) Where god and science meet: how brain and evolutionary studies alter our understanding of religion, vol 2. The neurology of religious experience. Praeger, Westport, pp 33–54

Azari NP, Nickel J, Wunderlich G et al (2001) Neural correlates of religious experience. Eur J Neurosci 13(8):1649–1652

Bacon F (1626; 1996) The New Atlantis. In: Vickers B (ed) Bacon F, A critical edition of the major works. Oxford University Press, Oxford, pp 457–489

Badmington N (2000) Posthumanism. Palgrave Macmillan, Basingstoke

Baily K (1995) The sociopath: cheater or warrior hawk? Behav Brain Sci 18:542–543

Bailey R (2006) Overpopulation does not threaten the environment or humanity. In: Gerdes LI (ed) Humanity's future. Greenhaven Press, Detroit, pp 148–156

Baillie HW, Casey TK (eds) (2004) Is human nature obsolete? Genetics, bioengineering, and the future of the human condition. MIT Press, Cambridge

Bainbridge WS (ed) (2010) Online worlds: convergence of the real and the virtual. Springer, London

Baird DT, Van Look PFA, Grimes DA (eds) (1995) Modern methods of inducing abortion. Blackwell Science, London

Bajema CJ (1971) The genetic implications of population control. Bioscience 21(2):71–75

Bajema CJ (ed) (1976) Eugenics. Then and now. Dowden, Hutchinson and Ross, Stroudsburg

Bajema CJ (1978) Differential transmission of genetic and cultural information about the environment: a cybernetic view of genetic and cultural evolution in animal species. In: Meier R, Otten Ch, Hameed A (eds) Evolutionary modes and studies in human diversity. Mouton, The Hague

Bajos N, Ferrand M, l'équipe GINE (2002) De la contraception à l'avortement: Sociologie des grossesses non prévues. (From contraception to abortion: sociology of unplanned pregnancies). Collection Questions en santé publique. Inserm, Paris

Bakar O (1987) Critique of evolutionary theory: a collection of essays. Islamic Academy of Science, Kuala Lumpur

Bakar O (2003) The nature and extent of criticism of evolutionary theory. In: Zarandi MM (ed) Science and the myth of progress. World Wisdom, Bloomington

Baker J (2008) An investigation of the sociological patterns of prayer frequency and content. Sociol Relig 69(2):169–185

Baker R, Bellis MA (1995) Human sperm competition. Copulation, masturbation and infidelity. Chapman and Hall, London

Baker TA, Wachsmann N, Caplan J (eds) (2010) Concentration camps in Nazi Germany: the new histories. Routledge, London

Bakermans-Kranenburg MJ, van IJzendoorn MH (2006) Gene-environment interaction of the Dopamine D4 receptor (DRD4) and observed maternal insensitivity predicting externalizing behavior in preschoolers. Dev Psychobiol 48(5):406–409

Bakre OM (2008) Looting by the ruling elites, multinational corporations and the accountants: the genesis of indebtedness, poverty and underdevelopment of Nigeria. School of Accounting, Finance and Management, University of Essex, Colchester
Baldassarri D, Grossman G (2011) Centralized sanctioning and legitimate authority promote cooperation in humans. PNAS 108(27):11023–11027
Balliet D, Van Lange PAM (2011) Trust, punishment, and cooperation across 18 societies: a meta-analysis. Perspect Psychol Sci 8(4):363–379
Baltes PB, Reuter-Lorenz PA, Rösler F (2006) Lifespan development and the brain: the perspective of biocultural co-constructivism. Cambridge University Press, Cambridge
Baltimore D, Berg P, Botchan M et al (2015) A prudent path forward for genomic engineering and germline gene modification. Science 348(6230):36–38
Bamshad M, Kivisild T, Watkins WS (2001) Genetic evidence on the origins of Indian caste populations. Genome Res 11:994–1004
Bandelt HJ, Richards M, Macaulay V (2010) Human mitochondrial DNA and the evolution of Homo sapiens (Nucleic acids and molecular biology). Springer, Berlin
Bankard J (2015) Training emotion cultivates morality: how loving-kindness meditation hones compassion and increases prosocial behavior. J Relig Health 54(6):2324–2343
Banquet JP (1972) EEG and meditation. Electrocephalography Clin Neurophysiol 33:454
Banton M (1987) Racial theories. Cambridge University Press, Cambridge
Baofu P (2010) The Future of post-human war and peace: a preface to a new theory of aggression and pacificity. Cambridge Scholars Publishing, Newcastle upon Tyne
Barak A, Fisher WA, Belfry S et al (1999) Sex, guys, and cyberspace: effects of internet pornography and individual differences on men's attitudes toward women. J Psychol Hum Sex 11(1):63–91
Barash D (1977) Sociobiology and behavior. Elsevier, New York
Barash D (2003) The survival game: how game theory explains the biology of cooperation and competition. Henry Holt, New York
Barash D (2012) Homo mysterious: evolutionary puzzles of human nature. Oxford University Press, New York
Barash D, Lipton JE (2001) The myth of monogamy: fidelity and infidelity in animals and people. W.H Freeman, New York
Barber N (1995) The evolutionary psychology of physical attractiveness: sexual selection and human morphology. Ethol Sociobiol 16(5):395–424
Barber B (1957) Social stratification. A comparative analysis of structure and process. Harcourt, Brace and Company, New York
Barbour IG (1990) Religion in an age of science: the Gifford lectures 1989–1991, vol 1. SCM, London
Barbour IG (1997) Religion and science: historical and contemporary issues. HarperSanFrancisco, San Francisco
Baril D (2006) La grande illusion. Comment la selection naturelle a créé l'idée de Dieu. (The great illusion. How natural selection has created the idea of God). MultiMondes, Québec
Baril D (2007) Déclin de la religion et persistance du religieux. (Decline of religion and religious persistence). Altérités 4(2):79–90
Baril D (2013) Does the shamanic model support the adaptive hypothesis of the origin of religion? Stud Relig-Sci Religieuses 42(3) SI:364–381
Barkow JH (1989) Darwin, sex and status: biological approaches to mind and culture. University of Toronto Press, Toronto
Barkow JH, Cosmides L, Tooby J (eds) (1992) The adapted mind: evolutionary psychology and the generation of culture. Oxford University Press, New York
Barna G (1999) Christians are more likely to experience divorce than are non-Christians. Barna Research Group, 1999-DEC-21. http://www.barna.org/cgi-bin/
Barnea M, Schwartz SH (1998) Values and voting. Polit Psychol 19(1):17–40

Barnes NJ (1987) Buddhism. In: Sharma A (ed) World religions. State University of New York Press, Albany, pp 105–134

Baron-Cohen S (ed) (1997) The maladapted mind: classic readings in evolutionary psychopathology. Psychology Press, Hove

Baron-Cohen S (2003) The essential difference: the truth about the male and female brain. Basic Books, New York

Baron-Cohen S (2011) Zero degrees of empathy. A new theory of human cruelty. Allen Lane, London

Barr KN, Quinsey VL (2004) Is psychopathy a pathology or a life strategy? Implications for social policy. In: Crawford C, Salmon C (eds) Evolutionary psychology, public policy, and personal decisions. Erlbaum, Hillsdale, pp 293–317

Barrett JL (2000) Exploring the natural foundations of religion. Trends Cogn Sci 4(1):29–34

Barrett JL (2001) Do children experience god as adults do? In: Andresen J (ed) Religion in mind: cognitive perspectives on religious belief, ritual, and experience. Cambridge University Press, Cambridge, pp 173–190

Barrett JL (2004) Why would anyone believe in God? AltaMira Press, New York

Barrett L, Dunbar R, Lycett J (2002) Human evolutionary psychology. Princeton University Press, Princeton

Barrett L, Henzi SP, Lusseau D (2012) Taking sociality seriously: the structure of multi-dimensional social networks as a source of information for individuals. Philos Trans R Soc B-Biol Sci 367 (SI 1599):2108–2118

Barta Z, McNamara JM, Huszar DB et al (2011) Cooperation among non-relatives evolves by state-dependent generalized reciprocity. Proc R Soc B-Biol Sci 278(1707):843–848

Bartels A, Zeki S (2000) The neural basis of romantic love. NeuroReport 11(17):3829–3834

Barthes J, Godelle B, Raymond M (2013) Human social stratification and hypergyny: toward an understanding of male homosexual preference. Evol Hum Behav 34(3):155–163

Barthold JA, Myrskylä M, Jones OR (2012) Childlessness drives the sex difference in the association between income and reproductive success of modern Europeans. Evol Hum Behav 33(6):628–638

Baruch S, Kaufman D, Hudson K (2008) Genetic testing of embryos: practices and perspectives of US IVF clinics. Fertil Steril 89(5):1053–1058

Baschetti R (2006) Evolution and morality. Lancet 367(9527):1975–1976

Baschetti R (2007) Evolutionary, neurobiological, gene-based solution of the ideological "puzzle" of human altruism and cooperation. Med Hypotheses 69(2):241–249

Baschetti R (2007) The dawn of science-based moral reasoning. Med Hypotheses 68(1):4–8

Basharat T (2009) The contemporary Hindu women of India: an overview. South Asian Stud. Res J South Asian Stud 24(2):242–249

Bassett AS, Bury A, Hodgkinson KA et al (1996) Reproductive fitness in familial schizophrenia. Schizophr Res 21(3):151–160

Basten S, Lutz W, Scherbov S (2013) Very long range global population scenarios to 2300 and the implications of sustained low fertility. Demogr Res 28(39):1145–1166

Batson CD (1983) Sociobiology and the role of religion in promoting prosocial behavior: an alternative view. J Pers Soc Psychol 45:1380–1385

Batson CD (1991) The altruism question: toward a social psychological answer. Lawrence Erlbaum Associates, Hillsdale

Batson CD, Ventis WL (1982) The religious experience: a social-psychological perspective. Oxford University Press, New York

Batten M (1994) Sexual strategies. How females choose their mates. G.P. Putnam's Sons, New York

Bauch CT, McElreath R (2016) Disease dynamics and costly punishment can foster socially imposed monogamy. Nat Commun 7:11219

Bauer U, Bittlingmayer UH, Richter M (2008) Health inequalities: Determinanten und Mechanismen gesundheitlicher Ungleichheit (Gesundheit und Gesellschaft) (German and, English edn. Verlag für Sozialwissenschaften, Wiesbaden

Baum SD (2015) The far future argument for confronting catastrophic threats to humanity: practical significance and alternatives. Futures 72:86–96

Baumard N (2016) The origins of fairness: how evolution explains our moral nature. Oxford University Press, Oxford

Baumard N, Andre JB, Sperber D (2013) A mutualistic approach to morality: the evolution of fairness by partner choice. Behav Brain Sci 36(1):59–122

Baumgardner J, Richards A (2000) Manifesta: young women, feminism, and the future. Farrar, Straus and Giroux, New York

Baumrind D (1971) Current patterns of parental authority. Devel Psych 4 (1):1–103

Bauserman R (1996) Sexual aggression and pornography: a review of correlational research. Basic Appl Soc Psychol 18(4):405–427

Bawer B (2006) While Europe slept: how radical Islam is destroying the West from within. Doubleday, New York

Baxter B (1999) Ecologism: an introduction. Edinburgh University Press, Edinburgh

Baylis F, Robert JS (2004) The inevitability of genetic enhancement technologies. Bioethics 18 (1):1–26

Baylis J, Smith S, Owens P (2011) The globalization of world politics: an introduction to international relations. Oxford University Press, Oxford

Beauregard M, O'Leary D (2008) The spiritual brain: a neuroscientist's case for the existence of the soul. HarperOne, New York

Beauregard M, Paquette V (2006) Neural correlates of a mystical experience in Carmelite nuns. Neurosci Lett 405:186–190

Beaver KM, Connolly EJ, Schwartz JA et al (2013) Genetic and environmental contributions to stability and change in levels of self-control. J Crim Just 41(5):300–308

Beaver KM, Gibson CL, Jennings WG et al (2009) A gene x environment interaction between DRD2 and religiosity in the prediction of adolescent delinquent involvement in a sample of males. Biodemogr Soc Biol 55(1):71–81

Beaver KM, Shutt JE, Boutwell BB et al (2009) Genetic and environmental influences on levels of self-control and delinquent peer affiliation: results from a longitudinal sample of adolescent twins. Crim Justice Behav 36(1):41–60

Bebel A (1879) Die Frau und der Sozialismus. (Woman under socialism). Volksbuchhandlung, Zürich

Bechara A (2002) The neurology of social cognition. Brain 125(8):1673–1675

Beck B (2015) Conceptual and practical problems of moral enhancement. Bioethics 29(4):233–240

Beck U, Beck-Gernsheim E (2002) Individualisation: institutionalized individualism and its social and political consequences. Polity Press, Cambridge

Becker K, Laucht M, El-Faddagh M et al (2005) The dopamine D4 receptor gene exon III polymorphism is associated with novelty seeking in 15-year-old males from a high-risk community sample. J Neur Trans 112(6):847–858

Becker GS, Lewis HG (1973) On the interaction between the quantity and quality of children. J Polit Econ 81(2), Part 2: New Econ Approaches Fertil S279–S288

Beekman T (2012) Door Spinoza's lens. (Through Spinoza's lens). Pelckmans/Klement, Kalmthout

Beer JM, Arnold RD, Loehlin JC (1998) Genetic and environmental influences on MMPI factor scales: joint model fitting to twin and adoption data. J Pers Soc Psychol 74(3):818–827

Behe MJ (1996) Darwin's black box: the biochemical challenge to evolution. The Free Press, New York

Behuniak SM (2011) Death with "dignity": the wedge that divides the disability rights movement from the right to die movement. Polit Life Sci 30(1):17–32

Beit-Hallahmi B (2010) Morality and immorality among the irreligious. In: Zuckerman P (ed) Atheism and secularity, vol 1. Issues, concepts, and definitions. Praeger, Santa Barbara, pp 113–148
Bekoff M, Pierce J (2009) Wild justice: the moral lives of aminals. University of Chicago Press, Chicago
Beland J-P, Patenaude J (2013) Risk and the question of the acceptability of human enhancement: the humanist and transhumanist perspectives. Dialogue-Can Philos Rev 52(2):377–394
Bell P (2002) Would you believe it? Mensa Magazine 12–13 Feb
Bell AP, Weinberg MS (1978) Homosexualities: A study of diversity among men and women. The Macmillan Company of Australia, South Melbourne
Bellah RN (2011) Religion in human evolution. From the Paleolithic to the axial age. The Belknap Press, Cambridge
Bellow A (2003) In praise of nepotism: a history of family enterprise from King David to George W. Bush. Doubleday, New York
Belva F, Bonduelle M, Roelants M et al (2016) Semen quality of young adult ICSI offspring: the first results. Hum Reprod. https://doi.org/10.1093/humrep/dew245
Benagiano G, Bastianelli C, Farris M (2007) Contraception: a social revolution. Eur J Contracept Reprod Health Care 12(1):3–12
Benedek E, Brown C (1999) No excuses: televised pornography harms children. Harvard Rev Psychiatry 7(4):236–240
Benjamin J, Ebstein RP, Belmaker RH (eds) (2002) Molecular genetics and the human personality. American Psychiatric Publications, Washington
Benjamin J, Li L, Patterson C et al (1996) Population and familial association between the D4 dopamine receptor gene and measures of novelty seeking. Nat Genet 12(1):81–84
Bennion J (2004) Desert patriarchy: Mormon and Mennonite communities in the Chihuahua Valley. University of Arizona Press, Tuczon
Benson H, Malhotra MS, Goldman RF et al (1990) Three case reports of the metabolic and electroencephalographic changes during advanced Buddhist meditation techniques. Behav Med 16(2):90–95
Benson P, Spilka B (1973) God image as a function of self-esteem and locus of control. J Sci Study Relig 12(3):297–310
Bensyl DM, Iuliano AD, Carter M et al (2002) Contraceptive use—United States and territories, Behavioral Risk Factor Surveillance System. http://www.cdc.gov/mmwr/
Bentham J (1834) Deontology or the science of morality. Longman, London
Bentley GR, Goldberg T, Jasieńska G (1993) The fertility of agricultural and non-agricultural traditional societies. Popul Stud 47(2):269–281
Bentley GR, Mascie-Taylor CG (eds) (2001) Infertility in the modern world: present and future prospects. Cambridge University Press, Cambridge
Bentley RA, Hahn MW, Shennan SJ (2004) Random drift and cultural change. Proc R Soc B 271:1443–1450
Bereczkei T, Csanaky A (1996) Mate choice, marital success, and reproduction in a modern society. Ethol Sociobiol 17(1):17–35
Berger B (2013) Altruistic reciprocity. Theoretical considerations and experimental evidence. Kölner Z für Soziologie und Sozialpsychologie 65(1):31–48
Berger B, Berger P (1983) The war over the family. Hutchinson, London
Berger PL, Luckmann T (1966) The social construction of reality: A treatise in the sociology of knowledge. Doubleday & Company, New York
Berger LR et al (2015) Homo naledi, a new species of the genus Homo from the Dinaledi Chamber, South Africa. eLife 4:e09560
Berger P (1999) The desecularization of the world: resurgent religion and world politics. W.B. Eerdmans, Washington
Berger P (2001) Le réenchantement du monde. (The re-enchantment of the world). Bayard, Paris

Bergman A, Siegal ML (2003) Evolutionary capacitance as a general feature of complex gene networks. Nature 424(6948):549–552

Bergman J (2001) Darwin's influence on ruthless Laissez-faire capitalism. http://www.icr.org/i/pdf/imp/imp-333.pdf

Bergstrom CT (Rapporteur) (2003) Group report: interspecific mutualism. In: Hammerstein P (ed) Genetic and cultural evolution of cooperation. MIT Press, Cambridge, pp 241–256

Bering JM (2006) The cognitive psychology of belief in the supernatural: a byproduct of the ability to reason about the minds of others may offer evolutionary advantage. Am Sci 94(2):142–149

Berkman LF, Kawachi I (2000) Social epidemiology. Oxford University Press, Oxford

Berlinski D (2008) The devil's delusion: atheism and its scientific pretentions. Crown Forum, New York

Berman E, Iannaccone LR, Ragusa G (2007) From empty pews to empty cradles: fertility decline among European catholics. http://econ.ucsd.edu/elib/pews.pdf

Bernhard H, Fischbacher U, Fehr E (2006) Parochial altruism in humans. Nature 442(7105):912–915

Bernhardt BA, Biesecker BB, Mastromarino CL (2000) Goals, benefits, and outcomes of genetic counseling: client and genetic counselor assessment. Am J Med Genet 94(3):189–197

Bernhardt BC, Singer T (2012) The neural basis of empathy. Ann Rev Neurosc 35(1):1–23

Berns G, Bell E, Capra CM et al (2012) The price of your soul: neural evidence for the non-utilitarian representation of sacred values. Philos Trans R Soc Lond B Biol Sci 367 (1589):754–762

Berryessa C, Cho MK (2013) Ethical, legal, social, and policy implications of behavioral genetics. Ann Rev Genomics Hum Genet 14:515–534

Bertka CM (2009) Exploring the origin, extent, and future of life: philosophical, ethical, and theological perspectives. Cambridge University Press, Cambridge

Bess M (2010) Enhanced humans versus "normal people": elusive definitions. J Med Philos 35 (6):641–655

Béteille A (1992) Caste and family: in representations of Indian society. Anthropol Today 8(1):13–18

Betzig L (1986) Despotism and differential reproduction: a Darwinian view of history. Aldine, Hawthorne

Betzig L (1989) Causes of conjugal dissolution: a cross-cultural study. Current Anthropol 30 (5):654–676

Betzig L (1995) Medieval monogamy. J Family Hist 20(2):181–215

Betzig L (2005) Politics as sex: the old testament case. Evol Psychol 3:326–346

Bhanji JP, Delgado MR (2014) The social brain and reward: social information processing in the human striatum. Wiley Interdisc Rev-Cogn Sci 5(1):61–73

Bhasin MK, Walter H (2001) Genetics of castes and tribes of India. Kamla-raj Enterprises, New Delhi

Bhaskarananda S (2002) The essentials of Hinduism: a comprehensive overview of the world's oldest religion. Viveka Press, Seattle

Bialasiewicza L (2006) 'The death of the West': Samuel Huntington, Oriana Fallaci and a new 'moral' geopolitics of births and bodies. Geopolitics 11(4):701–724

Billari F, Kohler H-P (2004) Patterns of low and lowest-low fertility in Europe. Popul Stud: J Demogr 58(2):161–176

Birdsell JB (1986) Some predictions for the Pleistocene based on equilibrium systems among recent hunter gatherers. In: Lee R, DeVore I (eds) Man the hunter. Aldine Publishing, New York, pp 229–240

Bishop KM, Wahlsten D (1997) Sex differences in the human corpus callosum: myth or reality? Neurosci Biobehav Rev 21(5):581–601

Black D (2000) On the origin of morality. J Conscious Stud 7:107–1191

Black DW, Lindon LC (2013) Bad boys, bad men: confronting antisocial personality disorder. Oxford University Press, New York

Blackburn S (2001) Ethics: a very short introduction. Oxford University Press, Oxford
Blackburn S (2014) Human nature and science: a cautionary essay. Behaviour 151(2–3):229–244
Blacker CP (1952) Eugenics: Galton and after. Harvard University Press, Cambridge
Blackford R (2010) Genetically engineered people: autonomy and moral virtue. Polit Life Sci 29 (1):82–84
Blackmore S (1993) Dying to live: near-death experiences. Prometheus Books, Amherst
Blackmore S (1999) The meme machine. Oxford University Press, New York
Blair J, Mitchell D, Blair K (2005) The psychopath: emotion and the brain. Blackwell Publishing, Malden
Blair RJ (1997) A cognitive developmental approach to morality: investigating the psychopath. In: Baron-Cohen S (ed) The maladapted mind: classic readings in evolutionary psychopathology. Psychology Press, Hove, pp 85–114
Blair RJR (2006) The emergence of psychopathy: implications for the neuropsychological approach to developmental disorders. Cognition 101(2):414–442
Blair RJR, Colledge E, Mitchell D (2001) Somatic markers and response reversal: Is there orbitofrontal cortex dysfunction in boys with psychopathic tendencies? J Abnorm Child Psychol 29:499–511
Blake C, Molloy C, Shakespeare S (2012) Beyond human: from animality to transhumanism. Continuum, New York
Blake R, Avis W, Mouton J (1966) Corporate Darwinism. Gulf Pub, Houston
Blakers M (2001) The global greens. Global Greens, Canberra
Blanc L (1839) Organisation du travail. (The organisation of labour). Bureau de la Société de l'Industrie Fraternelle, Paris
Blanchard R (1997) Birth order and sibling sex ratio in homosexual versus heterosexual males and females. Ann Rev Sex Res 8:27–68
Blanchard R (2011) Fertility in the mothers of firstborn homosexual and heterosexual men. Arch Sex Behav 41:551–556
Blanchard R, Klassen P (1997) H-Y antigen and homosexuality in men. J Theor Biol 185(3):373–378
Blancke S (2011) Towards an integrated understanding of creationism in Europe. Historical, philosophical and educational perspectives. Universiteit Gent, Faculteit Letteren en Wijsbegeerte, Gent
Blancke S, Hjermitslev HH, Kjærgaard PC (eds) (2014) Medicin, science, and religion in historical context. The Johns Hopkins University Press, Baltimore
Blancke S, Hjermitslev HH, Braeckman J, Kjærgaard PC (2013) Creationism in Europe: facts, gaps, and prospects. J Am Acad Relig.https://doi.org/10.1093/jaarel/lft034
Blane D, Davey Smith G, Bartley M (2008) Social selection: what does it contribute to social class difference in health? Sociol Health Illn 15(1):1–15
Blum D (1998) Sex on the brain: the biological differences between men and women. Penguin, New York
Blume M (2009) The reproductive benefits of religious affiliation. In: Voland E, Schiefenhövel W (eds) The biological evolution of religious mind and behavior. Springer, New York, pp 117–126
Blurton Jones NG (1987) Bushman birth spacing: direct tests of some simple predictions. Ethol Sociobiol 8:183–203
Bly R (2004) Iron John. A book about men. Da Capo Press, Cambridge
Bocherens H, Drucker DG, Billiou D et al (2006) Bears and humans in Chauvet Cave (Vallon-Pont-d'Arc, Ardèche, France): Insights from stable isotopes and radiocarbon dating of bone collagen. J Hum Evol 50(3):370–376
Bocquet-Appel J-P (2011) The agricultural demographic transition during and after the agriculture inventions. Curr Anthropol 52(S4):S497–S510
Boddy CR (2011) The corporate psychopaths theory of the global financial crisis. J Bus Ethics 102 (2):255–259

Boddy CRP, Ladyshewsky R, Galvin P (2010) Leaders without ethics in global business: corporate psychopaths. J Public Aff Special Issue Unethical Leadersh 10(3):121–138

Bodzsar EB, Susanne C (eds) (1998) Secular growth changes in Europe. Eotvos University Press, Budapest

Boehm C (1993) Egalitarian behavior and reverse dominance hierarchy. Curr Anthropol 34 (3):227–254

Boehm C (1996) Emergency decisions, cultural-selection mechanics, and group selection. Curr Anthropol 37(5):763–793

Boehm C (1999) Hierarchy in the forest: the evolution of egalitarian behavior. Harvard University Press, Cambridge

Boehm C (2008) Purposive social selection and the evolution of human altruism. Cross-Cult Res 24(4):319–352

Boehm C (2012) Moral origins: the evolution of virtue, altruism, and shame. Basic Books, New York

Boehm C (2014) The moral consequences of social selection. Behaviour 151 (2–3): SI 167–183

Boehm C (1997) Impact of the human egalitarian syndrome on Darwinian selection mechanics. Am Nat 150(S1):S100–S121

Bogin B, Bragg J, Kuzawa C (2014) Humans are not cooperative breeders but practice biocultural reproduction. Ann Hum Biol 41(4):368–380

Bonduriansky R, Chenoweth SF (2009) Intralocus sexual conflict. Trends Ecol Evol 24(5):280–288

Bone JE, Raihani NJ (2015) Human punishment is motivated by both a desire for revenge and a desire for equality. Evol Hum Behav 36(4):323–330

Bongaarts J (1978) A framework for analyzing the proximate determinants of fertility. Popul Dev Rev 4(1):105–132

Bongaarts J (2002) The end of the fertility transition in the developed world. Popul Dev Rev 28 (3):419–443

Bongaarts J, Potter RG (1983) Fertility, biology, and behavior. An analysis of the proximate determinants. Academic Press, New York

Boniolo G, De Anna G (eds) (2006) Evolutionary ethics and contemporary biology. Cambridge University Press, Cambridge

Bonneux L, Barendregt JJ, Nusselder WJ et al (1998) Preventing fatal diseases increases healthcare costs: cause elimination life table approach. BMJ 316(7124):26–29

Book AS, Quinsey VL (2004) Psychopaths: cheaters or warrior-hawks? Personality Individ Differ 36(1):33–35

Boomsma DI, Geus EJC, Van Baal GCM et al (1999) A religious upbringing reduces the influence of genetic factors on disinhibition: evidence for interaction between genotype and environment on personality. Twin Res 2(2):115–125

Boone JL (1998) The evolution of magnanimity: when is it better to give than to receive? Hum Nat 9(1):1–21

Boone JL, Kessler KL (1999) More status or more children? Social status, fertility reduction, and long-term fitness. Evol Hum Behav 20(4):257–277

Boorman SA, Levitt PR (1973) Group selection on the boundary of a stable population. Theor Popul Biol 4(1):85–128

Booth A, Carver K, Granger DA (2000) Biosocial perspectives on the family. J Marr Fam 62 (4):1018–1034

Booth A, Dabbs J (1993) Testosterone and men's marriages. Soc Forces 72(2):463–477

Borg J, Andree B, Soderstrom H et al (2003) The serotonin system and spiritual experiences. Am J Psychiatry 160(11):1965–1969

Borg JS, Hynes C, Van Horn J et al (2006) Consequences, action, and intention as factors in moral judgments: an FMRI investigation. J Cogn Neurosci 18(5):803–817

Borgerhoff Mulder M (1990) Kipsigis women's preferences for wealthy men: evidence for female choice in mammals? Behav Ecol and Sociobiol 27(4):255–264

Borgerhoff Mulder M (1998) Demographic transition: are we any closer to an evolutionary explanation? Trends Ecol Evol 13(7):266–270
Borgerhoff Mulder M, Schacht R (2012) Human behavioral ecology. Evol Hum Behav 24:99–112
Borrello ME (2005) The rise, fall and resurrection of group selection. Endeavour 29(1):43–47
Bossard JHS, Boll ES (1975) The Large family system: an original study in the sociology of family behaviour. Greenwood press, Westport
Bosshard G, Fischer S, Bar W (2002) Open regulation and practice in assisted dying. Swiss Med Weekly 132(37–38):527–534
Bostrom N (2002) Existential risks: analyzing human extinction scenarios and related hazards. J Evol Technol 9(1):1–33
Bostrom N (2003) The transhumanist FAQ: v 2.1. World Transhumanist Association. http://transhumanism.org/index.php/WTA/faq/
Bostrom N (2004) Human genetic enhancements: A transhumanist perspective. J Value Inq 37 (4):493–506
Bostrom N (2004) The future of human evolution. In: Tandy C (ed) Death and anti-death: two hundred years after Kant, fifty years after Turing. Ria University Press, Palo Alto, pp 339–371
Bostrom N (2005) A history of transhumanist thought. J Evol Technol 14(1):1–25
Bostrom N (2005) Transhumanist values. Review of contemporary. Philosophy 4(1–2):87–101
Bostrom N (2007) Technological revolutions: ethics and policy in the dark. In: de S. Cameron NM (ed) Nanotechnology and society. Wiley, Hoboken
Bostrom N (2008) Letter from Utopia. Stud Ethics Law Technol 2(1):1–7
Bostrom N (2009) The future of humanity. In: Olsen JKB, Pedersen SA, Hendricks VF (eds) Companion to philosophy of technology. Wiley-Blackwell, Oxford, pp 551–558
Bostrom N (2014) Superintelligence. Paths, dangers, strategies. Oxford University Press, Oxford
Bostrom N, Cirkovic M (eds) (2011) Global catastrophic risks. Oxford University Press, Oxford
Bostrom N, Roache R (2008) Ethical issues in human enhancement. In: Ryberg J, Petersen T, Wolf C (eds) New waves in applied ethics. Palgrave Macmillan, Basingstoke, pp 120–152
Bostrom N, Sandberg A (2007) Cognitive enhancement: methods, ethics, regulatory challenges. Sci Eng Ethics 15(3):311–341
Bostrom N, Savulescu J (2009) Human enhancement. Oxford University Press, Oxford
Botero CA, Gardner B, Kirby KR et al (2014) The ecology of religious beliefs. Proc Nat Acad Sci 111(47):16784–16789
Bouchard F, Hunemann P (2013) From groups to individuals. Evolution and emerging individuality. The MIT Press, Cambridge
Bouchard TJ Jr, Lublin JC (2001) Genes, evolution, and personality. Behav Genet 31(3):243–273
Bouchard TJ Jr, McGue M, Lykken DT et al (1999) Intrinsic and extrinsic religiousness: genetic and environmental influences and personality correlates. Twin Res 2(2):88–98
Boudry M (2014) Beste gematigde moslim, gebruik de Koran niet om de barbarij van IS af te wijzen. (Dear moderate Muslim, don't use the Quran to reject the barbarism of IS). De Morgen, 13 Sept 2014
Boudry M (2015) Illusies voor gevorderden. Of waarom waarheid altijd beter is. (Illusions for the advanced or why truth is always better). Polis, Kalmthout
Bourgeois-Pichat J (1988) Du XXe au XXIe siècle: l'Europe et sa population après l'an 2000. (From the twentieth to the twenty-first century: Europe and its population after 2000). Population 43(1):9–44
Bovens M, Dekker P, Tiemeijer W (eds) (2014) Gescheiden werelden? Een verkenning van sociaal-culturele tegenstellingen in Nederland. (Separate worlds? An exploraton of socio-cultural antagonisms in the Netherlands). Sociaal en Cultureel Planbureau; Wetenschappelijke Raad voor het Regeringsbeleid, Den Haag
Bowlby J (1969) Attachment and loss, attachment, vol I. Basic Books, New York
Bowler PJ (1984) Evolution. The history of an idea. University of California Press, Berkeley
Bowles S (2006) Group competition, reproductive leveling, and the evolution of human altruism. Science 314(5805):1569–1572

Bowles S (2009) Did warfare among ancestral hunter-gatherers affect the evolution of human social behaviors? Science 324(5932):1293–1298
Bowles S (2012) Warriors, levelers, and the role of conflict in human social evolution. Science 336 (6083):876–879
Bowles S, Choi J-K, Hopfensitz A (2003) The co-evolution of individual behaviors and social institutions. J Theor Biol 223(2):135–147
Bowles S, Gintis H (2011) A cooperative species: human reciprocity and its evolution. Princeton University Press, Princeton
Boxer MJ, Quataert JH (1978) Socialist women: European socialist feminism in the nineteenth and early twentieth centuries. Greenwood Publishing Group, Westport
Boyanowski B (2002) Fascism. http://departments.kings.edu/history/20c/fascism.html
Boyd R, Gintis H, Bowles S et al (2003) The evolution of altruistic punishment. PNAS 100 (6):3531–3535
Boyd R, Richerson PJ (1985) Culture and the evolutionary process. The University of Chicago Press, Chicago
Boyd R, Richerson PJ (1989) The evolution of indirect reciprocity. Soc Netw 11:213–236
Boyd R, Richerson PJ (1990) Group selection among alternative evolutionarily stable strategies. J Theor Biol 145:331–342
Boyd R, Richerson PJ (2005) Solving the puzzle of human cooperation. In: Levinson S (ed) Evolution and culture. MIT Press, Cambridge, pp 105–132
Boyd R, Richerson PJ (2007) Group selection: a tale of two controversies. In: Gangestad SW, Simpson JA (eds) The evolution of mind: fundamental questions and controversies. Guilford Press, New York, pp 221–225
Boyer P (2001) Religion explained: the evolutionary origins of religious thought. Basic Books, New York
Boyer P (2003) Religious thought and behaviour as by-products of brain function. Trends Cogn Sci 7(3):119–124
Boyle J (1989) Sanctity of life and suicide: tensions and development within common morality. In: Brody BA (ed) Suicide and euthanasia. Kluwer, Dortrecht, pp 221–250
Bozon M, Kontula O (1997) Initiation sexuelle et genre: comparaison des evolutions de douze pays européens. (Sexual initiation and gender: comparison of developments in twelve European countries). Population 52(6):1367–1400
Bozon M, Kontula O (1998) Sexual initiation and gender in Europe: a cross-cultural analysis of trends in the twentieth century. In: Hubert M, Bajos N, Sandfort Th (eds) Sexual behaviour and HIV/AIDS in Europe. UCL Press, London, pp 37–67
Brace C (1973) Sexual dimorphism in human evolution. Yearb Phys Anthropol 16:31–49
Brace C, Ryan A (1980) Sexual differences and human tooth size differences. J Hum Evol 9 (5):417–436
Bradie M (1994) The secret chain. Evolution and ethics. State University of New York, New York
Bradley JR (2012) After the Arab spring: how Islamists hijacked the Middle East revolts. Palgrave Macmillan, Basingstoke
Bradshaw M, Ellison CG (2008) Do genetic factors influence religious life? Findings from a behavior genetic analysis of twin siblings. J Sci Study Relig 47(4):529–544
Bradshaw J, Finch N, Mayhew E et al (2006) Child poverty in large families. Policy Press, Bristol
Braidotti R (2013) The posthuman. Polity Press, Bristol
Brassington I (2010) Enhancing evolution and enhancing evolution. Bioethics 24(8):395–402
Bravo G (2010) Voluntary contribution to public goods in mutual-help forums: reciprocity or group attachment? Soc-Econ Rev 8(4):709–733
Bregman R (2013) De geschiedenis van de vooruitgang. (The History of Progress). De Bezige Bij, Amsterdam
Breheny M, Stephens C (2010) Youth or disadvantage? The construction of teenage mothers in medical journals. Culture Health Sex 12(3):307–322

Brewer MB (1999) The psychology of prejudice: in-group love or out-group hate. J Soc Issues 55 (3):429–444
Britt L (2003) Fascism anyone? Free Inq 23(2):20
Brizendine L (2006) The female brain. Bantam Books, London
Broca P (1872) Les sélections, la descendance de l'homme, la sélection sexuelle de Darwin et la sélection naturelle de Wallace. (Selections, the descent of man, sexual selection of Darwin and natural selection of Wallace). Bull de la Société d'Anthropologie I:683–710
Brock DW, MacLean D (1993) Life and death: philosophical essays in biomedical ethics. Cambridge studies in philosophy and public policy. Cambridge University Press, New York
Brodie R (1996) Virus of the mind: the new science of the meme. Integral Press, Seattle
Bronstein J (2010) Objecting to the genetic virtue program premises, tradeoffs, and science. Polit Life Sci 29(1):85–87
Bronstein C (2011) Battling pornography: the American Feminist anti-pornography movement, 1976–1986. Cambridge University Press, New York
Broom DM (2004) The evolution of morality and religion. Cambridge University Press, Cambridge
Broom DM (2006) The evolution of morality. Appl Anim Behav Sci 100(1–2):20–28
Brosnan K (2010) Do the evolutionary origins of our moral beliefs undermine moral knowledge? Biol Philos 26(1):51–79
Brosnan SF (2011) An evolutionary perspective on morality. J Econ Behav Organ 77(1):23–30
Brosnan SF, de Waal FBM (2003) Monkeys reject unequal pay. Nature 425(6955):297–299
Brosnan SF, Salwiczek L, Bshary R (2010) The interplay of cognition and cooperation. Philos Trans R Soc B 365(1553):2699–2710
Brown DE (1991) Human universals. Temple University Press, Philadelphia
Brown CM (2012) Hindu perspectives on evolution: Darwin, dharma, and design. Routledge, London
Brown Travis C (ed) (2003) Evolution, gender, and rape. The MIT Press, Cambridge
Brown WS, Murphy NC, Malony HN (1998) Whatever happened to the soul? Scientific and theological portraits of human nature. Fortress Press, Minneapolis
Brownmiller S (1975) Against our will: men, women and rape. Simon and Schuster, New York
Bruce S (2002) God is dead: secularization in the West. Blackwell Publishers, Malden
Bruess CE, Greenberg JS (2008) Sexuality education: theory and practice. Jones and Bartlett Publishers, Sudury
Brugger P (2007) Das Glaubige Gehirn. In: Matthiesen S, Rosenzweig R (eds) Von Sinnen. (About senses). Mentis, Paderborn
Brunner HG, Nelen M, Breakefield XO et al (1993) Abnormal behavior associated with a point mutation in the structural gene for monoamine oxidase A. Science 262(5133):578–580
Buchanan A (2011) Beyond humanity? The ethics of biomedical enhancement. Oxford University Press, Oxford
Buchanan A, Brock DW, Daniels N et al (2000) From chance to choice: genetics and justice. Cambridge University Press, Cambridge
Buckholtz JW, Callicott JH, Kolachana B et al (2008) Genetic variation in MAOA modulates ventromedial prefrontal circuitry mediating individual differences in human personality. Mol Psychiatry 13:313–324
Buckholtz JW, Marois R (2012) The roots of modern justice: cognitive and neural foundations of social norms and their enforcement. Nat Neurosci 15(5):655–661
Buckman R (2000) Can we be good without God?. Viking, Toronto
Bucy EP (2010) Reconsidering genetic virtue. Polit Life Sci 29(1):72
Budin SL (2008) The myth of sacred prostitution in antiquity. Cambridge University Press, Cambridge
Buikhuisen W (1979) Kriminologie in biosociaal perspectief. (Criminology in Biosocial Perspective). Kluwer, Deventer

Bulbulia J (2004) Religious costs as adaptations that signal altruistic intention. Evol Cogn 10 (1):19–38
Bulbulia J (2012) Spreading order: religion, cooperative niche construction, and risky coordination problems. Biol Philos 27(1):1–27
Bulbulia J, Sosis R, Genet C et al (eds) (2008) The evolution of religion: studies, theories, and critiques. Collins Foundation Press, Santa Margarita
Bulkeley K (2004) The wondering brain: thinking about religion with and beyond cognitive neuroscience. Routledge, New York
Bulkeley K (2005) Soul, psyche, brain: new directions in the study of religion and brain-mind science. Palgrave Macmillan, New York
Bull J, Rice W (1991) Distinguishing mechanisms for the evolution of co-operation. J Theor Biol 149(1):63–74
Bullough VL, Bullough B (1987) Women and prostitution: a social history. Prometheus Books, Buffalo
Bulmer MG (1989) Structural stability in models of sexual selection. Theor Popul Biol 35:195–206
Burgess RL, Garbarino J (1983) Doing what comes naturally? An evolutionary perspective on abuse. In: Finkelhor D, Gelles RJ, Hotaling GT et al (eds) The dark side of families. Current family violence research. Sage, Beverly Hills, pp 88–101
Burkart JM, Allon O, Amici F et al (2014) The evolutionary origin of human hyper-cooperation. Nat Commun 5:4747
Burkert W (1996) Creation of the sacred: tracks of biology in early religions. Harvard University Press, Cambridge
Burley N, Symanski R (1981) Women without: an evolutionary and cross-cultural perspective on prostitution. In: Symanski R (ed) The immoral landscape: female prostitution in western societies. Butterworth, Toronto, pp 239–274
Burnham TC, Johnson DDP (2005) The biological and evolutionary logic of human cooperation. Anal Kritik 27:113–135
Burnham T, Phelan J (2000) Mean genes: from sex to money to food: taming our primal instincts. Perseus, Cambridge
Burt C (1961) Intelligence and social mobility. Br J Statist Psychol 14(1):3–24
Burton R (2008) A problem-oriented policing approach to tackling youth crime and antisocial behaviour on London's buses. Transport for London, London
Buss DM (1989) Sex differences in human mate preferences: evolutionary hypotheses tested in 37 cultures. Behav Brain Sci 12:1–49
Buss DM (1994) The evolution of desire. Basic Books, New York
Buss DM (1994) The strategies of human mating. Am Sci 82(3):238–249
Buss DM (1996) Sexual conflict: evolutionary insights into feminism and the "Battle of the sexes". In: Buss DM, Malamuth N (eds) Sex, power, conflict: evolutionary and feminist perspectives. Oxford University Press, New York, pp 296–318
Buss DM (1999) 2007) Evolutionary psychology. The new science of the mind. Allyn and Bacon, Boston
Buss DM (2002) Human mate guarding. Neuroendocrinol Lett 23(Suppl. 4):23–29
Buss DM (2002) The dangerous passion: why jealousy is as neccessary as love and sex. Free Press, New York
Buss DM (2007) The evolution of human mating strategies. In: Gangestad SW, Simpson JA (eds) The evolution of mind: fundamental questions and controversies. Guilford Press, New York, pp 375–382
Buss DM (ed) (2016) The handbook of evolutionary psychology, vol 1: Foundation. Wiley, Hoboken
Buss DM, Malamuth N (eds) (1996) Sex, power, conflict: evolutionary and feminist perspectives. Oxford University Press, New York

Buss DM, Schmitt DP (1993) Sexual strategies theory: an evolutionary perspective on human mating. Psychol Rev 100(2):204–232

Butler R (1969) Ageism: another form of bigotry. Gerontol 9(3):243–246

Button TMM, Stallings MC, Rhee SH et al (2011) The etiology of stability and change in religious values and religious attendance. Behav Genet 41(2):201–210

Byars SG, Ewbank D, Govindaraju DR et al (2010) Natural selection in a contemporary human population. PNAS 107(S1):1787–1792

Byrne R, Whiten A (eds) (1988, 1997) Machiavellian intelligence. Social expertise and the evolution of intellect in monkeys, apes, and humans. Clarendon Press, Oxford preview/mmwrhtml/ss5406a1.htm

Cabezón I (1992) Buddhism, sexuality, and gender. State University of New York Press, Albany

Cadoret RJ, Langbehn D, Caspers K et al (2003) Associations of the serotonin transporter promoter polymorphism with aggressivity, attention deficit, and conduct disorder in an adoptee population. Compr Psychiatry 44(2):88–101

Camperio Ciani AS, Fontanesi L, Iemmola F et al (2012) Factors associated with higher fecundity in female maternal relatives of homosexual men. J Sex Med 9(11):2878–2887

Caspi A, McClay J, Moffitt TE et al (2002) Role of genotype in the cycle of violence in maltreated children. Science 297(5582):851–854

Chanana K (2001) Hinduism and female sexuality: social control and education of girls in India. Sociol Bull 50(1):37–63

Chandra R (2005) Identity and genesis of caste system in India. Gyan Books, New Delhi

Chaudhuri S (2013) A life course model of human rights realization, female empowerment, and gender inequality in India. World Dev 52:55–70

Ching J (1993) Chinese religions. Orbis Books, Maryknoll

Chen Y (2012) Confucianism as religion: controversies and consequences. Brill, Leiden

Cherlin AJ, Scabini, Rossi G (2016) Still in the nest: delayed home leaving in Europe and the United States. J Fam Iss 18(6):572–575

Chesnais J-C (1998) Below-replacement fertility in the European Union (EU-15): Facts and policies, 1960–1997. Rev Pop Soc Pol 7:83–101

Cheverud JM, Dow MM, Leutenegger W (1985) The quantitative assessment of phylogenetic constraints in comparative analyses: sexual dimorphism in body weight among primates. Evolution 39(6):1335–1351

Clark KJ, Wang RR (2004) A confucian defense of gender equity. J Am Acad Relig 72(2):395–422

Coleman C (2017) Hinduism: adopting Hinduism as a way of life. CreateSpace Independent Publishing Platform, Colorado Springs CO

Cafaro P, Crist E (2012) Life on the brink: environmentalists confront overpopulation. University of Georgia Press, Athens

Cahill L, Haier RJ, White NS et al (2001) Sex-related difference in amygdala activity during emotionally influenced memory storage. Neurobiol Learn Mem 75(1):1–9

Caiazza J (2005) The evolution versus religion controversy: how two mystiques devolved into politics. Mod Age 47(2):104–113

Cain M (1982) Perspectives on family and fertility in developing countries. Popul Stud 36:159–175

Caldeira K, Kasting JF (1992) The life span of the biosphere revisited. Nature 360(6406):721–723

Caldwell BK, Caldwell JC (2003) Below-replacement fertility: determinants and prospects in South Asia. J Popul Res 20(1):19–34

Caldwell C (2009) Reflections on the revolution in Europe: immigration, Islam, and the West. Doubleday, New York

Caldwell JC (1982) Theory of fertility decline. Academic Press, London

Callahan D (2004) The cheating culture: why more Americans are doing wrong to get ahead. Houghton Mifflin Harcourt, Boston

Calman KC (2004) Evolutionary ethics: can values change? J Med Ethics 30(4):366–370

Calot G (1990) La fécondité en Europe: evolutions passées et perspectives d'avenir. (Fertility in Europe: past developments and future prospects). Bevolking en Gezin 1990(3):56–82
Camerer C, Loewenstein G, Prelec D (2005) Neuroeconomics. J Econ Lit 43(1):9
Campbell A (2006) Feminism and evolutionary psychology. In: Barkow JH (ed) Missing the revolution: Darwinism for social scientists. Oxford University Press, Oxford, pp 63–99
Campbell A (2008) Attachment, aggression and affiliation: the role of oxytocin in female social behavior. Biol Psychol 77(1):1–10
Campbell B (ed) (1972) Sexual selection and the descent of man. Aldine, Chicago
Campbell DT (1965) Variation and selective retention in sociocultural evolution. In: Barringer HR, Blanksten GI, Mack RW (eds) Social change in developing areas: a reinterpretation of evolutionary theory. Schenkman, Cambridge, pp 19–49
Campbell DT (1975) On the conflicts between biological and social evolution and between psychology and moral tradition. Am Psychol 30(12):1103–1126
Campbell DT (1975) The conflict between social and biological evolution and the concept of original sin. Zygon 10(3):234–249
Campbell DT (1979) Comments on the sociobiology of ethics and moralizing. Behav Sci 24 (1):37–45
Campbell H, Walker M (2005) Religion and transhumanism: introducing a conversation. J Evol Technol 14(2):i–xiv
Campbell R (1996) Can biology make ethics objective? Biol Philos 11(1):21–31
Campbella SM, Wahlerta L (2015) Is disability conservationism rooted in status quo bias? Am J Bioethics 15(6):20–22
Camperio-Ciani A, Battaglia U, Zanzotto G (2015) Human homosexuality: a paradigmatic arena for sexually antagonistic selection? Cold Spring Harbor Perspect Biol: 017657
Camperio-Ciani A, Corna F, Capiluppi C (2004) Evidence for maternally inherited factors favouring male homosexuality and promoting female fecundity. Proc R Soc Lond Ser B Biol Sci 271:2217–2221
Canli T (2008) Toward a "molecular psychology" of personality. In: John OP, Robins RW, Pervin LA (eds) Handbook of personality: theory and research. Guilford Press, New York, pp 311–327
Cannon WB (1939) The wisdom of the body. Norton, New York
Caplan A (2006) Is it wrong to try to improve human nature? In: Miller P, Wilsdon J (eds) Better humans. The politics of human enhancement and life extension. Demos, London, pp 31–39
Caporael L, Dawes RM, Orbell JM et al (1989) Selfishness examined: cooperation in the absence of egoistic incentives. Behav Brain Sci 12:683–739
Caprara GV, Schwartz S, Capanna C et al (2006) Personality and politics: values, traits, and political choice. Polit Psychol 27(1):1–28
Capraro V, Smyth C, Mylona K et al (2014) Benevolent characteristics promote cooperative behaviour among humans. PLOS ONE 9(8):e102881
Carbonell E, Mosquera M (2006) The emergence of a symbolic behaviour: the sepulchral pit of Sima de los Huesos, Sierra de Atapuerca, Burgos, Spain. Hum Palaeontol Prehist Comptes Rendus Palevol 5(1–2):155–160
Carey JR (2003) Longevity. The biology and demography of life span. Princeton University Press, Princeton
Carle R (2006) Demise of Dutch multiculturalism. Society 43(3):68–74
Carlsen E, Giwercman A, Keiding N et al (1992) Evidence of decreasing quality of semen during past 50 years. Br Med J 305(6854):609–613
Carlson EA (1973) Eugenics revisited: the case of germinal choice. Stadler symposium vol 5, pp 13–34. University of Missouri, Columbia
Carlson A, Lintott S (2008) Nature, aesthetics, and environmentalism. From beauty to duty. Columbia University Press, New York
Carnegie A (1920) Autobiography of Andrew Carnegie. Edited Van Dyke JC, reprint. Northeastern University Press, Boston, p 327

Carneiro RL (1992) The role of natural selection in the evolution of culture. Cult Dyn 5:113–140
Carson R (1962) Silent spring. Houghton Mifflin, Boston
Carroll L (1872) Through the looking-glass, and what Alice found there. Macmillan, London
Carter JA, Gordon EC (2015) On cognitive and moral enhancement: a reply to Savulescu and Persson. Bioethics 29(3):153–161
Carter CS, Grippo AJ, Pournajafi-Nazarloo H et al (2008) Oxytocin, vasopressin and sociality. Prog Brain Res 170:331–336
Carter S (2016) Could moral enhancement interventions be medically indicated? Health Care Anal. https://doi.org/10.1007/s10728-016-0320-8
Cartwright J (2000) Evolution and human behavior. Darwinian perspectives on human nature. Macmillan Press, Houndmills
Cartwright J (2010) Naturalising ethics: the implications of Darwinism for the study of moral philosophy. Sci Educ 19(4):407–443
Casavant L, Robertson JR (2007) The evolution of pornography law in Canada. Library of Parliament. Law and Government Division, Ottawa
Casebeer WD (2003) Natural ethical facts: evolution, connectivism, and moral cognition. MIT Press, Cambridge
Cassill D, Watkins A (2010) The evolution of cooperative hierarchies through natural selection processes. J Bioecon 12(1):29–42
Castellani C, Picci L, Tamanini A et al (2009) Association between carrier screening and incidence of cystic fibrosis. JAMA 302(23):2573–2579
Caton H (2006) Evolutionary constraints on democratic nation building. Evol Psychol 4(1):142–148
Cattell RB (1933) Psychology and social progress. Daniel, London
Cattell RB (1937) The fight for our national intelligence. King, London
Cattell RB (1938) Psychology and the religious quest. Nelson, New York
Cattell RB (1944) The place of religion and ethics in a civilization based on science. In: Wulsin R (ed) A reevaluation of our civilization. Argus Press, Albany
Cattell RB (1950) The scientific ethics of "Beyond". J Soc Issues 6(4):21–27
Cattell RB (1972) A new morality from science: beyondism. Pergamon, Elmsford
Catton W (1980) Overshoot: the ecological basis of revolutionary change. University of Illinois Press, Urbana
Cavalli-Sforza LL, Bodmer WF (2013) The genetics of human populations. Dover Publications, Mineola
Cavalli-Sforza LL, Feldman MW (1981) Cultural transmission: a quantitative approach. Princeton University Press, Princeton
Cavalli-Sforza LL, Feldman MW (2003) The application of molecular genetic approaches to the study of human evolution. Nat Genet 33 (Supplement):266–275
Cavalli-Sforza LL, Menozzi P, Piazza A (1994) History and geography of human genes. Princeton University Press, Princeton
Cela-Conde CJ (1987) On genes, gods and tyrants. The biological causation of morality. D. Reidel Publishing Company, Dordrecht
Cela-Conde CJ, Burges L, Nadal M et al (2010) Altruism and fairness: unnatural selection? Comptes Rendus Biol 333(2):SI 174–180
Cesarini D, Johannesson M, Lichtenstein P et al (2010) Genetic variation in financial decision-making. J Finan 65(5):1725–1754
Cetin I, Cozzi V, Antonazzo P (2003) Fetal development after assisted reproduction—a review. Placenta 24:S104–S113
Chagnon NA (1990) Reproductive and somatic conflicts of interest in the genesis of violence and warfare among tribesmen. In: Haas J (ed) The anthropology of war. Cambridge University Press, Cambridge
Chagnon NA (1997) Yanomamö. Harcourt Brace College Publishers, Fort Worth

Chagnon NA, Irons WG (eds) (1979) Evolutionary biology and human social behavior: an anthropological perspective. Duxbury Press, North Scituate
Chaisson EJ (2001) Cosmic evolution: the rise of complexity in nature. Harvard University Press, Cambridge
Chaladze G (2016) Heterosexual male carriers could explain persistence of homosexuality in men: Individual-based simulations of an x-linked inheritance model. Arch Sex Behav. https://doi.org/10.1007/s10508-016-0742-2
Chalfant HP, Beckley RE, Palmer CE (1994) Religion in contemporary society. F.E. Peacock, Itasca
Chamberlain HS (1911) The foundations of the nineteenth century. John Lane, London
Chan DK (2015) The concept of human dignity in the ethics of genetic research. Bioethics 29 (4):274–282
Chancer LS (1992) Sadomasochism in everyday life: the dynamics of power and powerlessness. Rutgers University Press, New Brunswick
Chapais B (2008) Primeval kinship: how pair bonding gave birth to human society. Harvard University Press, Cambridge
Chapais B (2013) Monogamy, strongly bonded groups, and the evolution of human social structure. Evol Anthropol 22(2):52–65
Chapais B (2015) Competence and the evolutionary origins of status and power in humans. Hum Nat 26(2):161–183
Chapman AR (2004) The social and justice implications of extending the human life span. In: Post S, Binstock R (eds) The fountain of youth. Oxford University Press, Oxford, pp 340–361
Chapman MJ (2004) Hominid failings: an evolutionary basis for sin in individuals and corporations. In: Clayton P, Schloss J (eds) Evolution and ethics: human morality in biological and religious perspective. Eerdmans, Grand Rapids, pp 101–113
Charbonneau H (1979) Les régimes de fécondité naturelle en Amérique du Nord: Bilan et analyse des observations. (Natural fertility regimes in North America: Review and analysis of observations). In: Léridon H, Menken J (eds) Natural fertility—Fécondité naturelle. Ordina Editions Liège, pp 441–491
Chatters LM (2000) Religion and health: public health research and practice. Ann Rev Public Health 21:335–367
Chauchard P (1959) Biologie et morale. (Biology and morality). Mame, Tours
Chauvet J-M, Brunel Deschamps É, Hillaire C (1995) La Grotte Chauvet à Vallon-Pont-d'Arc. (The Chauvet Cave in Vallon-Pont-d'Arc). Éditions du Seuil, Paris
Chaves M (2011) American religion: contemporary trends. Princeton University Press, Princeton
Cheng YW, Tzeng OJL, Decety J et al (2006) Gender differences in the human mirror system: a magnetoencephalography study. NeuroReport 17(11):1115–1119
Cherry M, Brown R (2009) Speaking freely about religion: religious freedom, defamation and blasphemy. International Humanist and Ethical Union policy paper. IHEU, London (www.iheu.org/UN-blasphemy-report)
Cherry MJ (2009) The normativity of the natural human goods, human virtues, and human flourishing. Springer, Dordrecht
Chesnais JC (1986) La transition démographique, étapes, formes, implications économiques. (Demographic transition, stages, forms, economic implications). Travaux et Documents, Cahier no 113, Institut National d'Etudes Démographiques. Presses Universitaires de France, Paris
Chetverikov SS (1926) On certain aspects of the evolutionary process from the standpoint of modern genetics. Translation by Barker M and Lerner IM (1961). Proc Am Philos Soc 105:167–195
Chew SC (2001) World ecological degradation: accumulation, urbanization, and deforestation, 3000BC-AD2000. Altamira press, New York
Chew SC (2008) Ecological futures: what history can teach us (Trilogy on world ecological degradation). Altamira press, New York

Chiang Y-S, Takahashi N (2011) Network homophily and the evolution of the pay-it-forward reciprocity. PLOS ONE 6(12):e29188
Chiao JY, Blizinsky K (2010) Culture-gene coevolution of individualism-collectivism and the serotonin transporter gene. Proc R Soc B-Biol Sci 277(1681):529–537
Chiesa A, Serretti A (2010) A systematic review of neurobiological and clinical features of mindfulness meditations. Eur Psychiatry 25(S1):1044
Choi J-K, Bowles S (2007) The coevolution of parochial altruism and war. Science 318 (5850):636–640
Chopra D (2006) Life after death: the burden of proof. Harmony, New York
Christ O, Schmid K, Lolliot S et al (2014) Contextual effect of positive intergroup contact on outgroup prejudice. PNAS 111(11):3996–4000
Christen M, Regard M (2012) The "immoral patient": analyzing the role of patients with brain lesions in moral research. Nervenheilkunde 31(4):209–214
Christensen FM (2008) Pornography: The other side. Praeger, New York
Chudek M, Henrich J (2011) Culture-gene coevolution, norm-psychology and the emergence of human prosociality. Trends Cogn Sci 15(5):218–226
Church Catholic (1993) Catechism of the Catholic Church. Libreria Editrice Vaticana, Citta del Vaticano
Churchland P (2002) Brain-Wise: studies in neurophilosophy. MIT Press, Cambridge
Churchland P (2011) Braintrust. What neuroscience tells us about morality. Princeton University Press, Princeton
Ciani AC, Fontanesi L, Iemmola F et al (2012) Factors associated with higher fecundity in female maternal relatives of homosexual men. J Sex Med 9(11):2878–2887
Ciaramelli E, Braghittoni D, di Pellegrino G (2012) It is the outcome that counts! Damage to the ventromedial prefrontal cortex disrupts the integration of outcome and belief information for moral judgment. J Int Neuropsychol Soc 18(6):962–971
CIBA Foundation Symposium (1996) Genetics of criminal and antisocial behaviour—Symposium no. 194. Wiley, Chichester
Cicirelli VG (1991) Attachment theory in old age: protection of the attached figure. In: Pillemer K, McCartney K (eds) Parent-child relations throughout life. Erlbaum, Hillsdale, pp 25–42
Cieri RL, Churchill SE, Franciscus RG et al (2014) Craniofacial feminization, social tolerance, and the origin of behavioral modernity. Curr Anthropol 55(4):417–443
Claidière N, André J-B (2012) The transmission of genes and culture: a questionable analogy. Evol Biol 39(1):12–24
Claidiere N, Scott-Phillips TC, Sperber D (2014) How Darwinian is cultural evolution? Philos Trans R Soc B-Biol Sci 369(1642):SI 20130368
Clark W, Schellenberg G (2008) Who's religious? http://www.statcan.gc.ca/pub/11-008-x/2006001/9181-eng.htm
Clarkson F (1997) Eternal hostility: the struggle between theocracy and democracy. Common Courage Press, Monroe
Clausen J, Levy N (eds) (2015) Handbook of neuroethics. Springer, Dordrecht
Clavien C (2010) Je t'aide moi non plus: biologique, comportemental ou psychologique, l'altruisme dans tous ses états. (I don't help you either: biological, behavioral or psychological altruism in all its forms). Vuibert, Paris
Clavien C (2015) Evolution, society and ethics: social Darwinism versus evolutionary ethics. In: Heams T, Huneman P, Lecointre G et al (eds) Handbook of evolutionary thinking in the sciences. Springer, Dordrecht, pp 725–745
Clavien C, Chapuisat M (2013) Altruism across disciplines: one word, multiple meanings. Biol Philos 28(1):125–140
Clavien C, El Bez C (eds) (2007) Morale et évolution biologique: Entre déterminisme et liberté. (Morality and biological evolution: Between determinism and freedom). Presses polytechniques et universitaires romandes, Lausanne
Clayton JN (2002) Darwin and adoption. Does God Exist? 29(5)

Clayton P (2013) Evolution, altruism and God: why the levels of emergent complexity matter. In: Nowak MA, Coakley S (eds) Evolution, games, and God. The principle of cooperation. Harvard University Press, Cambridge, pp 343–361

Clayton P, Schloss J (eds) (2004) Evolution and ethics: human morality in biological and religious perspective. Eerdmans, Grand Rapids

Cleckley HM (1984) The mask of sanity: an attempt to reinterpret the so-called psychopathic personality. http://www.quantumfuture.net/store/sanity_1.PdF

Cline A (2016) Is atheism an ism? Atheism is no religion, philosophy, ideology, belief system. http://atheism.about.com/od/definitionofatheism/p/AtheismReligion.htm

Cliquet R (1987) The consideration of below-replacement fertility in a long-term evolutionary perspective. C.B.G.S. Werkdocument 51. Centrum voor Bevolkings- en Gezinsstudiën, Ministerie van de Vlaamse Gemeenschap, Brussel

Cliquet R (1961) De zedelijke waardeëlementen in de eugenetiek in het licht van de ontwikkeling van de moderne cultuur. (The moral value elements in eugenics in the light of the development of modern culture). Tijdschr voor Sociale Wet 6:51–71

Cliquet R (1965) Antropologie: Herkomst, wezen en toekomst van de mens. (Anthropology: origin, present, and future of humankind). Onze Horizon. De Sikkel, Antwerpen

Cliquet R (1968) Social mobility and the anthropological structure of populations. Hum Biol 40:17–43

Cliquet R (1984) The relevance of sociobiological theory for emancipatory feminism. J Hum Evol 13:117–127

Cliquet R (1991) The second demographic transition: fact or fiction? Population studies no 23. Council of Europe, Strasbourg

Cliquet R (1996–1997) The demographic future of the human species: from quantity to quality? In: Deven F, Van den Brekel H (eds) Population and family in the low countries. NIDI/CBGS Publications, Brussels/The Hague, vol 33, pp 15–42

Cliquet R (2003) Major trends affecting families in the new millennium: Western Europe and North America. In: United Nations, major trends affecting families. A background document. prepared by the programme on the family. United Nations, New York, pp 1–26

Cliquet R (2010) Biosocial interactions in modernisation. Masaryk University Press, Brno

Cliquet R, Avramov D (1998) The future of the family: a sociobiological approach. In: Matthijs K (ed) The family. Contemporary perspectives and challenges. Festschrift in Honor of Wilfried Dumon. Leuven University Press, Leuven, pp 159–180

Cliquet R, Balcaen J (1983) Quantitative and qualitative aspects of family-size variation. Results of fertility research in Flanders. NIDI/CBGS Publications, vol 11. NIDI and CBGS, Voorburg/Brussels

Cliquet R, Callens M (eds) (1993) Gezinsvorming in Vlaanderen: Hoe en wanneer? Resultaten van de Enquête Gezinsontwikkeling 1991 (NEGO V). (Family formation in Flanders: how and when? Results of the Survey Family Development 1991—NEGO V). CBGS-Monografie, 1993, 1. Centrum voor Bevolkings- en Gezinsstudiën. Ministerie van de Vlaamse Gemeenschap, Brussel

Cliquet R, Delmotte G (1984) Possible effects of differences in family size on the genetic and social composition and structure of Western populations. Population and family in the low countries, NIDI/CBGS publications, vol 12. NIDI and CBGS, Voorburg/Brussels, pp 73–100

Cliquet R, Lodewijckx E (1986) The contraceptive transition in Flanders. Eur J Popul 2:71–74

Cliquet R, Maelstaf H (1977) Levensbeschouwing en gezinsvorming in Vlaanderen. (Worldview and family formation in Flanders). Bevolking en Gezin 3:313–345

Cliquet R, Schoenmaeckers R (1976) From incidental to planned parenthood. A NIDI/C.B.G.S. Publication 2. Martinus Nijhoff Social Sciences Division, Leiden

Cliquet R, Thienpont K (1995) Population and sustainability. The message from the Cairo Conference. European studies of population, volume 3. Kluwer Academic Publishers, Dordrecht

Cliquet R, Thienpont K (2002) Biologische variabiliteit bij de mens. Inleiding tot de biologische antropologie. Boekdeel 1. Determinanten en evolutie. (Biological variability in humans. Introduction to biological anthropology, vol 1. Determinants and evolution). Academia Press, Gent

Cliquet R, Thienpont K (2005) Biologische variabiliteit bij de mens. Inleiding tot de biologische antropologie. Boekdeel 2. Biologische variabiliteit. (Biological variability in humans. Introduction to biological anthropology, vol 2. Biological variability). Academia Press, Gent

Cliquet R, Thiery M (eds) (1972) Abortus provocatus. Een interdisciplinaire studie met betrekking tot beleidsalternatieven in België. (Induced abortion. An interdisciplinary study with regard to policy alternatives in Belgium). Studies en Documenten 2, Centrum voor Bevolkings- en Gezinsstudiën. De Nederlandsche Boekhandel/De Sikkel, Antwerpen/Utrecht

Cliquet R, Van De Velde (1985) De wereldbevolking op een keerpunt? Resultaten van de Internationale Bevolkingsconferentie, Mexico, 6-14 augustus 1984. (The World population at a turning point? Results of the International Conference on Population, Mexico, 6-14 August 1984). C.B.G.S. Monografie 1985/3. Centrum voor Bevolkings- en Gezinstudie, Brussel

Cliquet R, Veys D (1974) Naar een wereldbevolkingspolitiek? De Wereldbevolkingsconferentie in Boekarest, 19/30 augustus 1974. (Towards a world population politics? The World Population Conference in Bucharest, 19 August 30, 1974). Studies en Documenten 4. De Nederlandsche Boekhandel/De Sikkel, Kapellen

Cliteur P (2007), Moreel Esperanto. (Moral Esperanto). De Arbeiderspers, Amsterdam

Cliteur P (2010) The Secular outlook. In defense of moral and political secularism. Wiley-Blackwell, Chichester

Cloninger CR, Svrakic DM, Przybeck TR (1993) A psychobiological model of temperament and character. Arch Gen Psychiatry 50(12):975–990

Closson D (2006) Technological innovation must be limited to protect humanity. In: Gerdes LI (ed) Humanity's future. Greenhaven Press, Detroit, pp 29–36

Clutton-Brock TH, Harvey PH (1980) Primates, brains, and ecology. J Zool 190(3):309–323

Clutton-Brock TH, Harvey PH, Rudder B (1977) Sexual dimorphism, socionomic sex ratio and body weight in Primates. Nature 269(5631):797–800

Clutton-Brock TH, Parker GA (1995) Punishment in animal societies. Nature 373(6511):209–216

Coakley S (2013) Evolution, cooperation and divine providence. In: Nowak MA, Coakley S (eds) Evolution, games, and God. The principle of cooperation. Harvard University Press, Cambridge, pp 375–386

Coall DA, Hertwig R (2010) Grandparental investment: past, present, and future. Behav Brain Sci 33:1–19

Cochran G, Harpending H (2009) The 10,000 year explosion: how civilization accelerated human evolution. Basic Books, New York

Cochrane SH (1979) Fertility and education. What do we really know? The Johns Hopkins University Press, Baltimore

Cohen H, Buskila D, Neumann L et al (2002) Confirmation of an association between fibromyalgia and serotonin transporter promoter region (5-HTTLPR) polymorphism, and relationship to anxiety-related personality traits. Arthritis Rheum 46(3):845–847

Cohen JE (1995) The earth's carrying capacity. Sci J AAAS 269(5222):341–346

Cohen JE (1996) How many people can the earth support?. Norton, New York

Cole-Turner R (1998) Human cloning: religious responses. Westminster/John Knox Press, Louisville

Cole-Turner R (2011) Transhumanism and transcendence: Christian hope in an age of technological enhancement. Georgetown University Press, Washington

Coleman D (1994) Trends in fertility and intermarriage among immigrant populations in Western Europe as measures of integration. J Biosoc Sci 26(1):107–136

Coleman D (2004) Why we don't have to believe without doubting in the "Second Demographic Transition"—some agnostic comments. Vienna Yearbook of Population Research. Austrian Academy of Sciences, Vienna, pp 11–24

Coleman D (2006) Immigration and ethnic change in low-fertility countries: a third demographic transition. Popul Dev Rev 32(3):401–446
Coleman D, Rowthorn R (2011) Who's afraid of population decline? A critical examination of its consequences. Popul Dev Rev 37(S1):217–248
Collaer ML, Hines M (1995) Human behavioral sex differences: a role for gonadal hormones during early development? Psychol Bull 118(1):55–107
Collins FS (2006) The language of God: a scientist presents evidence for belief. Free Press, New York
Collins C, Alagiri P, Summers T, et al (2002) Abstinence only vs. comprehensive sex education: what are the arguments? What is the evidence? Policy Monograph Series. http://ari.ucsf.edu/science/reports/abstinence.pdf
Colman AM, Wilson JC (1997) Antisocial personality disorder: an evolutionary game theory analysis. Legal Criminol Psychol 2:23–34
Colson CW, de Cameron SNM (eds) (2004) Human dignity in the biotech century: a christian vision for public policy. InterVarsity Press, Downers Grove
Comfort A (1956) 1979) The biology of senescence. Churchill Livingstone, Edinburgh
Comfort A (1964) The process of ageing. New American Library, New York
Comfort A (2003) The joy of sex: fully revised and completely updated for the 21st century. Crown Archetype, New York
Comings D (2008) Did man create God? Is your spiritual brain at peace with your thinking brain?. Hope Press, Duarte
Comings DE (2010) The neurobiology, genetics and evolution of human spirituality. The central role of the temporal lobes. Neuroquantology 8(4):478–494
Comings DE, Gonzales N, Saucier G et al (2000) The DRD4 gene and the spiritual transcendence scale of the character temperament index. Psychiatr Genet 10(4):185–189
Commissie voor de behandeling van klachten wegens seksueel misbruik in een pastorale relatie (2010) Verslag activiteiten Commissie voor de behandeling van klachten wegens seksueel misbruik in een pastorale relatie. (Report Activities Commission for the Treatment of Complaints of Sexual Abuse in a Pastoral Relationship). Brussel. http://www.standaard.be/extra/static/pdf/eindrappport_adriaenssens.pdf
Commission European (2005) Special Eurobarometer: Europeans: science and technology. European Commission, Brussels
Commission European (2011) Gender equality in the European Union. Publications Office of the European Union, Luxembourg
Commission of the European Communities (1999) Communication from the Commission: towards a Europe for all ages. Promoting prosperity and intergenerational solidarity. Com (1999) 221 final. Brussels
Commission to Inquire Into Child Abuse (CICA) (2009) The Commission report. (http://www.childabusecommission.ie)
Congregation for the Doctrine of the Faith (1987) Instruction on respect for human life in its origin and on the dignity of procreation. Replies to certain questions of the day. The Vatican
Connell RW (1995) Masculinities. Polity Press, Cambridge
Connelly M (2008) Fatal misconception: the struggle to control world population. Belknap Press of Harvard University Press, Cambridge
Conrad-Martius H (1955) Utopien der Menschenzüchtung. Der Sozialdarwinismus und seine Folgen. (Utopias of human breeding. Social Darwinism and its consequences). Kösel-Verlag, Munich
Conte RL (2008) May the marriage bed be immaculate. www.catholicplanet.com
Coogan MD (2005) Eastern religions: Hinduism, Buddism, Taoism, Confucianism, Shinto. Oxford University Press, New York
Coontz S (2005) Marriage, a history: from obedience to intimacy, or how love conquered marriage. Viking, New York
Cordes C (2008) A potential limit on competition. J Bioecon 10(2):127–144

Corea G (1986) The mother machine. Reproductive technologies from artificial insemination to artificial wombs. Harpercollins, New York

Corijn M, Klijzing E (2001) Transitions to adulthood in Europe. Kluwer Academic Publishers, Dordrecht

Cornell D (2000) Feminism and pornography. Oxford readings in Feminism. Oxford University Press, New York

Corning P (1983) The synergism hypothesis: a theory of progressive evolution. McGraw-Hill, New York

Corning P (1996) The cooperative gene: on the role of synergy in evolution. Evol Theory 11:183–207

Corning P (1997) Holistic Darwinism. Synergistic selection and the evolutionary process. J Soc Evol Syst 20(4):363–400

Corning P (2000) Biological adaptation in human societies: a 'basic needs' approach. J Bioecon 2 (1):41–86

Corning P (2003) Evolutionary ethics: an idea whose time has come? An overview and an affirmation. Polit Life Sci 22(1):50–59

Corning P (2003) The basic problem is still survival, and an evolutionary ethics is indispensable. Complexity, ethics and creativity conference, London School of Economics, 17–18 September 2003

Corning P (2005) Holistic Darwinism: synergy, cybernetics, and the bioeconomics of evolution. University of Chicago Press, Chicago

Corning P (2008) Holistic Darwinism. Polit Life Sci 27(1):22–54

Corning P (2010) The biology of fairness and the social contract: beyond capitalism and socialism. APSA 2010 annual meeting paper. Available at SSRN: http://ssrn.com/abstract=1643438

Corning P (2011) The fair society: the science of human nature and the pursuit of social justice. University of Chicago Press, Chicago

Corning PA, Szathmáry E (2015) Synergistic selection: a Darwinian frame for the evolution of complexity. J Theor Biol 371:45–58

Cornish E (1977) The study of the future. An introduction to the art and science of understanding and shaping tomorrow's world. World Future Society, Washington

Cornish E (1979) The future of the family: intimacy in an age of loneliness. Futurist 13:45–58

Cornog M (2003) Big book of masturbation: from angst to zeal. Down There Press, San Francisco

Cornwell J (2007) Darwin's angel: an angelic response to "The Dod delusion". Profile Books, London

Correa S (1997) From reproductive health to sexual rights: achievements and future challenges. Reprod Health Matters 5(10):107–116

Correns C (1900) G. Mendels Regel über das Verhalten der Nachkommenschaft der Rassenbastarde. (Mendel's rule about the behavior of the offspring of racial bastards). Ber der Dtsch Botanischen Ges 18:158–168

Cortes Barragan R, Dweck CS (2014) Rethinking natural altruism: simple reciprocal interactions trigger children's benevolence. PNAS 111(48):17071–17074

Coşgel MM (2016) The family in Utopia: celibacy, communal child rearing, and continuity in a religious commune. J Fam Hist 25(4):491–503

Cosmides L, Tooby J (1992) Cognitive adaptations for social exchange. In: Barkow J, Cosmides L, Tooby J (eds) The adapted mind: evolutionary psychology and the generation of culture. Oxford University Press, New York

Costa PT, McCrae RR (2008) The Revised NEO Personality Inventory (NEO-PI-R). In: Boyle GJ, Matthews G, Saklofske DH (eds) The SAGE handbook of personality theory and assessment: Vol. 2, Personality measurement and testing. SAGE Publications, Thousand Oaks, pp 179–198

Costa PT, McCrae RR (1992) Five-factor model of personality and job satisfaction: a meta-analysis. J Appl Psych 87:530–541

Costanza R (ed) (2000) Forum: the ecological footprint. Ecol Econ 32(3):341–394

Cott NF (1987) The grounding of modern feminism. Yale University Press, New Haven

Cottler LB, Price RK, Compton WM et al (1995) Subtypes of adult antisocial behavior among drug abusers. J Nerv Mental Dis 183(3):154–161

Council for Biotechnology Policy (2002) The sanctity of life in a brave new world: a manifesto on biotechnology and human dignity. http://www.thecbc.org/manifesto.html

Council of Europe (1950, 2010) Convention for the protection of human rights and fundamental freedoms as amended by protocols no. 11 and no. 14. Council of Europe Strasbourg

Council of Europe (1999) Conventon for the protection of human rights and dignity of the human being with regard to the application of biology and medicine. CETS No. 164. Council of Europe, Strasbourg

Council of Europe (2000) European convention on the exercise of children's rights. CETS No. 160. Council of Europe, Strasbourg

Council of Europe (2007) The dangers of creationism in education. Parliamentary assembly, resolution 1580. Council of Europe, Strasbourg

Courtois CA (2010) Healing the incest wound. Adult survivors in therapy. Norton, New York

Cowell JM, Decety J (2015) Precursors to morality in development as a complex interplay between neural, socio-environmental, and behavioral facets. PNAS 112(41):12657–12662

Coyne JA (2009) Why evolution is true. Viking Adult, New York

Coyne JA (2012) Science, religion, and society: the problem of evolution in America. Evolution 66(8):2654–2663

Cozic CP (1995) Sexual values: opposing viewpoints. Greenhaven Press, San Diego

Crabtree S (2009) Analyst insights: religiosity around the world. Retrieved 14 Sept 2009 from http://www.gallup.com/video/114694/Analyst-Insights-Religiosity-Around-World.aspx

Crabtree S, Pelham B (2009) What Alabamians and Iranians have in common. A global perspective on Americans' religiosity offers a few surprises. http://www.gallup.com/poll/114211/alabamians-iranians-common.aspx

Cracraft J, Bybee RW (eds) (2005) Evolutionary science and society: Educating a new generation. BSCS, Colorado Springs

Craig R (2012) Cloning around: the ethics of human cloning and stem cell research. ThoughtWare Australia Pty Ltd, Helensvale

Craig IW, Halton KE (2009) Genetics of human aggressive behaviour. Hum Genet 126:101–113

Crawford C, Krebs DL (eds) (2008) Foundations of evolutionary psychology. Taylor and Francis, New York

Crawford C, Salmon C (2002) Psychopathology or adaptation? Genetic and evolutionary perspectives on individual differences and psychopathology. Neuroendocrinol Lett 23(S4):39–45

Crean T (2007) God is no delusion: a refutation of Richard Dawkins. Ignatius Press, San Franscico

Crespi B, Summers K (2014) Inclusive fitness theory for the evolution of religion. Anim Behav 92:313–323

Crew FAE, Darlington CD, Haldane JBS et al (1939) Social biology and population improvement. Nature 144(3646):521–522

Crippen T (1994) Toward a Neo-Darwinian sociology: its nomological principles and some illustrative applications. Sociol Perspect 37(3):309–335

Crippen T, Machalek R (1989) The evolutionary foundations of the religious life. Int Rev Sociol 3 (3):61–84

Crittenden P (1992) Learning to be moral. Humanity Books, Amherst

Crockett MJ, Apergis-Schoute A, Herrmann B et al (2013) Serotonin modulates striatal responses to fairness and retaliation in humans. J Neurosci 33(8):3505–3513

Crockett MJ, Clark L, Hauser MD et al (2010) Serotonin selectively influences moral judgment and behavior through effects on harm aversion. PNAS 107(40):17433–17438

Crockett MJ, Rini RA (2015) Neuromodulators and the (in)stability of moral cognition. In: Decety J, Wheatley T (eds) The moral brain. A multidisciplinary perspective. The MIT Press, Cambridge, pp 221–235

Croddy EA, Wirtz JJ, Larsen J (2004) Weapons of mass destruction: an encyclopedia of worldwide policy, technology, and history. ABC-CLIO, Santa Barbara

Cronin H (1993) The ant and the peacock: altruism and sexual selection from Darwin to today. Cambridge University Press, Cambridge

Cronin H (2006) The battle of the sexes revisited. In: Grafen A, Ridley M (eds) Richard Dawkins: how a scientist changed the way we think. Reflections by scientist writers and philosophers. Oxford University Press, Oxford, pp 14–26

Cronk L (1994) Evolutionary theories of morality and the manipulative use of signals. Zygon 29 (1):81–101

Cronk L (1999) That complex whole: culture and the evolution of human behavior. Westview Press, Boulder

Cronk L, Chagnon N, Irons W (2000) Adaptation and human behavior. An anthropological perspective. Aldine De Gruyter, New York

Crook JH (1980) The evolution of human consciousness. Clarendon Press, Oxford

Crook JH, Crook SJ (1988) Tibetan polyandry: problems of adaptation and fitness. In: Betzig L, Borgerhoff Mulder M, Turke P (eds) Human reproductive behavior. Cambridge University Press, Cambridge, pp 97–114

Crook P (1994) Darwinism, war and history: the debate over the biology of war from the 'Origin of Species' to the First World War. Cambridge University Press, Cambridge

Crow JF (1966) The quality of people: human evolutionary changes. BioScience 16(12):863–867

Crutzen PJ, Stoermer EF (2000) The anthropocene. IGBP Newslett 41:17–18

Cruz ER (2013) Transhumanism and the fate of natality: an introduction. Zygon 48(4):916–935

Crysdale CSW (2003) Playing God? Moral agency in an emerging world. J Soc Christ Ethics 23 (2):243–259

Cunningham MK (2007) God and evolution. A reader. Routledge, Oxon

Cunningham C (2010) Darwin's pious idea: why the ultra-Darwinists and creationists both get it wrong. B. Eerdmans Publishing Company, Grand Rapids

Cunningham WA (2010) In defense of brain mapping in social and affective neuroscience. Soc Cogn 28(6):717–722

Curry O (2006) Who's afraid of the naturalistic fallacy? Evol Psychol 4:234–247

Curry O, Roberts SGB, Dunbar RIM (2013) Altruism in social networks: evidence for a 'kinship premium'. Br J Psychol 104(2):283–295

Cziko G (1995) Without miracles. Universal selection theory and the Second Darwinian Revolution. The MIT Press, Cambridge

d'Aquili E, Newberg AB (1999) The mystical mind: probing the biology of religious experience. Fortress Press, Minneapolis

D'Onofrio BM, Eaves LJ, Murrelle L et al (1999) Understanding biological and social influences on religious affiliation, attitudes and behavior: a behavior-genetic perspective. J Pers 67 (6):953–984

D'Souza D (2007) What is so great about Christianity?. Regnery, Washington

D'Souza D (2009) Life after death: the evidence. Regnery, Washington

Dabbs JM Jr, Morris R (1990) Testosterone, social class, and antisocial behavior in a sample of 4462 men. Psychol Sci 1(3):209–211

Dacey JS, Packer AJ (1992) The nurturing parent: how to raise creative, loving, responsible children. Fireside, Simon and Schuster, New York

Daecke SM, Schakenberg J (2000) Gottesglaube—ein Selektionsvorteil? (Faith in god—a selective advantage?). Kaiser/Gütersloher Verlaghaus, Gütersloh

Dahlberg G (1948) Mathematical methods for population genetics. Karger, Basle

Dahlberg F (1981) Woman the gatherer. Yale University Press, New Haven

Daly M (1978) Gyn/ecology: the metaethics of radical feminism. Beacon Press, Boston

Daly M, Wilson M (1978) Sex, evolution, and behaviour. Adaptation for reproduction. Duxbury Press, North Scituate

Daly M, Wilson M (1988) Homicide. Aldine, New York

Daly M, Wilson M, Weghorst SJ (1982) Male sexual jalousy. Ethol Sociobiol 3(1):11–27
Damasio AR, Tranel D, Damasio H (1990) Individuals with sociopathic behavior caused by frontal damage fail to respond to autonomically social stimuli. Behav Brain Res 41(2):81–94
Daniels D (1983) The evolution of concealed ovulation and self-deception. Ethol Sociobiol 4 (2):69–87
Darling D (2002) Life everywhere. Basic Books, New York
Darlington CL (2009) The female brain. CRC Press, Boca Raton
Darragh M, Buniak L, Giordano J (2015) A four-part working bibliography of neuroethics: part 2—neuroscientific studies of morality and ethics. Philos Ethics Humanit Med. https://doi.org/10.1186/s13010-015-0022-0
Darroch JE (2013) Trends in contraceptive use. Contraception 87(3):259–263
Darroch JE, Singh S (2013) Trends in contraceptive need and use in developing countries in 2003, 2008, and 2012: an analysis of national surveys. Lancet 381(9879):1756–1762
Darwin C (1859) On the origin of species by means of natural selection, or the preservation of favoured races in the struggle for life. Murray, London
Darwin C (1871) The descent of man and selection in relation to sex. Murray, London
Davidson J, Davidson B, Va:nderkam L (2005) Genius denied: how to stop wasting our brightest young minds. Simon and Schuster, New York
Davis K (1986) Low fertility in evolutionary perspective. Popul Dev Rev 12:48–56
Davis H (2009) Caveman logic: the persistence of primitive thinking in de modern world. Prometheus Books, Amherst
Davis JJ (2016) A brief introduction to the neurobiology of spiritual values, morality, sovereignty and governance. Sci GOD J 7(5):316–327
Davis J, Daly M (1997) Evolutionary theory and the human family. Q Rev Biol 72(4):407–435
Davis P, Kenyon DH (1993) Of pandas and people. Foundation for Thought and Ethics, Richardson
Dawes CT, Fowle JH, Johnson T et al (2007) Egalitarian motives in humans. Nature 446 (7137):794–796
Dawes CT, Loewen PJ, Schreiber D et al (2012) Neural basis of egalitarian behavior. PNAS 109 (17):6479–6483
Dawkins R (1976) The selfish gene. Oxford University Press, New York
Dawkins R (1983) Universal Darwinism. In: Bendfall DS (ed) Evolution from molecules to man. Cambridge University Press, Cambridge, pp 403–425
Dawkins R (1986) The blind watchmaker: why the evidence of evolution reveals a universe without design. Norton, New York
Dawkins R (2000) Unweaving the rainbow: science, delusion and the appetite for wonder. Houghton Mifflin, Boston
Dawkins R (2003) A devil's chaplain: reflections on hope, lies, science, and love. Houghton Mifflin, Boston
Dawkins R (2006) The god delusion. Houghton Mifflin, Boston
Dawkins R (2010) The greatest show on earth: the evidence for evolution. Free Press, New York
Dawkins R, Krebs JR (1979) Arms races between and within species. Proc R Soc Lond B 205:489–511
Dawson DA, Meny DJ, Ridley JC (1980) Fertility control in the United States before the contraceptive revolution. Family Plann Perspect 12:76–86
De Beauvoir S (1949) Le deuxième sexe. Gallimard, Paris
De Dreu CKW, Greer LL, Van Kleef GA et al (2011) Oxytocin promotes human ethnocentrism. PNAS 108(4):1262–1266
De Duve C (2009) Génétique du péché originel. Le poids du passé sur l'avenir de la vie. Odile Jacob, Paris. See also: de Duve C, Patterson N (2010) Genetics of original sin: The impact of natural selection on the future of humanity. Yale University Press, New Haven
De Duve C (2011) De Jésus à Jésus… en passant par Darwin. (From Jesus to Jesus… via Darwin). Odile Jacob, Paris

De Garis H (2005) The artilect war: cosmists vs. terrans: a bitter controversy concerning whether humanity should build godlike massively intelligent machines. ETC Publications, Palm Springs

De Geus M (2003) The end of over-consumption: towards a lifestyle of moderation and self-restraint. International Books, Utrecht

De Gobineau A (1853–1855) (1967) Essai sur l'inégalité des races humaines. (Essay on the inequality of human races). Éditions Pierre Belfond, Paris

De Grey A (2007) Ending aging: the rejuvenation breakthroughs that could reverse human aging in our lifetime. St. Martin's Press, New York

De Guibert-Lantoine C, Léridon H (1998) La contraception en France: un bilan après 30 ans de libéralisation. (Contraception in France: a review after 30 years of liberalization). Population 53(4):785–812

De Jonge CJ, Barratt CL (eds) (2002) Assisted reproductive technology. Cambridge University Press, Cambridge

De Lanessan J-L (1908) La morale naturelle. (Natural morality). Félix Alcan, Paris

De Lapouge GV (1896) Sélections sociales. (Social selections). Thorin and Fils, Paris

De Oliveira-Souza R, Harec RD, Bramatia IE et al (2008) Psychopathy as a disorder of the moral brain: fronto-temporo-limbic grey matter reductions demonstrated by voxel-based morphometry. NeuroImage 40(3):1202–1213

De Quervain DJ-F, Fischbacher U, Treyer V et al (2004) The neural basis of altruistic punishment. Science 305(5688):1254–1258

De Rus E (2006) Humanisme et transhumanisme: l'Homme en question. Observatoire de la génétique, n°26. http://www.omics-ethics.org/observatoire/cadrages/cadr2006/c_no26_06/c_no26_06_02.html

De Saint Simon C-H (1817) L'Industrie ou discussions politiques, morales et philosophiques dans l'intérêt de tous les hommes livrés à des travaux utiles et indépendans. (Industry or political, moral and philosophical discussions in the interest of all men engaged in useful and independent work). Au bureau de l'administration, Paris

De Tarde G (1984) Darwinisme naturel et darwinisme social. Rev Philosophique 17:607–637

De Vries H (1900) Sur la loi de disjonction des hybrides. (On the law of segregation of hybrids). C R l'Academie des Sci 130:845–847

De Vries H (1904, 1910) Species and varieties: their origin by mutation. MacDougal, Chicago

De Waal F (1982) Chimpansee politics. Power and sex among apes. John Hopkins University Press, Baltimore

De Waal F (1996) Good natured: the origins of right and wrong in humans and other animals. Harvard University Press, Cambridge

De Waal F (2006) Primates and philosophers. Princeton University Press, Princeton

De Waal F (2007) Putting the altruism back into altruism: the evolution of empathy. Ann Rev Psychol 59:4.1–4.22

De Waal F (2014) Natural normativity: the 'is' and 'ought' of animal behavior. Behaviour 151:185–204

Deady DK, Law Smith MJ, Kent JP et al (2006) Is priesthood an adaptive strategy? Evidence from a historical Irish population. Hum Nat 17(4):393–404

Deane-Drummond C (2012) Christ and evolution: a drama of wisdom? Zygon 47(3):524–541

Decety J (2010) The neurodevelopment of empathy in humans. Dev Neurosci 32(4):257–267

Decety J, Cacioppo S (2012) The speed of morality: a high-density electrical neuroimaging study. J Neurophysiol 108(11):3068–3072

Decety J, Cowell JM (2015) Empathy, justice, and moral behavior. AJOB Neurosci 6(3):3–14

Decety J, Cowell JM (2015) The equivocal relationship between morality and empathy. In: Decety J, Wheatley T (eds) The moral brain. A multidisciplinary perspective. The MIT Press, Cambridge, pp 279–302

Decety J, Cowell JM, Lee K et al (2015) The negative association between religiousness and children's altruism across the world. Curr Biol 25(22):2951–2955

Decety J, Howard LH (2013) The role of affect in the neurodevelopment of morality. Child Dev Perspect 7(1):49–54
Decety J, Porges EC (2011) Imagining being the agent of actions that carry different moral consequences: An fMRI study. Neuropsychologia 49(11):2994–3001
Decety J, Svetlova M (2012) Putting together phylogenetic and ontogenetic perspectives on empathy. Dev Cogn Neurosci 2(1):1–24
Decety J, Wheatley T (2015) The moral brain. A multidisciplinary perspective. The MIT Press, Cambridge
Declerck CH, Boone C, Emonds G (2013) When do people cooperate? The neuroeconomics of prosocial decision making. Brain Cogn 81(1):95–117
Decraemer W, Coomans A, Geraert E (2010) Life and work of Prof. Dr. Lucien De Coninck: Biologist, humanist and freemason. J Story-Scientia, Gent
Defleur A, White T, Valensi P et al (1999) Neanderthal cannibalism at Moula-Guercy, Ardeche, France. Science 286(5437):128–131
Dein S (2011) Religion and psychosis: a common evolutionary trajectory? Transcult Psychiatry 48 (3):318–335
De Jong A, de Wert GMWR (2015) Prenatal screening: an ethical agenda for the near future. Bioethics 29(1):46–55
De La Croix D, Mariani F (2015) From polygyny to serial monogamy: a unified theory of marriage institutions. Rev Econ Stud 82(2):565–607
Delamontagne RG (2010) High religiosity and societal dysfunction in the United States during the first decade of the twenty-first century. Evol Psychol 8(4):617–657
Deleeck H (1992) De architectuur van de welvaartstaat. (The Architecture of the Welfare State). Acco, Leuven
Delhez C (2015) Quel homme pour demain? Science, ethique et christianisme. (Which human for tomorrow? Science, ethics and christianism). Editions jésuites, Namur
Deliens L, van der Wal G (2003) The euthanasia law in Belgium and the Netherlands. Lancet 362 (9391):1239–1240
Delmotte G, Cliquet R (1983) Bio-medische vruchtbaarheidsfactoren en gezinsgrootte. Onderzoekresultaten NEGO III (1975–1976). (Bio-medical fertility traits and family size. Research results of NEGO III, 1975-1976). C.B.G.S. Werkdocument, 2. C.B.G.S./Ministerie van de Vlaamse Gemeenschap, Brussel
Delton AW, Cosmides L, Guemo M et al (2012) The psychosemantics of free riding: dissecting the architecture of a moral concept. J Pers Soc Psychol 102(6):1252–1270
Delton AW, Krasnow MM (2015) Adaptationist approaches to moral psychology. In: Decety J, Wheatley T (eds) The moral brain. A multidisciplinary perspective. The MIT Press, Cambridge, pp 19–34
Delton AW, Krasnow MM, Cosmides L et al (2011) Evolution of direct reciprocity under uncertainty can explain human generosity in one-shot encounters. PNAS 108(32):13335–13340
Del Valle A (2016) Les vrais enemies de l'occident. Du rejet de la Russie à l'islamisation des sociétés ouvertes. (The true enemies of the West. From the rejection of Russia to the Islamization of open societies). L'artilleur, Paris
Demaerel I (2015) Wie heeft er vandaag nog een morele autoriteit? (Who has today still a moral authority?) Knack, 25 Oct 2015
Dembski W (1998) Mere creation. InterVarsity Press, Downer's Grove
Dembski W (2003) Three frequently asked questions about intelligent design. http://www.designinference.com
Dembski W (2004) The design revolution: answering the tough questions about intelligent design. Intervarsity Press, Downer's Grove
Dembski W, Witt J (2010) Intelligent design uncensored: an easy-to-understand guide to the controversy. InterVarsity Press, Downer's Grove

De Melo-Martin I, Salles A (2015) Moral bioenhancement: much ado about nothing? Bioethics 29 (4):223–232

Dennett DC (1995) Darwin's dangerous idea: evolution and the meanings of life. Simon and Schuster, New York

Dennett DC (2007) Breaking the spell: religion as a natural phenomenon. Viking, New York

Denniston GC, Hodges FM, Milos MF (eds) (1999) Male and female circumcision: medical, legal, and ethical considerations in pediatric practice. Kluwer Academic/Plenum, New York

Denniston GC, Hodges FM, Milos MF (2010) Genital autonomy: protecting personal choice. Springer, New York

Denno DW (1990) Biology and violence. Cambridge University Press, Cambridge

Denno DW (1998–1999) Evolutionary biology and rape. Jurimetrics 39:243–254

De Oliveira-Souza R, Zahn R, Moll J (2015) Neural correlates of human morality. In: Decety J, Wheatley T (eds) The moral brain. A multidisciplinary perspective. The MIT Press, Cambridge, pp 183–195

Depew DJ, Weber BH (1995) Darwinism evolving: systems dynamics and the genealogy of natural selection. MIT Press/A Bradford Book, Cambridge

Derosas R, van Poppel F (2006) Religion and the decline of fertility in the western world. Springer, Dordrecht

Desai S (1995) When are children from large families disadvantaged? Evidence from cross-national analyses. Popul Stud 49(2):195–210

Deschner K (1986–2013) Kriminalgeschichte des Christentums. Rowohlt Verlag, Reinbek bei Hamburg

DeSalle R, Tattersall I (2008) Human origins: what bones and genomes tell us about ourselves. Texas A&M University Press, College Station

DesAutels P (2010) Sex differences and neuroethics. Philos Psychol 23(1):95–111

DeScioli P, Krishna S (2013) Giving to whom? Altruism in different types of relationships. J Econ Psychol 34:218–228

Deven F (1996) Relatiepatronen en gezinsdynamiek. (Relational patterns and family dynamics). In: Cliquet R (ed) Gezinnen in de verandering. Veranderende gezinnen. CBGS Monografie 1996/2. Centrum voor Bevolkings- en Gezinsstudie, Brussel, pp 59–82

Deven F, Cliquet R (1986) One-parent families in Europe. A NIDI/CBGS Publication, The Hague/Brussels

Devore D (1992) Genetic screening and ethics: an overview. The Woodrow Wilson Biology Institute, Washington

DeVore I, Washburn SL (1963) Baboon ecology and human evolution. In: Howell FC, Bourliere F (eds) African ecology and human evolution. Viking Fund Publications in Anthropology vol 36, pp 335–367

Dewaraja LS (1994) The position of women in Buddhism. http://www.accesstoinsight.org/lib/authors/dewaraja/wheel280.html

Dewaraja LS (2009) The position of women in Buddhism. The Wheel Publication, pp 1–10

Dewey J (1898) Evolution and ethics. Monist 8(3):321–341

Dewey J (1922) Human nature and conduct. H. Holt and Co, New York

Dewey J (1934) A common faith. Yale University Press, New Haven

Dewhurst K, Beard AW (1970) Sudden religious conversions in temporal lobe epilepsy. Br J Psychiatry 117(540):497–507

Dhammananda KS (2005) A happy married life. A Buddhist perspective. http://www.accesstoinsight.org/lib/authors/dhammananda/marriage.html

Di Lampedusa GT (1958) Il Gattopardo. (The Leopard). Feltrinelli, Milano

Diamond I, Orenstein GF (eds) (1990) Reweaving the world: the emergence of ecofeminism. Sierra Club books, San Francisco

Diamond J (1992) The third chimpanzee: evolution and the future of the human animal. HarperCollins, New York

Diamond M (1993) Homosexuality and bisexuality in different populations. Arch Sex Behav 22 (4):291–310
Diamond J (1997) Guns, germs, and steel: the fates of human societies. Jonathan Cape, London
Diamond LM (2004) Emerging perspectives on distinctions between romantic love and sexual desire. Curr Dir Psychol Sci 13(3):116–119
Diamond J (2005) Collapse: how societies choose to fail or succeed. Viking Penguin, New York
Diamond M (2009) Pornography, public acceptance and sex related crime: a review. Int J Law Psychiatry 32(5):304–314
Diamond J (2012) The world until yesterday. What can we learn from traditional societies? Allen Lane, London
DiCarlo C (2002/03) The evolution of morality. Humanist Can 143:12–19
Dickemann M (1979) Female infanticide, reproductive strategies, and social stratification: a preliminary method. In: Chagnon N, Irons W (eds) Evolutionary biology and human social behaviour: an anthropological perspective. Duxbury Press, North Scituate, pp 321–368
Dijker AJM (2011) Physical constraints on the evolution of cooperation. Evol Biol 38(2):124–143
Dilley SC (ed) (2013) Darwinian evolution and classical liberalism: theories in tension. Lexington Books, Lanham
Dines G, Jensen B, Russo A (1997) Pornography: the production and consumption of inequality. Routledge, New York
Dionne G, Tremblay R, Boivin M et al (2003) Physical aggression and expressive vocabulary in 19-month-old twins. Dev Psychol 39(2):261–273
Ditch J, Barnes H, Bradshaw J et al (1995) European Observatory on National Family Policies. A synthesis of national family policies 1994. University of York, Heslington
Distelmans W (2012) Als de dood voor de dood. De zorgwekkende stijging van palliatieve sedatie. (As the death for death. The worrying rise of palliative sedation). Het Vrije Woord 57(3):22–23
Divale WT (1972) System population control in the middle and upper Paleolithic: inferences based on contemporary hunter-gatherers. World Archeol 4:222–243
Dobbelaere K (2002) Secularization: an analysis at three levels (Gods, humans and religion). Peter Lang Publishing, Brussels
Dobson A (1990) Green political thought: an introduction. Harper Collins Academic, London
Dobzhansky Th (1937) Genetics and the origin of species. Columbia University Press, New York
Dobzhansky Th (1956) The biological basis of human freedom. Columbia University Press, New York
Dobzhansky Th (1962) Mankind evolving: the evolution of the human species. Yale University Press, New Haven
Dobzhansky Th (1967) The biology of ultimate concern. New American Library, New York
Dobzhansky Th (1973) Nothing in biology makes sense except in the light of evolution. Am Biol Teach 35(3):125–129
Dodds WK (2008) Humanity's footprint: momentum, impact, and our global environment. Columbia University Press, NewYork
Doherty E (1999) The Jesus puzzle: did Christianity begin with a mythical Christ?. Canadian Humanist Publications, Ottawa
Dombrowski SC, Gischlar KL, Durst T (2007) Safeguarding young people from cyber pornography and cyber sexual predation: a major dilemma of the internet. Child Abuse Rev 16(3):153–170
Donaldson ZR, Young LJ (2008) Oxytocin, vasopressin, and the neurogenetics of sociality. Science 322(5903):900–904
Dong X, Milholland B, Vijg J (2016) Evidence for a limit to human lifespan. Nature http://dx.doi.org/10.1038/nature19793
Dooghe G, Vanden Boer L, Vanderleyden L (1988) De leefsituatie van bejaarden. (The living Conditions of Elderly). CBGS Mongrafie 1988/1. Centrum voor Bevolkings- en Gezinsstudie, Brussel

Dorbritz J, Schwarz K (1996) Kinderlosigkeit in Deutschland—ein Massenphänomen? Analysen zu Erscheinungsformen und Ursachen. (Childlessness in Germany—a mass phenomenon? Analyses of forms and causes). Z für Bevölkerungswissenschaft 21(3):231–261

Dorbritz J (2008) Germany: family diversity with low actual and desired fertility. Demographic Res 19(17):557–597

Dos Santos M, Rankin DJ, Wedekind C (2013) Human cooperation based on punishment reputation. Evolution 67(8):2446–2450

Doudna JA, Charpentier E (2014) The new frontier of genome engineering with CRISPR-Cas9. Science 346(6213). https://doi.org/10.1126/science.1258096

Doudna JA, Sternberg SH (2017) A Crack in creation: gene editing and the unthinkable power to control evolution. Houghton Mifflin Harcourt, Boston

Douglas T (2008) Moral enhancement. J Appl Philos 25(3):228–245

Douglas T (2015) The morality of moral neuroenhancement. In: Clausen J, Levy N (eds) Handbook of neuroethics. Springer, Dordrecht, pp 1227–1246

Douglas T, Devolder K (2013) Procreative altruism: beyond individualism in reproductive selection. J Med Philos 38(4):400–419

Dow J (2008) Is religion an evolutionary adaptation? J Artif Soc Soc Simul 11(2.2):2–6

Dowbiggin I (2005) A concise history of euthanasia: life, death, God and medicine. Rowman and Littlefield, Lanham

Dowd M (2009) Thank god for evolution: how the marriage of science and religion will transform your life and our World. Plume, New York

Doyal L, Gough I (1991) A theory of human need. Macmillan Education Ltd, London

Doyle D (2005) Ritual male circumcision. A brief history. J R Coll Phys Edinb 35(3):279–285

Draper E (1991) Risky business: genetic testing and exclusionary practices in the hazardous workplace. Cambridge University Press, Cambridge

Dreber A, Rand DG, Fudenberg D et al (2008) Winners don't punish. Nature 452:348–351

Drexler KE (1986) Engines of creation: the coming era of nanotechnology. Anchor Books, New York

Drexler KE, Peterson C, Pergamit G (1991) Unbounding the future: the nanotechology revolution. Quill William Morrow, New York

Du Pasquier R (1992) Unveiling Islam. Islamic Texts Society, Cambridge

Duberman L, Azumi K (1975) Sexism in Nepal. J Marriage Family 37(4):1013–1021

Dubessy J, Lecointre G, Silberstein M (2004), Les matérialismes (et leurs détracteurs). (The materialisms—and their critics). Syllepse, Paris

Ducros A (1981) Biosociologie et évolution. In: Ferembach D (ed), Les procesus de l'hominisation. L'évolution humaine. Les faits. Les modalités, Colloques internationaux du CNRS, 599, éditions du CNRS, Paris pp 229–232

Dugatkin L (1995) Partner choice, game theory and social behavior. J Quant Anthropol 5(1):3–14

Dugatkin L (2006) The altruism equation. Princeton University Press, Princeton

Dugatkin L, Mesterton-Gibbons M, Houston AI (1992) Beyond the prisoner's dilemma: towards models to discriminate among mechanisms of cooperation in nature. Trends Ecol Evol 7:202–205

Dugatkin L, Reeve HK (1994) Behavioral ecology and levels of selection—dissolving the group selection controversy. Adv Study Behav 23:101–133

Dulitzki M, Soriano D, Schiff E et al (1998) Effect of very advanced maternal age on pregnancy outcome and rate of cesarean delivery. Obstet Gynecol 92(6):935–939

Duménil G, Lévy D (2011) The crisis of neoliberalism. Harvard University Press, Cambridge

Dunbar R (1992) Neocortex size as a constraint on group size in Primates. J Hum Evol 6:469–493

Dunbar R (1998) The social brain hypothesis. Evol Anthropol 6(5):178–190

Dunbar R (2003) The social brain: mind, language, and society in evolutionary perspective. Ann Rev Anthropol 32(3):163–181

Dunbar R (2004) The human story: a new history of mankind's evolution. Faber and Faber, London

Dunbar R (2007) Evolution and the social sciences. Hist Hum Sci 20(2):29–50

Dunbar R, Barrett L (2009) Oxford handbook of evolutionary psychology. Oxford University Press, Oxford

Dunbar R, Spoors M (1995) Social networks, support cliques, and kinship. Hum Nat 6(3):273–290

Dunswort HM, Warrener AG, Deacon T et al (2012) Metabolic hypothesis for human altriciality. PNAS 109(38):15212–15216

Duntley J, Shackelford TK (2008) Evolutionary forensic psychology. Darwinian foundations of crime and law. Oxford University Press, New York

Duntley J, Buss DM (2005) The plausibility of adaptations for homicide. In: Carruthers P, Laurence S, Stich SP (eds) The innate mind: structure and contents. Oxford University Press, New York

Dupoux E, Jacob P (2007) Universal moral grammar: a critical appraisal. Trends Cogn Sci 11 (9):373–378

Durham WH (1991) Coevolution: genes, culture and human diversity. Stanford University Press, Stanford

Durrant R (2011) Collective violence: an evolutionary perspective. Aggression Violent Behav 16 (5):SI 428–436

Durrant R, Ward T (2012) The role of evolutionary explanations in criminology. J Theor Philos Criminol 4(1):1–37

Duschek K-J, Wirth H (2005) Kinderlosigkeit von Frauen im Spiegel des Mikrozensus. Eine Kohortenanalyse der Mikrozenses 1987 bis 2003. (Childlessness of women in the micro census. A cohort analysis of the micro censes 1987–2003). Wirtschaft und Statistik 8:800–820

Duster T (2003) Backdoor to eugenics. Routledge, New York

Dutton K (2012) The wisdom of psychopaths. William Heinemann, London

Duvold EM (1995) The Norwegian family: scenarios for 2020. Future consequences at the micro level of present-day family patterns. In: Lundh C (ed) Demography, economy and welfare. Lund University Press, Lund, pp 28–44

Duyndam J, Poorthuis M, De Wit T (2005) Humanisme en religie: Controverses, bruggen, perspectieven. (Humanism and religion: Controversies, bridges, perspectives). Eburon Uitgeverij, Delft

Dworkin A (1989) Pornography: men possessing women. E.P. Dutton, New York

Dvorsky G (2008) Better living through transhumanism. J Evol Technol 19(1):62–66

Dwyer S (2008) How not to argue that morality isn't innate/ comments on Prinz. In: Sinnott-Armstrong W (ed) Moral psychology, vol 1. The evolution of morality: adaptations and innateness. MIT Press, Cambridge, pp 407–418

Dyson FJ (2004) Infinite in all directions: Gifford lectures given at Aberdeen, Scotland April–November 1985. Harper Perennial, New York

Earp BD, Sandberg A, Savulescu J (2015) The medicalization of love. Cambridge Q Healthc Ethics 24(3):323–336

Easterlin RA (1975) An economic framework for fertility analysis. Stud Family Plann 6(3):54–63

Eaton JW, Mayer AJ (1953) The social biology of very high fertility among the Hutterites: the demography of a unique population. Hum Biol 25(3):206–264

Eaves L, Heath A, Martin N et al (1999) Comparing the biological and cultural inheritance of personality and social attitudes in the Virginia 30,000 study of twins and their relatives. Twin Res 2(2):62–80

Eaves LJ, D'Onofrio BM, Russell R (1999) Transmission of religion and attitudes. Twin Res 2 (2):59–61

Eaves LJ, Martin NG, Heath AC (1990) Religious affiliation in twins and their parents: testing a model of cultural inheritance. Behav Genet 20(1):1–22

Eaves LJ, Hatemi PK, Prom-Womley EC et al (2008) Social and genetic influences on adolescent religious attitudes and practices. Soc Forces 6(4):1621–1646

Ebenstein A, Ebenstein W, Fogelman E (1999) Today's ISMS: socialism, capitalism, fascism, communism, and libertarianism. Prentice Hall, Upper Saddle River

Eberl JTA (2014) Thomistic appraisal of human enhancement technologies. Theor Med Bioethics 35(4):289–310

Eberstadt N (1997) World population implosion? Public Interest 129:3–22

Ebstein RP (2006) The molecular genetic architecture of human personality: beyond self-report questionnaires. Mol Psychiatry 11(5):427–445

Ebstein RP, Novick O, Umansky R et al (1996) Dopamine D4 receptor (D4DR) exon III polymorphism associated with the human personality trait of Novelty seeking. Nat Genet 12:78–80

Ebstein RP, Israel S, Lerer E et al (2009) Arginine vasopressin and oxytocin modulate human social behavior. Ann New York Acad Sci 1167:87–102

Ebstein RP, Israel S, Chew SH et al (2010) Genetics of human social behavior. Neuron 65(6):831–844

Eckersley R (2001) Doomsday scenarios: how the world may go on without us. Futurist 35(6):20–23

Ecklund E, Scheitle C (2007) Religion among academic scientists: distinctions, disciplines, and demographics. Soc Prob 54(2):289–307

Edgell P, Gerteis J, Hartmann D (2006) Atheists as "other": moral boundaries and cultural membership in American society. Am Sociol Rev 71(2):211–234

Edgerton RB (1992) Sick societies. Sci Educ 18(6):885–903

Edis T (2002) The ghost in the universe: god in light of modern science. Prometheus, Buffalo

Edis T (2006) Grand themes, narrow constituency. In: Young M, Edis T (eds) Why intelligent design fails: a scientific critique of the new creationism. Rutgers University Press, New Brunswick, pp 9–19

Edis T (2007) An illusion of harmony: science and religion in Islam. Prometheus Books, Amherst

Edlund L, Korn E (2002) A theory of prostitution. J Polit Econ 110(1):181–214

Egas M, Riedl A (2008) The economics of altruistic punishment and the maintenance of cooperation. Proc R Soc B 275(1637):871–878

Egas M, Kats R, van der Sar X et al (2013) Human cooperation by lethal group competition. Sci Rep 3:1373

Ehrlich PR (2000) Human natures. Genes, cultures, and the human prospect. Penguin Books, New York

Ehrlich PR (2001) Intervening in evolution: ethics and actions. PNAS 98(10):5477–5481

Ehrlich PR, Ehrlich A, Holdren JP (1977) Ecoscience. Population, resources, environment. Freeman and Co, San Francisco

Ehrlich PR, Ehrlich AH (1981) Extinction: the causes and consequences of the disappearance of species. Random House, New York

Ehrlich PR, Ehrlich AH (1990) The population explosion. Hutchinson, London

Ehrlich PR, Ehrlich AH (2008) The dominant animal: human evolution and the environment. Island Press, Washington

Ehrlich PR, Holdren JP (1971) Impact of population growth. Science 171(3977):1212–1217

Ehrlich R (2003) Eight preposterous propositions. From the genetics of homosexuality to the benefits of global warming. Princeton University Press, Princeton

Ehrman BD (2008) God's problem: how the Bible fails to answer our most important question—why we suffer. HarperOne, New York

Eibl-Eibesfeldt I (1971) Love and hate. Holt, Rinehart and Winston, New York

Eibl-Eibesfeldt I (1979) The biology of peace and war: men, animals, and aggression. Viking, New York

Eibl-Eibesfeldt I (1982) Warfare, man's indoctrinability, and group selection. Z für Tierpsychologie 67:177–198

Eibl-Eibesfeldt I (1998) Us and the others. The familial roots of ethnonationalism. In: Eibl-Eibesfeldt I, Salter FK (eds), Ethnic conflict and indoctrination. Altruism and identity in evolutionary perspective. Berghahn Books, New York, pp 21–53

Eibl-Eibesfeldt I, Salter FK (eds) (1998) Ethnic conflict and indoctrination. Altruism and identity in evolutionary perspective. Berghahn Books, New York
Eibl-Eibesfeldt I, Salter FK (eds) (1998) Indoctrinability, ideology and warfare. Berghahn, Oxford
Eichhammer P, Sand PG, Stoertebecker P et al (2005) Variation at the DRD4 promoter modulates extraversion in Caucasians. Mol Psychiatry 10(6):520–522
Eimer T (1898) On orthogenesis and the impotence of natural selection in species formation. The Open Court Publishing Company, Chicago
Eisenberg DTA, Campbell B, MacKillop J et al (2007) Polymorphisms in the dopamine D2 and D4 receptor genes and reproductive, sexual and life history behaviors. Evol Psychol 5:696–715
Eisenberg E (2001) The adoption paradox. Discover 22(1):80–89
Eisenberg N, Fabes R (1998) Prosocial development. In: Damon W, Eisenberg N (eds) Handbook of child psychology, vol 3. Social, emotional, and personality development. Wiley, New York, pp 701–778
Eisenberg N, Guthrie IK, Cumberland A et al (2002) Prosocial development in early adulthood: a longitudinal study. J Person Soc Psychol 82(6):993–1006
Eisenstadt SN (2000) Multiple modernities. Daedalus 129(1):1–29
Ekelund CK, Jørgensen FS, Petersen OB et al (2008) Impact of a new national screening policy for down's syndrome in Denmark: population based cohort study. BMJ 337:a2547
El Mouden C, André JB, Morin O et al (2014) Cultural transmission and the evolution of human behaviour: a general approach based on the price equation. J Evol Biol 27(2):231–241
Elger B, Harding T (2003) Huntington's disease: do future physicians and lawyers think eugenically? Clin Genet 64(4):327–338
Elgin D (1993) Awakening Earth: exploring the evolution of human culture and consciousness. William Morrow, New York
Eliade M (1961) The sacred and the profane: the nature of religion. Harper Torchbooks, New York
Elliot KA (ed) (1997) Corruption and the global economy. Institute for International Economics, Washington
Elliot A, Lemert CC (2005) The new individualism: the emotional costs of globalisation. Routledge, New York
Ellis EC (2013) Overpopulation is not the problem. New York Times, 14 Sept 2013, A19
Ellis L (1987) Criminal behavior and r/K selection: an extension of gene-based evolutionary theory. Deviant Behav 8(2):149–176
Ellis L (1989) Theories of rape: inquiries into the cause of sexual aggression. Taylor and Francis, New York
Ellis L (1990) The evolution of collective counterstrategies to crime: from the primate control role to the criminal justice system. In: Ellis L, Hoffman H (eds) Crime in biological, social and moral contexts. Praeger, Westport, pp 81–99
Ellis L (1991) Monoamine oxidase and criminality: identifying an apparent biological marker for antisocial behavior. J Res Crime Delinquency 28(2):227–251
Ellis L (1993) Social stratification and socioeconomic inequality. Praeger, New York
Ellis L (1996) A discipline in peril: sociology's future hinges on curing its biophobia. Am Sociol 27(2):21–41
Ellis L (1998) Neodarwinian theories of violent criminality and antisocial behavior: photographic evidence from nonhuman animals and a review of the literature. Aggression Violent Behav 3 (1):61–110
Ellis L (2003) Genes, criminality and the evolutionary neuroandrogenic theory. In: Walsh A, Ellis L (eds) Biosocial criminology: challenging environmentalism's supremacy. Nova Science, New York, pp 13–34
Ellis L (2011) Evolutionary neuroandrogenic theory and universal gender differences in cognitive behavior. Person Individ Diff 51(5):552–561
Ellis L, Hoffman H (eds) (1991) Crime in biological, social and moral contexts. Praeger, New York
Ellis H, Symonds JA (1897) Sexual inversion. Wilson and Macmillan, Bedford Row

Ellsworth RM (2009) Evolution and religion: theory, definitions, and the natural selection of religious behavior. A thesis presented to the Faculty of the Graduate School at the University of Missouri in partial fulfillment of the requirements for the degree masters of arts by Dr. Craig T. Palmer, Thesis Committee Chair. University of Missouri, Columbia

Elqayam S, Evans JStBT (2011) Subtracting "ought" from "is": descriptivism versus normativism in the study of human thinking. Behav Brain Sci 34(5):233–248

Ember C (1978) Myths about hunter-gatherers. Ethnology 17(4):439–448

Emberger G (1994) Theological and scientific explanations for the origin and purpose of natural evil. Perspect Sci Christ Faith 46(2):150–158

Emery N, Clayton N, Frith C (2007) Social intelligence: from brain to culture. Oxford University Press, Oxford

Emlen ST (1995) An evolutionary theory of the family. PNAS 92:8092–8099

Engel D, Woolley AW, Jing LX et al (2014) Reading the mind in the eyes or reading between the lines? Theory of mind predicts collective intelligence equally well online and face-to-face. PloS One 9(12):e115212

Engelhardt HT (2006) Global bioethics: the collapse of consensus. M and M Scrivener Press, Salem

Engelke M (2014) Christianity and the anthropology of secular humanism. Curr Anthropol 55 (S10):S292–S301

Engels F (1878) Herrn Eugen Dühring's Umwälzung der Wissenschaft. (Mr Eugen Dühring's revolution in science). Druck und Verlag der Genossenschafts Buchdruckerei, Leipzig

Epstein E, Guttman R (1984) Mate selection in man: evidence, theory, and outcome. Soc Biol 31 (3–4):243–278

Epstein GM (2010) Good without God: what a billion nonreligious people do believe. HarperCollins, New York

Erdal D, Whiten A (1994) On human egalitarianism: an evolutionary product of Machiavellian status escalation? Curr Anthropol 35(2):175–178

Eriksson K, Strimling P (2012) The hard problem of cooperation. PLOS ONE 7:e40325

Erlenmeyer-Kimling L, Paradowski W (1966) Selection and schizophrenia. Am Nat 100:651–665

Escher DF (2010) Revisiting risk: sensation-seeking and the sex difference in religiosity. University of Notre Dame, Notre Dame

Eshel I, Cavalli-Sforza LL (1982) Assortment of encounters and the evolution of cooperativeness. PNAS 79(4):1331–1335

Eslinger PJ, Grattan LM, Damasio H et al (1992) Developmental consequences of childhood frontal lobe damage. Arch Neurol 49(7):764–769

Espenshade TJ, Guzman JC, Westoff CF (2003) The surprising global variation in replacement fertility. Popul Res Policy Rev 22(5/6):575–583

Esping-Andersen G (1990) The three worlds of welfare capitalism. Princeton University Press, Princeton

Essock-Vitale SM, McGuire MT (1980) Predictions derived from the theories of kin selection and reciprocation assessed by anthropological data. Ethol Sociobiol 1(3):233–243

Etkin W (1963) Social behaviour factors in the emergence of man. Hum Biol 35(3):299–310

Ettinger RCW (1965) The prospect of immortality. Sidgwick and Jackson, London

Ettinger RCW (1972) Man into superman. The startling potential of human evolution—and how to be part of it. St. Martin's Press, New York

Ettinger RCW, Tandy C, Perry RM (2005) The prospect of immortality. Ria University Press, Ann Arbor

European Greens (2006) A green future for Europe. Adopted as amended by the congress of the European Green Party. Geneva 14th Oct 2006. http://www.globalgreens.org/parties

European Parliament (1996) Resolution on the protection of human rights and dignity with regard to the application of biology and medicine. Off Journal of the European communities, no C320

European Police Office (Europol) (2016) European Union terrorism situation and trend report (TE-SAT) 2016. Europol, The Hague

Evans EM (2000) The emergence of beliefs about the origin of species in school-age children. Merrill Palmer Q 46(2):221–254
Evans PD, Gilbert SL, Mekel-Bobrov N et al (2005) Microcephalin, a gene regulating brain size, continues to evolve adaptively in humans. Science 309(5741):1717–1720
Everitt N (2004) The non-existence of god. Routledge, London
Everitt N (2006) The argument from imperfection: a new proof of the non-existence of God. Philo 9(2):113–130
Evers-Kiebooms G (1987) Genetic risk perception and decision making. Wiley, New York
Ewing B, Moore D, Goldfinger S et al (2010) Ecological footprint atlas 2010. Global Footprint Network, Oakland
Eysenck HJ (1977) Crime and personality. Routledge, Oxford
Eysenck HJ, Gudjonsson GH (1989) The causes and cures of criminality. Plenum, New York
Faderman L (1998) Surpassing the love of men: romantic friendship and love between women from the renaissance to the present. Harper Paperbacks, New York
Fairbanks DJ (2010) Relics of Eden: the powerful evidence of evolution in human DNA. Prometheus Books, Amherst
Falcicchio G, Barbiero G (2015) Loving openness towards nature: Aldo Capitini and the moral value of biophilia. Vis Sustain 3:05–15
Falger VSE, Falger ELF (2003) The cultural evolution of dying: euthanasia in the Netherlands. In: Somit A, Peterson SA (eds) Human nature and public policy: an evolutionary approach. Palgrave Macmillan, Basingstoke, pp 77–96
Family Safe (2008) Pornography statistics. http://familysafemedia.com/pornography_statistics.html
Fanon F (2004) The wretched of the Earth. Grove, New York
Faraone SV, Doyle AE, Mick E et al (2001) Meta-analysis of the association between the 7-repeat allele of the dopamine D(4) receptor gene and attention deficit hyperactivity disorder. Am J Psychiatry 158(7):1052–1057
Farber PL (1998) The temptations of evolutionary ethics. University of California Press, Berkeley
Farley M (2004) Bad for the body, bad for the heart: prostitution harms women even if legalized or decriminalized. Violence Against Women 10(10):1087–1125
Farmer S (2006) Neurobiology, primitive gods and textual traditions: from myth to religions and philosophies. Cosmos 22(2):55–119
Farrell W (1993) The myth of male power: why men are the disposable sex. Simon and Schuster, New York
Farrow T (ed) (2007) Empathy disorders in psychopathy. Oxford University Press, New York
Faundes A, Barzelatto JS (2006) The human drama of abortion: a global search for consensus. Vanderbilt University Press, Nashville
Fehl K, van der Post DJ, Semmann D (2011) Co-evolution of behaviour and social network structure promotes human cooperation. Ecol Lett 14(6):546–551
Fehr E, Bernhard H, Rockenbach B (2008) Egalitarianism in young children. Nature 454:1079–1083
Fehr E, Fischbacher U (2003) The nature of human altruism. Nature 425(6960):785–791
Fehr E, Fischbacher U, Gächter S (2002) Strong reciprocity, human cooperation, and the enforcement of social norms. Hum Nat 13(1):1–25
Fehr E, Gächter S (2000) Cooperation and punishment in public goods experiments. Am Econ Rev 90(4):980–994
Fehr E, Gächter S (2002) Altruistic punishment in humans. Nature 415(6868):137–140
Fehr E, Henrich J (2003) Is strong reciprocity a maladaption? On the evolutionary foundations of human altruism. In: Hammerstein P (ed) Genetic and cultural origins of cooperation. The MIT Press, Cambridge, pp 55–82
Fehr E, Rockenbach B (2004) Human altruism: economic, neural and evolutionary perspectives. Curr Opin Neurobiol 14(6):784–790

Fehr E, Schmidt KM (1999) A theory of fairness, competition, and cooperation. Q J Econ 114 (3):817–868

Fehr E, Schneider F (2010) Eyes are on us, but nobody cares: are eye cues relevant for strong reciprocity? Proc R Soc B-Biol Sci 277(1686):1315–1323

Feierman JR (ed) (2009) The biology of religious behavior: the evolutionary origins of faith and religion. ABC-CLIO, LLC, Santa Barbara

Feigin A, Kaplitt MG, Tang C et al (2007) Modulation of metabolic brain networks after subthalamic gene therapy for Parkinson's disease. PNAS 104(49):19559–19564

Feinberg J (1992) Freedom and fulfillment. Princeton University Press, Princeton

Feldman MW, Laland KN (1996) Gene-culture coevolutionary theory. Trends Ecol Evol 11 (11):453–457

Feldman R, Gordon I, Influs M (2013) Parental oxytocin and early caregiving jointly shape children's oxytocin response and social reciprocity. Neuropsychopharmacology 38:1154–1162

Feldman W, Feldman E, Goodman JT et al (1991) Is childhood sexual abuse really increasing in prevalence. An analysis of the evidence. Pediatrics 88(1):29–33

Ferguson ARB (1999) The logical foundations of ecological footprints. Environ Dev Sustain 1 2:149–156

Ferguson CJ (2008) An evolutionary approach to understanding violent antisocial behavior: diagnostic implications for duel-process etiology. J Forensic Psychol Pract 8(4):321–343

Ferguson CJ (2010) Genetic contributions to antisocial personality and behavior: a meta-analytic review from an evolutionary perspective. J Soc Psychol 150(2):160–180

Ferguson CJ, Beaver KM (2009) Natural born killers: the genetic origins of extreme violence. Aggression Violent Behav 14(5):286–294

Ferguson CJ, Hartley RD (2009) The pleasure is momentary…the expense damnable? The influence of pornography on rape and sexual assault. Aggression Violent Behav 14(5):323–329

Ferguson MA, Nielsen JA, King JB et al (2016) Reward, salience, and attentional networks are activated by religious experience in devout Mormons. Soc Neurosci. https://doi.org/10.1080/17470919.2016.1257437

Ferguson RB (1984) Warfare, culture, and environment. Academic Press, Orlando

Ferrari PF (2014) The neuroscience of social relations. A comparative-based approach to empathy and to the capacity of evaluating others' action value. Behaviour 151 (2–3):SI 297–313

Fessler DMT, Haley KJ (2003) The strategy of affect. Emotions in human cooperation. In: Hammerstein P (ed) Genetic and cultural evolution of cooperation. MIT Press, Cambridge, pp 7–36

Fetchenhauer D (2009) Evolutionary perspectives on religion—what they can and what they cannot explain (yet). In: Voland E, Schiefenhövel W (eds) The biological evolution of religious mind and behavior. Springer Publishers, New York, pp 275–291

Feuerbach L (1841) Das Wesen des Christenthums. (The essence of Christianity). O. Wigand, Leipzig

Fieder M, Huber S (2007) The effects of sex and childlessness on the association between status and reproductive output in modern society. Evol Hum Behav 28(6):392–398

Field AJ (2008) Why multilevel selection matters. J Bioecon 10(3):203–238

Filiu J-P (2011) The Arab revolution: ten lessons from the democratic uprising. Oxford University Press, Oxford

Filsinger EE (1988) Biosocial perspectives on the family. Sage, Newbury Park

Finch CE (1990) Longevity, senescence and the genome. University of Chicago Press, Chicago

Fincher CL, Thornhill R, Murray DR et al (2008) Pathogen prevalence predicts human cross-cultural variability in individualism/collectivism. Proc R Soc B-Biol Sci 275 (1640):1279–1285

Finer LB, Zolna MR (2011) Unintended pregnancy in the United States: incidence and disparities, 2006. Contraception 84(5):478–485

Fink B, Trivers R (2014) Cognitive simplicity and self-deception are crucial in martyrdom and suicide terrorism. Behav Brain Sci 37(4):366–367

Finkel D, Swartwout P, Sosis R (2010) The socio-religious brain: a developmental model. Proc Br Acad 158:287–312
Fischer I, Frid A, Goerg SJ et al (2013) Fusing enacted and expected mimicry generates a winning strategy that promotes the evolution of cooperation. PNAS 110(25):10229–10233
Fisher RA (1915) The evolution of sexual preference. Eugenics Rev 7:184–192
Fisher RA (1930) The genetical theory of natural selection. Clarendon Press, Oxford
Fisher H (1992) Anatomy of love. The natural history of monogamy, adultery, and divorce. Norton, New York
Fisher H (2004) Why we love: the nature and chemistry of romantic love. Henry Holt, New York
Fisher H, Aron A, Brown LL (2005) Romantic love: an fMRI study of a neural mechanism for mate choice. J Comp Neurol 493(1):58–62
Fisher W, Barak A (2001) Internet pornography: a social psychological perspective on internet sexuality. J Sex Res 38(4):312–323
Fiske J (1874) Outlines of cosmic philosophy. Houghton Mifflin and CO, Boston
Flack JC, de Waal FBM (2000) Any animal! Whatever. Darwinian building blocks of morality in monkeys and apes. J Conscious Stud 7(1–2):1–29
Flank L (2006) Creationism/ID. A short legal history. http://www.talkreason.org/articles/HistoryID.cfm
Flannelly KJ, Galek K (2010) Religion, evolution, and mental health: attachment theory and ETAS theory. J Relig Health 49(3):337–350
Flannelly KJ, Galek K, Ellison CG et al (2010) Beliefs about God, psychiatric symptoms, and evolutionary psychiatry. J Relig Health 49(2):246–261
Fleagle JG, Shea JJ, Grine FE et al (2010) Out of Africa I: the first hominin colonization of Eurasia. Springer, Dordrecht
Fleck RK (2011) Natural selection and the problem of evil: an evolutionary model with application to an ancient debate. Zygon 46(3):561–587
Fleischman DS, Fessler DMT, Cholakians AE (2014) Testing the affiliation hypothesis of homoerotic motivation in humans: the effects of progesterone and priming. Arch Sex Behav. https://doi.org/10.1007/s10508-014-0436-6
Fletcher J (1974) The ethics of genetic control: ending reproductive roulette. Prometheus books, New York
Flew A (1967) Evolutionary ethics. Macmillan, London
Flew A, Varghese RA (2007) There is a God. How the world's most notorious atheist changed his mind. Harper Collins, New York
Flinn M, Coe K (2007) The linked red queens of human cognition, coalitions, and culture. In: Gangestad SW, Simpson JA (eds) The evolution of mind: fundamental questions and controversies. Guilford press, New York, pp 339–356
Flinn MV, Geary DC, Ward CV (2005) Ecological dominance, social competition, and coalitionary arms races: why humans evolved extraordinary intelligence. Evol Hum Behav 26(1):10–46
Flint P, VanderKam J (2005) The meaning of the Dead Sea scrolls: their significance for understanding the Bible, Judaism, Jesus, and Christianity. Continuum International Publishing Group, New York
Flynn JR (1987) Massive IQ gains in 14 nations: what IQ tests really measure. Psychol Bull 101 (2):171–191
FM-2030 (Esfandiary FM) (1970) Optimism one: the emerging radicalism. Norton, New York
FM-2030 (Esfandiary FM) (1973) Up-wingers: a futurist manifesto. The John Day Company, New York
FM-2030 (Esfandiary FM) (1989) Are you a transhuman? Monitoring and stimulating your personal rate of growth in a rapidly changing world. Warner Books, New York
Fodor JA (1983) The modularity of mind. MIT Press, Bradford Books, Cambridge
Fog A (1999) Cultural selection. Kluwer Academic Publishers, Boston

Fomby P, Cherlin AJ (2016) Family instability and child well-being. Am Sociol Rev 72(2):181–204
Forber P, Smead R (2015) Evolution and the classification of social behavior. Biol Philos 30 (3):405–421
Forlini C, Hall W (2015) The is and ought of the ethics of neuroenhancement: mind the gap. Front Psychol 6:1998. https://doi.org/10.3389/fpsyg.2015.01998
Ford K (1979) Contraceptive utilisation, United States. Vital and health statistics, Series 23, N° 2. DHEW Publication N° (PHS) 79-1978
Fordham G (2006) As if they were watching my body: a study of pornography and the development of attitudes toward sex and sexual behavior among Cambodian youth. World Vision, Phnom Penh, Cambodia
Forrest B, Gross PR (2004) Creationism's Trojan Horse: the wedge of intelligent design. Oxford University Press, Oxford
Forthun LF, Bell NJ, Peek CW et al (1999) Religiosity, sensation seeking, and alcohol/drug use in denominational and gender contexts. J Drug Issues 29(1):75–90
Fortunato L, Archetti M (2010) Evolution of monogamous marriage by maximization of inclusive fitness. J Evol Biol 23(1):149–156
Foster L (1984) Religion and sexuality: the Shakers, the Mormons, and the Oneida Community. University of Illinois Press, Urbana
Four books and five classics of Confucianism. (http://chineseliterarycannon.weebly.com/-confucian-cannon-ldquofour-books-and-five-classics.html)
Fowers BJ (2015) The evolution of ethics: human sociality and the emergence of ethical mindedness. Palgrave Macmillan, Basingstoke
Fowler JH (2005) Altruistic punishment and the origin of cooperation. PNAS 102:7047–7049
Fowler JH, Schreiber D (2008) Perspective. Biology, politics, and the emerging science of human nature. Science 322(5903):912–914
Francis C, Pirkis J, Dunt D et al (2001) Mental health and illness in the media: a review of the literature. Commonwealth Department of Health and Aged Care, Canberra
Francis LJ (1997) The psychology of gender differences in religion: a review of empirical research. Religion 27(1):81–96
Frank RH (1988) Passions within reason: the strategic role of the emotions. Norton, New York
Frank SA (1995) George price's contributions to evolutionary genetics. J Theor Biol 175(3):373–388
Frank SA (1998) Foundations of social evolution. Princeton University Press, Princeton
Frank SA (2003) Repression of competition and the evolution of cooperation. Evolution 57 (4):693–705
Frank SA (2006) Social selection. In: Fox CW, Wolf JB (eds) Evolutionary genetics: concepts and case studies. Oxford University Press, New York, NY, pp 350–363
Frank RH (2010) Luxury fever: weighing the cost of excess. Princeton University, Princeton
Frank SA (2012) Natural selection. IV. The Price equation. J Evolut Biol 25(6):1002–1019
Franz M, Schuelke O, Ostner J (2013) Rapid evolution of cooperation in group-living animals. BMC Evol Biol 13:235. http://www.biomedcentral.com/1471-2148/13/235
Frayer DW (1980) Sexual dimorphism and cultural evolution in the late Pleistocene and Holocene of Europe. J Human Evol 9(5):399–416
Frayer DW, Wolpoff MH (1985) Sexual dimorphism. Annu Rev Anthropol 14:429–473
Freeman D (1983) Margaret Mead and Samoa: the making and unmaking of an anthropological myth. Harvard University Press, Cambridge
Freeman J (2001) Gifted children grown up. David Fulton Press, London
Freedman E (2003) No turning back: the history of feminism and the future of women. Ballantine Books, New York
Freedman E (2007) The essential feminist reader. Modern library, New York
Freedman R, Whelpton PK, Campbell AA (1959) Family planning, sterility and population growth. McGraw-Hill, New York

Freese J (2004) Risk preferences and gender differences in religiousness: evidence from the World Values Survey. Rev Religious Res 46(1):88–91

Frejka T (2008) Birth regulation in Europe: completing the contraceptive revolution. Demographic Res 19(5):73–84

Frejka T, Ross J (2001) Paths to subreplacement fertility: the empirical evidence. Population and Development Review, 27 (Supplement: Global Fertility Transition): 213–254

Frejka T, Sobodka, T, Hoem J, et al. (eds) (2008) Childbearing trends and policies in Europe. Demographic research, special collection 7. Max Planck Institute for Demographic Research, Rostock

Frejka T, Westoff CF (2008) Religion, religiousness and fertility in the US and Europe. Eur J Population 24(1):5–31

Frey U, Rusch H (2012) An evolutionary perspective on the long-term efficiency of costly punishment. Biol Philos 27(6):811–831

Friedan B (1963) The feminine mystique. Penguin, Harmondsworth

Friedmann Th (ed) (1998) The development of human gene therapy. Cold Spring Harbor Laboratory Press, Cold Spring Harbor NY

Fries JF (1980) Aging, natural death and the compression of morbidity. New England J Med 303 (3):130–135

Frimer JA, Walker LJ (2008) Towards a new paradigm of moral personhood. J Moral Educ 37:333–356

Fry DP (2006) Reciprocity: the foundation stone of morality. In: Killen M, Smetana J (eds) Handbook of moral development. Erlbaum, Mahwah, pp 399–422

Fu F, Tarnita CE, Christakis NA et al (2012) Evolution of in-group favoritism. Sci Rep 2:460

Fukuyama F (2002) Our posthuman future: consequences of the biotechnology revolution. Farrar Straus and Giroux, New York

Fuller R (2006) Wonder and the religious sensibility: a study on religion and emotion. J Rel 86 (3):364–384

Fuller S (2007) Science vs religion? Intelligent design and the problem of evolution. Polity Press, Cambridge

Fuller S (2011) Humanity 2.0: What it means to be human past, present and future. Palgrave Macmillan, Basingstoke

Fuller M (2016) Religion and the brain: what can science tell us about belief in God? Mod Believing 57(2):163–173

Fuller RW, Gerloff PA (2008) Dignity for all: how to create a world without rankism. Berrett-Koehler Publishers, San Francisco

Fuller JL, Thompson WR (1978) Foundations of behavior genetics. Mosby, Saint Louis MO

Fulton JJ, Marcus DK, Payne KT (2010) Psychopathic personality traits and risky sexual behavior. Pers Individ Differ 49(1):29–33

Fulton JJ, Marcus DK, Zeigler-Hill V (2014) Psychopathic personality traits, risky sexual behavior, and psychological adjustment among college-age women. J Soc Clin Psychol 33 (2):143–168

Fumagalli M, Priori A (2012) Functional and clinical neuroanatomy of morality. Brain 135:2006–2021

Furnham A, Swami V (2008) Body beautiful: evolutionary and socio-cultural perspectives. Palgrave Macmillan, Basingstoke

Gaard GC (2011) Ecofeminism revisited: rejecting essentialism and re-placing species in a material feminist environmentalism. Feminist Formations 23(2):26–53

Gächter S, Herrmann B (2006) Human cooperation from an economic perspective. In: Kappeler PM, van Schaik CP (eds) Cooperation in primates and humans. Springer, New York, pp 275–302

Gächter S, Herrmann B, Thöni C (2008) Antisocial punishment across societies. Science 319 (5868):1362–1367

Gailliot MT, Stilman TF, Schmeichel BJ et al (2008) Mortality salience increases adherence to salient norms and values. Pers Soc Psychol Bull 34(7):993–1003

Gaitan Torres A (2012) The evolution of morality. Teorema 31(1):109–127

Gallese V (2003) The roots of empathy: the shared manifold hypothesis and the neural basis of intersubjectivity. Psychopathology 36(4):171–180

Gallup GG, Suarez SD (1983) Homosexuality as a by-product of selection for optimal heterosexual strategies. Perspect Biol Med 26(2):315–322

Gallus A, Blacking J, Cohen E et al (1972) A biofunctional theory of religion. Current Anthropol 13(5):543–568

Galor O, Moav, O (2005) Natural Selection and the Evolution of Life Expectancy CEPR Discussion Paper No. 5373. http://ssrn.com/abstract=885956

Galton F (1865) Hereditary talent and character. Macmillan's Magazine 12:157–166; 318–327

Galton F (1869) Hereditary genius. Macmillan, London

Galton F (1883) Inquiries into human faculty and its development. Dent, London

Galton F (1889) Natural inheritance. Macmillan, London

Galton F (1905) Eugenics: Its definition, scope and aims. Sociological Papers 1: 45–50. Paper read before the sociological society at a meeting in the school of economics and political science, London University, 16 May 1904.)

Gamble C, Gowlett J, Dunbar R (2014) Thinking big: how the evolution of social life shaped the human mind. Thames and Hudson, London

Gangestad SW (2011) Human adaptations for mating: Frameworks for understanding patterns of family formation and fertility. In: Booth A, McHale SM, Landale NS (eds) Biosocial foundations of family processes. Springer, New York, pp 117–148

Gangestad SW, Buss DM, (1993) Pathogen prevalence and human mate preferences. Eth Sociobiol 14(2):89–96

Gangestad SW, Simpson JA (2007) Whither science of the evolution of mind? In: Gangestad SW, Simpson JA (eds) The evolution of mind: Fundamental questions and controversies. Guilford Press, New York, pp 397–437

Gangestad SW, Thornhill R (1997) The evolutionary psychology of extrapair sex: the role of fluctuating asymmetry. Evol Hum Behav 18(2):69–88

Gangoli G, Westmarland N (eds) (2011) International approaches to rape. The Policy Press, Bristol

Garcia JR, MacKillop J, Aller EL et al (2010) Associations between dopamine D4 receptor gene variation with both infidelity and sexual promiscuity. PLoS One 5(11):e14162

Garcia J, Traulsen A (2012) The structure of mutations and the evolution of cooperation. PLoS One 7(4):e35287

Gardner H (1983) 1993) Frames of mind: the theory of multiple intelligences. Basic Books, New York

Gardner A (2008) The Price equation. Curr Biol 18(5):R198–R202

Gardner A (2013) Adaptation of individuals and groups. In: Bouchard F, Hunemann P (eds) From groups to individuals. Evolution and emerging individuality. The MIT Press, Cambridge, pp 99–116

Gardner DK (2014) Confucianism: a very short introduction. Oxford University Press, Oxford

Gardner A (2015) The genetical theory of multilevel selection. J Evol Biol 28:305–319

Gardner DK, Weissman A, Howles CM et al (2004) Textbook of assisted reproductive techniques: laboratory and clinical perspectives. Taylor and Francis, London

Garland-Thomson R (2002) Integrating disability, transforming feminist theory. Nat Women's Stud Assoc J 14(3):1–32

Garland-Thomson R (2012) The case for conserving disability. J Bioethical Inquiry 9(3):339–355

Garner S (2005) Transhumanism and Christian social concern. J Evol Technol 14(2):89–103

Garreau J (2006) Radical evolution. The promise and peril of enhancing our minds, our bodies and what it means to be human. Broadway, New York

Gasenbeek B, Gogineni B (eds) (2002) International humanist and ethical union 1952–2002—past, present and future. De Tijdstroom, Utrecht

Gasman D (1998) Haeckel's Monism and the birth of fascist ideology. Peter Lang Publishing, New York
Gastmans C, Lemiengre J, van der Wal G et al (2006) Prevalence and content of written ethics policies on euthanasia in Catholic healthcare institutions in Belgium (Flanders). Health Policy Year 76(2):169–178
Gat A (2006) War in human civilization. Oxford University Press, New York
Gat A (2009) So why do people fight? Evolutionary theory and the causes of war. Eur J Int Relat 15(4):571–599
Gatens M (1996) Imaginary bodies. Ethics, power and corporeality. Routledge
Gauchet M (1997) The disenchantment of the world: a political history of religion. Princeton University Press, Princeton
Gaulin SJC, Schlegel A (1980) Paternal confidence and paternal investment: A cross-cultural test of a sociobiological hypothesis. Ethology Sociobiol 1(4):301–309
Gaus G (2015) The egalitarian species. Soc Philos Policy 31(02):1–27
Gautier E (1880) Le Darwinisme social. Derveaux, Paris
Gauthier AH, Smeeding TM, Furstenberg FF (2004) Are parents investing less time in children? Trends in selected industrialized countries. Pop Dev Rev 30(4):647–672
Gavrilets S (1997) Evolution and speciation on holy adaptive landscapes. Trends Ecologiccal. Evol 12:307–312
Gavrilets S (2012) On the evolutionary origins of the egalitarian syndrome. PNAS 109(35):14069–14074
Gavrilets S, Fortunato L (2014) A solution to the collective action problem in between-group conflict with within-group inequality. Nat Commun 5:3526
Gazzaniga MS (1985) The social brain. Psychol Today 19(11):29–38
Gazzaniga MS (2005) The ethical brain. Dana Press, New York
Geary DC (1998) Male, female: The evolution of human sex differences. American Psychological Association, Washington
Geary DC (2006) Coevolution of paternal investment and cuckoldry in humans. In: Shackelford TK, Platek S (eds) Female infidelity and paternal uncertainty. Cambridge University Press, New York, pp 14–34
Geary DC (2007) The motivation to control and the evolution of general intelligence. In: Gangestad SW, Simpson JA (eds) The evolution of mind: fundamental questions and controversies. Guilford Press, New York, pp 305–312
Geeraerts K (2010) Geschiedenis van het vak niet-confessionele zedenleer [History of the course non-confessional ethics]. http://users.skynet.be/inge.tordeur/ovm/zedenleer.htm
Geertz AW (2009) When cognitive scientists become religious, science is in trouble: on neurotheology from a philosophy of science perspective. Religion 39(4):319–324
Gellhorn E, Kiely WF (1972) Mystical states of consciousness: Neurophysiological and clinical aspects. J Nerv Mental Dis 154(6):399–405
Gensler HJ (2012) Ethics and the golden rule. Routledge, New York
Gerhardt V, Nida-Rümelin J (eds) (2010) Evolution in Natur und Kultur. de Gruyter, Berlin
Gervais WM (2014) Everything is permitted? People intuitively judge immorality as representative of atheists. PLoS One 9(4):e92302
Gesang B (2006) Enhancement und Gerechtigkeit. [Enhancement and justice]. In: Sorgner SL, Birx HJ, Knoepffler N (eds) Eugenik und die Zukunft. Verlag Karl Alber, Freiburg, pp 127–149
Gewirth A (1993) How ethical is evolutionary ethics? In: Nitecki MH, Nitecki DV (eds) evolutionary ethics. State University of New York, Albany, pp 241–256
Ghafouri-Fard S, Akrami SM (2011) Man evolution: An Islamic point of view. Eur J Sci Theol 7 (3):17–28
Ghesquiere J, Martin RD, Newcombe F (eds) (1985) Human sexual dimorphism. Taylor and Francis, London

Ghiglieri MP (1999) The dark side of man: tracing the origins of male violence. Perseus Books, Reading
Ghiselin MT (1974) The economy of nature and the evolution of sex. University of California Press, Berkeley
Ghurye GS (1996) Caste and race in India. Popular Prakashan, Mumbai
Giampetro M, Pimentel D (1993) The tightening conflict: population, energy use, and the ecology of agriculture. NPG Forum, Negative Population Growth, Teaneck
Giannelli F, Anagnostopoulos T, Green PM (1999) Mutation rates in humans. II. Sporadic mutation-specific rates and rate of detrimental human mutations inferred from hemophilia B. Am J Human Genet 65(6):1580–1587
Giberson K (2008) Saving Darwin: how to be a christian and believe in evolution. HarperOne, New York
Gibson MA, Lawson DW (2015) Applying evolutionary anthropology. Evol Anthropol 24(1):3–14
Giesen KG (2004) Transhumanisme et génétique humaine. [Transhumanism and human genetics]. l'Observatoire de la Génétique, Nr. 16
Gignoux CR, Henn BM, Mountain JL (2011) Rapid, global demographic expansions after the origins of agriculture. PNAS 108:6044–6049
Gilbert M (1986) The Holocaust: a history of the Jews of Europe during the Second World War. Holt, Rinehart, and Winston, New York
Gilbert P (1999) Family values and the nation-state. In: Jagger G, Wright C (eds) Changing family values. Edward Elgar, Cheltenham, pp 136–149
Gilbert SF, Opitz JM, Raff RA (1996) resynthesizing evolutionary and developmental biology. Developmental Biology 173(2):357–372
Giles H, Reid SA, Harwood J (eds) (2010) The dynamics of intergroup communication. Peter Lang, New York
Gillis S, Howie G, Munford R (2007) Third wave feminism: a critical exploration. Palgrave, Basingstoke
Gingerich O (1993) The eye of heaven: Ptolemy, Copernicus, Kepler. American Institute of Physics, New York
Ginges J, Atran S (2011) War as a moral imperative (not just practical politics by other means). Proc R Soc B Biol Sci 278(1720):2930–2938
Gingrich A, Banks M (2006) Neo-nationalism in Europe and beyond: perspectives from social anthropology. Berghahn Books, Oxford
Gintis H (2000) Game theory evolving. Princeton University Press, Princeton
Gintis H (2000) Strong reciprocity and human sociality. J Theoret Biol 206:169–179
Gintis H (2010) Reflections on Arnhart's Darwinian Liberalism. http://www.cato-unbound.org/archives/july-2010-darwin-and-politics/
Gintis H (2011) Gene-culture coevolution and the nature of human sociality. Philos Trans R Soc B 366(1566):878–888
Gintis H (2014) Inclusive fitness and the sociobiology of the genome. Biol Philos 29(4):477–515
Gintis H (2014) Thinking big: how the evolution of social life shaped the human mind. Nature 509 (7500):284–285
Gintis H (2015) Modeling Homo socialis: a reply to critics. Rev Behav Econ 2(1–2):211–237
Gintis H, Helbing D (2015) Homo socialis: an analytical core for sociological theory. Rev Behav Econ 2(1–2):1–59
Gintis H, Henrich J, Bowles S et al (2008) Strong reciprocity and the roots of human morality. Soc Justice Res 21(2):241–253
Giovannoli FD (1999) The Biology of belief: how our biology biases our beliefs and perceptions. Rosetta Press, New York
Giroud C, Augsburger M, Horisberger B et al (1999) Exit association-mediated suicide: toxicologic and forensic aspects. Am J Forensic Med Pathol 20(1):40–44

Giubilini A (2015) Normality, therapy, and enhancement. What should bioconservatives say about the medicalization of love? Cambridge Q Healthc Ethics 24(3):347–354

Giubilini A, Sanyal S (2015) The ethics of human enhancement. Philosophy Compass 10(4):233–243

Giuliani C, Bacalini MG, Sazzini M, Pirazzini C, Franceschi C, Garagnani P, Luiselli D (2014) The epigenetic side of human adaptation: hypotheses, evidences and theories. Ann Hum Biol 42(1):1–9

Giustiniani V (1985) Homo, humanus, and the meanings of humanism. J Hist Ideas 46(2):167–195

Glad J (2003) Future human evolution. Mankind Q XLIII 4:409–454

Glad J (2006) Future human evolution: eugenics in the twenty-first century. Hermitage Publishers, Schuylkill Haven

Glaeser EL, Sacerdote BI (2008) Education and religion. University of Chicago Press, Chicago

Glannon W (2001) Genes and future people. Philosophical issues in human genetics. Westview Press, Cambridge

Glannon W (2007) Bioethics and the brain. Oxford University Press, New York

Glass B (1966) The effect of changes in the physical environment on genetic change. In: Roslansky JD (ed) Genetics and the future of man. A discussion at the Nobel Conference organized by Gustavus Adolphus College, St. Peter, Minnesota, 1965. North-Holland Publishing Company, Amsterdam, pp 23–48

Glass JD (2007) The power of faith: mother nature's gift. Donington Press, Corona del Mar

Glenn EN (1987) Gender and the family. In: Hess B, Ferree MM (eds) Analyzing gender. Sage, Beverly Hills

Glenn AL, Kurzban R, Raine A (2011) Evolutionary theory and psychopathy. Aggression Violent Behav 16(5):371–380

Global Ethic Foundation (1993) Towards a global ethic. Statement by the 1993 Parliament of the World's Religions, 28 August-5 September 1993, Chicago

Glover J (1984) What sort of people should there be?. Penguin, Harmondsworth

Glover J (2003) Choosing children: genes, design, and disability. Oxford University Press, Oxford

Glover J (2006) Choosing children: the ethical dilemmas of genetic intervention. Oxford University Press, Oxford

Glover V (2011) Annual research review: Prenatal stress and the origins of psychopathology: an evolutionary perspective. J Child Psychol Psychiatry 52(4):356–367

Gneezy A, Fessler DMT (2012) Conflict, sticks and carrots: War increases prosocial punishments and rewards. Proc Royal Soc B-Bio Sci 279(1727):219–223

Godfrey LR (ed) (1983) Scientists confront creationism. W.W. Norton, New York NYGogineni B (2006) Humanism for the world—The IHEU in a nutshell. IHEU London

Goldberg S (1973/74) The inevitability of patriarchy. Why the biological difference between men and women always produce male domination. W. Morrow and Co, New York

Goldberg S (1993) Why men rule. A theory of male dominance. Open Court Publishing Company, Chicago

Goldin PR (2011) Confucianism. University of California Press, Oakland

Goldstein J, Lutz W, Testa MR (2003) The emergence of sub-replacement family size ideals in Europe. Population Research and Policy Review 22(5–6):479–496

Goldstein JR, Sobotka T, Jasilioniene A (2009) The end of "lowest-low" fertility? Population Dev Rev 35(4):663–699

Goldstone JA (1987) Cultural orthodoxy, risk and innovation: The divergence of east and west in the early modern world. Sociol Theor 5(2):119–135

Gómez JM, Verdú M, González-Megías A, Méndez M (2016) The phylogenetic roots of human lethal violence. Nature. https://doi.org/10.1038/nature19758

Gonzaga GC, Turner RA, Keltner D et al (2006) Romantic love and sexual desire in close relationships. Emotion 6(2):163–179

Gonzalez S (2009) Panorámica del debate creacionismo-evolucionismo en los últimos cien años en USA. [An outlook on the debate between creationism and evolutionism in the USA over the last 100 years]. Anuario de historia de la Iglesia 18:41–53
Goodman RE (1998) The paraphilias: an evolutionary and developmental perspective. In: Freeman H, Pullen I, Stein G et al (eds) Seminars in psychosexual disorders. RCPsych Publications, London, pp 142–155
Goodman N (2002) The serotonergic system and mysticism: Could LSD and the nondrug-induced mystical experience share common neural mechanisms? J Psychoactive Drugs 34(3):263–272
Goodman J (2014) Altruism and the golden rule. Zygon 49(2):381–395
Goodin E, Headey B, Muffels R et al (1999) The real worlds of welfare capitalism. Cambridge University Press, Cambridge
Goodman A, Koupil I (2009) Social and biological determinants of reproductive success in Swedish males and females born 1915–1929. Evol Hum Behav 30(5):329–341
Goodman A, Koupil I (2010) The effect of school performance upon marriage and long-term reproductive success in 10,000 Swedish males and females born 1915–1929. Evol Hum Behav 31(6):425–435
Goodman M, Porter CA, Czelusniak J et al (1998) Toward a phylogenetic classification of Primates based on DNA evidence complemented by fossil evidence. Mol Phylogenet Evol 9 (3):585–598
Gordijn B, Chadwick R (eds) (2010) Medical enhancements and posthumanity. Routledge, New York
Gordon JW (1999) Gene enhancement in humans. Science 283(5410):2023–2024
Gordon TJ, Gerjuoy H, Anderson H (1979) Life-extending technologies. A technological assessment. Pergamon Press, New York
Gorelik G, Shackelford TK (2014) Evolutionary awareness. Evolutionary Psychology 12(4):783–813
Gorelik G, Shackelford TK, Salmon CA (2010) New horizons in the evolutionary science of the human family. Rev Gen Psychol 14(4):330–339
Gorelik G, Shackelford TK, Weekes-Shackelford VA (2012) Human violence and evolutionary consciousness. Rev Gen Psychol 16(4):343–356
Gornick JC, Meyers MK (2003) Families that work: policies for reconciling parenthood and employment. Russell Sage Foundation, New York
Gorsuch NM (2006) The future of assisted suicide and euthanasia. Princeton University Press, Princeton
Gosselin D, Tindemans B (2016) Thinking Futures. Strategy at the Edge of Complexity and Uncertainty, LannooCampus
Gottfredson LS (1997) Why g matters: the complexity of everyday life. Intelligence 24(1):79–132
Gottfredson LS (2004) Life, death, and intelligence. J Cogn Educ Psychol 4(1):23–46
Gottfredson MR, Hirschi T (1990) A general theory of crime. Stanford University Press, Stanford
Gottschalk M, Ellis L (2009) Evolutionary and genetic explanations of violent crime. In: Ferguson C (ed) Violent: clinical and social implications. Sage, Thousand Oaks
Gottschall J (2004) Explaining wartime rape. J Sex Res 41(2):129–136
Goudie A (2005) The human impact on the natural environment: Past, present, and future. Wiley, Oxford
Gough K (1971) The origin of the family. J Marr Fam 33(4):760–771
Gould SJ (1977) Ontogeny and phylogeny. Harvard University Press, Cambridge
Gould SJ (1980) The panda's thumb. More reflections in natural history, Norton, New York
Gould SJ (1997) Nonoverlapping magisteria. Natural History 106:16–22
Gould SJ (1999) Rocks of ages: Science and religion in the fullness of life. The Ballantine Publishing Group, New York
Gould SJ (2008) On replacing the idea of progress with an operational notion of directionality. In: Nitecki MH (ed) Evolutionary progress. The University of Chicago Press, Chicago, pp 319–338

Gould SJ, Lewontin RC (1979) The spandrels of San Marco and the panglossian paradigm: a critique of the adaptationist programme. Proc Roy Soc London Ser B 205(1161):581–598

Gould SJ, Vrba E (1982) Exaptation: a missing term in the science of form. Paleobiology 8(1):4–15

Gove WR (1985) The effect of age and gender on deviant behavior: a biopsychosocial perspective. In: Rossi AS (ed) Gender and the life course. Aldine, New York NY, pp 210–234

Gowaty PA (1997) Feminism and evolutionary biology: boundaries, intersections, and frontiers. Springer, New York

Gowlett J, Gamble C, Dunbar R (2012) Human evolution and the archaeology of the social brain. Current Anthropology 53(6):693–722

Grafen A (2000) Developments of the Price equation and natural selection under uncertainty. Proc Roy Soc London B 267(1449):1223–1227

Graffin GW, Provine WB (2007) Evolution, religion and free will. The most eminent evolutionary scientists have surprising views on how religion relates to evolution. Am Sci 95(4):294–298

Graham RK (1970) The future of man. Christopher Publishing, North Quincy

Graham J, Haidt J (2010) Beyond beliefs: Religions bind individuals into moral communities. Pers Soc Psychol Rev 14(1):140–150

Grammer K, Fink B, Møller AP et al (2003) Darwinian aesthetics: sexual selection and the biology of beauty. Biol Rev 78(3):385–407

Granqvist P (2006) Religion as a by-product of evolved psychology: The case of attachment and implications for brain and religious research. In: McNamara P (ed) Where God and science meet: how brain and evolutionary studies alter our understanding of religion, vol 2. The neurology of religious experience. Praeger, Westport, pp 105–150

Granqvist P, Mikulincer M, Shaver P (2010) Religion as attachment: normative processes and individual differences. Pers Soc Psychol Rev 14(1):49–59

Grant L (1992) Elephants in the Volkswagen: facing the tough questions about our overcrowded country. WH Freeman, New York

Grant L (1996) Juggernaut: growth on a finite planet. Seven Locks Press, Santa Ana

Grant L (2000) Too many people: the case for reversing growth. Seven Locks Press, Santa Ana

Grassie W (2010) The new sciences of religion: exploring spirituality from the outside in and bottom up. Palgrave Macmillan, New York

Gray J (2002) Straw dogs: thoughts on humans and other animals. Granta Books, London

Gray PB, Garcia JR (2013) Evolution and human sexual behavior. Harvard University Press, Cambridge

Grayling AC (2014) The God argument: the case against religion and for humanism. Bloomsbury, London

Grayson DK (2007) Deciphering North American Pleistocene extinctions. J Anthropol Res 63 (2):185–213

Green TH (1884) Prolegomena to ethics. Clarendon, Oxford

Green K (1989) Prostitution, exploitation and taboo. Philosophy 64(250):525–534

Green RM (2013) Transhumanism and transcendence: christian hope in an age of technological enhancement. Hastings Cent Rep 43(4):45–47

Greenberg J, Cohen RL (1982) Equity and justice in social behavior. Academic Press, New York

Greene PJ (1980) Paternity and the avunculate. Am Anthropologist New Ser 82(2):381–382

Greene JD (2003) From neural 'Is' to moral 'Ought': what are the moral implications of neuroscientific moral psychology? Nat Rev Neurosci 4(10):846–849

Greene JD (2009) The cognitive neuroscience of moral judgment. In: Gazzaniga MS (ed) The cognitive neurosciences. MIT Press, Cambridge

Greene JD (2015) The cognitive neuroscience of moral judgment and decision making. In: Decety J, Wheatley T (eds) The moral brain. A multidisciplinary perspective. The MIT Press, Cambridge, pp 197–220

Greene JD, Sommerville RB, Nystrom LE et al (2001) An fMRI investigation of emotional engagement in moral judgment. Science 293(5537):2105–2107

Greenfield PM (2004) Inadvertent exposure to pornography on the Internet: Implications of peer-to-peer file-sharing networks for child development and families. J Appl Dev Psychol 25 (6):741–750
Greenhalgh S (2003) Science, modernity, and the making of China's one-child policy. Popul Dev Rev 29(2):163–196
Gregory AM, Light-Häusermann JH, Rijsdijk F et al (2009) Behavioral genetic analyses of prosocial behavior in adolescents. Dev Sci 12(1):165–174
Gregory MS, Silvers A, Sutch D (eds) (1978) Sociobiology and human nature. Jossey-Bass, San Francisco
Greksa LP (2002) Population growth and fertility patterns in an Old Order Amish settlement. Ann Hum Biol 29(2):192–201
Griffiths J, Weyers H, Adams M (2008) Euthanasia and law in Europe: with special reference to the Netherlands and Belgium. Hart Publishing, Oxford
Grinde B (1996) Darwinian happiness: Biological advice on the quality of life. J Soc Evol Sys 19 (3):249–260
Grinde B (1998) The biology of religion: a Darwinian gospel. J Soc Evol Syst 21(1):19–28
Grine FE, Fleagle JG, Leakey RE (2009) The first humans: origin and early evolution of the genus Homo. Springer, Dordrecht
Gross R (1993) Buddhism after patriarchy: a feminist history, analysis, and reconstruction of Buddhism. State University of New York Press, Albany
Gross R (1996) Child and family in Buddhism. Religious dimensions of child and family life: Reflections on the UN Convention of the Rights of the Child. University of Victoria, Canada
Gross R (2013) Buddhist perspectives on gender issues. In: Emmanuel SM (ed) A companion to Buddhist philosophy. Wiley, Oxford
Grove W (1990) Heritability of substance abuse and antisocial behaviour: A study of monozygotic twins reared apart. Biol Psychiatry 27(12):1293–1304
Grueter CC, Chapais B, Zinner D (2012) Evolution of multilevel social systems in nonhuman primates and humans. Int J Primatol 33(5):1002–1037
Gruman GJ (2003) A history of ideas about the prolongation of life. Springer, New York
Grundy E (2008) The challenges of ageing: Prospects for the family support of older people in 21st century Europe. In: Surkyn J, Deboosere P, Van Bavel J (eds) Demographic challenges for the 21st century. A state of the art in demography. VUBPress, Brussel, pp 223–246
Grundy E, Tomassini C (2005) Fertility history and health in later life: a record linkage study in England and Wales. Elsevier, Amsterdam
Grusky DB (1994) Contours of social stratification. In: Grusky DB (ed) Social stratification: class, race and gender in sociological perspective. Westview Press, Boulder CO
Gruter M, Masters RD (eds) (1986) Ostracism as a social and biological phenomenon. Elsevier, New York
Guala F (2012) Reciprocity: Weak or strong? What punishment experiments do (and do not) demonstrate. Behav Brain Sci 35(1):1–15
Guilaine J, Zammit J (2005) The origins of war: violence in prehistory. Blackwell, Malden
Guinote A, Cotzia I, Sandhu S et al (2015) Social status modulates prosocial behavior and egalitarianism in preschool children and adults. PNAS 112(3):731–736
Guldi M (2008) Fertility effects of abortion and birth control pill access for minors. Demography 45(4):817–827
Gunderson M (2008) Genetic engineering and the consent of future persons. J Evol Technol 18 (1):86–93
Guo G (2006) Genetic similarity shared by best friends among adolescents. Twin Res Human Genet 9(1):113–121
Gürerk Ö, Irlenbusch B, Rockenbach B (2006) The competitive advantage of sanctioning institutions. Science 312(5770):108–111
Gurven M, Kaplan H (2007) Longevity among hunter-gatherers: A cross-cultural examination. Popul Dev Rev 33(2):321–365

Gurven M, von Rueden C (2006) Hunting, social status and biological fitness. Biodemography Soc Biol 53(1):81–99

Guthrie S (1993) Faces in the clouds: a new theory of religion. Oxford University Press, New York

Gutiérrez G (1971) Teología de la liberación: Perspectivas. [A theology of liberation: History, politics, and salvation]. CEP, Lima

Gutman I (ed) (1990) Encyclopedia of the holocaust. Macmillan Publishing Company, New York

Guttmacher AE, Collins FS (2003) Welcome to the genomic era. N Engl J Med 349(10):996–998

Gyngell C (2015) The ethics of human life extension: The second argument from evolution. J Med Philos 40(6):696–713

Gyngell C, Douglas T (2016) Selecting against disability: the liberal eugenic challenge and the argument from cognitive diversity. J Appl Philos. https://doi.org/10.1111/japp.12199

Gyngell C, Easteal S (2015) Cognitive diversity and moral enhancement. Camb Q Healthc Ethics 24(01):66–74

Haacke W (1893) Gestaltung und Vererbung. Eine Entwicklungsmechanik der Organismen. [Design and Inheritance. A Developmental Mechanics of the Organisms]. Weigl, Leipzig

Habermas J (1987) The philosophical discourse of modernity. Polity Press, Cambridge (Translation from: Das philosophische Diskurs der Moderne: Zwölf Vorlesungen, 1985, Suhrkamp Verlag, Frankfurt-am-Main)

Habermas J (2001) Die Zukunft der menschlichen Natur. Auf dem Weg zu einer liberalen Eugenik? [The future of human nature. Towards a liberal eugenics?] Suhrkamp, Frankfurt-am-Main

Habermas J (2003) The future of human nature. Blackwell, Oxford

Hacker JS, Pierson P (2010) Winner-take-all politics: how Washington made the rich richer—and turned its back on the middle class. Simon and Schuster, New York

Haddad YY (2002) Muslims in the West: from sojourners to citizens. Oxford University Press, New York

Haddow JE, Palomaki GE (1996) Similarities in women's decision making. Prenat Diagn 16 (12):1161–1162

Haeckel E (1866) Generelle Morphologie der Organismen: Allgemeine Grundzüge der organischen Formen-Wissenschaft, mechanisch begründet durch die von C. Darwin reformirte Decendenz-Theorie. [General morphology of organisms: General principles of organic forms-science, mechanical founded by Darwin's reformed descendence theory]. Georg Reimer, Berlin

Haeckel E (1868) Naturliche Schopfungsgeschichte. [Natural creation history]. Georg Reimer, Berlin

Haeckel E (1905) Die Welträthsel: Gemeinverständliche Studien über monistische Philosophie. [The riddle of the universe: Community understandable studies of monistic philosophy]. Kröner, Stuttgart

Hagan J, Gillis AR, Simpson J (1985) The class structure of gender and delinquency: Toward a power-control theory of common delinquent behavior. Am J Soc 90(6):1151–1178

Hagan J, Gillis AR, Simpson J (1990) Clarifying and extending power-control theory. Am J Soc 95(4):1024–1037

Häggström O (2016) Here be dragons: science, technology and the future of humanity. Oxford University Press, Oxford

Haidt J (2001) The emotional dog and its rational tail: a social intuitionist approach to moral judgment. Psychol Rev 108(4):814–834

Haidt J (2007) The new synthesis in moral psychology. Science 316(5827):998–1002

Haidt J, Craig J (2007) The moral mind: How five sets of innate intuitions guide the development of many culture-specific virtues, and perhaps even modules. In: Carruthers P, Laurence S, Stich S (eds) The innate mind, Vol 3: Foundations and the future. Oxford University Press, New York

Hald GM, Malamuth NM (2007) Self-perceived effects of pornography consumption. Arch Sex Behav 37(4):614–625

Haldane JBS (1924) Daedalus; or, science and the future. K. Paul, Trench, Trubner and Co, London
Haldane JBS (1932) The causes of evolution. Longmans Green, London
Haley J (2003) Death and dying: opposing viewpoints. Greenhaven Press, Farmington Hills
Hall R (ed) (1982) Sexual dimorphism in Homo sapiens. Praeger, New York
Hall DE, Koenig HG, Meador KG (2008) Measuring religiousness in health research: Review and critique. J Relig Health 47(2):134–163
Hallam A, Wignall PB (1997) Mass Extinctions and their aftermath. Oxford University Press, Oxford
Haller M, Jowell R, Smith TW (eds) Charting the globe: The International Social Survey Programme, 1984–2009. Routledge, London
Halman L, Draulans V (2006) How secular is Europe? Br J Sociol 57(2):263–288
Halman L, Inglehart R, Díez-Medrano J et al (2007) Changing values and beliefs in 85 countries: Trends from the Values Surveys from 1981 to 2004 (European Values Studies). Brill, Leiden
Halman L, Riis O (eds) (2003) Religion in a secularizing society: The Europeans' religion at the end of the 20th century. Brill, Leiden
Halpern CT, Udry JR, Campbell B (1994) Testosterone and religiosity as predictors of sexual attitudes and activity among adolescent males: A biosocial model. J Biosoc Sci 26(2):217–234
Ham K (1987) The lie: evolution. Masterbooks, El Cajon
Ham BJ, Kim YH, Choi MJ (2004) Serotonergic genes and personality traits in the Korean population. Neurosci Lett 354(1):2–5
Hameed S (2010) Evolution and creationism in the Islamic world. In: Dixon T, Cantor G, Pumfrey S (eds) Science and religion: new historical perspectives. Cambridge University Press, Cambridge, pp 133–152
Hamer DH (2005) The God gene: how faith is hardwired into our genes. Anchor, New York
Hamer D, Copeland P (1995) The science of desire: the search for the gay gene and the biology of behavior. Touchstone, New York
Hamilton WD (1963) The evolution of altruistic behavior. Am Nat 97:354–356
Hamilton WD (1964) The genetical evolution of social behavior, I and II. J Theor Biol 7:1–32
Hamilton WD (1966) The moulding of senescence by natural selection. J Theoret Biol 12:12–45
Hamilton WD (1975) Innate social aptitudes of man: An approach from evolutionary genetics. In: Fox R (ed) Biosocial anthropology. Malaby Press, London, pp 133–155
Hamilton WD (1977) Review of E.O. Wilson: Sociobiology: The new synthesis. J Anim Ecol 46:975–983
Hamilton WD (1996) Narrow roads of gene land. Oxford University Press, Oxford
Hamlin JK, Wynn K, Bloom P et al (2011) How infants and toddlers react to antisocial others. PNAS 108(50):19931–19936
Hammerstein P (ed) (2003) Genetic and cultural evolution of cooperation. MIT Press, Cambridge
Hammerstein P, Hagen EH (2005) The second wave of evolutionary economics in biology. Trends in Ecology and Evolution 20(11):604–609
Hamplova D (2011) Religion and gender: why are women more religious than men? Sociologicky Casopis-Czech Sociol Rev 47(2):297–323
Hatemi PK, Hibbing JR, Medland SE et al (2010) Not by twins alone: using the extended family design to investigate genetic influence on political beliefs. Am J Pol Sc 54(3):798–814
Handwerker P (1986) The modern demographic transition: An analysis of subsistence choices and reproductive consequences. Am Anthropologist 88(2):400–417
Hansell GR, Grassie W (eds) (2011) H+/-: Transhumanism and its critics. Prometheus Books, Amherst
Hanson FA (2001) Donor insemination: eugenic and feminist implications. Med Anthropol Q 15 (3):287–311
Harari YN (2017) Homo deus: a brief history of tomorrow. Harper, New York
Harbaugh W, Mayr U, Burghart D (2007) Neural responses to taxation and voluntary giving reveal motives for charitable donations. Science 316(5831):1622–1625

Hardin G (1968) The tragedy of the commons. Science 162(3859):1243–1247
Hardin G (1977) The limits of altruism: an ecologist's view of survival. Indiana University Press, Bloomington
Hardin G (1993) Living within limits. Oxford University Press, New York
Hardin G (1994) The tragedy of the unmanaged commons. Trends in Ecology and Evolution 9 (5):199
Hardy GH (1908) Mendelian proportions in a mixed population. Science 28(706):49–50
Hare RD (1999) Without conscience: the disturbing world of the psychopaths among us. Guilford Press, New York
Hare RD, Neumann CS (2008) Psychopathy as a clinical and empirical construct. Ann Rev Clin Psychol 4:217–246
Harenski CL, Antonenko O, Shane MS et al (2008) Gender differences in neural mechanisms underlying moral sensitivity. Soc Cogn Affect Neurosci 3(4):313–321
Harenski CL, Harenski KA, Shane MS et al (2010) Aberrant neural processing of moral violations in criminal psychopaths. J Abnorm Psychol 119(4):863–874
Harman O (2011) The price of altruism: George Price and the search for the origins of kindness. Norton, New York
Harman O (2013) Altruism: a saga. Isr J Ecol Evol 59(2):SI 117–124
Harman O (2014) A history of the altruism-morality debate in biology. Behaviour 151(2–3):SI 147–165
Harms W (2000) The evolution of cooperation in hostile environments. In: Katz LD (ed) Evolutionary origins of morality: cross-disciplinary perspectives. Imprint Academic, Thorverton, pp 308–313
Harpending H, Draper P (1988) Antisocial behavior and the other side of cultural evolution. In: Moffitt TE, Mednick SA (eds) Biological contributions to crime causation. Martinus Nijhoff, Dordrecht
Harpending HC, Rogers AR (1990) Fitness in stratified societies. Ethol Sociobiol 11:497–509
Harpending HC, Sobus J (1987) Sociopathy as an adaptation. Ethol Sociobiol 8:63S–72S
Harrington A (1969) The immortalist: an approach to the engineering of man's divinity. Granada, London
Harris M (1974) Cows, pigs, wars and witches: the riddles of culture. Random House, New York
Harris M (1977) Cannibals and kings: the origins of cultures. Random House, New York
Harris G (1995) Can evolutionary psychology explain psychopathy? http://www.mhcp-research.com/en2.htm
Harris S (2004) The end of faith: religion, terror, and the future of reason. Norton, New York
Harris FR (ed) (2006) The baby bust: who will do the work? Who will pay the taxes?. Rowman and Littlefield Publishers, Lanham
Harris J (2007) Enhancing evolution: the ethical case for making better people. Princeton University Press, Princeton
Harris L (2007) The suicide of reason: radical Islam's threat to the enlightenment. Basic books, New York
Harris J (2009) Enhancements are a moral obligation. In: Savulescu J, Bostrom N (eds) Human enhancement. Oxford University Press, Oxford, pp 131–154
Harris S (2010) The moral landscape. How science can determine human values. Free Press, New York
Harris S (2014) Waking up. A guide to spirituality without religion. Bantam Press, London
Harris J (2016) How to be good: the possibility of moral enhancement. Oxford University Press, Oxford
Harris E, McNamara P (2008) Is religiousness a biocultural adaptation? In: Bulbulia J, Sosis R, Genet R et al (eds) The evolution of religion: studies, theories, and critiques. Collins Foundation Press, Santa Margarita, pp 69–75
Harris G, Rice ME, Quinsey VL (1994) Psychopathy as a taxon: evidence that psychopaths are a discrete class. J Consult Clin Psychol 62(2):387–397

Harrison JE (1909) The influence of Darwinism on the study of religion. In: Seward AC (ed) Darwin and modern science: essays in commemoration of the centenay of the birth of Charles Darwin and of the fiftieth anniversary of the publication of the Origin of Species. Cambridge University Press, Cambridge, pp 494–511

Harrison E (1974) Origin and evolution of the universe. Encycl Britannica Macropaedia 1974:1007

Hartmann T (1995) ADD success stories. Underwood Books, Grass Valley

Hartmann WK (2015) Chelyabinsk, Zond IV, and a possible first-century fireball of historical importance. Meteorit Planet Sci 50(3):368–381

Hartmann T, Palladino LJ (2005) The Edison gene: ADHD and the gift of the hunter child. Park Street Press, New York

Hartung J (1995) Love thy neighbor: the evolution of in-group morality. Skeptic 3(4):86–99

Harvey D (2007) A Brief history of neoliberalism. Oxford University Press, New York

Harvey M (2014) Early humans' egalitarian politics runaway synergistic competition under an adapted veil of ignorance. Hum Nat 25(3):299–327

Hashemi N (2010) The multiple histories of secularism: Muslim societies in comparison. Philos Soc Criticism 36(3–4):325–338

Hassan FA (1975) Determination of the size, density, and growth rate of hunting-gathering populations. In: Polgar S (ed) Population, ecology, and social evolution. Mouton, The Hague, pp 27–78

Hastings PD, Zahn-Waxler C, McShane KE (2006) We are, by nature, moral creatures: biological bases of concern for others. In: Killen M, Smetana J (eds) Handbook of moral development. Erlbaum, New York, pp 483–516

Hauert C, Traulsen A, Brandt H et al (2007) Via freedom to coercion: the emergence of costly punishment. Science 316(5833):1905–1907

Haught JF (2000) God after Darwin. A theology of evolution. Westview Press, Boulder

Haught JA (2002) Holy horrors: an illustrated history of religious murder and madness. Prometheus Books, Amherst

Haught JA (2010) Fading faith: the rise of the secular age. Gustav Broukal Press, Cranford

Haught JF (2004) Deeper than Darwin: the prospect for religion in the age of evolution. Westview Press, Boulder

Haught JF (2008) God and the new atheism: a critical response to Dawkins, Harris, and Hitchens. Westminster John Knox Press, Louisville

Haught JF (2010) Making sense of evolution: Darwin, God, and the drama of life. Westminster John Knox Press, Louisville

Haught JF (2013) Transhumanism, theology and the promise of nature. Universitas-Mon Rev Philos Cult 40(4):21–38

Hauser M (2006) Moral minds. How nature designed our universal sense of right and wrong. HarperCollins, New York

Hauser M, Singer P (2006) Morality without religion. Free Inq 26(1):18–19

Hauspie RC, Vercauteren M, Susanne C (1996) Secular changes in growth. Hormon Res 45 (S2):8–17

Havely N, Weisel O, Bornstein G (2012) "In-group love" and "out-group hate" in repeated interaction between groups. J Behav Decis Mak 25(2):188–195

Hawking S, Mlodinow L (2010) The grand design. Bantam, London

Hawks J (2016) Still evolving. New Sci 231(3083):18–19

Hawks J, Wang ET, Cochran GM et al (2007) Recent acceleration of human adaptive evolution. PNAS 104(52):20753–20758

Hawley P, Little T, Rodkin P (eds) (2007) Aggression and adaptation: the bright side to bad behavior. Lawrence Erlbaum, Mahwah

Hawley P, Vaughn B (2003) Aggression and adaptive function: the bright side to bad behavior. Merrill-Palmer Q 49(3):239–242

Hayden B (2003) Shamans, sorcerers, and saints. Smithsonian Institution Books, Washington

Hayek FA (1960) The constitution of liberty. University of Chicago Press, Chicago
Hayek FA (1979) The political order of a free people. University of Chicago Press, Chicago
Hayflick L (1996) How and why we age. Ballantine Books, NewYork
Hayflick L (2000) The future of ageing. Nature 408(6809):37–39
Hayles K (2011) Wrestling with transhumanism. In: Hansell GR, Grassie W (eds) H+/-: Transhumanism and its critics. Prometheus Books, Amherst, pp 215–226
Haynes J (2007) Religion and development: conflict or cooperation?. Palgrave Macmillan, Basingstoke
Haynes J (2013) An introduction to international relations and religion. Routledge, New York
Hazleton L (2014) The first Muslim: the story of Muhammad. Riverhead Trade, New York
Heard E, Martienssen RA (2014) Transgenerational epigenetic inheritance: myths and mechanisms. Cell 157(1):95–109
Hefner P (1993) The human factor: evolution, culture and religion. Fortress, Minneapolis
Hefner P (1999) Going as far as we can go: the Jesus proposal for stretching genes and cultures. Zygon 34(3):485–500
Heie N (2004) Ernst Haeckel and the morphology of ethics. J Can Hist Assoc 15(1):1–27
Heilbroner RL (1995) Visions of the future: the distant past, yesterday, today, tomorrow. Oxford University Press, New York
Heimlich J (2011) Breaking their will: shedding light on religious child maltreatment. Prometheus Books, Amherst
Hein G, Morishima Y, Leiberg S et al (2016) The brain's functional network architecture reveals human motives. Science 351(6277):1074–1078
Heinemann B, Wörmann V (2010) Psychische Beanspruchung, Lebenszufriedenheit und Religiosität: Beeinflusst Religiosität den Zusammenhang zwischen psychischer Beanspruchung und Lebenszufriedenheit bei Patienten mit psychischen Erkrankungen? (Mental stress, life satisfaction and religiosity: Affects religiosity the relationship between psychological stress and life satisfaction in patients with mental illness?) Internationale Zeitschrift für Philosophie und Psychosomatik, 1/2010
Heinrichs M, Domes G (2008) Neuropeptides and social behaviour: effects of oxytocin and vasopressin in humans. Prog Brain Res 170:337–350
Heltne PG (2012) Commonsense Darwinism: evolution, morality, and the human condition. Zygon 47(1):243–245
Henrich J (2004) Cultural group selection, coevolutionary processes and large-scale cooperation. J Econ Behav Organ 53(1):3–35
Henrich J (2008) A cultural species. In: Brown MJ (ed) Explaining culture scientifically. University of Washington Press, Seattle, pp 184–210
Henrich J, Boyd R (2001) Why people punish defectors: weak conformist transmission can stabilize costly enforcement of norms I. Cooperative dilemmas. J Theoret Biol 208:78–89
Henrich J, Boyd R, Bowles S et al (2001) In search of homo economicus: behavioral experiments in 15 small-scale societies. Am Econ Rev 91(2):73–78
Henrich J, Boyd R, Bowles S et al (eds) (2004) Foundations of human sociality: economic experiments and ethnographic evidence from fifteen small-scale societies. Oxford University Press, New York
Henrich J, Boyd R, Richerson PJ (2008) Five misunderstandings about cultural evolution. Hum Nat 19(2):119–137
Henrich J, Boyd R, Richerson PJ (2012) The puzzle of monogamous marriage. Philos Trans R Soc B 367(1585):657–669
Henrich J, Ensminger J, McElreath R et al (2010) Markets, religion, community size, and the evolution of fairness and punishment. Science 327(5972):1480–1484
Henrich J, Heine SJ, Norenzayan A (2010) The weirdest people in the world? Behav Brain Sci 33 (2–3):61–83
Henrich N, Henrich J (2007) Why humans cooperate: cultural and evolutionary explanation. Oxford University Press, Oxford

Henrich J, McElreath R (2007) Dual inheritance theory: the evolution of human cultural capacities and cultural evolution. In: Dunbar R, Barrett L (eds) Oxford handbook of evolutionary psychology. Oxford University Press, Oxford, pp 555–570
Henrich J, McElreath R, Barr A et al (2006) Costly punishment across human societies. Science 312(5781):1767–1770
Henry L (1961) Some data on natural fertility. Soc Biol 8(2):81–91
Henry A (2004) Not my mother's sister: generational conflict and third-wave feminism. Indiana University Press, Bloomington
Henshaw SK, Singh S, Haas T (1999) The incidence of abortion worldwide. Int Family Plan Perspect 25:S30–S38
Hepach R, Vaish A, Tomasello M (2012) Young children are intrinsically motivated to see others helped. Psychol Sci 23(9):967–972
Herder JG (1774–1787) Auch eine Philosophie der Geschichte zur Bildung der Menschheit. (This Too a Philosophy of History for the Formation of Humanity). Johann Gottfried Herder Werke 4. Schriften zu Philosophie, Literatur, Kunst und Altertum, hg. von Jürgen Brummack und Martin Bollacher. Frankfurt a.M. 1994
Herman-Giddens ME, Smith JB, Mittal M et al (2003) Newborns killed or left to die by a parent. A population-based study. JAMA 289(11):1425–1429
Herman A, Philbeck JW, Vasilopoulos NL et al (2003) Serotonin transporter promoter polymorphism and differences in alcohol consumption behaviour in a college student population. Alcohol 38(5):446–449
Hern WM (1990) Why are there so many of us? Description and diagnosis of a planetary ecopathological process. Popul Environ 12(1):9–39
Hern WM (1993) Has the human species become a cancer on the planet? A theoretical view of population growth as a sign of pathology. Curr World Leaders 36(6):1089–1124
Herrmann B, Thöni C, Gächter S (2008) Antisocial punishment across societies. Science 319 (5868):1362–1367
Hertig AT (1967) The overall problem in man. In: Benirschke K (ed) Comparative aspects of reproductive failure. Springer, New York
Hesketh T, Xing ZW (2006) Abnormal sex ratios in human populations: causes and consequences. PNAS 103(36):13271–13275
Hetzer M, Sornette D, Johnson N (2013) The co-evolution of fairness preferences and costly punishment. PLoS ONE 8(3):e54308
Hewitt JA (2002) The architecture of thought: a new look at human evolution. Holmhurst House Press, Bedfordshire
Hewlett BS (1991) Demography and childcare in preindustrial societies. J Anthropol Res 47(1):1–37
Hewlett SA (2002) Creating a life: professional women and the quest for children. Talk Miramax Books, New York
Heywood L, Drake J (eds) (1997) Third wave agenda: being feminist, doing feminism. University of Minnesota Press, Minneapolis
Hiatt LR (1989) On cuckoldry. J Soc Biol Struct 12(1):53–72
Hick J (1966) Evil and the God of love. Harper and Row, New York
Hilberg R (2003) The destruction of the European Jews. Yale University Press, New Haven
Hildebrand AR, Penfield GT, Kring DA et al (1991) Chicxulub Crater: a possible cretaceous/tertiary boundary impact crater on the Yucatán Peninsula, Mexico. Geology 19 (9):867–871
Hill J (1984) Prestige and reproductive success in man. Ethol Sociobiol 5(2):77–95
Hill K (1993) Life history theory and evolutionary anthropology. Evol Anthropol 2(3):78–88
Hill E (1999) Lineage interests and non-reproductive strategies: an evolutionary approach to medieval religious women. Hum Nat 10(2):109–134
Hill K, Hurtado AM (1991) The evolution of premature reproductive senescence and menopause in human females: an evaluation of the "grandmother hypothesis". Hum Nat 2(4):313–350

Hill PC, Hood RW (1999) Measures of religiosity. Religious Education Press, Birmingham
Hill PC, Pargament KI (2003) Advances in the conceptualization and measurement of religion and spirituality: implications for physical and mental health research. Am Psychol 58(1):64–74
Hill PC, Pargament KI, Hood RW et al (2000) Conceptualizing religion and spirituality: points of commonality, points of departure. J Theory Soc Behav 30(1):51–77
Hills P, Francis LJ, Argyle M et al (2004) Primary personality trait correlates of religious practice and orientation. Person Individ Diff 36(1):61–73
Himes NE (1936) A medical history of contraception. Williams and Wilkins, Baltimore
Hinde RA (1974) Biological bases of human social behaviour. McGraw-Hill, New York
Hinde RA (1999) Why gods persist: a scientific approach to religion. Routledge, London
Hinde RA (2002) Why good is good: the sources of morality. Routledge, London
Hinde RA (2011) Society needs morality. Eur Rev 19(1):105–117
Hirschi T, Hindelang MJ (1977) Intelligence and delinquency: a revisionist review. Am Sociol Rev 42(4):571–587
Hirshleifer J (1978) Competition, cooperation, and conflict in economics and biology. Am Ec Rev 68(2):238–243
Hirsi Ali A (2015) Heretic. Why Islam must change to join the modern world. Harper, New York
Hitchens C (2008) God the failed hypothesis. How science shows that God does not exist. Stenger, Amherst
Hitchens C (2009) God is not great: how religion poisons everything. Twelve, New York
Hitler A (1925–1926) Mein Kampf. (My fight). Franz Eher Verlag, München
Hobgood-Oster L (2005) Ecofeminism - historic and international evolution. In: Taylor B (ed) Encyclopedia of religion and nature. Continuum, London
Hockett CF (1973) Man's place in nature. McGraw-Hill, New York
Hočevar DK, Černič Istenič M (2010) Dom in delo na kmetijah. Raziskava odnosov med generacijami in spoloma (Home and work on farms. The study of relations between the generations and the sexes). Založba ZRC, ZRC SAZU, Ljubljana
Hodgson GM (2007) Taxonomizing the relationship between biology and economics: a very long engagement. J Bioecon 9(2):169–185
Hodgson GM (2014) The evolution of morality and the end of economic man. J Evolut Econ 24 (1):83–106
Hoem J (1993) Public policy as the fuel of fertility: effects of a policy reform on the pace of childbearing in Sweden in the 1980s. Acta Sociol 36(1):19–31
Hoffer P, Hull NEH (1981) Murdering mothers: infanticide in England and America, 1558-1803. New York University Press, New York
Hofferth SL (2003) The American family: changes and challenges for the 21st century. In: Wallace H, Green G, Jaros K (eds) Health and welfare for families in the 21st century. Jones and Bartlett, Sudbury, pp 71–79
Hoffman ML (2000) Empathy and moral development. Cambridge University Press, Cambridge
Hofmann W, Wisneski DC, Brandt MJ et al (2014) Morality in everyday life. Science 345 (6202):1340–1343
Hoffman M, Yoeli E, Nowak MA (2015) Cooperate without looking: why we care what people think and not just what they do. PNAS 112(6):1727–1732
Hoffmann-Nowotny HJ (1987) The future of the family. In: European Population Conference 1987, Plenaries. Central Statistical Office of Finland, Helsinki, pp 113–200
Hofstadter R (1944) 1955) Social Darwinism in American thought. The Beacon Press, Boston
Hogan R (2004) The superstitions of everyday life. Behav Brain Sci 27(6):737–738
Høgh-Olesen H (ed) (2010) Human morality and society: evolutionary and comparative perspectives. Palgrave Macmillan, Basingstoke
Holcomb HR (1993) Sociobiology, sex, and science. New York Press, New York
Holland J (1975/1992) Adaptation in natural and artificial systems. University of Michigan Press, Ann Arbor

Holloway RL (1981) Culture, symbols, and human brain evolution: a synthesis. Dialect Anthropol 5(4):287–303

Holloway R (1999) Godless morality: keeping religion out of ethics. Canongate, Edinburgh

Holland S (2003) Bioethics. A philosophical introduction. Polity Press, Cambridge

Holloway C (2008) The right Darwin? Evolution, religion, and the future of democracy. Spence Publishing Company, Dallas

Holloway SM, Smith C (1975) Effects of various medical and social practices on the frequency of genetic disorders. Am J Hum Genet 27(5):614–627

Hönekopp J, Bartholomé T, Jansen G (2004) Facial attractiveness, symmetry, and physical fitness in young women. Hum Nature 15(2):147–167

Honeywill R (2016) The man problem. Destructive masculinity in western culture. Palgrave Macmillan, New York

Hoorens S, Clift J, Staetsky L et al (2011) Low fertility in Europe. RAND Corporation, Santa Monica

Hopcroft R (2006) Sex, status, and reproductive success in the contemporary United States. Evol Hum Behav 27(2):104–120

Hopcroft RL (2015) Sex differences in the relationship between status and number of offspring in the contemporary U.S. Evol Hum Behav 36(2):146–151

Hopkins PD (2005) Transcending the animal: how transhumanism and religion are and are not alike. J Evol Technol 14(1):11–26

Hopkins PD (2008) A moral vision for transhumanism. J Evol Technol 19(1):3–7

Hoptman MJ (2003) Neuroimaging studies of violence and antisocial behavior. J Psychiatr Pract 9 (4):265–278

Horgan J (1995) The new social Darwinists. Sci Am 273(4):150–157

Houkes W (2012) Population thinking and natural selection in dual-inheritance theory. Biol Philos 27(3):401–417

House BR, Henrich J, Brosnan SF et al (2012) The ontogeny of human prosociality: behavioral experiments with children aged 3 to 8. Evol Hum Behav 33(4):291–308

House BR, Silk JB, Henrich J et al (2013) Ontogeny of prosocial behavior across diverse societies. PNAS 110(36):14586–14591

Hout M, Greeley A, Wilde MJ (2001) The demographic imperative in religious change in the United States. Am J Sociol 107(2):468–500

Howard T, Rifkin J (1977) Who should play God? The artificial creation of life and what it means for the future of the human race. Delacorte Press, New York

Howe M (2001) Genius explained. Cambridge University Press, Cambridge

Howell N (1979) Demography of the Dobe !Kung. Academic Press, New York

Howell RR, Stevenson RE (1971) The offspring of phenylketonuric women. Soc Biol 18(1):19–29

Hoyle C, Bradford A, Frenett R (2015) Becoming Mulan? Female western migrants to ISIS. Institute for Strategic Dialogue, London

Hrdy SB (1981) The woman that never evolved. Harvard University Press, Cambridge

Hrdy SB (1999) Mother nature: the history of mothers, infants, and natural selection. Pantheon, New York

Hrdy S (2011) Mothers and others: the evolutionary origins of mutual understanding. The Belknap Press, Cambridge

Hruschka DJ, Henrich J (2013) Economic and evolutionary hypotheses for cross-population variation in parochialism. Front Hum Neurosci 7:559. https://doi.org/10.3389/fnhum.2013.00559

Hu X, Lipsky RH, Zhu G et al (2006) Serotonin transporter promoter gain-of-function genotypes are linked to obsessive-compulsive disorder. Am J Hum Genet 78(5):815–826

Hubbeling D (2011) The moral brain: essays on the evolutionary and neuroscientific aspects of morality. Evol Psychol 9(3):430–437

Huber R (1971) The American idea of success. McGraw-Hill, New York

Huber S, Bookstein F (2010) Socioeconomic status, education, and reproduction in modern women: an evolutionary perspective. Am J Hum Biol 22(5):578–587
Hublin JJ, Ben-Ncer A, Bailey SE et al (2017) New fossils from Jebel Irhoud, Morocco and the pan-African origin of *Homo sapiens*. Nature 546:289–292
Hughes W (1986) Richards' defense of evolutionary ethics. Biol Philos 1:306–315
Hughes J (2004) Citizen cyborg: why democratic societies must respond to the redesigned human of the future. Westview Press, Cambridge
Hughes J (2010) Contradictions from the enlightenment roots of transhumanism. J Med Philos 35 (6):622–640
Hughes J (2012) The politics of transhumanism and the techno-millennial imagination. Zygon 47 (4):757–776
Hughes J (2015) Moral enhancement requires multiple virtues toward a posthuman model of character development. Camb Q Healthc Ethics 24(1):86–95
Hughes SM, Gallup GG (2003) Sex differences in morphological predictors of sexual behavior: shoulder to hip and waist to hip ratios. Evol Hum Behav 24(3):173–178
Hume D (1739/1969) A treatise of human nature. Penguin Books, London
Humm M (1992) Feminisms. Harvester Wheatsheaf, New York
Humphrey NK (1976) The social function of intellect. In: Bateson P, Hinde RA (eds) Growing points in ethology. Cambridge University Press, Cambridge, pp 303–317
Humphrey N (1995) Soul searching: human nature and supernatural belief. Chatto and Windus, London
Hunt L (1996) The invention of pornography: obscenity and the origins of modernity, 1500-1800. Zone Books, New York
Hunt M (1998) The new know-nothings: the political foes of the scientific study of human nature. Transaction Publishers, New Brunswick
Hunter C (2003) Darwin's proof: the triumph of religion over science. Brazos, Grand Rapids
Hunter I (2015) Secularization: the birth of a modern combat concept. Mod Intellect Hist 12(1):1–32
Huntington SP (1996) The clash of civilizations and the remaking of world order. Simon and Schuster, New York
Hur Y-M, Rushton JP (2007) Genetic and environmental contributions to prosocial behaviour in 2- to 9-year-old South Korean twins. Biol Lett 3(6):664–666
Hurd JP (2006) The shape of high fertility in a traditional Mennonite population. Ann Hum Biol 33(5–6):557–569
Hurd JP (2011) Continuing high fertility in a culturally-embedded Mennonite farming society. Ann Hum Biol 38(1):69–75
Hurlbut JB, Tirosh-Samuelson H (eds) (2016) Perfecting human futures. Transhuman visions and technological imaginations. Springer, Wiesbaden
Hurst R (1999) Training manual for media and disability rights. Disability Awareness in Action, Hullavington
Husén T (1972) Equality as an objective of educational and social policy. In: Husén T (ed) Social background and educational careers. Organization for Economic Cooperation and Development, Paris
Hutchinson GE (1959) A speculative consideration of certain possible forms of sexual selection in man. Am Nat 93(869):81–91
Hutchinson G, Bhugra D, Mallett R et al (1999) Fertility and marital rates in first-onset schizophrenia. Soc Psychiatry Psychiatr Epidemiol 34(12):617–621
Hutchinson J, Smith A (1995) Nationalism. Oxford University Press, Oxford
Huxley TH (1894) Evolution and ethics. Macmillan, London
Huxley JS (1927) Religion without revelation. Ernest Benn, London
Huxley A (1932) Brave new world. Harper Perennial, New York
Huxley JS (1936) Eugenics and society. Eugenics Rev 28(1):11–31
Huxley JS (1942) Evolution: the modern synthesis. Allen and Unwin, London

Huxley JS (1957) Knowledge, morality and destiny. New American Library, New York
Huxley JS (1957) Transhumanism. In: Huxley J (ed) New bottles for New Wine. Chatto and Windus, London, pp 13–17
Huxley JS (1962) Eugenics in evolutionary perspective. Eugenics Rev 54:123–141
Huxley JS (1964) Essays of a humanist. Chatto and Windus, London
Huyse L (1987) De verzuiling voorbij. (Beyond pillarisation). Kritak, Leuven
Huyse L (1970) Passiviteit, pacificatie en verzuiling in de Belgische politiek. Een sociologische studie. (Passivity, pacification and compartmentalization in Belgian politics. A sociological study). Standaard Wetenschappelijke Uitgeverij, Antwerpen/Utrecht
Hwang S-H, Bowles S (2012) Is altruism bad for cooperation? J Econ Behav Organ 83(3):330–341
Iemmola F, Camperio CA (2009) New evidence of genetic factors influencing sexual orientation in men: female fecundity increase in the maternal line. Arch Sex Behav 38(3):393–399
Ihara Y (2011) Evolution of culture-dependent discriminate sociality: a gene-culture coevolutionary model. Philos Trans R Soc Lond B Biol Sci 366(1566):889–900
Illies C (2009) Die Bedeutung von Anthropologie und Evolutionswissenschaften für die politische Philosophie. (The importance of anthropology and evolutionary science for political philosophy). In: Jörke D, Ladwig B (eds) Politische Anthropologie. Geschichte, Gegenwart, Möglichkeiten. Nomos, Baden-Baden, pp 197–231
Illies C (2009) Philosophische Anthropologie im biologischen Zeitalter. Zur Konvergenz von Moral und Natur. (Philosophical anthropology in biological age. On the convergence of morality and nature). Suhrkamp, Frankfurt a.M
Illies C (2010) Biologie statt Philosophie? Evolutionäre Kulturerklärungen und ihre Grenzen. (Biology instead of philosophy? Evolutionary explanations of culture and their limits). In: Gerhardt V, Nida-Rümelin J (eds) Evolution in Natur und Kultur. De Gruyter, Berlin, pp 15–38
Immerman RS, Mackey WC (2003) Religion and fertility. Mankind Q 43(4):377–408
Inglehart R (1990) Culture shift in advanced industrial society. Princeton University Press, Princeton
Inglehart R, Basanez M, Diez-Medrano J et al (2004) Human beliefs and values: a cross-cultural sourcebook based on the 1999–2002 value surveys. Siglo Veintiuno Editores, Buenos Aires
Inglehart R, Welzel C (2005) Modernization, cultural change, and democracy: the human development sequence. Cambridge University Press, New York
InterAcademy Panel on International Issues (IAP) (2006) IAP statement on the teaching of evolution. http://www.interacademies.net/Object.File/Master/6/150/Evolution%20statement.pdf
International Humanist and Ethical Union (2012) Freedom of thought 2012. Global report on discrimination against humanists, atheists and the nonreligious. www.iheu.org
International Summit on Human Gene Editing (2015) On human gene editing: International summit statement. The national Academies of Sciences, Engineering, Medicine. 30 Dec 2015
Inzlicht M, McGregor I, Hirsh JB et al (2009) Neural markers of religious conviction. Psychol Sci 20(3):385–392
IPCC (2013) Climate change 2013: The physical science basis. Contribution of working group I to the fifth assessment report of the intergovernmental panel on climate change. (Stocker TF, Qin D, Plattner G-K et al (eds)). Cambridge University Press, Cambridge
Irin L, Méndez A, Dairén A et al (2014) Assessing the possibility of biological complexity on other worlds, with an estimate of the occurrence of complex life in the milky way galaxy. Challenges 5(1):159
Irons W (1979) Cultural and biological success. In: Chagnon NA, Irons W (eds) Evolutionary biology and human social behavior: an anthropological perspective. Duxbury, North Scituate, pp 284–302
Irons W (1996) Morality as an evolved adaptation. In: Hurd JP (ed) Investigating the biological foundations of morality. Edwin Mellon Press, Lewiston, pp 1–34

Irons W (1998) Adaptively relevant environments versus the environment of evolutionary adaptedness. Evol Anthrop: Issues, News, and Reviews 6(6):194–204
Irons W (2001) Religion as a hard-to-fake sign of commitment. In: Nesse R (ed) Evolution and the capacity for commitment. Russell Sage Foundation, New York, pp 292–309
Isaac G (1978) The food-sharing behavior of protohuman hominids. Sci Am 238(4):90–108
Isaacson S (2012) Science refutes religion. An essay concerning how and what it means to prove God does not exist. Kindle Edition
Isakov A, Rand DG (2012) The evolution of coercive institutional punishment. Dyn Games Appl 2 (1) SI:97–109
Isler K, van Schaik CP (2012) How our ancestors broke through the gray ceiling: comparative evidence for cooperative breeding in early Homo. Curr Anthropol 53(S6):S453–S465
Israel S, Hasenfratz L, Knafo-Noam A (2015) The genetics of morality and prosociality. Curr Opin Psychol 6:55–59
Israel S, Lerer I Shalev et al (2008) Molecular genetic studies of the arginine vasopressin 1a receptor (AVPR1a) and the oxytocin receptor (OXTR) in human behaviour: from autism to altruism with some notes in between. Prog Brain Res 170:435–449
Israel S, Lerer E, Shalev I et al (2009) The oxytocin receptor (OXTR) contributes to prosocial fund allocations in the dictator game and the social value orientations task. PLoS ONE 4(5):e5535
Iwagami A, Masuda N (2010) Upstream reciprocity in heterogeneous networks. J Theoret Biol 265 (3):297–305
Izaka M, Mansell S, Fuller T (2015) Introduction: between no future and business-as-usual: exploring futures of capitalism. Futures 68:1–4
Jablensky A (2000) Epidemiology of schizophrenia. In: Gelder MG, Lopez-Ibor JJ, Andreasen N (eds) New Oxford textbook of psychiatry. Oxford University Press, Oxford, pp 585–598
Jablonka E, Lamb MJ (2005) Evolution in four dimensions: genetic, epigenetic, behavioral, and symbolic variation in the history of life. MIT Press, Cambridge
Jack AI, Friedman JP, Boyatzis RE et al (2016) Why do you believe in God? Relationships between religious belief, analytic thinking, mentalizing and moral concern. PLoS One 11(3): e0149989
Jackson S (1978) On the social construction of female sexuality. Women's Research and Resources. Centre Publications, London
Jackson J (2008) The amorality of preference: a response to the enemies of enhancement. J Evol Technol 19(1):42–50
Jacobs T (ed) (2004) Op latere leeftijd. De leefsituatie van 55-plussers in Vlaanderen. (Later in life. The living situation of people over 55 in Flanders). CBGS Publicaties 2004/1, Brussel
Jacobstein N (1999) Foresight guidelines for responsible nanotechnology development. Foresight Institute and IMM, Palo Alto
Jaeggi AV, Gurven M (2013) Reciprocity explains food sharing in humans and other primates independent of kin selection and tolerated scrounging: a phylogenetic meta-analysis. Proc R Soc B-Biol Sci 280(1768):20131615
Jaggar AM (1983) Feminist politics and human nature. The Harvester Press, Sussex
Jalajel DS (2009) Islam and biological evolution: exploring classical sources and methodologies. University of the Western Cape, Bellville
James W (1902, 1997) The varieties of religious experience. Touchstone, New York
James SM (2011) An introduction to evolutionary ethics. Wiley-Blackwell, Chichester
Jang KL (2005) The behavioral genetics of psychopathology: a clinical guide. Lawrence Erlbaum Associates, Mahwah
Jankowiak WR (1995) Romantic passion: a universal experience?. Columbia University Press, New York
Jankowiak WR, Fischer EF (1992) A cross-cultural perspective on romantic love. Ethnology 31 (2):149–155
Jansen JJG (1997) The dual nature of Islamic fundamentalism. Cornell University Press, Ithaca

Janssen F, Nusselder WJ, Looman CWN et al (2003) Stagnation in mortality decline among elders in The Netherlands. Gerontologist 43(5):722–734
Jasienska G, Ziomkiewicz A, Ellison PT et al (2004) Large breasts and narrow waists indicate high reproductive potential in women. Proc Biol Sci 271(1545):1213–1217
Jaspers K (1949) Vom Ursprung und Ziel der Geschichte. (The origin and goal of history). Artemis Verlag, Zürich
Jausovec N, Jausovec K (2005) Sex differences in brain activity related to general and emotional intelligence. Brain Cogn 59(3):277–286
Jean-Baptiste P (2003) La biologie de Dieu. (The biology of God). Agnès Viénot, Paris
Jebari K (2012) Moderate bioconservatism and moral enhancement. Doktorand i Praktisk Filosofi. Kungliga Tekniska Högskolan. http://utilitarian-essays.com/ModerateBioconservatismAndMoralEnhancement.pdf
Jeeves M, Brown WS (2009) Neuroscience, psychology and religion: illusions, delusions, and realities about human nature. Templeton Press, Radnor
Jeffrey J (2011) Defamation of religion? Or incitement to hatred? Statement on behalf of the International Humanist and Ethical Union (IHEU) at the UN Human Rights Council on Tuesday, 15 Mar 2011. United Nations, Geneva
Jensen R (2007) Getting off: pornography and the end of masculinity. South End Press, Cambridge
Jensen K (2010) Punishment and spite, the dark side of cooperation. Philos Trans R Soc B-Biol Sci 365(1553):2635–2650
Jensen PS, Mrazek D, Knapp PK et al (1997) Evolution and revolution in child psychiatry: ADHD as a disorder of adaptation. J Am Acad Child Adolesc Psychiatry 36(12):1672–1679
Jensen K, Vaish A, Schmidt MFH (2014) The emergence of human prosociality: aligning with others through feelings, concerns, and norms. Front Psychol 5:822. https://doi.org/10.3389/fpsyg.2014.00822
Jerison HJ (1973) Evolution of the brain and intelligence. Academic Press, New York
Jiang Y, Chew SH, Ebstein RP (2013) The role of D4 receptor gene exon III polymorphisms in shaping human altruism and prosocial behavior. Front Hum Neurosci 7:195. https://doi.org/10.3389/fnhum.2013.00195
Jogan M (1989) Patriarchy, misogyny and religion. Anthropos 20(5–6):417–428
Johansson SR (1987) Status anxiety and demographic contraction of privileged populations. Popul Dev Rev 13(3):349–470
Johansson S, Nygren O (1991) The missing girls of China: a new demographic account. Popul Dev Rev 17(1):35–51
Johnson PE (1991) Darwin on trial. Regnery Gateway, Washington
Johnson EA (1996) Does God play dice? Divine providence and chance. Theol Stud 57(1):3–18
Johnson PE (1997) Defeating Darwinism. InterVarsity Press, Downers Grove
Johnson DC (1997) Formal education vs. religious belief: Soliciting new evidence with multinomial logit modeling. J Sci Study Relig 36(2):231–246
Johnson PE (2002) Wedge of truth: splitting the foundations of naturalism. InterVarsity Press, Downers Grove
Johnson DDP (2005) God's punishment and public goods: a test of the supernatural punishment hypothesis in 186 world cultures. Hum Nat 16(4):410–446
Johnson DDP, Bering J (2006) Hand of God, mind of man: punishment and cognition in the evolution of cooperation. Evol Psychol 4:219–233
Johnson BR, Lam SK (2010) Self-organization, natural selection, and evolution: cellular hardware and genetic software. BioScience 60(11):879–885
Johnson DDP (2013) The uniqueness of human cooperation: cognition, cooperation, and religion. In: Nowak MA, Coakley S (eds) Evolution, games, and God. The principle of cooperation. Harvard University Press, Cambridge, pp 168–185
Johnson DDP, Kruger O (2004) The good of wrath: supernatural punishment and the evolution of cooperation. Polit Theol 5(2):159–176

Johnson W, Penke L, Spinath FM (2011) Heritability in the era of molecular genetics: Some thoughts for understanding genetic influences on behavioural traits. Eur J Person Special Issue: Eur Person Rev 25(4):254–266

Johnstone B, Bodling A, Cohen D et al (2012) Right parietal lobe-related "selflessness" as the neuropsychological basis of spiritual transcendence. Int J Psychol Relig 22(4):267–284

Joly Y, Knoppers BM, Godard B (2003) Genetic information and life insurance: a 'real' risk? Eur J Hum Genet 11:561–564

Jones G (1980) Social Darwinism and English thought. The interaction between biological and social theory. The Harvester Press, Brighton

Jones A (1994) Postfeminism, feminist pleasures, and embodied theories of art. In: Frueh J, Langer CL, Raven A (eds) New feminist criticism: art, identity, action. HarperCollins, New York, pp 16–41

Jones OD (1999) Sex, culture, and the biology of rape: toward explanation and prevention. Calif Law Rev 87(4):827–941

Jones JW (2005) Brain, mind, and spirit—a clinician's perspective, or why I am not afraid of dualism. In: Bulkeley K (ed) Soul, psyche, brain: new directions in the study of religion and brain-mind science. Palgrave Macmillan, New York, pp 36–60

Jones GW (2007) Delayed marriage and very low fertility in Pacific Asia. Popul Dev Rev 33 (3):453–478

Jones EF, Forrest JD, Goldman N et al (1986) Teenage pregnancy in industrialized countries. Yale University Press, New Haven

Jones E, Reynolds V (eds) (1995) Survival and religion. Biological evolution and cultural change. John Wiley, Chichester

Joravsky D (1970) The Lysenko affair. Harvard University Press, Boston

Jordan GE (2008) The invention of man: a response to C.S. Lewis's the abolition of man. J Evol Technol 19(1):35–41

Jordan JJ, McAuliffe K, Warneken F (2014) Development of in-group favoritism in children's third-party punishment of selfishness. PNAS 111(35):12710–12715

Jorde LB, Wooding SP (2004) Genetic variation, classification and 'race'. Nat Genet 36:S28–s33

Joseph R (2000) The transmitter to God: the limbic system, the soul, and spirituality. University Press California, San Jose

Joseph R (2001) The limbic system and the soul: evolution and the neuroanatomy of religious experience. Zygon 36(1):105–136

Joseph R (ed) (2002) NeuroTheology: brain, science, spirituality, religious experience. University of California Press, San Jose

Josephson M (1934) The robber barons: the great american capitalists. Harcourt Brace and Company, New York

Jotterand F (2010) At the roots of transhumanism: from the enlightenment to a post-human future. J Med Philos 35(6):617–621

Jotterand F (2014) Questioning the moral enhancement project. Am J Bioethics 14(4):1–3

Joy B (2000) Why the future doesn't need us: our most powerful 21st-century technologies—robotics, genetic engineering, and nanotech—are threatening to make humans an endangered species. Wired 8(4):238–264

Joyce R (2006) The evolution of morality. MIT Press, Cambridge

Jozifkova E, Flegr J (2006) Dominance, submissivity (and homosexuality) in general population. Testing of evolutionary hypothesis of sadomasochism by internet-trap-method. Neuroendocrinol Lett 27(6):711–718

Judge DS, Carey JR (2000) Postreproductive life predicted by primate patterns. J Gerontol: Biol Sci 55A(4):B201–B209

Juergensmeyer M (2001) Terror in the mind of God: the gobal rise of religious violence. University of California Press, Berkeley and Los Angeles

Julian T, McKenry PC (1989) Relationship of testosterone to men's family functioning in mid-life. Aggressive Behav 15(4):281–289

Jun KH (2005) The transition to sub-replacement fertility in South Korea: implications and prospects for population policy. Jpn J Popul 3(1):26–57
Kabasenche WP (2012) Moral enhancement worth having: thinking holistically. AJOB Neurosci 3 (4):18–20
Kachigan SK (1990) The sexual matrix. Boy meets girl on the evolutionary scale. Radius Press, New York
Kahane G (2011) Evolutionary debunking arguments. Noûs 45(1):103–125
Kaiser J-C, Nowak K, Schwartz M (1992) Eugenik, sterilisation, "Euthanasie": Politische biologie in Deutschland 1895-1945. Buchverlag Union, Halle
Kajiyama Y (1982) Women in Buddhism. East Buddhist 15(2):53–70
Kanazawa S (2010) Why liberals and atheists are more intelligent. Soc Psychol Q 73(1):33–57
Kanazawa S (2016) Possible evolutionary origins of human female sexual fluidity. Biol Rev. https://doi.org/10.1111/brv.12278
Kanazawa S, Still MC (1999) Why monogamy? Soc Forces 78(1):25–50
Kandel E, Freed D (1989) Frontal lobe dysfunction and antisocial behavior: a review. J Clin Psychol 45(3):404–413
Kandler C, Lewis GJ, Feldhaus LH et al (2015) The genetic and environmental roots of variance in negativity toward foreign nationals. Behav Genet 45(2):181–199
Kandler C, Riemann R (2013) Genetic and environmental sources of individual religiousness: the roles of individual personality traits and perceived environmental religiousness. Behav Genet 43(4):297–313
Kaplan H (1996) A theory of fertility and parental investment in traditional and modern human societies. Yearb Phys Anthropol 39:91–135
Kaplan HS, Lancaster JB (2000) The evolutionary economics and psychology of the demographic transition to low fertility. In: Cronk L, Irons W, Chagnon N (eds) Human behavior and adaptation: an anthropological perspective. Aldine de Gruyter, Hawthorne, pp 238–322
Kaplan H, Lancaster JB, Tucker WT et al (2002) Evolutionary approach to below replacement fertility. Am J Hum Biol 14(2):233–256
Kapogiannis D, Barbey AK, Su M et al (2009) Cognitude and neural foundations of religious belief. PNAS 106(12):4876–4881
Kapogiannis D, Barbey AK, Su M et al (2009) Neuroanatomical variability of religiosity. PLoS ONE 4(9):e7180
Kapp M (2001) Economic influences on end-of-life care: empirical evidence and ethical speculation. Death Stud 25(3):251–263
Kardong KV (2010) Beyond God: evolution and the future of religion. Humanity Books, Amherst
Karen R (1994) Becoming attached: first relationships and how they shape our capacity to love. Oxford University Press, New York
Karimi-Boosherhi F, Rasouli-Nia A (1988) The role of ideology and religion in psychosexual development of women. The sociobiology of sex differences. Abstracts of the 11th meeting of the European sociobiological Society, Held in the Veterinary College, Oslo: 2–4 Sept 1988
Karraker MW, Grochowski JR (2005) Families with futures: a survey of family studies for the 21st century. Lawrence Erlbaum, Philadelphia
Kasper C, Borgerhoff Mulder M (2015) Who helps and why? Cooperative networks in Mpimbwe. Curr Anthropol 56(5):701–732
Kass L (2002) Life, liberty, and the defense of dignity: the challenge for bioethics. Encounter Books, San Francisco
Kasting JF (1988) Runaway and moist greenhouse atmospheres and the evolution of earth and Venus. Icarus 74(3):472–494
Kastleman MB (2007) The drug of the new millennium—the brain science behind internet pornography use. PowerThink Publishing, Provo
Kato-Shimizu M, Onishi K, Kanazawa T et al (2013) Preschool children's behavioral tendency toward social indirect reciprocity. PLOS ONE 8(8):e70915
Katz SH (1999) Toward a new concept of global morality. Zygon 34(2):237–252

Katz L (2000) Toward good and evil. Evolutionary approaches to aspects of human morality. In: Katz L (ed) Evolutionary origins of morality: cross-disciplinary perspectives. Imprint Academic, Exeter, pp ix–xvi
Katz L (ed) (2000) Evolutionary origins of morality: cross-disciplinary perspectives. Imprint Academic, Exeter
Kauffman SA (1993) The origins of order: self-organization and selection in evolution. Oxford University Press, Oxford
Kaufman SA (1995) At home in the universe. The search for the laws of self-organization and complexity. Oxford University Press, New York
Kaufman GD (1997) The epic of evolution as a framework for human orientation in life. Zygon 32 (2):175–189
Kaufman GD (2001) On thinking of God as serendipitous creativity. J Am Acad Relig 69(2):409–425
Kaufmann EP (2011) Shall the religious inherit the earth? Demography and politics in the twenty-first century. Profile Books, London
Kauth MR (2007) Handbook of the evolution of human sexuality. Routledge, New York
Kautsky K (1918) Die Diktatur des Proletariats. (The dictatorship of the proletariat). Verlag Ignaz Band, Vienna
Kay AC, Gaucher D, McGregor I et al (2010) Religious belief as compensatory control. Person Soc Psychol Rev 14(1):37–48
Kearsey MJ, Pooni HS (1998) The genetical analysis of quantitative traits. Chapman and Hall, London
Keeley LH (1996) War before civilization: the myth of the peaceful savage. Oxford University Press, Oxford
Keith A (1946) Evolution and ethics. G.P. Putman's Sons, New York
Keizer R, Dykstra PA, Jansen MD (2007) Pathways into childlessness: evidence of gendered life-course dynamics. J Biosoc Sci 40(6):863–878
Kelemen D (2004) Are children "intuitive theists"? Reasoning about purpose and design in nature. Psychol Sci 15(5):295–301
Kellehear A (1996) Experiences near death: beyond medicine and religion. Oxford University Press, Oxford
Keller AG (1915) Societal evolution: a study of the evolutionary basis of the science of society. Macmillan Company, New York
Keller G (2014) Deccan volcanism, the Chicxulub impact and the end-Cretaceous mass extinction: coincidence? Cause and effect? In: Keller G, Kerr A (eds) Volcanism, impacts and mass extinctions: causes and effects. Geological Society of America Special Paper 505
Kellert SR (1996) The value of life: biodiversity and human society. Island Press, Washington
Kelly RC (2005) The evolution of lethal intergroup violence. PNAS 102:15294–15298
Kelly RL (2016) The fifth beginning: what six million years of human history can tell us about our future. University of California Press, Oakland
Keltner D, Kogan A, Piff PK et al (2014) The sociocultural appraisals, values, and emotions (SAVE) framework of prosociality: core processes from gene to meme. Ann Rev Psychol 65:425–460
Kemkes-Grottenthaler A (2003) Postponing or rejecting parenthood? Results of a survey among female academic professionals. J Biosoc Sci 35(2):213–226
Kendall TD (2007) Pornography, rape, and the internet. Clemson University, Clemson
Kendler KS, Gardner CO, Prescott CA (1997) Religion, psychopathology, and substance use and abuse: a multimeasure, genetic-epidemiologic study. Am J Psychiatry 154(3):322–329
Kendler KS, Myers J (2009) A developmental twin study of church attendance and alcohol and nicotine consumption: a model for analyzing the changing impact of genes and environment. Am J Psychiatry 166(10):1150–1155
Kendler KS, Liu X-Q, Gardner CO et al (2003) Dimensions of religiosity and their relationship to lifetime psychiatric and substance use disorders. Am J Psychiatry 160(3):496–503

Kennedy H (2007) The great Arab conquests: how the spread of Islam changed the world we live in. Da Capo, New York

Keown D (2013) Buddhism: a very short introduction. Oxford University Press, Oxford

Kertzer DJ (2006) Religion and the decline of fertility: conclusions. In: Derosas R, van Poppel F (eds) Religion and the decline of fertility in the western world. Springer, Dordrecht, pp 259–269

Ketting E, Van Praag PH (1983) Abortus provocatus. Wet en praktijk. (Abortion. Law and practice). Nisso, Zeist

Kevles DJ (1985) In the name of eugenics. Genetics and the use of human heredity. University of California Press, Berkeley and Los Angeles

Kevles DJ, Hood L (1992) The code of codes. Scientific ad social issues in the human genome project. Harvard University Press, Cambridge

Keys A, Brozek J (eds) (1950) The biology of human starvation. The University of Minnesota Press, Minneapolis

Khlat M, Khoury M (1991) Inbreeding and diseases: demographic, genetic, and epidemiologic perspectives. Epidemiol Rev 13:28–41

Khosrokhavar F (2005) Suicide bombers: Allah's new martyrs. Pluto Press, London

Kieffer G (1979) Bioethics. A textbook of issues. Wesley-Addison, Menlo Park

Kiehl K, Laurens K, Liddle P et al (2006) Brain potentials implicate temporal lobe abnormalities in criminal psychopaths. J Abnorm Psychol 115(3):443–453

Kiehl KA, Smith AM, Hare RD et al (2001) Limbic abnormalities in affective processing by criminal psychopaths as revealed by functional magnetic resonance imaging. Biol Psychiatry 50:677–684

Kiernan K (1993) The future of partnership and fertility. In: Cliquet R (ed) The future of Europe's population. Population Studies N° 26. Strasbourg: Council of Europe, pp 23–44

Kiernan K (1989) Who remains childless? J Biosoc Sci 21:387–398

Killen M, Smetana J (2007) The biology of morality: human development and moral neuroscience. Hum Dev 50(5):241–244

Kilner J, Mitchell CB (2003) Remaking humans: the new utopians versus a truly human future. The center for bioethics and human dignity. Routledge, New York

Kilpatrick DG (2000) Rape and sexual assault. http://www.musc.edu/vawprevention/research/sa.shtml

Kilpatrick DG, Edmunds C, Seymour A (1992) Rape in America: a report to the nation. National Victim Center and the Crime Victims Research and Treatment Center, Medical University of South Carolina, Charleston

Kimbrell A (1998) The human body shop. The cloning, engineering and marketing of life. Regnery Publishing, Washington

Kinfu Y (2000) Below-replacement fertility in Tropical Africa? Some evidence from Addis Ababa. J Popul Res 17(1):63–82

King BJ (2007) Evolving God: a provocative view on the origins of religion. Doubleday Religion, New York

King AJ, Johnson DDP, Van Vugt M (2009) The origins and evolution of leadership. Curr Biol 19:R911–R916

Kinsey AC, Martin CE, Pomeroy WB et al (1953) Sexual behavior in the human female. WB Saunders, Philadelphia

Kinsey A, Pomeroy W, Martin C (1948) Sexual behavior in the human male. W.B. Saunders, Philadelphia

Kirk D (1969) The genetic implications of family planning. Biol Family 44(11):80–83

Kirk KM (2001) Natural selection and quantitative genetics of life-history traits in western women: a twin study. Evolution 55(2):432–435

Kirk R (2002) Homo superior. DS Press

Kirk KM, Eaves LJ, Martin NG (1999) Self-transcendence as a measure of spirituality in a sample of older Australian twins. Twin Res 2(2):81–87

Kirk KM, Maes HH, Heath AC et al (1999) Frequency of church attendance in Australia and the United States: models of family resemblance. Twin Res 2(2):99–107
Kirkpatrick M (1982) Sexual selection and the evolution of female choice. Evolution 36(1):1–12
Kirkpatrick LA (1999) Towards an evolutionary psychology of religion and personality. J Person 67:921–949
Kirkpatrick RC (2000) The evolution of human homosexual behavior. Curr Anthropol 41(3):385–413
Kirkpatrick LA (2005) Attachment, evolution, and the psychology of religion. Guilford Publications, New York
Kirkpatrick LA (2006) Religion is not an adaptation. In: McNamara P (ed) Where God and science meet: how brain and evolutionary studies alter our understanding of religion. Praeger, Westport, pp 159–179
Kirkpatrick LA (2008) Religion is not an adaptation. Some fundamental issues and arguments. In: Bulbulia J, Sosis R, Genet C et al (eds) The evolution of religion: studies, theories, and critiques. Collins Foundation Press, Santa Margarita, pp 61–66
Kirkpatrick LA, Shaver PR (1992) An attachment-theoretical approach to romantic love and religious belief. Person Soc Psychol Bull 18(3):266–275
Kirsch JAW, Rodman JE (1982) Selection and sexuality: the Darwinian view of homosexuality. In: Paul W, Weinrich JD, Gonsiorek JC et al (eds) Homosexuality: social, psychological, and biological issues. Sage, Beverly Hills, pp 183–195
Kirschner M, Gerhart J (1998) Evolvability. PNAS 95(15):8420–8427
Kitcher P (1982) Abusing science: the case against creationism. MIT Press, Cambridge
Kitcher P (1985) Vaulting ambition: sociobiology and the quest for human nature. MIT Press, Cambridge
Kitcher P (1996) The lives to come: the genetic revolution and human possibilities. Penguin Books, London
Kitcher P (2006) Ethics and evolution. How to get here from there. In: de Waal F (ed) Primates and philosophers. Princeton University Press, Princeton, pp 120–139
Kitcher P (2007) Living with Darwin. Evolution, design, and the future of faith. Oxford University Press, Oxford
Kitcher P (2014) Is a naturalized ethics possible? Behaviour 151(2–3):245–260
Kivlahan C, Ewigman N (2010) Rape as a weapon of war in modern conflicts. Br Med J 340: c3270
Klarsfeld A, Revah F, Brady L (2003) The Biology of death: origins of mortality. Comstock Publishing, Sacramento
Klee E (1983) Euthanasie im NS-Staat. Die Vernichtung lebensunwerten Lebens. (Euthanasia in the Nazi State. The destruction of life unworthy of life). Fischer Verlag, Frankfurt
Klein C (2011) The dual track theory of moral decision-making: a critique of the neuroimaging evidence. Neuroethics 4(2):SI 143–162
Klein N (2014) This changes everything: capitalism vs. the climate. Simon and Schuster, New York
Kleinplatz PJ, Moser C (2006) Sadomasochism: powerful pleasures. Routledge, New York
Klerkx G (2006) The transhumanists as tribe. In: Miller P, Wilsdon J (eds) Better humans. The politics of human enhancement and life extension. Demos, London, pp 59–66
Klingorova K, Havlicek T (2015) Religion and gender inequality: the status of women in the societies of world religions. Moravian Geogr Rep 23(2):2–11
Kluger AN, Siegfried Z, Ebstein RP (2002) A meta-analysis of the association between DRD4 polymorphism and novelty seeking. Mol Psychiatry 7(7):712–717
Knafo A, Israel S (2009) Genetic and environmental influences on prosocial behavior. In: Mikulincer M, Shaver PR (eds) Prosocial motives, emotions, and behavior: the better angels of our nature. APA Books, Washington
Knafo A, Plomin R (2006) Prosocial behavior from early to middle childhood: genetic and environmental influences on stability and change. Dev Psychol 42(5):771–786

Knafo A, Zahn-Waxler C, Davidov M et al (2009) Empathy in early childhood. Ann New York Acad Sci 1167:103–114

Knapp A (1989) Soziobiologie und Moraltheologie. Kritik der ethischen Folgerungen moderner Biologie. (Sociobiology and moral theology. Critique of the ethical implications of modern biology). VCH Verlagsgesellschaft, Weinheim

Knauft B (1991) Violence and sociality in human evolution. Curr Anthropol 32(4):391–428

Knauft B (2000) Symbols, sex and sociality in the evolution of human morality. In: Katz L (ed) Evolutionary origins of morality: cross-disciplinary perspectives. Imprint Academic, Exeter, pp 130–139

Kneebone E, Raphael S (2011) City and suburban crime trends in metropolitan America. Brookings, Washington

Knickmeyer CR, Baron-Cohen S (2006) Fetal testosterone and sex differences. Early Hum Dev 82 (12):755–760

Knight LU (2001) The voluntary human extinction movement. http://www.vhemt.org/

Knoch D, Pascual-Leone A, Meyer K et al (2006) Diminishing reciprocal fairness by disrupting the right prefrontal cortex. Science 314(5800):829–832

Knoepffler N, Savulescu J (eds) (2009) Der neue Mensch. Enhancement und Genetik. (The new man. Enhancement and genetics). Alber Verlag, Freiburg i. B

Knott K (2016) Hinduism: a very short introduction. Oxford University Press, Oxford

Ko D, Haboush JH, Piggott J (2003) Women and Confucian cultures in premodern China, Korea, and Japan. University of California press, Oakland

Koch T (2010) Enhancing who? Enhancing what? Ethics, bioethics, and transhumanism. J Med Philos 35(6):685–699

Kochanska G (2002) Mutually responsive orientation between mothers and their young children: a context for the early development of conscience. Curr Dir Psychol Sci 11(6):191–195

Koedt A, Levine E, Rapone A (eds) (1973) Radical feminism. Times Books, New York

Koenig LB, Bouchard TJ Jr (2006) Genetic and environmental influences on the traditional moral values triad—authoritarianism, conservatism and religiousness—as assessed by quantitative behavior genetic methods. In: McNamara P (ed) Where God and science meet: how brain and evolutionary studies alter our understanding of religion, vol 1. Praeger, Westport, pp 31–60

Koenig H, Harvey G, Cohen J (2002) The link between religion and health: psycho-neuroimmunology and the faith factor. Oxford University Press, Oxford

Koenig H, McCullough ME, Larson DB (2000) Handbook of religion and health. Oxford University Press, Oxford

Koenig LB, McGrue M, Krueger RF et al (2005) Genetic and environmental influences on religiousness: findings for retrospective and current religiousness ratings. J Person 73(2):471–488

Koenig LB, McGue M, Iacono WG (2008) Stability and change in religiousness during emerging adulthood. Dev Psychol 44(2):532–543

Koenigs M, Young L, Adolphs R et al (2007) Damage to the prefrontal cortex increases utilitarian moral judgements. Nature 446(7138):908–911

Koerper HC, Stickel EG (1980) Cultural drift: a primary process of culture change. J Anthropol Res 36(4):463–469

Kohl M (1992) Altruistic humanism and voluntary beneficent euthanasia. Issues Law Med 8 (3):331–341

Kohler H-P, Billari FC, Ortega JA (2002) The emergence of lowest-low fertility in Europe during the 1990s. Popul Dev Rev 28(4):641–680

Kohler P, Lafferty W (2008) Abstinence-only and comprehensive sex education and the initiation of sexual activity and teen pregnancy. J Adolesc Health 42(4):344–351

Kohn RR (1978) Principles of mammalian ageing. Prentice Hall, New York

Kohn A (1987) The case against competition. Houghton Mifflin, Boston

Kolbert E (2014) The sixth extinction: an unnatural history. Henry Hold, New York

Komter A (2010) The evolutionary origins of human generosity. Int Sociol 25(3):443–464

Konietzka D, Kreyenfeld M (eds) (2007) Ein Leben ohne Kinder: Kinderlosigkeit in Deutschland. (A life without children: childlessness in Germany). VS Verlag für Sozialwissenschaften, Wiesbaden

Konner M (2015) Women after all: sex, evolution, and the end of male supremacy. WW Norton, New York

Konrad KA, Morath F (2012) Evolutionarily stable in-group favoritism and out-group spite in intergroup conflict. J Theoret Biol 306:61–67

Kontula O, Haavio-Mannila E (1995) Sexual pleasures: enhancement of sex life in Finland, 1971–1992. Dartmouth Publishing Company, Aldershot

Koons DN, Holmes RR, Grand JB (2007) Population inertia and its sensitivity to changes in vital rates or initial conditions. Ecology 88(11):2857–2867

Koopmans R (2013) Fundamentalism and out-group hostility. Muslim immigrants and Christian natives in Western Europe. WZB Mitt 142:21–25

Kopenhaver B (2010) The psychopath: a new subspecies of Homo Sapiens. www.Sott.net

Kortright C (2008) The biology of capitalism: scarcity, poverty and population (Part 1). http://isgreaterthan.net/2008/03/the-biology-of-capitalism-scarcity-poverty-and-population/

Kosfeld M, Heinrichs M, Zak P et al (2005) Oxytocin increases trust in humans. Nature 435 (7042):673–676

Koslowsky P (ed) (1999) Sociobiology and economics: the theory of evolution in biological and economic theory. Springer, Berlin

Kosmin BA, Keysar A (2009) American nones: the profile of the no religion population: a report based on the American religious identification survey 2008. Institute for the Study of Secularism in Society and Culture, Hartford

Koukl G (2012) Monkey morality: can evolution explain ethics? http://www.str.org/articles/monkey-morality-can-evolution-explain-ethics#.UZUTC7W1ka0

Kovacs M, Ott J, Merikangas KR (2009) Interaction between the serotonin transporter gene (5-HTTLPR), stressful life events, and risk of depression: a meta-analysis. JAMA 301:2462–2471

Kozinszky Z, Zadori J, Orvos H et al (2003) Obstetric and neonatal risk of pregnancies after assisted reproductive technology: a matched control study. Acta Obstet et Gynecol Scand 82 (9):850–856

Kramer KL, Otárola-Castillo E (2015) When mothers need others: the impact of hominin life history evolution on cooperative breeding. J Hum Evol 84:16–24

Krantz GS (1970) Human activities and megafaunal extinctions. Am Sci 58(2):164–170

Krasnow MM, Delton AW, Tooby J et al (2013) Meeting now suggests we will meet again: implications for debates on the evolution of cooperation. Sci Rep 3:1747. https://doi.org/10.1038/srep01747

Krause N (2005) God-mediated control and psychological well-being in late life. Res Aging 27 (2):136–164

Krauss LM (2012) A universe from nothing: why there is something rather than nothing. Free Press, New York

Kravdal Ö (1992) The emergence of a positive relation between education and third-birth rates in Norway with supportive evidence from the United States. Popul Stud 46(6):459–475

Krawczynski M, Walkowiak J, Krzyzaniak A (2003) Secular changes in body height and weight in children and adolescents in Poznan, Poland, between 1880 and 2000. Acta Paediatr 92(3):277–282

Krebs D (1983) Commentary and critique: sociobiological approaches to prosocial development. In: Bridgeman DL (ed) The nature of prosocial development. Academic Press, New York, pp 61–69

Krebs D (2011) The origins of morality: an evolutionary account. Oxford University Press, New York

Krimsky S, Sloan K (2011) Race and the genetic revolution: science, myth, and culture. Columbia University Press, New York

Krippner S (2005) Psychoneurological dimensions of anomalous experience in relation to religious belief and spiritual experience. In: Bulkeley K (ed) Soul, psyche, brain: new directions in the study of religion and brain-mind science. Palgrave Macmillan, New York, pp 61–92

Krolokke C, Sorensen AS (2005) Three waves of feminism: from suffragettes to grrls. Gender communication theories and analyses: from silence to performance. Sage, Thousand Oaks

Kropotkin P (1902) Mutual aid: a factor in evolution. McClure, Phillips and Co, New York

Krosch AR, Amodio DM (2014) Economic scarcity alters the perception of race. PNAS 111 (25):9079–9084

Krupp DB, DeBruine LM, Jones BC (2011) Apparent health encourages reciprocity. Evol Hum Behav 32(3):198–203

Krzywicki L (1951) Rozwój moralnośsci (Development of morality). Studia Socjologiczne, 233–295

Kübler-Ross E (2012) On life after death. EKR Family Limited Partnership, Scottsdale

Kuemmerli R (2011) A test of evolutionary policing theory with data from human societies. PLOS ONE 6(9):e24350

Kuhle BX (2013) Born both ways: the alloparenting hypothesis for sexual fluidity in women. Evol Psychol 11(2):304–323

Kuhlmeier VA, Dunfield KA, O'Neill AC (2014) Selectivity in early prosocial behavior. Front Psychol 5:836. https://doi.org/10.3389/fpsyg.2014.00836

Küng H (1990) Projekt Weltethos. (Project global ethic). Piper, München

Küng H (1996) Yes to a global ethic. Continuum Publishing Company, New York

Küng H (1997) A global ethic for global politics and economics. SCM Press, London

Küng H, Kuschel KJ (eds) (1993) A global ethic: the declaration of the world's religions. SCM Press, London

Küntzel M (2007) Jihad and Jew-Hatred: Islamism, Nazism and the roots of 9/11. Telos Press, New York

Kümmerli R, Burton-Chellew MN, Ross-Gillespie A et al (2010) Resistance to extreme strategies, rather than prosocial preferences, can explain human cooperation in public goods games. PNAS 107(22):10125–10130

Kurtz P (1992) The case for euthanasia: a humanistic perspective. Issues Law Med 8(3):309–316

Kurtz P (2003) Science and religion: are they compatible?. Prometheus Books, Amherst

Kurtz P (2007) What is secular humanism?. Prometheus books, Amherst

Kurtz P, Koepsell D (eds) (2007) Science and ethics. Can science help us make wise moral judgments? Prometheus Books, Amherst

Kurtz P, Wilson EH (1973) Humanist Manifesto II. American Humanist Association, Washington

Kurzban R (2010) Grand challenges of evolutionary psychology. Front Psychol 1(3):1–3

Kurzban R, Aktipis CA (2007) Modularity and the social mind. Are psychologists too selfish? Person Soc Psychol Rev 11(2):131–149

Kurzban R, Leary MR (2001) Evolutionary origins of stigmatization: the functions of social exclusion. Psychol Bull 127(2):187–208

Kurzweil R (2005) The singularity is near. When humans transcend biology. Viking Press - Penguin Group, New York

Kutchinsky B (1973) The effect of easy availability of pornography on the incidence of sex crimes: the Danish experience. J Soc Issues 14(3):47–64

Kutchinsky B (1991) Pornography and rape: theory and practice? Evidence from crime data in four countries where pornography is easily available. Int J Law Psychiatry 14(1–2):47–64

Kwame A (2006) Cosmopolitanism: ethics in a world of strangers. Norton, New York

Laakso MP, Vaurio O, Koivisto E et al (2001) Psychopathy and the posterior hippocampus. Behav Brain Res 118:187–193

Lacasse L, Lee M, Ishikawa S et al (2003) Corpus callosum abnormalities in psychopathic antisocial individuals. Arch Gen Psychiatry 60(11):1134–1142

Lachmann PJ (2010) Religion—an evolutionary adaptation. FASEB J 24(5):1301–1307

LaCroix RR (1974) Unjustified evil and God's choice. Sophia 13(1):20–28

LaHaye T (1980) The battle for the mind. Fleming H. Revell, Old Tappan
Lahti DC (2004) "Have you heard…but I tell you…": a test of the adaptive significance of moral evolution. In: Clayton P, Schloss J (eds) Evolution and ethics: human morality in biological and religious perspective. Eerdmans, Grand Rapids, pp 132–150
Lahti DC (2009) The correlated history of social organization, morality, and religion. In: Voland E, Schiefenhövel W (eds) The biological evolution of religious mind and behavior. Springer Publishers, New York, pp 67–88
Lai-wan CC, Eric B, Celia Hoi-yan C (2006) Attitudes to and practices regarding sex selection in China. Prenat Diagn 26(7):610–613
Laland KN, Brown GR (2002) Sense and nonsense: evolutionary perspectives on human behaviour. Oxford University Press, Oxford
Laland KN, Odling-Smee FJ, Myles S (2010) How culture has shaped the human genome: bringing genetics and the human sciences together. Nat Rev Genet 11:137–148
Lamarck J-B (1809) Philosophie zoologique ou exposition des considérations relatives à l'histoire naturelle des animaux. (Zoological philosophy: exposition with regard to the natural history of animals). Dentu, Paris
Lamba S, Mace R (2011) Demography and ecology drive variation in cooperation across human populations. PNAS 108(35):14426–14430
Lamm C, Majdandžić J (2015) The role of shared neural activations, mirror neurons, and morality in empathy—a critical comment. Neurosci Res 90:15–24
Lamont C (1996) The philosophy of humanism. Humanist Press, Amherst
Lampert A (1997) The evolution of love. Praeger, Westport
Lancaster J (1985) Evolutionary perspectives on sex differences in the higher primates. In: Rossi AS (ed) Gender and the life course. Aldine, Hawthorne, pp 3–27
Lancaster JB, Altmann J, Rossi A et al (1987) Parenting across the life-span: biosocial dimensions. de Gruyter, New York
Landa JT, Ghiselin MT (1999) The emerging discipline of bioeconomics: aims and scope of the journal of bioeconomics. J Bioecon 1(1):5–12
Landa JT, Wilson DS (2008) Group selection: theory and evidence. An introduction to the special issue. J Bioecon 10(3):199–202
Lande R (1981) Models of speciation by sexual selection of polygenic traits. PNAS 78(6):3721–3725
Landry A (1934) La révolution démographique: études et essais sur les problèmes de la population. (The demographic revolution: Studies and tests on population problems). Sirey, Paris
Lane A, Byrne M, Mulvany F et al (1995) Reproductive behaviour in schizophrenia relative to other mental disorders: evidence for increased fertility in men despite decreased marital rate. Acta Psychiatr Scand 91(4):222–228
Langdridge D, Barker M (eds) (2007) Safe, sane and consensual: contemporary perspectives on sadomasochism. Palgrave Macmillan, Basingstoke
Langergraber K, Schubert G, Rowney C et al (2011) Genetic differentiation and the evolution of cooperation in chimpanzees and humans. Proc R Soc B: Biol Sci 278(1717):2546–2552
Langhoff C (2007) Pre-transitional populations: natural fertility. GRIN Verlag, München
Langton R (2009) Sexual solipsism: philosophical essays on pornography and objectification. Oxford University Press, New York
Laor Y, Jablonka E (2013) The evolution and development of culture. Hist Theory 52(2):290–299
LaPierre D, Braun CMJ, Hodgins S (1995) Ventral frontal deficits in psychopathy: neuropsychological test findings. Neuropsychologia 33(2):139–151
Laqueur W (2007) The last days of Europe: Epitaph for an old continent. Thomas Dunne, New York
Larson EJ (1985) Trial and error: the American controversy over creation and evolution. Oxford University Press, New York

Larson EF (2002) The meaning of human gene testing for disability rights. Univ Cincinnati Law Rev 70:913–938

Larson EJ (2006) Evolution: the remarkable history of a scientific theory. Modern Library, New York

Larson EL, Witham L (1998) Leading scientists still reject God. Nature 394(6691):313

Larsson H, Andershed A, Lichtenstein P (2006) A genetic factor explains most of the variation in the psychopathic personality. J Abnorm Psychol 115(2):221–230

Laslett P (1965) The world we have lost further explored. Methen and Co, London

Latané B, Draley JM (1970) The unresponsive bystander: why doesn't he help?. Appleton-Century-Croft, New York

Latham E (1906) Famous sayings and their authors. Swan Sonnenschein, London

Laumann EO, Gagnon JH, Michael RT et al (1994) The social organization of sexuality. Sexual practices in the United States. The University of Chicago Press, Chicago

Laurin K, Shariff AF, Henrich J et al (2012) Outsourcing punishment to God: beliefs in divine control reduce earthly punishment. Proc R Soc B-Biol Sci 279(1741):3272–3281

Layden MA, Eberstadt M (2010) The social costs of pornography: a statement of findings and recommendations. The Witherspoon Institute, Princeton

Lazare D (2002) False testament: archaeology refutes the Bible's claim to history. Harper's March 2002:39–47

Leake CD, Romanell P (1950) Can we agree? A scientist and a philosopher argue about ethics. University of Texas Press, Austin

Leakey R, Lewin R (1995) The sixth extinction: patterns of life and the future of humankind. Doubleday, New York

Leben D (2011) Cognitive neuroscience and moral decision-making: guide or set aside? Neuroethics 4(2):SI 163–174

Leblanc SA (2003) Constant battles: the myth of the noble savage and a peaceful past. St Martin's Press, New York

Lecointre G, Wells J, Johnson P et al (2004) Les formes instruites de l'anti-science. (Educated forms of anti-science). In: Dubessy J, Lecointre G, Silberstein M (2004) Les matérialismes (et leurs détracteurs). Syllepse, Paris, pp 511–543

Lederberg J (1960) A view of genetics. Science 131(3396):269–276

Lederer (ed) (1980) Take back the night: women on pornography. William Morrow, New York

Lee RD (1997) Intergenerational relations and the elderly. In: Wachter KW, Finch C (eds) Between Zeus and the Salmon. The biodemography of longevity. National Academy Press, Washington, pp 212–233

Lee RD (2003) The demographic transition: three centuries of fundamental change. J Econ Perspect 17(4):167–190

Lee AJ, Dubbs SL, Von Hippel W et al (2014) A multivariate approach to human mate preferences. Evol Hum Behav 35(3):193–203

Lee HJ, Macbeth AH, Pagani JH et al (2009) Oxytocin: the great facilitator of life. Prog Neurobiol 88(2):127–151

Leeming DA (2009) Creation myths of the world. ABC-CLIO, Santa Barbara

Lehmann L, Feldman MW (2008) War and the evolution of belligerence and bravery. Proc R Soc Lond B 275(1653):2877–2885

Leibniz G (1710) Essais de théodicée sur la bonté de Dieu, la liberté de l'homme et l'origine du mal. (Essays of theodicy on the goodness of God, the freedom of man and the origin of evil). I. Troyel, Amsterdam

Leigh EG (2010) The group selection controversy. J Evol Biol 23(1):6–19

Lemos J (2008) Commonsense Darwinism: evolution, morality, and the human condition. OpenCourt, Chicago

Lenington S (1981) Child abuse: the limits of sociobiology. Ethol Sociobiol 2(1):17–29

Lenski G (1984) Power and privilege. A theory of social stratification. The University of North Carolina Press, Chapel Hill

Lenz F (1936) In: Baur E, Fischer E, Lenz F (eds) Grundriss der menschliche Erblehre und Rassenhygiene. Lehmanns Verlag, München
Leo PPXIII (1891) Rerum novarum. Tablet 77(2663):805–807
Léridon H (1973) Aspects biométriques de la fécondité humaine. (Biometric aspects of human fertility). INED, Travaux et Documents 65. Presses Universitaires de France, Paris
Léridon H (1987) Natural fertility: patterns and determinants of natural fertility. Ordina, Liège
Léridon H, Menken JA (1979) Natural fertility: patterns and determinants of natural fertility: proceedings of a seminar on natural fertility held in Paris in March 1977. Organized by the International Union for the Scientific Study of Population. IUSSP, Liège
Lerner G (1986) The creation of patriarchy. Oxford University Press, New York
Lesch KP (2007) Linking emotion to the social brain. The role of the serotonin transporter in human social behaviour. EMBO Rep 8(S1):S24–S29
Lesch KP, Bengel D, Heils A et al (1996) Association of anxiety-related traits with a polymorphism in the serotonin transporter gene regulatory region. Science 274(5292):1527–1531
Leslie J (1996) The end of the world: the science and ethics of human extinction. Routledge, London
Lesthaeghe R (1995) The second demographic transition in western countries: an interpretation. In: Oppenheim Mason K, Jensen AM (eds) Gender and family change in industrialised countries. Clarendon Press, Oxford, pp 17–62
Lesthaeghe R (ed) (2002) Meaning and choice. Value orientation and life course decisions. NIDI/CBGS Publications, Brussels/The Hague
Lesthaeghe R (2010) The unfolding story of the second demographic transition. University of Michigan, PSC Research Report No, pp 10–696
Lesthaeghe R (2014) The second demographic transition: a concise overview of its development. PNAS 111(51):18112–18115
Lesthaeghe R, Page H, Surkyn J (1988) Are immigrants substitutes for births? I.D.P.-working paper, 3, V.U.B., Brussel
Lesthaeghe R, Surkyn J (1991) Pratiques et appartenances: Résultats d'une enquête. (Practices and affiliations: results of a survey). In: Dassetto F (ed) Facettes de l'Islam belge. Louvain-la-Neuve, Academia-Bruylant, pp 6–46
Lesthaeghe R, Van de Kaa DJ (1986) Twee demografische transities? (Two demographic transitions?). In: Van de Kaa DJ, Lesthaeghe R (eds) Bevolking: groei en krimp. Van Loghum Slaterus, Deventer, pp 9–24
Leutenegger W (1982) Encephalization and obstetrics in primates with particular reference to human evolution. In: Armstrong E, Falk D (eds) Primate brain evolution. Plenum, New York, pp 85–95
Leutenegger W, Cheverud JM (1982) Correlates of sexual dimorphism in primates. Int J Primatol 3:387–402
LeVay S (1994) The sexual brain. MIT Press, Cambridge
Levie LH (1961) Seksuologie. Handleiding voor artsen. (Sexology. Manual for physicians). Stafleu, Leiden
Levin J (2001) God, faith, and health: exploring the spirituality-healing connection. Wiley, New York
Levin J (2010) Religion and mental health: theory and research. Int J Appl Psychoanal Stud 7 (2):102–115
Levy N (2004) What makes us moral? Crossing the boundaries of biology. Oneworld publications, Oxford
Levy D (2008) Love and sex with robots: the evolution of human-robot relationships. Harper Perennial, New York
Levy N (2010) The prospects for evolutionary ethics today. EurAmerica 40(3):529–571
Lewis CS (1943, 2001) Abolition of man. HarperOne, San Francisco
Lewis B (2003) The crisis of Islam: holy war and unholy terror. Modern Library, New York

Lewis CS (1952) Mere christianity. Barbour, Westwood
Lewis GJ, Bates TC (2010) Genetic evidence for multiple biological mechanisms underlying in-group favoritism. Psychol Sci 21(11):1623–1628
Lewis GJ, Kanai R, Bates TC et al (2012) Moral values are associated with individual differences in regional brain volume. J Cogn Neurosci 24(8):1657–1663
Lewis GJ, Kandler C, Riemann R (2014) Distinct heritable influences underpin in-group love and out-group derogation. Soc Psychol Person Sci 5(4):407–413
Lewis T, Amini F, Lannon R (2000) A General theory of love. Vintage, New York
Liddle JR, Machluf K, Shackelford TK (2010) Understanding suicide terrorism: premature dismissal of the religious-belief hypothesis. Evol Psychol 8(3):343–345
Lightcap JL, Kurland JA, Burgess RL (1982) Child abuse: a test of some predictions from evolutionary theory. Ethol Sociobiol 3(2):61–67
Lindsay RA (2008) Future bioethics: overcoming taboos, myths, and dogmas. Prometheus books, Amherst
Linquist S, Machery E, Griffiths PE et al (2011) Exploring the folkbiological conception of human nature. Philos Trans R Soc B: Biol Sci 366(1563):444–453
Littman DG (2003) Human rights and human wrongs. Sharia can't be an exception to international human-rights norms. http://www.nationalreview.com/articles/205577/human-rights-and-human-wrongs/david-g-littman
Liumbruno GM, Franchini M (2013) Beyond immunohaematology: the role of the ABO blood group in human diseases. Blood Transfus 11(4):491–499
Livesley WJ, Jang KL (2008) The behavioral genetics of personality disorder. Ann Rev Clin Psychol 4:247–274
Livi-Bacci M (1971) A century of Portuguese fertility. Princeton University Press, Princeton
Livingstone FB (1983) Do humans maximize their inclusive fitness? Behav Brain Sci 6(1):110–111
Lobaczewski AM, Knight-Jadczyk L (2007) Political ponerology (A science on the nature of evil adjusted for political purposes). Red Pill Press, Grande Prairie
Locke J (1690) Two treatises of government: in the former, the false principles and foundation of sir Robert Filmer, and his followers, are detected and overthrown. The latter is an essay concerning the true original, extent, and end of civil-government. Thomas Hollis, London
Lodewijckx E (2000) Vrijwillige sterilisatie in Vlaanderen. (Voluntary sterilization in Flanders). Bevolking en Gezin 29:59–89
Lodewijckx E (ed) (1988) Gezinsplanning in Vlaanderen. (Family planning in Flanders). C.B.G.S Monografie 1988/3, Centrum voor Bevolkings- en Gezinsstudie, Brussel
Lodewijckx E, Schoenmaeckers R (1994) Changes in fertility as a result of prior pregnancy-related experiences: an exploratory analysis. Population and family in the low countries, 30. NIDI/CBGS. Swets and Zeitlinger, Lisse
Loehlin JC, Nichols RC (1976) Heredity, environment, and personality: a study of 850 sets of twins. University of Texas Press, Austin
Logue B (1993) Last rights: death control and the elderly in America. New Lexington Press, New York
Longman P (2004) The empty cradle. Basic Books, New York
Lopreato J (1981) Toward a theory of genuine altruism in Homo sapiens. Ethol Sociobiol 2 (3):113–126
Lopreato J (1984) Human nature and biocultural evolution. Allen and Unwin, Winchester
Lopreato J (2001) Sociobiological theorizing: evolutionary sociology. In: Turner JH (ed) Handbook of sociological theory. Kluwer Academic/Plenum Publishers, New York, pp 405–433
Lopreato J, Crippen T (1999) Crisis in sociology: the need for Darwin. Transaction, London
Lopreato J, Yu M (1988) Human fertility and fitness optimization. Ethol Sociobiol 9(5):269–289
Lorenc T (2015) Artificial intelligence and the ethics of human extinction. J Conscious Stud 22(9–10):194–214

Lorenz RD, Lunine JI, McKay CP (1997) Titan under a red giant sun: a new kind of "habitable" moon. Geophys Res Lett 24(22):2905–2908
Louwe Kooijmans LP, Smirnov Y, Solecki RS et al (1989) On the evidence of Neanderthal burials. Curr Anthropol 30(3):322–330
Low BS (1990) Marriage systems and pathogen stress in human societies. Am Zool 30(2):325–340
Low BS (1993) An evolutionary perspective on war. In: Zimmerman W, Jacobson HK (eds) Behavior, culture and conflict in world politics. University of Michigan Press, An Arbor, pp 13–55
Low BS (1994) Men in demographic transition. Hum Nat 5(3):223–254
Low BS (1996) Behavioral ecology of conservation in traditional societies. Hum Nat 7(4):353–379
Low BS (2001) Why sex matters: a Darwinian look at human behavior. Princeton University Press, Princeton
Low A (2008) The origin of human nature: a Zen Buddhist looks at evolution. Sussex Academic Press, Eastbourne East Sussex
Low BS, Simon CP, Anderson KG (2002) An evolutionary ecological perspective on demographic transitions: modelling multiple currencies. Am J Hum Biol 14(2):149–167
Lowell CS (1967) The hidden wealth of the roman catholic church. Protestants and Other Americans United for Separation of Church and State, Washington
Loye D (1999) Can science help construct a new global ethic? The development and implications of moral transformation theory. Zygon 34(2):221–235
Lozada M, D'Adamo P, Angel Fuentes M (2011) Beneficial effects of human altruism. J Theoret Biol 289:12–16
Ludlow P (1998) Europe and the Mediterranean. Brassey's, London
Lüke U, Schnakenberg J, Souvignier G (2004) Darwin und Gott. Wissenschaftliche Buchgesellschaft, Darmstadt
Lukes S (1971) The meanings of "Individualism". J Hist Ideas 32(1):45–66
Lumsden CJ, Wilson EO (1981) Genes, mind and culture: the coevolutionary process. Harvard University Press, Cambridge
Luo Q, Nakic M, Wheatley T et al (2006) The neural basis of implicit moral attitude—an IAT study using event-related fMRI. Neuroimage 30(4):1449–1457
Lutz W (1987) Culture, religion, and fertility: a global view. Genus 43(3–4):15–35
Lutz W (2006) Fertility rates and future population trends: will Europe's birth rate recover or continue to decline? Int J Androl 29(1):25–33
Lutz W, Sanderson W, Scherbov S (2001) The end of world population growth. Nature 412 (6846):543–546
Luxemburg R (1918) The Russian revolution. Paul Levi, Berlin
Lykken DT (1995) The antisocial personalities. Erlbaum, Hillsdale
Lynch M, Walsh B (1998) Genetics and analysis of quantitaive traits. Sinauer Associates, Sunderland MA
Lynn R (2001) Eugenics. A reassessment. Praeger, Westport
Lynn R, Harvey J (2008) The decline of the world's IQ. Intelligence 36(2):112–120
Lynn R, Harvey J, Nyborg H (2009) Average intelligence predicts atheism rates across 137 nations. Intelligence 37(1):1–15
Lynn R, Van Court M (2004) New evidence of dysgenic fertility for intelligence in the United States. Intelligence 32(2):193–201
Lynn R, Vanhanen T (2002) IQ and the wealth of nations. Praeger, Westport
Lynn R, Vanhanen T (2006) IQ and global inequality. Washington Summit Books, Augusta
Lyon J, Corner P (1995) Altered fates: gene therapy and the retooling of human life. Norton, New York
Lyon BE, Montgomerie R (2012) Sexual selection is a form of social selection. Philos Trans R Soc B 367(1600):2266–2273

Lyons MN, Berlet C (1996) Too close for comfort: right wing populism, scapegoating, and fascist potentials in US politics. South End Press, Boston
Maccoby E, Jacklin C (1974) The psychology of sex differences. Stanford University Press, Palo Alto
MacDonald K (1990) Mechanisms of sexual egalitarianism in Western Europe. Ethol Sociobiol 11:195–238
MacDonald K (1995) Evolution, the 5-factor model, and levels of personality. J Person 63(3):525–567
MacDonald K (1995) The establishment and maintenance of socially imposed monogamy in Western Europe. Polit Life Sci 14:2–23
MacDonald K (1999) An evolutionary perspective on human fertility. Popul Environ 21(2):223–246
Mace R (2000) An adaptive model of human reproductive rate where wealth is inherited: why people have small families. In: Cronk L, Chagnon N, Irons W (eds) Adaptation and human behavior: an anthropological perspective. Aldine de Gruyter, New York, pp 261–281
Machalek R, Martin MW (2004) Sociology and the second Darwinian revolution: a metatheoretical analysis. Sociol Theory 22(3):455–476
Machery E, Mallon R (2010) Evolution of morality. In: Doris JM et al (eds) The moral psychology handbook. Oxford University Press, New York, pp 3–47
Macintosh KL (2012) Human cloning: four fallacies and their legal consequences. Cambridge University Press, Cambridge
MacIntyre F (2004) Was religion a kinship surrogate? J Am Acad Relig 72(3):653–694
MacIntyre F, Estep KW (1993) Sperm competition and the persistence of genes for male homosexuality. Biosystems 31(2–3):223–233
McKay R, Whitehouse H (2015) Religion and morality. Psychol Bull 141(2):447–473
MacKellar C, Bechtel C (eds) (2014) The ethics of the new eugenics. Berghahn, New York
MacKenzie D (2008) Will a pandemic bring down civilization? New Sci 2650:28–31
MacKenzie D (2008) Why the demise of civilisation may be inevitable. New Sci 2650:32–35
MacKinnon CA (1986) Pornography: not a moral issue. Women's Study Int Forum 9:63–78
MacKinnon CA (1993) Prostitution and civil rights. Mich J Gender Law 1:13–31
MacKinnon CA, Dworkin A (1998) In harm's way: the pornography civil rights hearings. Harvard University Press, Cambridge
MacKinnon B (2001) Human cloning: science, ethics and public policy. Cambridge University Press, Cambridge
Mackintosh NJ (2007) Book review. Race differences in intelligence: an evolutionary hypothesis. Intelligence 35(1):94–96
Macklon NS, Geraerdts JPM, Fauser BCJM (2002) Conception to ongoing pregnancy: the black box of early pregnancy loss. Hum Reprod Update 8(4):333–343
Macmillan J, Kofoed L (1984) Sociobiology and antisocial personality. J Nerv Mental Dis 172 (12):701–706
Maiello C (2007) Messung und Korrelate von Religiosität: Beziehungen zwischen Glaubensintensität und psychologisch, pädagogisch, soziologisch sowie medizinisch relevanten Variablen. (Measurement and correlates of religiosity: relations between faith intensity and psychological, educational, sociological, and medically relevant variables). Waxmann, Münster
Maienschein J, Ruse M (eds) (1999) Biology and the foundation of ethics. Cambridge University Press, New York
Makrushin AV (2016) On the expedience and prospects of immortality. Adv Gerontol 6(1):21–23
Malalasekera GP, Jayatilleke KN (2010) Buddhist studies: caste system. Buddhanet—Worldwide Buddhist Information and Education Network, Tullera
Malamuth NM (1980) Testing hypotheses regarding rape: exposure to sexual violence, sex differences, and the "normality" of rapists. J Res Person 14(1):121–137

Malamuth NM (1996) The confluence model of sexual aggression: feminist and evolutionary perspectives. In: Buss DM, Malamuth N (eds) Sex, power, conflict: evolutionary and feminist perspectives. Oxford University Press, New York, pp 269–295
Malamuth NM, Addison T, Koss M (2000) Pornography and sexual aggression: are there reliable effects and can we understand them? Ann Rev Sex Res 11:26–91
Malherbe JF (1981) Ethiek en menselijke erfelijkheid. (Ethics and human genetics). Streven, November 1981
Malinovsky B (1960) A scientific theory of culture and other essays. Oxford University Press, New York
Mameli M (2013) Meat made us moral: a hypothesis on the nature and evolution of moral judgment. Biol Philos 28(6):903–931
Mameli M, Bateson P (2006) Innateness and the sciences. Biol Philos 21(2):155–188
Manenschijn G (1987) Evolution and ethics. In: Anderson S, Peacocke A (eds) Evolution and creation: a European perspective. Aarhus University, Aarhus, pp 85–103
Manji I (2003) The trouble with Islam. St. Martin's, New York
Manner M, Gowdy J (2010) The evolution of social and moral behavior: evolutionary insights for public policy. Ecol Econ 69(4):753–761
Manson NA (ed) (2003) God and design: the teleological argument and modern science. Routledge, London
Marazziti D, Baroni S, Landi P et al (2013) The neurobiology of moral sense: facts or hypotheses? Ann Gen Psychiatry 12(6):https://doi.org/10.1186/1744-859x-12-6
March F (2009) The evolution of the mind and what it means to humanism. Humanist 69(1):16–20
Mardorossian CM (2004) Toward a new feminist theory of rape. Gend Stud 03(2004):243–275
Marean CW (2015) How Homo sapiens became the ultimate invasive species. Sci Am 313(2):32–39
Margulis L (1998) Human population growth in the gaian politic: the cancer analogy. American Anthropological Association, Washington
Marlowe F (2000) Paternal investment and the human mating system. Behav Process 51(1–3):45–61
Marlowe FW (2003) The mating system of foragers in the standard cross-cultural sample. Cross-Cult Res 37(3):282–306
Marlowe FW, Berbesque JC, Barr A et al (2008) More 'altruistic' punishment in larger societies. Proc R Soc B 275(1634):587–592
Marlowe FW, Berbesque JC, Barrett C et al (2011) The 'spiteful' origins of human cooperation. Proc R Soc B-Biol Sci 278(1715):2159–2164
Marmor J (ed) (1980) Homosexual behavior. A modern reappraisal. Basic Books, New York
Marsh AA, Stoycos SA, Brethel-Haurwitz KM et al (2014) Neural and cognitive characteristics of extraordinary altruists. PNAS 111(42):15036–15041
Marshall G, Swift A, Routh D et al (1999) What is and what ought to be popular beliefs about distributive justice in thirteen countries. Eur Sociol Rev 15(4):349–367
Martin D (2005) On secularization: towards a revised general theory. Ashgate, Aldershot
Martin LH (2008) Can religion really evolve? In: Bulbulia J, Sosis R, Genet C et al (eds) The evolution of religion: studies, theories, and critiques. Collins Foundation Press, Santa Margarita, pp 349–364
Martin M, Monnier R (2003) The impossibility of God. Prometheus Books, Amherst
Marx K (1844) Zur Kritik der Hegelschen Rechtsphilosophie (Towards a critique of Hegel's philosophy of right). Deutsch-Französische Jahrbücher, Februar
Marx K (1867) Das Kapital. Otto Meissner, Hamburg
Marx K (1875), Kritik des gothaer programms. (Critique of the Gotha Programme). Die Neue Zeit 18(1):1890–1891
Marx K, Engels F (1848) Manifest der Kommmunistischen Partei (Manifesto of the Communist Party). Burghard, London
Marx K, Engels F (1958) Über religion. Akademie Verlag, Berlin

Maryanski A (1998) Evolutionary sociology. Adv Hum Ecol 7:1–56
Maryanski A, Turner JH (1992) The social cage: human nature and the evolution of society. Stanford University Press, Stanford CA
Mascaro S, Korb KB, Nicholson AE (2001) Suicide as an evolutionarily stable strategy. In: Proceedings of the 6th European conference on advances in artificial life, 10–14 Sept 2001, pp 120–130
Masci D (2009) Darwin debated: religion vs. evolution. Pew Research Center's Forum on Religion and Public Life, Washington
Mascie-Taylor CG, Boyce AJ (eds) (1988) Human mating patterns. Cambridge University Press, Cambridge
Mashiah S, Ben-Rafael Z, Laufer N et al (1990) Advances in assisted reproductive technologies. Plenum Press, New York
Maslow A (1954) Motivation and personality. Harper, New York
Maslow A (1972) The farther reaches of human nature. Penguin, New York
Maslow A (1999) Toward a psychology of being. Wiley and Sons, New York
Mason DA, Frick PJ (1994) The heritability of antisocial behavior: a meta-analysis of twin and adoption studies. J Psychopathol Behav Assess 16(4):301–323
Masters R (1989) The nature of politics. Yale University Press, New Haven
Masuda N (2012) Ingroup favoritism and intergroup cooperation under indirect reciprocity based on group reputation. J Theor Biol 311:8–18
Mataré HF (1999) Bioethics: the ethics of evolution and genetic interference. Bergin and Garvey, Westport
Matejcek Z (2003) Is ADHD adaptive or non-adaptive behavoir? Neuroendocrinol Lett 24:148–150
Matheny JG (2007) Reducing the risk of human extinction. Risk Anal 27(5):1335–1344
Mather K, Jinks JL (1971) Biometrical genetics. Chapman and Hall, London
Mathew S, Boyd R (2011) Punishment sustains large-scale cooperation in prestate warfare. PNAS 108(28):11375–11380
Mathur VA, Harada T, Lipke T et al (2010) Neural basis of extraordinary empathy and altruistic motivation. Neuroimage 51(4):1468–1475
Matsumura M (ed) (1995) Voices for evolution. National Center for Science Education, Berkeley
May L, Clark A (1996) Mind and morals: essays on ethics and cognitive science. The MIT Press. A Bradford Book, Cambridge
May H, Striegnitz M, Hefner P (eds) (1989) Kooperation und Wettbewerb: zu Ethik und Biologie menschlichen Sozialverhaltens. (Cooperation and competition: ethics and the biology of human social behaviour). Evangelische Akademie Loccum, Rehburg-Loccum
Maynard Smith J (1964) Group selection and kin selection. Nature 201:1145–1146
Maynard Smith J (1972) On evolution. Edinburgh University Press, Edinburgh
Maynard Smith J (1978) The evolution of sex. Cambridge University Press, Cambridge
Maynard Smith J (1982) Evolution and the theory of games. Cambridge University Press, Cambridge
Maynard Smith J (1984) Game theory and the evolution of behaviour. Behav Brain Sci 7:95–125
Maynard Smith J (1989) Evolutionary genetics. Oxford University Press, Oxford
Maynard Smith J, Price GR (1973) The logic of animal conflict. Nature 246(5427):15–18
Maynard Smith J, Szathmáry E (1995) The major transitions in evolution. Freeman Press, Oxford
Maynard Smith J, Szathmáry E (1999) The origins of life: from the birth of life to the origin of language. Oxford University Press, Oxford
Mayr E (1942) Systematics and the origin of species. Columbia University Press, New York
Mayr E (1978) The nature of the Darwinian revolution. In: Washburn S, McCown R (eds) Human evolution. Biosocial perspectives. The Benjamin/Cummings Publ. Co., Menlo Park, pp 11–31
Mayr E (1988) Toward a new philosophy of biology. Belknap Press, Cambridge
Mazur A (2004) Biosociology of dominance and deference. Rowman and Littlefield, New York

McAllister P (2009) Manthropology: the science of the inadequate modern male. St. Martin's Press, New York
McAnulty RD, Burnette MM (2006) Sex and sexuality: sexual deviation and sexual offenses. Praeger, Westport
McCabe LL, McCabe ERB (2011) Down syndrome: coercion and eugenics. Genet Med 13 (8):708–710
McCall G, Shields N (2008) Examining the evidence from small-scale societies and early prehistory and implications for modern theories of aggression and violence. Aggression Violent Behav 13(1):1–9
McCauley RN (2000) The naturalness of religion and the unnaturalness of science. In: Keil FC, Wilson RA (eds) Explanation and cognition. Cambridge University Press, Cambridge, pp 61–85
McClelland RT (2011) A naturalistic view of human dignity. J Mind Behav 32(1):5–48
McClenon J (2002) Wondrous healing: Shamanism, human evolution, and the origin of religion. Northern Illinois University Press, DeKalb
McCloskey DN (2016) Bourgeois equality. How ideas, not capital or institutions, enriched the world. University of Chicago University Press, Chicago
McCrae RR (1996) Social consequences of experiential openness. Psychol Bull 120(3):322–337
McCullough ME, Tsang J, Brion S (2003) Personality traits in adolescence as predictors of religiousness in early adulthood: findings from the Terman longitudinal study. Person Soc Psychol Bull 29(8):980–991
McDermott R, Tingley D, Cowden J et al (2009) Monoamine oxidase a gene (MAOA) predicts behavioral aggression following provocation. PNAS 106(7):2118–2123
McDonald P (2000) Gender equity, social institutions, and the future of fertility. J Popul Res 17 (1):1–16
McDougall I, Brown FH, Fleagle JG (2005) Stratigraphic placement and age of modern humans from Kibish, Ethiopia. Nature 433(7027):733–736
McElreath R, Boyd R (2007) Mathematical models of social evolution: a guide for the perplexed. University of Chicago Press, Chicago
McElroy W (2002) Liberty for women: freedom and feminism in the 21st century. Ivan R, Dee, Chicago
McFague S (1993) The body of God: an ecological theology. Fortress Press, Minneapolis
McFaul TR (1978) Biomedical issues for the future. In: Fowles J (ed) Handbook of futures research. Greenwood Press, Westport, pp 653–674
McGrath AE (2004) Dawkins' God. Genes, memes and the meaning of life. Blackwell, Malden
McGrath B (2007) The new atheist crusaders and their unholy grail: the misguided quest to destroy your faith. Thomas Nelson, Nashville
McGrath AE (2010) The ideological uses of evolutionary biology in recent atheist apologetics. In: Alexander DE, Numbers RL (eds) Biology and ideology from Descartes to Dawkins. University of Chicago Press, Chicago, pp 329–351
McGrath AE, McGrath J (2007) The Dawkins delusion: atheist fundamentalism and the denial of the divine. InterVarsity Press, Downers Grove
McGuire M (2010) God's brain. Prometheus Books, Amherst
McGuire M, Gruter M (2003) Prostitution—an evolutionary perspective. In: Somit A, Peterson SA (eds) Human nature and public policy: an evolutionary approach. Palgrave Macmillan, New York, pp 29–40
McGuire MT, Marks I, Nesse R (1997) Evolutionary biology: a basic science for psychiatry? In: Baron-Cohen S (ed) The maladapted mind: classic readings in evolutionary psychopathology. Psychology Press, Hove, pp 23–38
McGuire MT, Tiger L (2009) The brain and religious adaptations. In: Feierman JR (ed) The biology of religious behavior: the evolutionary origins of faith and religion. ABC-CLIO, LLC, Santa Barbara, pp 125–140
McGuire MT, Troisi A (1998) Darwinian psychiatry. Oxford University Press, New York

McIntosh D (2008) Human, transhuman, posthuman: Implications of evolution-by-design for human security. J Hum Secur 4(3):4–20
McKenzie Alexander J (2007) The structural evolution of morality. Cambridge University Press, Cambridge
McKibben B (2003) Enough: genetic engineering and the end of human nature. Bloomsbury, London
McKinney LO (1994) Neurotheology. American Institute of Mindfulness, Cambridge
McKnight J (1997) Straight science? Homosexuality, evolution, and adaptation. Routledge, New York
McKown DB (1993) The mythmaker's magic: behind the illusion of "creation science". Prometheus Books, Buffalo
McLaren A (1990) A history of contraception. From antiquity to the present day. Basil Blackwell, Oxford
McNair B (1996) Mediated sex: pornography and postmodern culture. Arnold, London
McNally L, Brown SP, Jackson AL (2012) Cooperation and the evolution of intelligence. Proc R Soc B-Biol Sci 279(1740):3027–3034
McNamara P (2006) The frontal lobes and the evolution of cooperation and religion. In: McNamara P (ed) Where God and science meet: how brain and evolutionary studies alter our understanding of religion, vol 2. The neurology of religious experience. Praeger, Westport, pp 189–204
McNamara P (ed) (2006) Where God and science meet: how brain and evolutionary studies alter our understanding of religion. Volume 1: Evolution, genes and the religious brain. Volume 2: The neurology of religious experience. Volume 3: The psychology of religious experience. Praeger, Westport
McNamara P (2009) The neuroscience of religious experience. Cambridge University Press, Cambridge
McNamara P, Durso R, Brown A et al (2006) The chemistry of religiosity: evidence from patients with Parkinson's disease. In: McNamara P (ed) Where God and science meet: how brain and evolutionary studies alter our understanding of religion, vol 2. The neurology of religious experience. Praeger, Westport, pp 1–14
McQuaig L, Brooks N (2013) Billionaires' ball: Gluttony and hubris in an age of epic inequality. Beacon Press, Boston
McQuillan K (2004) When does religion influence fertility? Popul Dev Rev 30(1):25–56
McRobbie A (2004) Post-feminism and popular culture. Feminist Media Stud 4(3):255–264
McShea RJ (1990) Morality and human nature: a new route to ethical theory. Temple University Press, Philadelphia
McShea RJ, McShea D (1999) Biology and value theory. In: Maienschein J, Ruse M (eds) Biology and the foundation of ethics. Cambridge University, New York, pp 307–328
Mead M (1928) Coming of age in Samoa. Harper Collins, New York
Meadows DH, Randers J, Meadows DL (2004) Limits to growth: the 30-year update. Earthscan, London
Mealey L (1995) The sociobiology of sociopathy—an integrated evolutionary model. Behav Brain Sci 18(3):523–541
Medawar PB (1952) An unsolved problem of biology. HK Lewis, London
Mednick S, Christiansen KO (1964, 1977) Biosocial bases of criminal behavior. Gardner Press, New York
Mednick S, Gabrielli WF, Hutchins B (1984) Genetic influences in criminal convictions: evidence from an adoption cohort. Science 224(4651):891–894
Medora NP (2007) Strengths and challenges in the Indian family. Marriage Family Rev 41(1–2):165–193
Medvedev Z (1969) The rise and fall of T.D. Lysenko. Columbia University Press, New York
Meeks TW, Jeste DV (2009) Neurobiology of wisdom: a literature overview. Arch Gen Psychiatry 66(4):355–365

Meetoo V, Mirza HS (2007) There is nothing 'honourable' about honour killings: gender, violence and the limits of multiculturalism. Women's Stud Int Forum 30(3):187–200
Mehlman MJ (2003) Wondergenes. Genetic enhancement and the future of society. Indiana University Press, Bloomington
Meisenberg G (2006) Challenging the belief in a (genetically) just world. J Biosoc Sci 38:844–845
Meisenberg G (2011) Secularization and desecularization in our time. J Soc Polit Econ Stud 36 (3):318–359
Mekel-Bobrov N, Gilbert SL, Evans PD et al (2005) Ongoing adaptive evolution of ASPM, a brain size determinant in Homo sapiens. Science 309(5741):1720–1722
Melis AP, Semmann D (2010) How is human cooperation different? Philos Trans R Soc B-Biol Sci 365(1553):2663–2674
Mellen SLW (1981) The evolution of love. W.H. Freeman, Oxford
Meloni M (2013) Moralizing biology: the appeal and limits of the new compassionate view of nature. Hist Hum Sci 26(3):82–106
Meloni M (2016) Political biology. Science and social values in human heredity from eugenics to epigenetics. Palgrave Macmillan, London
Melotti U (1980) Towards a new theory on the origin of the family: some hypotheses on monogamy, polygyny, incest taboo, exogamy, and genetic altruism. Mankind Q XXI 2:99–134
Mendel GJ (1865) Versuche über Pflanzenhybriden. (Experiments on plant hybridization). Verh des Naturforschenden Ver Brünn IV 3:3–47
Mendez A (2009) Planetary habitability of the solar system. AAS/Division for planetary sciences meeting. http://adsabs.harvard.edu/abs/2009DPS....41.0813M
Mendez MF (2009) The neurobiology of moral behavior: review and neuropsychiatric implications. CNS Spectums 14(11):608–620
Mendez MF, Chen AK, Shapira JS et al (2005) Acquired sociopathy and frontotemporal dementia. Dement Geriatr Cogn Dis 20(2–3):99–104
Menelik GYI (2010) Europe: the future battleground of Islamic terrorism. GRIN Verlag, München
Mengzi. The book of Mengzi. (http://nothingistic.org/library/mencius/mencius01.html)
Mercadillo RE, Díaz JL, Pasaye EH et al (2011) Perception of suffering and compassion experience: brain gender disparities. Brain Cogn 76(1):5–14
Meredith HV (1974) Findings from Asia, Australia, Europe, and North America on secular change in mean height of children, and young adults. Am J Phys Anthropol 44(2):315–325
Mesoudi A (2011) Cultural evolution: how Darwinian theory can explain human culture and synthesize the social sciences. University of Chicago Press, Chicago
Mesoudi A (2016) Cultural evolution: integrating psychology, evolution and culture. Curr Opin Psychol 7:17–22
Metchnikoff E (1907) The prolongation of life. Optimistic studies. Putnam's Sons, New York
Micklethwait J, Wooldridge A (2009) God is Back. How the global revival of faith is changing the world. The Penguin Press, New York
Midgley M (1985) Evolution as a religion: strange hopes and stranger fears. Routledge, London
Midgley M (1994) The ethical primate. Humans, freedom and morality. Routledge, London
Mies M, Shiva V (1993) Ecofeminism. Zed Books, London
Mifune N, Hashimoto H, Yamagishi T (2010) Altruism toward in-group members as a reputation mechanism. Evol Hum Behav 31(2):109–117
Mikhail J (2007) Universal moral grammar: theory, evidence and the future. Trends Cogn Sci 11 (4):143–152
Mikolajczak M, Gross JJ, Lane A et al (2010) Oxytocin makes people trusting, not gullible. Psychol Sci 21(8):1072–1074
Milgram S (1974) Obedience to authority: an experimental view. Harper and Row, New York
Milinski M, Semmann D, Krambeck H et al (2006) Stabilizing the Earth's climate is not a losing game: supporting evidence from public goods experiments. PNAS 103(11):3994–3998
Mill JS (1863/1957) Utilitarianism. Bobbs-Merrill, Indianapolis

Millennium Ecosystem Assessment (MEA) (2005) Ecosystems and human well-being: synthesis. Island Press, Washington

Miller N (1992) Single parents by choice. A growing trend in family life. Plenum Press, New York

Miller G (1998) How mate choice shaped human nature: a review of sexual selection and human evolution. In: Crawford C, Krebs D (eds) Handbook of evolutionary psychology: ideas, issues, and applications. Lawrence Erlbaum, New York, pp 87–130

Miller KR (1999) Finding Darwin's God: a scientist's search for common ground between God and evolution. Harper Collins, New York

Miller G (2000) The mating mind: how sexual choice shaped the evolution of human nature. Doubleday, New York

Miller G (2007) Sexual selection for moral virtues. Q Rev Biol 82(2):97–125

Miller G (2008) Kindness, fidelity, and other sexually selected virtues. In: Sinnott-Armstrong W (ed) Moral psychology, vol 1. The evolution of morality: adaptations and innateness. MIT Press, Cambridge, pp 209–243

Miller G (2008) Neurobiology. The roots of morality. Science 320(5877):734–737

Miller A, Hoffmann J (1995) Risk and religion: an explanation of gender differences in religiosity. J Sci Study Relig 34(1):63–75

Miller AS, Stark R (2002) Gender and religiousness: can socialization explanations be saved? Am J Sociol 107(6):1399–1423

Miller JD, Scott EC, Okamoto S (2006) Public acceptance of evolution. Science 313(5788):765–766

Miller L, Bansal R, Wickramaratne P et al (2014) Neuroanatomical correlates of religiosity and spirituality a study in adults at high and low familial risk for depression. JAMA Psychiatry 71 (2):128–135

Miller P, Wilsdon J (eds) (2006) Better humans. The politics of human enhancement and life extension. Demos, London

Miller WR, Thoresen CE (2003) Spirituality, religion, and health: an emerging research field. Am Psychol 58(1):24–35

Millhouse T, Bush LS, Moss D (2015) The containment problem and the evolutionary debunking of morality. In: Shackelford TK, Hansen RD (eds) The evolution of morality. Springer, New York, pp 113–135

Millon T, Simonsen E, Davis RD et al (2002) Psychopathy: antisocial, criminal, and violent behavior. The Guilford Press, New York

Milner LS (2000) Hardness of heart/hardness of life: the stain of human infanticide. University Press of America, Lanham

Minichiello V, Browne J, Kendig H (2000) Perceptions and consequences of ageism: views of older people. Ageing Soc 20(3):253–278

Mirola WA (1999) A refuge for some: gender differences in the relationship between religious involvement and depression. Sociol Relig 60(4):419–437

Mishra RC (2006) Women in India: towards gender equality. Authorspress, New Delhi

Mix LJ, Masel J (2014) Chance, purpose, and progress in evolution and Christianity. Evolution 68 (8):2441–2451

Mizzoni J (2002) Ruse's Darwinian ethics and moral realism. The global spiral, http://www.metanexus.net/magazine/tabid/68/id/6008/Default.aspx

Modelski G, Devezas T, Thompson WR (2008) Globalization as evolutionary process. Modeling global change. Routledge, London

Modleski T (1991) Feminism without women: culture and criticism in a "postfeminist" age. Routledge, New York

Moffitt T (1993) Adolescence-limited and life-course-persistent antisocial behavior: a developmental taxonomy. Psychol Rev 100(4):674–701

Moir A, Jessel D (1992) Brain sex. The real difference between men and women. Dell Publishing, New York

Moir A, Jessel D (1995) A mind to crime. The controversial link between the mind and criminal behaviour. Penguin Books, London
Mokyr J (1990) The lever of riches. Oxford University Press, New York
Mol H (1985) The faith of Australians. George, Allen, and Unwin, Sydney
Moll J, de Oliveira-Souza R, Bramati IE et al (2002) Functional networks in emotional moral and nonmoral social judgments. Neuroimage 16(3 Pt 1):696–703
Moll J, de Oliveira-Souza R, Eslinger PJ et al (2002) The neural correlates of moral sensitivity: a functional magnetic resonance imaging investigation of basic and moral emotions. J Neurosci 22(7):2730–2736
Moll J, de Oliveira-Souza R, Eslinger P (2003) Morals and the brain; a working model. NeuroReport 14(3):299–305
Moll J, de Oliveira-Souza R, Zahn R (2008) The neural basis of moral cognition: sentiments, concepts, and values. Ann New York Acad Sci 1124(1):161–180
Moll J, de Oliveira-Souza R, Moll FT et al (2005) The moral affiliations of disgust: a functional MRI study. Cogn Behav Neurol 18(1):68–78
Moll H, Tomasello M (2007) Cooperation and human cognition: the Vygotskian intelligence hypothesis. Philos Trans R Soc B 362(1480):639–648
Moll J, Krueger F, Zahn R et al (2006) Human fronto-mesolimbic networks guide decisions about charitable donation. PNAS 103(42):15623–15628
Moll J, Schulkin J (2009) Social attachment and aversion in human moral cognition. Neurosci Biobehav Rev 33(3):456–465
Moll J, Zahn R, de Oliveira-Souza R et al (2005) Opinion: the neural basis of human moral cognition. Nat Rev Neurosci 6(10):799–809
Molleman L, Quinones AE, Weissing FJ (2013) Cultural evolution of cooperation: the interplay between forms of social learning and group selection. Evol Hum Behav 34(5):342–349
Monod J (1971) Le hazard et la néccesité. Essai sur la philosophie naturelle de la biologie moderne. (Hazard and neccesity. Essay on natural philosophy of modern biology). Editions du Seuil, Paris
Montagu A (1950) A consideration of the concept of race. Cold Spring Harbor Symp Quant Biol 15:315–336
Montagu A (1955) The direction of human development. Biological and social bases. Harper, New York
Moore GE (1903) Principia ethica. Cambridge University Press, Cambridge
Moore JA (2002) From Genesis to genetics. The case of evolution and creationism. University of California Press, Berkeley
Moore TM, Scarpa A, Raine A (2002) A meta-analysis of serotonin metabolite 5-HIAA and antisocial behavior. Aggressive Behav 28(4):299–316
Moorjani P, Thangaraj K, Patterson N et al (2013) Genetic evidence for recent population mixture in India. Am J Hum Genet 93(3):422–438
Moravec H (1988) Mind children. Harvard University Press, Cambridge
Moravec H (2000) Robot: mere machine to transcendent mind. Oxford University Press, New York
More M (1990) Transhumanism: toward a futurist philosophy. Extropy 6:6–12
More M (2010) The overhuman in the transhuman. J Evol Technol 21(1):1–4
More M, Vita-More N (2013) The transhumanist reader: classical and contemporary essays on the science, technology, and philosophy of the human. Wiley-Blackwell, Hoboken
Morgan TH (1903) Evolution and adaptation. Macmillan, New York
Morgan R (1980) Theory and practice: pornography and rape. In: Lederer L (ed) Take back the night: women on pornography. William Morrow, New York, pp 134–140
Morgan F (2002) Fascism in Europe, 1919–1945. Routledge, London
Morgan SP (2003) Is low fertility a twenty-first-century demographic crisis? Demography 40 (4):589–603

Moritz JM (2012) Human uniqueness, the other hominids, and "anthropocentrism of the gaps" in the religion and science dialogue. Zygon 47(1):65–96

Morris HM (ed) (1974, 1985) Scientific creationism. Master Books, Green Forest

Morris HM (1984) A history of modern creationism. Master Book Publishers, San Diego

Morris J (2007) The young earth: the real history of the earth—past, present, and future. Master Books, San Diego

Mosher WD, Bachrach CA (1988) Contraceptive use, United States, 1982. Vital and health statistics, Series 23, N° 12. DHHS Publication N° (PHS) 86-1988

Mosher WD, Westoff C (1982) Trends in contraceptive practice, United States: 1965–1976. Vital and health statistics, Series 23, N° 10

Motesharrei S, Rivas J, Kalnay E (2014) Human and nature dynamics (HANDY): modeling inequality and use of resources in the collapse or sustainability of societies. Ecol Econ 101:90–102

Moynihan DP, Smeeding TM, Rainwater L (eds) (2005) The future of the family. Russell Sage Foundation Publications, New York

Mueller E, Short K (1983) Effects of income and wealth on the demand of children. In: Bulatao RA, Lee RD (eds) Determinants of fertility in developing countries. Academic press, New York, pp 590–642

Mulder E, Milo T (2009) De omstreden bronnen van de Islam. (The controversial sources of Islam). Meinema, Zoetermeer

Mulder RT, Wells JE, Joyce PR et al (1994) Antisocial women. J Pers Dis 8(4):279–287

Mulhall D (2002) Our molecular future: how nanotechnology, robotics, genetics, and artificial intelligence will transform our world. Prometheus Books, Amherst

Mullatti L (1995) Families in India: beliefs and realities. J Comp Family Stud 26(1):11–25

Muller HJ (1934) The dominance of economics over eugenics. In: Perkins H et al (eds) A decade of progress in eugenics. Scientific papers third international congress of eugenics. The Williams and Wilkins Co., Baltimore, pp 138–144

Muller HJ (1935) Out of the night. A biologist's view of the future. The Vanguard press, New York

Muller HJ (1950) Our load of mutations. Am J Hum Genet 2:111–176

Muller HJ (1958) Human values in relation to evolution. Science 127(3299):625–629

Muller HJ (1960) The guidance of human evolution. Perspect Biol Med 3(1):1–43

Müller-Hill B (1984) Tötliche Wissenschaft. (Deadly science). Rowohlt Taschenbuch Verlag, Reinbeck bei Hamburg

Munro D (2005) A Chinese ethics for the new century. The Chinese University Press, Hong Kong

Murdock GP (1961) Outline of cultural material (OCM). HRAF, New Haven

Murdock DM (2007) Who was Jesus? Fingerprints of the Christ. Stellar House Publishing, Seattle

Murphy JG (1982) Evolution, morality, and the meaning of life. Rowman and Littlefield, Totowa

Murphy TF (2014) Genetic modifications for personal enhancement: a defence. J Med Ethics 40 (4):242–245

Murphy TF (2014) In defense of prenatal genetic interventions. Bioethics 28(7):335–342

Murray MJ (2008) Belief in God: a trick of our brain? In: Copan P, Craig WL (eds) Christianity's critics: answering the new atheists and other objectors. Broadman and Homan, Nashville

Muscarella F, Fink B, Grammer K et al (2001) Homosexual orientation in males: evolutionary and ethological aspects. Neuroendocrinol Lett 22(6):393–400

Mussweiler T, Ockenfels A (2013) Similarity increases altruistic punishment in humans. PNAS 110(48):19318–19323

Myers N (1997) The population/environment predicament: even more urgent than supposed. Polit Life Sci 16(2):211–213

Myers PZ (2010) Evolution is far freer than classical liberalism. http://www.cato-unbound.org/archives/july-2010-darwin-and-politics/

Mysterud I (2000) Group selection, morality, and environmental problems. J Conscious Stud 7(1–2):225–227

Na C (2016) Why is Confucianism not a religion? The impact of orientalism. Zygon 51(1):21–42
Naam R (2005) More than human: embracing the promise of biological enhancement. Broadway Books, New York
Nachman MW, Crowell SL (2000) Estimate of the mutation rate per nucleotide in humans. Genetics 156(1):297–304
Nachmani A (2009) Europe and its Muslim minorities: aspects of conflict, attempts at accord. Sussex Academic Press, Brighton
Naess A (1973) The shallow and the deep, long-range ecological movement. Inquiry 16(1–4):95–100
Nagle J (ed) (2001) Whores and other feminists. Routledge, New York
Nakamaru M, Iwasa Y (2007) The coevolution of altruism and punishment: role of the selfish punisher. J Theor Biol 245(1):193–196
Nakamura M, Masuda N (2012) Groupwise information sharing promotes ingroup favoritism in indirect reciprocity. BMC Evol Biol 12:213. https://doi.org/10.1186/1471-2148-12-213
Narvaez D (2008) Triune ethics: the neurobiological roots of our multiple moralities. New Ideas Psychol 26:95–119
Narvaez D (2010) The emotional foundations of high moral intelligence. New Dir Child Adolesc Dev 129:77–94
Narvaez D (2014) Neurobiology and the development of human morality. Evolution, culture, and wisdom. Norton, New York
Narvaez D, Lapsley DK (2009) Moral personality: themes, questions, futures. In: Narvaez D, Lapsley DK (eds) Personality, identity, and character: exploration in moral psychology. Cambridge University Press, New York, pp 441–448
Narvaez D, Vaydich JL (2008) Moral development and behaviour under the spotlight of the neurobiological sciences. J Moral Educ 37(3):289–312
Nasr SH (1989) Knowledge and the sacred. State University of New York Press, Albany
Nasr SH (1994) A young Muslim's guide to the modern world. KAZI Publications, Chicago
National Academy of Sciences (1998) Teaching about evolution and the nature of science. National Academy Press, Washington
National Academy of Sciences (1999) Science and creationism. National Academy Press, Washington
Neary D, Snowden J, Gustafson L et al (1998) Frontotemporal lobar degeneration: a consensus on clinical diagnostic criteria. Neurology 51(6):1546–1552
Neel JV (1970) Lessons from a "primitive" people. Science 1970(3960):815–822
Nelson-Pallmeyer J (2003) Is religion killing us? Violence in the Bible and the Quran. Trinity Press international, Harrisburg
Nesse R (1994) Why is group selection such a problem? Behav Brain Sci 17(4):633–634
Nesse R (2000) How selfish genes shape moral passions. In: Katz L (ed) Evolutionary origins of morality: cross-disciplinary perspectives. Imprint Academic, Exeter, pp 227–252
Nesse R (ed) (2001) Evolution and the capacity for commitment. Russell Sage, New York
Nesse R (2007) Runaway social selection for displays of partner value and altruism. Biol Theory 2 (2):143–155
Nesse R (2009) Runaway social selection for displays of partner value and altruism. In: Verplaetse J, de Schrijver J, Vanneste S et al (2009) The moral brain: essays on the evolutionary and neuroscientific aspects of morality. Springer, Berlin, pp 211–231
Nesse R, Williams G (1997) Are mental disorders diseases? In: Baron-Cohen S (ed) The maladapted mind: classic readings in evolutionary psychopathology. Psychology Press, Hove, pp 1–22
Nettle D (2002) Women's height, reproductive success and the evolution of sexual dimorphism in modern humans. Proc R Soc Lond B 269:1919–1923
Nettle D (2006) The evolution of personality variation in humans and other animals. Am Psychol 61(6):622–631
Nettle D, Pollet TV (2008) Natural selection on male wealth in humans. Am Nat 172(5):658–666

Newberg A (2006) Religious and spiritual practices: a neurochemical perspective. In: McNamara P (ed) Where God and science meet: how brain and evolutionary studies alter our understanding of religion, vol 2. The neurology of religious experience. Praeger, Westport, pp 15–31
Newberg A, D'Aquili E, Rause V (2001) Why God won't go away: brain science and the biology of belief. Ballantine Books, New York
Newberg A, Lee BY (2006) The relationship between religion and health. In: McNamara P (ed) Where God and science meet: how brain and evolutionary studies alter our understanding of religion, vol 3. The psychology of religious experience. Praeger, Westport, pp 35–66
Newberg A, Waldman MR (2009) How God changes your brain: breakthrough findings from a leading neuroscientist. Ballantine Books, New York
Newbigin E (2013) The Hindu family and the emergence of modern India: law, citizenship and community. Cambridge University Press, Cambridge
Newitz A (2013) Scatter, adapt, and remember. How humans will survive a mass extinction. Doubleday, New York
Newman SA, Muller GB (2001) Epigenetic mechanisms of character origination. In: Wagner GP (ed) The character concept in evolutionary biology. Academic Press, San Diego, pp 559–580
Newson L (2009) Cultural versus reproductive success: why does economic development bring new tradeoffs? Am J Hum Biol 21(4):464–471
Newson L, Postmes T, Lea SEG et al (2007) Influences on communication about reproduction: the cultural evolution of low fertility. Evol Hum Behav 28(3):199–210
Newson L, Richerson PJ (2009) Why do people become modern? A Darwinian explanation. Popul Dev Rev 35(1):117–158
Nicey C (2017) The darkening age: the Christian destruction of the classical world. Macmillan, London
Nichols S (1988) The posthuman manifesto. Games Monthly Magazine
Nichols DE, Chemel BR (2006) The neuropharmacology of religious experience: Hallucinogens and the experience of the divine. In: McNamara P (ed) Where God and science meet: how brain and evolutionary studies alter our understanding of religion, vol 3. The psychology of religious experience. Praeger, Westport, pp 1–34
Niedenzu HJ, Meleghy T, Meyer P (eds) (2008) The new evolutionary social science: human nature, social behavior, and social change. Paradigm Publishers, Boulder
Nieschlag E (1986) Reasons for abandoning immunization against FSH as an approach to male fertility regulation. In: Zatuchni GI, Goldsmith A, Spieler JM et al (eds) Male contraception: advances and future prospects. Harper and Row, Philadelphia, pp 395–400
Nietzsche F (1883–1885) Also sprach Zarathustra. (Thus spoke Zarathustra). Verlag von Ernst Schmeitzner, Chemnitz
Nietzsche F (1887) Zur Genealogie der Moral. Eine Streitschrift. (On the genealogy of morality. A polemic). Verlag von C.G. Neumann, Leipzig
Nietzsche F (1901) Der Wille zur Macht. (The will to power). Naumann. Leipzig
Nisbet RA (1980, 2009), History of the idea of progress. Transaction Publishers, New Brunswick
Nitecki MH (ed) (1988) Evolutionary progress. The University of Chicago Press, Chicago
Nitecki MH, Nitecki DV (eds) (1993) Evolutionary ethics. State University of New York Press, Albany
Noblett KL, Coccaro EF (2005) Molecular genetics of personality. Curr Psychiatry Rep 7(1):73–80
Nonaka K, Miura T, Peter K (1994) Recent fertility decline in Dariusleut Hutterites: an extension of Eaton and Mayer's Hutterite fertility study. Hum Biol 66(3):411–420
Noom M (1999) Adolescent autonomy. Characteristics and correlates. Eburon Academic Publishers, Delft
Norberg J (2016) Progress. Ten reasons to look forward to the future. Oneworld Publications, Woolloomooloo
Nordio E (2012) Could an antisocial behaviour be influenced by our genes? Roskilde University: International Basic Studies in Natural Sciences, Roskilde

Norenzayan A (2014) Does religion make people moral? Behaviour 151(2–3):SI 365–384
Norenzayan A, Shariff AF (2008) The origin and evolution of religious prosociality. Science 322 (5898):58–62
Norenzayan A, Shariff AF, Gervais WM et al (2016) The cultural evolution of prosocial religions. Behav Brain Sci 39: https://doi.org/10.1017/s0140525x14001356, e0
Norman KR (2003) The four noble truths. Indological and Buddhist studies. Volume for J.W. de Jong. K.R. Norman Collected Papers II, 10-223. Pali Text Society, Oxford
Norman R (2004) On humanism. Routledge, London
Norris P, Inglehart R (2004) Sacred and secular: religion and politics worldwide. Cambridge University Press, New York
Notaro SR (2012) health disparities among under-served populations: implications for research, policy and praxis. Emerald Group Publishing, Bingley
Notestein F (1945) Population: the long view. In: Schultz T (ed) Food for the World. University of Chicago Press, Chicago, pp 36–57
Notestein F (1954) Some demographic aspects of aging. Proc Am Philos Soc 98(1):38–45
Novacek MJ (2001) The biodiversity crisis. Losing what counts. The New Press, New York
Nowak MA (2006) Five rules for the evolution of cooperation. Science 314(5805):1560–1563
Nowak MA, Coakley S (eds) (2013) Evolution, games, and God. The principle of cooperation. Harvard University Press, Cambridge
Nowak MA, Highfield R (2011) Supercooperators: altruism, evolution, and why we need each other to succeed. Free Press, New York
Nowak MA, May RM (1992) Evolutionary games and spatial chaos. Nature 359(6398):826–829
Nowak MA, Sigmund K (1992) Tit for tat in heterogenous populations. Nature 355(6357):250–253
Nowak MA, Sigmund K (1993) A strategy of win-stay, lose-shift that outperforms tit-for-tat in the prisoner's dilemma game. Nature 364(6432):56–58
Nowak MA, Sigmund K (1994) The alternating prisoner's dilemma. J Theor Biol 168:219–226
Nowak MA, Sigmund K (1998) Evolution of indirect reciprocity by image scoring. Nature 393 (6685):573–577
Nowak MA, Sigmund K (1998) The dynamics of indirect reciprocity. J Theor Biol 194:561–574
Nowak MA, Sigmund K (2005) Evolution of indirect reciprocity. Nature 437(7063):1291–1298
Nowak MA, Tarnita CE, Wilson EO (2010) The evolution of eusociality. Nature 466(7310):1057–1062
Noziglia CM, Siegel J (2006) Criminal psychology and personality profiling. Mason Crest Publishers, Broomall
Nuland SB (1994) How we die: reflections on life's final chapter. Alfred A. Knoph, New York
Numbers R (2006) The creationists: from scientific creationism to intelligent design. Harvard University Press, Cambridge
Nunez M, Harris PL (1998) Psychological and deontic concepts: separate domains or intimate connection? Mind Lang 13(2):153–170
Nyborg H (2009) The intelligence-religiosity nexus: a representative study of white adolescent Americans. Intelligence 37(1):81–93
Nys H (2003) A presentation of the Belgian act on Euthanasia against the background of Dutch Euthanasia Law. Eur J Health Law 10(3):239–255
McCready Price G (1923) The new geology. Pacific Press, Mountain View
O'Connor LE, Berry JW, Lewis T (2007) Empathy and depression: the moral system on overdrive. In: Farrow T (ed) Empathy disorders in psychopathy. Oxford University Press, Oxford, pp 58–62
O'Brien M (1981) The politics of reproduction. Routledge and Kegan Paul, Boston
O'Connell Davidson J (1998) Prostitution, power and freedom. Polity Press, Cambridge
O'Donald P (1980) Genetic models of sexual selection. Cambridge University Press, Cambridge
O'Grady S (2013) And man created god: kings, cults and conquests at the time of Jesus. Atlantic Books, London

O'Hear A (1997) Beyond evolution: human nature and the limits of evolutionary explanation. Clarendon, Oxford
O'Malley PD (2006) New developments in sickle cell disease research. Nova Science Publishers, Hauppauge
O'Neill M (2001) Prostitution and feminism. Polity Press, Cambridge
Oakley K (1959) Man the toolmaker. University of Chicago Press, Chicago
Oakley A (1972) Sex, gender and society. Temple Smith, London
Obodovskiy I (2015) Fundamentals of radiation and chemical safety. Elsevier, New York
Oddone-Paolucci E, Genuis M, Violato C (2000) A meta-analysis of the published research on the effects of pornography. In: Violato C, Oddone-Paolucci E, Genuis M (eds) The changing family and child development. Ashgate Publishing, Aldershot, pp 48–59
Odegar O (1980) Fertility of psychiatric first admissions in Norway, 1936–1975. Acta Psychiatr Scand 62(3):212–220
Oderberg DS (2014) Could there be a superhuman species? South J Philos 52(2):206–226
Odeyemi A (2008) Female genital mutilation: reasons, practice, effects. VDM Verlag, Saarbrücken
Odling-Smee FJ (2003) Niche construction. Princeton University Press, Princeton
Ohlig KH, Puin GR (2009) The hidden origins of Islam. Prometheus Press, Amherst
Ohtsuki H, Hauert C, Lieberman E et al (2006) A simple rule for the evolution of cooperation on graphs and social networks. Nature 441:502–505
Okuyama Y, Ishiguro H, Nankai M et al (2000) Identification of a polymorphism in the promoter region of DRD4 associated with the human novelty seeking personality trait. Mol Psychiatry 5:64–69
Oldroyd DR (1980) Darwinian impacts. Open University Press, Milton Keynes
Olshansky J, Hayflick L, Carnes B (2002) Position statement on human aging. J Gerontol Ser A: Biol Sci Med Sci 57(8):B292–B297
Omark DR, Strayer FF, Freedman DG (eds) (1980) Dominance relations: an ethological view of human conflict and social interaction. Garland STPM Press, New York
OMIM (2017) https://www.ncbi.nlm.nih.gov/omim
Organization of the Islamic Conference (1990) The Cairo declaration on human rights in Islam. Cairo: http://www.oic-oci.org/english/article/human.htm
Orians GH, Heerwagen JH (1992) Evolved responses to landscapes. In: Barkow JH, Cosmides L, Tooby J (eds) The adapted mind: evolutionary psychology and the generation of culture. Oxford University Press, New York, pp 555–579
Osborn F (1940) Preface to eugenics. Harper, New York
Osborn F (1952) The eugenic hypothesis. (ii) Negative eugenics. Eugenics Rev 44(2):97–100
Osborn F (1968) The future of human heredity. An introduction to eugenics in modern society. Weybright and Talley, New York
Osborn F, Bajema C (1972) The eugenics hypothesis. Soc Biol 19(4):337–345
Ostow M (2006) Spirit, mind, and brain: a psychoanalytic examination of spirituality and religion. Columbia University Press, New York
Ott MA, Santelli JS (2007) Abstinence and abstinence-only education. Curr Opin Obstet Gynecol 19(5):446–452
Overhage P (1977) Die biologische Zukunft der Menschheit. (The biological future of humanity). Verlag Josef Knecht, Frankfurt am Main
Oviedo L (2009) Is religious behavior "internally guided" by religious feelings and needs? In: Feierman JR (ed) The biology of religious behavior: the evolutionary origins of faith and religion. ABC-CLIO, LLC, Santa Barbara, pp 141–156
Pahnke WN (1967) LSD and religious experience. In: DeBold RC, Leaf RC (eds) LSD, man and society. Wesleyan University Press, Middletown, pp 60–85
Pali Canon or Tipitaka. (http:// http://www.accesstoinsight.org/tipitaka/)
Palme J (2012) The future of Homo sapiens. The future of human evolution. http://web4health.info/en/aux/homo-sapiens-future.html
Palmore E (2001) The ageism survey. First findings. Gerontol 41(5):572–575

Paltiel AM (2013) Estimation of the size and vital rates of the Haredi (Ultra-orthodox) population in Israel for the purpose of long-range population projections. http://www.unece.org/fileadmin/DAM/stats/documents/ece/ces/ge.11/2013/7_3_Estimation_of__size_and_vital_rates_Haredim29112013.pdf
Parelius RJ, Parelius AP (1987) The sociology of education. Prentice-Hall, Englewood Cliffs
Parens E (ed) (1998) Enhancing human traits. Ethical and social implications. Georgetown University Press, Washington
Pargament KI, Smith BW, Koenig HG et al (1998) Patterns of positive and negative religious coping with major life stressors. J Sci Study Relig 37(4):710–724
Paridis JG, Williams GC (1989) Evolution and ethics. T.H. Huxley's evolution and ethics with new essays on its Victorian and sociobiological context. Princeton University Press, Princeton
Parkinson C, Sinnott-Armstrong W, Koralus PE et al (2011) Is morality unified? Evidence that distinct neural systems underlie moral judgments of harm, dishonesty, and disgust. J Cogn Neurosci 23(10):3162–3180
Parsons T (1964) Evolutionary universals in society. Am Sociol Rev 29(3):339–357
Parsons J (1999) Human population competition: a study of the pursuit of power through numbers. Edwin Mellin Press, Lewiston
Pascal B (1670) Penseés. (Thoughts). G. Desprez, Paris
Paterson C (2008) Assisted suicide and euthanasia: a natural law ethics approach. Ashgate, Aldershot
Paul D (1980) The Buddhist feminine ideal. Scholars' Press
Paul D (1985) Women in Buddhism. University of California Press, Oakland
Paul DB (1998) The politics of heredity. Essays on eugenics, biomedicine, and the nature-nurture debate. State University of New York Press, New York
Paul DB (2014) What was wrong with eugenics? Conflicting narratives and disputed interpretations. Sci Educ 23(2):259–271
Paul GS (2005) Cross-national correlations of quantifiable societal health with popular religiosity and secularism in the prosperous democracies: A first look. J Relig Soc 7:1–17
Paul P (2005) Pornified: how pornography is transforming our lives, our relationships, and our families. Time Books, New York
Paul GS (2009) The chronic dependence of popular religiosity upon dysfunctional psychosociological conditions. Evol Psychol 7(3):398–441
Paul GS (2010) The evolution of popular religiosity and secularism: how first world statistics reveal why religion exists, why it has been popular, and why the most successful democracies are the most secular. In: Zuckerman P (ed) Atheism and secularity, vol 1. Issues, concepts, and definitions. Praeger, Santa Barbara, pp 149–207
Paul GS, Zuckerman P (2007) Why the gods are not winning. www.edge.org/3rd_culture/paul07/paul07_index.html
Paulos JA (2008) Irreligion: a mathematician explains why the arguments for god just don't add up. Hill and Wang, New York
Pava ML (2009) The exaggerated moral claims of evolutionary psychologists. J Bus Ethics 85 (3):391–401
Paxton RO (2005) The anatomy of fascism. Vintage, New York
Peacocke A (1998) Biological evolution—a positive theological appraisal. In: Russell RJ, Stoeger WR, Ayala FJ (eds) Evolutionary and molecular biology: scientific perspectives on divine action. Center for Theology and the Natural Sciences and the Vatican Observatory, Berkeley and Rome, pp 357–376
Pearson K (1896) Regression, heredity and panmixia. Philos Trans A 187:253–318
Pearson K (1909) The groundwork of eugenics. Dulau and Co, London
Peccei JS (1995) The origin and evolution of menopause: the altriciality-lifespan hypothesis. Ethol Sociobiol 16(5):425–450
Pedersen CA (2004) How love evolved from sex and gave birth to intelligence and human nature. J Bioecon 6(1):39–63

Pedersen EJ, Kurzban R, McCullough ME (2013) Do humans really punish altruistically? A closer look. Proc R Soc B-Biol Sci 280(1758):2012–2723

Pena J, Pestelacci E, Berchtold A et al (2011) Participation costs can suppress the evolution of upstream reciprocity. J Theor Biol 273(1):197–206

Penn DJ, Mysterud I (eds) (2007) Evolutionary perspectives on environmental problems. Aldine transaction, New Brunswick

Penner LA, Dovidio JF, Piliavin JA et al (2005) Prosocial behavior: multilevel perspectives. Ann Rev Psychol 56:365–392

Pennock RT (1999) The tower of Babel: the evidence against the new creationism. MIT Press, Cambridge

Pennock RT (2003) Creationism and intelligent design. Ann Rev Genomics Hum Genet 4:143–163

Penny D (2015) Epigenetics, Darwin, and Lamarck. Genome Biol Evol 7(6):1758–1760

Penrose R (2010) Cycles of time: an extraordinary new view of the universe. The Bodley Head, London

Peoples HC, Marlowe FW (2012) Subsistence and the evolution of religion. Hum Nat 23(3):253–269

Perakh M (2003) Unintelligent design. Prometheus Books, Amherst

Perakh M, Young M (2006) Is intelligent design science? In: Young M, Edis T (eds) Why intelligent design fails: a scientific critique of the new creationism. Rutgers University Press, New Brunswick, pp 185–196

Perez RM (2004) Kings and untouchables: a study of the caste system in Western India. Orient Blackswan, Hyderabad

Perilloux C, Fleischman DS, Buss DM (2008) The daughter-guarding hypothesis: parental influence on, and emotional reactions to, offspring's mating behavior. Evol Psych 6(2):217–233

Perls TT, Fretts RC (2001) The evolution of menopause and human lifespan. Ann Hum Biol 28 (3):237–245

Perrett DI, Burt DM, Penton-Voak IS et al (1999) Symmetry and human facial attractiveness. Evol Hum Behav 20(5):295–307

Perry S (2015) Antinatalism in biological and cultural evolution: fertility and suicide. In: Shackelford TD, Hansen RD (eds) The evolution of morality. Springer, New York, pp 141–154

Persinger MA (1987) Neuropsychological bases of god beliefs. Praeger, New York

Persinger MA (2003) Experimental simulation of the god experience. In: Joseph R (ed) Neurotheology. University Press California, San Jose, pp 267–284

Persson I (2012) Could it be permissible to prevent the existence of morally enhanced people? J Med Ethics 38(11):692–693

Persson I, Savulescu J (2008) The perils of cognitive enhancement and the urgent imperative to enhance the moral character of humanity. J Appl Philos 25(3):162–177

Persson I, Savulescu J (2010) Moral transhumanism. J Med Philos 35(6):656–669

Persson I, Savulescu J (2012) Unfit for the future: the need for moral enhancement. Oxford University Press, Oxford

Pérusse D (1993) Cultural and reproductive success in industrial societies: testing the relationship at proximate and ultimate levels. Behav Brain Sci 16(2):267–322

Peters KE (1999) Reflections on the evolution of morality. Zygon 34(3):419–433

Peters KE (2001) Neurotheology and evolutionary theology: reflections on the mystical mind. Zygon 36(3):493–501

Peters T (2002) Playing God? Genetic discrimination and human freedom. Routledge, New York

Peters KE (2003) Pluralism and ambivalence in the evolution of morality. Zygon 38(2):333–355

Peterson GR (1999) The evolution of consciousness and the theology of nature. Zygon 34(2):283–306

Peterson GR (2000) God, genes and cognizing agents. Zygon 35(3):469–480

Peterson C, Seligman MEP (2004) Character strengths and virtues. Oxford University Press, Oxford

Petrinovich L (1995) Human evolution, reproduction, and morality. Plenum Press, New York

Petrinovich L, O'Neill P, Jorgensen M (1993) An empirical study of moral intuitions: towards an evolutionary ethics. J Person Soc Psychol 64(3):467–478

Pew Forum on Religion and Public Life (2009) Report on religious affiliation. Pew Research Center, Washington

Pew Research Center (2007) World publics welcome global trade—but not immigration. 47-Nation Pew Global Attitudes Survey. http://www.pewglobal.org/files/2007/10/Pew-Global-Attitudes-Report-October-4-2007-NOT-EMBARGOED.pdf

Pew Research Center (2015) The future of world religions: population growth projections, 2010–2050.http://www.pewforum.org/files/2015/03/PF_15.04.02_ProjectionsFullReport.pdf

Pfaff DW, Christen Y (2013) Multiple origins of sex differences in brain: neuroendocrine functions and their pathologies. Springer, Berlin

Phelan JC, Link BG, Feldman NM (2013) The genomic revolution and beliefs about essential racial differences. A backdoor to eugenics? Am Sociol Rev 78(2):167–191

Phelps S (2013) Emergence of social networks via direct and indirect reciprocity. Auton Agents Multi-Agent Syst 27(3):355–374

Philip JC, Cherian S (2013) Implications of human cloning. Philip Communications, Kochi

Philipov D, Berghammer C (2007) Religion and fertility ideals, intentions and behaviour: a comparative study of European countries. Vienna Yearb Popul Res 1:271–305

Philipse H (1995) Atheïstisch manifest. Drie wijsgerige opstellen over godsdienst en moraal. (Atheist Manifesto. Three Philosophical Essays on Religion and Morality). Prometheus, Amsterdam

Phillips PT (2007) One world, one faith: the quest for unity in Julian Huxley's religion of evolutionary humanism. J Hist Ideas 68(4):613–633

Philipse H (2012) God in the age of science? A critique of religious reason. Oxford University Press, Oxford

Phillips T, Ferguson E, Rijsdijk F (2010) A link between altruism and sexual selection: genetic influence on altruistic behaviour and mate preference towards it. Br J Psychol 101(4):809–819

Phillips T, Li J, Kendall G (2014) The effects of extra-somatic weapons on the evolution of human cooperation towards non-kin. PLOS ONE 9(5):e95742

Phillips RE, Pargament KI, Lynn QK et al (2004) Self-directing religious coping: a deistic God, abandoning God, or no God at all? J Sci Study Relig 43(3):409–418

Pianka ER (2000) Evolutionary ecology, Benjamin-Cummings. Addison-Wesley-Longman, San Francisco

Piazza A (1990) Migration and genetic differentiation in Italy. In: Adams J, Hermalin A, Lam D et al (eds) Convergent issues in genetics and demography. Oxford University Press, New York, pp 81–93

Pickel G, Müller O (2009) Church and religion in contemporary Europe. Results from empirical and comparative research. VS Verlag für Sozialwissenschaften, Wiesbaden

Piedmont RL (1999) Does spirituality represent the sixth factor of personality? Spiritual transcendence and the five factor model. J Person 67(6):985–1013

Piff PK, Stancato DM, Côté S et al (2012) Higher social class predicts increased unethical behavior. PNAS 109(11):4086–4091

Pigliucci M (2001) Phenotypic plasticity: beyond nature and nurture. Johns Hopkins University Press, Baltimore

Pigliucci M (2007) Do we need an extended evolutionary synthesis? Evolution 61(12):2743–2749

Pigliucci M, Müller GB (2010) Evolution: the extended synthesis. MIT Press, Cambridge

Piketty T (2013) Le capital au XXIe siècle. Seuil, Paris (see also: Piketty T (2014) Capital in the Twenty-First Century. Belknap Press, Cambridge)

Pilgrim D (2011) The child abuse crisis in the Catholic Church: international, national and personal policy aspects. Policy Polit 39(3):309–324

Pilnick A (2002) Genetics and society: an introduction. Open University Press, Philadelphia
Pimentel D, Bailey O, Kim P et al (1999) Will limits of the earth's resources control human numbers? Environ Dev Sustain 1:19–39
Pimentel D, Giampietro M (1994) Food, land, population and the US economy. Carrying Capacity Network, Washington
Pimentel D, Pimentel M (1991) Land, energy, and water: the constraints governing ideal human population size. NPG Forum, Teaneck
Pimentel D, Pimentel M (1997) Too many people for food resources and the environment. Polit Life Sci 16(2):217–218
Pimentel D, Pimentel M (2005) Global environmental resources versus world population growth. Ecol Econ 59(2):195–198
Pinker S (2002) The blank slate: the modern denial of human nature. Penguin, London
Pinker S (2004) Why nature and nurture won't go away. Daedalus 133(4):5–17
Pinker S (2007) A history of violence. New Republic. http://www.newrepublic.com/article/77728/history-violence
Pinker S (2008) The moral instinct. The New York Times Magazine, 13 Jan 2008
Pinker S (2011) The better angels of our nature: why violence has declined. Viking Adult, New York
Pinxten R (2007) De strepen van de zebra: voor een strijdend vrijzinnig humanisme. (The zebra stripes: For a militant secular humanism). Houtekiet, Antwerpen
Pinxten R (2010) The creation of God. Peter Lang, Brussels
Pitchford I (2001) The origins of violence: is psychopathy an adaptation? Hum Nat Rev 1:28–36
Pitman GR (2011) The evolution of human warfare. Philos Soc Sci 41(3):352–379
Platek SM, Shackelford TK (eds) (2006) Female infidelity and paternal uncertainty: evolutionary perspectives on male anti-cuckoldry tactics. Cambridge University Press, New York
Plavcan JM (2001) Sexual dimorphism in primate evolution. Yearb Phys Anthropol 44:25–53
Plavcan JM, van Schaik CP (1997) Interpreting hominid behavior on the basis of sexual dimorphism. J Hum Evol 32(4):345–374
Plomin R (1989) Environment and genes: determinants of behavior. Am Psychol 44(2):105–111
Plomin R, Asbury K (2005) Nature and nurture: genetic and environmental influences on behavior. Ann Am Acad Polit Soc Sci 600:86–98
Plomin R, DeFries JC, McClearn GE et al (2008) Behavioral genetics. Worth Publishers, New York
Plomin R, McGuffin P (2003) Psychopathology in the postgenomic era. Ann Rev Psychol 54:205–228
Plotkin HC, Odling-Smee FJ (1981) A multiple-level model of evolution and its implications for sociobiology. Behav Brain Scien 4(2):225–268
Polgar S (ed) (1975) Population, ecology, and social evolution. Mouton, The Hague
Pomiankowski AN (1987) The costs of choice in sexual selection. J Theor Biol 128:195–218
Poneli A (2012) Het toekomst manifest. Over duurzaamheid, zin, rechtvaardigheid en kwaliteit. (The future manifesto. About sustainability, sense, justice and quality). Aspekt, Soesterberg
Pontifical Council for Social Communications (1989) Pornography and violence in the media. Vatican City
Pope Benedict XVI (2006) Faith, reason and the university. Memories and reflections. Apostolic journey of His Holiness benedict XVI to München, Altötting and Regensburg (September 9–14, 2006). Meeting with the representatives of science. Lecture of the Holy Father. Aula Magna of the University of Regensburg, Tuesday, 12 Sept 2006
Pope Benedict XVI (2007) Encyclical letter Spe Salvi of the Supreme Pontiff Benedict XVI to the Bishops priests and deacons men and women religious and all the lay faithful on christian hope. Vatican City
Pope Benedict XVI (2007) Meeting of the Holy Father Benedict XVI with the clergy of the Dioceses of Belluno-Feltre and Treviso. Church of St Justin Martyr, Auronzo di Cadore Tuesday, 24 Jul 2007. Libreria Editrice Vaticana, Vatican City

Pope Benedict XVI (2009) To the participants of the Pontifical Academy for Life conference "New Frontiers of Genetics and the Danger of Eugenics". The Vatican

Pope John Paul II (1995) Evangelium vitae. To the Bishops priests and deacons men and women religious lay faithful and all people of good will on the value and inviolability of human life. The Vatican

Pope John Paul II (1995) The Gospel of life. Random house, New York

Pope John Paul II (1996) Message to the Pontifical Academy of Sciences on evolution. Magisterium is concerned with question of evolution for it involves conception of man. Message delivered to the Pontifical Academy of Sciences 22 Oct 1996

Pope Paul VI (1968) Humanae Vitae: encyclical of Pope Paul VI on the regulation of birth. The Vatican

Pope Pius XI (1930), Casti Connubii: encyclical of Pope Pius XI on Christian marriage. The Vatican

Pope S (1994) The evolution of altruism and the ordering of love. Georgetown University Press, Washington

Pope S (2007) Human evolution and Christian ethics. Cambridge University, New York

Porter D (1999) Health, civilization, and the state: a history of public health from ancient to modern times. Routledge, London

Portmann A (1944) Biologische Fragmente zu einer Lehre vom Menschen. (Biological fragments for a human science). Schwabe, Basel

Posner R (2005) Catastrophe. Risk and response. Oxford University Press, New York

Post RH (1971) Possible cases of relaxed selection in civilized populations. Hum Genet 13 (4):253–284

Post S, Binstock R (eds) (2004) The fountain of youth. Oxford University Press, Oxford

Potter S (2010) Designer genes: a new era in the evolution of man. Random House, New York

Poulin MJ, Holman EA, Buffone A (2012) The neurogenetics of nice: receptor genes for oxytocin and vaso-pressin interact with threat to predict prosocial behavior. Psychol Sci 23(5):446–452

Povinelli DJ, Preuss TM (1995) Theory of mind: evolutionary history of a cognitive specialization. Trends in Neurosciences 18(9):418–424

Powell R (2012) The evolutionary biological implications of human genetic engineering. J Med Philos 36(1):6–27

Powel R, Clarke S (2012) Religion as an evolutionary byproduct: a critique of the standard model. Br J Philos Sci 63(3):457–486

Powers ST, Lehmann L (2013) The co-evolution of social institutions, demography, and large-scale human cooperation. Ecol Lett 16(11):1356–1364

Powers ST, Taylor DJ, Bryson JJ (2012) Punishment can promote defection in group-structured populations. J Theor Biol 311:107–116

Poythress N (1975) Literal, antiliteral, and mythological religious orientations. J Sci Study Relig 14(3):271–284

Prabhat J, Kumar R, Dhingra N et al (2006) Low male-to-female sex ratio of children born in India: National survey of 1.1 million households. Lancet 367(9506):211–218

Prahalad CK, Hammond A (2002) Serving the world's poor, profitably. Harvard Bus Rev 80 (9):48–58

Prehn K, Wartenburger I, Meriau K et al (2008) Individual differences in moral judgment competence influence neural correlates of socio-normative judgments. Soc Cogn Affect Neurosci 3(1):33–46

Premack D, Woodruff G (1978) Does the chimpanzee have a theory of mind? Behav Brain Sc 1 (04):515

Premasiri PD (2015) Buddhist response to social conflict: some practical Buddhist suggestions for the resolution of the problem of social conflict. http://www.undv.org/vesak2015/-paper/-practical_Buddhist_suggestions.pdf President's Council on Bioethics (2003) Beyond therapy: Biotechnology and the pursuit of happiness. Regan Books, New York

President's Council on Bioethics (2003) Human cloning and human dignity: an ethical inquiry. National Bioethics Advisory Commission, Washington

Preston JL, Ritter RS (2013) Different effects of religion and god on prosociality with the ingroup and outgroup. Person Soc Psychol Bull 39(11):1471–1483

Previc F (2006) The role of the extrapersonal brain systems in religious activity. Conscious Cogn 15(3):500–539

Previc F (2009) The dopaminergic mind in human evolution and history. Cambridge University Press, New York

Price G (1995) The nature of selection. J Theor Biol 175(3):389–396

Price G (1970) Selection and covariance. Nature 227(5257):520–521

Price G (1972) Extension of covariance selection mathematics. Ann Hum Genet 35(4):485–490

Price RM (2000) Deconstructing Jesus. Prometheus Books, Amherst

Price ME (2008) The resurrection of group selection as a theory of human cooperation. Soc Justice Res 21(2):228–240

Pride M (1985) The way home/beyond feminism, back to reality. Crossway, Westchester

Prigent R (1955) Renouveau des idées sur la famille. (Renewing ideas about the family). Institut national d'études démographiques. Travaux et documents. Cahier n° 18. Presses Universitaires de France, Paris

Prigogine I, Stengers I (1984) Order out of chaos: man's new dialogue with nature. Flamingo, London

Prinz J (2008) Is morality innate? In: Sinnott-Armstrong W (ed) Moral psychology, vol 1. The evolution of morality: adaptations and innateness. MIT Press, Cambridge, pp 367–406

Prothero DR, Dott R (2009) Evolution of the Earth. McGraw-Hill Science/Engineering/Math, New York

Prinz J (2015) Is the moral brain ever dispassionate? In: Decety J, Wheatley T (eds) The moral brain: a multidisciplinary perspective. MIT Press, Cambridge, pp 51–67

Provine W (1988) Evolution and the foundation of ethics. Science 3(1):25–29

Provine W (1988) Progress in evolution and meaning of life. In: Nitecki MH (ed) Evolutionary progress. University of Chicago Press, Chicago, pp 49–74

Pruss A (2013) Altruism, normalcy, and God. In: Nowak MA, Coakley S (eds) Evolution, games, and God. The principle of cooperation. Harvard University Press, Cambridge, pp 329–342

Przepiorka W, Diekmann A (2013) Individual heterogeneity and costly punishment: a volunteer's dilemma. Proc R Soc B-Biol Sci 280(1759):20130247

Pugh GE (1978) The biological origin of human values. Routledge and Kegan Paul, London

Purdom G, Lisle J (2009) Morality and the irrationality of an evolutionary worldview. Phi Kappa Phi Forum 89(1):8–12

Purzycki BG, Apicella C, Atkinson QD et al (2016) Moralistic gods, supernatural punishment and the expansion of human sociality. Nature 530(7590):327–330

Purzycki BG, Sosis R (2008) The religious system as adaptive: cognitive flexibility, public displays, and acceptance. In: Voland E, Schiefenhövel W (eds) The biological evolution of religious mind and behavior. Springer Publishers, New York, pp 243–256

Putnam RD (2000) Bowling alone: America's declining social capital. In: Crothers L, Lockhart C (eds) Culture and politics. Palgrave Macmillan, New York, pp 223–234

Putkonen A, Cederwall A, Klier E et al (2009) Filicide in Austria and Finland—a register-based study on all filicide cases in Austria and Finland 1995–2005. BMC Psychiatry 9:74–83

Puurtinen M, Heap S, Mappes T (2015) The joint emergence of group competition and within-group cooperation. Evol Hum Behav 36(3):211–217

Pyysiäinen I (2001) How religion works: towards a new cognitive science of religion. Brill, Leiden

Pyysiäinen I (2001) Magic, miracles and religion: a scientist's perspective. William Heinnemann, London

Pyysiäinen I (2006) Religion and the evolution of the human mind. In: McNamara P (ed) Where God and science meet. How brain and evolutionary studies alter our understanding of religion. Praeger, Westport, pp 209–226

Pyysiäinen I, Hauser M (2010) The origins of religion: evolved adaptation or by-product? Trends Cogn Sci 14(3):104–109

Quayle AP, Bullock S (2006) Modelling the evolution of genetic regulatory networks. J Theor Biol 238:737–753

Quinsey VL (2002) Evolutionary theory and criminal behavior. Legal Criminol Psychol 7(1):1–13

Rabinowicz T, Dean DE, Petetot JM, de Courten-Myers GM (1999) Gender differences in the human cerebral cortex: more neurons in males; more processes in females. J Child Neurol 14 (2):98–107

Rachels J (1990) Created from animals: the moral implications of Darwinism. Oxford University Press, Oxford

Radman M, Carton D (2011) Au-delà de nos limites biologiques. Les secrets de la longévité. (Beyond our biological limits. The secrets of longevity). Plon, Paris

Rahman Q, Wilson GD (2003) Born gay? The psychobiology of human sexual orientation. Person Individ Diff 34(8):1337–1382

Raihani NJ, McAuliffe K (2012) Human punishment is motivated by inequity aversion, not a desire for reciprocity. Biol Lett 8(5):802–804

Raine A (1993) The psychopatholoy of crime: criminal behavior as a clinical disorder. Academic Press, San Diego

Raine A (2013) The anatomy of violence: the biological roots of crime. Pantheon Books, New York

Raine A, Lencz T, Bihrle S et al (2000) Reduced prefrontal gray matter volume and reduced autonomic activity in antisocial personality disorder. Arch Gen Psychiatry 57(2):119–127

Rainer T (2007) Religion, born again. Spiegel online international, 15 Jan 2007. www.spiegel.de/international/spiegel/0,1518,459500,00.html

Rakić V (2014) Voluntary moral enhancement and the survival-at-any-cost bias. J Med Ethics 40 (4):246–250

Rakić V (2015) We must create beings with moral standing superior to our own. Cambridge Q Healthc Ethics 24(01):58–65

Ramachandran VS, Blakeslee S (1998) Phantoms in the brain. William Morrow, New York

Rampino MR, Self S (1992) Volcanic winter and accelerated glaciation following the Toba super-eruption. Nature 359:50–52

Ramsey P (1970) Fabricated man: the ethics of genetic control. Yale University Press, New Haven

Ramsey G, Brandon R (2011) Why reciprocal altruism is not a kind of group selection. Biol Philos 26:385–400

Rancour-Laferriere D (1985) Signs of the flesh. An essay on the evolution of hominid sexuality. Mouton de Gruyter, Berlin

Rand DG, Armao JJ, Nakamaru M et al (2010) Anti-social punishment can prevent the co-evolution of punishment and cooperation. J Theor Biol 265(4):624–632

Rand DG, Dreber A, Ellingsen T et al (2009) Positive interactions promote public cooperation. Science 325:1272–1275

Rand DG, Greene JD, Nowak MA (2012) Spontaneous giving and calculated greed. Nature 489 (7416):427–430

Rand DG, Nowak MA (2013) Human cooperation. Trends Cogn Sci 17(8):413–425

Randolph-Seng B, Nielsen ME (2007) Honesty: one effect of primed religious representations. Int J Psychol Relig 17(4):303–315

Rauf FA (2004) What's right with Islam: a new vision for Muslims and the West. Harper, San Francisco

Ravat J (2015) Darwinian morality, moral Darwinism. In: Heams T, Huneman P, Lecointre G et al (eds) Handbook of evolutionary thinking in the sciences. Springer, Dordrecht, pp 747–759

Rawls J (1971) A theory of justice. Harvard University Press, Cambridge

Raymond J (1994) Women as wombs. Spinifex Press, Melbourne

Redmond G (2000) Children in large families: disadvantage or just different? Luxembourg income study. Working paper no. 225. www.lisproject.org/publications/liswps/225.pdf

Rée P (1877) Der Ursprung der moralischen Empfindungen. (The origin of the moral sentiments). Verlag von Ernst Schmeitzner, Chemnitz
Reed EW (1971) Mental retardation and fertility. Soc Biol 18:S42–S49
Rees M (2003) Our final hour: a scientist's warning: how terror, error, and environmental disaster threaten humankind's future in this century—on earth and beyond. Basic Books, New York
Rees M (2004) Our final century. Random House, London
Rees D, Rose S (2004) New brain sciences: perils and prospects. Cambridge University Press, Cambridge
Reher DS (2004) The demographic transition revisited as a global process. Popul Space Place 10 (1):19–41
Reher DS (2007) Towards long-term population decline: a discussion of relevant issues. Eur J Popul/Revue Européenne de Démographie 23(2):189–207
Reich D, Thangaraj K, Patterson N et al (2009) Reconstructing Indian population history. Nature 461(7263):489–494
Reilly P (1999) Eugenics and the misuse of genetic information to restrict reproductive freedom. Board of Directors of the American Society of Human Genetics. Am J Hum Genet 64(2):335–338
Reiss MJ (1999) What sort of people do we want? The ethics of changing people through genetic engineering. Notre Dame J Law Ethics Public Policy 13(1):63–92
Reiss MJ (2009) The relationship between evolutionary biology and religion. Evolution 63 (7):1934–1941
Reiss MJ, Straughan R (1996) Improving nature? The science and ethics of genetic engineering. Cambridge University Press, Cambridge
Remmel RJ, Glenn AL (2015) Immorality in the adult brain. In: Decety J, Wheatley T (eds) The moral brain. A multidisciplinary perspective. The MIT Press, Cambridge, pp 239–251
Rendu W, Beauval C, Crevecoeur I et al (2014) Evidence supporting an intentional Neandertal burial at La Chapelle-aux-Saints. Proc Nat Acad Sci (PNAS) 111(1):81–86
Relethford JH (2001) Genetics and the search for modern human origins. Wiley-Liss, New York
Retherford RD (1993) Demographic transition and the evolution of intelligence: theory and evidence. East-West Center, Honululu
Retherford RD, Sewell WH (1988) Intelligence and family size reconsidered. Soc Biol 35(1–2):1–40
Retz W, Retz-Junginger P, Supprian T et al (2004) Association of serotonin transporter promoter gene polymorphism with violence: relation with personality disorders, impulsivity, and childhood ADHD psychopathology. Behav Sci Law 22(3):415–425
Reuter M, Frenzel C, Walter NT et al (2011) Investigating the genetic basis of altruism: the role of the COMT Val158Met polymorphism. Soc Cogn Affect Neurosci 6(5):662–668
Reynolds V (1995) Secular change and religious inertia. In: Jones E, Reynolds V (eds) Survival and religion. Biological evolution and cultural change. Wiley, Chichester, pp 187–205
Reynolds V, Falger V, Vine I (1987) The sociobiology of ethnocentrism. evolutionary dimensions of xenophobia, discrimination, racism and nationalism. Croom Helm, London
Reynolds V, Tanner R (1983) The biology of religion. Longman, London
Reynolds V, Tanner R (1985) The effects of religion on human biology. In: Durant J (ed) Darwinism and divinity. Basil Blackwell, Oxford, pp 131–153
Reynolds V, Tanner R (1995) The social ecology of religion. Oxford University Press, New York
Rhee SH, Waldman ID (2002) Genetic and environmental influences on antisocial behavior: a meta-analysis of twin and adoption studies. Psychol Bull 128(3):490–529
Rich F (2006) The greatest story ever sold: the decline and fall of truth in Bush's America. Penguin, New York
Richards RJ (1986) A defense of evolutionary ethics. Biol Philos 1(3):265–293
Richards RJ (1987) Darwin and the emergence of evolutionary theories of mind and behavior. University of Chicago Press, Chicago

Richards RJ (1993) Birth, death, and resurrection of evolutionary ethics. In: Nitecki MH, Nitecki DV (eds) Evolutionary ethics. State University of New York Press, Albany, pp 113–131

Richards RJ (1999) Darwin's romantic biology: the foundation of his evolutionary ethics. In: Maienschein J, Ruse M (eds) Biology and the foundation of ethics. Cambridge University Press, Cambridge

Richards JR (2000) Human nature after Darwin. Routledge, London

Richards RJ (2008) The tragic sense of life: Ernst Haeckel and the struggle over evolutionary thought. University of Chicago Press, Chicago

Richards RJ (2010) Myth 19: That Darwin and Haeckel were complicit in Nazi biology. http://home.uchicago.edu/~rjr6/articles/Myth.pdf

Richards EH (2011) Euthenics, the science of controllable environment: a plea for better living conditions as a first step toward higher human efficiency. Nabu Press, Charleston

Richerson P, Baldini R, Bell A et al (2016) Cultural group selection plays an essential role in explaining human cooperation: a sketch of the evidence. Behav Brain Sci 39:e30

Richerson PJ, Boyd R (1997) The evolution of human ultra-sociality. In: Eibl-Eibesfeldt I, Salter F (eds) Ideology, warfare, and indoctrinability. Berghahn, Oxford, pp 71–95

Richerson PJ, Boyd R (2005) Not by genes alone. How culture transformed human evolution. University of Chicago Press, Chicago

Richerson PJ, Boyd R, Bettinger RL (2009) Cultural innovations and demographic change. Hum Biol 81(2–3):211–235

Richerson PJ, Boyd R, Henrich J (2003) Cultural evolution of human cooperation. In: Hammerstein P (ed) Genetic and cultural evolution of cooperation. MIT Press, Cambridge, pp 357–388

Richerson PJ, Boyd R, Henrich J (2010) Gene-culture coevolution in the age of genomics. PNAS 107(Supplement 2):8985–8992

Richman S (2008) Fascism. The concise encyclopedia of economics. http://www.econlib.org/LibraryEnc1/Fascism.html

Richter R, Grün R, Joannes-Boyau R et al (2017) The age of the hominin fossils from Jebel Irhoud, Morocco, and the origins of the Middle Stone Age. Nature 546:293–296

Riddle JM (1992) Contraception and abortion from the ancient world to the renaissance. Harvard University Press, Cambridge

Ridley M (1993) The Red queen. Sex and the evolution of human nature. Penguin Books, New York

Ridley M (1996) The origins of virtue: human instincts and the evolution of cooperation. Penguin, London

Rietveld MJH, Posthuma D, Dolan CV et al (2003) ADHD. Sibling interaction or dominance: an evaluation of statistical power. Behav Genet 33(3):247–255

Riexinger M (2010) Islamic opposition to the Darwinian theory of evolution. In: Lewis JR, Hammer O (eds) Handbook of religion and the authority of science. Brill, Leiden, pp 483–510

Rifkin J (1983) Algeny - New word, new world. Viking, New York

Rijsdijsk FV, Viding E, De Brito S et al (2010) Heritable variations in gray matter concentration as a potential endophenotype for psychopathic traits. Arch Gen Psychiatry 67(4):406–413

Rilling JK, Glenne AL, Jairama MR et al (2007) Neural correlates of social cooperation and non-cooperation as a function of psychopathy. Biol Psychiatry 61(11):1260–1271

Rilling JK, Gutman DA, Zeh T et al (2002) A neural basis for social cooperation. Neuron 35 (2):395–405

Rilling JK, Sanfey AG (2011) The neuroscience of social decision-making. Ann Rev Psychol 62:23–48

Rindfuss RR, Stephen EH (1990) Marital noncohabitation: separation does not make the heart grow fonder. J Marr Fam 52(1):259–269

Ritsner M, Sherina O, Ginath Y (1992) Genetic epidemiological study of schizophrenia: reproduction behaviour. Acta Psychiatr Scand 85(6):423–429

Rizzolatti G, Craighero L (2004) The mirror neuron system. Ann Rev Neurosci 27(1):169–192
Roach J, Pease K (2011) Evolution and the prevention of violent crime. Psychology 2(4):393–404
Roache R, Clarke S (2009) Bioconservatism, bioliberalism, and the wisdom of reflecting on repugnance. Monash Bioethics Rev 28(1):1–21
Roberts DE (1999) Killing the black body: race, reproduction and the meaning of liberty. Vintage Books, New York
Roberts G (2013) When punishment pays. PLOS ONE 8(3):e57378
Roberts BW, Jackson JJ (2008) Sociogenomic personality psychology. Personality 76(6):1523–1544
Robertson JA (1994) Children of choice: freedom, and the new reproductive technologies. Princeton University Press, Princeton
Robertson D, Snarey J, Ousley O et al (2007) Neural processing of moral sensitivity to issues of justice and care. Neuropsychologia 45(8):755–766
Robine JM, Vaupel JW (2002) Emergence of supercentenarians in low mortality countries. North Am Actuarial J 6(3):54–63
Robinson R (1964) An atheist's values. The Clarendon Press, Oxford
Robinson GE, Grozinger CM, Whitfield CW (2005) Sociogenomics: social life in molecular terms. Nat Rev Genet 6:257–270. https://doi.org/10.1038/nrg1575
Robinson WC, Ross JA (eds) (2007) The global family planning revolution: three decades of population policies and programs. The World Bank, Washington
Robson AJ (2001) The biological basis of economic behavior. J Econ Lit 39(1):11–33
Roccas S (2005) Religion and value systems. J Soc Issues 61(4):747–759
Rock PJ (1996) Eugenics and euthanasia: a cause for concern for disabled people, particularly disabled women. Disabil Soc 11(1):121–128
Rockenbach B, Milinski M (2006) The efficient interaction of indirect reciprocity and costly punishment. Nature 444(7120):718–723
Rockström J, Steffen W, Noone K et al (2009) Planetary boundaries: exploring the safe operating space for humanity. Ecol Soc 14(2):32
Rockström J, Steffen W, Noone K et al (2009) A safe operating space for humanity. Nature 461 (7263):472–475
Rocque M, Welsh BC, Raine A (2012) Biosocial criminology and modern crime prevention. J Crim Justice 40(4):306–312
Rodgers JL, Cleveland HH, van den Ord E et al (2000) Do large families make low-IQ children, or do low-IQ parents make large families? Am Psychol 55(6):599–612
Rodinson M (1996) Muhammad. Penguin Books, London
Rodrigues SM, Saslow LR, Garcia N et al (2009) Oxytocin receptor genetic variation relates to empathy and stress reactivity in humans. PNAS 106(50):21437–21441
Roduit JAR, Heilinger JC, Baumann H (2015) Ideas of perfection and the ethics of human enhancement. Bioethics 29(9SI):622–630
Roede MJ (1988) Women's biological inferiority: abuse of sociobiology by certain feminists. The sociobiology of sex differences. Abstracts of the 11th meeting of the European Sociobiological Society, Held in the Veterinary College, Oslo: 2–4 Sept 1988
Roele M (1993) Religious behavior as a utility and inclusive fitness optimizing strategy. Soc Sci Inf 32(3):387–417
Roes FL, Raymond M (2003) Belief in moralizing gods. Evol Hum Behav 24(2):126–135
Rogers AR (1990) Evolutionary economics of human reproduction. Ethol Sociobiol 11(6):479–495
Rogers SA, Paloutzian RF (2006) Schizophrenia, neurology, and religion: what can psychosis teach us about the evolutionary role of religion? In: McNamara P (ed) Where God and science meet: how brain and evolutionary studies alter our understanding of religion, vol 3. The psychology of religious experience. Praeger, Westport, pp 161–186
Rokeach M (1973) The nature of human values. Free Press, New York

Rolston H III (1999) Genes, genesis and God: values and their origins in natural and human history. Cambridge University Press, Cambridge
Ropelato J (2011) Internet pornography statistics. http://internet-filter-review.toptenreviews.com/internet-pornography-statistics.html
Rosas A (2010) Evolutionary game theory meets social science: is there a unifying rule for human cooperation? J Theor Biol 264(2):450–456
Rose S (2005) The 21st century brain. Jonathan Cape, London
Rosenfeld A, Wise N (2001) The over-scheduled child: avoiding the hyper-parenting Trap. St Martin's Press, New York
Rosenkranz P, Charlton BG (2013) Individual differences in existential orientation: empathizing and systemizing explain the sex difference in religious orientation and science acceptance. Arch Psychol Relig-Arch fur Relig Psychol 35(1):119–146
Rosenlee LL (2006) Confucianism and women. A philosophical interpretation. Suny series in Chinese philosophy and culture. State University of New York Press, Albany
Ross JA (1992) Sterilization: past, present, future. Stud Family Plann 23(3):187–198
Ross MW, Wells AL (2000) The modernist fallacy in homosexual selection theories: homosexual and homosocial exaptation in South Asian society. Psychology, Evolution & Gender 2(3):253–262
Rossano M (2007) The religious mind and the evolution of religious forms. The Global Spiral. http://www.metanexus.net/magazine/tabid/68/id/9771/Default.aspx
Rossano M (2010) Supernatural selection: how religion evolved. Oxford University Press, New York
Rossi A (1977) A biosocial perspective on parenting. Daedalus 106(2):1–32
Rossi A, Rossi P (1990) Of human bonding. Parent-child relationships across the life course. Aldine de Gruyter, New York
Rota M (2013) The problem of evil and cooperation. In: Nowak MA, Coakley S (eds) Evolution, games, and God. The principle of cooperation. Harvard University Press, Cambridge, pp 362–374
Rothman S, Rothman D (2003) The pursuit of perfection: the promise and perils of medical enhancement. Pantheon, New York
Rothschild ML (1990) Bionomics: the inevitability of capitalism. H. Holt, New York
Rothstein MA (1995) Genetic testing: employability, insurability, and health reform. J Nat Cancer Inst—Monogr 17:87–90
Rottschaefer WA (1998) The biology and psychology of moral agency. Cambridge University Press, Cambridge
Rottschaefer WA (2000) It was a pleasure, but that's not why I did it: are Sober and Wilson too generous toward their selfish brethren? In: Katz LD (ed) Evolutionary origins of morality: cross-disciplinary perspectives. Imprint Academic, Exeter, pp 239–243
Rottschaefer WA (2007) Mythic religious naturalism. Zygon 42(2):369–408
Roughgarden J (2006) Evolution and Christian faith: reflections of an evolutionary biologist. Island Press, Washongton
Roussel L (1989) La famille incertaine. (The uncertain family). Jacob, Paris
Rowe D (2002) Biology and crime. Roxbury, Los Angeles
Rowland DT (2007) Historical trends in childlessness. J Family Issues 28(10):1311–1337
Rowthorn R (2011) Religion, fertility and genes: a dual inheritance model. Proc R Soc B 278 (1717):2519–2527
Royal Society (2006) A statement by the royal society on evolution, creationism and intelligent design. http://royalsociety.org/
Royal Society (2012) Brain waves module 3: neuroscience, conflict and security. Royal Society Policy document 06/11, London
Rubin PH (2002) Darwinian politics. The evolutionary origin of freedom. Rutgers University Press, New Brunswick

Ruckenbauer HW (2002) Moralität zwischen Evolution und Normen: eine Kritik biologistischer Ansätze in der Ethik. (Morality between evolution and standards: a critique of biologically deterministic approaches in ethics). Königshausen and Neumann, Würzburg

Rueckert L, Naybar N (2008) Gender differences in empathy: the role of the right hemisphere. Brain Cogn 67(2):162–167

Ruether RR (1993) Ecofeminism: symbolic and social connections of the oppression of women and the domination of nature. In: Adams CJ (ed) Ecofeminism and the sacred. The Continuum Publishing Company, New York, p 13

Ruff C (2002) Variation in human body size and shape. Ann Rev Anthropol 31:211–232

Rummel RJ (1994) Power, genocide and mass murder. J Peace Res 31(1):1–10

Rummel RJ (1995) Democracies are less warlike than other regimes. Eur J Int Relat 1(4):457–479

Rummel RJ (2002) Power kills: democracy as a method of nonviolence. Transaction Publishers, New Brunswick

Rumney PNS (2006) False allegations of rape. Camb Law J 65(1):128–158

Rusch H (2014) The two sides of warfare: an extended model of altruistic behavior in ancestral human intergroup conflict. Hum Nat 25(3):359–377

Rusch H, Leunissen JM, van Vugt M (2015) Historical and experimental evidence of sexual selection for war heroism. Evol Hum Behav 36(5):367–373

Ruse M (1979) Sociobiology: sense or nonsense?. D. Reidel Publishing Co, Dordrecht

Ruse M (1986) Evolutionary ethics: a phoenix arisen. Zygon 21(1):95–112

Ruse M (1986) Taking Darwin seriously: a naturalistic approach to philosophy. Blackwell, Oxford

Ruse M (1989) The Darwinian paradigm: essays on its history, philosophy, and religious implications. Routledge, London

Ruse M (1993) The new evolutionary ethics. In: Nitecki MH, Nitecki DV (eds) Evolutionary ethics. State University of New York Press, Albany, pp 133–162

Ruse M (1994) Evolutionary naturalism: selected essays. Routledge, London

Ruse M (1996) Monad to man: the concept of progress in evolutionary biology. Harvard University Press, Cambridge

Ruse M (1999) Evolutionary ethics in the 20th century: Julian Sorell Huxley and George Gaylord Simpson. In: Maienschein J, Ruse M (eds) Biology and the foundation of ethics. Cambridge University Press, New York, pp 198–217

Ruse M (1999) Evolutionary ethics: what can we learn from the past? Zygon 34(3):435–452

Ruse M (2000) Can a Darwinian be a Christian? Sociobiological issues. Zygon 35(2):299–331

Ruse M (2001) Can a Darwinian be a Christian? The relationship between science and religion. Oxford University Press, New York

Ruse M (2008) Evolution and religion: a dialogue. Rowman and Littlefield Publishers, Lanham

Ruse M (2010) Science and spirituality: making room for faith in the age of science. Cambridge University Press, Cambridge

Ruse M (2015) Atheism: what everyone needs to know. Oxford University Press, Oxford

Ruse M, Wilson EO (1985) The evolution of ethics. New Sci 108(1478):50–52

Ruse M, Wilson EO (1986) Darwinism as applied science. Philosophy 61(236):173–192

Rushby AJ, Claire MW, Osborn H et al (2013) Habitable zone lifetimes of exoplanets around main sequence stars. Astrobiology 13(9):833–849

Rushdie S (2005) Do we have to fight the battle for the enlightenment over again? The Independent, 22 Jan 2005

Rushdie S (2005) The right time for an Islamic reformation. The Washington Post, 7 Aug 2005

Rushton JP (2004) Genetic and environmental contributions to pro-social attitudes: a twin study of social responsibility. Proc Biol Sci 271(1557):2583–2585

Rushton JP, Fulker DW, Neale MC et al (1984) Altruism and genetics. Acta Geneticae Med et Gemellologiae 33(2):265–271

Rushton JP, Fulker DW, Neale MC et al (1986) Altruism and aggression: the heritability of individual differences. J Person Soc Psychol 50(6):1192–1198

Russell B (1957) Why I am not a Christian. Simon and Schuster, New York

Russell B (1997) Religion and science. Oxford University Press, New York
Russell D (1977) Pornography. A feminist perspective. Russell Publications, Berkeley
Russell C (2010) The limits of liberal choice: racial selection and reprogenetics. South J Philos 48 (S1):97–108
Russian Academy of Science (1995) Human losses of the USSR in the period of WWII: collection of articles (In Russian). Saint-Petersburg. (Rossiiskaia Akademiia nauk. Liudskie poteri SSSR v period vtoroi mirovoi voiny: sbornik statei. Sankt-Peterburg 1995 ISBN 5-86789-023-6)
Rutherford SL, Lindquist S (1998) Hsp90 as a capacitor for morphological evolution. Nature 396:336–342
Rutter M (1972) Maternal deprivation reassessed. Penguin Books, Harmondsworth
Rutter M, Kim-Cohen J, Maughan B (2006) Continuities and discontinuities in psychopathology between childhood and adult life. J Child Psychol Psychiatry 47(3–4):276–295
Ruyle EE (1973) Genetic and cultural pools: some suggestions for a unified theory of biocultural evolution. Hum Ecol 1(3):201–215
Sääksvuori L, Mappes T, Puurtinen M (2011) Costly punishment prevails in intergroup conflict. Proc R Soc B: Biol Sci 278(1723):3428–3436
Saavedra S, Smith D, Reed-Tsochas F (2010) Cooperation under indirect reciprocity and imitative trust. PLOS ONE 5(10):e13475
Sachchidananda A, Lal K (1980) Elite and development. Naurang Rai, New Delhi
Sachdeva S, Singh P, Medin D (2011) Culture and the quest for universal principles in moral reasoning. Int J Psychol 46(3):161–176
Sacher GA (1959) Relation of lifespan to brain weight and body weight in mammals. In: Wolstenholme GEW, O'Connor M (eds) The lifespan of animals. CIBA Foundation Colloquia on Aging, vol 5. Churchill, London, pp 115–133
Sachsenmaier D, Eisenstadt SN, Riedel J (2002) Reflections on multiple modernities: European, Chinese, and other interpretations. Brill Academic Publications, Leiden
Sackmann I-J, Boothroyd AI, Kraemer KE (1993) Our sun. III. Present and future. Astrophys J 418:457–468
Sacred Congregation for the Doctrine of the Faith (1975) Persona humana. Declaration on certain questions concerning sexual ethics. The Vatican
Sacred Congregation for the Doctrine of the Faith (1980) Declaration on euthanasia. The Vatican
Safa RF (1997) Inside Islam: exposing and reaching the world of Islam. Charisma House, Peabody
Sagarin R, Taylor T (eds) (2008) Natural security: a Darwinian approach to a dangerous world. University of California Press, Berkeley
Sahlins M (1976, 1979) The use and abuse of biology. An anthropological critique of sociobiology. The University of Michigan Press, Ann Arbor
Sahni M, Verma N, Narula D et al (2008) Missing girls in India: infanticide, feticide and made-to-order pregnancies? Insights from hospital-based sex-ratio-at-birth over the last century. PLoS ONE 3(5):e2224
Salihu HM, Shumpert MN, Slay M et al (2003) Childbearing beyond maternal age 50 and fetal outcomes in the United States. Obstet Gynecol 102(5):1006–1014
Saller K (1961) Die Rassenlehre des Nationalsozialismus in Wissenschaft und Propaganda. (The racial theory of Nazism in science and propaganda). Progress-Verlag, Darmstadt
Salmon CA, Shackelford TK (eds) (2007) Family relationships: an evolutionary perspective. Oxford University Press, New York
Salomonsson M (2010) Group selection: the quest for social preferences. J Theor Biol 264(3):737–746
Salter F (2003) On genetic interests: family, ethny, and humanity in an age of mass migration. Peter Lang, Frankfurt
Salter F (2004) Welfare, ethnicity and altruism: new findings and evolutionary theory. Frank Cass, London

Salter F (2008) Class mobility, environment and genes. A Darwinian conflict analysis. In: Niedenzu HJ, Meleghy T, Meyer P (eds) The new evolutionary social science: human nature, social behavior, and social change. Paradigm Publishers, Boulder, pp 159–171

Salter F, Harpending H (2013) JP Rushton's theory of ethnic nepotism. Person Individ Diff 55(3): SI 256–260

Sánchez-Barricarte JJ, Fernández-Carro R (2007) Patterns in the delay and recovery of fertility in Europe. Eur J Popul 23(2):145–170

Sandberg A (2001), Morphological freedom—why we not just want it, but need it. Presented at Transvision 2001, 22–24 June 2001, Berlin. HTTP: http://www.nada.kth.se/~asa/Texts/MorphologicalFreedom.htm

Sandel M (2004) The case against perfection. What's wrong with designer children, bionic athletes, and genetic engineering. The Atlantic Monthly, April 2004. https://www.theatlantic.com/past/docs/issues/2004/04/sandel.htm

Sandel M (2007) The case against perfection. Ethics in the age of genetic engineering. Harvard University Press, Cambridge

Sanderson SK (2001) The evolution of human sociality. A Darwinian conflict perspective. Rowman and Littlefield, Lanham

Sanderson SK (2008) Adaptation, evolution, and religion. Religion 38(2):141–156

Sanderson SK, Roberts WW (2008) The evolutionary forms of the religious life: a cross-cultural, quantitative study. Am Anthropol 110(4):454–466

Sandu A (2015) The anthropology of immortality and the crisis of posthuman conscience. J Study Relig Ideologies 14(40):3–26

Santelli J, Ott MA, Lyon M et al (2006) Abstinence and abstinence-only education: a review of U. S. policies and programs. J Adolesc Health 38(1):72–81

Santos FP, Santos FC, Pacheco JM (2016) Social norms of cooperation in small-scale societies. PLoS Comput Biol 12(1):e1004709

Santow G (1978) A simulation approach to the study of human fertility. A NIDI/CBGS Publication. Martinus Nijhoff, Leiden

Sardon JP (2002) La conjoncture démographique: L'Europe et les pays développés d'outre-mer. (The demographic situation: Europe and the developed overseas countries). Population 57 (1):123–170

Saroglou V (2010) Religiousness as a cultural adaptation of basic traits: a five-factor model perspective. Person Soc Psychol Rev 14(1):108–125

Saroglou V, Muñoz-García A (2008) Individual differences in religion and spirituality: an issue of personality traits and/or values. J Sci Study Relig 47(1):83–101

Sartorius C (2003) An evolutionary approach to social welfare. Routledge, London

Sasaki JY, Kim HS, Mojaverian T et al (2013) Religion priming differentially increases prosocial behavior among variants of the dopamine D4 receptor (DRD4) gene. Soc Cogn Affect Neurosci 8(2):209–215

Sasaki JY, Kim HS, Xu J (2011) Religion and well-being: the moderating role of culture and the oxytocin receptor (OXTR) gene. J Cross-Cult Psychol 42(8):SI 1394–1405

Sasaki T, Uchida S (2013) The evolution of cooperation by social exclusion. Proc R Soc B 280 (1752):20122498

Saslow LR, John OP, Piff PK et al (2013) The social significance of spirituality: new perspectives on the compassion-altruism relationship. Psychol Relig Spirituality 5(3):201–218

Sauerteig L, Davidson R (eds) (2008) Shaping sexual knowledge: a cultural history of sex education in twentieth century Europe. Routledge, London

Saugstad LF (1989) Social class, marriage, and fertility in schizophrenia. Schizophr Bull 15(1):9–43

Savage M, Egerton M (1997) Social mobility, individual ability, and the inheritance of class inequality. Sociology 31(4):645–672

Saver JL, Rabin J (1997) The neural substrates of religious experience. J Neuropsychiatry 9 (3):498–510

Savulescu J (2001) Procreative beneficence: why we should select the best children. Bioethics 15 (5–6):413–426
Savulescu J (2009) Enhancing human capabilities. Wiley-Blackwell, Oxford
Savulescu J (2015) Procreative beneficence, diversity, intersubjectivity, and imprecision. Am J Bioethics 15(6):16–18
Savulescu J, Bostrom N (eds) (2009) Human enhancement. Oxford University Press, Oxford
Savulescu J, Kahane G (2009) Procreative beneficence and disability: is there a moral obligation to create children with the best chance of the best life? Bioethics 23(5):274–290
Savulescu J, Sandberg A (2008) Neuroenhancement of love and marriage: the chemicals between us. Neuroethics 1(1):33–44
Savulescu J, ter Meulen R, Kahane G (2011) Enhancing human capabilities. Wiley-Blackwell, Oxford
Sawaguchi T, Kudo H (1990) Neocortical development and social structure in primates. Primates 31(2):283–290
Saxton A (2009) The great God debate and the future of faith. Free Inq 29(1):37–43
Schaar J (1967) Equality of opportunity, and beyond. In: Pennock JR, Chapman JW (eds) Equality. Atherton Press, New York, pp 228–249
Schachter SC (2006) Religion and the brain. Evidence from temporal lobe epilepsy. In: McNamara P (ed) Where God and science meet: how brain and evolutionary studies alter our understanding of religion, vol 2. The neurology of religious experience. Praeger, Westport, pp 171–188
Schallmayer W (1903) Vererbung und Auslese im Lebenslauf der Völker. Eine staatswissenschaftliche Studie auf Grund der neueren Biologie. (Inheritance and selection in the life course of the peoples. A political science study based on the new biology). Verlag von Gustav Fischer, Jena
Scheidt W (1925) Allgemeine Rassenkunde. J.F.Lehmanns Verlag, München
Schelling TC (1960) The strategy of conflict. Oxford University Press, New York
Schieman S, Bierman A, Ellison CG (2010) Religious involvement, beliefs about God, and the sense of mattering among older adults. J Sci Study Relig 49(3):517–535
Schieve LA, Wilcox LS, Zeitz J, et al (2002) Assessment of outcomes for assisted reproductive technology: overview of issues and the US experience in establishing a surveillance system. In: Vayena E, Rowe PJ, Griffin PD (eds) Medical, ethical and social aspects of assisted reproduction. Current practices and controversies in assisted reproduction: report of a WHO Meeting. WHO Publications, Geneva, pp 363–376
Schilbach L, Wohlschlaeger AM, Kraemer NC et al (2006) Being with virtual others: neural correlates of social interaction. Neuropsychologia 44(5):718–730
Schillebeeckx E (1984) On christian faith: the spiritual, ethical and political dimension. Crossroad, New York
Schinka JA, Busch RM, Robichaux-Keene N (2004) A meta-analysis of the association between the serotonin transporter gene polymorphism (5-HTTLPR) and trait anxiety. Mol Psychiatry 9 (2):197–202
Schinka JA, Letsch EA, Crawford FC (2002) DRD4 and novelty seeking: results of meta-analyses. Am J Med Genet 114:643–648
Schirmann F (2013) Invoking the brain in studying morality: a theoretical and historical perspective on the neuroscience of morality. Theory Psychol 23(3):289–304
Schjoedt U, Geertz A, Stødkild-Jørgensen H et al (2009) Highly religious participants recruit areas of social cognition in personal prayer. Soc Cogn Affect Neurosci 4(2):199–207
Schjoedt U, Stødkilde-Jørgensen H, Geertz A et al (2010) The power of charisma: perceived charisma inhibits the attentional and executive systems of believers in intercessory prayer. Soc Cogn Affect Neurosci 4(2):199–207
Schloss JP (2004) Evolutionary ethics and Christian morality: surveying the issues. In: Clayton P, Schloss J (eds) Evolution and ethics: human morality in biological and religious perspective. Eerdmans, Grand Rapids, pp 1–24

Schloss JP (2013) Unpredicted outcomes in the games of life. In: Nowak MA, Coakley S (eds) Evolution, games, and God. The principle of cooperation. Harvard University Press, Cambridge, pp 201–219

Schloss JP, Murray MJ (2011) Evolutionary accounts of belief in supernatural punishment: a critical review. Relig Brain Behav 1(1):46–99

Schmeiser SR (2008) The ungovernable citizen: psychopathy, sexuality, and the rise of medico-legal reasoning. Yale J Law Humanit 20(2):163–240

Schmid J (1984) The background of recent fertility trends in the member states of the Council of Europe. Population studies no 15. Council of Europe, Strasbourg

Schmidt MFH, Sommerville JA (2011) Fairness expecations and altruistic sharing in 15-month-old human infants. PLOS ONE 6(10):e23223

Schmitt DP, Alcalay L, Allik J et al (2004) Patterns and universals of mate poaching across 53 nations: the effects of sex, culture, and personality on romantically attracting another person's partner. J Person Soc Psychol 86(4):560–584

Schmitt DP, Shackelford TK, Buss DM (2001) Are men really more 'oriented' toward short-term mating than women? A critical review of theory and research. Psychol Evol Gender 3(3):211–239

Schoenmaeckers R, Callens M (eds) (1999) Gezinsvorming in Brussel. Resultaten van de 'Fertility and Family Survey' (FFS) in het Brussels Hoofdstedelijk Gewest. (Family building in Brussels. Results from the Fertility and Family Survey (FFS) in the Brussels Metropolitan Region). CBGS-Publicaties. Centrum voor Bevolkings- en Gezinsstudie. Garant, Leuven-Apeldoorn

Schoenmaeckers R, Lodewijckx E (1999) Religious affiliation, education and number of children: patterns of change in flanders (Belgium). Eur J Popul 15(3):207–240

Schramm K, Skinner D, Rottenburg R (eds) (2012) Identity politics and the new genetics: re/creating categories of difference and belonging. Berghahn Books, New York

Schröder KP, Smith RC (2008) Distant future of the sun and earth revisited. Mon Not R Astron Soc 386(1):155–163

Schroeder GL (2001) The hidden face of God: how science reveals the ultimate truth. Free Press, New York

Schroeder E, Kuriansky J (eds) (2009) Sexuality education: past, present, and future. Praeger, Westport

Schroeder KB, McElreath R, Nettle D (2013) Variants at serotonin transporter and 2A receptor genes predict cooperative behavior differentially according to presence of punishment. PNAS 110(10):3955–3960

Schulte-Rüther M, Markowitsch HJ, Shah NJ et al (2008) Gender differences in brain networks supporting empathy. NeuroImage 42(1):393–403

Schultes RC, Hoffmann A, Rätsch C (2001) Plants of the Gods: their sacred, healing, and hollucinogenic powers. Healing Arts Press, Rochester

Schwartz G, Kim RM, Kolundzija AB et al (2010) Biodemographic and physical correlates of sexual orientation in men. Arch Sex Behav 39(1):93–109

Schwidetzky I (1950) Grundzüge der Völkerbiologie. (Principles of ethnobiology). Ferdinand Enke Publisher, Stuttgart

Scott EC (2005) Evolution vs. creationism: an introduction. Greenwood, Westport

Scourfield JJB, Martin N, McGuffin P (2004) The development of prosocial behaviour in children and adolescents: a twin study. J Child Psychol Psychiatry 45(5):927–935

Scully JL (2008) Disability bioethics: moral bodies, moral difference. Rowman and Littlefield, Lanham

Sebastian-Enesco C, Hernandez-Lloreda MV, Colmenares F (2013) Two and a half-year-old children are prosocial even when their partners are not. J Exp Child Psychol 116(2):SI 186–198

Sedikides C (2010) Why does religiosity persist? Person Soc Psychol Rev 14(1):3–6

Sedikides C, Gebauer JE (2010) Religiosity as self-enhancement: a meta-analysis of the relation between socially desirable responding and religiosity. Person Soc Psychol Rev 14(1):17–36

Segerstraele U (1988) Sociobiology and feminism—enemies or allies? The sociobiology of sex differences. Abstracts of the 11th meeting of the European Sociobiological Society, Held in the Veterinary College, Oslo, 2–4 Sept

Seitz RJ, Nickel J, Azari NP (2006) Functional modularity of the medial prefrontal cortex: involvement in human empathy. Neuropsychology 20(6):743–751

Selgelid MJ (2014) Moderate eugenics and human enhancement. Med Health Care Philos 17(1):3–12

Sellars RW, Bragg R (1933) The humanist Manifesto. American Humanist Association, Washington

Sen S, Burmeister M, Ghosh D (2004) Meta-analysis of the association between a serotonin transporter promoter polymorphism (5-HTTLPR) and anxiety-related personality traits. Am J Med Genet 127B(1):85–89

Seto MC, Maric A, Barbaree HE (2001) The role of pornography in the etiology of sexual aggression. Aggression Violent Behav 6(1):35–53

Seung S (2013) Connectome: how the brain's wiring makes us who we are. Mariner Books, Boston

Sevinc G, Spreng RN (2014) Contextual and perceptual brain processes underlying moral cognition: a quantitative meta-analysis of moral reasoning and moral emotions. PLOS ONE 9 (2):e87427

Shackelford T, Hanson R (eds) (2014) The evolution of human sexuality. Springer, New York

Shackelford TK, Hansen RD (eds) (2015) The evolution of morality. Springer, New York

Shackelford TK, Larsen RJ (1999) Facial attractiveness and physical health. Evol Hum Behav 20 (1):71–76

Shackelford TK, Pound N (eds) (2006) Sperm competition in humans: classic and contemporary readings. Springer, New York

Shaftesbury AAC (1711/1999) Characteristics of men, manners, opinions, times. Cambridge University Press, Cambridge

Shanavas TO (2005) Evolution and/or creation: an Islamic perspective. Xlibris Corporation, Philadelphia

Shanks N (2004) God, the devil, and Darwin: a critique of intelligent design. Oxford University Press, Oxford

Sharabany R, Mayseless O, Edri G (2001) Ecology, childhood experiences, and adult attachment styles of women in the kibbutz. Int J Behav Dev 25(3):214–225

Shariff AF, Norenzayan A (2007) God is watching you: priming God concepts increases prosocial behavior in an anonymous economic game. Psychol Sci 18(9):803–880

Shariff AF, Willard AK, Andersen T et al (2016) Religious priming: a meta-analysis with a focus on prosociality. Person Soc Psychol Rev 20(1):27–48

Sharma A (1987) Women in world religions. State University of New York Press, New York

Sharpe KJ (1991) Science and religion: from warfare over sociobiology to a working alliance. Curr Contents 23:5–13

Shatz SM (2008) IQ and fertility: a cross-national study. Intelligence 36(2):109–111

Shaw D (2006) Genetic morality. Peter Lang, Bern

Shaw RP, Wong Y (1989) Genetic seeds of warfare: evolution, nationalism, and patriotism. Unwin Hyman, London

Sheils WJ, Wood D (1987) The church and wealth: papers read at the 1986 summer meeting and the 1987 winter meeting of the ecclesiastical history Society. Boydell and Brewer, Martlesham

Shelley-Tremblay JF, Rosén LA (1996) Attention deficit hyperactivity disorder: an evolutionary perspective. J Genet Psychol 157(4):443–453

Shenk MK, Towner MC, Kress HC et al (2013) Model comparison approach shows stronger support for economic models of fertility decline. PNAS 110(20):8045–8050

Shepher J (1983) Incest: a biosocial view. Academic Press, New York

Shepher J, Reisman J (1985) Pornography: a sociobiological attempt at understanding. Ethol Sociobiol 6(2):103–114

Sherkat DE (2008) Beyond belief: atheism, agnosticism, and theistic certainty in the United States. Sociol Spect 28(5):438–459
Sherkat DE (2011) Religion and scientific literacy in the United States. Soc Sci Q 92(5):1134–1150
Sherkat DE, Ellison CG (1999) Recent developments and current controversies in the sociology of religion. Ann Rev Sociol 25(1):363–394
Shermer M (1997) Why people believe weird things: pseudoscience, superstition and other confusions of our time. W.H. Freeman, New York
Shermer M (2003) How we believe: science, skepticism and the search for God. AW Freeman/Owl Book, New York
Shermer M (2004) The science of good and evil: why people cheat, gossip, care, share, and follow the golden rule. Holt, New York
Shermer M (2007) Why Darwin matters: the case against intelligent design. Holt Paperbacks, New York
Shermer M (2011) The believing brain: from ghosts and gods to politics and conspiracies—how we construct beliefs and reinforce them as truths. Times Books, New York
Shichor D, Decker DL, O'Brien RM (1979) Population density and criminal victimization. Some unexpected findings in central cities. Criminology 17(2):184–193
Shimao H, Nakamaru M (2013) Strict or graduated punishment? Effect of punishment strictness on the evolution of cooperation in continuous public goods games. PLOS ONE 8(3):e59894
Shoemaker WJ (2012) The social brain network and human moral behaviour. Zygon 47(4):806–820
Shook JR (2012) Neuroethics and the possible types of moral enhancement. AJOB Neurosci 3 (4):3–14
Short RV, Potts M (2009) The impact of population growth on tomorrow's world. Philos Trans R Soc B 364:1532
Shults FL (2015) How to survive the anthropocene: adaptive atheism and the evolution of Homo deiparensis. Religions 6(2):724–741
Shutters ST (2012) Punishment leads to cooperative behavior in structured societies. Evol Comput 20(2):301–319
Sibly RM, Curnow RN (2012) Evolution of discrimination in populations at equilibrium between selfishness and altruism. J Theor Biol 313:162–171
Sidanius J, Pratto F (1999) Social dominance: an intergroup theory of social hierarchy and oppression. Cambridge University Press, New York
Siddle R, Haddock G, Tarrier N et al (2002) Religious delusions in patients admitted to hospital with schizophrenia. Soc Psychiatry Psychiatr Epidemiol 37(3):130–138
Siderits M (2007) Buddhism as philosophy: an introduction. Hackett Pub Co, Cambridge
Sidgwick H (1876) The theory of evolution and its application in practice. Mind 1(1):52–67
Siegel DJ (2001) Toward an interpersonal neurobiology of the developing mind: attachment relationships, 'mindsight' and neural integration. Infant Ment Health 22(1–2):67–94
Sigmund K (2012) Moral assessment in indirect reciprocity. J Theor Biol 299, Special Issue:SI 25–30
Sigvardsson S, Cloninger R, Bohman R et al (1982) Predisposition to petty criminality in Swedish adoptees: sex differences and validation of the male typology. Arch Gen Psychiatry 39 (11):1248–1253
Siker JS (2007) Homosexuality and religion: an encyclopedia. Greenwood Press, Westport
Silberman NA, Finkelstein I (2002) The Bible unearthed: archaeology's new vision of ancient Israel and the origin of its sacred texts. Free Press, New York
Silk JB (1990) Human adoption, evolutionary perspective. Hum Nat 1(1):25–52
Silk JB, Brosnan SF, Henrich J et al (2013) Chimpanzees share food for many reasons: the role of kinship, reciprocity, social bonds and harassment on food transfers. Anim Behav 85(5):941–947

Silk JB, House BR (2011) Evolutionary foundations of human prosocial sentiments. PNAS 108 (S2):10910–10917
Silver L (1998) Remaking Eden: cloning and beyond in a Brave New World. Avon, New York
Silverman I (2003) Confessions of a closet sociobiologist: personal perspectives on the Darwinian movement in psychology. Evol Psychol 1(1):1–9
Silverman I, Case D (1998) Ethnocentrism vs. pragmatism in the conduct of human affairs. In: Eibl-Eibesfeldt I, Salter FK (eds) Indoctrinability, ideology, and warfare: evolutionary perspectives. Berghahn Books, New York, pp 389–406
Silverstein M, Long JD (1998) Trajectories of grandparents' perceived solidarity with adult grandchildren: a growth curve analysis over 23 years. J Marriage Family 60(4):912–923
Sim DC (1997) Christianity and ethnicity in the gospel of Matthew. In: Brett MG (ed) Ethnicity and the Bible. Brill Academic Publishers, Leiden, pp 171–195
Simon J (1981, 1998) The ultimate resource. Princeton University Press, Princeton
Simon J, Kahn H (eds) (1984) The resourceful earth: a response to "Global 2000". Basil Blackwell, Oxford
Simpson GG (1944) Tempo and mode in evolution. Columbia University Press, New York
Simpson GG (1949) The meaning of evolution: a study of the history of life and of its significance for man. Yale University Press, New Haven
Simpson EH (1951) The interpretation of interaction in contingency tables. J R Stat Soc B 13 (2):238–241
Simpson GG (1964) This view of life. Harcourt, brace, and world, New York
Simpson GG (1969) Biology and ethics. In: Simpson GG (ed) Biology and man. Harcourt brace and world, New York, pp 130–148
Simpson JA, Kenrick DT (eds) (1997) Evolutionary social psychology. Lawrence Erlbaum Associates, Hillsdale
Sinclair D, Dangerfield P (1998) Human growth after birth. Oxford University Press, New York
Singh D (1993) Adaptive significance of female physical attractiveness: role of waist-to-hip ratio. J Person Soc Psychol 65(2):293–307
Singer P (1981) Ethics and sociobiology. Philos Public Aff 11(1):40–64
Singer P (1981) The expanding circle. Farrar. Straus and Giroux, New York
Singer P (1984) Ethics and sociobiology. Zygon 19(2):141–158
Singer P (1993) How are we to live? Ethics in an age of self-interest. Text Publishing Company, Melbourne
Singer P (1999) A Darwinian left: politics, evolution and cooperation. Weidenfeld and Nicholson, London
Singer P (2002) One world: the ethics of globalization. Yale University Press, New Haven
Singer P (2009) Parental choice and human improvement. In: Savulescu J, Bostrom N (eds) Human enhancement. Oxford University Press, Oxford, pp 277–289
Singer T, Seymour B, O'Doherty JP et al (2006) Empathic neural responses are modulated by the perceived fairness of others. Nature 439(7075):466–469
Sinnott-Armstrong W (ed) (2008) Moral psychology, volume 1: the evolution of morality: adaptations and innateness. Volume 2: the cognitive science of morality: Intuition and diversity. Volume 3: the neuroscience of morality: Emotion, brain disorders, and development. MIT Press, Cambridge
Sipe AWR (1990) A secret world: sexuality and the search for celibacy. Brunner/Mazel, New York
Sirimanne CR (2016) Buddhism and women—the Dhamma has no gender. J Int Women's Stud 18 (1):273–299
Sjoberg RL, Ducci F, Barr CS et al (2008) A non-addictive interaction of a functional MAO-A VNTR and testosterone predicts antisocial behavior. Neuropsychopharmacology 33(2):425–430
Skaine R (2005) Female genital mutilation: legal, cultural and medical issues. McFarland and Company, Jefferson

Skene L, Thompson J (eds) (2001) The sorting society. The ethics of genetic screening and therapy. Cambridge University Press, Cambridge

Skirbekk V (2008) Fertility trends by social status. Demographic Res 18(5):145–180

Skinner MK (2015) Environmental epigenetics and a unified theory of the molecular aspects of evolution: a neo-Lamarckian concept that facilitates neo-Darwinian evolution. Genome Biol Evol 7(5):1296–1302

Sklar RL (1979) The nature of class domination in Africa. J Mod Afr Stud 17(4):531–552

Skybreak A (2006) The science of evolution and the myth of creationism: knowing what's real and why it matters. Insight Press, Chicago

Slater E, Hare EH, Price JS (1971) Marriage and fertility of psychiatric patients compared with national data. Soc Biol 18:S60–S73

Slavich GM, Cole SW (2013) The emerging field of human social genomics. Clin Psych Sc 1 (3):331–348

Slembrouck J (ed) (2010) Onze waarden? Welke toekomst voor het humanisme? (Our values? What future for humanism?). University Press Antwerp, Brussel

Sloan RP, Bagiella E (2002) Claims about religious involvement and health outcomes. Ann Behav Med 24(1):14–21

Slocum S (1980) Woman the gatherer: male bias in anthropology. Issues in feminism: a first course in women's studies. Houghton Mifflin, Boston

Smail JK (1997) Averting the 21st century's demographic crisis: can human numbers be reduced by 75%? Popul Environ 18(6):565–580

Smail JK (1997) Beyond population stabilization: the case for dramatically reducing global human numbers. Polit Life Sci 16(2):183–192

Smail JK (1997) Population growth seems to affect everything but is seldom held responsible for anything. Polit Life Sci 16(2):231–236

Smail JK (2002) Confronting a surfeit of people: reducing global human numbers to sustainable levels. An essay on population two centuries after Malthus. Environ Dev Sustain 4(1):21–50

Smaldino PE (2014) The cultural evolution of emergent group-level traits. Behav Brain Sci 37 (3):243–254

Smaldino PE, Schank JC, McElreath R (2013) Increased costs of cooperation help cooperators in the long run. Am Nat 181(4):451–463

Smead R (2010) Indirect reciprocity and the evolution of "moral signals'. Biol Philos 25(1):33–51

Smirnov O, Arrow H, Kennett D et al (2007) Ancestral war and the evolutionary origins of heroism. J Polit 69(4):927–940

Smith A (1759) (2006) The theory of moral sentiments. Dover, Mineola

Smith A (1776) An inquiry into the nature and causes of the wealth of nations. W. Strahan, London

Smith A (1991) National identity. University of Nevada Press, Reno

Smith A (1995) Nations and nationalism in a global era. Polity Press, Cambridge

Smith A (1998) Nationalism and modernism: a critical survey of recent theories of nations and nationalism. Routledge, London

Smith MJ (1998) Ecologism: toward ecological citizenship. Open University Press, Buckingham

Smith DL (2007) The most dangerous animal: human nature and the origins of war. St. Martin's Press, New York

Smith P (2007) Why has aggression been thought of as maladaptive? In: Hawley P, Little T, Rodkin P (eds) Aggression and adaptation: the bright side to bad behavior. Lawrence Erlbaum, Mahwah, pp 65–83

Smith EA (2010) Communication and collective action: language and the evolution of human cooperation. Evol Hum Behav 31(4):231–245

Smith SS, Arnett PA, Newman JP (1992) Neuropsychological differentiation of psychopathic and nonpsychopathic criminal offenders. Person Individ Diff 13(11):1233–1243

Smith EA, Borgerhoff Mulder M, Bowles S et al (2010) Production systems, inheritance, and inequality in premodern societies: conclusions. Curr Anthropol 51(1):85–94

Smith DA, Visher CA (1980) Sex and involvement in deviance/crime: a quantitative review of the empirical literature. Am Sociol Rev 45(4):691–701

Snarey J (1996) The natural environment's impact upon religious ethics: a cross-cultural study. J Sci Study Relig 35(2):85–96

Sober E, Wilson DS (1998) Unto others: the evolution and psychology of unselfish behavior. Harvard University Press, Cambridge

Sobotka T (2009) Does persistent low fertility threaten the future of European populations? In: Surkyn J, Deboosere P (eds) Demographic challenges for the 21st century: a state of the art in demography. ASP—Academic and Scientific Publishers, Brussels

Sobotka T, Martin MA, Jensen TK et al (2008) The contribution of assisted reproduction to completed fertility: an analysis of Danish data. Popul Dev Rev 34(1):79–101

Society for Aesthetic Plastic Surgery (2010) Cosmetic surgery national data bank statistics. Aesthetic Surg J, 1–18

Soler M (2008) Commitment costs and cooperation: evidence from Cadomble, an AfroBrazilian religion. In: Bulbulia R, Sosis E, Harris C et al (eds) The evolution of religion: studies, theories, and critiques. Collins Foundation Press, Santa Margarita, pp 167–173

Soltis J, Boyd R, Richerson PJ (1995) Can group-functional behaviors evolve by cultural group selection? An empirical test. Curr Anthropol 36(3):473–494

Somit A (1976) Biology and politics. Recent explorations. Mouton, The Hague

Somit A, Peterson SA (1997) Darwinism, dominance, and democracy. The biological bases of authoritarianism. Praeger, Westport

Sommers CH (1994) Who stole feminism? How women have betrayed women. Simon and Schuster, New York

Sorgner S (2009) Nietzsche, the overhuman, and transhumanism. J Evol Technol 20(1):29–42

Sorgner S (2010) Beyond humanism: reflections on trans- and posthumanism. J Evol Technol 21 (2):1–19

Sorgner S, Birx HJ, Knoepffler N (eds) (2006) Eugenik und die Zukunft. (Eugenics and the future). Alber Verlag, Freiburg i. B

Sorokin P (1927) Social mobility. Harper and Brothers, New York

Sosis R (2000) Religion and intragroup cooperation: preliminary results of a comparative analysis of utopian communities. Cross-Cult Res 34(1):70–87

Sosis R (2003) Why aren't we all Hutterites? Costly signaling theory and religious behavior. Hum Nat 14(2):91–127

Sosis R (2006) Religious behaviors, badges, and bans: signaling theory and the evolution of religion. In: McNamara P (ed) Where God and science meet: how brain and evolutionary studies alter our understanding of religion, vol 1. Evolution, genes, and the religious brain. Praeger, Westport, pp 61–86

Sosis R (2009) The adaptationist-byproduct debate on the evolution of religion: five misunderstandings of the adaptationist program. J Cogn Cult 9(3):315–332

Sosis R, Alcorta CS (2008) Militants and martyrs. Evolutionary perspectives on religion and terrorism. In: Sagarin R, Taylor T (eds) Darwinian security: perspectives from ecology and evolution. University of California Press, Berkeley, pp 105–124

Sosis R, Ruffle BJ (2004) Ideology, religion, and the evolution of cooperation. Field experiments on Israeli Kibbutzim. Res Econ Anthropol 23:89–117

Southgate C (2008) The groaning of creation: god, evolution, and the problem of evil. Westminster John Knox Press, Louisville

Soyfer VN (1994) Lysenko and the tragedy of Soviet science. Rutgers University Press, New Brunswick

Sparrow R (2011) Liberalism and eugenics. Australas J Philos 89(3):499–517

Sparrow R (2015) Imposing genetic diversity. Am J Bioethics 15(6):2–10

Spector J (2006) Prostitution and pornography: philosophical debate about the sex industry. Stanford University Press, Stanford

Spencer H (1851–1888) Social statics. D. Appleton, New York

Spencer H (1862–1898) First Principles. Williams and Norgate, London
Spencer H (1864) Principles of biology. Williams and Norgate, London
Spencer H (1879) The data of ethics. Hurst and Co, New York
Spencer H (1892) The principles of ethics. Williams and Norgate, London
Sperber D (1985) On anthropological knowledge: three essays. Cambridge University Press, Cambridge
Spilka B, Hood RW Jr, Hunsberger B et al (2003) The psychology of religion: an empirical approach. Guilford Press, New York
Spira A, Bajos N et le groupe ACSF (1993) Les comportements sexuels en France. (Sexual behavior in France). La Documentation Française, Paris
Spiro PJ (2007) Non-governmental organizations and civil society. In: Bodansky D, Brunnée J, Hey E (eds) The Oxford handbook of international environmental law. Oxford University Press, Oxford, pp 771–790
Spong JS (2005) The sin of the scripture. HarperOne; HarperCollins, New York
Sprinkle RA (2010) Moral suasion, installed. Polit Life Sci 29(1):88–89
Sproul BC (1979) Primal myths: creating the world. Harper and Row, New York
Staatscommissie voor de Ethische Problemen (1976), Rapport aangaande de anticonceptie. (Report regarding contraception). Ministerie van Justitie, Brussel
Staatscommissie voor de Ethische Problemen (1977) Abortus: Aalyse van de twee verslagen van de Staatscommssie voor de Ethische Problemen ingediend bij de Eerste-minister op 15 juni 1976. (Abortion: analysis of the two reports of the State commission for the ethical problems submitted to the Prime Minister on June 15, 1976). Reinaert Uitgaven, Brussel
Stacey J (1983) The new conservative feminism. Feminist Stud 9(3):559–583
Stambler I (2010) Life extension—a conservative enterprise? Some fin-de-siècle and early twentieth century precursors of transhumanism. J Evol Technol 21(1):13–26
Stamos DN (2008) Evolution and the big questions: sex, race, religion, and other matters. Wiley-Blackwell, Malden
Stanley M (2014) Huxley's church and Maxwell's demon. From theistic science to naturalistic science. University of Chicago Press, Chicago
Stapledon O (1930) Last and first men: a story of the near and far future. Methuen, London
Stapledon O (1935) Odd John: a story between jest and earnest. Methuen, London
Stark R (1997) The rise of christianity: how the obscure, marginal Jesus movement became the dominant religious force in the western world in a few centuries. HarperCollins, San Francisco
Stark R (2001) One true God: historical consequences of monotheism. Princeton University Press, Princeton
Stark R (2002) Physiology and faith: addressing the 'universal' gender difference in religious commitment. J Sci Study Relig 41(3):495–507
Stark R (2003) For the glory of God: how monotheism led to reformations, science, witch-hunts, and the end of slavery. Princeton University Press, Princeton
Stark R, Finke R (2000) Acts of faith. Explaining the human side of religion. University of California Press, Berkeley
Steadman LB, Palmer CT (2008) The supernatural and natural selection: religion and evolutionary success. Paradigm Publishers, Boulder
Stearns SC (2007) Are we stalled part way through a major evolutionary transition from individual to group ? Evol 61(10):2275–2280
Stearns SC, Byars SG, Govindaraju DR et al (2010) Measuring selection in contemporary human populations. Nat Rev Genet 11(9):611–622
Stebbing LS (1941) Ideals and illusions. Watts, London
Steckel RH, Wallis J (2009) Stones, bones, and states. A new approach to the neolithic revolution. http://www.nber.org:confer/2007/daeso7.steckel.pdf
Steerneman AGM, Van Der Meulen EA, Schaafsma W et al (1992) Testing some hypotheses on human evolution and sexual dimorphism. Hum Evol 8(1):25–31

Steffen W, Grinevald J, Crutzen P et al (2011) The anthropocene: conceptual and historical perspectives. Philos Trans R Soc A 369(1938):842–867

Steffensmeier DJ (1980) Sex differences in patterns of adult crime, 1965-1977: a review and assessment. Soc Forces 58(4):1080–1108

Steger MB, Roy RK (2010) Neoliberalism: a very short introduction. Oxford University Press, New York

Steinbock B (2008) Designer babies: choosing our children's genes. Lancet 372(9646):1294–1295

Steinem G (1980) Erotica and pornography: a clear and present difference. In: Lederer L (ed) Take back the night: women on pornography. William Morrow, New York, pp 35–39

Stenger V (2003) Has science found god? The latest results in the search for purpose in the universe. Prometheus Books, Amherst

Stenger V (2008) God: the failed hypothesis. How science shows that god does not exist. Prometheus Books, Amherst

Stenger V (2009) The new atheism: taking a stand for science and reason. Prometheus Books, Amherst

Stenger V (2011) The fallacy of fine-tuning: why the universe is not designed for us. Prometheus Books, Amherst

Stenger V (2012) God and the folly of faith: the incompatibility of science and religion. Prometheus Books, Amherst

Stengers J (1971) Les pratiques anticonceptionnelles dans le mariage au XIXe et au XXe siècle: problèmes humains et attitudes religieuses. (The contraceptive practices in marriage in the nineteenth and twentieth century: human problems and religious attitudes). Belgisch Tijdschrift voor Filologie en Geschiedenis 49:403–481; 1119–1174

Stenmark M (2001) Scientism: science, ethics and religion. Ashgate, Aldershot

Stent GS (1969) The coming of the golden age: a view of the end of progress. Doubleday, New York

Stent GS (1976) The promise of structural ethics. Hastings Cent Rep 6(6):2–40

Stent GS (ed) (1980) Morality as a biological phenomenon: the presuppositions of sociobiological research. University of California Press, Berkeley

Stephen L (1882–1907) The science of ethics. Smith, Elder, London

Stevens A, Price J (1996) Evolutionary psychiatry: a new beginning. Routledge, London

Stevenson CL (1963) Facts and values. Yale University Press, New Haven

Stewart J (2008) Evolutionary manifesto. Our role in the future evolution of life. http://www.evolutionarymanifesto.com

Stewart AJ, Plotkin JB (2014) Collapse of cooperation in evolving games. PNAS 111(49):17558–17563

Stewart AL, Reynolds EO, Lipscomb AP (1981) Outcome for infants of very low birthweight: survey of world literature. Lancet 9, 1, 8228:1038–1040

Stiglitz JE (2003) Globalization and its discontents. Norton, New York

Stock G (2002) Redesigning humans: our inevitable genetic future. Houghton Mifflin, Boston

Stock G, Campbell J (2000) Engineering the human germline: an exploration of the science and ethics of altering the genes we pass to our children. Oxford University Press, New York

Stonawski M, Skirbekk V, Kaufmann E et al (2015) The end of secularisation through demography? Projections of Spanish religiosity. J Contemp Relig 30(1):1–21

Stone L, Lurquin PF, Cavalli-Sforza LL (2007) Genes, culture, and human evolution: a synthesis. Wiley-Blackwell, Malden

Strachan T, Read A (2010) Human molecular genetics. Garland Science, Oxford

Strassmann BI, Kurapati NT, Hug BF et al (2012) Religion as a means to assure paternity. PNAS 109(25):9781–9785

Streeter SA, McBurney DH (2003) Waist-hip ratio and attractiveness: new evidence and a critique of "A Critical Test". Evol Hum Behav 24(2):88–98

Strickland SS, Shetty PS (1998) Human biology and social inequality. Cambridge University Press Cambridge

Stringer C (2012) Lone survivors: how we came to be the only humans on earth. Times Books, New York

Stringer C, Andrews P (2005) The complete world of human evolution. Thames and Hudson, London

Strobel A, Zimmermann J, Schmitz A et al (2011) Beyond revenge: neural and genetic bases of altruistic punishment. Neuroimage 54(1):671–680

Stroeykens S (2016) Het einde van de wereld. (The end of the world). Pelckmans, Kalmthout

Strong J (1885) Our country: its possible future and its present crisis. Baker and Taylor, New York

Strossen N (2000) Defending pornography: free speech, sex, and the fight for women's rights. New York University Press, New York

Struthers WM (2009) Wired for intimacy: how pornography hijacks the male brain. IVP Books, Downers Grove

Stump E, Murphy MJ (eds) (1999) Philosophy of religion: the big questions. Basil Blackwell, Oxford

Sturgeon N (2016) Ecofeminist natures: race, gender, feminist theory and political action. Routledge, London

Suarez E, Gadalla TM (2010) Stop blaming the victim: a meta-analysis on rape myths. J Interpers Violence 25(11):2010–2035

Sugirtharajah S (2002) Hinduism and feminism—some concerns. J Feminist Stud Relig 18(2):97–104

Sul S, Tobler PN, Hein G et al (2015) Spatial gradient in value representation along the medial prefrontal cortex reflects individual differences in prosociality. PNAS 112(25):7851–7856

Sullins DP (2006) Gender and religion: deconstructing universality, constructing complexity. Am J Sociol 112(3):838–880

Sumner WG (1883) What social classes owe to each other. Harper and Bros, New York

Sumner WG (1914) The challenge of facts and other essays. Yale University Press, New Haven

Surovell TA (2008) Extinctions of big game. Elsevier, Amsterdam

Sussman RW, Cloninger CR (eds) (2011) Origins of altruism and cooperation. Springer, New York

Sutter J (1950) L'Eugénique. Problèmes, méthodes, résultats. Travaux et Documents, Cahier n° 11. Institut national d'études démographiques. Presses Universitaires de France, Paris

Sutton WS (1903) The chromosomes in heredity. Biol Bull 4:231

Sutton M (2009) Lowest-low fertility in Japan: consequences for a once-great nation. Bull Geogr Socio-econ Ser 11:5–14

Sutton A (2012) Germ-line gene therapy could prove a two-edged tool. Christ Bioethics 18 (2):145–155

Suziedelis A, Potvin RH (1981) Sex differences in factors affecting religiousness among catholic adolescents. J Sci Study Relig 20:38–50

Suzuki S, Kimura H (2013) Indirect reciprocity is sensitive to costs of information transfer. Sci Rep 3:1435. https://doi.org/10.1038/srep01435

Svensson E, Calsbeek R (eds) (2012) The adaptive landscape in evolutionary biology. Oxford University Press, Oxford

Swaab DF (2004) Sexual differentiation of the human brain: relevance for gender identity, transsexualism and sexual orientation. Gynecol Endocrinol 19(6):301–312

Swaab DF (2007) Sexual diffentiation of the brain and behavior. Best Pract Res Clin Endocrinol Metab 21:431–444

Swami Nikhilananda (2017) Ethics. An essay on Hindu ethics. Ramakrishna—Vivekananda Center, New York: Burton Watson. Columbia University Press, New York (http://www.hinduism.co.za/-ethics.htm)

Swanson GE (1960) The birth of the gods. University of Michigan, Ann Arbor

Sykes B (2003) Adam's curse. A future without men. Corgi Books, London

Sylwester K, Roberts G (2013) Reputation-based partner choice is an effective alternative to indirect reciprocity in solving social dilemmas. Evol Hum Behav 34(3):201–206

Symons D (1979) The evolution of human sexuality. Oxford University Press, New York
Tainter J (1988) The collapse of complex societies. Cambridge University Press, Cambridge
Tahir E, Yazgan Y, Cirakoglu B et al (2000) Association and link of DRD4 and DRD5 with attention deficit hyperactivity disorder (ADHD) in a sample of Turkish children. Mol Psychiatry 5(4):396–404
Tajfel H, Turner JC (1986) An integrative theory of intergroup conflict. In: Worchel S, Austin WG (eds) Psychology of intergroup relations. Nelson-Hall, Chicago, pp 2–24
Takahashi H, Takano H, Camerer CF et al (2012) Honesty mediates the relationship between serotonin and reaction to unfairness. PNAS 109(11):4281–4284
Talbott WJ (2015) How could a "blind" evolutionary process have made human moral beliefs sensitive to strongly universal, objective moral standards? Biol Philos 30(5):691–708
Tan S-L, Lu J-H, Yu X-H et al (2013) Evolution and maintenance of cooperation via inheritance of neighborhood relationship. Chin Sci Bull 58(28–29):3491–3498
Tancredi L (2005) Hardwired behavior: what neuroscience reveals about morality. Cambridge University Press, New York
Tang T, Ye H (2016) The blessing of sexuality: evolution of altruism with mating preference. J Artif Soc Soc Simul 19(2):2. https://doi.org/10.18564/jasss.3058
Tanghe KB (2013) The non-mendelian revolution. A conceptual reinterpretation of the genetic revolution. Faculteit Wijsbegeerte en Letteren, Universiteit Gent, Gent
Tankersley D, Stowe CJ, Huettel SA (2007) Altruism is associated with an increased neural response to agency. Nat Neurosci 10(2):150–151
Tanner JM (1962) Growth at adolescence. Blackwell, Oxford
Tanner JM (1978) Foetus into man. Physical growth from conception to maturity. Open Books, London
Tapp RB (ed) (2002) Ecohumanism. Prometheus books, Amherst
Tarnita CE, Antal T, Ohtsuki H et al (2009) Evolutionary dynamics in set structured populations. PNAS 106(21):8601–8604
Tattersall I (2012) Masters of the planet: the search for our human origins. Palgrave Macmillan, New York
Tava VI (2013) The role of non-governmental organisations, peoples and courts in implementing international environmental laws. In: Shawkat A, Bhuiyan MJH, Chowdhury TMR et al (eds) Routledge handbook of international environmental law. Routledge, London, pp 123–135
Taylor L (1984) Born to crime: the genetic causes of criminal behavior. Greenwood Press, Westport
Taylor SE (2002) The tending instinct: how nurturing is essential to who we are and how we live. Times Books, New York
Taylor C (2007) A secular age. Harvard University Press, Cambridge
Taylor M, Quayle E (2003) Child pornography: an internet crime. Brunner-Routledge, Hove
Teehan J (2006) The evolutionary basis of religious ethics. Zygon 41(3):747–775
Teehan J (2009) The evolved brain: understanding religious ethics and religious violence. In: Verplaetse J, de Schrijver J, Vanneste S et al (2009) The moral brain: essays on the evolutionary and neuroscientific aspects of morality. Springer, Dordrecht, pp 233–254
Teehan J (2010) In the name of God. The evolutionary origins of religious ethics and violence. Wiley-Blackwell, Chichester
Teehan J (2013) The cognitive bases of the problem of evil. Monist 96(3):325–348
Teehan J, DiCarlo C (2004) On the naturalistic fallacy: a conceptual basis for evolutionary ethics. Evol Psychol 2:32–46
Teilhard de Chardin P (1956) Le phénomène humain. (The human phenomenon). Les Éditions du Seuil, Paris
Teitelbaum MS (1972) Some genetic implications of population policies. In: Bajema CJ (ed) Eugenics, then and now. Hutchinson and Ross, Stroudsburg, pp 321–331
Teitelbaum MS (1999) Sustained below-replacement fertility in Europe. Popul Bull Unit Nat 40 (41):161–183

Temkin LS (2008) Is living longer living better? J Appl Philos 25(3):193–210
Temkin LS (2013) What's wrong with enhancements? J Med Ethics 39(12):729–731
Tennant N (1983) Evolutionary versus evolved ethics. Philosophy 58(225):289–302
Tenniso MN (2012) Moral transhumanism: the next step. J Med Philos 37(4):405–416
Tennison MN, Moreno JD (2012) Neuroscience, ethics, and national security: the state of the art. PLoS Biol 10(3):e1001289
Tennov D (1999) Science and policy. The dilemma of racial science. In: Thienpont K, Cliquet R (eds) Ingroup-outgroup behaviour in modern societies. An evolutionary perspective. NIDI/CBGS Publications 34, Brussels/The Hague, pp 259–270
Tennyson AL (1908) Poetical works. Macmillan, London
Terhune KW (1974) A review of the actual and expected consequences of family size. Calspan Report nr. DP.5333-G-1, US Department of Health, Education, and Welfare, Public Health Service, Publication nr NIH 75-779
Terman R, Women Living Under Muslim Laws (2007) The stop stoning forever campaign: a report. http://www.meydaan.com/UserFiles/File/Terman_stoning-2.pdf
Thagard P (2005) The emotional coherence of religion. J Cogn Cult 5(1–2):58–74
Thayer BA (2008) Causes of and solutions to Islamic fundamentalist terrorism. In: Sagarin R, Taylor T (eds) Natural security: a Darwinian approach to a dangerous world. University of California Press, Berkeley, pp 125–140
The Holy Bible. English Standard version. Containing the Ol and new Testaments. Crossway Ill
The Qur'an. A new translation by MAS Abdel Haleem. Oxford University Press, Oxford
The Vedas or Books of Holy Knowledge. http://e-libdigital.com/download/the-vedas-or-books-of-holy-knowledge.pdf
The World Bank (2010) Unmet need for contraception. The World Bank, Washington
Thibault O (1972) L'Home inachevé. Biologie et promotion humaine. (The unfinished human. Biology and human development). Casterman, Tournai
Thienpont K (2005) Over daders. Een basisstudie in de biosociale criminology. (About perpetrators. A basic study in biosocial criminology). Academia Press, Gent
Thienpont K, Cliquet R (eds) (1999) Ingroup-outgroup behaviour in modern societies. An evolutionary perspective. NIDI/CBGS Publications, vol 34, Brussels/The Hague
Thiessen D (1996) Bitter-sweet destiny: the stormy evolution of human behavior. Transaction, New Brunswick
Thiessen D, Young RK, Burrough R (1993) Lonely hearts advertisements reflect sexually dimorphic mating strategies. Ethol Sociobiol 14(3):209–229
Thomas B (1984) Evolutionary stability: states and strategies. Theor Popul Biol 26(1):49–67
Thompson P (1990) Evolutionary ethics, Darwinian ethics and ethical naturalism. Hum Evol 5 (2):133–138
Thompson EH (1991) Beneath the status characteristic: gender variations in religiousness. J Sci Study Relig 30(4):381–394
Thompson CB (1993) Socialism vs. capitalism: which is the moral system. On Principle, v1n3 October
Thompson P (ed) (1995) Issues in evolutionary ethics. State University of New York Press, Albany
Thompson P (1999) Evolutionary ethics: its origin and contemporary face. Zygon 34(3):473–484
Thompson TL (2000) The mythic past: biblical archaeology and the myth of Israel. Basic Books, New York
Thomson AM (1959) Maternal stature and reproductive efficiency. Eug Rev 51(3):157–162
Thompson EH, Remmes KR (2002) Does masculinity thwart being religious? An examination of older men's religiousness. J Sci Study Relig 41(3):521–532
Thompson JN, Thoday JM (eds) (1979) Quantitative genetic variations. Academic Press, New York
Thornburgh D, Lin H (2002) Youth, pornography and the internet. National Academies Press, Washington

Thornhill R (1998) Darwinian aesthetics. In: Crawford C, Krebs D (eds) Handbook of evolutionary psychology. L. Erlbaum, Mahway, pp 543–572
Thornhill R, Grammer K (1999) The body and face of woman: one ornament that signals quality? Evol Hum Behav 20(2):105–120
Thornhill NW, Thornhill R (1983) Human rape: an evolutionary analysis. Ethol Sociobiol 4:137–173
Thornhill R, Palmer CT (2000) A natural history of rape. Biological bases of sexual coercion. The MIT Press, Cambridge
Tiberius V (2008) The nativism debate and moral philosophy: comments on Prinz. In: Sinnott-Armstrong W (ed) Moral psychology, vol 1. The evolution of morality: adaptations and innateness. MIT Press, Cambridge, pp 419–426
Tietze C (1981) Induced abortion. A world review 1981. The Population Council, New York
Tiffin L (1994) Creationism's upside-down pyramid: how science refutes fundamentalism. Prometheus Books, Amherst
Tiger L (1999) The decline of males. Golden Books Publishing Company, New York
Tiger L (2010) Much work left to be done in connecting politics and evolution. http://www.cato-unbound.org/archives/july-2010-darwin-and-politics/
Tiger L, Fox R (1971) The imperial animal. Holt, Rinehart and Winston, New York
Tiger L, McGuire M (2010) God's brain. Prometheus Books, Amherst
Tiihonen J, Hodgins S, Vaurio O et al (2000) Amygdaloid volume loss in psychopathy. Society for Neuroscience Abstracts, 2017
Timio M, Verdecchia P, Venanzi S et al (1988) Age and blood pressure changes a 20-year follow-up study in nuns in a secluded order. Hypertension 12(4):457–461
Tirosh-Samuelson H (2011) Engaging transhumanism. In: Hansell GR, Grassie W (eds) H+/-: Transhumanism and its critics. Prometheus Books, Amherst, pp 19–54
Tirosh-Samuelson H (2012) Transhumanism as a secularist faith. Zygon 47(4):710–734
Tishkoff SA, Kidd KK (2004) Implications of biogeography of human populations for 'race' and medicine. Nat Genet 36(11):S21–S27
Tomasello M (2009) Why we cooperate. MIT Press, Cambridge
Tomasello M (2016) A natural history of human morality. Harvard University Press, Cambridge
Tomasello M, Melis AP, Tennie C et al (2012) Two key steps in the evolution of human cooperation the interdependence hypothesis. Curr Anthropol 53(6):673–692
Tonkens R (2015) 'My child will never initiate Ultimate Harm': an argument against moral enhancement. J Med Ethics 41(3):245–251
Tooby J, Cosmides L (1990) The past explains the present: adaptations and the structure of ancestral environments. Ethol Sociobiol 11:375–424
Tooby J, Cosmides L (1992) The psychological foundation of culture. In: Barkow JH, Cosmides L, Tooby J (eds) (1992) The adapted mind. Evolutionary psychology and the generation of culture. Oxford University Press, New York, pp 119–136
Torr JD (2000) Euthanasia: opposing viewpoints. Greenhaven Press, San Diego
Tort P (1992) Darwinisme et société. (Darwinism and Society). Presses Universitaires de France, Paris
Tost H, Kolachana B, Hakimi S et al (2010) A common allele in the oxytocin receptor gene (OXTR) impacts prosocial temperament and human hypothalamic-limbic structure and function. PNAS 107(31):13936–13941
Toumey CP (1993) Evolution and secular humanism. J Am Acad Relig 61(2):275–301
Traulsen A, Nowak MA (2006) Evolution of cooperation by multilevel selection. PNAS 130 (29):10952–10955
Traulsen A, Röhl T, Milinski M (2012) An economic experiment reveals that humans prefer pool punishment to maintain the commons. Proc R Soc B 279(1743):3716–3721
Traulsen A, Sengupta AM, Nowak MA (2005) Stochastic evolutionary dynamics on two levels. J Theor Biol 235:393–401

Travison TG, Araujo AB, O'Doneell AB et al (2007) A population-level decline in serum testosterone levels in American men. J Clin Endocrinol Metab 92(1):196–202
Tremlin T (2006; 2010) Minds and Gods: the cognitive foundations of religionOxford University Press, Oxford
Tricomi E, Rangel A, Camerer CF et al (2010) Neural evidence for inequality-averse social preferences. Nature 463(7284):1089–1091
Trimble MR (2007) The soul in the brain: the cerebral basis of language, art, and belief. Johns Hopkins University Press, Baltimore
Trinkaus E (2005) Early modern humans. Ann Rev Anthropol 34(1):207–230
Trinkhaus E, Shipman P (1993) The Neandertals: changing the image of mankind. Knopf, New York
Trivers RL (1971) The evolution of reciprocal altruism. Q Rev Biol 46(1):35–57
Trivers RL (1972) Parental investment and sexual selection. In: Campbell B (ed) Sexual selection and the descent of man. Aldine, Chicago, pp 136–179
Trivers RL (1974) Parent offspring conflict. Am Ecol 14(1):249–264
Trivers RL (1985) Social evolution. The Benjamin/Cummings P.C, Menlo Park
Trivers RL (2000) The elements of a scientific theory of self-deception. Ann New York Acad Sci 907:114–131
Trivers RL (2002) Natural selection and social theory: selected papers of Robert Trivers. Oxford University Press, New York
Trivers RL (2007) Reciprocal altruism: 30 years later. In: Kappeler PK, van Schaik CP (eds) Cooperation in primates and humans: mechanisms and evolution. Springer, Berlin, pp 67–85
Trivers RL (2011) The folly of fools: the logic of deceit and self-deception in human life. Basic Books, New York
Trivers RL, Willard DE (1973) Natural selection of parental ability to vary the sex ratio of offspring. Science 179(4068):90–92
Trost J (1998) LAT relationships now and in the future. In: Matthijs K (ed) The family. Contemporary perspectives and challenges. Festschrift in honor of Wilfried Dumon. Leuven University Press, Leuven, pp 209–220
Truett KR, Eaves LJ, Walters EE et al (1994) A model system for analysis of family resemblance in extended kinships of twins. Behav Genet 24(1):35–49
Trussell J, Wynn LL (2008) Reducing unintended pregnancy in the United States. Contraception 77(1):1–5
Tsai SJ, Hong CJ, Cheng CY (2002) Serotonin transporter genetic polymorphisms and harm avoidance in the Chinese. Psychiatr Genet 12:165–168
Tsui AO, McDonald-Mosley Burke AE (2010) Family planning and the burden of unintended pregnancies. Epidemiol Rev 32(1):152–174
Tucker ME (2015) World religions, ethics and the earth charter for a sustainable future. In: Rozzi R, Chapin FS III, Callicott JB et al (eds) Earth stewardship: linking ecology and ethics in theory and practice. Springer, Cham, pp 395–405
Tucker JB, Zilinskas RA (2006) The promise and perils of synthetic biology. New Atlantis 12:25–45
Tullberg J, Tullberg BS (1999) Separatism—choice or desperation? In: Thienpont K, Cliquet R (eds) Ingroup-outgroup behaviour in modern societies. An evolutionary perspective. NIDI/CBGS Publications, vol 34 Brussels/The Hague, pp 189–209
Turchin P (2011) Warfare and the evolution of social complexity: a multilevel selection approach. Struct Dyn 4(3):1–37
Turchin P, Currie TE, Turner EAL et al (2013) War, space, and the evolution of old world complex societies. PNAS 110(41):16384–16389
Turda M (ed) (2013) Crafting humans: from genesis to eugenics and beyond. V&R Unipress Göttingen
Turke PW (1989) Evolution and the demand for children. Popul Dev Rev 15(1):61–90

Turke PW (1990) Which humans behave adaptively, and why does it matter? Eth Sociobiol 11 (4):305–339

Turner J, Machalek R, Maryanski A (2015) Handbook of evolution and society: toward an evolutionary social science. Paradigm Press, Boulder

Turner J, Maryanski A (2005) Incest: origins of the taboo. Paradigm Publishers, Boulder

Turner J, Maryanski A (2008) On the origin of societies by natural selection. Paradigm Press, Boulder

Turner J, Maryanski A (2015) Evolutionary sociology: a cross-species strategy for discovering human nature. In: Turner J, Machalek R, Maryanski A (eds) Handbook of evolution and society: toward an evolutionary social science. Paradigm Press, Boulder

Tybur JM (2012) Nothing in morality makes sense except in the light of evolution. A review of Dennis Krebs, the origins of morality: an evolutionary account. Oxford University Press: New York, 2011. Evol Psychol 10(1):35–38

Udry JR (1988) Biological predispositions and social control in adolescent sexual behavior. Am Sociol Rev 53(5):709–722

Udry JR (2000) Biological limits of gender construction. Am Sociol Rev 65(3):443–457

UNDP (2010) Human development report 2010. UN Development Programme. United Nations, New York

UNDP (2013) Human development report 2013 the rise of the south: human progress in a diverse world. UNDP, New York

UNESCO/UN (1997) Universal declaration on the human genome and human rights. UNESCO, Paris

United Nations (1948) Universal declaration of human rights. United Nations, New York

United Nations (1975) Report on the United Nations world population conference, 1974, Bucharest 19–30 Aug 1974. United Nations, New York

United Nations (1975) World conference of the international women's year. Mexico City. United Nations, New York

United Nations (1980) World conference of the United Nations decade for women: equality, development and peace, Copenhagen. United Nations, New York

United Nations (1984), Report on the United Nations international conference on population, 1984, Mexico City, 6–14 Aug 1984. United Nations, New York

United Nations (1985) World conference to review and appraise the achievements of the United Nations decade for women: equality, development and peace, Nairobi. United Nations, New York

United Nations (1989) Convention on the rights of the child. General assembly resolution 25 session 44, 20 Nov 1989. United Nations, New York

United Nations (1992) United Nations framework convention on climate change. United Nations, New York, p 1992

United Nations (1992) Agenda 21. United Nations Environment Programe. United Nations, New York

United Nations (1994) Report on the international conference on population and development, Cairo: 5–13 Sept 1994. United Nations, New York

United Nations (1995) Beijing declaration and platform for action. United Nations, New York

United Nations (1995) Fourth world conference on women, Beijing, China. Action for equality, development and peace. United Nations, New York

United Nations (2002) International plan of action on ageing 2002. Report of the second world assembly on ageing, Madrid, 8–12 Apr 2002. United Nations, New York

United Nations (2006) Convention on the rights of persons with disabilities. United Nations, New York

United Nations (2009) World water development report 2009. Water in a changing world. UNESCO, Paris/Earthscan, London

United Nations (2011) The millennium development goals report 2011. United Nations, New York

United Nations Development Programme (2004) Human development report 2004. Cultural Liberty in Today's Diverse World. United Nations, New York
United Nations Population Division (2004) World population 2300. United Nations, New York
United Nations Population Division (2015) World population prospects. The 2015 revision. United Nations, New York
United Nations Population Division (2011) World contraceptive use 2010. (POP/DB/CP. Rev2010). United Nations, New York
University of Michigan (2006) Human population: a historical sociological prospective. www.globalchange.umich.edu/globalchange2/.../pop_socio.html
Urbanek A (1993) Evolutionary origin of moral principles. In: Nitecki MH, Nitecki DV (eds) Evolutionary ethics. State University of New York Press, Albany, pp 325–335
Ussher J (1650) Annals of the world: James Ussher's classic survey of world history (Modern English republication, ed. Larry and Marion Pierce), Master Books, Green Forest AR, 2003
Vaas R, Blume M (2009) Gott, Gene, und Gehirn. Warum Glaube nützt die evolution der Religiosität. (God, genes and brain). Hirzel, Stuttgart
Vail K III, Rothchild Z, Weise D et al (2010) A terror management analysis of the psychological functions of religion. Person Soc Psychol Rev 14:84–94
Vaillancourt T (2013) Do human females use indirect aggression as an intrasexual competition strategy? Philos Trans R Soc B 368:20130080
Vaknin S (2009) After the rain—how the West lost the East. Narcissus Publications in association with Central Europe Review/CEENMI, Prague and Skopje
Valdecasas AG, Boto L, Correas AM (2013) There is no common ground between science and religion. J Biosci 38(2):181–187
Van Anders SM, Steiger J, Goldey KL (2015) Effects of gendered behavior on testosterone in women and men. PNAS 112(45):13805–13810
Van Bavel J (2008) Subreplacement fertility in the West before the baby boom. Current and contemporary perspectives. Interface demography working paper 2008-1. VUB, Brussel
Van Bavel J (2010) Choice of study discipline and the postponement of motherhood in Europe: the impact of expected earnings, gender composition, and family attitudes. Demography 47 (2):439–458
Van Dongen W (2009), Towards a democratic division of labour in Europe? The combination model as a new integrated approach to professional and family life. The Policy Press, Bristol
Vance T, Maes HH, Kendler KS (2010) Genetic and environmental influences on multiple dimensions of religiosity: a twin study. J Nerv Men Dis 198(10):755–761
Van Dam P (2015) Constructing a modern society through "depillarization". Understanding post-war history as gradual change. J Hist Sociol 28(3):291–313
Van de Kaa DJ (1978) Towards a population policy for Western Europe. In: Council of Europe (ed) Population decline in Europe: implications of a declining or stationary population. Edward Arnold, London, pp 215–233
Van de Kaa DJ (1987) Europe's second demographic transition. Popul Bull 42:1–57
Van De Kaa DJ (2010) Universal history and population change. Demográfia 53(5):5–20
Van den Berg F (2009) Hoe komen we van religie af? Een ongemakkelijke liberale paradox. (How do we get away from religion? An uncomfortable liberal paradox). Houtekiet/Atlas, Antwerpen
Van den Berghe PL (1975) Man in society: a biosocial view. Elsevier, New York
Van den Berghe PL (1988) The family and the biological base of human sociality. In: Filsinger EE (ed) Biosocial perspectives on the family. Sage, Newbury Park, pp 39–60
Van Den Berghe PL (1979) Human family systems. Elsevier, New York
Van den Berghe PL (1980) Incest and exogamy: a sociobiological reconsideration. Ethol Sociobiol 1:151–162
Van den Berghe PL (1990) Why most sociologists don't (and won't) think evolutionarily. Sociol Forum 5(2):173–185

Van den Berghe PL (1999), Racism, ethnocentrism and xenophobia: In our genes or in our memes? In: Thienpont K, Cliquet R (eds) (1999) Ingroup-outgroup behaviour in modern societies. An evolutionary perspective. NIDI/CBGS Publications, vol 34, Brussels/The Hague, pp 21–33

Van den Bogaert G (1976), Profiel van de vrouw in België: Een doorlichting van de relatie- en gezinsvorming en de maatschappelijke achtergrond van de vrouw. (Profile of women in Belgium: a review of the partnership and family formation and the social background of women). De Sikkel/De Nederlansche Boekhandel, Antwerpen

Van den Bos W, Van Dijk E, Westenberg M et al (2009) What motivates repayment? Neural correlates of reciprocity in the Trust Game. Soc Cogn Affect Neurosci 4(3):294–304

van den Broek A, van Campen C, de Haan J, Roeters A, Turkenburg M, Vermeij L (eds) (2016) De toekomst tegemoet: Nederland richting 2050. Leren, werken, zorgen, samenleven en consumeren in het Nederland van later. (Towards the future: Netherlands towards 2050. Learning, working, carrying, living together and consume in future Netherlands). Sociaal en cultureel planbureau, Den Haag

Van Der Dennen JMG (1995) The origin of warT the evolution of a male-coalitional reproductive strategy. 2 volumes. Origin Press, Groningen

van der Steen WJ (1999) Methodological problems in evolutionary biology. XII. Against evolutionary ethics. Acta Biotheor 47(1):41–57

Van der Valk JC, Verhulst FC, Stroet TM et al (1998) Quantitative genetic analysis of internalizing and externalizing problems in a large sample of 3-year-old twins. Twin Res 1:25–33

Van Dijk E, Feith SW (2010) The neuropeptide oxytocin regulates parochial altruism in intergroup conflict among humans. Science 328(5984):1408–1411

Van Duppen D, Hoebeke J (2016) De supersamenwerker. (The super cooperator). EPO, Berchem

Van Heek F (1956) Roman-catholicism and fertility in the Netherlands. Demographic aspects of minority status. Popul Stud 10(2):125–138

van Niekerk AA (2012) Biomedical enhancement: makeability or disenchantment? Tydskrif vir Geesteswetenskappe 52(4):581–595

Van Peer C (ed) (2007) De impact van een (echt)scheiding op kinderen en ex-partners. (The impact of divorce/separation on children and ex-partners). Studiedienst van de Vlaamse Regering, Studies 2007/03, Brussel

Van Praag Ph (1979) Het bevolkingsvraagstuk in België: Ontwikkeling van standpunten en opvattingen (1900–1977). (The population question in Belgium: development of views and beliefs, 1900–1977). Studies en Documenten 12. De Sikkel/De Nederlandsche Boekhandel, Antwerpen/Amsterdam

Van Rooy W (2008) De malaise van de multiculturaliteit. (The malaise of multiculturalism). Acco, Leuven

Van Rooy S, Van Rooy W (eds) (2010) De Islam. Kritische essays over een politieke religie. (Islam. Critical essays on political religion). Academic and Scientific Publishers, Brussel

Van Schaik CP (2015) The primate origins of human nature. Wiley-Blackwell, Hoboken

Van Schaik C, Michel K (2016) The good book of human nature. An evolutionary reading of the Bible. Basic Books, New York

Van Till H (1998) The creation: intelligently designed or optimally equipped? Theol Today 55 (3):344–364

Van Valen L (1973) A new evolutionary law. Evol Theory 1:1–30

Van Waeleghem K, De Clercq N, Vermeulen L et al (1996) Deterioration of sperm quality in young healthy Belgian men. Hum Reprod 11(2):325–329

Vanden Boer L (1999) Over wonen-met-zorg van ouderen. Serviceflats en rusthuis: (g)een paar apart? (On live-in care of the elderly. serviced residences and nursing home: A/no pair apart?). Centrum voor Bevolkings- en Gezinsstudies (CBGS), Vlaamse Wetenschappelijke Instellling. Garant, Leuven

VanderLaan DP, Ren Z, Vasey PL (2013) Male androphilia in the ancestral environment. Hum Nat 24(4):375–401
Vandermassen G (2005) Who's afraid of Charles Darwin? Debating feminism and evolutionary theory. Rowman and Littlefield, Lanham
Vannelli R (2001) Evolutionary theory and human nature. Kluwer Academic Publishers, Boston
Vasey PL, VanderLaan DP (2014) Evolving research on the evolution of male androphilia. Can J Hum Sex 23(3):137–147
Vatsyayana (400 BCE—200 CE?) The Kama Sutra. (Translator: Burton RF). CreateSpace Independent Publishing Platform, Colorado Springs, CO
Veenhoven R (2005) Is life getting better? How long and happily do people live in modern society? Eur Psychol 10(4):330–343
Ventafridda V (2006) According to the 2002 WHO definition of palliative care. Palliat Med 20 (3):159–160
Venter C (2013) Life at the speed of light. from the double helix to the dawn of digital life. Viking, New York
Verbeek P (2006) Everyone's monkey: primate moral roots. In: Killen M, Smetana J (eds) Handbook of moral development. Erlbaum, Mahwah, pp 423–459
Vercauteren M, Susanne C (1985) The secular trend of height and menarche in Belgium: are there any signs of a future stop? Eur J Pediatr 144(4):306–309
Verdoux P (2009) Transhumanism, progress and the future. J Evol Technol 20(2):49–69
Verhage F (1964) Intelligence and religious persuasion. Ned Tijdschr voor de Psychol en haar Grensgebieden 19:247–254
Verhagen E, Sauer PJJ (2005) The Groningen protocol—Euthanasia in severely ill newborns. New Engl J Med 352(10):959–962
Verhofstadt D (2013) Atheïsme als basis voor de moraal. (Atheism as a basis for morality). Houtekiet, Antwerpen
Vermeersch E (1988) De ogen van de panda. Een milieufilosofisch essay. (The eyes of the panda. An environmental philosophical essay). Uitgeverij Marc Van de Wiele, in cooperation with Stichting Leefmilieu, Brugge
Vermeersch E (2011) De multiculturele samenleving. (The multicultural society). Luster, Antwerpen
Vermeersch E (2016) Over god. (About god). Uitgeverij Vrijdag, Antwerpen
Vermes G (2004) The authentic gospel of Jesus. Penguin Books, London
Verplaetse J (2008) Het morele instinct. Over de natuurlijke oorsprong van onze moraal. (The moral instinct. About the natural origins of our morality). Nieuwezijds, Amsterdam
Verplaetse J, de Schrijver J, Vanneste S et al (2009) The moral brain: essays on the evolutionary and neuroscientific aspects of morality. Springer, Dordrecht
Victoroff J (2005) The mind of the terrorist: a review and critique of psychological approaches. J Confl Resolut 49(1):3–42
Viding E, Blair RJR, Moffitt TE et al (2005) Evidence for substantial genetic risk for psychopathy in 7-year-olds. J Child Psychol Psychiatry 46(6):592–597
Vigilant L, Stoneking M, Harpending H et al (1991) African populations and the evolution of human mitochondrial DNA. Science 253(5027):1503–1507
Viney W (2008) Religion and science in Christendom: a history of intellectual warfare and accommodation. Midwest Q 49(4):343–358
Vinge V (1993) The coming technological singularity: how to survive in the post-human era. In: Landis GA (ed) Vision-21: interdisciplinary science and engineering in the era of cyberspace. NASA Publication CP-10129, pp 11–22
Vining D (1986) Social versus reproductive success: the central theoretical problem of human sociobiology. Behav Brain Sci 9(1):167–216
Virkkunen M, Goldman D, Linnoila M (1996) Serotonin in alcoholic violent offenders. Ciba Found Symp 194:68–77

Vivekjivandas S, Dave J (eds) (2013) Hinduism: sacred texts—Vedas, Upanishads, Agamas, Puranas. Swaminarayan Aksharpith, Ahmedabad India

Voland E, Schiefenhövel W (eds) (2009) The biological evolution of religious mind and behavior. Springer, New York

Voland E (1993) Grundriss der Soziobiologie. (Principles of sociobiology). Fischer, Stuttgart

Voland E (1998) Evolutionary ecology of human reproduction. Ann Rev Anthropol 27:347–374

Voland E (2009) Evaluating the evolutionary status of religiosity and religiousness. In: Voland E, Schiefenhövel W (eds) The biological evolution of religious mind and behavior. Springer, New York, pp 9–24

Voland E (2010) Die evolution der Religiosität. (Evolution of religiosity). Biol unserer Zeit 40 (1):29–35

Voland E (2013) Soziobiologie: die evolution von Kooperation und Konkurrenz. (Sociobiology: the evolution of cooperation and competition). Springer Spektrum, Berlin

Voland E, Söling C (2004) Aus der Sicht der Soziobiologie: die biologische Basis der Religiosität in Instinkten—Beiträge zu einer evolutionären Religionstheorie. (From the perspective of sociobiology: the biological basis of religiosity in instincts—contributions to evolutionary theory of religion). In: Lüke U, Schnakenberg J, Souvignier G (eds) Darwin und Gott. Darmstadt, Wissenschaftliche Buchgesellschaft, pp 47–65

Volbrecht MM, Lemery-Chalfant K, Aksan N et al (2007) Examining the familial link between positive affect and empathy development in the second year. J Genet Psychol 168(2):105–129

Volk AA, Atkinson JA (2013) Infant and child death in the human environment of evolutionary adaptation. Evol Hum Behav 34(3):182–192

Volken T (2003) IQ and the wealth of nations: a critique of Richard Lynn and Tatu Vanhanen's recent book. Eur Sociol Rev 19(4):411–412

Vollan B (2012) Weird reciprocity? A 'within-culture across-country' trust experiment and methodological implications. J Inst Econ 8(3):371–397

von Hippel T (1792) Über die bürgerliche Verbesserung der Weiber. (On the civic improvement of women). Verlag, in der Vossischen Buchhandlung, Berlin

Von Schilcher F, Tennant N (1984) Philosophy, evolution and human nature. Routledge and Kegan Paul, London

Wachsmann N, Caplan J (2010) Concentration camps in Nazi Germany: The new histories. Routledge, Oxon

Wachter KW, Finch C (eds) (1997) Between Zeus and the Salmon. The biodemography of longevity. National Academy Press, Washington

Wackernagel M, Yount JD (2000) Footprints for sustainability: the next step. Environ Dev Sustain 2(1):23–44

Wackernagel M, Rees W (1996) Our ecological footprint: reducing human impact on the earth. New Society Publishers, Gabriola Island

Waddington CH (1941) The relations between science and ethics. Nature 148(3749):270–274

Waddington CH (1960) The ethical animal. Allen and Unwin, London

Wade MJ (1978) A critical review of models of group selection. Q Rev Biol 53(2):101–114

Wade N (2009) The faith instinct. How religion evolved and why it endures. The Penguin Press, New York

Wager TD, Phan KL, Liberzon I et al (2003) Valence, gender, and lateralization of functional brain anatomy in emotion: a metaanalysis of findings from neuroimaging. NeuroImage 19:513–531

Wagner GP, Altenberg L (1996) Complex adaptations and the evolution of evolvability. Evolution 50:967–976

Wahlund K, Kristiansson M (2009) Aggression, psychopathy and brain imaging—review and future recommendations. Int J Law Psychiatry 32(4):266–271

Walby K, Carrier N (2010) The rise of biocriminology: capturing observable bodily economies of 'criminal man'. Criminol Crim Justice 10(3):261–285

Waldron-Perrine B, Rapport LJ, Hanks RA et al (2011) Religion and spirituality in rehabilitation outcomes among individuals with traumatic brain injury. Rehabil Psychol 56(2):107–116

Walker R (1996) To be real: telling the truth and changing the face of feminism. Anchor Books, London

Walker M (2009) Enhancing genetic virtue: a project for twenty-first century humanity? Polit Life Sci 28(2):27–47

Walker M (2009) Ship of fools: why transhumanism is the best bet to prevent the extinction of civilization. http://www.metanexus.net/essay/h-ship-fools-why-transhumanism-best-bet-prevent-extinction-civilization

Walker M (2010) In defense of the genetic virtue program. Polit Life Sci 29(1):90–96

Wall D (2010) The rise of the Green Left. Inside the worldwide ecosocialist movement. Pluto, London

Wallace AR (1900) Studies: scientific and social. Macmillan, London

Waller NG, Kojetin BA, Bouchard TJ Jr et al (1990) Genetic and environmental influences on religious interests, attitudes, and values: a study of twins reared apart and together. Psychol Sci 1(2):138–142

Walsh A (1995) Biosociology. An emerging paradigm. Greenwood Publishing, New York

Walsh A (2002) Biosocial criminology. Introduction and integration. Anderson, Cincinnati

Walsh A, Ellis L (eds) (2003) Biosocial criminology: challenging environmentalism's supremacy. Nova Science, New York

Walter N (1997) Humanism—what's in the word. Rationalist Press Association, London

Walter A (2006) The anti-naturalistic fallacy: evolutionary moral psychology and the insistence of brute facts. Evol Psychol 4(1):33–48

Walter T, Davie G (1998) The religiosity of women in the modern West. Br J Sociol 49(4):640–660

Walter NT, Montag C, Markett S et al (2012) Ignorance is no excuse: moral judgments are influenced by a genetic variation on the oxytocin receptor gene. Brain Cogn 78(3):268–273

Walters L, Palmer JG (1997) The ethics of human gene therapy. Oxford University Press, New York

Walton JH (2012) Human origins and the Bible. Zygon 47(4):875–889

Walum H, Wesstberg L, Henningsson S et al (2008) Genetic variation in the vasopressin receptor 1a gene (AVPR1A) associates with pair-bonding behavior in humans. PNAS 105(37):14153–14156

Wang Y (2013) Fairness is a virtue in cooperation. Eur J Sci Theol 9(1):61–70

Wang J, Suri S, Watts DJ (2012) Cooperation and assortativity with dynamic partner updating. PNAS 109(36):14363–14368

Ward P (2001) Future evolution. W.H. Freeman, New York

Ward PD, Brownlee D (2000) Rare earth: why complex life is uncommon in the universe. Copernicus Press, New York

Warneken F, Tomasello M (2007) Helping and cooperation at 14 months of age. Infancy 11 (3):271–294

Warneken F, Tomasello M (2013) The emergence of contingent reciprocity in young children. J Exp Child Psychol 116(2):SI 338–350

Warraq I (1998) The origins of the Koran: classic esays on Islam's holy book. Prometheus Books, Amherst

Warraq I (2009) Democracy vs. theocracy—Islamic human rights and the universal declaration. International Humanist News, February 2009. http://www.iheu.org/democracy-vs-theocracy-islamic-human-rights-and-universal-declaration

Washburn SL (1959) Speculations on the interrelations of the history of tools and biological evolution. Hum Biol 31(1):21–31

Washburn SL (1960) Tools and human evolution. Sci Am 203(3):3–16

Wasserman D (2015) Disability, diversity, and preference for the status quo: bias or justifiable preference? Am J Bioethics 15(6):11–12

Watanabe T, Takezawa M, Nakawake Y et al (2014) Two distinct neural mechanisms underlying indirect reciprocity. PNAS 111(11):3990–3995

Watson JD, Crick FHC (1953) Molecular structure of nucleic acids: a structure for desoxyribose nucleic acid. Nature 171(4356):964–967

Wattenberg BJ (1987) The birth dearth. Pharos Books, New York

Waytz A, Zaki J, Mitchell JP (2012) Response of dorsomedial prefrontal cortex predicts altruistic behavior. J Neurosci 32(22):7646–7650

Weaver TD (2012) Did a discrete event 200,000–100,000 years ago produce modern humans? J Hum Evol 63(1):121–126

Weber S, Habel U, Amunts K et al (2008) Structural brain abnormalities in psychopaths—a review. Behav Sci Law 26(1):7–28

Webster GD (2007) What's in a name: is "evolutionary psychology" eclipsing "sociobiology" in the scientific literature? Evol Psychol 5(4):683–695

Wedekind C, Milinski M (1996) Human cooperation in the simultaneous and the alternating prisoner's dilemma: Pavlov versus generous tot-for-tat. PNAS 93(7):2686–2689

Wedekind C, Milinski M (2000) Cooperation through image scoring in humans. Science 288 (5467):850–852

Weeden J, Abrams MJ, Green MC et al (2006) Do high-status people really have fewer children? Education, income, and fertility in the contemporary US. Hum Nat 17(4):377–392

Weeden J, Cohen AB, Kenrick DT (2008) Religious attendance as reproductive support. Evol Hum Behav 29(5):327–334

Weeden J, Kurzban R (2013) What predicts religiosity? A multinational analysis of reproductive and cooperative morals. Evol Hum Behav 34(6):440–445

Weekes-Shackelford V, Shackelford T (eds) (2014) Evolutionary perspectives on human sexual psychology and behavior. Springer, New York

Weenink AW (2015) Behavioral problems and disorders among radicals in police files. Perspect Terrorism 9(2):17–33

Weidenreich F (1943) The skull of Sinanthropus Pekinensis. Geological Survey of China, Pehpei, Chungking

Weikart R (2004) From Darwin to Hitler. Palgrave Macmillan, New York

Weikart R (2009) Hitler's ethic: the Nazi pursuit of evolutionary progress. Palgrave Macmillan, New York

Weinberg W (1908) Über den Nachweis der Vererbung beim Menschen. (About the proof of inheritance in humans). Jahreshefte des Ver für vaterländische Naturkunde in Wurttemberg 64:368–382

Weinrich JD (1978) Nonreproduction, homosexuality, transsexuality, and intelligence: a systematic literature search. J Homosex 3(3):275–289

Weisbord A (2011) Offshoots of Liberalism—Feminism. http://www.weisbord.org/Feminism.htm

Weismann A (1868), Über die Berechtigung der Darwin'schen Theorie: Ein akademischer Vortrag gehalten am 8. Juli 1868 in der Aula der Universität zu Freiburg im Breisgau. (About the legitimacy of Darwinian theory: an academic lecture held on 8 July 1868 in the auditorium of the University of Freiburg im Breisgau). Engelmann, Leipzig

Weismann A (1892) Das Keimplasma: Eine Theorie der Vererbung. (The germ-plasm: a theory of heredity). Fischer, Jena

Weismann A (1902) Vorträge über Deszendenztheorie gehalten an der Universität zu Freiburg im Breisgau. (Lectures on descendence theory held at the University of Freiburg im Breisgau). Fischer, Jena

Weiss V (2002) Zur Vererbung der Intelligenz, zu Sozialstruktur und Familienpolitik: Eine Nachbetrachtung zum Bericht PISA 2000. (About the inheritance of intelligence, social structure and family policy: a review of the PISA report 2000). Veröff der Ges für Freie Publizistik 18:31–59

Weiss KM (2015) Life goes on. Are humans still evolving? Evol Anthropol 24(3):106–110

Weiss KM (2016) Finding a moral compass…without a compass. Evolution and ethics. Evol Anthropol 25(1):1–5

Weiss KM, Buchanan AV (2009) The mermaid's tale: four billion years of cooperation in the making of living things. Harvard University Press, Cambridge
Weitzer R (ed) (2009) Sex for sale: prostitution, pornography, and the sex industry. Routledge, New York
Wellings K, Field J, Johnson AM et al (1994) Sexual behaviour in Britain. The national survey of sexual attitudes and lifestyles. Penguin, London
Wells HG (1905) A modern Utopia. Chapman and Hall, London
Wells GA (1999) The Jesus myth. Open Court, Chicago
Wenegrat B (1990) The divine archetype. The sociobiology and psychology of religion. Lexington Books, Lexington
Wenneberg SR, Schneider RH, Walton KG et al (1997) A controlled study of the effects of the transcendental meditation program on cardiovascular reactivity and ambulatory blood pressure. Int J Neurosci 89(1–2):15–29
Westermarck E (1906) The origin and development of the moral ideas. Macmillan, London
Westermarck E (1921) The history of human marriage, 5th ed, rev. 3 vols. Allerton. New York
West-Eberhard M (1975) Evolution of social behaviour by kin selection. Q Rev Biol 50(1):1–33
West-Eberhard M (1979) Sexual selection, social competition, and evolution. Proc Am Philos Soc 51(4):222–234
West-Eberhard M (2003) Developmental plasticity and evolution. Oxford University Press, Oxford
West SA, El Mouden C, Gardner A (2011) Sixteen common misconceptions about the evolution of cooperation in humans. Evol Hum Behav 32(4):231–262
Westley SB, Choe MK, Retherford R (2010) Very low fertility in Asia: is there a problem? Can it be solved?. East-West Center, Honolulu
Westmarland N, Gangoli G (2012) International approaches to rape. Polity Press, Bristol
Westoff CF (2006) New estimates of unmet need and the demand for family planning. DHS comparative reports no. 14. Macro International Inc, Calverton
Westoff CF, Frejka T (2007) Religiousness and fertility among European Muslims. Popul Dev Rev 33(4):785–809
Wheale P, McNally R (1988) Genetic engineering. Catastrophe or Utopia? Wheatsheaf, Harvester
Westoff CF, Ryder NB (1977) The contraceptive revolution. The Office of Population Research. Princeton University Press, Princeton
Whelpton PK, Campbell A, Patterson J (1966) Fertility and family planning in the United States. Princeton University Press, Princeton
Whitcomb JC, Morris H (1961) The genesis flood: the Biblical record and its scientific implications. Presbyterian and Reformed Publishing Co, Philadelphia
White E (1972) Genetic diversity and political life: toward a populational interaction paradigm. J Polit 34(4):1203–1242
White S (2007) Equality. Polity Press, Cambridge
White TD, Asfaw B, DeGusta D et al (2003) Pleistocene Homo sapiens from Middle Awash, Ethiopia. Nature 423(6941):742–747
Whitehead JW, Conlan J (1978) The establishment of the religion of secular humanism and its first amendment implications. Texas Tech Law Rev 10(1):1–66
Whitehouse H (2008) Cognitive evolution and religion: cognition and religious evolution. In: Bulbulia J, Sosis R, Harris E et al (eds) The evolution of religion: studies, theories, and critiques. Collins Foundation Press, Santa Margarita, pp 31–41
Whitmeyer JM (1998) On the relationship between memes and genes: a critique of Dennett. Biol Philos 13(2):187–204
Wiebe RP (2004) Psychopathy and sexual coercion: a Darwinian analysis. Couns Clin Psychol J 1 (1):23–41
Wiebe D (2013) Pseudo-speciation of the human race: religions as hazard-precaution systems. Method Theory Study Relig 25(4–5):410–430

Wiederman MW, Allgeier ER (1992) Gender differences in mate selection criteria: sociobiological or socioeconomic explanation? Eth Sociob 13(2):115–124

Wiederman M, Kendall E (1999) Evolution, sex, and jealousy: investigation with a sample from Sweden. Evol Hum Behav 20(2):121–128

Wielemans W (1993) Voorbij het individu. Mensbeelden in wetenschappen. (Beyond the individual. Human representations in sciences). Garant, Leuven-Apeldoorn

Wielenberg EJ (2005) Value and virtue in a godless universe. Cambridge University Press, Cambridge

Wierzbicka A (2007) 'Moral sense'. J Soc Evol Cult Psych 1(3):66–85

Wijkman A, Rockström J (2012) Bankrupting nature. Denying our planetary boundaries. A report to the Club of Rome. Routledge, London

Wiker B (2002) Moral Darwinism. InterVarsity. Downers, Grove

Wilcox DL (2016) A proposed model for the evolutionary creation of human beings: from the image of god to the origin of sin. Perspect Sci Christ Faith 68(1):22–43

Wilcockson M (2008) Issues of life and death. Hodder Murray, London

Wilkins JF, Thurner S (2010) The Jerusalem Game: cultural evolution of the golden rule. Adv Complex Syst 13(5):635–641

Wilkinson W (2005) Capitalism and human nature. Cato Policy Report, XXVII, 1

Wilkinson R (2010) Unjustified discrimination: is the moratorium on the use of genetic test results by insurers a contradiction in terms? Health Care Anal 18(3):279–293

Willey D (1996) An optimum population for Europe. Optimum Population Trust, Manchester

Williams CM (1893) A review of the systems of ethics founded on the theory of evolution. Macmillan, London

Williams GC (1957) Pleiotropy, natural selection, and the evolution of senescence. Evolution 11 (4):398–411

Williams GC (1966) Adaptation and natural selection. Oxford University Press, New York

Williams GC (1975) Sex and evolution. Princeton University Press, Princeton

Williams GC (1988) Huxley's evolution and ethics in sociobiological perspective. Zygon 23 (4):383–407

Williams GC (1992) Natural selection: domains, levels and challenges. Oxford University Press, New York

Williams PA (1996) Christianity and evolutionary ethics: sketch toward a reconciliation. Zygon 31 (2):253–268

Williams ME (2000) Sex: opposing viewpoints. Greenhaven Press, San Diego

Williams PA (2001) Doing without Adam and Eve: sociobiology and original sin. Fortress Press, Minneapolis

Williams GC, Williams DC (1957) Natural selection of individually harmful social adaptations among sibs with special reference to social insects. Evolution 11:32–39

Williams J, Taylor E (2006) The evolution of hyperactivity, impulsivity and cognitive diversity. J R Soc Interface 3(8):399–413

Williamson L (1978) Infanticide: an anthropological analysis. In: Kohl M (ed) Infanticide and the value of life. Prometheus Books, Buffalo, pp 61–75

Williamson SH, Hubisz MJ, Clark AG et al (2007) Localizing recent adaptive evolution in the human genome. PLoS Genet 3(6):e90

Willis-Owen SA, Turri MG, Munafo MR et al (2005) The serotonin transporter length polymorphism, neuroticism, and depression: a comprehensive assessment of association. Biol Psychiatry 58:451–456

Willis E (1984) Radical feminism and feminist radicalism. In: Willis E (ed) No more nice girls: countercultural essays. Wesleyan University Press, Middletown, pp 117–150

Wilmoth JR (1997) In search of limits. In: Wachter KW, Finch C (eds) Between Zeus and the Salmon. The biodemography of longevity. National Academy Press, Washington, pp 38–64

Wilmoth JR (2000) Demography of longevity: past, present, and future trends. Exp Gerontol 35(9–10):1111–1129

Wilson C (2004) Fertility below replacement level. Science 304(5668):207–208

Willmott C (2016) Biological determinism, free will and moral responsibility. Insights from genetics and neuroscience, Springer, New York
Wilmoth JR, Lundström H (1996) Extreme longevity in five countries: presentation of trends with special attention to issues of data quality. Eur J Popul 12(1):63–93
Wilson DS (1975) A theory of group selection. PNAS 72(1):143–146
Wilson EO (1975) Sociobiology. The new synthesis. The Belknap Press of Harvard University Press, Cambridge
Wilson EO (1978) On human nature. Harvard University Press, Cambridge
Wilson EO (1984) Biophilia: the human bond with other species. Harvard University Press, Cambridge
Wilson EO (1992) The diversity of life. Harvard University Press, Cambridge
Wilson JQ (1993) The moral sense. Free Press, New York
Wilson DS (1997) Human groups as units of selection. Science 276(5320):1816–1817
Wilson EO (1998) Consilience: the unity of knowledge. Alfred A. Knopf, New York
Wilson EO (1998) The biological basis of morality. Atlantic Mon 281(4):53–66
Wilson DS (2002) Darwin's cathedral: evolution, religion, and the nature of society. University of Chicago Press, Chicago
Wilson DS (2003) The biological basis of morality. Med Ethics 10(2):1–2
Wilson DS (2007) The role of group selection in human psychological evolution. In: Gangestad SW, Simpson JA (eds) The evolution of mind: fundamental questions and controversies. Guilford press, New York, pp 213–220
Wilson DS (2010) Multilevel selection and major transitions. In: Pigliucci M, Müller GB (eds) Evolution: the extended synthesis. MIT Press, Cambridge, pp 81–94
Wilson EO (2012) The social conquest of earth. Liveright, New York
Wilson EO (2014) The meaning of human existence. Liveright, New York
Wilson L (2014) Buddhism and family. Relig Compass 8(6):188–198
Wilson DS (2015) Does altruism exist? Culture, genes, and the welfare of others. Yale University Press, New Haven
Wilson L (ed) (2013) Family in Buddhism. Suny Press, Albany
Wilson M, Daly M, Gordon S (1998) The evolved psychological apparatus of human decision-making is one source of environmental problems. In: Caro T (ed) Behavioral ecology and conservation biology. Oxford University Press, New York, pp 501–523
Wilson DB, Dolphin WD (1996) Did the devil make Darwin do it? Modern perspectives on the creation-evolution controversy. Iowa State University Press, Ames
Wilson DS, Dugatkin LA (1997) Group selection and assortative interactions. Am Nat 149 (2):336–351
Wilson JQ, Herrnstein RJ (1985) Crime and human nature. Simon and Schuster, New York
Wilson DS, Wilson EO (2007) Rethinking the theoretical foundation of sociobiology. Q Rev Biol 82(4):327–348
Wilson DS, Wilson EO (2008) Evolution "for the good of the group". Am Sci 96:380–389
Wilson G, Rahman Q (2008) Born gay: the psychobiology of sex orientation. Peter Owen Publishers, London
Win-Gallup International (2015) Losing our religion? Two thirds of people still claim to be religious. Press release, 13/04/2015. http://www.wingia.com/en/news/losing_our_religion_two_thirds_of_people_still_claim_to_be_religious/290/
Wink P, Ciciolla L, Dillon M et al (2007) Religiousness, spiritual seeking, and personality: findings from a longitudinal study. J Person 75(5):1051–1070
Winner L (1989) The whale and the reactor: a search for limits in an age of high technology. University of Chicago Press, Chicago
Winterhalder B, Baillargeon W, Cappelleto F et al (1988) The population dynamics of hunter-gatherers and their prey. J Anthropol Archaeol 7:289–328
Wiseman H (2016) The myth of the moral brain: the limits of moral enhancement. MIT Press, Cambridge

Wissenburg M (2007) Ecologie en hedendaags humanisme. (Ecology and contemporary humanism). Tijdschr voor Humanistiek 8(29):2–7
Wlodarski R, Pearce E (2016) The God allusion. Hum Nat 27(2):160–172
Wolbring G (2000) Science and the disadvantaged. http://www.edmonds-institute.orgwolbring.html
Wolbring G (2003) Disability rights approach toward bioethics? J Disabil Policy Stud 14(3):154–180
Wolbring G (2006) The unenhanced underclass. In: Miller P, Wilsdon J (eds) Better humans. The politics of human enhancement and life extension. Demos, London, pp 122–128
Wolf AP (1995) Sexual attraction and childhood association. Stanford University Press, Stanford
Wolf AP, Durham WH (2004) Inbreeding, incest, and the incest taboo: the state of knowledge at the turn of the century. Stanford University Press, Stanford
Wolfe C (2009) What is posthumanism?. University of Minnesota Press, Minneapolis
Wollstonecraft M (1792) A vindication of the rights of woman. Peter Edes, Boston
Wolpert L (2008) Six impossible things before breakfast: the evolutionary origins of belief. Faber and Faber, London
Woltmann L (1899) Die Darwinsche Theorie und der Sozialismus: Ein Beitrag zur Naturgeschichte der menschlichen Gesellschaft. (Darwinian theory and socialism: a contribution to the natural history of human society). H Michels, Düsseldorf
Women Radical (2001) The radical women manifesto: socialist feminist theory, program, and organizational structure. Red Letter Press, Seattle
Wongsurawat W (2006) Pornography and social ills: evidence from the early 1990s. J Appl Econ 9 (1):185–213
Wood JW (1990) Fertility in anthropological populations. Ann Rev Anthropol 19:211–242
Wood JW (1994) Dynamics of human reproduction: biology, biometry, demography. Aldine Transaction, Hawthorne
Wood D (2006) A middling introduction to better humans. J Evol Technol 15(1):109–110
Wood W, Eagly AH (2007) Social structural origins of sex differences in human mating. In: Gangestad SW, Simpson JA (eds) The evolution of mind: fundamental questions and controversies. Guilford press, New York, pp 383–390
Wood JL, Heitmiller D, Andreasen NC et al (2008) Morphology of the ventral frontal cortex: relationship to femininity and social cognition. Cereb Cortex 18(3):534–540
Wood JL, Murko V, Nopoulos P (2008) Ventral frontal cortex in children: morphology, social cognition and femininity/masculinity. Soc Cogn Affect Eurosci 3(2):168–176
Woolcock PG (1999) The case against evolutionary ethics. In: Maienschein J, Ruse M (eds) Biology and the foundation of ethics. Cambridge University Press, New York, pp 276–306
Woolley AW, Chabris CF, Pentland A et al (2010) Evidence for a collective intelligence factor in the performance of human groups. Science 330(6004):686–688
Worchel S, Austin WG (eds) (1986) Psychology of intergroup relations. Nelson-Hall, Chicago
World Health Organisation (2002) Active ageing. A policy framework. WHO, Geneva
Wrangham R (1999) Evolution of coalitionary killing. Yearb Phys Anthropol 42:1–30
Wrangham R, Peterson D (1996) Demonic males: apes and the origins of human violence. Houghton Mifflin, New York
Wray GA (2010) Integrating genomic into evolutionary theory. In: Pigliucci M, Müller GB (eds) Evolution: the extended synthesis. MIT Press, Cambridge
Wright S (1921) Systems of mating. I-V. Genetics 6:111–178
Wright S (1922) Coefficients of inbreeding and relationship. Am Nat 56:330–338
Wright S (1929) The evolution of dominance. Am Nat 63(689):556–561
Wright S (1931) Evolution in Mendelian populations. Genetics 16(2):97–159
Wright S (1945) Tempo and mode in evolution: a critical review. Ecology 26:415–419
Wright S (1965) The interpretation of population structure by F-statistics with special regard to system of mating. Evolution 19:395–420

Wright R (1994) The moral animal: the new science of evolutionary psychology. Pantheon, New York
Wright R (2000) Nonzero: the logic of human destiny. Vintage Books, New York
Wright R (2009) The evolution of God. Little. Brown and Company, New York
Wright C, Jagger G (1999) End of century, end of family? In: Jagger G, Wright C (eds) Changing family values. Edward Elgar, Cheltenham, pp 17–37
Wuerfel J, Krishnamoorthy ES, Brown RC et al (2004) Religiosity is associated with hippocampal but not amygdala volumes in patients with refractory epilepsy. J Neurol Neurosurg Psychiatry 75(4):640–642
Wright JP, Tibbetts SG, Daigle LE (2008) Criminals in the making: criminality across the life course. Sage, Thousand Oaks
Wuketits FM (1984) Evolution, Erkenntnis, Ethik—Folgerungen aus der modernen Biologie. (Evolution, cognition, ethics—implications of modern biology). Wissenschaftliche Buchgesellschaft, Sonderausgabe, Darmstadt
Wynne-Edwards VC (1962) Animal dispersion in relation to social behaviour. Oliver and Boyd, Edinburgh
Xia C-Yi, Zhang J-J, Wang Y-L et al (2011) Evolution of cooperation in public goods games. Commun Theor Phys 56(4):638–644
Xie Y (1990) What is natural fertility? The remodeling of a concept. Popul Index 56(4):656–663
Yahya H (1999) The evolution deceit: the collapse of Darwinism and its ideological background. Ta-Ha Publishers, London (first published in Turkey in 1997)
Yahya H (2006–2007) Atlas of creation, vol 1–3. Global Publishing Company, Istanbul
Yamagishi T, Horita Y, Mifune N et al (2012) Rejection of unfair offers in the ultimatum game is no evidence of strong reciprocity. PNAS 109(50):20364–20368
Yamamoto ME, Leitão M, Castelo-Branco R et al (2009) Religious behavior and cooperation. In: Feierman JR (ed) The biology of religious behavior: the evolutionary origins of faith and religion. ABC-CLIO, LLC, Santa Barbara, pp 219–240
Yang Y, Raine A, Narr KL et al (2009) Localization of deformations within the amygdala in individuals with psychopathy. Arch Gen Psychiatry 66(9):986–994
Yang Q, Yan L, Luo J et al (2013) Temporal dynamics of disgust and morality: an event-related potential study. PLOS ONE 8(5):e65094
Yao S, Långström N, Temrin H et al (2014) Criminal offending as part of an alternative reproductive strategy: investigating evolutionary hypotheses using Swedish total population data. Evol Hum Behav 35(6):481–488
Yarbrough RM (2005) Teaching bioeconomics. J Bioecon 7(1):1–38
Ye H, Tan F, Ding M et al (2011) Sympathy and punishment: evolution of cooperation in public goods game. JASSS 14(4):20
Yeoman I, Mars M (2012) Robots, men and sex tourism. Futures 44(4):365–371
Yilmaz O, Bahçekapili HG (2016) Supernatural and secular monitors promote human cooperation only if they remind of punishment. Evol Hum Behav 37(1):79–84
Yoder KJ, Decety J (2014) Spatiotemporal neural dynamics of moral judgment: a high-density ERP study. Neuropsychologia 60:39–45
Yokoyama S (1983) Social selection and evolution of human disease. Am J Phys Anthropol 62 (1):61–66
Yokoyama S (1983) Theories of social selection in human populations. Am J Hum Genet 35 (3):362–375
Young C (1999) Ceasefire! Why women and men must join forces to achieve true equality. Free Press, New York
Young KK (2002) Women and Hinduism. In: Sharma A (ed) Women in Indian religions. Oxford University Press, New Delhi, pp 3–37
Young S (2006) Designer evolution. A transhumanist manifesto. Prometheus Books, Amherst

Young L, Camprodon JA, Hauser M et al (2010) Disruption of the right temporoparietal junction with transcranial magnetic stimulation reduces the role of beliefs in moral judgments. PNAS 107(15):6753–6758
Young L, Dungan J (2012) Where in the brain is morality? Everywhere and maybe nowhere. Soc Neurosci 7(1):1–10
Young M, Edis T (2006) Why intelligent design fails: a scientific critique of the new creationism. Rutgers University Press, New Brunswick
Young L, Koenigs M (2007) Investigating emotion in moral cognition: A review of evidence from functional neuroimaging and neuropsychology. Br Med Bull 84:69–79
Yount L (2000) Physician-assisted suicide and euthanasia. Facts on File, New York
Ysseldyk Y, Matheson K, Hymie A (2010) Religiosity as identity: toward an understanding of religion from a social identity perspective. Person Soc Psychol Rev 14:60–71
Zagzebski L (2004) Divine motivation theory. Cambridge Uiversity Press, Cambridge
Zahavi A, Zahavi A (1997) The handicap principle: a missing piece of Darwin's puzzle. Oxford University Press, New York
Zahid HJ, Robinson E, Kelly RL (2016) Agriculture, population growth, and statistical analysis of the radiocarbon record. PNAS 113(4):931–935
Zaidel DW, Nadal M (2011) Brain intersections of aesthetics and morals perspectives from biology, neuroscience, and evolution. Perspect Biol Med 54(3):367–380
Zaidi ZF (2010) Gender differences in human brain: a review. Open Anat J 2:37–55
Zak PJ (2012) The moral molecule. Bantam Press, London
Zak PJ, Stanton AA, Ahmadi S (2007) Oxytocin increases generosity in humans. Public Libr Sci ONE 2:1–5
Zaki J, Mitchell JP (2013) Intuitive prosociality. Curr Dir Psychol Sci 22(6):466–470
Zanc I, Lupu I, Muresan O (2010) Bioethics and the challenges of science. Rev Rom de Bioetica 8 (4):121–132
Zarandi MM (ed) (2003) Science and the myth of progress. World Wisdom, Bloomington
Zeedyk S (2007) The science of rape: (mis)constructions of women's trauma in evolutionary theory. Feminist Rev 86:67–88
Zeiss F (1982) Is wetenschap seksistisch? (Is science sexist?) De Nieuwe Linie 20.01.1982:7–8
Zellner K, Jaeger U, Kromeyer-Hauschild K (2004) Height, weight and BMI of schoolchildren in Jena, Germany—are the secular changes levelling off? Econ Hum Biol 2:281–294
Zey MG (2001) Man's evolutionary path into the universe. Futurist 35(3):28–33
Zey MG (2006) Technology will enhance humanity's future. In: Gerdes LI (ed) Humanity's future. Greenhaven Press, Detroit, pp 19–28
Zhang L (2008) Religious affiliation, religiosity, and male and female fertility. Demographic Res 18:233–262
Zhen-Wang B, Cheng-Ye J (2005) Secular growth changes in body height and weight in children and adolescents in Shandong, China between 1939 and 2000. Ann Hum Biol 32(5):650–665
Zhuang Q, Wang D, Fan Y et al (2012) Evolution of cooperation in a heterogeneous population with influential individuals. Phys A-Stat Mech Appl 391(4):1735–1741
Zonabend F (1996) An anthropological perspective on kinship and the family. In: Burguiere A, Klapisch-Zuber C, Segalen M et al (eds) A history of the family. Volume One. Distant worlds, ancient worlds. Polity Press, Cambridge, pp 8–70
Zick CD, Mathews CJ, Roberts JS et al (2005) Genetic testing for Alzheimer's disease and its impact on insurance purchasing behavior. Health Aff 24(2):483–490
Zietsch B, Morley K, Shekar S et al (2008) Genetic factors predisposing to homosexuality may increase mating success in heterosexuals. Evol Hum Behav 29(6):424–433
Zietsch BP, Verweij KJH, Bailey JM et al (2010) Genetic and environmental influences on risky sexual behaviour and its relationship with personality. Behav Genet 40(1):12–21
Zietsch BP, Westberg L, Santtila P et al (2015) Genetic analysis of human extrapair mating: heritability, between-sex correlation, and receptor genes for vasopressin and oxytocin. Evol Hum Behav 36(2):130–136

Zillmann D, Bryant J (1988) Effects of prolonged consumption of pornography on family values. J Family Issues 9(4):518–544

Zillmann D, Bryant J (1988) Pornography's impact on sexual satisfaction. J Appl Soc Psychol 18 (5):438–453

Zindler FR (2003) The Jesus the Jews never knew: Sepher toldoth yeshu and the quest of the historical Jesus in Jewish sources. American Atheist Press, Cranford

Zinn MB, Dill BT (1996) Theorizing difference from multiracial feminism. Feminist Stud 22 (2):321–331

Zinnbauer BJ, Pargament KI (2005) Religiousness and spirituality. In: Paloutzian RF, Park CL (eds) Handbook of the psychology of religion and spirituality. Guilford Press, New York, pp 21–42

Zinnbauer BJ, Pargament KI, Scott AB (1999) The emerging meanings of religiousness and spirituality: problems and prospects. J Person 67(6):889–919

Zirkle C (1959) Evolution, Marxian biology, and the social scene. University of Pennsylvania Press, Philadelphia

Zuckerman P (2005) Atheism: contemporary rates and patterns. In: Martin M (ed) Cambridge companion to atheism. Cambridge University Press, Cambridge, pp 47–68

Zuckerman P (2008) Society without God: what the least religious nations can tell us about contentment. New York University Press, New York

Zuckerman P (ed) (2010) Atheism and secularity: volume 1: issues, concepts, and definitions. Volume 2: global expressions. Praeger Santa, Barbara

Zycinski J (2006) Ethics in medical technologies: the roman catholic viewpoint. J Clin Neurosci 13(5):518–523

Printed by Printforce, the Netherlands